Numerical Analysis of Variational Inequalities

STUDIES IN MATHEMATICS AND ITS APPLICATIONS

VOLUME 8

NORTH-HOLLAND PUBLISHING COMPANY – AMSTERDAM • NEW YORK • OXFORD

NUMERICAL ANALYSIS OF VARIATIONAL INEQUALITIES

ROLAND GLOWINSKI
Université Pierre et Marie Curie, Paris and INRIA

JACQUES-LOUIS LIONS
Collège de France, Paris and INRIA

RAYMOND TRÉMOLIÈRES
Université d'Aix-Marseille

English version edited, prepared and produced by TRANS-INTER-SCIENTIA P.O. Box, 16, Tonbridge, TN11 8DY, England

50
1931 N·H 1981
P C

NORTH-HOLLAND PUBLISHING COMPANY – AMSTERDAM • NEW YORK • OXFORD

ISBN 0 444 86199 8

Translation and revised edition of:
Analyse Numérique des Inéquations Variationnelles
(Tome 1 et 2)
©Bordas (Dunod), Paris, 1976

Publishers:
NORTH-HOLLAND PUBLISHING COMPANY
AMSTERDAM · NEW YORK · OXFORD

Sole distributors for the U.S.A. and Canada:
ELSEVIER NORTH-HOLLAND, INC.
52 VANDERBILT AVENUE, NEW YORK, N.Y. 10017

Library of Congress Cataloging in Publication Data

Glowinski, R.
Numerical analysis of variational inequalities.

(Studies in mathematics and its applications ; v. 8)
Revised translation of: Analyse numérique des inéquations variationelles.
Bibliography: p.
1. Differential inequalities. 2. Numerical analysis. I. Lions, Jacques Louis. II. Trémolières, Raymond. III. Title. IV. Series.
QA374.G5713 1981 515.3'6 81-38345
ISBN 0-444-86199-8 AACR2

PRINTED IN THE NETHERLANDS

INTRODUCTION

1. INTRODUCTION TO CHAPTERS 1-3 (VOL. I OF THE FRENCH EDITION): GENERAL THEORY AND FIRST APPLICATIONS

Numerous problems in Mechanics, Physics and Control Theory lead to the study of systems of partial differential inequalities, the solution of which leans heavily on the techniques of so-called *variational inequalities*.

1.1 We start by giving a very simple example, the physical motivation for which is discussed in the book by Duvaut-Lions /1/. [1]

On a bounded open domain Ω of $\mathbb{R}^n$ ($n = 2, 3$ in practical applications) with boundary Γ, we seek a real-valued function $x \to u(x)$ which in Ω satisfies the classical equation:

(1) $$-\Delta u + u = f, \quad f \text{ given in } \Omega,$$

$\Delta u = \sum_{i=1}^{n} \frac{\partial^2 u}{\partial x_i^2}$, with *boundary conditions in the form of inequalities*:

(2) $$u \geqslant 0, \quad \frac{\partial u}{\partial \nu} \geqslant 0, \quad u \frac{\partial u}{\partial \nu} = 0 \text{ on } \Gamma,$$

where $\partial/\partial\nu$ denotes differentiation along the outward normal to Γ.

A similar *time-dependent* problem consists of finding $u = u(x, t)$ defined for $x \in \Omega$ and $t > 0$, which is a solution of

(3) $$\frac{\partial u}{\partial t} - \Delta u = f, \quad f \text{ given in } \Omega \times \{t > 0\},$$

with *inequality conditions* identical to (2) on Γ (for $t > 0$) and with the initial condition

(4) $$u(x, 0) = u_0(x), \quad u_0 \text{ given in } \Omega.$$

---

(1) A number enclosed between slashes corresponds to the main list of references, which follows Chapter 6.

The problem (3), (2), (4) is an *inequality problem of parabolic type*. Other inequalities encountered in practical applications are of "hyperbolic type" or of "Petrowski type", together with numerous variants of these.

In fact problem (1), (2) is a very simple problem in the *calculus of variations with constraints*. Indeed, if we introduce the functional [1]

$$J(v) = \tfrac{1}{2} a(v, v) - (f, v) \tag{5}$$

in which

$$a(u, v) = \sum_{i=1}^{n} \int_{\Omega} \frac{\partial u}{\partial x_i} \frac{\partial v}{\partial x_i} \,\mathrm{d}x + \int_{\Omega} uv \,\mathrm{d}x , \qquad (f, v) = \int_{\Omega} fv \,\mathrm{d}x ,$$

and if we introduce the convex (closed) set K defined by

$$K = \{ v \mid v \geqslant 0 \ \text{ on } \Gamma \} , \tag{6}$$

then *problem* (1), (2) *is equivalent* to finding $u \in K$ such that

$$J(u) = \inf J(v) , \qquad v \in K . \tag{7}$$

Problem (7) *admits a unique solution* u, characterised by:

$$\begin{cases} u \in K , \\ a(u, v - u) \geqslant (f, v - u) \quad \forall v \in K . \end{cases} \tag{8}$$

The problem in the form (8) is called an *elliptic variational inequality problem*.

The formulation (8) allows problem (3), (2), (4) to be stated in a very convenient manner; denoting by $u(t)$ the function $x \to u(x, t)$, we need to determine $u(t)$ such that

$$\begin{cases} u(t) \in K , \\ \left(\dfrac{\partial u(t)}{\partial t} , v - u(t) \right) + a(u(t), v - u(t)) \geqslant (f(t), v - u(t)) \quad \forall v \in K , \\ u(0) = u_0 , \end{cases} \tag{9}$$

where now

$$a(u, v) = \sum_{i=1}^{n} \int_{\Omega} \frac{\partial u}{\partial x_i} \frac{\partial v}{\partial x_i} \,\mathrm{d}x .$$

---

(1) The classes (*Sobolev spaces*) in which these functionals must be taken will be defined in the text.

This is a *time-dependent variational inequality* problem. It may be shown that *problem* (9) *admits a unique solution* (see Lions-Stampacchia /1/; we shall give in Chapter 6 an existence proof based on *discretisation*).

1.2 Without claiming to be exhaustive, we indicate below some areas in which steady-state and time-dependent variational inequalities arise.

(i) *Elasticity with unilateral conditions* (see Signorini /1/, Fichera /1/ and Frémond /1/) and *elasticity with friction* (see Duvaut-Lions /1/ and Kalker /1/, /2/) lead to elliptic variational inequalities and in the time-dependent case to hyperbolic variational inequalities (see Duvaut-Lions /1/); similarly the theory of thin plates leads to inequality problems of a new type (see Duvaut-Lions /4/).

(ii) *Elasto-plastic materials* lead to constrained minimisation problems (see Koiter /1/, Mandel /1/, Prager /1/,..., /8/, and Prager and Hodge /1/) which have been studied comprehensively, using inequality techniques, in Brézis and Sibony /1/, Duvaut-Lions /1/, Lanchon /1/, Lewy and Stampacchia /1/, /2/, Moreau /2/, /3/ and Ting /1/, /2/, /3/.

(iii) Problems of *hydrodynamics in porous media* lead to steady state or parabolic inequalities (see Duvaut-Lions /1/).

(iv) Problems of the flow of certain non-Newtonian fluids of the *Bingham fluid* type (see Duvaut-Lions /1/, Mosolov and Miasnikov /1/) lead to inequalities which contain the Navier-Stokes equations as a special case.

(v) Various problems arising in *servo-control* and in *electromagnetism in unstable media* lead to time-dependent inequalities of various types.

(vi) Numerous situations in which *several of the above phenomena* are present simultaneously, such as for example thermal phenomena in plastic media or in Bingham fluids (see Prager /3/, Duvaut-Lions /2/, /3/), also lead to inequalities.

(vii) In the foregoing applications, inequalities (steady state or time-dependent) on the "*physical variable*" (displacement, pressure, temperature, etc.) are obtained.

Classical *free boundary* problems, however, are *not expressed* in the form of variational inequalities in the physical variables; nonetheless, Biaocchi, in his study of free-boundary seepage problems, has observed (see Biaocchi /1/) that *after a suitable* (and non-trivial!) *change of unknown function*, variational inequalities or problems using the theory of inequalities are obtained.

This has been developed in a series of investigations at the C.N.R. Laboratory in Pavia: see Baiocchi, Comincioli, Magenes and Pozzi /1/ and the bibliography of the latter.

(viii) Using methods inspired by that of Biaocchi /1/, classical "free-boundary" problems may be reduced to variational inequalities; these include steady flows (see Brézis and Stampacchia /1/), wakes (see Brézis and Duvaut /1/), problems of Stefan type (see Duvaut /1/) and cavitation problems (see Bayada /1/). For numerical details of wake problems solved using the methods of this book, see Bourgat and Duvaut /1/.

(ix) Similarly, *optimal stopping-time problems* may be reduced to time-dependent variational inequalities (see Bensoussan-Lions /1/ and Friedman /1/).

(x) *Impulse control* problems may be reduced to problems which call upon variational inequality techniques: these are the *quasi-variational inequalities* introduced in Bensoussan-Lions /1/, /2/ and Bensoussan, Goursat and Lions /1/, and will not be considered here; from a *numerical* point of view, however, their treatment is based, in particular, on techniques described in the present book: see Goursat /1/ and Bensoussan-Lions /1/.

(xi) Control Theory for distributed systems leads to numerous problems of similar type, notably in the consideration of "a priori feedback" (see Lions /2/).

1.3 The main objective of the present work is to describe the principal numerical methods for computing the solution of steady-state and time-dependent variational inequalities, and to present some of the numerical calculations which have been performed.

Most steady-state problems reduce to one of the two following cases:

(i) Minimisation of a *differentiable* functional J over a convex closed set K (called the "convex set *of the constraints*"); this is the case, for example, with problem (7) stated earlier.

(ii) Minimisation over the entire space (unconstrained) of a *non-differentiable* functional J.

The *principle* behind the approximation is clear: we "approximate" the functional $J(v)$ (or the form $(J'(u), v)$ in the differentiable case) by means of *discretisation*, using either a *finite difference* method or a *finite element* method (i.e. using the *classical methods of Numerical Analysis*).

The specific difficulties (which fundamentally distinguish the numerical analysis of *elliptic inequalities* from that of elliptic *equations*) are concerned with:

(j) In case (i), the *approximation* by discretisation of the *convex set* K;

(jj) *The solution of the discretised problem*; this is a (finite-dimensional) *nonlinear programming* problem.

In the case in which $J(v)$ is "approximated" by $J_h(v_h)$ (where h denotes the discretisation parameter - for example the mesh size) and the convex set K is approximated by the convex set K_h, the approximate problem (in case (i)) is:

$$\inf J_h(v_h)\,, \qquad v_h \in K_h\,, \tag{10}$$

and in case (ii) is:

(11) $\begin{cases} \inf J_h(v_h)\,, & \text{unconstrained, } J_h \text{ being nondifferentiable or} \\ \text{in any case the "limit" of } J_h \text{ not being differentiable.} \end{cases}$

The question of the *convergence* of the methods does not present any particular difficulty. The problem of the *error estimation* is less clear cut; see Falk /1/, Mosco and Strang /1/. The fundamental issue is then that of *effectively solving* (10), (11).

In our experience, direct application of the "standard" methods of nonlinear programming generally leads, in the cases considered here, to prohibitive computation times; this observation has prompted us to reconsider the question in a systematic manner, taking into account the *particular structure* of the functionals to be minimised and of the convex sets of the constraints.

1.4 The first three chapters of this book (which formed Volume I of the original French edition) describe the *general methods* relating to *steady state* problems, and studies in some depth a typical example taken from the theory of elasto-plasticity.

Chapter 1 describes the various ways in which the original problem may be approximated by a finite-dimensional problem. First, however, a brief résumé of inequalities is given, which should *enable the reader to tackle this book without having to possess any specialised knowledge of inequalities.*

Chapter 2 is devoted to the methods used in the solution of finite-dimensional problems. This chapter may be considered as part of a book on the methods of *nonlinear programming*.

Chapter 3 treats, by means of examples, the detailed implementation of the general methods, and gives the results of a fairly large number of numerical applications; this makes it possible for *comparisons* to be drawn between the various methods. The problem of the *elasto-plastic torsion of cylindrical bars* is studied in a relatively exhaustive manner. This was chosen as a "model" problem not only because it is of both physical and theoretical interest, but also because its numerical solution illustrates many of the *fundamental difficulties* of the theory; in fact, the convex set K is in this case defined by

$$K = \{ v \mid |\operatorname{grad} v(x)| \leqslant 1 \text{ a.e. in } \Omega, \ v = 0 \text{ on } \Gamma \} \tag{12}$$

and the *approximation* of K raises a number of interesting points.

2. INTRODUCTION TO CHAPTERS 4 - 6 (VOL. II OF THE FRENCH EDITION): APPLICATIONS TO STEADY-STATE AND TIME-DEPENDENT PHENOMENA

2.1 The previous three chapters presented the general methodology: Chapter 1 considered the infinite-dimensional case and Chapter 2 the finite-dimensional case, whilst Chapter 3 investigated a steady-state elasto-plasticity problem which illustrates some of the difficulties which are typical of this subject. Chapters 4 to 6 (which formed Vol. II of the original French edition) continue the study of the Numerical Analysis of Variational Inequalities, by analysing several classes of *steady-state* problems occurring in Mechanics and Physics, and a number of *time-dependent* problems.

2.2 The steady-state case is taken up in Chapters 4 and 5. These chapters investigate some problems which are of practical interest and which contain *a number of typical difficulties* in steady-state inequality problems, concerning in particular *non-differentiable functionals.*

It would appear that the following practical conclusions may be drawn: when the convex set K is of suitable form, the "optimal" methods (with regard to ease of programming, storage requirement, computational speed, etc,....) are of the *overrelaxation* type (see Chapter 2); this is true in particular for elliptic problems of order 2.

Duality methods (see Chapters 1 and 2) have a more general range of application (in particular, they are well suited to the solution of 4th order elliptic problems and of certain nondifferentiable problems) and have a high performance in terms of computational speed; however, all other things being equal, they are more difficult to implement and more costly in terms of storage requirement (and often of computation time) than overrelaxation methods when these are applicable.

In the problems considered, *penalty methods* (see Chapter 2) are, in general, more complicated to implement and more costly both in computation time and in storage requirement than the above methods. However, they should not be dismissed completely, since certain problems, particularly those arising in optimal control, lead to the idea that these methods could prove to be useful if they were employed *simultaneously* with overrelaxation and duality methods.

Methods based on decomposition and splitting of the constraints (see Lions-Témam /1/) will not be presented in the present book; a general description of these is given in Bensoussan-Lions-Témam /1/; their numerical application and, in particular, the search for "optimal" decompositions of the right-hand side, are described in Lavainne /1/.

2.3 *Time-dependent problems* form the subject of Chapter 6.

If, in the inequality (9), $\partial u(t)/\partial t$ is replaced by $(u^{n+1} - u^n)/\Delta t$, where u^n denotes an approximation of $u(n\,\Delta t)$, then it is possible to "approximate" (9) by using *semi-discretisation*:

$$(10)\qquad \begin{cases} \left(\dfrac{u^{n+1} - u^n}{\Delta t}, v - u^{n+1}\right) + a(u^{n+1}, v - u^{n+1}) \geqslant \\ \geqslant (f(n\,\Delta t), v - u^{n+1}) \quad \forall v \in K, \quad u^{n+1} \in K \qquad (u^0 = u_0)\,. \end{cases}$$

This is of the same form as a steady-state variational inequality, and is therefore amenable to the methods mentioned in 2.2 above.

Hence, by *discretisation* of (10), we achieve the complete discretisation of problem (9).

The choice leading to (10) is obviously the simplest, but it is not the only one possible! The following points require consideration:

(i) choice of the methods of discretisation in t
(ii) analysis of the stability and convergence
(iii) numerical applications allowing comparisons to be made.

From this viewpoint, Chapter 6 investigates time-dependent inequalities connected with operators which may or may not be linear:

- parabolic inequalities of type I and type II (see chapter 6);
- "hyperbolic" inequalities, or those corresponding to operators of the Petrowski type; then in Section 11, the analogous problems for nonlinear operators of the Bingham type are studied briefly - we refer the reader to Fortin /1/ and Bégis /1/ for further technical details.

2.4 There already exist in the literature numerous works devoted to the Numerical Analysis of problems in Mechanics, with constraints. This is especially true in the field of elasto-plasticity; here we may cite the works of Prager-Hodge /1/, Hérakovitch and Hodge /1/, Capurso /1/, /2/, /3/, Capurso and Maier /1/, Maier /1/, /2/ and works using finite elements in elasto-plasticity for which we refer to Argyris /1/, /2/, Argyris, Scharpf and Spooner /1/, Yamada /1/, Zienkiewicz /1/, Zienkiewicz, Valliapan and King /1/ and their associated bibliographies.

Some general methods for elliptic problems are described in Haugazeau /1/, /2/ and Aubin /1/. Herein we have primarily used the works of Céa and Glowinski /1/ and Céa, Glowinski, Nédélec /1/. In the case of time dependent problems, the first results were given in a paper by Lions /5/; herein we have primarily used the works of Trémolières /4/, Viaud /1/ and Fortin /1/.

We have also used, throughout the book, the results obtained by Bégis /1/, Bourgat /1/, Goursat /2/, Lemonnier /1/, Leroy /1/, Marrocco /1/, Mercier /1/, Thomasset /1/ to whom we are indebted.

For other methods relating to the steady-state case, we refer the reader to the articles by Auslender /1/ and Mosco /1/.

The following works amongst others may be consulted for further applications: Baiocchi, Comincioli, Magènes and Pozzi /1/, Comincioli, Guerri and Volpi /1/, Frémond /1/, Fuscardi, Mosco, Scarpini and Schiaffino /1/.

We would like to express our thanks to Messrs. Céa, Nédélec and Strang for several interesting discussions on the material contained within this book.

Since the final drafting of Volumes I and II of the original French version of the present work (June 1974) certain progress has been made in the Numerical Analysis of variational inequalities, principally in connection with the problems of *estimating the approximation errors* related to the use of the *finite element method*. We may cite here, amongst others, the works of Brezzi-Hager-Raviart, Falk, Falk-Mercier, Glowinski, Johnson, Johnson-Mercier, Mosco-Scarpini, Nitsche, etc. Some of the references corresponding to these investigations may be found in Glowinski /9/.

We should also emphasise the potential, confirmed by numerical experiments, of the *augmented Lagrangian* methods mentioned in Chapter 2, Section 6, in Chapter 3, Section 10 and in Chapter 5, Section 9. The reader may refer to Gabay-Mercier /1/, Glowinski-Marrocco /1/, /2/ and Mercier /1/ for some applications of these methods to nonlinear partial differential equations and variational inequalities.

3. SUPPLEMENTARY INTRODUCTION TO THE ENGLISH-LANGUAGE EDITION

Progress which has taken place since the publication of Volumes I and II of the original French edition of this book has led to developments under the following headings:

1. Introduction and investigation of new algorithms.
2. New applications of Variational Inequalities to free-boundary problems.
3. Variants of the Variational Inequalities: the Quasi-Variational Inequalities.
4. Optimal Control of systems governed by Variational Inequalities.
5. Application of the methodology of Variational Inequalities to the solution of other problems.

We shall now give a few brief comments on these points.

3.1 Introduction and investigation of new algorithms

This topic is the one most directly related to the theme of the present book. The new work has prompted us to produce Appendices 1 to 6, in which we present those contributions which seem to us to be the most important (see also the bibliography corresponding to these appendices ([1])).

([1]) The bibliography for the appendices follows Appendix 6. The corresponding references in the text are followed by a letter A.

3.2 New applications of Variational Inequalities to free-boundary problems

It is especially important here to point out the work of the team at the Numerical Analysis Laboratory of the University of Pavia, and in particular that of C. Baiocchi, V. Comincioli and their associates. Using the techniques of steady-state and *time-dependent* Variational Inequalities, the authors have solved progressively more complicated problems related to steady and time-dependent seepage.

Bibliographic references are cited in Appendices 1 and 6.

3.3 Variants of the Variational Inequalities: the Quasi-Variational Inequalities

The theory of *Impulse Control* brings in new problems, similar in some ways to the obstacle problem but with "*an obstacle which depends upon the solution*". We thus have "implicit" Variational Inequalities, which A. Bensoussan and J.L. Lions have termed *Quasi*-Variational Inequalities (abbreviated to Q.V.I.).

These problems give rise to some quite serious theoretical difficulties (in particular, it seems difficult to apply these techniques to operators which do not possess a maximum principle); fundamental difficulties also arise in connection with the *dimension* of the problems to be investigated.

A brief introduction to the investigation of this type of problem is given in Appendix 1, Section 7.

3.4 Optimal Control of systems governed by Variational Inequalities

Some recent and important applications, notably in the control of steel rolling mills [1], have led to a renewed interest in the problem (which was posed in a purely "academic" manner in Lions /2/) of *Variational Inequality Control* and also, more generally, in *free-boundary control*.

A number of recent theoretical investigations have been directed at this question, in particular those of V. Barbu and F. Mignot. A full account of these problems will be presented at a later date.

---

(1) See Ch. Saguez /1A/.

3.5 Application of the methodology of Variational Inequalities to the solution of other problems

The techniques introduced for the numerical solution of Variational Inequalities have also proved useful for, amongst other things, problems of the following type:

- *Nonlinear problems in Fluid Mechanics* (see in particular Appendix 4, Section 4)
- *Free-boundary problems*, even when it is not known whether these problems actually fall within the scope of Variational Inequalities.

Some interesting issues, not addressed in the present book, emerge in connection with all the headings outlined above; some of these have already given rise to books or monographs, and it is reasonable to conjecture that eventually the same will apply for those which are at present unresolved or not solved in full.

Finally, we would like to acknowledge our particular indebtedness to Mrs. F. Weber of INRIA who painstakingly typed the French version of the Appendices, to Mr. Barry Hunt and his colleagues in *Trans-Inter-Scientia* who translated, edited and typed this English edition, and to the North-Holland Publishing Company for agreeing to publish this translation in the series

Studies in Mathematics and its Applications

Paris, France
January 1981.

Roland Glowinski
Jacques-Louis Lions
Raymond Trémolières

TABLE OF CONTENTS

Chapter 1

GENERAL METHODS OF APPROXIMATION FOR STEADY-STATE INEQUALITY PROBLEMS

SYNOPSIS

After first giving some *examples* of *steady-state inequality* problems (referring to Duvaut-Lions /1/ for a detailed motivation) this chapter deals with the *general methods* for approximating these inequalities.

In the succeeding four chapters, these methods will then be applied, and the main technical difficulties which arise discussed in some detail, for three classes of examples.

The reader who simply requires an overall view of steady-state and *time-dependent* problems may proceed straight to Chapter 6 after reading Chapter 1; to be able to apply the methods investigated in this book effectively, however, it is essential to read Chapters 2 to 5.

1. EXAMPLES

1.1 Fluid mechanics problems in media with semi-permeable boundaries

Consider a medium Ω of $\mathbb{R}^n$ (n = 1, 2 or 3 in normal applications) with boundary Γ; we assume that Ω is occupied by a fluid for which the pressure, in the steady state, is $u(x)$ (1) and we assume that the boundary Γ allows fluid entering Ω to pass freely, but prevents fluid leaving. Suppose that we apply, on the boundary Γ surrounding Ω, a fluid pressure $h = h(x)$.

Under suitable conditions, it can be assumed that, in Ω, u satisfies:

$$(1.1) \qquad -\Delta u = f\,,$$

where $\quad \Delta = \sum_{i=1}^{n} \frac{\partial^2}{\partial x_i^2}, \qquad f = f(x) \quad$ given in Ω,

---

(1) We denote a point in Ω by $x = \{x_1, \dots, x_n\}$. We shall always put $\mathrm{d}x = \mathrm{d}x_1 \dots \mathrm{d}x_n$.

the boundary conditions being as follows: for $x \in \Gamma$ it is necessary to distinguish two situations:

(i) $u(x) > h(x)$;

the fluid thus has a tendency to leave Ω; the wall Γ prevents this, however, so that the flux is zero at x, i.e:

$$\frac{\partial u}{\partial n}(x) = 0 \qquad (1.2)$$ [1];

(ii) $u(x) \leqslant h(x)$.

The fluid tends to enter Ω; the wall Γ allows this and therefore the flux satisfies

$$\frac{\partial u}{\partial n}(x) \geqslant 0, \qquad (1.3)$$

where $\frac{\partial u}{\partial n}(x)$ is *finite*. Hence, u is continuous in the neighbourhood of x, which, if the wall Γ is infinitely thin (Duvaut-Lions /1/), implies that $u(x) = h(x)$.

Summing up, the boundary conditions are:

$$u - h \geqslant 0, \quad \frac{\partial u}{\partial n} \geqslant 0, \quad (u - h)\frac{\partial u}{\partial n} = 0 \text{ on } \Gamma. \qquad (1.4)$$ ■

Remark 1.1. There are thus a priori two regions Γ_1 and Γ_2 on Γ, for which $u = h$ and $\partial u/\partial n = 0$, respectively. However, the location of these two regions *is not known a priori*, and finding these regions is in fact *equivalent* to solving the problem; we thus have a "*free boundary*" *type problem* [2]. ■

Transformation of the problem.

We shall now transform the problem (1.1), (1.4) into the form of a (simple) problem in the *calculus of variations*. This will make it possible:

(i) to show that (under suitable assumptions) the problem is "*well posed*";

(ii) to lay a foundation for *constructive methods for approximating the solution.* ■

---

(1) $\partial/\partial n$ = the derivative normal to Γ, *directed towards the outside of* Ω (though this choice is arbitrary).

(2) This observation applies to *all* the problems investigated in this book.

We shall use several functional analysis tools; *the basic principles* behind these ideas may be found in Duvaut-Lions /1/, Chapter 1; here we simply recall the *definitions* (for a complete study of Sobolev spaces see for example Necas /1/, Lions-Magenes /1/).

We denote by $H^1(\Omega)$ *the Sobolev space* (of order 1) of functions v (1) such that (see Sobolev /1/):

$$(1.5) \qquad v \in L^2(\Omega), \qquad \frac{\partial v}{\partial x_i} \in L^2(\Omega), \qquad i = 1, \dots, n$$

where $L^2(\Omega)$ is the space of (classes of) square summable functions on Ω (2), and where in (1.5) the derivatives $\partial v/\partial x_i$ are taken in the sense of distributions on Ω. $H^1(\Omega)$ is a *Hilbert space* equipped with the inner product:

$$(1.6) \qquad (u, v)_{H^1(\Omega)} = \int_\Omega uv \, dx + \sum_{i=1}^n \int_\Omega \frac{\partial u}{\partial x_i} \frac{\partial v}{\partial x_i} \, dx .$$

It may be shown (see the references cited) that, if the boundary Γ of Ω is finite and sufficiently regular, we can uniquely define the *trace* γv of $v \in H^1(\Omega)$ on Γ, and $\gamma v \in L^2(\Gamma)$, the mapping $v \to \gamma v$ being linear continuous from $H^1(\Omega) \to L^2(\Gamma)$(3).

If the function h is given on Γ, then we can define the subset K of $H^1(\Omega)$ by:

$$(1.7) \qquad K = \{ v \mid v \in H^1(\Omega), \ \gamma v \geqslant h \text{ a.e. on } \Gamma \} .$$

K can be shown to be a *closed convex subset of* $H^1(\Omega)$.

We now introduce

$$(1.8) \qquad a(u, v) = \int_\Omega \sum_{i=1}^n \frac{\partial u}{\partial x_i} \frac{\partial v}{\partial x_i} \, dx ,$$

and the functional

$$(1.9) \qquad J(v) = \tfrac{1}{2} a(v, v) - (f, v) ,$$

where

$$(1.10) \qquad (f, v) = \int_\Omega f(x) \, v(x) \, dx .$$

---

(1) *All* the functions considered in this book are *real-valued*.

(2) Hence $\varphi \in L^2(\Omega) \Leftrightarrow \varphi$ measurable on Ω and $\int_\Omega \varphi(x)^2 \, dx < \infty$.

(3) Since the image of $H^1(\Omega)$ by γ is a smaller space than $L^2(\Gamma)$, namely $H^{1/2}(\Gamma)$, this result can be extended (see e.g. Lions-Magenes /1/, Chapter 1).

We shall now verify that *the problem* (1.1), (1.4) *is equivalent to the minimisation of* $J(v)$ *over* K. This is based on the following general observations:

1) The functional $v \to J(v)$ is *convex* and *differentiable*, the derivative $J'(u)$ ([1]), given by:

$$(1.11)\qquad (J'(u), v) = \frac{\mathrm{d}}{\mathrm{d}\lambda} J(u + \lambda v)\Big|_{\lambda = 0},$$

being expressed by:

$$(1.12)\qquad (J'(u), v) = a(u, v) - (f, v)\,.$$

2) If $u \in K$ constitutes a minimum of $J(v)$ over K, i.e.

$$(1.13)\qquad J(u) \leqslant J(v) \qquad \forall v \in K\,,$$

then

$$(1.14)\qquad (J'(u), v - u) \geqslant 0 \qquad \forall v \in K\,,$$

and conversely.

Therefore, using (1.12), it may be seen that the problem in the calculus of variations: $\inf J(v), v \in K$, *is equivalent to the variational inequality:*

$$(1.15)\qquad \begin{cases} u \in K\,, \\ a(u, v - u) \geqslant (f, v - u) \quad \forall v \in K\,. \end{cases}$$

We shall now show that (1.1), (1.4) *are equivalent to* (1.15).

If we denote by $\mathcal{D}(\Omega)$ the space of indefinitely differentiable functions with *compact* support in Ω, we see that if $\varphi \in \mathcal{D}(\Omega)$ then

$$(1.16)\qquad v = u \pm \varphi$$

is in K. The choice (1.16) in the inequality (1.15) therefore gives

$$a(u, \varphi) = (f, \varphi) \quad \forall \varphi \in \mathcal{D}(\Omega)$$

and hence equation (1.1) (by definition of the derivatives in the sense of distributions on Ω). However, by taking the inner product of (1.1) with $v - u$ and integrating by parts, this

---

([1]) In (1.11), $J'(u)$ denotes an element of the dual space $(H^1(\Omega))'$ of $H^1(\Omega)$ and $(J'(u), v)$ denotes the inner product between $(H^1(\Omega))'$ and $H^1(\Omega)$.

leads to

$$-\int_\Gamma \frac{\partial u}{\partial n}(v-u)\,\mathrm{d}\Gamma + a(u, v-u) = (f, v-u)\,,$$

so that (1.15) *is equivalent to* (1.1) *and*

$$\int_\Gamma \frac{\partial u}{\partial n}(v-u)\,\mathrm{d}\Gamma \geqslant 0 \quad \forall v \in K\,. \tag{1.17}$$

By taking $v = u + \psi$, $\psi \geqslant 0$ to be arbitrary (regular) on Γ, we then deduce that

$$\int_\Gamma \frac{\partial u}{\partial n}\psi\,\mathrm{d}\Gamma \geqslant 0 \quad \forall\psi \geqslant 0 \text{ on } \Gamma\,,$$

and therefore that

$$\frac{\partial u}{\partial n} \geqslant 0 \text{ on } \Gamma\,.$$

Taking successively, $v = h$ and $v = 2u - h$ on Γ (assuming h to be sufficiently regular on Γ), we deduce that

$$\frac{\partial u}{\partial n}(u-h) = 0 \text{ on } \Gamma\,,$$

and hence (1.4), and conversely (1.4) implies (1.17).

This therefore proves the stated equivalence. ■

To summarise: A physical problem gave rise to the inequality problem (1.1), (1.4) and we have just shown that this problem is *equivalent* to a problem in the calculus of variations. The latter itself has *two equivalent formulations:*

1) $\inf_{v \in K} J(v)$;

2) Find u in K which is a solution of

$$a(u, v-u) \geqslant (f, v-u) \quad \forall v \in K\,.$$

Remark 1.2. The above equivalences, do not, at the moment, show *the existence of a solution.* However, we shall give some indications on this point in Section 2, using the formulation corresponding to a problem in the calculus of variations. ■

Synopsis.

We shall now give *some examples,* formulated directly in variational form. ■

1.2 Model elasto-plastic problem

We shall now give a formulation of a model problem encountered in elasto-plasticity (see Annin /1/, Lanchon /1/, Duvaut-Lions /1/, Chapter 5 and the bibliography cited therein, and Mandel /1/), directly in the form of a problem in the calculus of variations.

We assume Ω to be an open domain, as in Section 1.1. We introduce

$$(1.18) \qquad H_0^1(\Omega) = \{ v \mid v \in H^1(\Omega) , \gamma v = 0 \} ,$$

which is a closed vector subspace of $H^1(\Omega)$.

We then introduce

$$(1.19) \qquad K = \{ v \mid v \in H_0^1(\Omega) , | \operatorname{grad} v(x) | \leqslant 1 \text{ a.e. in } \Omega \} .$$

It may easily be shown that

(1.20) K is a *closed convex set* in $H_0^1(\Omega)$.

$J(v)$ is defined by (1.9).

Hence, the problem becomes

$$(1.21) \qquad \inf J(v) , \qquad v \in K .$$

As in Section 1.1, we see that (1.21) is *equivalent* to seeking u such that

$$(1.22) \qquad \begin{cases} u \in K , \\ a(u, v - u) \geqslant (f, v - u) \quad \forall v \in K . \end{cases}$$

When Ω is *simply connected* (1) the problem (1.21) — or (1.22) — is an elasto-plasticity problem. ■

Remark 1.3 In Section 2 it will be shown that problem (1.21) admits a unique solution. ■

Remark 1.4 The problem can be interpreted as follows (2): the open domain Ω consists of two regions Ω_e and Ω_p (the elastic region and the plastic region, respectively). In Ω_e (situated

---

(1) In the case of a multiply-connected section, it is necessary to modify the formulations (1.21), (1.22) (see e.g. Lanchon /2/, Glowinski-Lanchon /1/).

(2) This will be dealt with in more detail in Chapter 3.

"in the interior" of Ω) we have the classical equation:

$$-\Delta u = f, \tag{1.23}$$

and in Ω_p we have:

$$|\operatorname{grad} u| = 1, \tag{1.24}$$

to which we add: $u = 0$ on Γ and some transmission conditions - which incidentally are somewhat awkward to express explicitly (see Brézis /1/) - at the interface between Ω_e and Ω_p.

These observations lead to two remarks:

(i) We once more encounter the "free-boundary" nature of this type of problem - as mentioned earlier in Remark 1.1 for the problem in Section 1.1;

(ii) The "*global*" formulation in the form (1.21) or (1.22) is *much more amenable to calculation* than the "*pointwise*" formulation using (1.23), (1.24) and transmission conditions at the interface (or "free boundary"). ■

Remark 1.5. Many other steady-state or time-dependent inequality problems arise in plasticity theory; in particular, we refer the reader to Duvaut-Lions /1/, Chapter 5. ■

1.3 Model friction problem

We now present a highly idealised case of one of the problems encountered in the theory of elasticity with friction, or with unilateral constraints (see Signorini /1/, Fichera /1/). Further examples of this type will be given in Chapter 4.

We again take the space $H^1(\Omega)$ introduced in Section 1.1 and define the functional

$$j(v) = \int_\Gamma g\,|v|\,\mathrm{d}\Gamma \tag{1.25}$$

where g is a constant > 0 [1].

We define

$$a(u, v) = \int_\Omega \left(uv + \sum_{i=1}^{n} \frac{\partial u}{\partial x_i} \frac{\partial v}{\partial x_i} \right) \mathrm{d}x, \tag{1.26}$$

$$J(v) = \tfrac{1}{2} a(v, v) - (f, v) + j(v). \tag{1.27}$$

The problem considered is then to find:

---

[1] The case where g is a function which, for example, is continuous on Γ and $\geqslant 0$, could equally well be considered.

(1.28) $\inf J(v)\,, \qquad v \in H^1(\Omega)\,.$

This problem differs from those discussed earlier in that we now seek the lower bound over *a vector space*, and not over a convex set K which is not a vector space.

Moreover, the problem (1.28) *does not reduce* to the "Euler equation"

$$J'(u) = 0$$

since the functional j (and hence also J) *is not differentiable*.

It may easily be shown that problem (1.28) is *equivalent* to seeking u such that

$$\text{(1.29)} \quad \begin{cases} u \in H^1(\Omega)\,, \\ a(u, v-u) + j(v) - j(u) \geqslant (f, v-u) \quad \forall v \in H^1(\Omega)\,. \end{cases}$$

Remark 1.6. Generally, if

$$\text{(1.30)} \quad J(v) = H(v) + j(v)\,,$$

where H is differentiable and convex, and where j is continuous and convex but not necessarily differentiable, the condition (1.13) is *equivalent* to

$$\text{(1.31)} \quad (H'(u), v-u) + j(v) - j(u) \geqslant 0 \quad \forall v \in H^1(\Omega)\,. \quad \blacksquare$$

Remark 1.7. The inequality (1.29) is a *variational inequality*. ■

It is possible to *interpret* (1.29).

By first taking $v = u \pm \varphi, \varphi \in \mathscr{D}(\Omega)$, we find

$$\text{(1.32)} \quad -\Delta u + u = f \quad \text{in } \Omega\,.$$

From (1.32) and Green's formula, we deduce that

$$-\int_\Gamma \frac{\partial u}{\partial n}(v-u)\,\mathrm{d}\Gamma + a(u, v-u) = (f, v-u)\,,$$

so that (1.29) is *equivalent* to (1.32) and

$$\text{(1.33)} \quad \int_\Gamma \frac{\partial u}{\partial n}(v-u)\,\mathrm{d}\Gamma + j(v) - j(u) \geqslant 0 \quad \forall v\,.$$

Replacing v by $\pm \lambda v, \lambda > 0$, we deduce that

$$\lambda \int_\Gamma \left(\pm \frac{\partial u}{\partial n} v + g\,|\,v\,|\right)\mathrm{d}\Gamma - \int_\Gamma \left(u \frac{\partial u}{\partial n} + g\,|\,u\,|\right)\mathrm{d}\Gamma \geqslant 0 \quad \forall \lambda \geqslant 0$$

and hence

$$\left| \int_\Gamma \frac{\partial u}{\partial n} v \, \mathrm{d}\Gamma \right| \leqslant \int_\Gamma g \, | v | \, \mathrm{d}\Gamma , \tag{1.34}$$

$$\int_\Gamma \left(u \frac{\partial u}{\partial n} + g \, | u | \right) \mathrm{d}\Gamma \leqslant 0 . \tag{1.35}$$

However (1.34) is *equivalent* to $| \partial u / \partial n | \leqslant g$ on Γ; hence

$$u \frac{\partial u}{\partial n} + g \, | u | \geqslant 0$$

which in combination with (1.35) shows that

$$u \frac{\partial u}{\partial n} + g \, | u | = 0 .$$

To summarise, (1.29) *is equivalent to* (1.32) with *the boundary conditions on* Γ:

$$\left| \frac{\partial u}{\partial n} \right| \leqslant g , \qquad u \frac{\partial u}{\partial n} + g \, | u | = 0 . \tag{1.36}$$ ■

Remark 1.8. Conditions (1.36) are equivalent to the following [1]:

$$\begin{cases} \left| \dfrac{\partial u}{\partial n} \right| < g \Rightarrow u = 0 , \\ \dfrac{\partial u}{\partial n} = g \Rightarrow u \leqslant 0 , \\ \dfrac{\partial u}{\partial n} = - g \Rightarrow u \geqslant 0 . \end{cases} \tag{1.37}$$ ■

Remark 1.9. Much more complicated problems, corresponding to realistic physical situations, may be found in the articles and literature already cited, and in Chapter 4 of the present book. ■

1.4 A flow problem

The laminar flow of a Bingham fluid in a cylindrical duct (see Duvaut-Lions /1/, Chapter 6; Mosolov and Miasnikov /1/) leads to the following problem.

On the space $H_0^1(\Omega)$ we define

$$j(v) = \int_\Omega g \, | \operatorname{grad} v | \, \mathrm{d}x , \qquad g > 0 , \tag{1.38}$$

---

(1) In this form the "free-boundary" nature of the problems considered is again evident.

(so that as in Section 1.3 above, j is a *non*-differentiable functional). We define $a(u, v)$ using (1.8) and, as in (1.27), we put

$$J(v) = \tfrac{1}{2} a(v, v) - (f, v) + j(v) . \tag{1.39}$$

The problem considered is therefore analogous to (1.28) with $H^1(\Omega)$ replaced by $H_0^1(\Omega)$:

$$\inf J(v) , \quad v \in H_0^1(\Omega) , \tag{1.40}$$

which is *equivalent* to finding u such that

$$\begin{cases} u \in H_0^1(\Omega) , \\ a(u, v - u) + j(v) - j(u) \geqslant (f, v - u) \quad \forall v \in H_0^1(\Omega) . \end{cases} \tag{1.41}$$ ■

The interpretation of this problem in "standard" terms is complicated; an interpretation using *multipliers* is possible - see Theorem 3.3, later.

1.5 Synopsis

In the following section (Section 2) we shall present *a general framework* for all the above problems which will then enable us (Sections 3 onwards) to develop systematic methods for approximating these problems.

2. GENERAL FORMULATION OF STEADY-STATE VARIATIONAL INEQUALITIES

2.1 The symmetric case

We consider a Hilbert space V (on the field of real numbers $\mathbb{R}$) we denote by $\| v \|$ the norm of v in V. On V we take *a continuous bilinear form* $u, \ v \to a(u, v)$, which thus satisfies

$$| a(u, v) | \leqslant c \, \| u \| \, \| v \| , \qquad c = \text{const.}, \qquad u, v \in V .$$

We assume that the form $a(u, v)$ *is symmetric*, and hence

$$a(u, v) = a(v, u) \quad \forall u, v \in V . \tag{2.1}$$

We then take $K \subset V$ with

$$K = \text{closed convex subset of } V . \tag{2.2}$$

If V' denotes the dual of V and (f, v) the inner product between $f \in V'$ and $v \in V$, we consider the functional

$$J(v) = \tfrac{1}{2} a(v, v) - (f, v) . \tag{2.3}$$

The problem considered is:

$$\inf J(v)\,, \qquad v \in K\,. \tag{2.4}$$

Remark 2.1. The examples considered in Sections 1.1 and 1.2 can be shown to come within this category. ■

As in Section 1, problem (2.4) is *equivalent* to the following:

$$\begin{cases} \text{find } u \in K \text{ such that} \\ a(u, v - u) \geqslant (f, v - u) \quad \forall v \in K\,. \end{cases} \tag{2.5}$$

Problem (2.5) is a general *variational inequality* problem.

We can also consider a functional $v \to j(v)$ such that

$$v \to j(v) \text{ is continuous from } V \to \mathbb{R}\,,\ \text{convex } (^1). \tag{2.6}$$

We then consider

$$J(v) = \tfrac{1}{2}\, a(v, v) - (f, v) + j(v)\,, \tag{2.7}$$

the problem being

$$\inf J(v)\,, \qquad v \in V\,. \tag{2.8}$$

Remark 2.2. The examples considered in Sections 1.3 and 1.4 can be shown to come within this category. ■

Problem (2.8) is equivalent to the *variational inequality* problem:

$$\begin{cases} \text{find } u \in V \quad \text{such that} \\ a(u, v - u) + j(v) - j(u) \geqslant (f, v - u) \quad \forall v \in V\,. \end{cases} \tag{2.9}$$ ■

Concerning the *existence* of solutions, we have the following classical result:

Theorem 2.1. *If the function* $v \to J(v)$ *satisfies*

$$J(v) \to +\infty \text{ when } \| v \| \to \infty\,,\ v \in K \qquad (^2) \tag{2.10}$$

then there exists a solution u of (2.5) *or* (2.9).

A *sufficient* condition for (2.10) to be satisfied is that

$$\begin{cases} \text{there exists } \alpha > 0 \text{ such that} \\ a(v, v) \geqslant \alpha \| v \|^2 \quad \forall v \in V\,. \end{cases} \tag{2.11}$$ ■

---

(1) In practice it is sufficient to assume that j is convex, *proper* and lower semi-continuous; j is said to be *proper* if $j \not\equiv +\infty, j(v) > -\infty\ \forall v \in V$.

(2) This condition is redundant if *K is bounded in V*, which is the case, for example, in the context of the example in Section 1.2.

Remark 2.3. Condition (2.11) is satisfied in the examples considered in Sections 1.2 to 1.4, *but not* in the example of Section 1.1. It can be shown that in this case (2.10) is satisfied if we assume that

(2.12) $$\int_\Omega f(x)\,\mathrm{d}x < 0\,. \quad \blacksquare$$

For the possible *uniqueness* of the solution, the following result is immediate:

Theorem 2.2. *If the function* $v \to J(v)$ *is strictly convex* [1] *problem* (2.5) *(or* (2.9)*) admits at most one solution.*

This condition is met when (2.11) is true, and this shows that the *problems considered in Sections* 1.2 *to* 1.4 *admit a unique solution.*

The uniqueness of the solution of the problem of Section 1.1 may be verified *directly*, under the assumption (2.12). Suppose u_1 and u_2 are two possible solutions of (1.15); by taking $v = u_2$ (resp. $v = u_1$), in the inequality associated with u_1 (resp. u_2) and adding, we get, putting $w = u_1 - u_2$:

$$-\,a(w, w) \geqslant 0\,,$$

hence $a(w, w) = 0$ and therefore w = constant = c [2].

If for simplicity we put $u_2 = u$, we thus have the solutions u and $(u + c)$. Hence

(2.13) $$-\,\Delta u = f\,,$$

(2.14) $$u - h \geqslant 0\,, \quad \frac{\partial u}{\partial n} \geqslant 0\,, \quad (u - h)\frac{\partial u}{\partial n} = 0 \quad \text{on } \Gamma\,,$$

(2.15) $$u + c - h \geqslant 0\,, \quad \frac{\partial u}{\partial n} \geqslant 0\,, \quad (u + c - h)\frac{\partial u}{\partial n} = 0 \quad \text{on } \Gamma\,.$$

From (2.13) and (2.12) we deduce that

$$\int_\Gamma \frac{\partial u}{\partial n}\,\mathrm{d}\Gamma > 0\,,$$

so that there exists a set $\Gamma_1 \subset \Gamma$, measure $(\Gamma_1) > 0$, with

$$\frac{\partial u}{\partial n} > 0 \text{ on } \Gamma_1\,.$$

However, (2.14) and (2.15) then give

$$u - h = 0 \text{ and } u + c - h = 0 \quad \text{on } \Gamma_1\,,$$

---

(1) i.e. $J((1-\theta)\,u + \theta v) < (1-\theta)\,J(u) + \theta J(v)$ if $0 < \theta < 1$, $u, v \in K$, $u \neq v$.

(2) Assuming Ω to be connected.

and hence $c = 0$, which proves the uniqueness. ■

Remark 2.4. It is possible to render the hypotheses considerably less stringent by assuming that V is a *reflexive Banach space* and that the form $u, v \to a(u, v)$ is continuous over V, linear in v, but *nonlinear* in u; then

$$a(u, v) = (A(u), v), \qquad A(u) \in V'. \tag{2.16}$$

If K is a closed convex subset of V, the problem can be considered *directly in the form* (2.5):

$$\begin{cases} \text{find } u \in K \text{ such that} \\ (A(u) - f, v - u) \geqslant 0 \quad \forall v \in K. \end{cases} \tag{2.17}$$

It may be shown (see Brézis /1/, /2/, Browder /1/, Lions /1/) that if A is monotone, i.e.

$$(A(u) - A(v), u - v) \geqslant 0 \quad \forall u, v \in V \tag{2.18}$$

and *coercive*, i.e. (with $\|v\|$ again denoting the norm of v in V):

$$\frac{1}{\|v\|}(A(v), v) \to +\infty \text{ if } \|v\| \to \infty, \qquad v \in K\,(^1) \tag{2.19}$$

then problem (2.17) *admits a solution.*

Clearly, we have *uniqueness* in the case where A is *strictly* monotone i.e. when we have the *strict* inequality in (2.18) for $u \neq v$.

We note that (2.17) is not necessarily a problem in the calculus of variations, since the assumption (2.18) does not require that A should be a gradient (on this point, see Rockafellar /1/).

For extensions and examples, see Lions /1/. ■

2.2 The nonsymmetric case

We again consider the situation where V is a Hilbert space and we consider a bilinear form $a(u, v)$ which is continuous on V, but *not necessarily symmetric* i.e.

$$a(u, v) \neq a(v, u) \text{ in general.} \tag{2.20}$$

The problem is then expressed *directly* in the form (2.5); we are thus no longer dealing with a problem in the calculus of variations.

---

(1) This hypothesis is redundant if K is bounded.

Remark 2.5. In the nonsymmetric case, problem (2.9) can also be expressed directly. ■

It may then be shown (a constructive proof of this will be given in Section 3.1 below) that *if* (2.11) *is satisfied, (where the form* a *is not symmetric) problem* (2.5) *(or problem* (2.9)) *admits a unique solution.*

Remark 2.6. (*Variant in the formulation of the problems*).

Problem (2.5) is *equivalent* to the following:

$$(2.21)\qquad \begin{cases} \text{find } u \in K \text{ such that} \\ a(v, v-u) \geqslant (f, v-u) \quad \forall v \in K. \end{cases}$$

(similarly (2.9) is equivalent to seeking $u \in V$ such that

$$(2.22)\qquad a(v, v-u) + j(v) - j(u) \geqslant (f, v-u) \quad \forall v \in V).$$

In fact, if (2.5) is satisfied then

$$a(v, v-u) - (f, v-u) = a(u, v-u) - (f, v-u) + a(v-u, v-u) \geqslant 0$$

and hence (2.21); conversely, using (2.21) and dividing by θ:

$$a((1-\theta)\,u + \theta w, w-u) - (f, w-u) \geqslant 0$$

and by making $\theta \to 0$ we deduce (2.5). ■

Remark 2.7. The following remark is due both to L. Nirenberg and to R. Trémolières; various variants and applications are given in Trémolières /4/.

We introduce, for $v, w \in V$:

$$(2.23)\qquad L(v, w) = a(v, v-w) - (f, v-w).$$

If u is the solution of (2.5) or (2.21) we have:

$$L(v, u) \geqslant 0 \quad \forall v \in K \quad (\text{from } (2.21)),$$

$$L(u, v) \leqslant 0 \quad \forall v \in K \quad (\text{from } (2.5)),$$

and therefore

$$(2.24)\qquad L(u, v) \leqslant L(u, u) = 0 \leqslant L(v, u) \quad \forall v \in K.$$

Hence

$$(2.25)\qquad \{u, u\} \textit{ is a saddle point of } L(v, w) \textit{ on } K \times K.$$

Conversely, if $\{u, \hat{u}\}$ is a saddle point of $L(v, w)$ on $K \times K$, then $u = \hat{u} =$ solution of (2.5). ■

2.3 Synopsis

We shall now study some constructive methods for the *infinite dimensional* approximation of the solution u of elliptic variational inequalities. This will form the subject of Section 3. We shall then go on to see (in Sections 4 and 5) how the equations obtained in the infinite-dimensional case can be approximated by means of *finite dimensional* problems.

3. INFINITE-DIMENSIONAL APPROXIMATION METHODS

3.1 Successive approximations

We direct our attention to the case of (2.5), in which $a(u,v)$ is not necessarily symmetric, and satisfies

(3.1) $$a(v, v) \geqslant \alpha \| v \|^2, \quad \alpha > 0, \quad \forall v \in V.$$

If K is a closed convex subset of V, let u be the solution in K of

(3.2) $$a(u, v - u) \geqslant (f, v - u) \quad \forall v \in K.$$

The idea is as follows: assume that a symmetric, continuous bilinear form $b(u,v)$ is known on V, such that

(3.3) $$b(v, v) \geqslant \beta \| v \|^2, \quad \beta > 0, \quad \forall v \in V$$

and for which the problem of finding $w \in K$, the solution of

(3.4) $$b(w, v - w) \geqslant (g, v - w) \quad \forall v \in K,$$

(g given in V') *admits one and only one solution, which is "easy to calculate"*. ([1]). ■

Remark 3.1. Since b is symmetric, problem (3.4) is equivalent to the minimisation over K of

(3.5) $$\tfrac{1}{2} b(v, v) - (g, v),$$

so that (Theorem 2.1) problem (3.4) does indeed admit a unique solution. ■

Now, for u given in K (or in V) and for $\rho > 0$ fixed but arbitrary for the moment, there exists a unique $w \in K$ such that

---

([1]) Naturally this should be defined more precisely.... see also Remark 3.2 below.

(3.6) $$b(w, v-w) \geqslant b(u, v-w) - \rho[a(u, v-w) - (f, v-w)] \quad \forall v \in K.$$

We therefore introduce a mapping

(3.7) $$u \to w = S(u) \ \text{from} \ K \to K$$

and we immediately see that (3.2) is *equivalent* to seeking a *fixed point* of (3.7), i.e. a u such that

(3.8) $$u = S(u).$$

Now it will be shown below (using the method given in Lions-Stampacchia /1/) that ρ can be chosen such that for all $u_1, u_2 \in K$ or V we have

(3.9) $$\| S(u_1) - S(u_2) \| \leqslant \delta \| u_1 - u_2 \|, \quad 0 < \delta < 1;$$

if for the moment we assume (3.9) to be true, this gives:

(i) the existence and uniqueness of the solution u of (3.2);
(ii) the possibility of using *successive approximations* for the approximation of u; starting from an arbitrary u^0 in K, we construct $u^{n+1} \in K$ as *the* solution in K of

(3.10) $$b(u^{n+1}, v - u^{n+1}) \geqslant b(u^n, v - u^{n+1}) - \\ - \rho[a(u^n, v - u^{n+1}) - (f, v - u^{n+1})] \quad \forall v \in K$$

and hence $u^n \to u$ in V.

Let us now verify (3.9). Since b is symmetric and satisfies (3.3) we can take $b(v, w)$ to be a *new inner product* in V and take

$$\| v \| = (b(v, v))^{1/2}.$$

We put $w_i = S(u_i)$, $i = 1, 2$ and take $v = w_2$ (resp. w_1) in the inequality (3.6) associated with w_1 (resp. w_2). Adding, we deduce:

(3.11) $$-\| w_1 - w_2 \|^2 \geqslant -b(u_1 - u_2, w_1 - w_2) + \rho a(u_1 - u_2, w_1 - w_2).$$

We take: $u_1 - u_2 = \varphi$, $w_1 - w_2 = \psi$, $b(\varphi, \psi) = (\varphi, \psi)_V$ and note that

(3.12) $$a(u, v) = (\mathscr{A} u, v)_V, \quad \mathscr{A} \in \mathscr{L}(V; V).$$

Then (3.11) is expressed as

(3.13) $$\| \psi \|^2 \leqslant ((I - \rho \mathscr{A}) \varphi, \psi)_V.$$

However

$$\| (I - \rho \mathscr{A}) \varphi \|^2 = \| \varphi \|^2 + \rho^2 \| \mathscr{A} \varphi \|^2 - 2 \rho (\mathscr{A} \varphi, \varphi)_V \leqslant \\ \leqslant (I + \rho^2 \| \mathscr{A} \|^2 - 2 \alpha \rho) \| \varphi \|^2$$

so that (3.13) gives

(3.14) $$\| \psi \| \leqslant (1 + \rho^2 \| \mathscr{A} \|^2 - 2\alpha\rho)^{1/2} \| \varphi \| .$$

Since $\alpha > 0$ we can choose ρ such that $(1 + \rho^2 \| \mathscr{A} \|^2 - 2\alpha\rho)^{1/2} < 1$ and hence, (3.9) follows. ■

Remark 3.2. We shall now give a *variant* of the above method; see also Brézis-Sibony /1/.

Let W be a Hilbert space with

(3.15) $W \subset V$, W dense in V with continuous injection,

(3.16) $$\begin{cases} K_0 = \text{closed convex set in } W, \text{ such that} \\ K = \text{closure of } K_0 \text{ in } V \end{cases}$$

and let $b(u,v)$ be a continuous bilinear form, symmetric or otherwise, on W such that

(3.17) $$b(v, v) \geqslant \beta \| v \|_W^2 , \quad \beta > 0 , \quad \forall v \in W .$$

Thus, for all $\varepsilon > 0$, there exists a unique $u_\varepsilon \in K_0$ such that

(3.18) $$\varepsilon b(u_\varepsilon, v - u_\varepsilon) + a(u_\varepsilon, v - u_\varepsilon) \geqslant (f, v - u_\varepsilon) \quad \forall v \in K_0 .$$

It may be shown (Huet /1/) that as $\varepsilon \to 0$, $u_\varepsilon \to u$ in V, where u is the solution of (2.5).

Using (3.18) it is then possible to construct an iterative scheme:

(3.19) $$\varepsilon_n b(u^n, v - u^n) \geqslant (f, v - u^n) - a(u^{n-1}, v - u^n) \quad \forall v \in K_0$$

which may offer some advantage in the case where (3.19) is "simpler" than the initial problem.

In the case of *equations*, the following is a useful choice ([1]) (Godunov-Prokopov /1/); we restrict our attention to an example in two dimensions. We assume that $V = H_0^1(\Omega)$, and that $a(u,v)$ is as in (1.8). We then introduce:

$$W = \left[v \mid v \in H_0^1(\Omega) , \frac{\partial^2 v}{\partial x_1 \partial x_2} \in L^2(\Omega) \right] ,$$

$$b(v, w) = \int_\Omega \left[vw + \frac{\partial v}{\partial x_1} \frac{\partial w}{\partial x_1} + \frac{\partial v}{\partial x_2} \frac{\partial w}{\partial x_2} + \frac{\partial^2 v}{\partial x_1 \partial x_2} \frac{\partial^2 w}{\partial x_1 \partial x_2} \right] \mathrm{d}x ,$$

$$v, w \in W .$$

([1]) At least when Γ is the union of segments parallel to the coordinate axes.

The above theory applies. Here, the advantage lies in the fact that the differential operator B corresponding to $b(v, w)$ can be *decomposed*:

$$B = \left(I - \frac{\partial^2}{\partial x_1^2}\right)\left(I - \frac{\partial^2}{\partial x_2^2}\right).$$

In the case of equations we therefore introduce the scheme:

$$b(u^n, v) = b(u^{n-1}, v) + \rho[(f, v) - a(u^{n-1}, v)] \quad \forall v \in W, \; u^n \in W. \tag{3.20}$$

If we assume that the solution u of $a(u, v) = (f, v) \quad \forall v \in V$ is an element of W (which in the applications will come from a *regularity* theorem) then

$$b(u, v) = b(u, v) + \rho[(f, v) - a(u, v)] \quad \forall v \in W.$$

Therefore, if we put

$$\varphi^n = u^n - u,$$

we deduce from (3.20) that

$$b(\varphi^n - \varphi^{n-1}, v) = -\rho a(\varphi^{n-1}, v). \tag{3.21}$$

By taking $v = \varphi^n - \varphi^{n-1}$ and then $v = \varphi^n + \varphi^{n-1}$ in (3.21), we deduce successively (by putting:

$$b(v, v)^{1/2} = |||\, v \,|||, \quad \text{the norm on } W \text{ equivalent to } \|v\|_W,$$

and using

$$|a(\varphi, \psi)| \leqslant M \|\varphi\| \|\psi\|, \qquad \|\varphi\| \leqslant c_1 |||\,\varphi\,||| \quad)$$

that:

$$|||\,\varphi^n - \varphi^{n-1}\,||| \leqslant \rho M c_1 \|\varphi^{n-1}\|,$$

$$|||\,\varphi^n\,|||^2 - |||\,\varphi^{n-1}\,|||^2 = -2\rho a(\varphi^{n-1}, \varphi^{n-1}) - \rho a(\varphi^{n-1}, \varphi^n - \varphi^{n-1}),$$

and hence

$$\begin{aligned} |||\,\varphi^{n-1}\,|||^2 - |||\,\varphi^n\,|||^2 &\geqslant 2\alpha\rho \|\varphi^{n-1}\|^2 - M\rho \|\varphi^{n-1}\| \|\varphi^n - \varphi^{n-1}\| \\ &\geqslant 2\alpha\rho \|\varphi^{n-1}\|^2 - M\rho c_1 \|\varphi^{n-1}\| \, |||\,\varphi^n - \varphi^{n-1}\,||| \\ &\geqslant 2\alpha\rho \|\varphi^{n-1}\|^2 - (M\rho c_1)^2 \|\varphi^{n-1}\|^2 \\ &= \rho(2\alpha - \rho M^2 c_1^2) \|\varphi^{n-1}\|^2. \end{aligned}$$

Thus, if we choose ρ such that

$$0 < \rho < \frac{2\alpha}{M^2 c_1^2}, \tag{3.22}$$

we have

$$||| \varphi^{n-1} ||| \geqslant ||| \varphi^n ||| \geqslant \cdots$$

and

$$\| \varphi^{n-1} \| \to 0 .$$

Consequently: *if we choose* ρ *in accordance with* (3.22) *scheme* (3.20) *is convergent in* V *towards the solution of the problem.*

3.2 Penalisation

We now consider the case of (3.2).

The idea of penalisation consists of approximating (3.2) by *equations* in which the relation appearing in (3.2), which expresses the fact that u belongs to K, is replaced by a *penalisation* term which becomes progressively larger as the solution "moves away" from K and thus forces the limit of the approximate solutions to belong to K.

Specifically, we introduce *a penalisation operator* β which has the following properties:

$$(3.23) \quad \begin{cases} \beta \text{ maps } V \text{ into } V', \ \beta \text{ is Lipschitz continuous } (^1) \\ \operatorname{Ker}(\beta) = \text{kernel of } \beta = K \\ \beta \text{ is monotone.} \end{cases}$$

It can be shown that there *exists* a β which has the properties (3.23). Indeed, there exists *an infinite number* of penalisation operators (for a given convex set K). ■

Example 3.1. We take Example (1.7) of Section 1.1.

For $v, w \in V = H^1(\Omega)$ we put

$$(3.24) \qquad (\beta(v), w) = -\int_\Gamma (v - h)^- \, w \, \mathrm{d}\Gamma$$

where in a general manner we have defined

$$(3.25) \qquad \psi^- = \sup(-\psi, 0) .$$

It may easily be shown that (3.23) is satisfied.

We may alternatively define β by

$$(3.26) \qquad (\beta(v), w) = -\int_\Gamma M(v - h)^- \, w \, \mathrm{d}\Gamma ,$$

where M is a continuous function > 0 on Γ. ■

Penalised equation.

For $\varepsilon > 0$ we consider *the equation:*

---

(1) Globally or locally.

(3.27) $$a(u_\varepsilon, v) + \frac{1}{\varepsilon}(\beta(u_\varepsilon), v) = (f, v) \quad \forall v \in V.$$

Equation (3.27) is *nonlinear*. The existence of a solution u_ε of the problem comes from the general theory of monotone operators (see Minty /1/, Browder /1/, Leray and Lions /1/).

Equation (3.27) is the *penalised equation* associated with problem (3.2). ■

Example 3.2. Using the choice (3.24) a penalised problem associated with (1.15) is:

(3.28) $$\begin{cases} -\Delta u_\varepsilon = f, \\ \dfrac{\partial u_\varepsilon}{\partial n} - \dfrac{1}{\varepsilon}(u_\varepsilon - h)^- = 0. \end{cases}$$ ■

We now show (see Lions /1/) that:

Theorem 3.1. *As* $\varepsilon \to 0$, $u_\varepsilon \to u$ *in* V, u *being the solution of* (3.2).

The *basic principle* behind the proof is as follows:

First, it is verified that u_ε resides in a bounded subset of V (take $v = u_\varepsilon - v_0$, $v_0 \in K$ in (3.27), introduce Av_0, use the coercivity and the continuity of A and

$$(\beta(u_\varepsilon), u_\varepsilon - v_0) = (\beta(u_\varepsilon) - \beta(v_0), u_\varepsilon - v_0) \geqslant 0).$$

It may then be deduced from (3.27) that

$$(\beta(u_\varepsilon), v) = \varepsilon[(f, v) - a(u_\varepsilon, v)],$$

and, hence

(3.29) $$\| \beta(u_\varepsilon) \|_{V'} = 0(\varepsilon).$$

A subsequence, also denoted by u_ε, can then be extracted from the sequence u_ε, such that:

(3.30) $u_\varepsilon \to u$ weakly in V.

From this it may be deduced (by using the monotonicity of β) that

$$\beta(u) = 0,$$

and, hence, (using (3.23)) that $u \in K$.

By replacing v in (3.27) by $v - u_\varepsilon$, $v \in K$, it may be deduced that:

$$a(u_\varepsilon, v - u_\varepsilon) - (f, v - u_\varepsilon) = \frac{1}{\varepsilon}(\beta(v) - \beta(u_\varepsilon), v - u_\varepsilon) \geqslant 0,$$

and, hence

(3.31) $$a(u_\varepsilon, v) - (f, v - u_\varepsilon) \geqslant a(u_\varepsilon, u_\varepsilon).$$

Since $\liminf_{\varepsilon \to 0} a(u_\varepsilon, u_\varepsilon) \geq a(u, u)$, we deduce from (3.31) that

$$a(u, v) - (f, v - u) \geq a(u, u) \quad \forall v \in K, \text{ i.e. } (3.2)\,. \ \blacksquare$$

Remark 3.3. Numerous variants of the above method are possible; we shall meet some examples of these in the following chapters. ■

Remark 3.4. If we express the penalty method in the context of the "calculus of variations" - by assuming that a is symmetric and that β is the gradient of a functional Ψ (which is permissible) - problem (3.27) amounts to the minimisation of the *penalised functional:*

$$(3.32) \qquad \frac{1}{2} a(v, v) - (f, v) + \frac{1}{\varepsilon} \Psi(v) = J_\varepsilon(v)\,.$$

In its present form, this method was introduced by Courant /1/.
This idea has prompted numerous investigations in the theory of mathematical programming. We shall come back to this point in the following chapters. ■

3.3 Regularisation

We now consider a problem of the type (2.9) where the functional j is *nondifferentiable*.
It is natural to "approximate" j by a family of functionals, j_ε which are *convex* and *differentiable*. ■

Example 3.3. Within the setting of Section 1.3 we may take

$$(3.33) \qquad j_\varepsilon(v) = \int_\Gamma g \varphi_\varepsilon(v)\, \mathrm{d}\Gamma\,,$$

where the functions $\lambda \to \varphi_\varepsilon(\lambda)$ are convex differentiable approximations of $\lambda \to |\lambda|$. For example, we can take

$$(3.34) \qquad \varphi_\varepsilon(\lambda) = \frac{1}{\varepsilon + 1} |\lambda|^{\varepsilon + 1}$$

or, as is preferable in numerical applications

$$(3.35) \qquad \begin{cases} \varphi_\varepsilon(\lambda) = \sqrt{\lambda^2 + \varepsilon^2} & \text{or} \\ \varphi_\varepsilon(\lambda) \text{ convex differentiable}, & \varphi_\varepsilon(\lambda) = |\lambda| \text{ if } |\lambda| \geq \varepsilon\,. \ \blacksquare \end{cases}$$

We then "approximate" problem (2.9) by

$$(3.36) \qquad a(u_\varepsilon, v - u_\varepsilon) + j_\varepsilon(v) - j_\varepsilon(u_\varepsilon) \geq (f, v - u_\varepsilon) \quad \forall v \in V\,.$$

However, since j_ε is differentiable (3.36) is equivalent to the equation

$$(3.37) \qquad a(u_\varepsilon, v) + (j'_\varepsilon(u_\varepsilon), v) = (f, v) \quad \forall v \in V\,.$$

This nonlinear equation admits a unique solution, by virtue of the general theory of monotone operators (since the operator $v \to j_\varepsilon'(v)$ is monotone from $V \to V'$).

It may be shown that

$$u_\varepsilon \to u \text{ in } V \text{ when } \varepsilon \to 0, \text{ where } u \text{ is the solution of (2.9).} \tag{3.38}$$

This is achieved by first showing that u_ε resides in a bounded subset of V; a subsequence, also denoted by u_ε, can then be extracted, such that $u_\varepsilon \to u$ weakly in V. We deduce from (3.36), (3.37) that

$$a(u_\varepsilon, v - u_\varepsilon) + j_\varepsilon(v) - j_\varepsilon(u_\varepsilon) - (f, v - u_\varepsilon) = \\ = j_\varepsilon(v) - j_\varepsilon(u_\varepsilon) - (j_\varepsilon'(u_\varepsilon), v - u_\varepsilon) \geqslant 0 \quad (^1)$$

therefore

$$a(u_\varepsilon, v) - (f, v - u_\varepsilon) + j_\varepsilon(v) \geqslant a(u_\varepsilon, u_\varepsilon) + j_\varepsilon(u_\varepsilon)\,. \tag{3.39}$$

However (and these are the axiomatic approximation assumptions which are satisfied, for example, in (3.33)):

$$\begin{cases} j_\varepsilon(v) \to j(v)\,, \\ \liminf j_\varepsilon(u_\varepsilon) \geqslant j(u)\,, \end{cases} \tag{3.40}$$

so that, since $\liminf a(u_\varepsilon, u_\varepsilon) \geqslant a(u, u)$, we deduce from (3.39) that

$$a(u, v) - (f, v - u) + j(v) \geqslant a(u, u) + j(u)$$

i.e. (2.9). ∎

Remark 3.5. It is easy to extend (3.33) to the other situations considered. Some examples of applications are given in the following chapters. ∎

3.4 Duality (I)

The theory of duality is open to a number of approaches, which lead to problems with greatly differing analytic expressions. In this section and the two sections which follow, we shall present the three ideas which strike us as being the most important in this direction.

The methods described in this section and the next relate to *one example* but we shall endeavour to extract from this example points which are of a general nature. Other examples and, in particular, numerous variants and numerical applications will be given in the following chapters. ∎

---

(1) Using the convexity of j_ε.

The basic concept behind all duality theories is that *a lower semi-continuous, proper, convex function is the upper envelope of the affine functions which are below it.*

Assuming the conditions of Section 1.4, the application of this idea amounts to saying that the functional j given by (1.38) satisfies

$$j(v) = \sup_{p \in \mathscr{P}} \int_\Omega gp.\mathrm{grad}\, v \, dx \,, \tag{3.41}$$

where the $\sup$ over p is taken for $p \in \mathscr{P}$ with

$$\mathscr{P} = \{ p \mid p = \{ p_1, \dots, p_n \} \,,\ |p| \leqslant 1 \text{ a.e. in } \Omega \} \,. \tag{3.42}$$

The problem (1.40) is therefore equivalent to

$$\inf_{v \in H_0^1(\Omega)} \sup_{p \in \mathscr{P}} \left[\frac{1}{2} a(v, v) - (f, v) + \int_\Omega gp.\mathrm{grad}\, v \, dx \right]. \tag{3.43}$$

We put ($\mathscr{L}$ is often termed a "Lagrangian"):

$$\mathscr{L}(v, p) = \frac{1}{2} a(v, v) - (f, v) + \int_\Omega gp.\mathrm{grad}\, v \, dx \,. \tag{3.44}$$

We then have:

$$\inf_{v \in H_0^1(\Omega)} \sup_{p \in \mathscr{P}} \mathscr{L}(v, p) = J(u) \,. \tag{3.45}$$

where u is *the* solution of (1.40).

From (3.45) we deduce (since we also have sup inf $\leqslant$ inf sup) that:

$$\sup_{p \in \mathscr{P}} \inf_{v \in H_0^1(\Omega)} \mathscr{L}(v, p) \leqslant J(u) \,. \tag{3.46}$$

Using a *direct* method (1), we shall now prove

<u>Theorem 3.2</u>. *We have*

$$\sup_{p \in \mathscr{P}} \inf_{v \in H_0^1(\Omega)} \mathscr{L}(v, p) = J(u) \,. \tag{3.47}$$

The proof lies in the following result, which is itself of interest:

<u>Theorem 3.3</u>. *If u is the solution of* (1.40) *there exists* $m = \{ m_1, \dots, m_n \}$ *with*

---

(1) It is also possible to use the general results on min-max theorems.

(3.48) $|m| \leqslant 1$ a.e. *in* Ω [1]

(3.49). $m . \operatorname{grad} u = |\operatorname{grad} u|$ a.e. *in* Ω,

(3.50) $-\Delta u - g \operatorname{div} m = f$ *in* Ω,

(3.51) $u = 0$ on Γ.

Conversely, if u,m satisfy (3.48),...,(3.51) *then u is the solution of* (1.40).

Remark 3.6. Theorem 3.3 gives an interpretation of the problem in Section 1.4. ■

Proof of theorem 3.3. By successively taking $v = 0$ and $v = 2u$ in (1.41) we obtain

(3.52) $a(u, u) - (f, u) + j(u) = 0$

then, with $v = \pm v$ and taking account of (3.52), we obtain

(3.53) $|a(u, v) - (f, v)| \leqslant j(v), \quad \forall v \in H_0^1(\Omega)$.

(it will be noted that (1.41) is equivalent to (3.52), (3.53)).

However, putting

(3.54) $r = \frac{1}{g}(\Delta u + f)$,

we see that (3.53) is equivalent to

(3.55) $|(r, v)| \leqslant \int_\Omega |\operatorname{grad} v| \, dx, \quad \forall v \in H_0^1(\Omega)$.

We introduce the space

$$\Phi = \{ \varphi \mid \varphi \in (L^1(\Omega))^n \}, \qquad \|\varphi\|_\Phi = \int_\Omega |\varphi| \, dx$$

and the mapping

$$v \to \pi v = \left\{ \frac{\partial v}{\partial x_1}, \ldots, \frac{\partial v}{\partial x_n} \right\}$$

from $H_0^1(\Omega) \to \Phi$ [2]. Thus (3.55) is equivalent to

---

[1] Therefore $m \in \mathscr{P}$ (defined in (3.42)).

[2] Ω is assumed to be *bounded*.

$$(3.56) \qquad |(r, v)| \leqslant \| \pi v \|_{\Phi}$$

and therefore, from the Hahn-Banach theorem, there exists $m \in \Phi' = (L^{\infty}(\Omega))^n$, such that

$$(3.57) \qquad \| m \|_{\Phi'} \leqslant 1 ,$$

$$(3.58) \qquad (r, v) = (m, \pi v) = \sum \int_{\Omega} m_i \frac{\partial v}{\partial x_i} \, dx ,$$

and (3.52) is equivalent to

$$(3.59) \qquad (r, u) = \int_{\Omega} | \operatorname{grad} u | \, dx .$$

However (3.57) is equivalent to (3.48), (3.58) to (3.50) and (3.59) to (3.49) (since $|m| \leqslant 1$ a.e.). Since $u \in H_0^1(\Omega)$ we therefore have (3.48) to (3.51).

The inverse property may be verified by means of a direct calculation. ■

Proof of theorem 3.2. We can *explicitly* calculate

$$\inf_{v \in H_0^1(\Omega)} \mathscr{L}(v, p)$$

which is met by the solution $v = v(p)$ of

$$(3.60) \qquad \begin{cases} -\Delta v = g \operatorname{div} p + f , \\ v = 0 \quad \text{on } \Gamma . \end{cases}$$

We then have, for $v = v(p)$,

$$(3.61) \qquad \mathscr{L}(v, p) = -\tfrac{1}{2} a(v, v) ,$$

and consequently

$$(3.62) \qquad \begin{cases} \sup\limits_{p \in \mathscr{P}} \inf\limits_{v \in H_0^1(\Omega)} \mathscr{L}(v, p) = \sup\limits_{p \in \mathscr{P}} -\tfrac{1}{2} a(v(p), v(p)) , \\ v(p) \text{ solution of } (3.60) . \end{cases}$$

If we take $v = u$, $p = m$, where u, m satisfy the conditions of Theorem 3.3, we have $p \in \mathscr{P}$ and

$$\sup_{p \in \mathscr{P}} \inf_{v \in H_0^1(\Omega)} \mathscr{L}(v, p) \geqslant -\tfrac{1}{2} a(u, u) = J(u) ,$$

which, in combination with (3.46), proves (3.47). ■

Remark 3.7. From the point of view of numerical applications, we therefore have the option of solving, instead of the initial (or primal) problem, the (dual) problem:

$$(3.63) \qquad \inf_{p \in \mathscr{P}} \tfrac{1}{2} a(v(p), v(p)) \qquad \text{where } v \text{ is a solution of } (3.60).$$

In the following chapters we shall present some *approximation algorithms* for (3.63). ■

Remark 3.8. The simultaneous solution of the primal problem and a dual problem gives, as in classical theory, *bounds* for $J(u)$

$$J(v) \leqslant J(u) \leqslant \tfrac{1}{2}\, a(v(p), v(p)) \quad \forall v, \forall p \in \mathscr{P}\,. \quad ■$$

Remark 3.9. We can consider (3.63) as an *optimal control problem for a system governed by partial differential equations* (see Lions /2/ for this theory). In fact, we consider p as the *control* (or the command). The *state* of the system $v = v(p)$ is given by the solution of (3.60) and the cost function, or criterion, is given by

$$J(p) = \tfrac{1}{2}\, a(v(p), v(p))\,. \tag{3.64}$$

The *optimal control problem* then consists of seeking

$$\inf J(p)\,, \qquad p \in \mathscr{P}\,. \quad ■ \tag{3.65}$$

3.5 Duality (II)

Two other ideas, which are also fundamental in duality, are the following:

(i) It is possible to "suppress constraints" by using "Lagrange multipliers";

(ii) The problem can be decomposed by the addition of "artificial constraints" which are then suppressed using method (i).

We shall now consider this in more explicit terms, using the same example as in Section 3.4. ■

Introduction of artificial constraints.

The problem (1.40) is clearly equivalent to seeking

$$\inf \ \frac{1}{2}\int_\Omega |\, w\,|^2\, \mathrm{d}x - (f, v) + \int_\Omega g\, |\, w\,|\, \mathrm{d}x \tag{3.66}$$

subject to the constraint

$$w = \operatorname{grad} v\,, \qquad v \in H_0^1(\Omega)\,. \quad ■ \tag{3.67}$$

Suppression of constraints by using multipliers.

We use the fact that:

$$\sup_p\, (p, w - \operatorname{grad} v)\,, \qquad p \in (L^2(\Omega))^n$$

is equal to $+\infty$ *unless* (3.67) is satisfied; this is an extension of the idea (3.41).

Hence we get the natural introduction of the problem:

$$\sup_{p} \inf_{v, w} \mathcal{M}(v, w, p) \tag{3.68}$$

where

$$\mathcal{M}(v, w, p) = \frac{1}{2} \int_{\Omega} | w |^2 \, dx - (f, v) + \int_{\Omega} g \, | w | \, dx + \\ + (p, w - \operatorname{grad} v) . \quad ■ \tag{3.69}$$

Remark 3.10. In (3.68) the functions v, w are *independent*; v covers $H_0^1(\Omega)$ and w covers $(L^2(\Omega))^n$. ■

We note (and we have done what is necessary for this) that:

$$\mathcal{M}(v, \operatorname{grad} v, p) = J(v) . \tag{3.70}$$

Hence

$$\inf_{v, w} \mathcal{M}(v, w, p) \leqslant \inf_{v} J(v) = J(u)$$

and consequently

$$\sup_{p} \inf_{v, w} \mathcal{M}(v, w, p) \leqslant J(u) . \tag{3.71}$$

We shall now prove

Theorem 3.5. *We have*

$$\sup_{p} \inf_{v, w} \mathcal{M}(v, w, p) = J(u) . \tag{3.72}$$

Proof. We have

$$\inf_{v, w} \mathcal{M}(v, w, p) = \inf_{v} \mathcal{M}_1(v, p) + \inf_{w} \mathcal{M}_2(w, p) ,$$

$$\mathcal{M}_1(v, p) = - (f, v) - (p, \operatorname{grad} v) ,$$

$$\mathcal{M}_2(w, p) = \frac{1}{2} \int_{\Omega} | w |^2 \, dx + \int_{\Omega} g \, | w | \, dx + (p, w) .$$

However

$$\inf_{v} \mathcal{M}_1(v, p) = \begin{cases} 0 & \text{if } \operatorname{div} p = f , \\ -\infty & \text{otherwise.} \end{cases} \tag{3.73}$$

Then

$$\inf_{w} \mathcal{M}_2(w, p) = -\frac{1}{2} \int_{\substack{\Omega \\ | p | \geqslant g}} (| p | - g)^2 \, dx . \tag{3.74}$$

In fact, this reduces to calculating

$$\inf_{\xi \in \mathbb{R}^n} \left[\tfrac{1}{2} | \xi |^2 + g \, | \xi | + p\xi\right] .$$

We immediately see that this lower bound is *zero* if $|p| \leqslant g$. If $|p| \geqslant g$, we note that

$$\inf_{\xi} [\tfrac{1}{2} |\xi|^2 + g|\xi| + p\xi] = - \tfrac{1}{2}(|p| - g)^2 + \\ + \inf_{\xi} [\tfrac{1}{2}(|\xi| - |p| + g)^2 + p\xi + |p||\xi|] = \\ = - \tfrac{1}{2}(|p| - g)^2$$

which is met by

$$\xi = - \frac{p}{|p|}(|p| - g).$$

Consequently we have (3.74) and

$$\text{(3.75)} \qquad \sup_p \inf_{v, w} \mathscr{M}(v, w, p) = - \inf_p \begin{cases} \dfrac{1}{2} \displaystyle\int_{\substack{\Omega \\ |p| \geqslant g}} (|p| - g)^2 \, dx, \\ \operatorname{div} p = f. \ (^1) \end{cases}$$

We now take $\{u, m\}$ which satisfy the conditions of Theorem 3.3 and let:

$$\text{(3.76)} \qquad p = - \operatorname{grad} u - gm.$$

Then, using (3.50), we have $\operatorname{div} p = f$; moreover, if $\operatorname{grad} u = 0$, then $|p| = g|m| \leqslant g$; hence we need only be concerned with the region where $\operatorname{grad} u \neq 0$. Now

$$\text{(3.77)} \qquad p = - \operatorname{grad} u \left(1 + g \frac{1}{|\operatorname{grad} u|}\right)$$

therefore

$$|p| = g + |\operatorname{grad} u| \geqslant g$$

and

$$\text{(3.78)} \qquad \frac{1}{2} \int_{\substack{\Omega \\ |p| \geqslant g}} (|p| - g)^2 \, dx = \frac{1}{2} \int_{\substack{\Omega \\ |p| \geqslant g}} |\operatorname{grad} u|^2 \, dx = \frac{1}{2} \int_{\Omega} |\operatorname{grad} u|^2 \, dx$$

(since $\operatorname{grad} u = 0$ in the region excluded from the initial integrations). Hence

$$\sup_p \inf_{v, w} \mathscr{M}(v, w, p) \geqslant - \frac{1}{2} \int_{\Omega} (|p| - g)^2 \, dx$$

for the choice (3.76) of p and therefore from (3.78):

---

(1) The following should be understood: $\operatorname{div} p = f$ in Ω and the integral is extended to the region where $|p(x)| \geqslant g$.

$$\sup_p \inf_{v,\, w} \mathcal{M}(v, w, p) \geqslant -\tfrac{1}{2} a(u, u) = J(u),$$

which, in combination with (3.71), proves the Theorem. ∎

Remark 3.11. In view of the extreme importance of the present subject matter to subsequent arguments we shall give a *further example* of the above method. We consider the Stokes problem: find a vector $u = \{u_1, \dots, u_n\}$ and a scalar p such that:

(3.79) $$-\mu \Delta u = f - \operatorname{grad} p,$$

(3.80) $$\operatorname{div} u = 0,$$

(3.81) $$u = 0 \quad \text{on} \quad \Gamma.$$

We are thus dealing with a problem of *equations* in which (3.80) is to be considered as a *constraint*.

We introduce

(3.82) $$V = \{ v \mid v \in (H_0^1(\Omega))^n,\ \operatorname{div} v = 0 \ \text{ in } \ \Omega \}$$

(3.83) $$a(u, v) = \mu \sum_{i,\, j=1}^{n} \int_\Omega \frac{\partial u_i}{\partial x_j} \frac{\partial v_i}{\partial x_j} \, dx, \qquad \mu > 0,$$

$$(f, v) = \sum_{i=1}^{n} \int_\Omega f_i v_i \, dx.$$

Thus the problem (3.79), (3.80), (3.81) is *equivalent* to seeking:

(3.84) $$\inf_{v \in V} J(v), \qquad J(v) = \tfrac{1}{2} a(v, v) - (f, v)$$

or, equally, to seeking $u \in V$ with

(3.85) $$a(u, v) = (f, v) \quad \forall v \in V.$$

Taking (3.80) as a constraint, we introduce, by analogy with the above:

(3.86) $$\mathcal{N}(v, p) = \frac{1}{2} a(v, v) - (f, v) - \int_\Omega p (\operatorname{div} v) \, dx$$

where

(3.87) $$v \in W = (H_0^1(\Omega))^n, \qquad p \in L^2(\Omega).$$

It may now be verified directly that

(3.88) $$\sup_p \inf_v \mathcal{N}(v, p) = J(u).$$

We calculate $\inf_v \mathcal{N}(v, p), v \in W$. For a given p, the lower bound is attained for $v = v(p)$, the solution of

$$(3.89) \qquad \begin{cases} -\mu \Delta v = f - \operatorname{grad} p, \\ v = 0 \text{ on } \Gamma, \end{cases}$$

and thus

$$\inf_v \mathcal{N}(v, p) = -\tfrac{1}{2} a(v, v).$$

We then have:

$$(3.90) \qquad -\inf_p \tfrac{1}{2} a(v, v) = J(u)$$

where $v = v(p)$ is the solution of (3.89) (1).

Algorithms for the approximation of (3.90) are given later in this volume. ■

Remark 3.12. The above methods are clearly not confined to *2nd order* operators. In Chapter 4 we shall meet some examples of their application for certain 4th order operators which appear in the theory of thin plates. ■

Remark 3.13. As we have just seen, the ideas of duality can lead to a formulation which - in appearance - differs widely from the initial formulation. Moreover, more generally, starting with a problem in the calculus of variations

$$\inf_v J(v)$$

for which the lower bound is not necessarily reached (*e.g. the theory of minimum surfaces*) we can introduce duality problems

$$\operatorname{Sup}_p \Phi(p)$$

such that

(i) $\sup_p \Phi(p) = \inf_v J(v)$;

(ii) there exists p which gives the sup of Φ.

For minimum surfaces, see Témam /1/. We thus have *a relaxation process* (2).

---

(1) We again have a "distributed-system optimal control" *type* problem; see Remark 3.9.

(2) Here, the term "relaxation" is to be taken in a completely different sense from that of the classical *relaxation methods* of numerical analysis which, incidentally, we shall also use later.

Also, see Remark 3.17 later.

Various relaxation processes, particularly for *non* convex functionals, are encountered in *optimal control* theory: see Ekeland /1/ and Bidaut /1/.

3.6 Duality (III)

Another possibility is offered by the use of *conjugate functions*; this classical idea, due originally to Legendre, is developed in Fenchel /1/, Mandelbrojt /1/, Hörmander /1/, Moreau /1/, and Rockafellar /2/.

As we shall now see, this *does not assume* that the form $a(u,v)$ is symmetric, as was the case in the two previous sections. ■

Over a Hilbert space V we say that a *convex* function $v \to \Phi(v)$ is *proper* if:

(i) Φ is lower semi-continuous, with values in $]-\infty, +\infty]$;
(ii) Φ is not identical to $+\infty$. ■

Example 3.4. If K is a closed convex subset of V, $K \neq \emptyset$, the function given by

$$\text{(3.91)} \qquad \begin{cases} \Phi_K(v) = 0 & \text{if } v \in K, \\ = +\infty & \text{if } v \notin K, \end{cases}$$

is a proper convex function. ■

Using this concept, we see that the inequality (3.2) is *equivalent* to seeking u *in* V such that

$$\text{(3.92)} \qquad a(u, v-u) - (f, v-u) + \Phi_K(v) - \Phi_K(u) \geqslant 0, \quad \forall v \in V.$$

We also note that the function $\Phi = j$ introduced in Section 2.1 is a proper convex function, so that *all* the inequalities mentioned so far come within the following general framework:

$$\text{(3.93)} \qquad \begin{cases} \text{find } u \in V \text{ with} \\ a(u, v-u) - (f, v-u) + \Phi(v) - \Phi(u) \geqslant 0 \quad \forall v \in V. \end{cases}$$

In order to simplify notation in the following, we introduce the operator A where:

$$\text{(3.94)} \qquad \begin{cases} A \in \mathscr{L}(V; V'), \\ a(u, v) = (Au, v). \end{cases}$$

Thus (3.93) is equivalent to

$$\text{(3.95)} \qquad (Au - f, v-u) + \Phi(v) - \Phi(u) \geqslant 0 \quad \forall v \in V. \ ■$$

We shall now transform (3.95), using *sub-differentials*.

If the function Φ is differentiable, then, from the convexity:

$$(3.96)\qquad \Phi(v) - \Phi(u) - (\Phi'(u), v - u) \geqslant 0 \quad \forall v \in V .$$

We now introduce, in a general manner, *the set of* $\xi \in V'$ such that

$$(3.97)\qquad \Phi(v) - \Phi(u) - (\xi, v - u) \geqslant 0 \quad \forall v \in V ,$$

a concept which is valid whether Φ is differentiable at *u or not.*

The set of ξ such that (3.97) is satisfied is denoted by $\partial\Phi(u)$ and called the *sub-differential of* Φ *at u* (see Moreau /1/).

We often write

$$(3.98)\qquad \Phi(v) - \Phi(u) - (\partial\Phi(u), v - u) \geqslant 0 \quad \forall v \in V$$

- the correct expression being (3.97) with $\xi \in \partial\Phi(u)$.

Clearly, if Φ is differentiable at u:

$$(3.99)\qquad \partial\Phi(u) = \{ \Phi'(u) \} .$$

Using this concept, we see that (3.95) *is equivalent to*

$$(3.100)\qquad -(Au - f) \in \partial\Phi(u) . \quad ■$$

Remark 3.14. We observe that (3.100) may be viewed as an equation related to a *multivalued operator* (sometimes called a *set-valued operator*). ■

We shall now transform (3.100) using the concept of a *conjugate convex function.*

If Φ is a proper convex function, we define *the conjugate function* Φ^* on the space V' by

$$(3.101)\qquad \Phi^*(v^*) = \sup_v \left[(v^*, v) - \Phi(v)\right] .$$

It may be shown (see Moreau /1/) that Φ^* is a proper convex function. ■

Example 3.5. We take $V = H_0^1(\Omega)$, K given by

$$(3.102)\qquad K = \{ v \mid v \in V,\ v \geqslant 0 \text{ a.e. in } \Omega \} ,$$

$$\Phi = \Phi_K \quad \text{given by (3.91)} .$$

We identify $L^2(\Omega)$ with its dual; thus $V \subset L^2(\Omega) \subset V' = H^{-1}(\Omega)$, where $H^{-1}(\Omega)$ is the space of distributions of the form

$$f = f_0 + \sum_{i=1}^{n} \frac{\partial f_i}{\partial x_i}, \qquad f_0, f_1, \ldots, f_n \in L^2(\Omega) .$$

Now the conjugate function of Φ_K is given by

$$\Phi^*(v^*) = \sup (v^*, v) , \qquad v \in K$$

and hence

(3.103) $$\Phi^*(v^*) = \begin{cases} 0 & \text{if } v^* \leqslant 0 \quad (^1) \\ +\infty & \text{otherwise.} \end{cases}$$

If we introduce

(3.104) $$K^* = \{ v^* \mid v^* \in V', v^* \leqslant 0 \} = \{ \text{cone of measures } \leqslant 0 \text{ in } H^{-1}(\Omega) \},$$

we then have, using the notation (3.91):

(3.105) $$\Phi^* = \Phi_{K^*}. \quad \blacksquare$$

The following equivalence can be verified immediately:

(3.106) $$u^* \in \partial\Phi(u) \Leftrightarrow u \in \partial\Phi^*(u^*).$$

This equivalence forms the basis of the duality in the present sense; here we follow the method described by Mosco /1/ (2). From (3.100) and (3.106) we deduce that

(3.107) $$u \in \partial\Phi^*(-(Au - f)).$$

We put:

(3.108) $$A^{-1} f = F,$$

(3.109) $$u^* = -Au + f.$$

Thus (3.107) is equivalent to

$$-A^{-1}(u^* - f) = -(A^{-1} u^* - F) \in \partial\Phi^*(u^*)$$

or

(3.110) $$(A^{-1} u^* - F, v^* - u^*) + \Phi^*(v^*) - \Phi^*(u^*) \geqslant 0, \quad \forall v^* \in V'.$$

Inequality (3.110) is *one* "*dual*" inequality of (3.95). ■

Example 3.6. (see Fusciardi, Mosco, Scarpini and Schiaffino /1/). We consider the context of Example 3.5, with

(3.111) $$a(u, v) = \int_\Omega \operatorname{grad} u . \operatorname{grad} v \, dx.$$

Suppose ψ to be given in $H_0^1(\Omega)$, and

(3.112) $$f = -\Delta\psi.$$

---

(1) $v^* \leqslant 0$ in the sense of distributions in Ω, which is equivalent to saying that v^* is a measure $\leqslant 0$. (Schwartz /1/).

(2) In which *nonlinear* operators $\mathcal{A}$ are also considered.

The inequality:

$$\begin{cases} u \in K, \\ a(u, v - u) \geqslant (f, v - u) \quad \forall v \in K, \end{cases}$$

therefore amounts to:

$$a(u - \psi, v - u) \geqslant 0 \quad \forall v \in K, \tag{3.113}$$

and hence is equivalent to finding *the projection* of ψ, *in the sense of the inner product* (3.111), *on the cone of* $H_0^1(\Omega)$ *of functions* $\geqslant 0$.

We put:

(3.114) $A^{-1} = G =$ Green's operator in Ω for the Dirichlet problem.

Thus, from (3.105) and by noting that $F = \psi$, problem (3.110) is equivalent to

$$\begin{cases} u^* \leqslant 0 \text{ on } \Omega, \\ (Gu^* - \psi, v^* - u^*) \geqslant 0 \quad \forall v^* \leqslant 0 \text{ on } \Omega. \end{cases} \tag{3.115}$$

Examples of numerical applications of (3.115) are given in Fusciardi, Mosco, Scarpini and Schiaffino /1/. ■

Remark 3.15. In actual applications, a difficulty in the implementation of the above method rests in the *explicit calculation* of the conjugate function Φ^* of Φ. ■

Remark 3.16. A slightly different approach is given in Cea /1/, which is also valid for *nonsymmetric* $a(u,v)$. See also the article by Cea /2/. ■

Remark 3.17. The function Φ is given on V. We have introduced the conjugate function *on* V'. However, identifying V with V', we can *also* introduce *the dual function on* V. Moreover, this concept of a dual function *depends on the inner product chosen*.

We *assume* that $a(u,v)$ is *symmetric* and take $a(u,v)$ as a new inner product on V; we then put

$$a(u, v) = [u, v]. \tag{3.116}$$

We introduce the "Green's operator" $G \in \mathscr{L}(V'; V)$ by means of:

$$(f, v) = [Gf, v] \quad \forall v \in V. \tag{3.117}$$

Thus (3.93) *is equivalent to*

$$[u - Gf, v - u] + \Phi(v) - \Phi(u) \geqslant 0 \quad \forall v \in V. \tag{3.118}$$

The conjugate function of Φ on V which must now be introduced is:

$$(3.119) \qquad \hat{\Phi}^*(v) = \sup_{\varphi \in V} ([v, \varphi] - \Phi(\varphi)) .$$

Then (3.118) is equivalent in turn to

$$- (u - Gf) \in \partial\Phi(u)$$
$$\Leftrightarrow u \in \partial\hat{\Phi}^*(- (u - Gf)) .$$

We put:

$$(3.120) \qquad w = - u + Gf .$$

Then (3.118) is equivalent to

$$- w + Gf \in \partial\hat{\Phi}^*(w)$$
$$\Leftrightarrow [w - Gf, v - w] + \hat{\Phi}^*(v) - \hat{\Phi}^*(w) \geqslant 0$$
$$(3.121) \qquad \Leftrightarrow a(w, v - w) - (f, v - w) + \hat{\Phi}^*(v) - \hat{\Phi}^*(w) \geqslant 0$$

this problem being in turn equivalent to the minimisation of the functional:

$$(3.122) \qquad \tfrac{1}{2} a(v, v) - (f, v) + \hat{\Phi}^*(v) , \qquad v \in V .$$

We thus have the very "symmetric" result: *the problem of the minimisation of the functional*

$$(3.123) \qquad \tfrac{1}{2} a(v, v) - (f, v) + \Phi(v) , \qquad v \in V$$

is "in duality" with (3.122); *if* u (resp. w) *denotes the solutions of* (3.123) (resp. (3.122)), *then we have* (3.120), i.e.

$$u + w = Gf .$$

(see also, for example, Lions /2/, Chapter 3, Section 12). ■

We now present *an application of this result*, taken from Brézis /2/, (Theorem 1.5, Chapter 1).

We take

$$V = H_0^1(\Omega) , \qquad a(u, v) = \sum_{i=1}^{n} \int_\Omega \frac{\partial u}{\partial x_i} \frac{\partial v}{\partial x_i} \, dx ,$$

$$K = \{ v \mid v \in V, \ | v(x) | \leqslant 1 \ \text{a.e.} \}$$

$$\Phi(v) = 0 \text{ if } v \in K , \quad + \infty \text{ if } v \notin K .$$

We take f *in* $L^2(\Omega)$ (and not in $H^{-1}(\Omega) = V'$). It may then be shown (Brézis, *loc. cit.*) that the solution u of (3.123) satisfies ([1]) (*regularity theorem*):

([1]) Assuming the boundary Γ of Ω to be sufficiently regular.

(3.124) $u \in H^2(\Omega) \cap H_0^1(\Omega)$.

However, if $f \in L^2(\Omega)$ we have: $Gf \in H^2(\Omega) \cap H_0^1(\Omega)$ and consequently the solution w of the dual problem (in the above sense) satisfies

(3.125) $w = Gf - u \in H^2(\Omega) \cap H_0^1(\Omega)$.

Hence, *the lower bound of the functional* (3.122) *is reached on* $H^2(\Omega) \cap H_0^1(\Omega)$, and therefore

(3.126)
$$\inf_{v \in H_0^1(\Omega)} [\tfrac{1}{2} a(v, v) - (f, v) + \hat{\Phi}^*(v)] = \inf_{v \in H^2(\Omega) \cap H_0^1(\Omega)} [\tfrac{1}{2} a(v, v) - (f, v) + \hat{\Phi}^*(v)].$$

The advantage of the second formulation in (3.126) is that $\hat{\Phi}^*(v)$ *immediately becomes explicit* on $H^2(\Omega) \cap H_0^1(\Omega)$, whereas the same is not true on $H_0^1(\Omega)$. In fact, if

$v \in H^2(\Omega) \cap H_0^1(\Omega)$,

then (3.119) gives

$$\hat{\Phi}^*(v) = \sup_{|\varphi| \leqslant 1} \int_\Omega (-\Delta v)\, \varphi \,\mathrm{d}x = \int_\Omega |\Delta v| \,\mathrm{d}x.$$

The dual problem is thus *to find the lower bound over* $H^2(\Omega) \cap H_0^1(\Omega)$ of

(3.127)
$$\frac{1}{2} a(v, v) - (f, v) + \int_\Omega |\Delta v| \,\mathrm{d}x = \mathcal{J}(v).$$

(The *noncoercive* nature of the functional (3.127) on $H^2(\Omega) \cap H_0^1(\Omega)$ may be noted; if we introduce the space X of $v \in H_0^1(\Omega)$, such that $\Delta v \in L^1(\Omega)$, then $\mathcal{J}(v)$ is well defined on $L^1(\Omega)$, but the space X *is not reflexive*; a *direct* solution of (3.127) is not easy). ■

3.7 Various remarks

Remark 3.18. Let us consider an inequality of the type (3.2) K_0 is assumed to be another closed convex subset of V and u_0 the solution in K_0 of

(3.128) $a(u_0, v - u_0) \geqslant (f, v - u_0) \quad \forall v \in K_0$.

Thus, it is clearly possible that *for certain* f we may have

(3.129) $u = u_0$.

In Chapter 3 we shall present an example (due to Brézis

and Sibony /2/) related to the situation in Section 1.2, K being given by (1.19), and where

$$(3.130) \qquad K_0 = \{ v \mid v \in H_0^1(\Omega),\ |v(x)| \leqslant \text{ distance from } x \text{ to } \Gamma \}.$$

It is now clear that the calculation of u will be *much more simple* using (3.128), (3.130) than by using the initial formulation. ■

Remark 3.19. In the spirit of Remark 3.18, we can *check the calculations* by using, in particular, certain "*comparison properties*" relating the solution u of (3.2) to solutions of more simple problems. On this subject see Haugazeau /1/, /2/. ■

Remark 3.20. Another method consists of approximating the steady-state problems by time-dependent inequality problems. We shall return to this point in Chapter 6. ■

Remark 3.21. For applications of duality theory to approximation theory see Laurent /1/. ■

3.8 Synopsis

In the previous sections we have given a number of *equivalent formulations* of variational inequalities, leading to *infinite-dimensional* approximation algorithms, i.e. the approximate solutions are themselves supplied by the solution of a problem supposedly "more simple" than the initial problem, but with infinite dimensions.

It is clear that the practical development of these algorithms presupposes that they may be reduced to *finite-dimensional* problems. Therefore, in the remainder of this chapter, we shall *provide the numerical tools which make this reduction to finite dimensions possible.*

For this there are two general types of method: *interior* approximations (Section 4) and *exterior* approximations (Section 5).

4. INTERIOR APPROXIMATIONS

4.1 Interior approximations of V. Finite elements

A family V_h of *finite-dimensional* vector spaces where h is a parameter $\in \mathbb{R}^m$ (or in the neighbourhood of the origin in $\mathbb{R}^m$) is said to be an *interior approximation* to V when the following conditions are satisfied:

(4.1) $\forall h, \quad V_h \subset V$ (1)

(4.2) $\left\{\begin{array}{l}\text{the } V_h \text{ "approximate" } V \text{ in the following sense: we can} \\ \text{find a space } \mathscr{V} \text{ which is dense in } V \text{ (2) such that}, \forall v \in \mathscr{V}, \\ \text{we can construct } v_h \in V_h \text{ with} \\ v_h \to v \text{ in } V \text{ as } h \to 0 \text{ in } \mathbb{R}^m . \ \blacksquare\end{array}\right.$

Example 4.1. (Galerkin method). Let $w_1, w_2, \ldots, w_j, \ldots$ be a "basis" of V in the following sense:

(4.3) $\forall j$, the $w_1, \ldots, w_j$ are linearly independent;

(4.4) the combinations $\sum_{\text{finite}} \lambda_j w_j$, $\lambda_j \in \mathbb{R}$, are dense in V (3).

Then putting

$h = 1/\mu$, where μ is an integer,

we define

(4.5) $V_h = [w_1, \ldots, w_\mu]$ = the space generated by $w_1, \ldots, w_\mu$.

It can be shown that (4.1), (4.2) are satisfied (with $\mathscr{V} = V$). ■

Example 4.2. (Finite elements). Let Ω be an open domain of $\mathbb{R}^2$, $V = H_0^1(\Omega)$.

We consider a function $h \to \varphi(h)$, which is real and continuous, such that

(4.6) $\varphi(h) \to 0$ if $h \to 0$.

We consider a triangulation $\mathscr{T}_h$ consisting of a *finite* family of triangles T which have the following properties:

(4.7) $T \in \mathscr{T}_h$, measure of $T \leqslant \varphi(h)$
and $\forall h$, the angles of the triangles are bounded below by a strictly positive number, independent of h;

(4.8) $T \in \mathscr{T}_h \Rightarrow T \subset \Omega$;

(4.9) $T, T' \in \mathscr{T}_h \Rightarrow T \cap T' = \varnothing$ or T and T' have a common side or T and T' have a common vertex;

---

(1) It is this condition which explains the terminology "*interior*"; this condition is not satisfied in the case of "*exterior*" approximations (see Section 5).

(2) The introduction of $\mathscr{V}$ is not absolutely essential. However, in order to *satisfy the hypotheses in practice*, it is clearly advantageous to have as weak as possible a definition of convergence.

(3) Such bases always exist if V is separable.

$$(4.10)\quad \begin{cases} \text{if } \Omega_h = \cup T,\ T \in \mathcal{T}_h,\ \Omega_h \to \Omega \text{ in the sense:} \\ \text{for all compact } E \subset \Omega \text{ we have: } \Omega_h \supset E \text{ for "sufficiently small" } h. \end{cases}$$

It is clear that such triangulations exist, for any φ satisfying (4.6).

We associate the space V_h with $\mathcal{T}_h$ in the following manner: let M be an "internal" vertex of $T \in \mathcal{T}_h$ (see Figure 1) with the adjacent vertices $P_1, \ldots, P_k$; we define

$$(4.11)\quad \begin{cases} w_M = \text{an affine function over each triangle with vertex } M, \text{ such that } w_M(M) = 1,\quad w_M(P_i) = 0,\quad i = 1, \ldots, k, \\ \text{and where } w_M \text{ is zero outside the triangles with vertex } M. \end{cases}$$

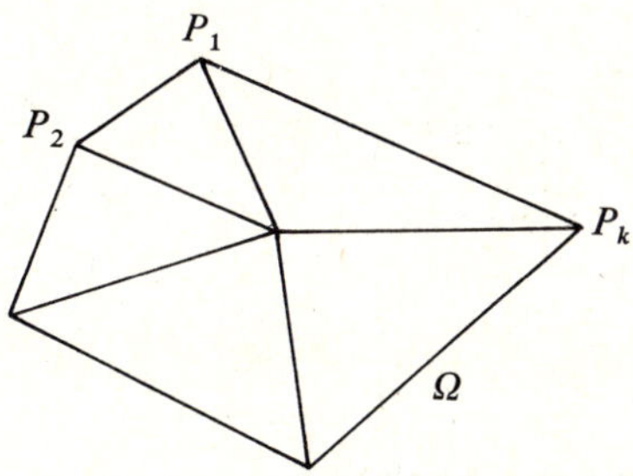

Fig. 1.

Now:

$$(4.12)\quad V_h = \text{the space generated by the } w_M.$$

Clearly (4.1) is satisfied. In order to satisfy (4.2) we take $v \in \mathcal{D}(\Omega)$ and define

$$(4.13)\quad v_h = \sum_M v(M)\, w_M .$$

The above construction is the most elementary of the *finite element* methods - a class of methods in which considerable developments have been made (use of basis functions of degree greater than 1, discretisation of the structure into rectangles, tetrahedra, etc...).

We shall come back to this later. We refer for the moment only to Ciarlet and Wagschal /1/, Fraeijs de Veubeke /1/, Zienkiewicz /1/, Zlamal /1/, /2/, Ph. Ciarlet and P.A. Raviart /1/ and the bibliographies therein. ■

4.2 Interior approximation schemes in the case of equations

The interior approximation schemes for the variational *equation*

(4.14) $$a(u, v) = (f, v) \quad \forall v \in V .$$

can be deduced immediately from the concepts of Section 4.1. Indeed, let us now introduce the problem: find $u_h \in V_h$ such that

(4.15) $$a(u_h, v_h) = (f, v_h) \quad \forall v_h \in V_h .$$

By assuming, as usual, that

(4.16) $$a(v, v) \geqslant \alpha \| v \|^2 , \quad \alpha > 0 , \quad \forall v \in V ,$$

we immediately get:

Theorem 4.1. *Using the hypotheses* (4.1), (4.2) *we have*:

(4.17) $$u_h \to u \qquad \text{in } V \text{ as } h \to 0.$$

Proof. It may be deduced from (4.15) that

$$\alpha \| u_h \|^2 \leqslant a(u_h, u_h) = (f, u_h) \leqslant \| f \|_{V'} \| u_h \|$$

and hence

(4.18) $$\| u_h \| \leqslant \frac{1}{\alpha} \| f \|_{V'} .$$

We can extract from u_h a sequence, also denoted by u_h, such that

(4.19) $$u_h \to w \qquad \text{weakly in } V \text{ as } h \to 0.$$

For $v \in \mathscr{V}$, we take v_h satisfying (4.2); since $v_h \to v$ *strongly* in V, for this choice of v_h we have:

$$a(u_h, v_h) \to a(w, v)$$

and hence, in the limit, (4.15) gives

$$a(w, v) = (f, v) \quad \forall v \in \mathscr{V} ;$$

since $\mathscr{V}$ is dense in V we deduce that $w = u$ and hence that $u_h \to u$ weakly in V without having to extract a subsequence.

It now remains to show that $u_h \to u$ *strongly* in V. For this we consider

$$X_h = a(u_h - u, u_h - u) .$$

From (4.15), $X_h = (f, u_h) - a(u, u_h - u) - a(u_h, u) \to (f, u) - a(u, u) = 0$, and therefore, $\| u_h - u \| \to 0$. ■

The problem is now *to extend these considerations to variational inequalities.*

4.3 Approximations of K

We consider, for each h, a set K_h with

$$K_h = \text{closed convex subset of } V_h, \tag{4.20}$$

$$\left\{\begin{array}{l} K_h \text{ "approximates" } K \text{ in the following sense:} \\ \text{(i) } \forall v \in K, \text{ we can find } v_h \in K_h \text{ with } v_h \to v \text{ in } V; \\ \text{(ii) If } u_h \in K_h,\ u_h \to u \text{ weakly in } V, \text{ then } u \in K. \ \blacksquare \end{array}\right. \tag{4.21}$$

Example 4.3. We consider the situation as in Example 4.2 and let

$$K = \{ v \mid v \in V,\ v \geqslant 0 \quad \text{a.e. in } \Omega \}. \tag{4.22}$$

Thus if we take

$$K_h = \{ v \mid v \in V_h,\ v \geqslant 0 \quad \text{a.e. in } \Omega \}, \tag{4.23}$$

where V_h is defined by (4.12), then (4.21) is satisfied. ■

Remark 4.1. We can weaken (4.21) by introducing $K \cap \mathscr{V}$ and assuming $K \cap \mathscr{V}$ to be dense in K. ■

Remark 4.2. In practical applications, *the construction of* K_h is *one* of the difficulties presented by inequality problems. We shall give numerous examples in the following chapters. (We shall also use *exterior* approximations, see Section 5 below). ■

Remark 4.3. A difficulty in terminology should be pointed out. The approximation V_h of V is *interior* since $V_h \subset V$; the approximation K_h of K has an interior character since $K_h \subset V$ but *it is not assumed that* $K_h \subset K$. ■

Remark 4.4. For a general study of "the approximation of convex sets", we refer the reader to the works of Mosco /2/, Joly /1/. See also Aubin /1/, Trémolières /4/.

4.4. Approximation schemes for the initial problem

We now consider *the inequality problems*

$$\left\{\begin{array}{l} u \in K, \\ a(u, v-u) \geqslant (f, v-u) \quad \forall v \in K, \end{array}\right. \tag{4.24}$$

$$a(u, v-u) + j(v) - j(u) \geqslant (f, v-u) \quad \forall v \in V. \tag{4.25}$$

The approximation schemes using the above concepts are:

$$(4.26)\qquad \begin{cases} u_h \in K_h\,, \\ a(u_h, v_h - u_h) \geqslant (f, v_h - u_h)\ \forall v_h \in K_h\,, \end{cases}$$

and

$$(4.27)\qquad a(u_h, v_h - u_h) + j(v_h) - j(u_h) \geqslant (f, v_h - u_h) \quad \forall v_h \in V_h\,,\quad u_h \in V_h\,,$$

respectively.

We have

Theorem 4.2. *We assume* (4.1), (4.2) *to be satisfied* (*and for problem* (4.26), *the conditions* (4.20), (4.21)). *Then if* u_h *is the solution of* (4.26) (resp. (4.27)) *we have*

$$(4.28)\qquad u_h \to u \quad \text{in } V,$$

where u is the solution of (4.24) (resp. (4.25)).

Proof. For fixed v in K, we take v_h satisfying (4.21) (i); with this choice of v_h in (4.26) we get:

$$(4.29)\qquad a(u_h, u_h) \leqslant a(u_h, v_h) - (f, v_h - u_h)$$

and hence

$$\alpha \| u_h \|^2 \leqslant c \| u_h \| \| v_h \| + \| f \|_{V'} (\| v_h \| + \| u_h \|)\,;$$

since $\| v_h \| \leqslant c$ we therefore have (with c denoting *various* constants)

$$\alpha \| u_h \|^2 \leqslant c(\| u_h \| + 1)$$

and hence

$$(4.30)\qquad \| u_h \| \leqslant c\,.$$

From u_h we can then extract a subsequence such that

$$u_h \to w \quad \text{weakly in } V.$$

From (4.21) (ii), $w \in K$ and from (4.29) we deduce:

$$a(w, w) \leqslant \liminf a(u_h, u_h) \leqslant a(w, v) - (f, v - w)\,,$$

therefore $w = u$, the solution of (4.24).

To demonstrate the *strong* convergence of u_h towards u we consider, as in the proof of Theorem 4.1,

$$X_h = a(u_h - u, u_h - u)\,.$$

From (4.26) we have:

$$X_h \leqslant a(u_h, v_h) - (f, v_h - u_h) - a(u_h, u) - a(u, u_h - u)$$

and by choosing v_h satisfying (4.21) (i) we deduce

$$\limsup X_h \leqslant a(u, v) - (f, v - u) - a(u, u) \; ; \tag{4.31}$$

in (4.31) v is fixed *arbitrarily* in K. Taking $v = u$ we thus have $\limsup X_h = 0$ and hence strong convergence. ■

The reasoning is the same in the case of (4.25), (4.27).

Remark 4.5. The problem has thus now been reduced to the *effective solution* of (4.26) or (4.27). The *algorithms* for this solution are studied in the following chapters. ■

Remark 4.6. The schemes (4.26), (4.27) *do not use* the concepts introduced in Section 3. *We are now going to see how these concepts can be used for the approximation*: in a general manner, which will later be defined more precisely, we shall no longer approximate the "initial" problem (4.24) or (4.25), but *one of the "transformed" problems* obtained in Section 3. ■

4.5. Approximation schemes for the penalised or regularised problems.

Penalised equation.

We shall now approximate (3.27) using the "interior" method. Thus:

1) We choose $\varepsilon = \varepsilon(h)$, $\varepsilon(h) \to 0$ if $h \to 0$; [1]

2) We define $u_h \in V_h$ as *the* solution of

$$a(u_h, v_h) + \frac{1}{\varepsilon(h)}(\beta(u_h), v_h) = (f, v_h) \quad \forall v_h \in V_h \, . \tag{4.32}$$

We then have:

Theorem 4.3. *As* $h \to 0$, *then*:

$$u_h \to u \quad \textit{in } V, \tag{4.33}$$

where u *is the solution of* (4.24).

The proof, which is based on principles presented earlier, will not be given here. ■

---

(1) The *practical* choice of $\varepsilon(h)$ poses certain difficulties.

Remark 4.7. We can consider the scheme (4.32) from a different viewpoint: *first* we consider scheme (4.26) *and then we penalise problem* (4.26). The two procedures can lead to certain technical differences. ■

Regularised equation.

We consider a family $j_{\varepsilon(h)}$ of regularised functionals (cf. (3.33), for example). We then discretise (3.37), which gives the scheme:

$$(4.34)\qquad a(u_h, v_h) + (j'_{\varepsilon(h)}(u_h), v_h) = (f, v_h) \quad \forall v_h \in V_h .$$

We again obtain the result (4.33). ■

Remark 4.8. This remark is analogous to 4.7; *first*, we can discretise as in (4.27) and then regularise scheme (4.27). ■

Remark 4.9. Algorithms for the effective solution of (4.32) or (4.34) will be given in the following chapters. ■

4.6 Approximation schemes for the dual problem

Formulation (I).

Here we use Section 3.4.
The initial problem has been transformed (see (3.63)) to:

$$(4.35)\qquad \inf_{p \in \mathscr{P}} \tfrac{1}{2} a(v(p), v(p)) \qquad (^1)$$

where $v = v(p)$ is the solution of

$$(4.36)\qquad -\Delta v = g \operatorname{div} p + f, \qquad v = 0 \quad \text{on} \quad \Gamma .$$

Discretisation of problem (4.35), (4.36).

As a first stage, we consider V_h = "interior approximation" of $V = H_0^1(\Omega)$ and we define, for $p \in \mathscr{P}$, the approximate solution $v_h \in V_h$ by

$$(4.37)\qquad a(v_h, \varphi_h) = (g \operatorname{div} p + f, \varphi_h) \quad \forall \varphi_h \in V_h .$$

We then consider

$$(4.38)\qquad \inf \tfrac{1}{2} a(v_h, v_h) .$$

A *second stage* consists of "discretising" $\mathscr{P}$. For this we

---

(1) $\mathscr{P}$ is defined in (3.42).

introduce a family $\mathscr{P}_\xi$ of subsets of $\mathscr{P}$, "approximating" $\mathscr{P}$ (1) and we consider $v_{h,\xi}$ to be the solution in V_h of

$$(4.39) \qquad a(v_{h,\xi}, \varphi_h) = (g \operatorname{div} p_\xi + f, \varphi_h) \quad \forall \varphi_h \in V_h$$

which gives the problem (2)

$$(4.40) \qquad \inf \tfrac{1}{2} a(v_{h,\xi}) .$$

We then have

$$(4.41) \qquad \inf_{p_\xi} \tfrac{1}{2} a(v_{h,\xi}) \to \inf_{p} \tfrac{1}{2} a(v)$$

as $h, \xi \to 0$. ■

Remark 4.10. *Algorithms* for the effective solution of problems of the type (4.40) are given in the following chapters. ■

Formulation (II).

We now consider the setting of Section 3.5.
The initial problem has been transformed (see (3.75)) to:

$$(4.42) \qquad \inf \frac{1}{2} \int_{\substack{\Omega \\ |p| \geq g}} (|p| - g)^2 \, dx ,$$

where

$$(4.43) \qquad \operatorname{div} p = f .$$

The method then consists of discretising (4.43), the *exterior* methods (see the next section) being simpler here. We then substitute the "discretised p" into (4.42), i.e. p_h.
As already pointed out in Remark 4.10, it now only remains to provide an algorithm for the calculation of

$$(4.44) \qquad \inf \frac{1}{2} \int_{\substack{\Omega \\ |p_h| \geq g}} (|p_h| - g)^2 \, dx .$$

Remark 4.11. For a numerical application of the concepts of Section 3.6, see Fusciardi, Mosco, Scarpini and Schiaffino /1/. ■

---

(1) In the examples we generally take $\xi = h$.

(2) We use the notation $a(v) = a(v, v)$.

4.7 Open problems. Estimates of the approximation error

In the above, a significant problem, and one which is still largely *open*, is that of *obtaining estimates of the error*. It would be interesting to see to what extent the error estimates obtained for equations (see Aubin /2/, Bramble and Schatz /1/, Strang and Fix /1/, Ciarlet and Raviart /1/, /2/, etc...) apply to *inequalities*.

It appears that there are very few general results in this area, although we should point out the work of Falk /1/, /2/ and Mosco-Strang /1/ relating to certain particular problems; error estimates for finite element approximations will also be found in the present book, in particular in Chapters 3, 4 and 5; these estimates are probably not optimal, however.

In order to conclude, for the time being, the discussion of interior approximations, we shall consider an example for which we can easily provide an error estimate which appears to be *of optimal order*.

We again consider Example 3.5 of Section 3.6, assuming Ω to be a bounded open domain of $\mathbb{R}^2$, with a sufficiently regular boundary Γ; we then have

$$V = H_0^1(\Omega) \tag{4.45}$$

$$K = \{ v \mid v \in V,\ v \geqslant 0 \text{ a.e. in } \Omega \} \tag{4.46}$$

and in order to simplify things slightly we shall restrict our attention to the case where the bilinear form a is defined by

$$a(u, v) = \int_\Omega \operatorname{grad} u . \operatorname{grad} v \, \mathrm{d}x . \tag{4.47}$$

Let $f \in L^2(\Omega)$; the variational inequality

$$\begin{cases} a(u, v - u) \geqslant \displaystyle\int_\Omega f(v - u)\, \mathrm{d}x & \forall v \in K \\ u \in K , \end{cases} \tag{4.48}$$

then admits *one and only one solution* and we recall (see Brézis-Stampacchia /2/) that

$$u \in H_0^1(\Omega) \cap H^2(\Omega) \tag{4.49}$$

with

$$\| u \|_{H^2(\Omega)} \leqslant C \| f \|_{L^2(\Omega)} , \tag{4.50}$$

where, in (4.50), C is independent of f and u. ∎

We shall now approximate the solution of (4.48) by means of a *finite element* method; the notation is as in Section 4.1, Example 4.2. In order to simplify the argument we shall assume that Ω is a *convex polygon*, so that relations (4.49), (4.50) still hold, and we say that $\mathcal{T}_h$ must satisfy,

$$\overline{\bigcup_{T \in \mathcal{T}_h} T} = \bar{\Omega}. \tag{4.51}$$

as well as (4.7)-(4.10).

We introduce

$$\Sigma_h = \{ M \mid M \in \bar{\Omega}, \quad M \text{ is a vertex of } T \in \mathcal{T}_h \} \tag{4.52}$$

and we define the "approximation" V_h of $H_0^1(\Omega)$ by

$$V_h = \{ v_h \mid v_h \in H_0^1(\Omega) \cap C^0(\bar{\Omega}),\ v_h|_T \in P_1 \ \forall T \in \mathcal{T}_h \} \tag{4.53}$$

where $v_h|_T$ denotes the restriction of v_h to T and P_1 denotes the space of polynomials in two variables of degree $\leqslant 1$. We then define K_h, an approximation of K, by

$$K_h = V_h \cap K = \{ v_h \mid v_h \in V_h, \quad v_h \geqslant 0 \text{ on } \bar{\Omega} \}\ ; \tag{4.54}$$

K_h is a *closed convex subset* of V_h and it is immediate, since $v_h|_T$ belongs to P_1, that

$$K_h = \{ v_h \mid v_h \in V_h,\ v_h(M) \geqslant 0 \ \forall M \in \Sigma_h \}\,. \tag{4.55}$$

We then define the *approximate problem* by

$$\begin{cases} a(u_h, v_h - u_h) \geqslant \displaystyle\int_\Omega f(v_h - u_h)\,\mathrm{d}x \quad \forall v_h \in K_h \\ u_h \in K_h \end{cases} \tag{4.56}$$

and (4.56) admits *one and only one solution*.

In connection with the approximation error $\| u_h - u \|_{H_0^1(\Omega)}$, we shall now prove

Proposition 4.1. *We assume that the angles of* $\mathcal{T}_h$ *are bounded below, uniformly in* h, *by* $\theta_0 > 0$; *we then have*

$$\| u_h - u \| \leqslant \| f \|_{L^2(\Omega)} \sqrt{h} \tag{4.57}$$

where C *is independent of* h, u, f *and where we recall that*

$$h = \max_{T \in \mathcal{T}_h} (\text{Area of } T).$$

Proof. In the following, C will denote various constants; we have

$$a(u, u_h - u) \geqslant \int_\Omega f(u_h - u)\,\mathrm{d}x$$

$$a(u_h, v_h - u_h) \geqslant \int_\Omega f(v_h - u_h)\,\mathrm{d}x \quad \forall v \in K_h\,,$$

and hence by addition

$$\begin{cases} \| u_h - u \|^2_{H^1(\Omega)} = a(u_h - u, u_h - u) \leqslant a(u_h, v_h - u) - \\ - \int_\Omega f(v_h - u) \, dx \quad \forall v_h \in K_h , \end{cases}$$

or

(4.58) $$\| u_h - u \|^2_{H^1(\Omega)} \leqslant a(u_h - u, v_h - u) + a(u, v_h - u) - \\ - \int_\Omega f(v_h - u) \, dx \quad \forall v_h \in K_h .$$

From (4.58) and the inequality

$$2 \, a(v, w) \leqslant \| v \|^2_{H^1_0(\Omega)} + \| w \|^2_{H^1_0(\Omega)} \quad \forall v, w \in H^1_0(\Omega) ,$$

we deduce

(4.59) $$\frac{1}{2} \| u_h - u \|^2_{H^1_0(\Omega)} \leqslant \frac{1}{2} \| v_h - u \|^2_{H^1_0(\Omega)} + a(u, v_h - u) - \\ - \int_\Omega f(v_h - u) \, dx \quad \forall v_h \in K_h .$$

We have $u \in H^1_0(\Omega) \cap H^2(\Omega)$ and hence $- \Delta u - f \in L^2(\Omega)$ and if we put

(4.60) $$\lambda = - \Delta u - f ,$$

we get, using (4.50),

(4.60') $$\begin{cases} \| \lambda \|_{L^2(\Omega)} \leqslant C \| f \|_{L^2(\Omega)} \\ C \text{ independent of } f . \end{cases}$$

From (4.60) we deduce that

(4.61) $$a(u, v) = \int_\Omega (f + \lambda) \, v \, dx \quad \forall v \in H^1_0(\Omega) ,$$

and from (4.59), (4.61)

$$\frac{1}{2} \| u_h - u \|^2_{H^1_0(\Omega)} \leqslant \frac{1}{2} \| v_h - u \|^2_{H^1_0(\Omega)} + \int_\Omega \lambda(v_h - u) \, dx \\ \forall v_h \in K_h ,$$

and hence from (4.60')

(4.62) $$\tfrac{1}{2} \| u_h - u \|^2_{H^1(\Omega)} \leqslant \tfrac{1}{2} \| v_h - u \|^2_{H^1(\Omega)} + C \| f \|_{L^2(\Omega)} \| v_h - u \|_{L^2(\Omega)} \\ \forall v_h \in K_h .$$

To estimate $u_h - u$ we shall use (4.62), choosing a suitable v_h; first we define the *interpolation* operator $\Pi_h : H^1_0(\Omega) \cap C^0(\Omega) \to V_h$ by

$$(4.63)\qquad \begin{cases} \Pi_h v \in V_h & \forall v \in H_0^1(\Omega) \cap C^0(\bar{\Omega}) \\ \Pi_h v(M) = v(M) & \forall M \in \Sigma_h , \end{cases}$$

and, from (4.53), (4.55) we have

$$(4.64)\qquad \Pi_h v \in K_h \quad \forall v \in K \cap C^0(\bar{\Omega}) .$$

In (4.62) we shall take $v_h = \Pi_h u$; this is meaningful owing to the fact that, as Ω is a *two-dimensional* domain with a *Lipschitz* boundary, we have $H^2(\Omega) \subset C^0(\bar{\Omega})$ and, from (4.64) $\Pi_h u \in K_h \quad \forall h$.

We recall that $h = \max_{T \in \mathcal{T}_h}$ (Area of T); thus from, for example, Fix-Strang /1/, Ciarlet-Raviart /1/, we have the result that under the specified condition on the angles of $\mathcal{T}_h$ we have

$$(4.65)\qquad \| \Pi_h u - u \|_{L^2(\Omega)} \leqslant C \| u \|_{H^2(\Omega)} h ,$$

$$(4.66)\qquad \| \Pi_h u - u \|_{H_0^1(\Omega)} \leqslant C \| u \|_{H^2(\Omega)} \sqrt{h}$$

with C independent of h,u.

The estimate (4.57) then follows trivially from (4.50), (4.62) with $v_h = \Pi_h u$, (4.65) and (4.66). ∎

Remark 4.12. The results and the proof of Proposition 4.1 are easily extended to the case where Ω is a convex polyhedron of $\mathbb{R}^3$ since in this case the inclusion $H^2(\Omega) \subset C^0(\bar{\Omega})$ still holds. ∎

Remark 4.13. In Chapter 2, Section 5 we shall study a problem similar to problem (4.48), using a finite difference method. ∎

5. EXTERIOR APPROXIMATIONS

5.1 An example

In applications not only are interior methods used, but also *exterior* methods.

Before going on to the "general axioms", which are somewhat involved, we shall point out the basic ideas using an example which has been chosen to be as simple as possible. ∎

We consider *the Dirichlet problem*:

$$(5.1)\qquad -\Delta u + u = f, \quad f \text{ given in } L^2(\Omega),$$

$$(5.2)\qquad u = 0 \quad \text{on } \Gamma .$$

We introduce

$$(5.3)\qquad V = H_0^1(\Omega)$$

and

(5.4) $$a(u, v) = \int_\Omega (uv + \text{grad } u . \text{grad } v) \, dx .$$

The problem (5.1), (5.2) is then equivalent to

(5.5) $$a(u, v) = (f, v) , \quad \forall v \in V .$$

Our aim is to approximate u using *characteristic functions*, i.e. functions which *do not belong to* $H^1(\Omega)$; the use of these characteristic functions corresponds to *finite difference* methods. More specifically: let $h \in \mathbb{R}^n$; we associate with it the grid R_h :

(5.6) $$R_h = \{ M \mid M \in \mathbb{R}^n, \ M = \{ m_1 h_1, \ldots, m_n h_n \}, \ m_i \in \mathbb{Z} \} .$$

With each node M of R_h, we associate *the panel* with centre M:

(5.7) $$\varpi_h^0(M) = \prod_{i=1}^n \left]\left(m_i - \frac{1}{2}\right) h_i, \left(m_i + \frac{1}{2}\right) h_i\right[$$

and *the cross* (with centre M):

(5.7') $$\varpi_h^1(M) = \bigcup_{i=1}^n \varpi_h^0\left(M \pm \frac{h_i}{2} e_i\right) \qquad (1)$$

where e_i denotes the i th unit vector in $\mathbb{R}^n$.

We then define

(5.8) $$\Omega_h = \{ M \mid \varpi_h^1(M) \subset \Omega \} ,$$

(5.9) $$\theta_h^M = \text{ characteristic function of } \varpi_h^0(M) ,$$

and

(5.10) $$V_h = \text{ space generated by } \theta_h^M , \quad M \in \Omega_h .$$

The functions θ_h^M *are not* in $H^1(\Omega)$, so that V_h *is not* a subspace of $H^1(\Omega)$. Hence, we cannot define $a(u,v)$ on V_h and we must therefore *modify* $a(u,v)$. For this we replace the derivatives by differential quotients. In general, we put

(5.11) $$\delta_i \varphi(x) = \frac{1}{h_i}\left[\varphi\left(x + \frac{h_i}{2} e_i\right) - \varphi\left(x - \frac{h_i}{2} e_i\right)\right] .$$

Then for $u_h, v_h \in V_h$ *we define*

---

(1) The indices "0" and "1" which appear in this notation correspond to the different *orders* of the approximation; we can introduce "crosses" of arbitrarily high order. See Aubin /2/.

$$a_h(u_h, v_h) = \int_\Omega \left(u_h v_h + \sum_{i=1}^n \delta_i u_h \delta_i v_h \right) dx . \tag{5.12}$$

The *approximate problem* is thus: find $u_h \in V_h$ such that

$$a_h(u_h, v_h) = (f, v_h) \quad \forall v_h \in V_h . \tag{5.13}$$

Since we have

$$a_h(v_h, v_h) = \sum_{i=1}^n \| \delta_i v_h \|^2_{L^2(\Omega)} + \| v_h \|^2_{L^2(\Omega)} , \tag{5.14}$$

it can be seen immediately that (5.13) admits a unique solution. The fundamental result is then

Theorem 5.1. *As $h \to 0$ we have:*

$$\left\{ \begin{array}{l} \{ u_h, \delta_1 u_h, \dots, \delta_n u_h \} \to \left\{ u, \dfrac{\partial u}{\partial x_1}, \dots, \dfrac{\partial u}{\partial x_n} \right\} \quad \textit{weakly in} \\ (L^2(\Omega))^{n+1} \qquad\qquad (^1). \end{array} \right. \tag{5.15}$$

Proof. We put

$$\| v_h \|_{V_h} = \left(\| v_h \|^2_{L^2(\Omega)} + \sum_{i=1}^n \| \delta_i v_h \|^2_{L^2(\Omega)} \right)^{1/2} . \tag{5.16}$$

Then, from (5.13), (5.14) we deduce that

$$\| u_h \|_{V_h} \leqslant c . \tag{5.17}$$

We can thus extract a subsequence, also denoted by u_h, such that

$$u_h \to w , \quad \delta_i u_h \to w_i \quad \text{weakly in } L^2(\Omega) . \tag{5.18}$$

We shall now verify the two following properties:

$$w_i = \frac{\partial w}{\partial x_i} , \tag{5.19}$$

$$w \in H^1_0(\Omega) . \tag{5.20}$$

In fact, for $\varphi \in \mathcal{D}(\Omega)$ and for sufficiently small $|h|$ we have:

$$(\delta_i u_h, \varphi) = - (u_h, \delta_i \varphi)$$

---

(1) We can in fact demonstrate strong convergence. However, (5.15) is adequate for our purposes.

and proceeding to the limit: $(w_i, \varphi) = -\left(w, \frac{\partial\varphi}{\partial x_i}\right)$ (since $\delta_i\varphi \to \frac{\partial\varphi}{\partial x_i}$ *strongly* in $L^2(\Omega)$), and hence (5.19).

In order to verify (5.20), we consider $\tilde{u}_h$ = extension of u_h to $\mathbb{R}^n$ by 0 outside Ω; since (5.8) is true, then:

$$(\widetilde{\delta_i u_h}) = \delta_i(\tilde{u}_h)$$

and $\tilde{u}_h \to \tilde{w}$, $\delta_i(\tilde{u}_h) \to \tilde{w}_i$ weakly in $L^2(\Omega)$.

However, as for (5.19), it may be verified that $\tilde{w}_i = \frac{\partial}{\partial x_i}(\tilde{w})$ and hence

$$\frac{\partial(\tilde{w})}{\partial x_i} \in L^2(\mathbb{R}^n) \quad \forall i ,$$

therefore $\tilde{w} = 0$ on Γ, which gives (5.20).

It now remains to show that $w = u$.

For this we take $v \in V$ and *define*

$$(5.21) \qquad r_h v = \sum_{M \in \Omega_h} \frac{1}{h_1, \ldots, h_n} (v, \theta_h^M)\, \theta_h^M .$$

It may be verified that:

$$(5.22) \qquad r_h v \to v , \qquad \delta_i r_h v \to \frac{\partial v}{\partial x_i} \qquad \text{strongly in } L^2(\Omega) , \quad \forall i .$$

Then, taking $v_h = r_h v$, we have $a_h(u_h, v_h) \to a(w, v)$ and hence (5.13) gives

$$a(w, v) = (f, v) \quad \forall v \in V ,$$

and therefore $w = u$, which proves the theorem. ■

We shall now "axiomatise" this proof [1], by also introducing "approximations" of K and j.

5.2 Exterior approximations for V, a, K, j.

Approximation of V.

We consider a Hilbert space F and an operator σ with

$$(5.23) \qquad \sigma \in \mathscr{L}(V; F) , \qquad \sigma \text{ injective from } V \text{ into } F.$$

For example, in the context of Section 5.1, $V = H_0^1(\Omega)$, $F = (L^2(\Omega))^{n+1}$ and σ is defined by

$$(5.24) \qquad \sigma v = \left\{ v, \frac{\partial v}{\partial x_1}, \ldots, \frac{\partial v}{\partial x_n} \right\} . \ ■$$

---

[1] In order to avoid repetition in the applications.

We next consider a family of finite-dimensional spaces V_h, equipped with (Hilbert) norms $\| v_h \|_{V_h}$. We assume *prolongation operators* $p_h \in \mathcal{L}(V_h; F)$ to be given, with

$$\| p_h \|_{\mathcal{L}(V_h;F)} \leqslant c . \tag{5.25}$$

For example, in the context of Section 5.1, V_h is defined by (5.10) and $\| v_h \|_{V_h}$ is defined by (5.16); p_h is defined by

$$p_h v_h = \{ v_h, \delta_1 v_h, \ldots, \delta_n v_h \} ; \tag{5.26}$$

we have:

$$\| p_h \|_{\mathcal{L}(V_h;F)} \leqslant 1 . \quad \blacksquare$$

We say that V_h constitutes an *exterior* approximation [1] of V if

$$v_h \in V_h , \qquad p_h v_h \text{ converges weakly towards } \xi \text{ in } F \Rightarrow \xi \in \sigma V \tag{5.27$_1$}$$

and if

$$\begin{cases} \forall v \in V , \ \exists v_h \in V_h \text{ with } p_h v_h \to \sigma V \text{ strongly in } F \text{ and} \\ \| v_h \|_{V_h} \leqslant C . \end{cases} \tag{5.27$_2$}$$

In the context of Section 5.1, we shall take $v_h = r_h v$ defined by (5.21); (5.22) then corresponds to $(5.27)_2$. ■

Approximation of a.

We now consider a bilinear form $a_h(u_h, v_h)$ on V_h [2] and we assume that

$$| a_h(u_h, v_h) | \leqslant c \| u_h \|_{V_h} \| v_h \|_{V_h} \tag{5.28}$$

$$a_h(v_h, v_h) \geqslant \alpha \| v_h \|^2_{V_h} , \quad \alpha > 0 , \quad \forall v_h \in V_h . \tag{5.29}$$

We say that the a_h *constitute an approximation of* a *if:*

$$\begin{cases} \text{as } p_h v_h \to \sigma v \text{ weakly in } F, \ p_h w_h \to \sigma w \text{ strongly in } F, \text{ we} \\ \text{have } \ a_h(v_h, w_h) \to a(v, w) , \quad a_h^*(v_h, w_h) \to a^*(v, w) \quad [3] \end{cases} \tag{5.30}$$

---

(1) Terminology justified by the fact that $V_h \not\subset V$.

(2) In fact, the whole of this may be extended to certain *nonlinear* problems.

(3) Where, in general, $a^*(\varphi, \psi) = a(\psi, \varphi)$, $a_h^*(\varphi, \psi) = a_h(\psi, \varphi)$. This assumption becomes redundant in the case where a is symmetric and where we then choose a_h to be symmetric (which is possible, if not essential).

and also if

(5.31) $$\begin{cases} \text{if} \quad p_h v_h \to \sigma v \quad \textit{weakly} \text{ in } F, \text{ then} \\ \liminf a_h(v_h, v_h) \geqslant a(v, v) . \end{cases}$$ ■

In the context of Section 5.1, a_h is defined by (5.12) and we have (5.30) and (5.31). ■

Approximation of K.

We take a family of sets K_h where

(5.32) $$K_h \text{ is a closed convex subset of } V_h .$$

We say that the K_h *constitute an approximation of* K if:

(5.33) $$v_h \in K_h , \quad p_h v_h \text{ converges weakly towards } \xi \text{ in } F \Rightarrow \xi \in \sigma K ,$$

and if

(5.34) $$\begin{cases} \text{for all } v \in K, \text{ we can find } v_h \in K_h \text{ such that} \\ p_h v_h \to \sigma v \quad \text{strongly in } F \text{ and } \| v_h \|_{V_h} \leqslant C . \end{cases}$$

Remark 5.1. If $K = V$, $K_h = V_h$, we recover $(5.27)_1$, $(5.27)_2$. ■

Remark 5.2. The *construction* of K_h *and the satisfaction of* (5.33), (5.34) form *one* of the main difficulties in applications. We shall give some examples later; we shall then see that the use of *exterior* approximations is often more convenient than that of *interior* approximations. ■

Approximation of j.

Let j be a proper convex function (see Section 3.6) on V.
We consider j_h , a proper convex function on V_h.
We say that the j_h *constitute an approximation of* j if:

(5.35) $$v_h \in V_h , \quad p_h v_h \to \sigma v \text{ weakly in } F \Rightarrow \liminf j_h(v_h) \geqslant j(v) ,$$

(5.36) $$v_h \in V_h , \quad v \in V , \quad p_h v_h \to \sigma v \text{ strongly in } F \Rightarrow j_h(v_h) \to j(v) .$$

Some *examples* will be given in Chapters 4 and 5. ■

Approximation of the right-hand side.

Let $f \in V'$. We consider the continuous linear form f_h on V_h and we say that f_h *constitutes an approximation of* f if

(5.37) $$| f_h(v_h) | \leqslant C \| v_h \|_{V_h} ,$$

(5.38) $$p_h v_h \to \sigma v \quad \text{weakly in } F \Rightarrow f_h(v_h) \to (f, v) .$$ ■

5.3 Exterior approximation schemes

Approximation of problem (4.24).

We seek $u_h \in K_h$ with

$$a_h(u_h, v_h - u_h) \geqslant f_h(v_h - u_h) \quad \forall v_h \in K_h . \quad \blacksquare \tag{5.39}$$

Approximation of problem (4.25).

We seek $u_h \in V_h$ with

$$a_h(u_h, v_h - u_h) + j_h(v_h) - j_h(u_h) \geqslant f_h(v_h - u_h) \quad \forall v_h \in V_h . \quad \blacksquare \tag{5.40}$$

We now have

Theorem 5.2. *Assume that the hypotheses* (5.23) *to* (5.34) *hold. Let* u_h *(resp.* u*) be the solutions of* (5.39) *or* (5.40), *(resp.* of (4.24) *or* (4.25)). *Then*

$$p_h u_h \to \sigma u \quad \text{strongly in } F. \tag{5.41}$$

Remark 5.3. Result (5.41) again gives (5.15). ■

Proof of Theorem 5.2. We shall perform the proof for the case of (5.39). The same reasoning applies to the case (5.40).

We take $v \in K$ and $v_h \in K_h$ which satisfy (5.34). Then, with this choice of v_h in (5.39) we have:

$$a_h(u_h, u_h) \leqslant a_h(u_h, v_h) - f_h(v_h - u_h) \tag{5.42}$$

and hence using (5.28), (5.29), (5.34) and (5.37) [1]

$$\alpha \| u_h \|_h^2 \leqslant C(1 + \| u_h \|_h)$$

and therefore

$$\| u_h \|_h \leqslant C .$$

From (5.25) we then have:

$$\| p_h u_h \|_F \leqslant C . \tag{5.44}$$

Hence, we can extract a subsequence, also denoted by u_h, such that

$$p_h u_h \to \xi \quad \text{weakly in } F$$

---

[1] From now on we put $\| v_h \|_{V_h} = \| v_h \|_h$.

which with (5.33) shows that $\xi = \sigma w, w \in K$ and hence:

(5.45) $p_h u_h \to \sigma w$ weakly in F, $w \in K$.

However, from (5.31), (5.45) gives

(5.46) $\liminf a_h(u_h, u_h) \geqslant a(w, w)$.

Moreover, since $p_h v_h \to \sigma v$ strongly in F and since we have (5.45) it results from (5.30) that

(5.47) $a_h(u_h, v_h) \to a(w, v)$,

and from (5.38) that

(5.48) $f_h(v_h - u_h) \to (f, v - w)$.

Using (5.46), (5.47), (5.48) in (5.42) we deduce that

$$a(w, w) \leqslant a(w, v) - (f, v - w) \quad \forall v \in K,$$

hence $w = u$ which shows that $p_h u_h \to \sigma u$.

In order to prove (5.41) we introduce $\tilde{u}_h \in K_h$ with

$$p_h \tilde{u}_h \to \sigma u \quad \text{strongly in } F$$

($\tilde{u}_h$ exists, from (5.34)) and we put

$$X_h = a_h(u_h - \tilde{u}_h, u_h - \tilde{u}_h).$$

We have:

$$X_h = a_h(u_h, u_h) - a_h(\tilde{u}_h, u_h) - a_h(u_h, \tilde{u}_h) + a_h(\tilde{u}_h, \tilde{u}_h).$$

From (5.39) we get

$$X_h \leqslant a_h(u_h, v_h) - f_h(v_h - u_h) - a_h(\tilde{u}_h, u_h) - a_h(u_h, \tilde{u}_h) + a_h(\tilde{u}_h, \tilde{u}_h).$$

We then use (5.30) (1) and assume that v_h is chosen as in (5.34). We then deduce that

(5.49) $\limsup X_h \leqslant a(u, v) - (f, v - u) - a(u, u) \quad \forall v \in K$.

Taking $v = u$ in (5.49) we deduce that

$$X_h \to 0$$

and hence, from (5.29), that

$$\| u_h - \tilde{u}_h \|_{V_h} \to 0$$

---

(1) The 2nd assumption in (5.30) is only involved in the case of *strong* convergence.

and hence, from (5.25),

$$p_h u_h - p_h \tilde{u}_h \to 0 \text{ strongly in } F \text{ which gives (5.41).} \quad \blacksquare$$

Remark 5.4. In applications it is sufficient that (5.34) be satisfied on *a dense subset of K*, i.e. that the following property be satisfied:

$$(5.50) \quad \begin{cases} \text{there exists } \mathscr{K} \subset K, \mathscr{K} \text{ dense in } K, \text{ such that } \forall v \in \mathscr{K} \text{ we} \\ \text{can find } v_h \in K_h \text{ with } p_h v_h \to \sigma v \text{ strongly in } F \text{ and } \| v_h \|_h \leqslant C . \end{cases}$$

In fact (5.34) results from (5.50); that is, $v \in K$ and $\varphi_\varepsilon \in \mathscr{K}$ with

$$\| \varphi_\varepsilon - v \| \leqslant \varepsilon , \quad \varepsilon \to 0 .$$

From (5.50) there exists h_ε and $\varphi_{h_\varepsilon} \in K$ with:

$$\| p_{h_\varepsilon} \varphi_{h_\varepsilon} - \sigma \varphi_\varepsilon \|_F \leqslant \varepsilon ;$$

thus:

$$\| p_{h_\varepsilon} \varphi_{h_\varepsilon} - \sigma V \|_F \leqslant \| p_{h_\varepsilon} \varphi_{h_\varepsilon} - \sigma \varphi_\varepsilon \|_F + \| \sigma \varphi_\varepsilon - \sigma V \|_F \leqslant C\varepsilon$$

giving the required result. ■

Remark 5.5. The systematic theory of exterior approximation was introduced by Cea /3/, the special case of approximations using "crosses" being introduced by J.L. Lions /3/. An investigation in the case of inequalities is made in Trémolières /4/.

Remark 5.6. Naturally, as for "interior" methods, it still remains to give *algorithms* for the solution of (5.39), (5.40). These algorithms, developed in the following chapters, will clearly be *the same* for the interior and exterior schemes.

Remark 5.7. We can now go back, *in the "exterior" case*, to the considerations of Sections 4.6, 4.7. It is not necessary to expand on this. Some examples will be given in the following chapters. ■

6. CONCLUSIONS

The numerical approximation of *steady-state* variational inequalities may be described as follows:

Stage 1.

Choice of the formulation: the *main* possibilities which we have mentioned are:

- formulation in the form of a problem in the calculus of variations;
- variational inequality formulations;
- penalised variational inequalities;
- regularised variational inequalities;
- dualised variational inequalities

though there are also *various possibilities* within each of these types of method.

Stage 2.

Choice of the discretisation. We have mentioned *interior* and *exterior* methods. Clearly a great number of variations is possible within *each* of these methods: the choice of finite element the choice of prolongation operators of various orders of approximation, etc...

Stage 3.

Choice of the algorithm. Having completed stages 1 and 2 (ar so far we have been developing the tools which enable us to arriv at this point), we are faced with a nonlinear programming problem or with a problem in *finite-dimensional* nonlinear equations. It is now necessary *to develop solution algorithms* and then *to choos* between the various possibilities. This will form the main objec ive of the next four chapters, first, for general cases and then for some concrete examples and numerical applications.

Chapter 2

OPTIMISATION ALGORITHMS

SYNOPSIS

In this chapter, we shall develop various optimisation algorithms - *concentrating especially on those suited to problems involving the inequalities (or functionals) encountered in Chapter 1.*

We shall illustrate the theory by using a number of very simple examples (or counter-examples). The fundamental examples will be developed in Chapters 3 to 5 (*for the steady-state case*).

1. THE RELAXATION METHOD

1.1 Description of the point relaxation method. Unconstrained case.

We consider a *continuous convex* functional $v \to J(v)$ from $\mathbb{R}^N \to \mathbb{R}$ and we shall first examine the so-called "*unconstrained*" case in which we seek

$$\inf J(v)\,, \qquad v \in \mathbb{R}^N\,. \tag{1.1}$$

We shall assume that

$$J(v) \to +\infty \quad \text{if} \quad \|v\| \to \infty\,, \tag{1.2}$$

$$J \text{ is } \textit{strictly convex}, \tag{1.3}$$

so that there *exists a unique u such that*

$$J(u) = \inf_v J(v)\,. \tag{1.4}$$

Starting from $u^0 = \{u_1^0, \ldots, u_N^0\}$ and assuming u^n to be known, we calculate u_i^{n+1} successively for $i = 1, \ldots, N$.

We define u_i^{n+1} as *the* solution of

$$\begin{aligned} &J(u_1^{n+1}, \ldots, u_{i-1}^{n+1}, u_i^{n+1}, u_{i+1}^n, \ldots, u_N^n) \leqslant \\ &\leqslant J(u_1^{n+1}, \ldots, u_{i-1}^{n+1}, v_i, u_{i+1}^n, \ldots, u_N^n)\,, \quad \forall v_i\,, \end{aligned} \tag{1.5}$$

for $i = 1, 2, \ldots, N$. ■

Before going on to *more general algorithms* (constrained

problems, block relaxation, etc...), we shall first study the problem of the (possible) *convergence* of algorithm (1.5).

Let us begin with a *counter-example*:

Counter-Example 1.1. The *point relaxation* method (1.5) may not be convergent to the solution of the problem. Suppose for example we take, with $N = 2$, the functional:

$$J(v_1, v_2) = v_1^2 + v_2^2 - 2(v_1 + v_2) + 2\,|\,v_1 - v_2\,| \tag{1.6}$$

which satisfies (1.2), (1.3) and which is *nondifferentiable* (due to the term $2\,|\,v_1 - v_2\,|$). We have

$$u = \{\,1, 1\,\}\,.$$

Let us now apply algorithm (1.5), starting from

$$u^0 = \{\,0, 0\,\}\,.$$

We thus define u_1^1 by the minimisation w.r.t. v_1 of

$$J(v_1, 0) = v_1^2 - 2\,v_1 + 2\,|\,v_1\,|$$

giving $u_1^1 = 0$. Next we define u_2^1 by the minimisation w.r.t. v_2 of $J(u_1^1, v_2) = J(0, v_2)$ giving $u_2^1 = 0$ and so on:

$$u^n = \{\,0, 0\,\} \quad \forall n$$

thus u^n does not converge to u. ∎

Remark 1.1. In Sections 4.3 and 4.4 we shall present various methods (in particular, the duality method) which allow us to minimise functionals of the type (1.6).

Furthermore, in Section 1.5 below, it will be seen that algorithm (1.5) is *convergent* for *certain nondifferentiable* functionals.

From the point of view of Counter-Example 1.1, it is reasonable to assume initially that the functional J is *differentiable*; we have:

Theorem 1.1. *We assume that the function J satisfies* (1.2), (1.3) *and is of class* C^1. *The point relaxation method* (1.5) *then converges to the solution u of problem* (1.4).

The proof lies in a number of lemmas:

Lemma 1.1. *If J is of class C^1 and strictly convex, then, for any $M > 0$, there exists a strictly increasing continuous function $\lambda \to \delta_M(\lambda)$ from $[0, 2\,M] \to \mathbb{R}^+$ with $\delta_M(0) = 0$, such that*

$$(J'(v) - J'(u), v - u) \geqslant \delta_M(\|\,v - u\,\|) \quad \forall u, v \tag{1.7}$$

$$\text{with } \|\,u\,\| \leqslant M\,, \quad \|\,v\,\| \leqslant M\,.$$

Proof. We denote by B_M the ball $B_M = \{\,v \mid \|\,v\,\| \leqslant M\,\}$ and for $\tau \in [0, 2\,M]$ we put

$$(1.8) \qquad \delta_M^0(\tau) = \inf_{\substack{\|v-u\|=\tau,\\ u,v\in B_M}} (J'(v) - J'(u), v - u)\,.$$

we have

$$\delta_M^0(0) = 0$$

and since J is *strictly* convex, J' is strictly monotone, and hence

$$\delta_M^0(\tau) > 0 \quad \text{if} \quad \tau > 0\,.$$

Since $B_M \times B_M$ is compact in $\mathbb{R}^N \times \mathbb{R}^N$ and since J' is of class C^0, for a given τ there exists a pair $u_0, v_0 \in B_M \times B_M$ such that

$$\| v_0 - u_0 \| = \tau\,, \quad (J'(v_0) - J'(u_0), v_0 - u_0) = \delta_M^0(\tau)\,.$$

We shall now show that the function δ_M^0 is strictly increasing. Consider τ_1, τ_2 with

$$0 < \tau_1 < \tau_2 \leqslant 2M\,.$$

There exists u_2, v_2 such that

$$\| v_2 - u_2 \| = \tau_2\,, \qquad \delta_M^0(\tau_2) = (J'(v_2) - J'(u_2), v_2 - u_2)\,.$$

However, since J' is strictly monotone,

$$(J'(v_2) - J'(u_2 + t(v_2 - u_2)), v_2 - u_2) > 0 \qquad \text{if} \qquad 0 \leqslant t < 1$$

therefore

$$(1.9) \qquad (J'(v_2) - J'(u_2), v_2 - u_2) > (J'(u_2 + t(v_2 - u_2)) - J'(u_2), v_2 - u_2)\,, \quad 0 \leqslant t < 1\,.$$

However if we introduce w by

$$w - u_2 = \frac{\tau_1}{\tau_2}(v_2 - u_2)\,,$$

we have

$$\| w - u_2 \| = \tau_1$$

and if, in (1.9), we take $t = \tau_1/\tau_2$ we get

$$\begin{aligned}\delta_M^0(\tau_2) &> \frac{\tau_2}{\tau_1}(J'(w) - J'(u_2), w - u_2) > \\ &> (J'(w) - J'(u_2), w - u_2) \geqslant \\ &\geqslant \inf_{\substack{\|v-u\|=\tau_1\\ u,v\in B_M}} (J'(v) - J'(u), v - u) = \delta_M^0(\tau_1)\,,\end{aligned}$$

which proves that δ_M^0 is strictly increasing.

From the definition of δ_M^0 in (1.8) we have

$$(1.10) \qquad (J'(v) - J'(u), v - u) \geqslant \delta_M^0(\| v - u \|)\,.$$

Applying (1.10) to the pair $\{u + t(v - u), u\}$, $t \in]0, 1]$, we get

(1.11) $$(J'(u + t(v - u)) - J'(u), v - u) \geqslant \frac{1}{t}\delta_M^0(t \| v - u \|),$$

and hence, with $\tau = t \| v - u \|$,

$$0 \leqslant \frac{1}{\tau}\delta_M^0(\tau) \leqslant \frac{1}{\| v - u \|}\left(J'\left(u + \tau \frac{v - u}{\| v - u \|}\right) - J'(u), v - u\right),$$

and hence

(1.12) $$\frac{1}{\tau}\delta_M^0(\tau) \to 0 \quad \text{if} \quad \tau \to 0.$$

Integrating (1.11) we deduce

(1.13) $$J(v) - J(u) \geqslant (J'(u), v - u) + \int_0^1 \delta_M^0(t \| v - u \|)\frac{dt}{t}.$$

By interchanging u and v and adding, we get

(1.14) $$(J'(v) - J'(u), v - u) \geqslant 2\int_0^1 \delta_M^0(t \| v - u \|)\frac{dt}{t} = \\ = 2\int_0^{\|v-u\|} \delta_M^0(\theta)\frac{d\theta}{\theta},$$

so that the function δ_M defined by

(1.15) $$\delta_M(\tau) = 2\int_0^\tau \delta_M^0(\theta)\frac{d\theta}{\theta}$$

meets the required properties. ■

Lemma 1.2. *Using the assumptions and notation of Lemma* 1.1, *we have:*

(1.16) $$J(v) - J(u) \geqslant (J'(u), v - u) + \tfrac{1}{2}\delta_M(\tfrac{1}{2}\| v - u \|), \\ \| u \|, \| v \| \leqslant M.$$

Proof. In (1.11) and (1.13) we can replace δ_M^0 by δ_M, and hence

$$J(v) - J(u) \geqslant (J'(u), v - u) + \int_0^1 \delta_M(t \| v - u \|)\frac{dt}{t},$$

which in particular, gives (1.16). ■

Proof of Theorem 1.1. Taking $v_i = u_i^n$ in (1.5) and successively setting $i = N, N - 1, \ldots, 1$, we have:

$$J(u^{n+1}) \leqslant J(u_1^{n+1}, \ldots, u_{N-1}^{n+1}, u_N^n) \leqslant J(u_1^{n+1}, \ldots, u_{N-2}^{n+1}, u_{N-1}^n, u_N^n) \leqslant \cdots \\ \cdots \leqslant J(u^n)$$

which gives

$$(1.17)\qquad J(u^{n+1}) \leqslant J(u^n) \leqslant \cdots \leqslant J(u^0).$$

Since (1.2) is satisfied, there then exists a constant M such that

$$(1.18)\qquad \| u^n \| \leqslant M \quad \forall n \geqslant 0.$$

From Lemma 1.2 we get

$$(1.19)\qquad \begin{aligned} & J(u_1^{n+1}, \dots, u_{i-1}^{n+1}, u_i^n, \dots, u_N^n) - J(u_1^{n+1}, \dots, u_{i-1}^{n+1}, u_i^{n+1}, u_{i+1}^n, \dots, u_N^n) \geqslant \\ & \geqslant (u_i^n - u_i^{n+1}) \frac{\partial J}{\partial u_i}(u_1^{n+1}, \dots, u_i^{n+1}, u_{i+1}^n, \dots, u_N^n) + \\ & + \frac{1}{2}\delta_M\left(\frac{1}{2} | u_i^{n+1} - u_i^n |\right) \end{aligned}$$

and since, *by definition of* u_i^{n+1}, we have

$$(1.20)\qquad \frac{\partial J}{\partial u_i}(u_1^{n+1}, \dots, u_i^{n+1}, u_{i+1}^n, \dots, u_N^n) = 0$$

we deduce from (1.19) that

$$(1.21)\qquad \begin{aligned} & J(u_1^{n+1}, \dots, u_{i-1}^{n+1}, u_i^n, \dots, u_N^n) - J(u_1^{n+1}, \dots, u_{i-1}^{n+1}, u_i^{n+1}, u_{i+1}^n, \dots, u_N^n) \geqslant \\ & \geqslant \tfrac{1}{2}\delta_M(\tfrac{1}{2} | u_i^{n+1} - u_i^n |), \end{aligned}$$

and hence

$$(1.22)\qquad J(u^n) - J(u^{n+1}) \geqslant \frac{1}{2}\sum_{i=1}^{N} \delta_M\left(\frac{1}{2} | u_i^{n+1} - u_i^n |\right).$$

However, from (1.17) $J(u^n) - J(u^{n+1}) \to 0$ so that, since δ_M is strictly increasing, (1.22) gives

$$(1.23)\qquad u^{n+1} - u^n \to 0.$$

We shall now show that $u^n \to u$, where u is the solution of problem (1.4) which therefore satisfies $J'(u) = 0$. From (1.18) we also have $\| u \| \leqslant M$ and, in (1.7), we can take $v = u^{n+1}$ and u, the solution of problem (1.4); this gives

$$(1.24)\qquad (J'(u^{n+1}), u^{n+1} - u) \geqslant \delta_M(\| u^{n+1} - u \|).$$

We put

$${}^i u^{n+1} = \{ u_1^{n+1}, \dots, u_i^{n+1}, u_{i+1}^n, \dots, u_N^n \},$$

so that (1.20) can also be expressed by

$$(1.20')\qquad \frac{\partial J}{\partial u_i}({}^i u^{n+1}) = 0.$$

Thus (1.24) is *equivalent* to

$$(1.25)\qquad \sum_{i=1}^{N}(u_i^{n+1}-u_i)\left(\frac{\partial J}{\partial u_i}(u^{n+1})-\frac{\partial J}{\partial u_i}({}^iu^{n+1})\right)\geqslant \delta_M(\| u^{n+1}-u\|).$$

Since $\| u^{n+1} - {}^iu^{n+1}\| \leqslant \| u^{n+1} - u^n\|$ and J' is uniformly continuous on bounded domains of $\mathbb{R}^N$, we see from (1.23) that the left-hand side of (1.25) tends to zero as $n \to \infty$; therefore

$$\delta_M(\| u^{n+1} - u\|) \to 0$$

which proves the theorem. ■

1.2 Point relaxation. Constrained case

We now consider the case where J is defined on K with

(1.26) K = *closed convex subset* of $\mathbb{R}^N$.

We assume that

$$(1.27)\qquad K=\prod_{i=1}^{N}K_i,$$

$$K_i = [a_i, b_i] \subset \mathbb{R} \ \ (a_i, b_i \text{ finite } \textit{or otherwise}).$$

We shall frequently state that under these circumstances K is defined by *local constraints*, since the membership of K is, at all points, expressed for each coordinate separately.

The formal extension of algorithm (1.5) is obvious: we define u_i^{n+1} as *the* solution *in* K_i of

$$(1.28)\qquad \begin{aligned} &J(u_1^{n+1},\dots,u_{i-1}^{n+1},u_i^{n+1},u_{i+1}^n,\dots,u_N^n)\leqslant\\ &\quad\leqslant J(u_1^{n+1},\dots,u_{i-1}^{n+1},v_i,u_{i+1}^n,\dots,u_N^n)\quad \forall v_i\in K_i,\\ &\quad i=1,\dots,N. \ ■\end{aligned}$$

If we assume that

$$(1.29)\qquad J(v)\to+\infty \text{ if } \|v\|\to+\infty,\qquad v\in K \qquad (^1)$$

and that (1.3) is satisfied, then *there exists a unique u in K such that*

$$(1.30)\qquad J(u) = \inf J(v),\quad v\in K.$$

We therefore have the following extension of Theorem 1.1:

---

(1) This assumption becomes redundant if K is bounded.

Theorem 1.2. *We assume that J is of class C^1 and satisfies* (1.29), (1.3). *Then the point relaxation algorithm* (1.28) *converges to the solution u of problem* (1.30).

Proof. The procedure is the same as in the proof of Theorem 1.1. First, we have (1.17) which gives (1.18), (1.19).

On this occasion, instead of (1.20) (or (1.20')) we have

$$(1.31) \qquad \frac{\partial J}{\partial u_i}({}^i u^{n+1})(v_i - u_i^{n+1}) \geqslant 0 \quad \forall v_i \in K_i\,,$$

so that (1.21) and (1.22) still hold, and (1.23) is satisfied. We deduce from (1.7) that

$$(J'(u^{n+1}) - J'(u), u^{n+1} - u) \geqslant \delta_M(\| u^{n+1} - u \|)\,,$$

and since

$$(1.32) \qquad (J'(u), v - u) \geqslant 0 \quad \forall v \in K\,,$$

we deduce (1.24), which, using (1.31), gives the inequality (1.25), and hence the same result as in Theorem 1.1. ■

A natural question at this point is to establish how far the method can be extended to convex sets K *which are not* of the form (1.27). First of all, it is necessary to determine *what meaning* is to be given to the method; it would appear reasonable to replace (1.28) by the same inequality, for:

$$\{ u_1^{n+1}, \ldots, u_{i-1}^{n+1}, v_i, u_{i+1}^n, \ldots, u_N^n \} \in K\,,$$
$$\{ u_1^{n+1}, \ldots, u_{i-1}^{n+1}, u_i^{n+1}, u_{i+1}^n, \ldots, u_N^n \} \in K\,.$$

However, the method then becomes unsatisfactory. To illustrate this, we may give a counter-example:

Counter-Example 1.2. Let J be strictly convex on K, represented in Figure 1.

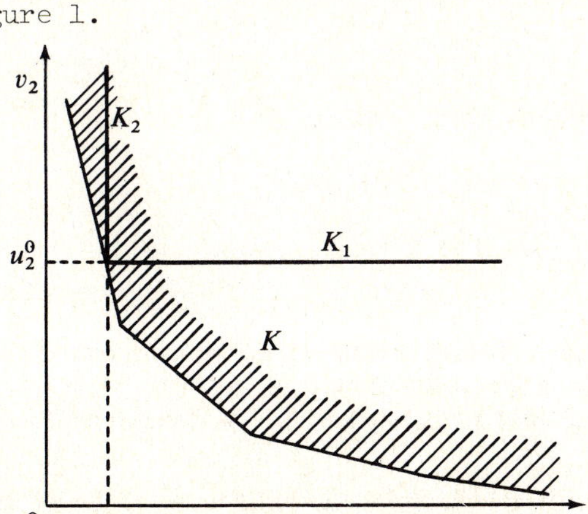

Fig. 1.

We assume J to be increasing w.r.t. each of the variables v_1, v_2. We start from

$$u^0 = \{ u_1^0, u_2^0 \} \in \partial K \quad (\partial K = \text{ boundary of } K).$$

Then

$$u_1^1, u_2^0 \in K_1 \quad \text{must satisfy} \quad J(u_1^1, u_2^0) \leqslant J(v_1, u_2^0)$$
$$\text{for} \quad \{ v_1, u_2^0 \} \in K_1$$

which gives

$$u_1^1 = u_1^0, \quad \text{and then} \quad u_2^1 = u_2^0, \quad \text{etc.}: \quad u^n = u^0 \quad \forall n.$$

The points on the boundary of K are thus "blockage" points of the algorithm, i.e. although convergence of the algorithm is implied by this last result, the limit is in general not the solution. ■

Remark 1.2. In Chapter 3, Section 7, in connection with the elasto-plastic torsion of a cylindrical bar (see Chapter 1, Section 1.2), we shall see that a minimisation problem has to be solved for a convex set K which does not satisfy (1.27) [1].

Nevertheless, it will be seen that the relaxation method converges to the solution, provided that, amongst other things, *a suitable* u^0 *is chosen.* ■

Remark 1.3. We can calculate u_i^{n+1} in two stages, as follows: First we define $u_i^{n+1/2}$ as *the* solution of

$$\frac{\partial J}{\partial u_i}(u_1^{n+1}, \ldots, u_{i-1}^{n+1}, u_i^{n+1/2}, u_{i+1}^n, \ldots, u_N^n) = 0, \tag{1.33}$$

and then define u_i^{n+1} by

$$u_i^{n+1} = P_{K_i}(u_i^{n+1/2}), \tag{1.34}$$

where P_{K_i} is the operator which projects onto K_i, i.e.

$$P_{K_i}(f) = \max(a_i, \min(b_i, f)). \quad ■$$

1.3 Block relaxation

The *decomposition* (1.27) of K is assumed to be performed *coordinate by coordinate*; it is clear that this is not necessary and may easily be extended to the case where we perform the decomposition using *blocks of coordinates*.

We put $V = \mathbb{R}^M$, and assume that V is decomposed as follows:

(1) We say: for a *nonlocal* convex set K.

$$V = \prod_{i=1}^{N} V_i ,$$
V_i of dimension m_i, $\quad m_1 + \cdots + m_N = M$.

Let K be a closed convex subset of V of the form

(1.35) $\quad K = \prod_{i=1}^{N} K_i$, K_i a closed convex subset of V_i.

Method (1.28) *is readily extended to this situation, as is Theorem* 1.2. ■

Remark 1.4. All this also extends to the case where the spaces V_i are uniformly convex Hilbert or Banach spaces, provided the properties of Lemmas 1.1 and 1.2 *are then either included in the hypotheses*, or else deduced from hypotheses *stronger* than (1.2), (1.3) and the fact that J is of class C^1.

For considerations of this kind, see Levitin and Polyak /1/, /2/, Polyak /1/, Glowinski /4/, and Cea and Glowinski /2/.

1.4 Over-relaxation and under-relaxation

Suppose A is a symmetric, positive-definite, (N,N) matrix

$$A = (a_{ij}), \qquad a_{ij} = a_{ji} \quad \forall i, j ,$$

and $\quad f \in \mathbb{R}^n$.

The equation

(1.36) $$Au = f$$

is equivalent to the unconstrained minimisation of

(1.37) $$J(v) = \tfrac{1}{2}\, a(v, v) - (f, v) ,$$

where

(1.38) $$a(u, v) = \sum_{i,j=1}^{N} a_{ij}\, u_j\, v_i .$$

Algorithm (1.5) then becomes

(1.39) $$u_i^{n+1} = -\frac{1}{a_{ii}} \left(\sum_{j=1}^{i-1} a_{ij}\, u_j^{n+1} + \sum_{j=i+1}^{N} a_{ij}\, u_j^n - f_i \right),$$

which is the classical Gauss-Seidel algorithm (see Varga /1/) which, as is well known, can often be generalised to advantage using the following algorithm:

(1.40) $$\begin{cases} u_i^{n+1/2} = -\dfrac{1}{a_{ii}} \left(\displaystyle\sum_{j=1}^{i-1} a_{ij}\, u_j^{n+1} + \sum_{j=i+1}^{N} a_{ij}\, u_j^n - f_i \right), \\ u_i^{n+1} = (1 - \omega)\, u_i^n + \omega\, u_i^{n+1/2} . \end{cases}$$

Algorithm (1.40) is termed *over-relaxation*, (resp. under-relaxation) if $\omega > 1$ (resp. $\omega < 1$), the idea naturally being *to accelerate the convergence* by choosing a suitable ω. ∎

We shall now extend the above algorithm to the *constrained* case, *where*

$$K = \prod_{i=1}^{N} K_i .$$

We note that the first equation in (1.40) is equivalent to the minimisation *over* $\mathbb{R}$ of

$$v_i \to J(u_1^{n+1}, \dots, u_{i-1}^{n+1}, v_i, u_{i+1}^n, \dots, u_N^n),$$

i.e.

$$(1.41)\quad \begin{cases} J(u_1^{n+1}, \dots, u_{i-1}^{n+1}, u_i^{n+1/2}, u_{i+1}^n, \dots, u_N^n) \leqslant \\ \qquad \leqslant J(u_1^{n+1}, \dots, u_{i-1}^{n+1}, v_i, u_{i+1}^n, \dots, u_N^n) \quad \forall v_i \in \mathbb{R} . \end{cases}$$

We next define

$$(1.42)\quad u_i^{n+1} = P_{K_i}((1-\omega)\, u_i^n + \omega u_i^{n+1/2}) .$$

We then have

Theorem 1.3. *J is assumed to be defined by* (1.37), (1.38), *where A is symmetric positive-definite. If it is assumed that*

$$(1.43)\quad 0 < \omega < 2$$

then u^n, *defined by* (1.41), (1.42), *converges to* u, *the element of K minimising J over K.*

Proof. First, we show that the sequence $J(u^n)$ is decreasing. For this, it will be sufficient to show that

$$(1.44)\quad X_i = J({}^{i-1}u^{n+1}) - J({}^{i}u^{n+1}) \geqslant \frac{1}{2}\left(\frac{2-\omega}{\omega}\right) a_{ii}\, | u_i^{n+1} - u_i^n |^2 ,$$

$$i = 1, \dots, N, \ (\text{using the convention } {}^{0}u^{n+1} = u^n)$$

from which, by summation, we get:

$$J(u^n) - J(u^{n+1}) \geqslant \frac{1}{2}\left(\frac{2-\omega}{\omega}\right) \sum_{i=1}^{N} a_{ii}\, | u_i^{n+1} - u_i^n |^2$$

and since $a_{ii} > 0$, $i = 1, \dots, N$, then with $\delta = \inf \{ a_{ii} \mid i = 1, \dots, N \}$, we have

$$J(u^n) - J(u^{n+1}) \geqslant \frac{1}{2}\left(\frac{2-\omega}{\omega}\right) \delta\, | u^{n+1} - u^n |^2 .$$

We have, with $i = 1$ (the other values of i having a similar role):

$$X_1 = \frac{1}{2} a_{11}(u_1^n)^2 + \sum_{j=2}^N a_{1j} u_1^n u_j^n - f_1 u_1^n - \frac{1}{2} a_{11}(u_1^{n+1})^2 -$$
$$- \sum_{j=2}^N a_{1j} u_1^{n+1} u_j^n + f_1 u_1^{n+1} = \frac{1}{2} a_{11}((u_1^n)^2 - (u_1^{n+1})^2) +$$
$$+ \left(\sum_{j=2}^N a_{1j} u_j^n \right) (u_1^n - u_1^{n+1}) - f_1(u_1^n - u_1^{n+1}) ,$$

so that, using the first equation (1.40) (which results from (1.41)), we obtain

$$X_1 = \tfrac{1}{2} a_{11}((u_1^n)^2 - (u_1^{n+1})^2) - a_{11} u_1^{n+1/2}(u_1^n - u_1^{n+1})$$
$$= a_{11}[- \tfrac{1}{2} (u_1^{n+1} - u_1^n)^2 + (u_1^n - u_1^{n+1}) (u_1^n - u_1^{n+1/2})] ,$$

from which we get (1.44) provided we ensure that

$$(u_1^n - u_1^{n+1}) (u_1^n - u_1^{n+1/2}) \geq \frac{1}{\omega} | u_1^{n+1} - u_1^n |^2 ,$$

which is equivalent to

$$(u_1^n - u_1^{n+1}) ((1 - \omega) u_1^n + \omega u_1^{n+1/2} - u_1^{n+1}) \leq 0 ,$$

the inequality which results from (1.42) (with $i = 1$).

Thus the sequence $J(u^n)$ is indeed decreasing, and since it is bounded below by $J(u)$ it converges and we have

$$(1.45) \qquad \| u^n \| \leq C , \qquad J(u^n) - J(u^{n+1}) \to 0 ,$$

$$(1.46) \qquad \| u^n - u^{n+1} \| \to 0 .$$

We have

$$(J'(u^{n+1}) - J'(u), u^{n+1} - u) \geq \lambda_m \| u^{n+1} - u \|^2 ,$$

where $\lambda_m > 0$ is the smallest eigenvalue of A.

If u is the solution minimising J over K, we have

$$(1.47) \qquad (J'(u^{n+1}), u^{n+1} - u) \geq \lambda_m \| u^{n+1} - u \|^2 .$$

We shall now prove that the left-hand side of this inequality tends to zero, which will give

$$\| u^{n+1} - u \| \to 0$$

which will prove the theorem.

Let $\bar{u}_i^n$ be the minimum over K_i of the functional

$$v_i \to J(u_1^{n+1}, \ldots, u_{i-1}^{n+1}, v_i, u_{i+1}^n, \ldots, u_N^n) ,$$

and

$${}^i\bar{u}^n = (u_1^{n+1}, \dots, u_{i-1}^{n+1}, \bar{u}_i^n, u_{i+1}^n, \dots, u_N^n).$$

We then write the left-hand side of (1.47) in the form [1]

$$(J'(u^{n+1}), u^{n+1} - u) = \sum_{i=1}^N (J_i'(u^{n+1}) - J_i'({}^iu^{n+1}))(u_i^{n+1} - u_i) +$$
$$+ \sum_{i=1}^N (J_i'({}^iu^{n+1}) - J_i'({}^i\bar{u}^n))(u_i^{n+1} - u_i) +$$
$$+ \sum_{i=1}^N J_i'({}^i\bar{u}^n)(u_i^{n+1} - \bar{u}_i^n) + \sum_{i=1}^N J_i'({}^i\bar{u}^n)(\bar{u}_i^n - u_i).$$

By definition of ${}^i\bar{u}^n$ the last term on the right-hand side is $\leqslant 0$. From (1.45) and (1.46) the first term on the right-hand side tends to zero.

It now remains to show that

$$(1.48) \qquad \| u^{n+1} - \bar{u}^n \| \to 0$$

with $\bar{u}^n = (\bar{u}_1^n, \dots, \bar{u}_N^n)$.

By definition, $\bar{u}_i^n$ is characterised by

$$\left(\sum_{j<i} a_{ij} u_j^{n+1} + a_{ii} \bar{u}_i^n + \sum_{j>i} a_{ij} u_j^n - f_i \right)(v_i - \bar{u}_i^n) \geqslant 0,$$
$$\forall v_i \in K_i, \quad \bar{u}_i^n \in K_i,$$

or, using the first equation in (1.40)

$$(\bar{u}_i^n - u_i^{n+1/2})(v_i - \bar{u}_i^n) \geqslant 0, \quad \forall v_i \in K_i$$

i.e. (which is obvious anyway)

$$\bar{u}_i^n = P_{K_i}(u_i^{n+1/2})$$

which is also equivalent to

$$\bar{u}_i^n = P_{K_i}((1-\omega)\bar{u}_i^n + \omega u_i^{n+1/2}).$$

Comparing with (1.42) we deduce that, since the projection P_{K_i} reduces the distances,

$$|\bar{u}_i^n - u_i^{n+1}| \leqslant |1-\omega|\,|\bar{u}_i^n - u_i^n| \leqslant$$
$$\leqslant |1-\omega|\,(|\bar{u}_i^n - u_i^{n+1}| + |u_i^{n+1} - u_i^n|),$$

and hence

$$(1 - |1-\omega|)\,|\bar{u}_i^n - u_i^{n+1}| \leqslant |1-\omega|\,|u_i^{n+1} - u_i^n|, \quad \forall i,$$

from which we deduce (1.48) since $0 < \omega < 2$. ■

Remark 1.5. In (1.42) we can *use variable parameters* ω [2]:

---

(1) With $J_i'(v) = \dfrac{\partial J}{\partial v_i}(v)$.

(2) We can also take $\omega_{i,n}$ dependent on both i and n.

$$u_i^{n+1} = P_{K_i}((1-\omega_n)\,u_i^n + \omega_n\,u_i^{n+1/2})\,. \tag{1.49}$$

If we assume that

$$0 < \varepsilon_1 \leqslant \omega_n \leqslant 2 - \varepsilon_2\,, \quad \varepsilon_i \text{ fixed,} \tag{1.50}$$

we again have the result of Theorem 1.3 (with the same proof). ■

Remark 1.6. Clearly, everything we have said can be extended to the case of *block relaxation* (as in Section 1.3); see Glowinski /4/, Cea and Glowinski /2/. ■

Remark 1.7. The proofs given, which are based on *estimates of the energy*, *can be extended to the case of Hilbert spaces*, see Glowinski /4/, Cea and Glowinski /2/. ■

Remark 1.8. In the unconstrained case, Theorem 1.3 is proved by means of different methods in Varga /1/. ■

Remark 1.9. In the constrained case, Auslender /1/, /2/, has proposed the scheme:

$$u_i^{n+1} = (1-\omega)\,u_i^n + \omega\bar{u}_i^{n+1}\,, \quad 0 < \omega \leqslant 1 \tag{1.51}$$

where $\bar{u}_i^n$ is the minimum over K_i of the functional

$$v_i \to J(u_1^{n+1}, \dots, u_{i-1}^{n+1}, v_i, u_{i+1}^n, \dots, u_N^n)\,.$$

This method, is identical to the one we propose, as long as the $u_i^{n+1/2}$ are all outside the K_i (and of course with $0 < \omega \leqslant 1$); it is however, very different when this is no longer the case. In fact (1.51) does not allow over-relaxation; in the applications which we have considered it is, however, important to be able to use $\omega > 1$ (this can be seen geometrically). ■

Remark 1.10. Certain variants of the above algorithms are useful in the solution of systems of equations. See Schechter /1/, /2/, /3/ and Ortega and Rheinboldt /1/, /2/.

1.5 A class of nondifferentiable functionals which can be minimised using relaxation

We have seen (Counter-Example 1.1) that, *in general*, the relaxation algorithm is not suitable for the minimisation of non-differentiable functions.

We shall however prove a result which is in fact valid for the class of functionals of the form:

$$\begin{cases} J(v) = J_0(v) + \sum_{i=1}^{N} \alpha_i\,|v_i|\,, \quad \alpha_i \geqslant 0\,, \\ \text{where } J_0 \text{ is } C^1, \text{ strictly convex, } J_0(v) \to +\infty \text{ if } \|v\| \to \infty\,. \end{cases} \tag{1.52}$$

Theorem 1.4. *If J is given by* (1.52), *the point relaxation algorithm* (1.5) *converges to the solution u of problem* (1.4).

We introduce *supplementary variables* (1) y_i, $i = 1, \ldots, N$, $y = \{ y_i \}$; we put:

$$J_1(v, y) = J_0(v) + \sum_{i=1}^{N} \alpha_i y_i , \tag{1.53}$$

which defines J_1, (not strictly) convex and of class C^1 on $\mathbb{R}^{2N}$; we introduce the closed convex subset of $\mathbb{R}^{2N}$:

$$\begin{cases} K = \prod_{i=1}^{N} K_i , \\ K_i = \{ v_i, y_i \mid y_i \geqslant | v_i | \} . \end{cases} \tag{1.54}$$

It is immediate that

$$\inf_K J_1(v, y) = J_1(u, | u |) = J(u) . \tag{1.55}$$

We introduce:

(i) The sequence u^n $(n \geqslant 0)$ corresponding to the relaxation algorithm for J;

(ii) The sequence z^n, corresponding to the (block) relaxation algorithm, with respect to

$J_1(z)$, $z = \{ v, y \}$, hence (2)

$$J_1(z_1^{n+1}, \ldots, z_{i-1}^{n+1}, z_i^{n+1}, z_{i+1}^n, \ldots, z_N^n) \leqslant \\ \leqslant J_1(z_1^{n+1}, \ldots, z_{i-1}^{n+1}, z_i, z_{i+1}^n, \ldots, z_N^n) \quad \forall z_i \in K_i . \tag{1.56}$$

It can be shown that if we start from

$$z^0 = \{ u^0, | u^0 | \} \tag{1.57}$$

then

$$z^n = \{ u^n, | u^n | \} , \tag{1.58}$$

and hence

$$J_1(z^n) = J(u^n) . \tag{1.59}$$

Since

$J(u^n) \geqslant J(u^{n+1})$ (this *does not assume* that J is differentiable),

---

(1) We put $J_1(v, y) = J_1(v_1, y_1, v_2, y_2, \ldots, v_N, y_N) = J_1(z_1, z_2, \ldots, z_N)$.

(2) So as to transform an "*unconstrained nondifferentiable*" problem into a "*constrained differentiable*" problem.

we have:

$$\| u^n \| \leqslant M . \tag{1.60}$$

If $z = \{ v, y \}$, $\zeta = \{ u, x \}$, we have:

$$(J_1'(z) - J_1'(\zeta), z - \zeta)_{\mathbb{R}^{2N}} = (J_0'(v) - J_0'(u), v - u)_{\mathbb{R}^N} \tag{1.61}$$

and therefore from Lemma 1.1 (for J_0 instead of J):

$$(J_1'(z) - J_1'(\zeta), z - \zeta)_{\mathbb{R}^{2N}} \geqslant \delta_M(\| v - u \|) \tag{1.62}$$

which, as in Lemma 1.2, gives

$$J_1(z) - J_1(\zeta) \geqslant (J_1'(\zeta), z - \zeta)_{\mathbb{R}^{2N}} + \tfrac{1}{2} \delta_M(\tfrac{1}{2} \| v - u \|) . \tag{1.63}$$

From (1.59), $J_1(z^n)$ is decreasing and $\geqslant J(u)$, hence

$$J_1(z^n) - J_1(z^{n+1}) \to 0 . \tag{1.64}$$

However

$$\begin{aligned} J_1(z^n) - J_1(z^{n+1}) &= \sum_{i=1}^{N} [J_1(z_1^{n+1}, \ldots, z_{i-1}^{n+1}, z_i^n, \ldots, z_N^n) - \\ &\quad - J_1(z_1^{n+1}, \ldots, z_{i-1}^{n+1}, z_i^{n+1}, z_{i+1}^n, \ldots, z_N^n)] \qquad \geqslant \text{ from } (1.63) \geqslant \\ &\geqslant \sum_{i=1}^{N} \frac{\partial J_0}{\partial u_i} (u_1^{n+1}, \ldots, u_i^{n+1}, u_{i+1}^n, \ldots, u_N^n) (u_i^n - u_i^{n+1}) + \\ &\quad + \alpha_i(y_i^n - y_i^{n+1}) \quad + \frac{1}{2} \sum_{i=1}^{N} \delta_M(| u_i^n - u_i^{n+1} |) \geqslant \frac{1}{2} \sum_{i=1}^{N} \times \\ &\quad \times \delta_M(| u_i^n - u_i^{n+1} |) \end{aligned} \tag{1.65}$$

(since we have $\dfrac{\partial J_0}{\partial u_i} (u_1^{n+1}, \ldots, u_i^{n+1}, u_{i+1}^n, \ldots, u_N^n) (u_i^n - u_i^{n+1}) + \alpha_i(y_i^n - y_i^{n+1}) \geqslant 0$ from the definition of z_i^{n+1}).

From (1.65) and (1.64) we deduce that:

$$u^{n+1} - u^n \to 0 . \tag{1.66}$$

It now remains to show that $u^n \to u$. The reasoning is the same as in the differentiable case (see (1.24)); we put

$$z^* = \{ u, | u | \}$$

and we start from

$$\begin{aligned} &(J_1'(z^{n+1}) - J_1'(z^*), z^{n+1} - z^*)_{\mathbb{R}^{2N}} = \\ &\quad = (J_0'(u^{n+1}) - J_0'(u), u^{n+1} - u)_{\mathbb{R}^N} \geqslant \delta_M(\| u^{n+1} - u \|) . \end{aligned}$$

Since

$$(J_1'(z^*), z^{n+1} - z^*)_{\mathbb{R}^{2N}} \geqslant 0 ,$$

we deduce

$$(1.67) \qquad (J_1'(z^{n+1}), z^{n+1} - z^*)_{\mathbb{R}^{2N}} \geqslant \delta_M(\| u^{n+1} - u \|),$$

which, by writing out the left-hand side explicitly, gives

$$(1.68) \qquad \sum_{i=1}^{N} \left(\frac{\partial J_0}{\partial u_i}(u^{n+1})(u_i^{n+1} - u_i) + \alpha_i(y_i^{n+1} - |u_i|) \right) \geqslant \geqslant \delta_M(\| u^{n+1} - u \|)$$

or

$$(1.69) \qquad \sum_{i=1}^{N} \left(\frac{\partial J_0}{\partial u_i}(u^{n+1}) - \frac{\partial J_0}{\partial u_i}(u_1^{n+1}, \dots, u_i^{n+1}, u_{i+1}^n, \dots, u_N^n) \right)(u_i^{n+1} - u_i) + + \sum_{i=1}^{N} \left[\frac{\partial J_0}{\partial u_i}(u_1^{n+1}, \dots, u_i^{n+1}, u_{i+1}^n, \dots, u_N^n)(u_i^{n+1} - u_i) + + \alpha_i(y_i^{n+1} - |u_i|) \right] \geqslant \delta_M(\| u^{n+1} - u \|).$$

From the definition of $\{ u_i^{n+1}, y_i^{n+1} \}$ and since $\{ u_i, |u_i| \} \in K_i$, the second term on the left-hand side of (1.69) is $\leqslant 0$, and hence

$$(1.70) \qquad \sum_{i=1}^{N} \left(\frac{\partial J_0}{\partial u_i}(u^{n+1}) - \frac{\partial J_0}{\partial u_i}(u_1^{n+1}, \dots, u_i^{n+1}, u_{i+1}^n, \dots, u_N^n) \right)(u_i^{n+1} - u_i) \geqslant \geqslant \delta_M(\| u^{n+1} - u \|).$$

Since $u_i^{n+1} - u_i^n \to 0$, we deduce that $u^n \to u$. ■

Remark 1.11. Theorem 1.4 supplements a result due to Auslender /1/, /2/ where the convergence of the algorithm towards a "critical point" (1), which can be distinct from the solution u, is shown. ■

Remark 1.12. The algorithms given in this section are often considered as coming within the class of so-called "direct" methods, i.e. those not using the expression for the derivative of the functional - at least in *the writing* of the algorithm. There are many other "direct" methods: the coordinate rotation method, Rosenbrock's method, the method of local variations or the Hooke and Jeeves method; see, for example, Cea /2/. ■

Remark 1.13. These algorithms can be modified in such a way that they can be used on parallel processors (see Morice /1/). ■

---

(1) "Blockage point" in the sense of the Counter-Examples 1.1 and 1.2.

SYNOPSIS

We shall now give a brief review of the methods which use the derivative of the functional to be minimised.

2. METHODS OF THE GRADIENT AND GRADIENT PROJECTION TYPE

Here we shall only present the algorithms. For the proofs we refer the reader to Cea /2/ (see, in particular, pp. 70-109 and pp. 118-156) and the bibliography therein.

2.1 General remarks

Assuming u^n to have been calculated, we seek u^{n+1} using:

$$u^{n+1} = u^n - \rho_n w^n , \tag{2.1}$$

where $\rho_n \geqslant 0$ is to be chosen, as is the direction w^n.

Assuming J to be differentiable and using formal reasoning we have:

$$J(u^n - \rho_n w^n) = J(u^n) - \rho_n(J'(u^n), w^n) + \cdots \tag{2.2}$$

so that it would be of advantage to take

$$(J'(u^n), w^n) \geqslant 0 . \tag{2.3}$$

The most natural choice is to take

$$w^n = J'(u^n) \tag{2.4}$$

and this leads to methods of the gradient type.

2.2 Methods of the gradient type (unconstrained case)

- *Optimal step method.*

We choose ρ_n so that

$$J(u^n - \rho_n J'(u^n)) = \inf_{\rho} J(u^n - \rho J'(u^n)) . \tag{2.5}$$

Assuming

(2.6) J strictly convex, of class C^1, $J(v) \to +\infty$ if $\| v \| \to \infty$, the method is convergent. ■

- *Fixed step method.*

We choose

(2.7) $\rho_n = \rho$ fixed.

The method is clearly easier to apply, but the conditions for convergence are fairly strict; for example, if we assume J to be of class C^2 and if we put

$$\mu_1 = \sup_{J(v) \leqslant J(v^0)} \| J''(v) \| \tag{2.8}$$

the method converges if:

$$0 < \delta \leqslant \rho \leqslant 2\rho_1 - \delta , \qquad \rho_1 \leqslant 1/\mu_1 . \tag{2.9}$$ ■

- *Variable step method.*

We seek ρ a priori in the form

$$\begin{cases} \rho = a^k \rho_0 , \quad k \in Z , \quad \rho_0 > 0 , \\ a > 0 \text{ and } k \text{ to be adjusted in the course of the iterations.} \end{cases} \tag{2.10}$$

We choose ρ *of the form* (2.10) such that

$$J(u^n - \rho J'(u^n)) < J(u^n) \tag{2.11}$$

(we assume $J'(u^n) \neq 0$).
The method is again convergent *if* ρ_0 *is sufficiently small.* ■

- *Divergent series method.*

We consider a sequence ρ_n such that $\rho_n > 0$, $\rho_n \to 0$ and $\sum_{n=0}^{\infty} \rho_n = +\infty$.

We then define

$$\begin{cases} u^{n+1} = u^n \quad \text{if} \quad J'(u^n) = 0 , \\ u^{n+1} = u^n - \rho_n \dfrac{J'(u^n)}{\| J'(u^n) \|} \quad \text{if} \quad J'(u^n) \neq 0 . \end{cases} \tag{2.12}$$

The choice of a suitable sequence ρ_n is very tricky, and this method generally converges very slowly.

2.3 Methods of the conjugate gradient type (unconstrained case)

Consider the *quadratic case*

$$J(v) = \tfrac{1}{2}(Av, v) - (f, v) , \tag{2.13}$$

where A is a symmetric positive-definite (N,N) matrix.
The directions $\{ w^0, \ldots, w^{N-1} \}$ of $\mathbb{R}^N$ are said to be *conjugate* with

respect to A if

$$(Aw^i, w^j) = 0 \text{ if } i \neq j, \neq 0 \text{ otherwise.} \tag{2.14}$$

Starting from u^0, we assume u^n to have been calculated and search for u^{n+1} by *displacement* in the direction w^n, thus

$$u^{n+1} = u^n + \lambda w^n, \tag{2.15}$$

λ being calculated in such a way as to minimise

$$J(u^{n+1}) = J(u^n) + \frac{\lambda^2}{2}(Aw^n, w^n) + \lambda((Au^n, w^n) - (f, w^n)),$$

and hence

$$\lambda = \lambda_n = \frac{(f, w^n) - (Au^n, w^n)}{(Aw^n, w^n)}. \tag{2.16}$$

After N iterations, we thus arrive at

$$u^N = u^0 + \sum_{i=0}^{N-1} \frac{(f, w^i) - (Au^0, w^i)}{(Aw^i, w^i)} w^i. \tag{2.17}$$

However a direct calculation shows that, $\forall g \in \mathbb{R}^N$, we have

$$A^{-1} g = \sum_{i=0}^{N-1} \frac{(g, w^i)}{(Aw^i, w^i)} w^i, \tag{2.18}$$

so that (2.17) gives

$$u^N = A^{-1} f = u = \text{ solution of the problem.} \tag{2.19}$$

Hence the algorithm converges in a finite number of iterations. ■

The problem now becomes, first of all, to discover how to generate the conjugate directions, $w^0, \ldots, w^{N-1}$.

Starting from u^0, we take as the first direction the direction of steepest descent at u^0, hence

$$w^0 = -J'(u^0) \tag{2.20}$$

i.e. in the quadratic case

$$w^0 = -Au^0 + f. \tag{2.21}$$

We then seek

$$u^1 = u^0 + \lambda w^0,$$

calculated so as to minimise $J(u^0 + \lambda w^0)$; we then seek w^1 in the form

$$w^1 = -J'(u^1) + \mu w^0, \tag{2.22}$$

where μ is chosen in such a way that w^0 and w^1 are conjugate, i.e.

$$(w^1, Aw^0) = 0\,,$$

and so on. *We hence arrive at the following algorithm*:

We assume u^n, r^n, w^n to be known (u^0 arbitrary, $r^0 = -J'(u^0), w^0 = r^0$) and we define:

$$u^{n+1} = u^n + \mu_n w^n\,, \qquad \mu_n = \frac{(w^n, r^n)}{(Aw^n, w^n)}, \tag{2.23}$$

then

$$r^{n+1} = -J'(u^{n+1})\,, \tag{2.24}$$

$$w^{n+1} = r^{n+1} + \lambda_{n+1} w^n\,, \qquad \lambda_{n+1} = -\frac{(r^{n+1}, Aw^n)}{(w^n, Aw^n)} = \frac{\| r^{n+1} \|^2}{\| r^n \|^2}\,. \tag{2.25}$$ ■

In the *non-quadratic* case, the algorithm is adapted as follows. We define:

$$u^{n+1} = u^n + \mu_n w^n\,, \tag{2.26}$$

where μ_n is chosen so as to minimise $J(u^n + \mu w^n), \mu \in \mathbb{R}$; then

$$r^{n+1} = -J'(u^{n+1})\,.$$

In choosing w^{n+1} it is necessary to take certain precautions. We put

$$\tilde{w}^{n+1} = r^{n+1} + \lambda_{n+1} w^n\,, \qquad \lambda_{n+1} = \frac{\| r^{n+1} \|^2}{\| r^n \|^2}\,. \tag{2.27}$$

We check whether the vectors r^{n+1} and $\tilde{w}^{n+1}$ form an angle $\leqslant \frac{\pi}{2} - \delta$, with sufficiently small $\delta > 0$. If so, we put

$$w^{n+1} = \tilde{w}^{n+1} \tag{2.28}$$

otherwise, the direction $\tilde{w}^{n+1}$ must be discarded and we then take

$$w^{n+1} = r^{n+1}\,. \tag{2.29}$$

The algorithm is convergent, by virtue of the classical assumptions (1.2), (1.3) and with J of class C^1.

From a numerical point of view, it is advantageous to take the following additional precaution: if, for N successive iterations, $j = i, i+1, \ldots, i+N-1$, we have

$$w^j \neq r^j$$

we put

$$w^{i+N} = r^{i+N}\,.$$ ■

Remark 2.1. The method described above is the so-called conjugate gradient method. There are various methods which allow the construction of N conjugate directions. In particular, we mention the method of Fletcher and Powell which also allows the inverse matrix of $J''(u)$ to be obtained.

An application of the so-called conjugate direction method to the problem of the minimisation of

$$J(v) = \tfrac{1}{2}(Av, v) - (f, v)$$

subject to the constraints $Bv = b$, will be given later (see Remark 4.10).

2.4 Constrained case

We now consider the system

$$\inf J(v)\,, \quad v \in K\,, \quad K = \text{closed convex subset of } \mathbb{R}^N. \tag{2.30}$$

Here, we shall adapt the above methods by using the operator P_K projecting onto K.

- *Point projection method*

Starting from $u^0 \in K$, we define u^{n+1} from u^n using

$$u^{n+1} = P_K(u^n - \rho_n J'(u^n))\,. \tag{2.31}$$

Theorem 2.1. *Consider the case* (2.13) (*quadratic functional*). *We can now choose* ρ_0 *and* ρ_1 *so that with*

$$0 < \rho_0 \leqslant \rho_n \leqslant \rho_1 \tag{2.32}$$

the method (2.31) *converges to the solution* u *of problem* (2.30).

Proof. If u is the solution of the problem we have:

$$u = P_K(u - \rho_n J'(u)) \quad \forall \rho_n \tag{2.33}$$

since (2.33) is equivalent to

$$(u - \rho_n J'(u) - u, v - u) \leqslant 0 \quad \forall v \in K\,, \quad \text{i.e. } (J'(u), v - u) \geqslant 0 \quad \forall v \in K\,.$$

By putting $w^n = u^n - u$ and using the fact that P_K is a *contraction* we deduce from (2.31), (2.33) that

$$\| w^{n+1} \| \leqslant \| w^n - \rho_n(J'(u^n) - J'(u)) \| \tag{2.34}$$

and hence

$$\| w^{n+1} \|^2 \leqslant \| w^n \|^2 - 2\,\rho_n(w^n, J'(u^n) - J'(u)) + \\ + \rho_n^2 \| J'(u^n) - J'(u) \|^2\,. \tag{2.35}$$

However, $J'(u^n) - J'(u) = A(u^n - u) = Aw^n$ so that (2.35) gives

$$\| w^{n+1} \|^2 \leqslant (1 - 2\,\alpha\rho_n + C\rho_n^2) \| w^n \|^2 . \tag{2.36}$$

We can choose ρ_0 and ρ_1 in such a way that (2.32) gives

$$1 - 2\,\alpha\rho_n + C\rho_n^2 \leqslant \beta < 1$$

and then (2.36) shows that $\| w^n \| \to 0$. ■

Remark 2.2. The above proof is readily extended to the non-quadratic case with Lipschitz J'. ■

- *Gradient projection method.*

In the unconstrained iteration method of (2.31), the points were projected onto K. We are now going to *project the gradients*. We assume:

$$\left\{ \begin{array}{l} \text{the convex set } K \text{ is bounded and defined by a finite} \\ \text{number of affine linear constraints} \\ K = \{ v \mid G_j(v) = 0,\ 1 \leqslant j \leqslant l,\ G_k(v) \leqslant 0,\ l + 1 \leqslant k \leqslant m \} \end{array} \right. \tag{2.37}$$

and that for all points $v \in K$, the vectors

$$G_j'(v)\,, \quad j \in I(v)\,, \quad I(v) = \{ j \mid 1 \leqslant j \leqslant m,\ G_j(v) = 0 \}\,,$$

are linearly independent.

Then the projection onto K of the straight line proceeding from v in the direction $-J'(v)$, i.e. the set of points w such that

$$w = P_K(v - \rho J'(v))\,, \qquad \rho \geqslant 0 \tag{2.38}$$

is a segmented line, the first segment being denoted by $[v, v^+]$. The algorithm is thus as follows: having obtained $u^n \in K$, we define u^{n+1} by:

$$J(u^{n+1}) \leqslant J(v)\,, \quad \forall v \in [u^n, (u^n)^+]\,. \tag{2.39}$$

The direction $[v, v^+]$ is obtained using an operator which projects onto an affine linear manifold.

If the manifold is defined by

$$S = \{ v \mid Av = b \}$$

the projection of a vector a onto S is given by

$$P_s(a) = [I - A^*(AA^*)^{-1} A]\, a + A^*(AA^*)^{-1}\, b\,.$$

where A^* is the transpose of the matrix A.

The method is convergent (e.g. using the assumption (2.37) and with J is strictly convex and of class C^1) (see Rosen /1/, /2/ and Canon, Cullum and Polak /1/).

Remark 2.3. Apart from the *gradient projection* method, it is appropriate to mention the *reduced gradient* method introduced by Wolfe /1/ in the context of the minimisation of *convex* functions subject to *linear constraints*, by extension of the *simplex* method. This method has been generalised to the case of *non-linear constraints* by Abadie-Carpentier /1/ (see also Fletcher /1/). A method somewhat similar to the above is that of Frank-Wolfe /1/. ■

3. PENALTY METHODS AND VARIANTS

3.1 General remarks

The idea of penalisation (already encountered in Chapter 1, Section 3.2; see in particular Remark 3.4) consists of "approximating" the problem,

$$\text{(3.1)} \qquad \begin{cases} \inf J(v)\,, \\ v \in K\,, \end{cases}$$

K = a closed convex subset of $\mathbb{R}^N$, by *unconstrained* problems (as will be the case in the *exterior* method described in Section 3.3 below), or by problems with *passive constraints*, i.e. where the inf is attained *in the interior of* the set of constraints (as will be the case in the *interior* method (Section 3.2 below) and in the method of centres with variable truncation described in Section 3.4 below) [1]. ■

In the following we shall assume that K is *defined* by

$$\text{(3.2)} \qquad \begin{cases} K = \{\, v \mid G_j(v) \geqslant 0,\ 1 \leqslant j \leqslant m \,\}\,, \\ G_j = \text{a concave continuous function on } \mathbb{R}^N. \end{cases}$$

We shall assume that J satisfies:

(3.3) J is defined and continuous on $\mathbb{R}^N$, strictly convex. ■

3.2 Interior methods

We introduce the functional

$$\text{(3.4)} \qquad I(v, \varepsilon) = J(v) + \varepsilon \sum_{j=1}^{m} \frac{1}{G_j(v)}, \quad \varepsilon > 0 \qquad \text{(which will tend towards 0)}$$

We make the assumptions:

---

(1) Clearly it *then* becomes necessary to choose an algorithm for *the solution of the penalised problem*.

$$\text{(3.5)} \quad \begin{cases} K \text{ is bounded [1], with a nonempty interior } \mathring{K}, \text{ given by} \\ \mathring{K} = \{ v \mid G_j(v) > 0,\ j = 1, \dots, m \} . \end{cases}$$

This last condition is assured when there exists v_0 such that

$$\text{(3.6)} \qquad G_j(v_0) > 0 , \qquad j = 1, \dots, m .$$

The function $v \to I(v, \varepsilon)$ is therefore not identical to $+\infty$ or to $-\infty$ on $\mathring{K}$. In fact,

Lemma 3.1. *Using the hypotheses* (3.2), (3.3), (3.5), *there exists a unique* u_ε *such that:*

$$\text{(3.7)} \qquad u_\varepsilon \in \mathring{K} ,$$

$$\text{(3.8)} \qquad I(u_\varepsilon, \varepsilon) \leqslant I(v, \varepsilon) \quad \forall v \in \mathring{K} .$$

Proof. With the element v_0 chosen as in (3.6), we introduce the set

$$\text{(3.9)} \qquad S_\varepsilon = \{ v \mid v \in K,\ I(v, \varepsilon) \leqslant I(v_0, \varepsilon) \} ,$$

which is nonempty since $v_0 \in S_\varepsilon$. It can be shown that S_ε is *closed* (and hence compact since $S_\varepsilon \subset K$, where K is bounded); since $I(v, \varepsilon)$ is bounded on S_ε the $G_j(v)$ remain > 0 on S_ε and $I(v, \varepsilon)$ is *continuous* on $S_{\varepsilon'}$, which gives the required result. Furthermore,

$$\text{(3.10)} \qquad S_\varepsilon \subset \mathring{K} ,$$

and hence the lemma. ■

We can now state and prove:

<u>Theorem 3.1</u> *Under the assumptions* (3.2), (3.3), (3.5) *if* u_ε *is the element defined by* (3.7), (3.8) *we have*

$$\text{(3.11)} \qquad u_\varepsilon \to u \quad \text{in } \mathbb{R}^N$$

where u *is the solution of*

$$\text{(3.12)} \qquad u \in K , \qquad J(u) = \inf_{v \in K} J(v) .$$

Remark 3.1. The penalty method thus defined is said to be *interior* since $u_\varepsilon \in \mathring{K}$. ■

---

[1] The assumption "*K* bounded" is introduced only with a view to simplifying the proofs. In practice, it is sufficient to assume that the minimum of $J(v)$ over K is reached for a point u at a finite distance, which is the case if, for example, J satisfies (1.2).

Proof. Since $u_\varepsilon \in \mathring{K}$ and K is bounded, we can extract a subsequence, also denoted by u_ε, such that

$$u_\varepsilon \to w \text{ in } \mathbb{R}^N, \quad w \in K \tag{3.13}$$

We have

$$J(u_\varepsilon) \leqslant J(u_\varepsilon) + \varepsilon \sum_{j=1}^{m} \frac{1}{G_j(u_\varepsilon)} \leqslant J(v) + \varepsilon \sum_{j=1}^{m} \frac{1}{G_j(v)} \quad \forall v \in \mathring{K},$$

which, in the limit, gives

$$J(w) \leqslant J(v) \quad \forall v \in \mathring{K},$$

and hence

$$J(w) \leqslant J(v) \quad \forall v \in K$$

and since J is strictly convex, $w = u$, giving the required result. ■

Remark 3.2. We have the following additional property:

The function $\varepsilon \to J(u_\varepsilon)$ is *decreasing* towards $J(u)$ as $\varepsilon \to 0$. To show this, let $0 < \eta < \varepsilon$.

We put:

$$\frac{1}{G(v)} = \sum_{j=1}^{m} \frac{1}{G_j(v)}.$$

From the definitions of u_ε and u_η we have:

$$J(u_\varepsilon) + \varepsilon \frac{1}{G(u_\varepsilon)} \leqslant J(u_\eta) + \varepsilon \frac{1}{G(u_\eta)}$$

and

$$J(u_\eta) + \eta \frac{1}{G(u_\eta)} \leqslant J(u_\varepsilon) + \eta \frac{1}{G(u_\varepsilon)} \tag{3.15}$$

so that, by addition,

$$(\varepsilon - \eta)\left(\frac{1}{G(u_\varepsilon)} - \frac{1}{G(u_\eta)}\right) \leqslant 0$$

and hence

$$\frac{1}{G(u_\varepsilon)} - \frac{1}{G(u_\eta)} \leqslant 0.$$

Thus (3.15) then gives

$$J(u_\eta) - J(u_\varepsilon) \leqslant \eta\left(\frac{1}{G(u_\varepsilon)} - \frac{1}{G(u_\eta)}\right) \leqslant 0. \quad ■$$

3.3 Exterior methods

We introduce the functional

$$E(v, \varepsilon) = J(v) + \frac{1}{\varepsilon} G(v)^- \tag{3.16}$$

where by definition

$$G(v)^- = \sum_{j=1}^{m} G_j(v)^- = \sum_{j=1}^{m} \sup(-G_j(v), 0). \tag{3.17}$$

We note (cf. Chapter 1, Section 3.2) that

$$G(v)^- = 0 \Leftrightarrow v \in K. \tag{3.18}$$

We do not make any assumption analogous to (3.5). Since, in particular, K is not assumed to be bounded, it is necessary to assume that

$$J(v) \to +\infty \text{ if } \|v\| \to \infty, \quad v \in \mathbb{R}^N. \tag{3.19}$$

Then, since

$$E(v, \varepsilon) \geqslant J(v), \tag{3.20}$$

we have, a fortiori, $E(v, \varepsilon) \to +\infty$ if $\|v\| \to \infty$ and hence, with $E(v, \varepsilon)$ strictly convex in v,

$$\begin{cases} \text{there exists a unique } u_\varepsilon \text{ in } \mathbb{R}^N \text{ such that} \\ E(u_\varepsilon, \varepsilon) = \inf_{v \in \mathbb{R}^N} E(v, \varepsilon). \end{cases} \tag{3.21}$$

We then have:

Theorem 3.2. *We assume that* (3.2), (3.3), (3.19) *are satisfied. Then, if* u_ε *is defined by* (3.21) *and* u *by* (3.12) *we have*:

$$u_\varepsilon \to u \quad \textit{in} \quad \mathbb{R}^N. \tag{3.22}$$

Remark 3.3. The element approximating u_ε *is not necessarily in* K; hence the terminology: *exterior* method. ■

Proof. We have:

$$J(u_\varepsilon) \leqslant E(u_\varepsilon, \varepsilon) \leqslant \inf_{v \in K} E(v, \varepsilon) = \inf_{v \in K} J(v) = J(u) \tag{3.23}$$

from which it results that u_ε lies in a bounded subset of $\mathbb{R}^N$. We can therefore extract a subsequence, also denoted by u_ε, such that $u_\varepsilon \to w$ in $\mathbb{R}^N$ and we deduce from (3.23) that

$$J(w) \leqslant J(u). \tag{3.24}$$

Moreover from (3.23)

$$\frac{1}{\varepsilon} G(u_\varepsilon)^- \leqslant J(u) - J(u_\varepsilon)$$

therefore

$$G(w)^- = \lim_{\varepsilon \to 0} G(u_\varepsilon)^- = 0$$

and hence (see (3.18): $w \in K$ which, along with (3.24) shows that $w = u$. ■

Remark 3.4. It should be emphasised that the exterior method still holds if $\overset{\circ}{K} = \emptyset$. This allows this method to be used for the solution of problems involving *linear equality constraints*: the penalisation term associated with a constraint $G_j(v) = 0$ then takes the form

$$\frac{1}{\varepsilon} \mid G_j(v) \mid . \quad ■$$

Remark 3.5. In (3.17) we can also define $G(v)^-$ by

$$G(v)^- = \sum_{j=1}^{m} [\sup(-G_j(v), 0)]^\alpha \text{ with } \alpha \geqslant 1 . \quad ■$$

Remark 3.6. In the context of exterior methods, the following variant can also be used: with each constraint

$$G_j(v) \geqslant 0 \ (j = 1, m)$$

we associate a variable $t_j \geqslant 0$ (called a slack variable) from which we get the equivalent constraint pair $\{ G_j(v) - t_j = 0, t_j \geqslant 0 \}$ and the penalised functional

$$F(v, t, \varepsilon) = J(v) + \frac{1}{\varepsilon} \sum_{j=1}^{m} (G_j(v) - t_j)^2$$

where $t = (t_1, \ldots, t_m) \in R_+^m$; the penalised problem then consists of minimising $F(v, t, \varepsilon)$ over $R^n \times R_+^m$.

The advantage of this approach over that described above is that the order of differentiation of J and of the G_j is conserved at the cost of introducing the vector t (the condition that the t_j must be positive does not introduce any additional practical difficulties).

The results of Section 3.3, concerning $E(v, \varepsilon)$ as defined by (3.16), are easily adapted (see, for example, Cea /2/).

An application relating to the model elasto-plastic problem defined in Chapter 1, Section 1.2 is given in Chapter 3, Section 8.2.2. ■

3.4 Method of centres with variable truncation

General remarks.

We introduce, again for problem (3.1):

$$(3.25) \qquad K(\lambda) = \{ v \mid v \in K, J(v) \leqslant \lambda \}$$

which defines a nonempty closed convex set if:

$$\lambda \geqslant J(u) \ \ (\text{ and } K(J(u)) = \{ u \}) .$$

The *formal* idea is then to construct an algorithm which determines:

(i) a decreasing sequence of λ_n $(\to J(u))$

(ii) a "centre" of $K(\lambda_n)$ associated with each λ_n (in fact λ_{n+1} is chosen using the "centre" of $K(\lambda_n)$). ■

"Centre of a convex set".

The concept of the "centre" of a convex set *is not intrinsic*, and numerous basically equivalent concepts can be defined.

Here we shall give the two concepts which seem to be the most natural and the most useful.

In general, let $\mathscr{L}$ be a closed convex subset of $\mathbb{R}^N$ defined by

$$\mathscr{L} = \{ v \mid L_j(v) \geqslant 0,\ 0 \leqslant j \leqslant m \} \tag{3.26}$$

where, $\forall j$:

(3.27) L_j is a continuous concave function.

We then consider the "centre" of $\mathscr{L}$ to be *either* the point in $\mathscr{L}$, in fact in $\mathscr{L}^0$, where

$$\prod_{j=0}^{m} L_j(v) \quad \textit{is maximum} \tag{3.28}$$

or the point in $\mathscr{L}$, in fact in $\mathscr{L}^0$, where

$$\sum_{j=0}^{m} \frac{1}{L_j(v)} \quad \text{is } \textit{minimum}. \ ■ \tag{3.29}$$

Application of the concepts (3.28), (3.29) *to* $K(\lambda)$.

We use the above concepts for $K(\lambda)$, with

$$L_0(v) = \lambda - J(v)\,, \quad L_j(v) = G_j(v)\,, \quad 1 \leqslant j \leqslant m\,.$$

In order to simplify the notation, we put:

$$D(v, \lambda) = (\lambda - J(v)) \prod_{j=1}^{m} G_j(v)\,, \tag{3.30}$$

$$F(v, \lambda) = \frac{1}{\lambda - J(v)} + \sum_{j=1}^{m} \frac{1}{G_j(v)}\,. \ ■ \tag{3.31}$$

Algorithm. We consider $v_0 \in K$. We take:

$$\lambda_1 = J(v_0)\,.$$

Then we define v_1 to be the "centre" of $K(\lambda_1)$ using method (3.28) [1], i.e. :

---

(1) If we use (3.29), v_1 (different from the above) is defined by

$$F(v_1, \lambda_1) = \inf_{v \in K(\lambda_1)} F(v, \lambda_1)\,.$$

$$D(v_1, \lambda_1) = \sup_{v \in K(\lambda_1)} D(v, \lambda_1) . \quad (3.32)$$ [1]

We next define $w_1 \in K(\lambda_1)$ (see Remark 3.7 below) such that

$$J(w_1) \leqslant J(v_1) \quad (3.33)$$

and then define λ_2 by

$$\lambda_2 = \lambda_1 - \rho_1(\lambda_1 - J(w_1)) \qquad 0 < \rho_1 < 1 . \quad (3.34)$$

Assuming v_{n-1} and λ_n to be known we define v_n by

$$D(v_n, \lambda_n) = \sup_{v \in K(\lambda_n)} D(v, \lambda_n) . \quad (3.35)$$

We define $w_n \in K(\lambda_n)$ such that

$$J(w_n) \leqslant J(v_n) \quad (3.36)$$

and we choose

$$\lambda_{n+1} = \lambda_n - \rho_n[\lambda_n - J(w_n)] \quad (3.37)$$

where

$$0 < \delta \leqslant \rho_n \leqslant 1 . \quad (3.38)$$

It may be noted that

$$v_n \in \overline{\overset{\circ}{K(\lambda_n)}} . \quad (3.39)$$

We have [2]

Theorem 3.3. *We assume that* (3.2), (3.3), (3.5) *are satisfied. Then the algorithm of the method of centres with variable truncation defined by* (3.35), (3.36), (3.37) *is convergent in the following sense:*

$$\lambda_n \to J(u) , \quad (3.40)$$

$$v_n \to u . \quad (3.41)$$

where u *is the solution of problem* (3.1).

---

(1) A similar algorithm results using (3.29) instead of (3.28).

(2) There is a similar result (and proof) if we use (3.29).

Proof. From the construction of the algorithm, λ_n and $J(v_n)$ are decreasing sequences and from (3.37) and (3.36) we have:

$$\lambda_n - \lambda_{n+1} = \rho_n[\lambda_n - J(w_n)] \geqslant \rho_n[\lambda_n - J(v_n)]$$

and hence

$$\lambda_n - J(v_n) \leqslant \frac{1}{\delta}(\lambda_n - \lambda_{n+1}) \to 0 .$$

Therefore

$$\lim \lambda_n = \lim J(v_n) = \mu \geqslant J(u) . \tag{3.42}$$

We shall now show that

$$\mu = J(u) \tag{3.43}$$

(which proves the theorem since v_n is then a minimising sequence and, since the function J is strictly convex, we have (3.41)).

Let us assume that $\mu > J(u)$. Then $\overline{\overset{\circ}{K}(\mu)} \neq \varnothing$, so that there exists v_0 such that

$$D(v_0, \mu) > 0 .$$

However, since $\lambda_n \geqslant \mu$ we have $D(v_0, \lambda_n) \geqslant D(v_0, \mu)$ and, by definition of v_n we have:

$$D(v_n, \lambda_n) \geqslant D(v_0, \lambda_n)$$

hence

$$D(v_n, \lambda_n) \geqslant D(v_0, \mu) > 0 . \tag{3.44}$$

However

$$D(v_n, \lambda_n) = (\lambda_n - J(v_n)) \prod_{j=1}^{m} G_j(v_n) \leqslant C(\lambda_n - J(v_n)) \to 0$$

which contradicts (3.44). ■

Remark 3.7. (On the choice of the w_n).

1) First we can take $w_n = v_n$:

- if $\rho_n = 1$, this is the *method of centres* (Huard /1/),
- if $\rho_n < 1$, we can consider this method to be an under-relaxation method.

2) The convergence is accelerated significantly if we take w_n such that

$$J(w_n) < J(v_n) .$$

In practice, we can seek $w_n \in K(\lambda_n)$ either on the straight line proceeding from v_n with direction vector $-J'(v_n)$, or on the segment $[v_{n-1}, v_n]$.

The method is interpreted here as an over-relaxation method. ■

4. DUALITY METHODS

4.1 General remarks

We now return to the concepts introduced in Chapter 1, Sections 3.4 and 3.5 and from them we shall derive algorithms which play an important part in the remainder of the book.

We consider *a Hilbert space* V — which in the case of numerical applications will always be *finite dimensional*: $V = \mathbb{R}^N$ — and the functional

$$J_0(v) = \tfrac{1}{2}\, a(v, v) - (f, v) \tag{4.1}$$

where $a(u, v) = a(v, u)\ \forall u, v \in V$, is a continuous bilinear form on V satisfying:

$$a(v, v) \geqslant \alpha \, \| v \|^2 , \quad \alpha > 0 , \quad \forall v \in V \tag{4.2}$$

(where $\| v \|$ = the norm of v in V), and where in (4.1) $v \to (f, v)$ is a continuous linear form on V.

We also take:

(4.3) M = closed convex subset of V.

(4.4) $\begin{cases} L = \text{Hilbert space, and } \Phi \text{ a function from } V \to L \text{ or from} \\ \quad M \to L, \text{ linear } \textit{or otherwise}. \end{cases}$

(4.5) Λ = closed convex subset of L.

If $(\ ,\ )_L$ denotes the inner product in L, we assume for all functions $q \in L$ that:

(4.6) $\begin{cases} v \to (q, \Phi(v))_L \text{ is lower semi-continuous and} \\ \text{convex on } V. \end{cases}$

We then consider the problem

$$\inf_{v \in M} \left[J_0(v) + \sup_{q \in \Lambda} (q, \Phi(v))_L \right] . \tag{4.7}$$ ■

Remark 4.1. Thus, *by definition*, we consider the problem in a form adapted to duality (see Chapter 1, Section 3.4). We shall see in Section 4.2 below how a number of important problems come within the framework of (4.7).

Remark 4.2. The function

$$v \to \sup_{q \in \Lambda} (q, \Phi(v))_L$$

is convex, lower semi-continuous for the weak topology of V; this is therefore also the case for J defined by

$$J(v) = J_0(v) + \sup_{q \in \Lambda} (q, \Phi(v))_L . \quad \blacksquare \tag{4.8}$$

4.2 Examples

Example 4.1. Using the notation (1.18), Chapter 1, we take:

$$V = H_0^1(\Omega), \qquad M = V,$$
$$a(u, v) = \int_\Omega \text{grad}\, u . \text{grad}\, v \,\mathrm{d}x, \quad (f, v) = \int_\Omega fv \,\mathrm{d}x, \quad f \in L^2(\Omega),$$
$$L = (L^2(\Omega))^n, \quad \Phi v = \text{grad}\, v \quad \text{hence} \quad \Phi \in \mathscr{L}(V; L),$$
$$\Lambda = \{ q \mid q \in L, \ |q(x)| \leqslant g \ \text{ a.e. in } \Omega \}.$$

We have (see Chapter 1, Section 3.4 above):

$$\sup_{q \in \Lambda} (q, \Phi(v)) = g \int_\Omega |\,\text{grad}\, v\,| \,\mathrm{d}x .$$

Then (4.7) becomes *the flow problem* considered in Chapter 1, Section 1.4, which has already been considered from the present standpoint in Chapter 1, Section 3.4. ■

Example 4.2. We take $V, M, a, (f,v)$ as in Example 4.1:

$$L = L^2(\Omega), \quad \Lambda_1 = [q \mid q \in L, \ q \geqslant 0 \ \text{ a.e.}]$$

and we define $\Phi_1 : V \to L$ by:

$$\Phi_1(v) = |\,\text{grad}\, v\,| - 1 .$$

The function Φ_1 is Lipschitz on V.
This comes within the context of Section 4.1 and we have:

$$\sup_{q \in \Lambda_1} (q, \Phi_1(v)) = \begin{cases} 0 & \text{if } |\,\text{grad}\, v\,| \leqslant 1 \ \text{ a.e.} \\ +\infty & \text{otherwise.} \end{cases}$$

In this example, (4.7) is thus the elasto-plasticity problem considered in Chapter 1, Section 1.2.

We note that Φ_1 is *nondifferentiable*, and since this is a disadvantage (from the numerical point of view [1]) we introduce $\Phi_2 : H_0^1(\Omega) \to L^1(\Omega)$ and Λ_2 defined by

---

(1) Easily surmountable: see Chapter 5, Section 8.

$$\Phi_2(v) = |\operatorname{grad} v|^2 - 1\,,$$

$$\Lambda_2 = L^\infty_+(\Omega) = [q \mid q \in L^\infty(\Omega),\ q \geq 0 \ \text{a.e.}]\,;$$

indeed we then have:

$$\sup_{q \in \Lambda_2} \int_\Omega q\Phi_2(v)\,dx = \begin{cases} 0 & \text{if } |\operatorname{grad} v| \leq 1 \ \text{a.e.} \\ +\infty & \text{otherwise.} \end{cases}$$

but, amongst other difficulties, the *Hilbert space* setting of Section 4.1 is not appropriate for Φ_2, in the *infinite dimensional* case [1], since $\Phi_2(v) \in L^1(\Omega)$.

We note that Φ_2 is indefinitely Frechet-differentiable (and hence locally Lipschitz) on $H^1_0(\Omega)$.

4.3 A saddle-point search algorithm

Let us now introduce the Lagrangian

$$\mathscr{L}(v, q) = J_0(v) + (q, \Phi(v))_L\,. \tag{4.9}$$

We make the *following assumptions*:

(4.10) there exists a saddle point of $\mathscr{L}(v, q)$ on $M \times \Lambda$, i.e. a point $\{u, p\} \in M \times \Lambda$ such that $\mathscr{L}(u, q) \leq \mathscr{L}(u, p) \leq \mathscr{L}(v, p)$, $\forall v \in M$, $\forall q \in \Lambda$;

(4.11) the function $\Phi : M \to \Lambda$ is Lipschitz, i.e. there exists a constant C_1 such that $\| \Phi(u) - \Phi(v) \|_L \leq C_1 \| u - v \| \quad \forall u, v \in M$.

A variant of (4.11) which is useful in applications is as follows:

(4.11') $\forall q \in \Lambda$, the solution of $\inf_{v \in M} \mathscr{L}(v, q)$ lies in a bounded set independent of q, and Φ is *locally* Lipschitz on V (in the sense of being Lipschitz on all bounded subsets of V).

Under these conditions *we define the following algorithm* [2]: Starting from p^0, we calculate u^0, and then p^1, etc. The general rule is

(4.12) with p^n known, $(\in \Lambda)$, we define u^n as the element of M which minimises $J_0(v) + (p^n, \Phi(v))_L$

---

1) We shall return to this in Chapter 3.

2) This is essentially an algorithm of the Uzawa type /1/.

(this certainly defines u^n in a unique manner; see Remark 4.3 below); we next define

(4.13) $$p^{n+1} = P_\Lambda(p^n + \rho_n \Phi(u^n))$$

where

(4.14) $P_\Lambda =$ projection operator from $L \to \Lambda$,

(4.15) $\rho_n > 0$ is to be chosen appropriately (see below). ■

Remark 4.3. The function $v \to J_0(v) + (p^n, \Phi(v))_L$ is *strictly convex* and it is "infinite at infinity"; in fact, from (4.11) where we take u to be fixed arbitrarily in M, we have

$$\| \Phi(v) \|_L \leqslant C(\| v \| + 1)$$

and hence

$$J_0(v) + (p^n, \Phi(v))_L \geqslant \frac{\alpha}{2} \| v \|^2 - C(\| v \| + 1)$$

(where the C denote various constants). ■

Remark 4.4. (Motivation of the algorithm). If $\{u, p\}$ is a saddle point then

(4.16) $$J_0(u) + (p, \Phi(u))_L \leqslant J_0(v) + (p, \Phi(v))_L \quad \forall v \in M ,$$

which gives (4.12), and

(4.17) $$(q - p, \Phi(u)) \leqslant 0 \quad \forall q \in \Lambda , \quad p \in \Lambda$$

which is equivalent to

(4.18) $$p = P_\Lambda(p + \rho\Phi(u)) \quad \forall \rho > 0$$

and (4.18) explains (4.13). ■

Remark 4.5. In fact the algorithm is not completely defined since in (4.15) we have *the choice of the algorithm for calculating* u^n. In Section 4.4 we shall meet a variant in which this choice is specified. ■

Convergence of the algorithm.

We shall now prove:

Theorem 4.1. *We assume that* (4.3),...,(4.6) *are true, along with* (4.10) *and* (4.11) (*or* (4.11')). *Thus the algorithm defined by* (4.12), (4.13) *is convergent in the sense that*

(4.19) $u^n \to u$ *strongly in* V,

where u *is the solution of* (4.7), *when*

(4.20) $0 < \alpha_0 \leqslant \rho_n \leqslant \alpha_1$, *with suitable* α_0 *and* α_1 ([1])

Proof. Properties (4.12), (4.16) are equivalent to

(4.21) $$(J_0'(u^n), v - u^n) + (p^n, \Phi(v) - \Phi(u^n))_L \geqslant 0 \quad \forall v \in M,$$

(4.22) $$(J_0'(u), v - u) + (p, \Phi(v) - \Phi(u))_L \geqslant 0 \quad \forall v \in M;$$

taking $v = u$ (resp. $v = u^n$) in (4.21) (resp. (4.22)) we then deduce

(4.23) $$a(u^n - u, u^n - u) + (p^n - p, \Phi(u^n) - \Phi(u))_L \leqslant 0.$$

We put

(4.24) $$p^n - p = r^n.$$

From (4.13), (4.18) and the fact that P_Λ is *a contraction*, we deduce that

(4.25) $$\| r^{n+1} \|_L^2 \leqslant \| r^n + \rho_n(\Phi(u^n) - \Phi(u)) \|_L^2 = \| r^n \|_L^2 + \\ + 2\rho_n(r^n, \Phi(u^n) - \Phi(u))_L + \rho_n^2 \| \Phi(u^n) - \Phi(u) \|_L^2.$$

However, using (4.23) and (4.11) (or (4.11')), we deduce from (4.25) that

(4.26) $$\| r^{n+1} \|_L^2 \leqslant \| r^n \|_L^2 - 2\rho_n a(u^n - u, u^n - u) + C_1^2 \rho_n^2 \| u^n - u \|^2 \leqslant \\ \leqslant \| r^n \|_L^2 - (2\alpha\rho_n - C_1^2 \rho_n^2) \| u^n - u \|^2.$$

We then choose α_0 and α_1 in such a way that (4.20) implies that

(4.27) $$2\alpha\rho_n - C_1^2 \rho_n^2 \geqslant \beta > 0.$$

Then (4.26) gives

(4.28) $$\| r^{n+1} \|_L^2 + \beta \| u^n - u \|^2 \leqslant \| r^n \|_L^2.$$

From (4.28) we get that the sequence $n \to \| r^n \|_L^2$ is decreasing and hence tends to a limit, so that

$$\beta \| u^n - u \|^2 \to 0 \qquad \text{and hence (4.19)}$$

Remark 4.6. For some general results on the existence of saddle points, see Ekeland-Temam /1/. ■

---

1) Estimates of α_0 and α_1 are supplied in the proof which follows.

4.4 A second saddle-point search algorithm

We shall now supplement Remark 4.5, by assuming that

$$\text{(4.29)} \qquad \begin{cases} \Phi \in \mathscr{L}(V; L), \\ M = V. \end{cases}$$

Then in (4.12) u^n is defined by

$$J_0'(u^n) + \Phi^* p^n = 0,$$

i.e. ($A \in \mathscr{L}(V, V')$ is defined by $a(u, v) = (Au, v)$) :

$$\text{(4.30)} \qquad Au^n + \Phi^* p^n - f = 0$$

and if we introduce an iterative algorithm for the solution of (4.30) we naturally arrive at the following algorithm: with u^n and p^n assumed to have been calculated, we define u^{n+1} by (1):

$$\text{(4.31)} \qquad u^{n+1} = u^n - \rho_1 S^{-1}(Au^n + \Phi^* p^n - f)$$

where: S = identity if $V' = V$ = finite-dimensional space (or alternatively S = arbitrary symmetric positive-definite matrix), S = the duality operator from $V \to V'$ if V is infinite-dimensional (2), and we then define p^{n+1} by:

$$\text{(4.32)} \qquad p^{n+1} = P_\Lambda(p^n + \rho_2 \Phi u^{n+1}),$$

where in (4.31) and (4.32) ρ_1 and ρ_2 are two parameters > 0 to be suitably chosen. We have:

Theorem 4.2. *We assume that* (4.3),...,(4.6), (4.29) *are true. We can then choose* $\rho_1, \rho_2 > 0$ (3) *in such a way that algorithm* (4.31), (4.32) *is convergent in the sense that*

$$\text{(4.33)} \qquad u^n \to u \quad \textit{strongly in } V$$

where u *is the solution of* (4.7).

Proof. If $\{u, p\}$ denotes a saddle point of $\mathscr{L}(v, q)$ we have:

$$\text{(4.34)} \qquad u = u - \rho_1 S^{-1}(Au + \Phi^* p - f),$$

---

(1) This is essentially an algorithm of the Arrow-Hurwicz type.

(2) If we know that the solution u of the problem is in a reflexive space $W \subset V$, we can take an isomorphism from $W \to W'$ for S.

(3) Estimates of ρ_1 and ρ_2 result from the proof below.

(4.35) $\qquad p = P_\Lambda(p + \rho_2 \Phi u)$.

Putting $w^n = u^n - u$ and using the notation (4.24) we get:

(4.36) $\qquad w^{n+1} = w^n - \rho_1 S^{-1}(Aw^n + \Phi^* r^n)$,

(4.37) $\qquad \| r^{n+1} \|_L^2 \leqslant \| r^n \|_L^2 + 2\rho_2(r^n, \Phi w^{n+1})_L + \rho_2^2 \| \Phi w^{n+1} \|_L^2$.

We deduce from (4.36) that

$$Sw^{n+1} = Sw^n - \rho_1(Aw^n + \Phi^* r^n)$$

and hence, taking the inner product of both sides with w^{n+1} in the duality between V' and V:

$$\| w^{n+1} \|^2 = ((S - \rho_1 A) w^n, w^{n+1}) - \rho_1(\Phi^* r^n, w^{n+1}).$$

Since $A^* = A, (Av, v) \geqslant \alpha \| v \|^2$, we have:

$$\| (S - \rho_1 A) v \| \leqslant \beta \| v \|, \quad 0 < \beta < 1, \quad \text{for} \quad 0 < \rho_1 \leqslant \rho_0.$$

Thus:

(4.38) $\qquad \| w^{n+1} \|^2 \leqslant \beta \| w^n \| \| w^{n+1} \| - \rho_1(\Phi^* r^n, w^{n+1})$.

We seek ρ_1, ρ_2 in the form:

$$\rho_1 = \rho, \qquad \rho_2 = \rho C.$$

Using (4.38) in (4.37) we get

$$\begin{aligned} \| r^{n+1} \|_L^2 &\leqslant \| r^n \|_L^2 + 2C(\beta \| w^n \| \| w^{n+1} \| - \| w^{n+1} \|^2) + \\ &+ \rho^2 C^2 \| \Phi \|^2 \| w^{n+1} \|^2 \leqslant \| r^n \|_L^2 + \\ &+ 2C\beta(\| w^n \| \| w^{n+1} \| - \| w^{n+1} \|^2) + \\ &+ [\rho^2 C^2 \| \Phi \|^2 - 2C(1 - \beta)] \| w^{n+1} \|^2 \leqslant \| r^n \|_L^2 + \\ &+ C\beta(\| w^n \|^2 - \| w^{n+1} \|^2) + \\ &+ [\rho^2 C^2 \| \Phi \|^2 - 2C(1 - \beta)] \| w^{n+1} \|^2. \end{aligned}$$

However

$$2C(1 - \beta) - \rho^2 C^2 \| \Phi \|^2 \geqslant \gamma > 0 \quad \text{if} \quad 0 < \rho \leqslant \rho_0^*,$$

so that, if $0 < \rho \leqslant \min(\rho_0, \rho_0^*)$:

(4.39) $\qquad \| r^n \|_L^2 + C\beta \| w^n \|^2 - (\| r^{n+1} \|^2 + C\beta \| w^{n+1} \|^2) \geqslant \gamma \| w^{n+1} \|^2$.

Hence

$\| r^n \|_L^2 + C\beta \| w^n \|^2$ decreases with n, and therefore converges,

which in turn means that the left-hand side of (4.39) → 0 and hence $\| w^{n+1} \| \to 0$. ■

Remark 4.7. The behaviour of p^n depends on the characteristics of the problem in hand. We can show, for all the examples solved by means of the above methods, that the sequence p^n remains bounded and that all of the closure values of this sequence are such that $\{u, p\}$ is a saddle point of $\mathscr{L}(v, q)$ on $M \times \Lambda$. ■

Remark 4.8. The first method provides an estimate from below of

$$\inf_{v \in M} \left[J_0(v) + \sup_{q \in \Lambda} (q, \Phi(v))_L \right].$$

In fact, by definition of u^n,

$$J(u^n) + (p^n, \Phi(u^n))_L \leqslant J(u) + (p^n, \Phi(u))_L \leqslant$$
$$\leqslant J(u) + \sup_{q \in \Lambda} (q, \Phi(u))_L = \inf_{v \in M} \left(J_0(v) + \sup_{q \in \Lambda} (q, \Phi(v))_L \right). \quad ■$$

Remark 4.9. The above methods apply to the minimisation of nondifferentiable functionals, and in particular to all the problems of this category studied in this book, such as that of Example 4.1 which can be written explicitly as follows:

$$(4.40) \qquad \inf_{v \in H_0^1(\Omega)} \left[\frac{1}{2} \int_\Omega |\operatorname{grad} v|^2 \, dx + g \int_\Omega |\operatorname{grad} v| \, dx - (f, v) \right]$$

and which will be studied in Chapter 5. In the following we shall show that the discretisation of (4.40) leads to functionals of the *type* found in Counter-Example 1.1 (although clearly more complex). ■

Remark 4.10. The solution of the problem

$$\min J(v) = \tfrac{1}{2}(Av, v) - (f, v)$$

where A is a symmetric positive-definite $N \times N$ matrix, subject to the constraints

$$Bv = b$$

where B is a $M \times N$ matrix of *rank* $M \leqslant N$, reduces, after introduction of the Lagrange variables $p \in \mathbb{R}^M$, to the solution of the system

$$\begin{bmatrix} A & B^* \\ B & 0 \end{bmatrix} \begin{bmatrix} u \\ p \end{bmatrix} = \begin{bmatrix} f \\ b \end{bmatrix},$$

which is a system of the form $\mathscr{A} z = d$, where $\mathscr{A}$ is a symmetric, but not generally positive definite matrix. In particular, the conjugate gradient method cannot be applied in this case. It will be noted however that the problem

$$\mathscr{A}^* \mathscr{A} z = \mathscr{A}^* d$$

is equivalent to the above and that the matrix $\mathscr{A}^* \mathscr{A}$ is symmetric and positive definite. The application of the conjugate gradient method is therefore possible and convergence is achieved, in a finite-dimensional space, in a finite number of iterations. From the numerical point of view, the use of $\mathscr{A}^* \mathscr{A}$ introduces rounding errors, due to a deterioration in the conditioning of the matrix, and this can adversely affect the convergence of the algorithm. This has led a number of authors (see Beckmann /1/, Luenberger /2/) to adapt the conjugate gradient method directly to the case in which it is required to solve a system $\mathscr{A}z = d$, with $\mathscr{A}$ symmetric but not necessarily positive definite. ■

5. APPLICATION OF RELAXATION AND DUALITY METHODS TO THE NUMERICAL ANALYSIS OF A MODEL VARIATIONAL PROBLEM.

5.1 General remarks

We shall now apply the results of Sections 1 and 4, and those of Chapter 1, Section 5, to the solution of a variational problem which arises in *lubrication* theory (1); this is a variant of the problem considered in Chapter 1, Example 3.6. This problem is solved by a variety of methods in Fusciardi, Mosco, Scarpini and Schiaffino /1/ and Marzulli /1/ (2).

5.2 The continuous problem

5.2.1 *Statement of the continuous problem*

Let $\Omega =]0, 1[\times]0, b[$, $\Gamma_1 =]0, 1[\times \{ 0 \}$, $\Gamma_2 =]0, 1[\times \{ b \}$ and

$$V = [v \mid v \in H^1(\Omega),\ v|_{\Gamma_1} = 0,\ v|_{\Gamma_2} = 0,\ v(0, x_2) = v(1, x_2) \text{ a.e. on }]0, b[]$$

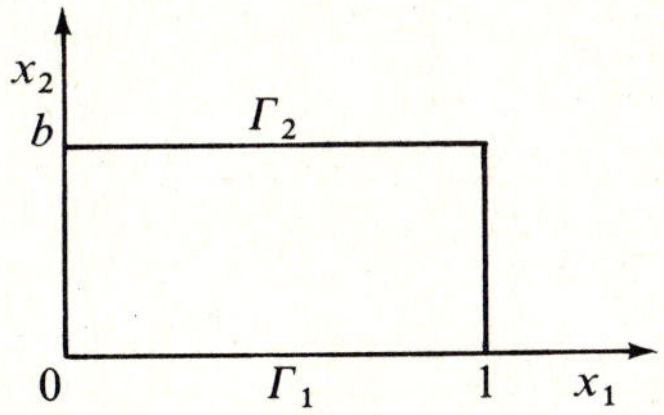

Fig. 5.1.

(1) For the physical motivation, see Marzulli /1/ and the bibliography therein.

(2) Cryer /2/ gives an application of the method of over-relaxation with projection to a lubrication problem of the same type as that considered in this section, the domain Ω considered being one-dimensional.

be a Hilbert space for the inner product and the norm:

$$(5.1) \qquad (u, v)_V = \int_\Omega \operatorname{grad} u . \operatorname{grad} v \, dx ,$$

$$(5.2) \qquad \| u \|_V = \left(\int_\Omega | \operatorname{grad} u |^2 \, dx \right)^{1/2} .$$

Given $f \in L^2(\Omega)$ (which is sufficient for the applications which we have in mind), we define on V the functional J, strictly convex and continuous, by:

$$(5.3) \qquad J(v) = \frac{1}{2} \int_\Omega | \operatorname{grad} v |^2 \, dx - \int_\Omega f v \, dx$$

and we consider the problem (P) :

$$(P) \quad \begin{cases} \min_{v \in K} J(v) , \\ K = [v \mid v \in V, \; v \geqslant 0 \text{ a.e. on } \Omega] , \end{cases}$$

since K is convex and closed in V and $\lim_{\|v\|_V \to +\infty} J(v) = +\infty$, (P) admits a single optimal solution; we denote this solution by u. ■

5.2.2 *Regularity results*

According to Brézis and Stampacchia /2/, if $f \in L^p(\Omega)$ $(2 \leqslant p < +\infty)$ we have

$u \in V \cap W^{2,p}(\Omega)$ [1] ■

5.2.3 *Characterisation of the solution*

It follows from Chapter 1, Section 2 that the optimal solution u is characterised by:

$$(5.4) \qquad \int_\Omega \operatorname{grad} u . \operatorname{grad} (v - u) \, dx \geqslant \int_\Omega f (v - u) \, dx \quad \forall v \in K , \quad u \in K .$$

In view of the regularity (see 5.2.2 above) we can integrate by parts in (5.4), which leads to the characterisation:

---

[1] By definition and for $p \geqslant 1$

$$W^{2,p}(\Omega) = \left[v \mid v, \frac{\partial v}{\partial x_i}, \frac{\partial^2 v}{\partial x_i \, \partial x_j} \in L^p(\Omega), i, j = 1, n \right] .$$

$$(5.5)\qquad \begin{cases} u \in V, \\ -\Delta u - f \geqslant 0 \quad \text{in} \quad \Omega, \quad \displaystyle\int_\Gamma v \frac{\partial u}{\partial n}\, d\Gamma = 0 \quad \forall v \in V, \\ u \geqslant 0 \quad \text{a.e.} \\ (-\Delta u - f)\, u = 0 \quad \text{a.e. on } \Omega \end{cases}$$

K being a cone with apex O.

The characterisation (5.5) is fundamental since it introduces $\lambda = -\Delta u - f$ ($\lambda \in L^2(\Omega)$ from 5.2.2) as a *Lagrange multiplier* for the problem (P); we shall return to the above in Section 5.6. ■

5.3 The approximate problem

Remark 5.1. It is convenient, bearing in mind the periodicity in x_1 of the elements of V, to identify Ω with $\tilde{\Omega} = \gamma \times]0, b[$ where γ is the circle of radius $1/2\pi$ (torus of dimension 1) orientated in the usual sense (the physical problem is itself expressed in such a domain); the boundary $\tilde{\Gamma}$ of $\tilde{\Omega}$ is thus

$$\gamma \times \{0\} \cup \gamma \times \{b\}$$

and hence

$$V = \{ v \mid v \in H^1(\tilde{\Omega}),\ v|_{\tilde{\Gamma}} = 0 \}.$$

(P) is therefore a (constrained) Dirichlet problem on a cylindrical manifold. ■

5.3.1 *Exterior approximation of* (P)

The considerations of Chapter 1, Section 5.1, adapt readily to (P); taking account of Remark 5.1 and taking N_1, N_2 to be positive integers which will tend towards infinity, where

$$h_1 = 1/N_1,\ h_2 = b/N_2,\ h = (h_1, h_2),$$

we define:

$$(5.6)\qquad R_h = \{ M \mid M \in \gamma \times R,\ M = (m_1 h_1, m_2 h_2),\ 0 \leqslant m_1 \leqslant N_1 - 1,\ m_2 \in \mathbb{Z} \}.$$

With each node of R_h we associate the "panel" with centre M:

$$(5.7)\qquad \varpi_h^0(M) = \gamma_{m_1} \times](m_2 - \tfrac{1}{2})\, h_2, (m_2 + \tfrac{1}{2})\, h_2[\,,$$

$$(5.8)\qquad \varpi_h^1(M) = \varpi_h^0(M) \cup \varpi_h^0\left(M \pm \frac{h_2}{2} e_2\right),$$

with γ_{m_1} $(0 \leqslant m_1 \leqslant N_1 - 1)$ defined by:

$$(5.9)\qquad \begin{cases} \gamma_0 =]1 - h_1/2, 1] \cup [0, h_1/2[\,, \\ \gamma_{m_1} =](m_1 - \tfrac{1}{2})\, h_1, (m_1 + \tfrac{1}{2})\, h_1[\quad \text{if} \quad 1 \leqslant m_1 \leqslant N_1 - 1\,. \end{cases}$$

By analogy with Chapter 1, Section 5.1, we next define:

(5.10) $$\tilde{\Omega}_h = \{ M \mid M \in R_h,\ \varpi_h^1(M) \subset \tilde{\Omega} \},$$

(5.11) $$V_h = \{ v_h \mid v_h = (v_M)_{M \in \tilde{\Omega}_h},\ v_M \in \mathbb{R} \}$$

and the mapping q_h [1]: $V_h \to L^2(\gamma \times \mathbb{R})$ by:

(5.12) $$q_h v_h = \sum_{M \in \tilde{\Omega}_h} v_M \theta_h^M$$

where: (see (5.9), Chapter 1):

(5.13) $$\theta_h^M = \text{characteristic function of } \varpi_h^0(M).$$

Once again putting,

(5.14) $$\delta_i \varphi(x) = \frac{1}{h_i}\left[\varphi\left(x + \frac{h_i}{2} e_i\right) - \varphi\left(x - \frac{h_i}{2} e_i\right)\right]$$ [2]

we "approximate" J by $J_h : V_h \to \mathbb{R}$ defined by:

(5.15) $$J_h(v_h) = \sum_{i=1}^{2} \frac{1}{2} \int_{\tilde{\Omega}} | \delta_i q_h v_h |^2 \, dx - \int_{\tilde{\Omega}} f q_h v_h \, dx$$

which gives the approximate problem:

(5.16) $$(P_h) \begin{cases} \min\limits_{v_h \in K_h} J_h(v_h), \\ K_h = [v_h \mid v_h \in V_h,\ v_M \geqslant 0\ \forall M \in \tilde{\Omega}_h]. \end{cases}$$

It may be noted that $q_h v_h \geqslant 0\ \forall v_h \in K_h$. ■

5.3.2 *Solvability of* (P_h)

It is immediate that

$$v_h \to \left(\sum_{i=1}^{2} \int_{\tilde{\Omega}} | \delta_i q_h v_h |^2 \, dx \right)^{1/2}$$

---

(1) We could, as in Chapter 1, Section 5.2, define $p_h : V_h \to (L^2(\gamma \times$ using q_h and the δ_i defined in (5.14); however, in view of the elementary nature of the convex set K considered in this section, this would not be particularly useful.

(2) In (5.14), for $i = 1$, $x \pm \frac{h_1}{2} e_1$ is defined *modulo* 1 such that it belongs to $[0, 1[$.

defines a norm on V_h, which will be denoted by $\| v_h \|_h$; the functional J_h is therefore continuous and strictly convex on V_h with

$$\lim_{\| v_h \|_h \to +\infty} J_h(v_h) = +\infty ,$$

and since K_h is closed we have the existence and uniqueness of an optimal solution for (P_h), denoted by u_h. ■

5.3.3 *Explicit formulation of* (P_h)

With a view to the numerical solution of (P_h), it is necessary to make (5.15), (5.16) explicit; if $M \in \tilde{\Omega}_h$, it is convenient to identify M by a double subscript (k, l) which gives $M_{kl} = (kh_1, lh_2)$ with $0 \leqslant k \leqslant N_1 - 1, 1 \leqslant l \leqslant N_2 - 1$. We can then write (5.15) in explicit form, as follows:

$$(5.17)\quad \begin{cases} 2\, J_h(v_h) = \dfrac{h_1 h_2}{2} \displaystyle\sum_{M_{kl} \in \tilde{\Omega}_h^*} \left[\left(\frac{v_{k+1\,l} - v_{kl}}{h_1} \right)^2 + \left(\frac{v_{k-1\,l} - v_{kl}}{h_1} \right)^2 + \right. \\ \qquad \left. + \left(\dfrac{v_{kl+1} - v_{kl}}{h_2} \right)^2 + \left(\dfrac{v_{kl-1} - v_{kl}}{h_2} \right)^2 \right] - 2 h_1 h_2 \displaystyle\sum_{M_{kl} \in \tilde{\Omega}_h} f_{kl} v_{kl} , \\ \\ f_{kl} = \dfrac{1}{h_1 h_2} \displaystyle\int_{\varpi_h^0(M_{kl})} f(x)\, \mathrm{d}x \quad (^1) \end{cases}$$

where $\tilde{\Omega}_h^* = \tilde{\Omega}_h \cup (\tilde{\Gamma} \cap R_h)$; the explicit form of (5.16) is:

$$(5.18)\quad \begin{cases} \min\limits_{v_h \in K_h} J_h(v_h) , \\ K_h = [v_h \mid v_h \in V_h,\ v_{kl} \geqslant 0 \ \forall M_{kl} \in \tilde{\Omega}_h] . \quad ■ \end{cases}$$

Remark 5.2. In (5.17) it is necessary to take $v_{kl} = 0$ if $M_{kl} \notin \tilde{\Omega}_h$, and in order to take account of the periodicity in x_1, $v_{-1l} = v_{N_1 - 1l}$, $v_{N_1 l} = v_{0l}\ \forall l$. ■

5.4 Convergence of the approximate solution as $h \to 0$

Preliminary remark. The convergence of u_h (in fact of $q_h u_h$) to the solution of problem (P) is a direct consequence of the results of Chapter 1, Section 5.2. However, since the problem considered in this section is the first example of the numerical solution of a variational inequality encountered in this book we have chosen to detail the various points of the proof. ■

We have:

Theorem 5.1. *As* $h \to 0$ *we have*

---

(1) If f is continuous, we take $f_{kl} = f(M_{kl})$.

(5.19) $$\{ q_h u_h, \delta_1 q_h u_h, \delta_2 q_h u_h \} \to \left\{ u, \frac{\partial u}{\partial x_1}, \frac{\partial u}{\partial x_2} \right\} \text{ strongly in } (L^2(\tilde{\Omega}))^3$$

(5.20) $$J_h(u_h) \to J(u) .$$

Proof. The norm $\| v_h \|_h$ defined in 5.3.2 satisfies the discrete Poincaré inequality:

(5.21) $$\| q_h v_h \|_{L^2(\tilde{\Omega})} \leqslant \alpha \left(\sum_{i=1}^{2} \int_{\tilde{\Omega}} | \delta_i q_h v_h |^2 \, dx \right)^{1/2}$$
$$(\alpha > 0) , \quad \forall h, \forall v_h \in V_h .$$

For this result we refer the reader to Céa /3/, Lions /3/.

Since $0 \in K_h$ we have $J_h(u_h) \leqslant J_h(0) = 0$ so that, taking account of (5.21):

(5.22) $$\begin{cases} \| q_h u_h \|_{L^2(\tilde{\Omega})} \leqslant C \quad \forall h , \\ \| \delta_i q_h u_h \|_{L^2(\tilde{\Omega})} \leqslant C \quad \forall h \quad (i = 1, 2) , \end{cases}$$

and hence, by proceeding as in the proof of Theorem 5.1, Chapter 1, Section 5.1, it can be shown that for a subsequence, also denoted by u_h, we have:

(5.23) $$\begin{cases} q_h u_h \to w \quad \text{weakly in } L^2(\tilde{\Omega}) \\ \delta_i q_h u_h \to \dfrac{\partial w}{\partial x_i} \quad \text{weakly in } L^2(\tilde{\Omega}) \qquad (i = 1, 2) , \end{cases}$$

with:

(5.24) $$w \in V .$$

Moreover $q_h u_h \geqslant 0$ a.e. and since the set $[v \mid v \in L^2(\tilde{\Omega}), v \geqslant 0 \text{ a.e.}]$ is convex and closed in $L^2(\tilde{\Omega})$ and therefore weakly closed, we deduce:

(5.25) $$w \geqslant 0 \quad \text{a.e.} \quad \Rightarrow w \in K .$$

It now remains to show that $w = u$; for this, we define $r_h : V \to V_h$ by:

(5.26) $$(r_h v)_M = \frac{1}{h_1 h_2} \int_{\varpi^0(M)} v(x) \, dx , \qquad M \in \tilde{\Omega}_h ,$$

and if $v \geqslant 0$ a.e. we clearly have $(r_h v)_M \geqslant 0 \ \forall M \in \tilde{\Omega}_h \Rightarrow r_h v \in K_h$ which gives:

(5.27) $$v \in K \Rightarrow r_h v \in K_h .$$

It can be shown that $\forall v \in V$ we have:

$$(5.28)\qquad \begin{cases} q_h\, r_h\, v \to v & \text{strongly in } L^2(\tilde{\Omega}) \\ \delta_i q_h\, r_h\, v \to \dfrac{\partial v}{\partial x_i} & \text{strongly in } L^2(\tilde{\Omega}) \quad (i = 1, 2)\,, \end{cases}$$

from which we deduce

$$(5.29)\qquad J_h(r_h\, v) \to J(v)\,.$$

In view of (5.27), since u_h is optimal on K_h we then have:

$$(5.30)\qquad J_h(u_h) \leqslant J_h(r_h\, v) \quad \forall v \in K$$

and in the limit, by weak lower semi-continuity:

$$(5.31)\qquad J(w) \leqslant \liminf J_h(u_h) \leqslant J(v) \quad \forall v \in K\,,$$

and from (5.24), (5.25), (5.27), (5.31), we deduce that $w = u$.

It may be noted that the uniqueness of u implies the convergence of the *entire* sequence $q_h\, u_h$.

To prove (5.20) it is sufficient to set $v = u$ in (5.30), (5.31); from this the strong convergence is deduced; in fact:

$$\sum_{i=1}^{2} \int_{\tilde{\Omega}} |\, \delta_i q_h\, u_h \,|^2 \,\mathrm{d}x - \sum_{i=1}^{2} \int_{\tilde{\Omega}} |\, \delta_i q_h\, r_h\, u \,|^2 \,\mathrm{d}x = 2\, J_h(u_h) -$$

$$- 2\, J_h(r_h\, u) + 2\left[\int_{\tilde{\Omega}} f q_h\, u_h \,\mathrm{d}x - \int_{\tilde{\Omega}} f q_h\, r_h\, u \,\mathrm{d}x \right],$$

and in the limit, from (5.20), (5.23), (5.28), (5.29), (5.31):

$$(5.32)\qquad \sum_{i=1}^{2} \int_{\tilde{\Omega}} |\, \delta_i q_h\, r_h\, u \,|^2 \,\mathrm{d}x \to \int_{\tilde{\Omega}} |\, \mathrm{grad}\, u \,|^2 \,\mathrm{d}x\,,$$

$$(5.33)\qquad J_h(u_h) \to J(u)\,,$$

$$(5.34)\qquad J_h(r_h\, u) \to J(u)\,,$$

$$(5.35)\qquad \int_{\tilde{\Omega}} f q_h\, u_h \,\mathrm{d}x \to \int_{\tilde{\Omega}} f u \,\mathrm{d}x\,,$$

$$(5.36)\qquad \int_{\tilde{\Omega}} f q_h\, r_h\, u \,\mathrm{d}x \to \int_{\tilde{\Omega}} f u \,\mathrm{d}x\,,$$

and hence:

$$(5.37)\qquad \sum_{i=1}^{2} \int_{\tilde{\Omega}} |\, \delta_i q_h\, u_h \,|^2 \,\mathrm{d}x \to \int_{\tilde{\Omega}} |\, \mathrm{grad}\, u \,|^2 \,\mathrm{d}x\,,$$

which proves convergence of the norm in $(L^2(\tilde{\Omega}))^3$, and hence, with (5.23), the strong convergence. ■

5.5 Solution of the approximate problem by point over-relaxation with constraints

We use the algorithm (1.41), (1.42) of Section 1.4 which, taking account of (5.17), (5.18), takes the form:

$$(5.38)\quad \begin{cases} u_{kl}^0 \geqslant 0 \quad \forall M_{kl} \in \tilde{\Omega}_h\,, \\ u_{kl}^{n+1/2} = (1-\omega)\,u_{kl}^n + \dfrac{\omega h_1^2 h_2^2}{2(h_1^2+h_2^2)} \times \\ \qquad \times \left[f_{kl} + \dfrac{1}{h_1^2}\left(u_{k+1\,l}^n + u_{k-1\,l}^{n+1}\right) + \dfrac{1}{h_2^2}\left(u_{kl+1}^n + u_{kl-1}^{n+1}\right)\right], \\ u_{kl}^{n+1} = \max\left(0, u_{kl}^{n+1/2}\right), \\ M_{kl} \in \tilde{\Omega}_h\,, \end{cases}$$

if we iterate with k, l increasing.

Remark 5.2 is valid for (5.38).
Since Theorem 1.3, Section 1.4 applies to (P_h) we have:

Proposition 5.1. *The sequence* $u_h^n \to u_h$ *if* $0 < \omega < 2$. ■

Remark 5.3. If $u_h^0 = 0$ and if $0 < \omega \leqslant 1$, we immediately get from relations (5.38) that the sequence $(u_{kl}^n)_n$ is increasing $\forall M_{kl} \in \tilde{\Omega}_h$; the convergence is thus monotone and the coordinates u_{kl} of u_h are approached from below.

5.6 Solution by a duality method

Problem (P) comes within the setting of Section 4 if we take:

$$J_0(v) = J(v)\,,\quad M = V\,,\quad \Phi(v) = v\,,\quad L = L^2(\Omega)\,,$$
$$\Lambda = L_-^2(\Omega) = \{\, q \mid q \in L^2(\Omega),\ q \leqslant 0 \quad \text{a.e.}\ \}\,.$$

In fact, we then have

$$\sup_{q\in\Lambda}\,(q, \Phi(v))_L = \begin{cases} 0 & \text{if} \quad v \in K\,, \\ +\infty & \text{if} \quad v \notin K\,. \end{cases}$$

It is somewhat simpler if we use only $L_+^2(\Omega)$, replacing q by $-\mu$ problem (P) then takes the form:

$$(5.39)\qquad \inf\left[J(v) + \sup_{\mu\in L_+^2(\Omega)} (-\mu, v)\right].$$

We therefore introduce the Lagrangian:

$$(5.40)\qquad \mathscr{L}(v,\mu) = J(v) - (\mu, v)\,.$$

Let u be the solution of (P), or equivalently, of (5.5).
We put

$$\lambda = -\Delta u - f. \tag{5.41}$$

In order to show *that the conditions of Theorem* 4.1 *are satisfied*, it is sufficient to prove the analogue of (4.10). Now, more precisely, we have:

Theorem 5.2. *The pair* $\{u, \lambda\}$ (λ defined by (5.41)) *is the unique saddle-point of* $\mathscr{L}(v, \mu)$ *on* $V \times L^2_+(\Omega)$.

Proof. 1) From 5.2.2 and (5.5) we have $\lambda \in L^2_+(\Omega)$ and $\lambda u = 0$ a.e.; furthermore:

$$u \geqslant 0 \Rightarrow \int_\Omega \mu u \, \mathrm{d}x \geqslant 0 \quad \forall \mu \in L^2_+(\Omega),$$

and hence:

$$\mathscr{L}(u, \mu) \leqslant \mathscr{L}(u, \lambda) \, (= J(u)) \quad \forall \mu \in L^2_+(\Omega) ; \tag{5.42}$$

2) With λ fixed, we consider the optimisation problem:

$$\min_{v \in V} \mathscr{L}(v, \lambda) . \tag{5.43}$$

The unique optimal solution u_λ of (5.43) is characterised by:

$$\begin{cases} -\Delta u_\lambda = f + \lambda, & \displaystyle\int_\Gamma v \frac{\partial u_\lambda}{\partial n} \mathrm{d}\Gamma = 0 \quad \forall v \in V \\ u_\lambda \in V, \end{cases} \tag{5.44}$$

from which $u_\lambda = u$, and hence:

$$\mathscr{L}(u, \lambda) \leqslant \mathscr{L}(v, \lambda) \quad \forall v \in V \tag{5.45}$$

which, with (5.42), shows that $\{u, \lambda\}$ is a saddle point.

We observe that:

$$\int_\Omega (\mu - \lambda) \, u \, \mathrm{d}x \geqslant 0 \quad \forall \mu \in L^2_+(\Omega) . \tag{5.46}$$

3) In order to prove the uniqueness, we note that any saddle-point (u^*, λ^*) of $\mathscr{L}$ on $V \times L^2_+(\Omega)$ satisfies a relation of the type (5.41), from which we easily deduce the characterisation:

$$
(5.47)\qquad \begin{cases} -\Delta u^* = f + \lambda^* \quad \text{on } \Omega\,, & \displaystyle\int_\Gamma v \frac{\partial u^*}{\partial n}\,\mathrm{d}\Gamma = 0 \quad \forall v \in V\,, \\ u^* \in V\,, \\ \displaystyle\int_\Omega (\mu - \lambda^*)\, u^*\,\mathrm{d}x \geqslant 0 \quad \forall \mu \in L^2_+(\Omega)\,, \\ \lambda^* \in L^2_+(\Omega)\,. \end{cases}
$$

We then set $\mu = \lambda^*$ in (5.46), and $\mu = \lambda$ in the third relation in (5.47), which by addition gives:

$$
(5.48)\qquad \int_\Omega (\lambda^* - \lambda)\,(u^* - u)\,\mathrm{d}x \leqslant 0\,;
$$

now

$$
(5.49)\qquad \lambda^* - \lambda = -\,\Delta(u^* - u)\,;
$$

we therefore deduce from (5.47), (5.48) that:

$$
(5.50)\qquad -\int_\Omega \Delta(u^* - u)\,(u^* - u)\,\mathrm{d}x = \int_\Omega |\operatorname{grad}(u^* - u)|^2\,\mathrm{d}x \leqslant 0\,,
$$

which gives $\| u^* - u \|_V^2 \leqslant 0$ and hence $u^* = u$ and $\lambda^* = \lambda$. ■

Remark 5.4. It is easy to prove an analogous result for the approximate problem (P_h), using the approximate Lagrangian $\mathscr{L}_h$ defined by:

$$
(5.51)\qquad \mathscr{L}_h(v_h, \mu_h) = J_h(v_h) - h_1\, h_2 \sum_{M_{kl} \in \tilde{\Omega}_h} \mu_{kl}\, v_{kl}\,,
$$

with

$$
\mu_h = \{ \mu_{kl} \}_{M_{kl} \in \tilde{\Omega}_h}\,. \quad ■
$$

5.6.2 *Solution of the problem (P) by a dual method*

The algorithm of Section 4.3 applied to the problem (P) takes the form:

$$
(5.52)\qquad \lambda^0 \in L^2_+(\Omega) \qquad \text{chosen arbitrarily (e.g. zero)},
$$

$$
(5.53)\qquad \begin{cases} -\Delta u^{n+1} = f + \lambda^n \quad \text{in } \Omega, & \displaystyle\int_\Gamma v \frac{\partial u^{n+1}}{\partial n}\,\mathrm{d}\Gamma = 0 \quad \forall v \in V\,, \\ u^{n+1} \in V\,, \end{cases}
$$

(5.54) $\lambda^{n+1} = (\lambda^n - \rho_n u^{n+1})^+ =$
$= \text{positive part of } \lambda^n - \rho_n u^{n+1}, \quad \rho_n > 0.$

Since the canonical injection from $V \to L^2(\Omega)$ is continuous, there exists $C > 0$ such that:

(5.55) $\| v \|_{L^2(\Omega)} \leqslant C \| v \|_V \quad \forall v \in V$

(this gives the Lipschitz constant of Φ; see Section 4.3).
Theorem 4.1 gives

(5.56) $u^n \to u$ strongly in V (1)

under the condition

(5.57) $0 < \alpha_0 \leqslant \rho_n \leqslant \alpha_1 < 2/C^2,$

where the constant C is that of (5.55).
In the case of problem (P) it may easily be shown that the optimal value of C (i.e. the smallest possible), say C_0, is the reciprocal of the square root of the smallest eigenvalue, λ_0, associated with the sequence of eigenfunctions of the operator $-\Delta$ in V, this sequence being given by:

(5.58) $\left\{\cos 2\pi m x_1 \sin n\pi \frac{x_2}{b}, \sin 2\pi m x_1 \sin n\pi \frac{x_2}{b}; m \geqslant 0, n \geqslant 1\right\}$

and the sequence of corresponding eigenvalues by:

(5.59) $\left\{\pi^2\left[4m^2 + \frac{n^2}{b^2}\right]; m \geqslant 0, n \geqslant 1\right\}$

which gives $\lambda_0 = \pi^2/b^2$, $C_0 = b/\pi$ and, for ρ_n the condition:

(5.60) $0 < \alpha_0 \leqslant \rho_n \leqslant \alpha_1 < 2\pi^2/b^2$

which defines (5.57) more precisely. ■

5.6.3 *Solution of the problem* (P_h) *by a dual method*

The analogue of the above algorithm for the problem (P_h) is given by:

(5.61) $\lambda_{kl}^0 \geqslant 0 \quad \forall M_{kl} \in \tilde{\Omega}_h$, chosen *arbitrarily* (e.g. zero)

---

(1) In fact we have strong convergence in $H^{2-\varepsilon}(\Omega) \cap V$, for arbitrarily small $\varepsilon > 0$: indeed, the proof of Theorem 4.1 shows that λ^n is bounded in $L^2(\Omega) \Rightarrow \Delta u^n$ is bounded in $L^2(\Omega) \Rightarrow u^n$ is bounded in $H^2(\Omega) \cap V$, which gives the required result. For the definition of the spaces $H^s(\Omega)$, for non-integer s, see Lions-Magenes /1/.

(5.62)
$$\begin{cases} -\left[\dfrac{u_{k+1l}^{n+1}+u_{k-1l}^{n+1}-2\,u_{kl}^{n+1}}{h_1^2}+\dfrac{u_{kl+1}^{n+1}+u_{kl-1}^{n+1}-2\,u_{kl}^{n+1}}{h_2^2}\right]= \\ \quad = f_{kl}+\lambda_{kl}^n, \quad M_{kl}\in\tilde{\Omega}_h\,, \end{cases}$$

(5.63)
$$\lambda_{kl}^{n+1}=\max\,(0,\lambda_{kl}^n-\rho_n\,u_{kl}^{n+1})\,,$$

with Remark 5.2 holding for (5.62).

Under a condition on ρ_n of the type (5.57),

$$u_h^n=\{\,u_{kl}^n\,\}_{M_{kl}\in\tilde{\Omega}_h}$$

converges to u_h, the solution of the problem (P_h)(1).

5.7 Analysis of the numerical results

All the calculations were carried out for:

(5.64)
$$f(x_1,x_2)=4\,\pi^2\sin 2\,\pi x_1\,,$$

using the CII 10070 computer at IRIA.

5.7.1 *Numerical values of the parameters*

For b we used the three values:

$$b = 0.2, \quad 0.34, \quad 0.69:$$

for h_1:

$$h_1 = 1/10, \quad 1/30, \quad 1/50;$$

for h_2:

$$h_2 = b/10.$$

With the termination criterion chosen as:

(5.65)
$$\sum_{M_{kl}\in\tilde{\Omega}_h}|\,u_{kl}^{n+1}-u_{kl}^n\,|\leqslant\varepsilon$$

we took:

$$\varepsilon = 10^{-4} \text{ for } b = 0.69,\ \varepsilon = 5.10^{-5} \text{ for } b = 0.34,$$
$$\varepsilon = 10^{-5} \text{ for } b = 0.2. \qquad \blacksquare$$

---

(1) And $\lambda_h^n=\{\,\lambda_{kl}^n\,\}_{M_{kl}\in\tilde{\Omega}_h}$ converges to $\lambda_h=\{\,\lambda_{kl}\,\}_{M_{kl}\in\tilde{\Omega}_h}$ where λ_{kl} is defined by:

$$\lambda_{kl}=-\left[\frac{u_{k+1l}+u_{k-1l}-2\,u_{kl}}{h_1^2}+\frac{u_{kl+1}+u_{kl-1}-2\,u_{kl}}{h_2^2}\right]-f_{kl}\quad\forall M_{kl}\in\tilde{\Omega}_h\,.$$

5.7.2 *Solution using over-relaxation*

The results in Table 5.1 correspond to the solution of the problem (P_h) using algorithm (5.38) with:

$$(5.66)\quad \begin{cases} \omega = 1.6 \text{ for } h_1 = 1/10\,, \quad 1.7 \text{ for } h_1 = 1/30\,, \\ \qquad 1.8 \text{ for } h_1 = 1/50\,, \\ u_h^0 = 0\,, \end{cases}$$

the parameters h_1, h_2, b, ε having the values given in 5.7.1, and the termination criterion being given by (5.65).

b	h_1	Number of iterations	CII 10070 computation time in minutes
0.2	1/10	17	0.025
	1/30	25	0.08
	1/50	30	0.19
0.34	1/10	17	0.025
	1/30	30	0.10
	1/50	35	0.22
0.69	1/10	17	0.025
	1/30	35	0.12
	1/50	50	0.31

Table 5.1

Fig. 5.2 shows, for $b = 0.69$ and $h_1 = 1/50$, the number of iterations as a function of ω, the termination criterion in all cases being given by (5.65); also shown on this figure is the curve giving the number of iterations as a function of ω (again using the same termination criterion) for the *unconstrained* problem (1), i.e.

$$(5.67)\quad \begin{cases} -\Delta u = f \quad \text{in } \Omega\,, \quad \displaystyle\int_\Gamma \frac{\partial u}{\partial n} v \,\mathrm{d}\Gamma = 0\,, \\ u \in V\,. \end{cases}$$

(1) In fact for the discretised form of (5.67).

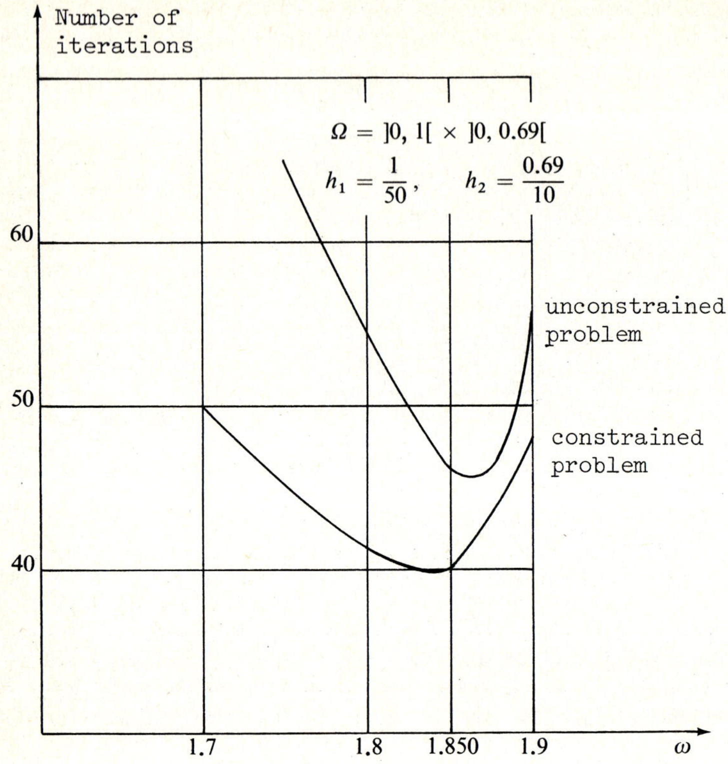

Fig. 5.2. Influence of the parameter ω on the rate of convergence.

We can therefore make the following remarks:

Remark 5.5. For a given ω, the over-relaxation method is more rapid, in terms of the number of iterations, for the constrained problem than for the unconstrained problem. ■

Remark 5.6. It may be noted that the maximal rate of convergence is achieved for values of ω much greater than 1; this justifies Remark 1.9 of Section 1.4. ■

Remark 5.7. In the constrained case, the optimal value of ω is a function of f (in contrast to the unconstrained case). More precisely, if we write $\Omega^+ = [x \mid x \in \Omega,\ u(x) > 0]$, the value of ω is essentially that which corresponds to the approximate solution of the Dirichlet problem: $-\Delta u = f$ on Ω^+, the discretisation parameter

h_1, h_2 remaining the same. Clearly Ω^+ is not known a priori, so that the above remark is of theoretical interest only. ■

5.7.3 *Solution using Uzawa's algorithm*

The problem (P_h) was solved using the algorithm (5.61), (5.62), (5.63) with a *constant* parameter ρ and $\lambda_h^0 = 0$, the termination criterion again being given by (5.65).

The approximate Dirichlet sub-problems (5.62) were solved by point over-relaxation with $\omega = 1.6$ for $h_1 = 1/10$, 1.7 for $h_1 = 1/30$ and 1.8 for $h_1 = 1/50$.

Table 5.2 shows the number of iterations and the computation times for algorithm (5.61), (5.62), (5.63), the parameter ρ having its optimal value (determined experimentally).

Figure 5.3 shows, for $h_1 = 1/50$ and $b = 0.69$, the variation of the number of iterations as a function of ρ, using the termination criterion (5.65); the following remarks may be made:

Remark 5.8. The total number of iterations required to solve (P_h), with a given ρ, is more or less independent of the accuracy with which the approximate Dirichlet sub-problems (5.62) are solved. ■

Remark 5.9. For $b = 0.69$, we have $2\pi^2/b^2 \simeq 42.5$; the estimate (5.60) (1) is thus *very realistic* for the problem (P_h) since for $\rho = 42.5$ convergence is attained in 522 iterations, for $\rho = 44$ in 698 iterations, and for $\rho = 45$ the solution actually diverges.

Remark 5.10. In view of the above results, it appears that in the case of problem (P) the over-relaxation method is quicker than the duality method, at least when the Dirichlet sub-problems (5.62) are solved by over-relaxation.

It has also been established that the adjustment of ρ to obtain satisfactory convergence is more critical than that of ω. It may be noted, moreover, that the duality method is more demanding in terms of storage requirements than the over-relaxation method, since it is necessary to store the components of λ_h.

The rate of convergence of the duality method could clearly be improved by solving the Dirichlet problems (5.62) by a method which is quicker than point over-relaxation for this type of problem (alternating direction, block over-relaxation) or by a direct method (e.g. Gauss or Choleski, etc....); however, this would also generally increase the computer storage requirement.

The above is also valid if we replace the constraint

$$v(x) \geqslant 0 \quad \text{a.e. by} \quad \Psi_1(x) \leqslant v(x) \leqslant \Psi_2(x) \qquad ■$$

---

(1) Relating to (P) rather than (P_h).

b	h_1	Optimal ρ	Number of iterations	CII 10070 computation time in minutes
0.2	1/10	45	30	0.045
	1/30	41	35	0.105
	1/50	30	50	0.3
0.34	1/10	16	30	0.04
	1/30	13	40	0.12
	1/50	10	50	0.3
0.69	1/10	13	40	0.09
	1/30	8	70	0.27
	1/50	5	100	0.6

Table 5.2

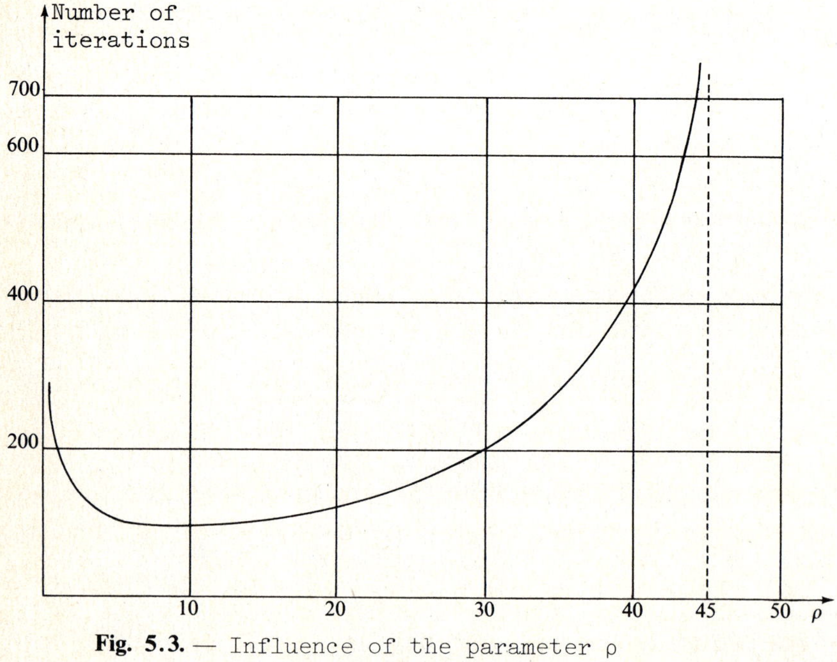

Fig. 5.3. — Influence of the parameter ρ on the rate of convergence.

5.7.4 *Shape of the free surface*

We have shown in Figure 5.4, for $b = 0.69$ and using the solution of problem (P_h) for $h_1 = 1/50$, $h_2 = 0.69/10$, the region where $u = 0$, a part of the region where $u > 0$ (i.e. Ω^+) and the boundary between these two regions which was one of the unknown of (P) considered as a *free boundary* problem.

We recall that in Ω^+, we have $-\Delta u = f$. ■

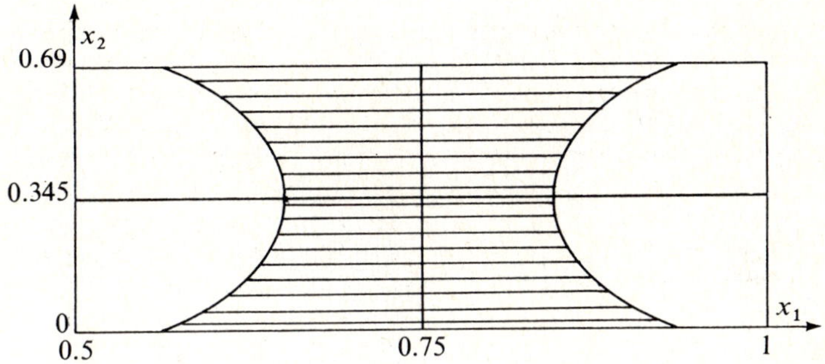

Fig. 5.4. The hatched region corresponds to $u = 0$.

6. DISCUSSION

We should emphasise the fact that this Chapter does not aim to give an exhaustive investigation of minimisation algorithms. For more comprehensive studies of this subject, the following may be consulted:

- *for unconstrained optimisation*: Daniel /2/, Kowalik and Osborne /1/, Spang /1/, Vainberg /1/;

- *for constrained optimisation*: Abadie /1/, /2/, Balakrishnan and Neustadt /1/, Box /1/, Bram and Saaty /1/, Cannon, Cullum and Polak /1/, Céa /2/, Fletcher /1/, Graves and Wolfe /1/, Hadley /1/, Kuhn and Tucker /1/, Kunzi, Krelle and Oettli /1/, Kunzi, Tzschach and Zehnder /1/, Lasdon /1/, Levitin and Polyak /1/, Luenberger /1/, Polak /1/, Trémolières /3/, Varayia /1/, Zangwill /1/, Zoutendijk /1/.

Ideas similar to those introduced in Section 1 have been investigated by a number of authors: Crouzeix /1/, Elkin /1/, Céa and Glowinski /2/, Glowinski /4/, /5/, Mezlyakov /1/, Miellou /1/, Motzkin and Schoenberg /1/, Ortega and Rheinboldt /1/, /2/, /3/, Ortega and Rockoff /1/, Petrischyn /1/, /2/, Schechter /1/, /2/, /3/, Southwell /1/, Stiefel /1/, Comincioli /1/, Varga /1/, Cryer /1/, /2/.

- *for various properties of convex functions*, see Levitin and Polyak /1/.

The concepts introduced in Section 2 are of a classical nature. For the unconstrained case we refer the reader to Blum /1/, Cauchy /1/, Crockett and Chernoff /1/, Curry /1/, Goldstein /1/, /2/, Greenstadt /1/, Polyak /1/.

The convergence of the method of divergent series, due to Ermolyev, is given in Polyak /2/ and for the general unconstrained case in Trémolières /2/. A transposition of this method to variational inequalities is made in Auslender-Gourand-Guillet /1/.

The so-called conjugate gradient method, due to Hestenes /1/, has been studied by Antosiewicz and Rheinboldt /1/, Beckman /1/, Durand /1/, Fletcher and Reeves /1/, Hestenes and Stiefel /1/. This method has been generalised to infinite dimensions by Daniel /1/, /2/. For methods of the same type, consult Davidon /1/, Fletcher and Powell /1/, Pearson /1/. The extension of the method of conjugate directions to the case of constrained minimisation has been carried out by Goldfarb /1/, Goldfarb and Lapidus /1/, Luenberger /2/ and Trémolières /4/.

For methods of the point projection or gradient projection type see Altman /1/, Demyanov /1/, Goldstein /1/, Levitin and Polyak /1/, /2/, Polyak /3/, /4/, Rosen /1/, /2/, Sibony /1/.

The penalisation of constraints of the type $G(v) = 0$ using the function

$$J(v) + \frac{1}{\varepsilon} G(v)^2 , \quad \varepsilon \to 0 ,$$

was introduced by Courant /1/. In the case of inequality constraints, the exterior penalty function $J(v) + \frac{1}{\varepsilon} \sum_{j=1}^{m} \max [- G_j(v), 0]$ is due to Ablow and Brigham /1/. The interior penalty function $J(v) + \varepsilon \sum_{j=1}^{m} \frac{1}{G_j(v)}$ is due to Carroll /1/.

The method of centres using the distance-function

$$(\lambda - J(v)) \prod_{j=1}^{m} G_j(v)$$

was introduced by Huard /1/ and extended to topological spaces by Bui-Trong-Lieu and Huard /1/.

The second distance function $\frac{1}{\lambda - J(v)} + \sum_{j=1}^{m} \frac{1}{G_j(v)}$ is that of Fiacco and McCormick /1/. The idea of varying the truncation errors and the possibility of accelerating this method by over-relaxation comes from Trémolières /1/.

For a general presentation of penalty methods, see Fiacco and McCormick /1/ and Lootsma /1/.

The concepts of Section 4 are less classical in nature. For the general theorems on the existence of saddle-points see Berge /1/, Browder /1/, Ky-Fan /1/, Sion /1/, /2/, Ekeland-Témam /1/. For duality theorems in mathematical programming see Arrow, Hurwicz, and Uzawa /1/, Bensoussan, Lions and Témam /1/, Mangasarian /1/, Rockafellar /3/ and Varayia /2/.

Some investigations of the two principal duality algorithms may be found in Arrow, Hurwicz and Uzawa /1/, Céa, Glowinski and Nédelec /2/, Céa and Glowinski /1/ and in Trémolières /3/.

In order to bring this chapter to a close, we should mention the *augmented Lagrangian* methods which are obtained by combining *penalisation and duality*; we shall meet some applications of this in Chapter 3, Section 10 and in Chapter 5, Section 9.

This approach appears particularly interesting for two fundamental reasons:

1) the possibility of accelerating the convergence of the duality methods,
2) the possibility of improving the conditioning of the functional to be minimised.

For the theoretical aspects, we refer to Hestenes /2/, Powell /1/, Rockafellar /4/, /5/, etc... Also, some applications of this methodology to the solution of nonlinear boundary-value problems are given in Glowinski-Marrocco /1/, /2/, Mercier /1/.

Chapter 3

NUMERICAL ANALYSIS OF THE PROBLEM OF THE ELASTO-PLASTIC TORSION OF A CYLINDRICAL BAR

INTRODUCTION

In this chapter we shall mainly study the application of the results and methods of Chapters 1 and 2 to the numerical analysis of the model elasto-plastic problem considered in Chapter 1, Section 1.2.

1. STATEMENT OF THE CONTINUOUS PROBLEM. PHYSICAL MOTIVATION. SYNOPSIS

1.1 Statement of the continuous problem

We again adopt the formalism of Chapter 1, Section 1.2; with Ω a *bounded* open domain in $\mathbb{R}^n$, with boundary Γ, we introduce

(1.1) $$H_0^1(\Omega) = \{ v \mid v \in H^1(\Omega),\ \gamma v = 0 \},$$

which is a closed vector subspace of $H^1(\Omega)$ and a Hilbert space for the inner product and the norm defined by

(1.2) $$a(u, v) = \int_\Omega \sum_{i=1}^n \frac{\partial u}{\partial x_i} \frac{\partial v}{\partial x_i} \, dx = \int_\Omega \operatorname{grad} u . \operatorname{grad} v \, dx$$

(1.3) $$\| v \| = \left(\int_\Omega | \operatorname{grad} v |^2 \, dx \right)^{1/2},$$

the norm (1.3) being equivalent to the norm induced by $H^1(\Omega)$. We then introduce

(1.4) $$K_0 = \{ v \mid v \in H_0^1(\Omega),\ | \operatorname{grad} v(x) | \leqslant 1 \quad \text{a.e. in} \quad \Omega \}$$

a *closed convex* set in $H_0^1(\Omega)$, and the functional

(1.5) $$J(v) = \tfrac{1}{2} a(v, v) - L(v)$$

where L is a continuous linear form on $H_0^1(\Omega)$.

We then consider the variational problem (P_0) defined by:

$$(1.6) \qquad (P_0) \operatorname*{Min}_{v \in K_0} J(v).$$

Since the functional J is continuous and strictly convex on $H_0^1(\Omega)$, with $\lim\limits_{\|v\| \to +\infty} J(v) = +\infty$, Theorem 2.1 of Chapter 1, Section 2.1 applies, and this implies the existence and uniqueness of a solution for (P_0). We denote this solution by u which is also the unique solution of the variational inequality

$$(1.7) \qquad \begin{cases} a(u, v-u) \geqslant L(v-u) \quad \forall v \in K_0 \\ u \in K_0, \end{cases}$$

which is equivalent to (P_0). ∎

Certain properties of the solution u, for *particular* forms L, are detailed in Section 2.

1.2 Physical motivation

With the set Ω a *simply connected*, *bounded* open domain in $\mathbb{R}^2$, we consider a cylindrical bar of cross-section Ω, made of an elastic - perfectly plastic material for which the threshold of plasticity (i.e. the yield stress) is defined by the von Mises criterion (see eg. Mandel /1/).

This bar is then subjected to an increasing torque, starting from an unstressed initial state, the torsion being characterised by the twist angle per unit length C. For each value of C, and with a suitable system of units, we can, using the Haar-Karman principle, reduce the determination of the stress field to the solution of the variational problem (where v denotes a "stress potential"):

$$(1.8) \qquad \begin{cases} \operatorname*{Min}_{v \in K_0} \left[\dfrac{1}{2} \displaystyle\int_\Omega |\operatorname{grad} v|^2 \, dx - C \int_\Omega v(x) \, dx \right], \\ K_0 = [v \mid v \in H_0^1(\Omega), \ |\operatorname{grad} v| \leqslant 1 \quad \text{a.e.} \ ], \end{cases}$$

this is a special case of (P_0) in which the form L is thus given by:

$$(1.9) \qquad L(v) = C \int_\Omega v(x) \, dx. \quad ∎$$

Remark 1.1. In the case of a *multiply-connected* section Ω it is convenient to employ a modified version of formulation (1.8); see Glowinski-Lanchon /1/, H. Lanchon /2/.

In the case where the initial state is not stress-free, and/or the torque is not increasing, formulation (1.8) is incorrect for the physical problem considered; a suitable approach is then to use time-dependent, or more precisely quasi-static, mathematical models (see Duvaut-Lions /1/, Chapter 5, for a detailed study of quasi-static models in plasticity). ■

Remark 1.2. We can supplement Remark 1.4 of Chapter 1, Section 1.2 by pointing out that several authors, in particular Shaw /1/, have carried out the numerical solution of the problem of the elasto-plastic torsion of a cylindrical bar, considered as a *free boundary* problem, by using the point formulation given in the above-mentioned remark, ie.:

$$-\Delta u = C \quad \text{in} \quad \Omega_e \tag{1.10}$$

$$|\operatorname{grad} u| = 1 \quad \text{in} \quad \Omega_p \tag{1.11}$$

with

$$\Omega_e = [x \mid x \in \Omega, \ |\operatorname{grad} u(x)| < 1] \tag{1.12}$$

$$\Omega_p = \Omega - \Omega_e ; \tag{1.13}$$

it is appropriate to add to the above relations

$$\begin{cases} u|_\Gamma = 0 \\ + \text{ the continuity of } \operatorname{grad} u \text{ at the interface between } \Omega_e \text{ and } \Omega_p . \end{cases} \tag{1.14}$$

Numerical methods based on the use of (1.10), ..., (1.14) give satisfactory results (see, for example, Hodge-Stout-Herakovich /1/) but appear to be much more complicated to apply than the majority of those given in this chapter, based on the variational formulation (1.8). ■

1.3 Synopsis

In Section 2 we shall give a number of properties of the solution of problem (1.6) (ie. (P_0)), and of the particular case (1.8), these properties having important consequences in the Numerical Analysis of the problem. In particular (see Remark 3.18, Chapter 1, Section 3.7) the optimal solution of (1.8) is also the solution of

$$(P_1)\begin{cases} \underset{v \in K_1}{\text{Min}} \left[\frac{1}{2}\int_\Omega |\operatorname{grad} v|^2 \, dx - C\int_\Omega v(x)\, dx\right] \\ K_1 = [v \mid v \in H_0^1(\Omega), \ |v(x)| \leqslant \delta(x,\Gamma) \ \text{a.e.}\ ] \end{cases} \tag{1.15}$$

where $\delta(x,\Gamma)$ = distance from x to Γ.

Problem (P_1), which is the subject of Section 3, is very similar to the lubrication problem studied in Chapter 2, Section 5.

From the point of view of applications to mechanics, it

would appear that in view of the equivalence of (1.8), (1.15), we could restrict our attention to (1.15); however, in view of possible applications to other situations, we have considered it worthwhile to investigate the numerical analysis of (P_0) with $\Omega \subset \mathbb{R}^2$ and $f \in L^2(\Omega)$, *by working directly with* K_0.

Sections 4, 5 and 6 will be devoted to the approximation of (P_0), and Sections 7, 8 and 9 to the solution of the approximate problems by relaxation, penalisation and duality, respectively. ■

2. SOME PROPERTIES OF THE SOLUTION OF PROBLEM (P_0)

2.1 Regularity results

We assume, in (1.6), that the linear form L can be written:

$$L(v) = \int_\Omega f(x)\, v(x)\, \mathrm{d}x\,. \tag{2.1}$$

Denoting the solution of (P_0) by u, it is shown in Brézis-Stampacchia /2/, Section 3 that:

1) If Ω is a *convex*, bounded, open domain in $\mathbb{R}^n$, with *Lipschitz* boundary Γ, and if $f \in L^p(\Omega) \cap H^{-1}(\Omega)$(1), $p > 1$, we then have:

$$\begin{cases} \Delta u \in L^p(\Omega) & (1 < p < +\infty) \\ u \in W^{2,p}(\Omega) & (1 < p < +\infty)\,. \end{cases} \tag{2.2}$$

2) If Ω is a bounded open domain in $\mathbb{R}^n$ with a sufficiently regular boundary Γ(2) and if $f \in L^p(\Omega) \cap H^{-1}(\Omega)$, $p > 1$, we again have (2.2).

Remark 2.1. It will be seen in Section 2.3 that the above properties are optimal, at least in Sobolev spaces of integer order. ■

2.2 An equivalent variational problem

In Brézis-Sibony /1/, it is shown that if Ω is a bounded open domain in $\mathbb{R}^n$ with a sufficiently regular boundary Γ, and if L is of the form

---

(1) This is automatically satisfied if $f \in L^p(\Omega)$ with $p \geq 2$.

(2) We refer to Brézis-Stampacchia, *loc cit.*, for the definition of a regularity condition for Γ which is sufficient to ensure (2.2)

$$(2.3) \qquad L(v) = C \int_{\Omega} v(x)\,dx,$$

then the solution of (P_0) is also a solution of (P_1), defined in (1.15), and of the equivalent variational inequality:

$$(2.4) \qquad \begin{cases} a(u, v - u) \geqslant C \int_{\Omega} (v - u)\,dx & \forall v \in K_1 \\ u \in K_1 = [v \mid v \in H_0^1(\Omega),\ |v(x)| \leqslant \delta(x, \Gamma) \ \text{a.e.}], \end{cases}$$

the bilinear form being defined as in (1.2). ■

Remark 2.2. In fact, it is easy to show that the solution of (P_1) has the same sign as C; in the case $C > 0$, K_1 can therefore be replaced, in (1.15) and (2.4), by the even simpler convex set

$$K_1' = [v \mid v \in H_0^1(\Omega),\ 0 \leqslant v(x) \leqslant \delta(x, \Gamma) \quad \text{a.e.}]. \ ■$$

Remark 2.3. From the numerical point of view, the solution of (P_1) is easier than that of (P_0); the numerical analysis of (P_1), which is a variant of that of the variational problem of Chapter 2, Section 5, will form the subject of Section 3. ■

2.3 Some particular cases where the solution is known(1)

2.3.1. *A first example*

We take $\Omega =]0, 1[$ and $L(v) = C \int_0^1 v(x)\,dx$ with $C > 0$, so that (P_0) is defined by:

$$(2.5) \qquad \begin{cases} \underset{v \in K_0}{\text{Min}} \left[\int_0^1 v'^2\,dx - 2C \int_0^1 v\,dx \right], \\ K_0 = [v \mid v \in H_0^1(0, 1),\ |v'(x)| \leqslant 1 \ \text{a.e.}] \end{cases}$$

with $v' = dv/dx$.

It can be verified that the solution u of (P_0)(and of (P_1)) is given by:

$$(2.6) \qquad u(x) = \frac{C}{2} x(1 - x) \qquad \text{if } C < 2$$

---

(1) We restrict our attention here to *very simple* examples, designed to act as test cases for the verification of the methods developed in this chapter. For various explicit solutions see Mandel /1/.

and if $C \geqslant 2$ by:

$$(2.7)\qquad \begin{cases} x \ \text{ if } 0 \leqslant x \leqslant \frac{1}{2} - \frac{1}{C} \\ \frac{C}{2}\left[x(1-x) - \left(\frac{1}{2} - \frac{1}{C}\right)^2\right] \ \text{ if } \frac{1}{2} - \frac{1}{C} \leqslant x \leqslant \frac{1}{2} + \frac{1}{C} \\ 1 - x \ \text{ if } \frac{1}{2} + \frac{1}{C} \leqslant x \leqslant 1 . \ \blacksquare \end{cases}$$

Figure 2.1 shows the solution of (P_0) (and (P_1)) corresponding to $C = 4$ (we thus have $1/2 - 1/C = 1/4$).

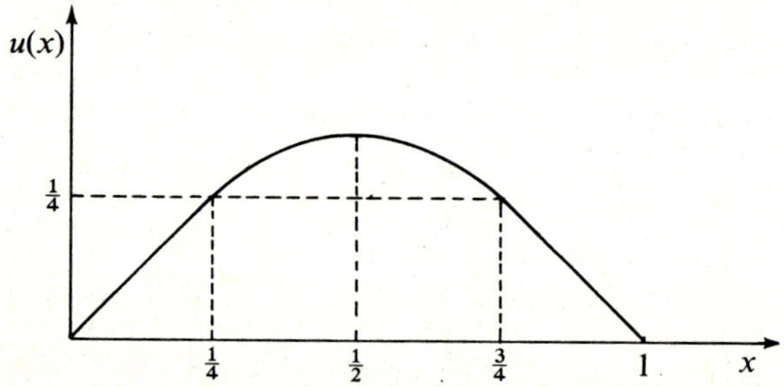

Fig. 2.1. Solution of (P_0) for $C = 4$.

2.3.2 *A second example*

We take $n = 2$,

$$\Omega = [x \mid x = (x_1, x_2),\ x_1^2 + x_2^2 < R^2],$$

$$L(v) = C \int_\Omega v(x)\,\mathrm{d}x, \quad C > 0;$$

the solution of (P_0) (and (P_1)) is then given by:

$$(2.8)\qquad u(x) = (C/4)(R^2 - r^2) \quad \text{if} \quad C \leqslant 2/R,$$

where

$$r = \sqrt{x_1^2 + x_2^2}, \ \text{ and if } C \geqslant 2/R \text{ by}$$

$$(2.9)\qquad u(x) = \begin{cases} R - r & \text{if } \ R' \leqslant r \leqslant R \\ -Cr^2/4 + (R - 1/C) & \text{if } \ 0 \leqslant r \leqslant R', \end{cases}$$

with

$$(2.10)\qquad R' = 2/C . \ \blacksquare$$

Remark 2.4. We can now supplement Remark 2.1; in the two above examples, for which the data are very regular, we have, for sufficiently large C,

$$u \in C^1(\overline{\Omega}) \cap H^2(\Omega) \cap H_0^1(\Omega) \quad (^1)$$

but

$$u \notin C^2(\overline{\Omega}), \quad u \notin H^3(\Omega) . \quad \blacksquare$$

3. NUMERICAL ANALYSIS OF PROBLEM (P_1)

3.1 Synopsis

In this section we shall apply the results of Chapters 1 and 2 to problem (P_1), with the open domain Ω simply-connected in $\mathbb{R}^2$; we shall consider only the *exterior approximation* of (P_1) (i.e. of the "finite difference" type) and no additional difficulties arise, relative to the variational problem of Chapter 2, Section 5, other than the convergence of the approximate solution as $h \to 0$, which is rather more awkward to prove.

With regard to algorithms, we shall restrict our attention to the over-relaxation method, which appears to be the best suited to the type of constraints met in this problem. ■

3.2 Exterior approximation of problem (P_1)

3.2.1. *Formulation of the approximate problem*

We recall that (P_1) is defined by

$$(3.1)\quad \begin{cases} \displaystyle\operatorname*{Min}_{v \in K_1} \left[J(v) = \frac{1}{2} \int_\Omega |\operatorname{grad} v|^2 \, dx - C \int_\Omega v \, dx \right], \quad (C > 0), \\ K_1 = [v \mid v \in H_0^1(\Omega), \ |v(x)| \leqslant \delta(x, \Gamma) \ \text{a.e.}] . \end{cases}$$

Suppose $h > 0$, intended to approach zero; proceeding as in Chapter 1, Section 5.1 and Chapter 2, Section 5, we define the grid R_h by:

$$(3.2)\qquad R_h = \{ M_{ij} \mid M_{ij} \in \mathbb{R}^2, \ M_{ij} = \{ ih, jh \}, \ i, j \in \mathbb{Z} \} .$$

With each node M_{ij} of R_h we associate the panel with centre M_{ij}

$$(3.3)\qquad \varpi_h^0(M_{ij}) =](i - \tfrac{1}{2})\, h, (i + \tfrac{1}{2})\, h[\times](j - \tfrac{1}{2})\, h, (j + \tfrac{1}{2})\, h[,$$

---

(1) In fact in the case of these two examples we have rather more, since

$$u \in W^{2,\infty}(\Omega) = \left[v \mid v, \frac{\partial v}{\partial x_i}, \frac{\partial^2 v}{\partial x_i \, \partial x_j} \in L^\infty(\Omega) \right] .$$

and then the *cross* (with centre M_{ij}):

(3.4) $$\varpi_h^1(M_{ij}) = \varpi_h^0\left(M_{ij} \pm \frac{h}{2}e_1\right) \cup \varpi_h^0\left(M_{ij} \pm \frac{h}{2}e_2\right).$$

We next define

(3.5) $$\Omega_h = \{ M_{ij} \mid \varpi_h^1(M_{ij}) \subset \Omega \}$$

(3.5') $$\begin{cases} \Gamma_h = \{ M_{ij} \mid M_{ij} \in R_h,\ M_{ij} \notin \Omega_h, \text{ at least one of the 4 points} \\ \qquad M_{ij} \pm he_1,\ M_{ij} \pm he_2 \in \Omega_h \} \end{cases}$$

(3.5") $$\bar{\Omega}_h = \Omega_h \cup \Gamma_h$$

(3.6) $$V_h = \{ v_h \mid v_h = (v_{ij})_{M_{ij} \in \Omega_h},\ v_{ij} \in \mathbb{R} \},$$

and the mapping $q_h : V_h \to L^2(\mathbb{R}^2)$ (1) by

(3.7) $$q_h v_h = \sum_{M_{ij} \in \Omega_h} v_{ij}\, \theta_h^{ij},$$

where

(3.8) $$\theta_h^{ij} = \text{characteristic function of } \varpi_h^0(M_{ij}).$$

Having put (as in Chapter 1, Section 5.1 and in Chapter 2, Section 5), for $k = 1, 2$

(3.9) $$\delta_k \varphi(x) = \frac{1}{h}\left[\varphi\left(x + \frac{h}{2}e_k\right) - \varphi\left(x - \frac{h}{2}e_k\right)\right]$$

we "approximate" J by $J_h : V_h \to \mathbb{R}$ defined by

(3.10) $$J_h(v_h) = \frac{1}{2}\sum_{k=1}^{2}\int_\Omega \left| \delta_k q_h v_h \right|^2 dx - C\int_\Omega q_h v_h\, dx,$$

and we take as the approximate problem

(3.11) $$(P_{1h})\begin{cases} \min\limits_{v_h \in K_{1h}} J_h(v_h), \\ K_{1h} = \{ v_h \mid v_h \in V_h,\ |v_{ij}| \leqslant \delta(M_{ij}, \Gamma),\ \forall M_{ij} \in \Omega_h \}. \end{cases}$$

It may be noted that K_{1h} is bounded and closed in V_h. ■

3.2.2 *Solvability of* (P_{1h})

We proceed as in Chapter 2, Section 5.3.2: the mapping

---

(1) We shall use the same notation for the mapping $V_h \to L^2(\Omega)$ defined by $v_h \to$ restriction to Ω of $q_h v_h$.

$$v_h \to \left(\sum_{k=1}^{2} \int_{\Omega} \left| \delta_k q_h v_h \right|^2 dx \right)^{1/2}$$

defines a norm over V_h, denoted by $\| v_h \|_h$; the functional J_h is therefore continuous and strictly convex on V_h and since K_{1h} is bounded and closed in V_h we have the existence and uniqueness of an optimal solution for (P_{1h}), say u_h. ■

3.2.3. *Explicit formulation of* (P_{1h})

We can write (3.10) in explicit form as follows:

$$\text{(3.12)} \quad \begin{cases} J_h(v_h) = \dfrac{h^2}{4} \displaystyle\sum_{M_{ij} \in \bar{\Omega}_h} \left[\left| \dfrac{v_{i+1j} - v_{ij}}{h} \right|^2 + \left| \dfrac{v_{i-1j} - v_{ij}}{h} \right|^2 + \right. \\ \left. \quad + \left| \dfrac{v_{ij+1} - v_{ij}}{h} \right|^2 + \left| \dfrac{v_{ij-1} - v_{ij}}{h} \right|^2 \right] - h^2 C \displaystyle\sum_{M_{ij} \in \Omega_h} v_{ij} . \end{cases} \quad ■$$

Remark 3.1. In (3.12) it is necessary to take $v_{pq} = 0$ if $M_{pq} \notin \Omega_h$. ■

3.3 Convergence of the approximate solution as $h \to 0$

Preliminary remark. In this section we shall prove that

$$\{ q_h u_h, \delta_1 q_h u_h, \delta_2 q_h u_h \} \to \left\{ u, \frac{\partial u}{\partial x_1}, \frac{\partial u}{\partial x_2} \right\} \text{ strongly in } (L^2(\Omega))^3$$

where u is the optimal solution of (P_1). This convergence result can be proved directly by proceeding as in Chapter 2, Section 5, though there are some additional technical difficulties; however, we have preferred to use the more general approach of Chapter 1, Section 5 even though this makes the working somewhat more complicated. ■

3.3.1. *Reduction of* (P_{1h}) *to an equivalent variational inequality in* K_{1h}

In order to get back to the setting of Chapter 1, Section 5, it is convenient to formulate (P_1) and (P_{1h}) in terms of variational inequalities; this gives

$$\text{(3.13)} \quad \begin{cases} \displaystyle\int_{\Omega} \operatorname{grad} u . \operatorname{grad} (v - u) \, dx \geq C \int_{\Omega} (v - u) \, dx \quad \forall v \in K_1 \\ u \in K_1 , \end{cases}$$

$$(3.14)\qquad \begin{cases} \sum_{k=1}^{2} \int_{\Omega} \delta_k q_h u_h \, \delta_k q_h(v_h - u_h)\, dx \geqslant C \int_{\Omega} q_h(v_h - u_h)\, dx \quad \forall v_h \in K_{1h}, \\ u_h \in K_{1h}. \end{cases}$$

In (1.2) we have put

$$\int_{\Omega} \operatorname{grad} u . \operatorname{grad} v \, dx = a(u, v)$$

and we define $a_h : V_h \times V_h \to \mathbb{R}$ by

$$(3.15)\qquad a_h(u_h, v_h) = \sum_{k=1}^{2} \int_{\Omega} \delta_k q_h u_h \, \delta_k q_h v_h \, dx . \quad \blacksquare$$

3.3.2. *Exterior approximation for* $H_0^1(\Omega), a, K_1, L$

We shall now verify point by point that the properties required in Chapter 1, Section 5.2 are satisfied:

Approximation of $H_0^1(\Omega)$.

In view of Chapter 1, Section 5.1, we take

$$(3.16)\qquad V = H_0^1(\Omega)$$

$$(3.17)\qquad F = (L^2(\Omega))^3$$

and $\sigma : V \to F$, *injective*, defined by:

$$(3.18)\qquad \sigma v = \left\{ v, \frac{\partial v}{\partial x_1}, \frac{\partial v}{\partial x_2} \right\} . \quad \blacksquare$$

In Section 3.2.1 we defined a family of finite-dimensional spaces V_h which are Hilbert spaces for the norms $\| \cdot \|_h$ defined by:

$$(3.19)\qquad \| v_h \|_h = (a_h(v_h, v_h))^{1/2} = \left(\sum_{k=1}^{2} \int_{\Omega} | \delta_k q_h v_h |^2 \, dx \right)^{1/2} ;$$

we recall (see Céa /3/, Lions /3/) that the norm $\| \cdot \|_h$ satisfies the discrete Poincaré inequality:

$$(3.20)\qquad \| q_h v_h \|_{L^2(\Omega)} \leqslant \alpha \| v_h \|_h \quad (\alpha > 0), \ \forall h, \quad \forall v_h \in V_h . \quad \blacksquare$$

Finally, we define a family of *prolongation operators* $p_h \in \mathscr{L}(V_h ; F)$ derived from the operator $q_h : V_h \to L^2(\Omega)$, defined in Section 3.2.1, relation (3.7), by:

$$(3.21)\qquad p_h v_h = \{ q_h v_h, \ \delta_1 q_h v_h, \ \delta_2 q_h v_h \} ,$$

and in view of (3.20), we indeed have:

$$\| p_h \|_{\mathcal{L}(V_h,F)} \leqslant C. \quad \blacksquare \tag{3.22}$$

This being the case, we shall now prove:

Proposition 3.1. *The family* $(V_h)_h$ *constitutes an exterior approximation of* V.

Proof. It is necessary for us to verify that (5.27) of Chapter 1, Section 5 is satisfied. We define $r_h : V \to V_h$ by:

$$(r_h v)_{M_{ij}} = \frac{1}{h^2} \int_{\varpi^0(M_{ij})} v(x)\, dx, \quad \forall M_{ij} \in \Omega_h. \tag{3.23}$$

Thus (see Chapter 1, Section 5.1), $\forall v \in V = H_0^1(\Omega)$

$$p_h r_h v \to \left\{ v, \frac{\partial v}{\partial x_1}, \frac{\partial v}{\partial x_2} \right\} \quad \text{strongly in} \quad (L^2(\Omega))^3 \tag{3.24}$$

and hence, taking account of (3.19),

$$\| r_h v \|_h \leqslant c \quad \text{independent of } h. \quad \blacksquare \tag{3.25}$$

Approximation of a

The form $a_h : V_h \times V_h \to \mathbb{R}$ defined in (3.15) satisfies, firstly

$$| a_h(u_h, v_h) | \leqslant \| u_h \|_h \| v_h \|_h \tag{3.26}$$

$$a_h(v_h, v_h) = \| v_h \|_h^2 , \tag{3.27}$$

and secondly (this may be proved without difficulty), (5.30) and (5.31) of Chapter 1, Section 5.2, so that we have:

Proposition 3.2. *The family* $(a_h)_h$ *constitutes an approximation of* a. $\blacksquare$

Approximation of K_1

With the family K_{1h} defined as in (3.11), we have:

Proposition 3.3. *The family* K_{1h} *constitutes an approximation of* K_1.

Proof. It is necessary to prove (see (5.33) and (5.34) Chapter 1, Section 5.2):

$$\text{if } v_h \in K_{1h} \text{ with } \lim_{h \to 0} p_h v_h = \xi \quad \text{weakly in } F, \text{ then } \xi \in \sigma K_1 \tag{3.28}$$

and

(3.29) $\forall v \in K_1 , \quad \exists v_h \in K_{1h}$ such that $p_h v_h \to \sigma v$ strongly in F and $\| v_h \|_h \leqslant c$.

Verification of (3.28).

We put $\xi = (\xi_0, \xi_1, \xi_2)$; from Chapter 1, Section 5.1 we then obtain that

$$\left.\begin{cases} \xi_0 \in H_0^1(\Omega) = V \\ \xi_k = \dfrac{\partial \xi_0}{\partial x_k} \ (k = 1, 2) \end{cases}\right\} \Leftrightarrow \xi \in \sigma V . \tag{3.30}$$

Moreover, since the function $x \to \delta(x, \Gamma)$ is continuous on $\bar{\Omega}$, it is uniformly continuous and thus, by definition of K_{1h}, gives

$$v_h \in K_{1h} \Rightarrow | q_h v_h(x) | \leqslant \delta(x, \Gamma) + \eta(h) \quad \text{a.e.} \tag{3.31}$$

with

$$\eta(h) \ \text{independent of } x, \quad \lim_{h \to 0} \eta(h) = 0 . \tag{3.32}$$

Since the set $[v \mid v \in L^2(\Omega), \ | v(x) | \leqslant \delta(x, \Gamma) + \varepsilon \text{ a.e.}]$ is, $\forall \varepsilon > 0$, convex and closed in $L^2(\Omega)$, and therefore weakly closed, we have from (3.31) and (3.32)

$$| \xi_0(x) | \leqslant \delta(x, \Gamma) + \varepsilon \ \text{ a.e., } \forall \varepsilon > 0 \Leftrightarrow | \xi_0(x) | \leqslant \delta(x, \Gamma) \quad \text{a.e.}, \tag{3.33}$$

and we indeed have $\xi \in \sigma K_1$. ■

Verification of (3.29).

It is sufficient (see Chapter 1, Section 5.3, Remark 5.4) *to verify that* (3.29) *is satisfied for a subset* χ_1 *of* K_1, *which is dense in* K_1 :

Let $v \in K_1$; given $\varepsilon > 0$, we define $v_\varepsilon = \tau_\varepsilon(v)$ by:

$$\tau_\varepsilon(v) = v_\varepsilon = (v - \varepsilon)^+ - (v + \varepsilon)^- . \tag{3.34}$$

Since $\varphi \to \varphi^+, \varphi^-$ map $H^1(\Omega) \to H^1(\Omega)$, $\tau_\varepsilon(v) \in H^1(\Omega)$ and

$$v_\varepsilon(x) \begin{cases} = 0 & \text{if } \delta(x, \Gamma) \leqslant \varepsilon \\ \leqslant \delta(x, \Gamma) - \varepsilon & \text{otherwise,} \end{cases} \tag{3.35}$$

as may easily be verified.

Hence

$$\tau_\varepsilon(v) \in K_1 \quad \forall \varepsilon > 0 . \tag{3.36}$$

We then define a subset χ_1 of K_1 by

$$\chi_1 = \bigcup_{\varepsilon > 0} \tau_\varepsilon K_1 . \tag{3.37}$$

We have

$$\bar{\chi}_1 = K_1 . \tag{3.38}$$

In fact $(v - \varepsilon)^+$ (resp. $(v + \varepsilon)^-$) tends strongly towards v^+ (resp. v^-) in $H^1(\Omega)$ as $\varepsilon \to 0$, therefore $\tau_\varepsilon(v) \to v$ strongly in $H^1(\Omega)$.

Variant.

The above procedure shows that the set of functions v of K_1, which are zero in a neighbourhood of Γ and bounded above in modulus by $\delta(x, \Gamma) - \varepsilon_v$ is *dense in* K_1. This fact may also be demonstrated by proceeding as follows : let $v \in K_1$ and let θ_n be a sequence of functions of $C^1(\bar{\Omega})$, $|\theta_n(x)| \leqslant 1$, $\theta_n = 0$ in the neighbourhood of Γ and $\theta_n(x) \to 1$ uniformly over all compact subsets of Ω,

$$\left| \frac{\partial \theta_n}{\partial x_i}(x)\, \delta(x, \Gamma) \right| \leqslant \text{ constant; thus } \theta_n v \to v \text{ in } H^1(\Omega) \text{ (since } v \in H_0^1(\Omega))$$

and $|\theta_n v(x)| \leqslant |v(x)| \leqslant \delta(x, \Gamma)$. Thus we are already able to assume that v is zero in the neighbourhood of Γ. However, if $0 < \lambda < 1, \lambda \to 1$, we have : $\lambda v \to v$ in $H^1(\Omega)$ and $|\lambda v(x)| \leqslant \lambda\, \delta(x, \Gamma)$, which combined with the fact that v is zero in the neighbourhood of Γ shows that $|\lambda v(x)| \leqslant \delta(x, \Gamma) - \varepsilon_v$, and hence the required result. ■

This being the case, let $v \in \chi_1$; with the operator $r_h : V \to V_h$ defined as in (3.23), if we put $v_h = r_h v$, we have, taking account of (3.35) and the continuity of $x \to \delta(x, \Gamma)$ on Ω :

$$\begin{cases} |v_{ij}| = \dfrac{1}{h^2} \left| \displaystyle\int_{\varpi_h^0(M_{ij})} v(x)\, \mathrm{d}x \right| \leqslant \dfrac{1}{h^2} \displaystyle\int_{\varpi_h^0(M_{ij})} |v(x)|\, \mathrm{d}x \leqslant \delta(M_{ij}, \Gamma) = \delta_{ij}, \\ \forall h > 0 \text{ sufficiently small} \quad \Leftrightarrow v_h \in K_{1h} \\ \forall h > 0, \text{sufficiently small.} \end{cases} \tag{3.39}$$

From (3.24) and (3.25), $p_h r_h v \;(= p_h v_h) \to \sigma v$ strongly in F and thus condition (3.29) is indeed satisfied in χ_1. ■

Remark 3.2. Following Raviart /1/, Chapter 0, we can verify (3.28) by using the fact that if $p_h v_h \to \xi$ weakly in F, then we have for a subsequence also denoted by v_h

(3.40) $\quad q_h v_h \to \xi_0$ strongly in $L^2(\Omega)$ and a.e. in Ω. ■

Remark 3.3. Basing our argument on (3.35), we shall now prove that $\mathscr{D}(\Omega) \cap K_1 = K_1$; since the set $\chi_1 \subset K_1$ is dense in K_1 it is sufficient to show this for χ_1; we shall proceed by *regularisation*:

Consider then a regularising sequence $(\rho_n)_n$ (i.e. $\rho_n \in \mathscr{D}(\mathbb{R}^2)$, $\rho_n \geqslant 0$, $\rho_n \to \delta(0)$, $\int_{\mathbb{R}^2} \rho_n(x)\,\mathrm{d}x = 1$, support $(\rho_n) \to 0)$; if $v \in \chi_1$, we extend v to $\tilde{v}$ by 0 outside Ω; $\tilde{v} \in H^1(\mathbb{R}^2)$, and we define $\tilde{v}_n \in \mathscr{D}(\mathbb{R}^2)$ by

(3.41) $$\tilde{v}_n = \tilde{v} * \rho_n \Leftrightarrow \tilde{v}_n(x) = \int_{\mathbb{R}^2} \tilde{v}(y)\,\rho_n(x-y)\,\mathrm{d}y\,.$$

We put

(3.42) $\quad v_n =$ restriction of $\tilde{v}_n$ to Ω.

We then have

(3.43) $\quad \tilde{v}_n \to \tilde{v}$ strongly in $H^1(\mathbb{R}^2)$

and hence, by restriction to Ω,

(3.44) $\quad v_n \to v$ strongly in $H^1(\Omega)$.

From (3.35) and for a sufficiently large n, we have

(3.45) $\quad$ support $(\tilde{v}_n) \subset \Omega \Rightarrow v_n \in \mathscr{D}(\Omega)$.

It now remains to show that $v_n \in K_1$ for a sufficiently large n. Suppose $x \in \Omega$; we then have:

(3.46) $$v_n(x) = \int_{\mathbb{R}^2} \tilde{v}(y)\,\rho_n(x-y)\,\mathrm{d}y\,;$$

we put:

(3.47) $$\omega_n(x) = [y \mid y \in \mathbb{R}^2,\ x - y \in \text{support}\,(\rho_n)]\,,$$

then $x \in \omega_n(x)$ and

(3.48) $\quad$ diameter $(\omega_n(x)) =$ diameter support $(\rho_n) \to 0$ as $n \to +\infty$.

We therefore have, $\forall x \in \Omega$,

$$(3.49) \quad \begin{cases} | v_n(x) | \leqslant \int_{\mathbb{R}_2} | \tilde{v}(y) | \rho_n(x-y) \, dy \leqslant \| \tilde{v}(y) \|_{L^\infty[\omega_n(x)]} \times \\ \qquad \times \int_{\mathbb{R}_2} \rho_n(y) \, dy = \| \tilde{v}(y) \|_{L^\infty[\omega_n(x)]} . \end{cases}$$

It can readily be deduced from (3.35) together with the definition of $\tilde{v}$ and (3.48) and (3.49), that

$$(3.50) \quad | v_n(x) | \leqslant \delta(x, \Gamma) \quad \forall x \in \Omega \text{ provided } n \text{ is sufficiently large,}$$

and hence $v_n \in K_1$ provided n is sufficiently large. ■

Approximation of L.

In the case considered, where $L(v) = C \int_\Omega v(x) \, dx$, the verification of (5.37), (5.38) of Chapter 1, Section 5.2 is immediate for $L_h : V_h \to \mathbb{R}$ defined by:

$$L_h(v_h) = C \int_\Omega q_h \, v_h \, dx . \quad ■$$

3.3.3. *A result concerning strong convergence*

In view of the properties demonstrated in Section 3.3.2 for exterior approximations of $H_0^1(\Omega)$, a, K_1, L we can apply to problem (P_{1h}), and more precisely to the equivalent formulation (3.14), Theorem 5.2 of Chapter 1, Section 5.3; this gives:

Theorem 3.1. *Let* u_h *be the solution of problem* (P_{1h}); *as* $h \to 0$, *we have*:

$$(3.51) \quad \begin{cases} p_h \, u_h \to \sigma u \quad \textit{strongly in } F \\ J_h(u_h) \to J(u) \end{cases}$$

where u *is the solution of problem* (P_1). ■

3.4 Solution of the approximate problem by point over-relaxation with projection.

We use algorithm (1.41), (1.42) of Chapter 2, Section 1.4; taking account of (3.11), (3.12) and iterating with increasing i, j this takes the form:

$$(3.52) \quad \begin{cases} u_h^0 \in K_{1h} \quad \text{given} \\ u_{ij}^{n+1/2} = (1-\omega) \, u_{ij}^n + \dfrac{\omega}{4} \left(u_{i+1j}^n + u_{ij+1}^n + u_{i-1j}^{n+1} + u_{ij-1}^{n+1} + h^2 \, C \right) \\ u_{ij}^{n+1} = \max \left(-\delta_{ij}, \min \left(u_{ij}^{n+1/2}, \delta_{ij} \right) \right) \\ M_{ij} \in \Omega_h . \end{cases}$$

We again use the convention $u_{kl} = 0$ if $M_{kl} \notin \Omega_h$.

Since Theorem 1.3 of Chapter 2, Section 1.4 applies to (P_{1h}) we have:

Theorem 3.2. *The sequence u_h^n tends towards u_h as $n \to +\infty$ if* $0 < \omega < 2$. ■

Remark 3.4. If $u_h^0 = 0$, and if $0 < \omega \leqslant 1$, it results from (3.52) that the sequence $(u_{ij}^n)_n$ is increasing $\forall M_{ij} \in \Omega_h$; the convergence is therefore monotone and the coordinates u_{ij} of u_h are approached from below. ■

Remark 3.5. We could also solve (P_{1h}) by other methods and in particular by a duality method similar to that used in Chapter 2, Section 5.6; although the application of such a method to problem (P_{1h}) does not present any difficulties, we have been content in this example, to use the over-relaxation method, which seems the best adapted to the problem considered. ■

3.5. Applications. Example 1.

We take $\Omega =]0, 1[\times]0, 1[$ with $C = 10$.

3.5.1. *Numerical values of the parameters*

Mesh size : $h = 1/40$.

Initialisation of algorithm (3.52) : $u_h^0 = 0$.

Termination criterion for algorithm (3.52) :

$$(3.53) \qquad \sum_{M_{ij} \in \Omega_h} |u_{ij}^{n+1} - u_{ij}^n| < 10^{-4}. \quad ■$$

3.5.2. *Analysis of the numerical results*

Table 3.1 shows the rate of convergence as a function of ω:

ω	1	1.4	1.5	1.6	1.7	1.8	1.9	2
Number of iterations	290	138	118	98	93	108	188	Divergence

Table 3.1

The execution time on an IBM 360/91 is 7.33 sec. for $\omega = 1$ and 2.51 sec. for $\omega = 1.7$. ■

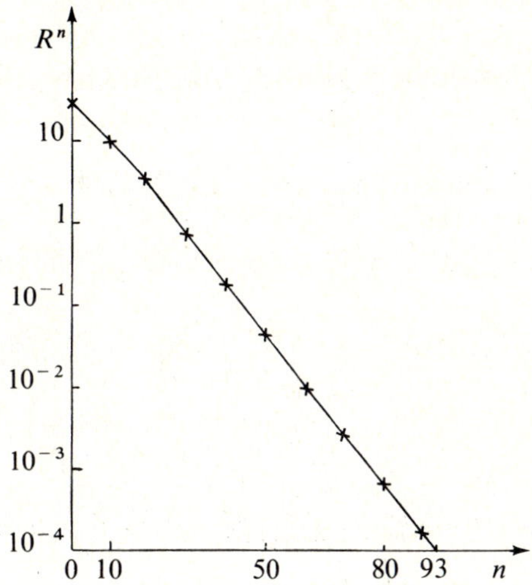

Fig. 3.1. Decrease of R^n as a function of the number of iterations n.
The case where $C = 10$ (log-linear scale).

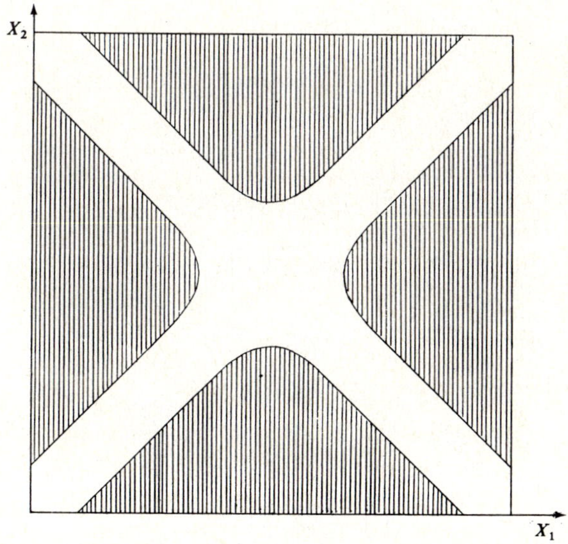

Fig. 3.2. The plastic regions are shown hatched ($C = 10$).

Remark 3.6. We note that once again the rate of convergence is greatest when ω is considerably larger than 1. ■

Figure 3.1 shows the decrease of the quantity

$$R^n = \sum_{M_{ij} \in \Omega_h} | u_{ij}^{n+1} - u_{ij}^n | ,$$

as a function of the number of iterations, for $\omega = 1.7$; Figure 3.2 shows the plastic $(| u(x) | = \delta(x, \Gamma))$ (1) and elastic

$(| u(x) | < \delta(x, \Gamma))$ (2)

regions and relating to the optimal value of ω we can state:

Remark 3.7. For the unconstrained Dirichlet problem on $\Omega =]0, 1[\times]0, 1[$, discretised with a mesh size $h = 1/40$, the optimal value of ω is very close to 1.85, whilst for problem (P_{1h}) with $C = 10$ it is close to 1.7; this value is essentially that which corresponds to the solution of the Dirichlet problem $-\Delta u = C$ in the elastic region for a discretisation with mesh size h; this results from the experimentally observed fact that the *plastic region*, i.e. the region where the constraints are satisfied with equality, *is located very quickly*, and that the computation time is spent mainly in solving $-\Delta u = C$ in the elastic region. The above comments corroborate Remark 5.7 of Chapter 2, Section 5.7.2. ■

Figure 3.3 shows, using the numerical results, the functions $x \to u(x, 0.2)$ and $x \to u(x, 0.5)$, where u is the solution of problem (P_1).

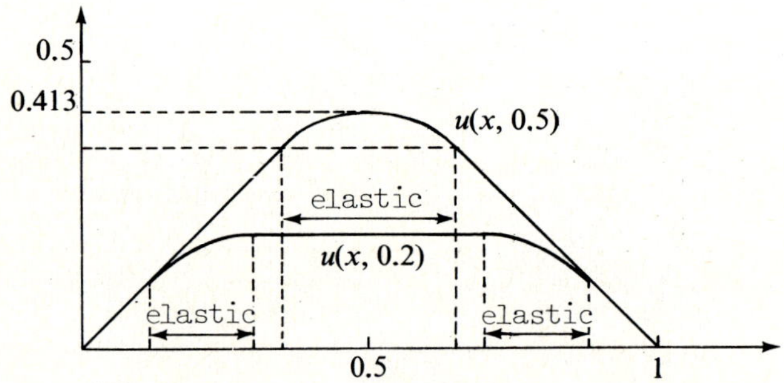

Fig. 3.3. $\Omega =]0, 1[\times]0, 1[$, $C = 10$.
Representation of $u(x, 0.2)$ and $u(x, 0.5)$.

(1) i.e. $| \text{grad}\, u | = 1$.

(2) i.e. $| \text{grad}\, u | < 1, \; -\Delta u = C$.

3.6. Applications. Example 2.

We take for Ω the domain represented in Figure 3.4, with $C = 5$ and 10.

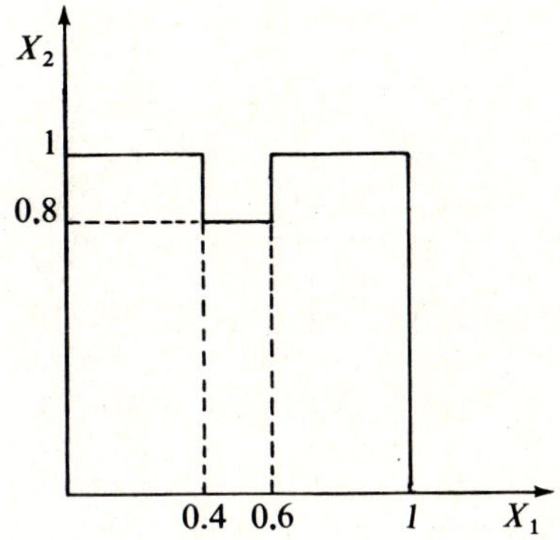

Fig. 3.4.

3.6.1. *Numerical values of the parameters*

Mesh size : $h = 1/20$.

Initialisation of algorithm (3.52) : $u_h^0 = 0$.

Termination criterion for algorithm (3.52) : we again use condition (3.53).

3.6.2. *Analysis of the numerical results*

The results in Table 3.2. show the rate of convergence as a function of ω.

C	5	5	10	10
ω	1	1.5	1	1.5
Number of iterations	185	81	64	38

Table 3.2

The computation time for $C = 10$, $\omega = 1$ is 0.88 sec. on an IBM 360/91.

These results confirm the advantage of using $\omega > 1$ in algorithm (3.52); moreover, since the *plastic region* is larger for

$C = 10$ than for $C = 5$, more constraints are satisfied *with equality* in the former case, which explains the faster rate of convergence of algorithm (3.52).

Figure 3.5 shows the *elastic* and *plastic* regions corresponding to $C = 10$.

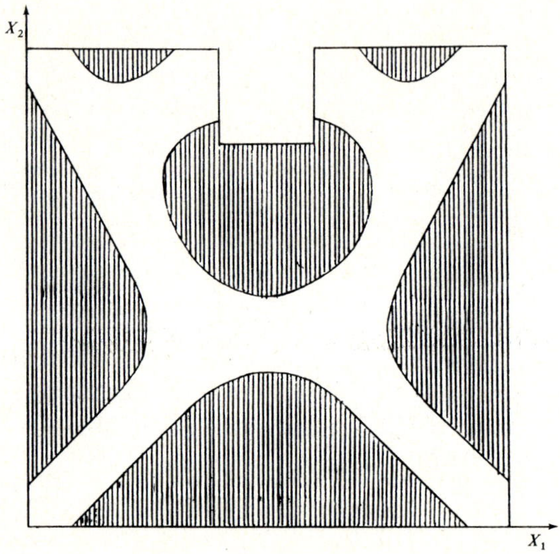

Fig. 3.5. Representation of the elastic and plastic regions for $C = 10$ (the plastic regions are shown hatched).

4. INTERIOR APPROXIMATIONS OF (P_0)

Synopsis.

In the previous section we studied problem (P_1); we shall now consider problem (P_0), for which the discretisation of the functional J *is treated in the same way* whereas the application of the constraints (membership of K_0) *poses new problems* which vary widely with the mode of approximation used for J. In this and the following sections we shall therefore carry out a systematic study of the approximation of J and *the corresponding approximation of the convex set* K_0.

In this section we shall formulate two types of interior approximation in the sense of Chapter 1, Section 4.

In Section 4.1 we shall consider a *finite element* method with approximation by affine linear functions over triangles (see Chapter 1, Section 4, Example 4.2), and in Section 4.2, a *Galerkin* approximation (see Chapter 1, Section 4, Example 4.1) using the *eigenfunctions* of the operator $-\Delta$ in $H_0^1(\Omega)$ as a basis.

The convergence of these approximations will be studied in

Section 6. ■

4.1. A finite element method

4.1.1 *Triangulation of* Ω. *Definition of* V_h

We now go back to Example 4.2 of Chapter 1, Section 4; suppose $\mathcal{T}_h$ is a *triangulation* generated by a *finite* number of triangles with the following properties:

$$T \in \mathcal{T}_h \Rightarrow T \subset \bar{\Omega} \tag{4.1}$$

$$\begin{cases} T, T' \in \mathcal{T}_h \Rightarrow T \cap T' = \phi \text{ or } T \text{ and } T' \text{ have a common side} \\ \text{or } T \text{ and } T' \text{ have a common vertex;} \end{cases} \tag{4.2}$$

for φ (see relation (4.6) of Chapter 1, Section 4, Example 4.2) we take

$$\varphi(h) = h = \operatorname*{Max}_{T \in \mathcal{T}_h} (\text{Measure of } T) \tag{4.3}$$

which satisfies the relations (4.6), (4.7) of Chapter 1, Section 4.

We define the sets

$$\bar{\Omega}_h = \bigcup_{T \in \mathcal{T}_h} T \tag{4.4}$$

$$\Omega_h = \mathring{\bar{\Omega}}_h \tag{4.5}$$

$$\Gamma_h = \partial\Omega_h \tag{4.6}$$

$$\Sigma_h = \text{set of vertices of } \mathcal{T}_h \tag{4.7}$$

$$\mathring{\Sigma}_h = [M \mid M \in \Sigma_h,\ M \notin \Gamma_h]\,. \tag{4.8}$$

If $M \in \mathring{\Sigma}_h$, we denote by $P_1, \ldots, P_k$ the neighbouring vertices of M (see Figure 4.1);

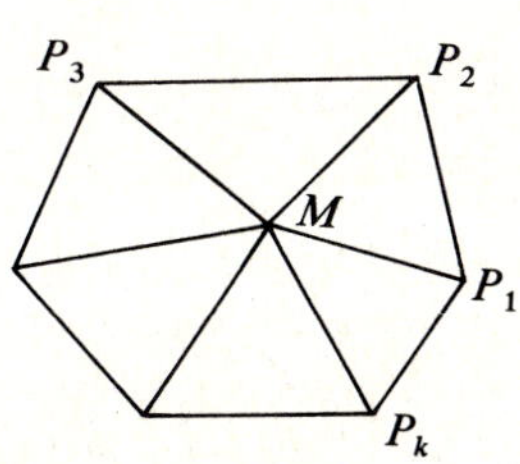

Fig. 4.1.

We define:

$$(4.9)\qquad \begin{cases} w_M = \text{affine continuous function over each triangle} \\ \text{with vertex } M, \text{ such that } w_M(M) = 1, w_M(P_i) = 0, i = 1, \dots, k \\ \text{and with } w_M \text{ zero on } \Omega \text{ outside the triangles with} \\ \text{vertex } M, \end{cases}$$

then:

$$(4.10)\qquad V_h = \text{ the space spanned by the } w_M,\ M \in \mathring{\Sigma}_h\,,$$

and we have:

Proposition 4.1. *V_h is a subspace of $H_0^1(\Omega)$ with*

$$\dim(V_h) = N = \text{Card}(\mathring{\Sigma}_h)\,.\ \blacksquare$$

Remark 4.1. Since the functions $v_h \in V_h$ are affine over T, $\text{grad}\, v_h$ is constant over $T, \forall T \in \mathcal{T}_h$. ■

4.1.2. *Definition of the approximate problem*

For (P_0), we restrict our attention to the case where

$$(4.11)\qquad L(v) = \int_\Omega fv \,dx\,dy,\quad f \in L^2(\Omega)$$

and hence, with $a(u, v) = \int_\Omega \text{grad}\, u . \text{grad}\, v \,dx\,dy$,

$$(4.12)\qquad J(v) = \frac{1}{2} a(v, v) - \int_\Omega fv \,dx\,dy$$

and

$$(P_0)\begin{cases} \underset{v \in K_0}{\text{Min}}\, J(v) \\ K_0 = [v \mid v \in H_0^1(\Omega),\ |\text{grad}\, v| \leqslant 1 \text{ a.e.}]\,. \end{cases}$$

We then define the approximate problem (P_{0h}) by:

$$(4.13)\qquad (P_{0h})\ \underset{v_h \in K_0 \cap V_h}{\text{Min}}\ J(v_h)\,.\ \blacksquare$$

Remark 4.2. The condition $v_h \in K_0 \cap V_h \Rightarrow |\text{grad}\, v_h|^2 \leqslant 1$ a.e. ; if we put (see (4.10)):

$$(4.14)\qquad v_h = \sum_{M \in \mathring{\Sigma}_h} v_M\, w_M\,,\quad v_M \in R$$

then in view of Remark 4.1 the approximate problem (4.13) reduces to an optimisation problem in $\mathbb{R}^N$ with N' $(= \text{Card}(\mathcal{T}_h))$ quadratic constraints with respect to the v_M. ■

4.1.3. *Solvability of the approximate problem*

Since the functional J has the required properties (see Chapter 1, Section 2.1) we have:

Proposition 4.2. *The approximate problem* (4.13) *admits a unique solution* u_h *characterised by*:

$$\begin{cases} a(u_h, v_h - u_h) \geqslant \int_\Omega f(v_h - u_h)\,dx\,dy \quad \forall v_h \in K_0 \cap V_h\,, \\ u_h \in K_0 \cap V_h\,. \quad \blacksquare \end{cases} \tag{4.15}$$

4.1.4. *Explicit formulation of the approximate problem*

Let M_1, M_2, M_3 be the vertices of a triangle $T \in \mathcal{T}_h$ with coordinates (x_i, y_i), $i = 1, 2, 3$; the affine linear function ϕ defined by:

$$\phi(M_i) = v_{M_i}, \quad i = 1, 2, 3 \tag{4.16}$$

is expressed as follows:

$$\phi(x, y) = \frac{1}{\delta} \sum_{i=1}^{3} (\alpha_i x + \beta_i y + \gamma_i)\, v_{M_i} \tag{4.17}$$

with

$$\begin{cases} \alpha_1 = y_2 - y_3\,, & \beta_1 = x_3 - x_2\,, & \gamma_1 = x_2 y_3 - x_3 y_2\,, \\ \alpha_2 = y_3 - y_1\,, & \beta_2 = x_1 - x_3\,, & \gamma_2 = x_3 y_1 - x_1 y_3\,, \\ \alpha_3 = y_1 - y_2\,, & \beta_3 = x_2 - x_1\,, & \gamma_3 = x_1 y_2 - x_2 y_1\,, \end{cases} \tag{4.18}$$

and

$$\begin{cases} \delta = \sum_{i=1}^{3} \gamma_i\,, \\ |\delta| = 2\ \text{Measure (Triangle}\ M_1 M_2 M_3)\,. \quad \blacksquare \end{cases} \tag{4.19}$$

In the following we shall use the notation $m(T)$ = measure of T, and, for simplification, $v_{M_i} = v_i$.

If we consider $v_h \in V_h$, we have (see (4.14)):

$$v_h(x, y) = \sum_{M \in \mathring{\Sigma}_h} v_M\, w_M(x, y)\,. \tag{4.20}$$

It is also possible (and useful) to define v_h using the characteristic functions of the triangles $T \in \mathcal{T}_h$ which, taking account of (4.17) and (4.19) and using the obvious notation, gives:

$$v_h(x, y) = \frac{1}{2} \sum_{T \in \mathcal{T}_h} \frac{1}{m(T)} \left(\sum_{i=1}^{3} (\alpha_{iT}\, x + \beta_{iT}\, y + \gamma_{iT}) v_{iT} \right) \theta_T(x, y)\,, \tag{4.21}$$

with

(4.22) θ_T = characteristic function of T.

From this we deduce

$$\frac{\partial v_h}{\partial x}(x, y) = \frac{1}{2} \sum_{T \in \mathcal{T}_h} \frac{1}{m(T)} \left(\sum_{i=1}^{3} \alpha_{iT} v_{iT} \right) \theta_T(x, y) \tag{4.23}$$

$$\frac{\partial v_h}{\partial y}(x, y) = \frac{1}{2} \sum_{T \in \mathcal{T}_h} \frac{1}{m(T)} \left(\sum_{i=1}^{3} \beta_{iT} v_{iT} \right) \theta_T(x, y) \tag{4.24}$$

$$| \operatorname{grad} v_h(x, y) |^2 = \frac{1}{4} \sum_{T \in \mathcal{T}_h} \frac{1}{(m(T))^2} \left[\left(\sum_{i=1}^{3} \alpha_{iT} v_{iT} \right)^2 + \left(\sum_{i=1}^{3} \beta_{iT} v_{iT} \right)^2 \right] \theta_T(x, y) , \tag{4.25}$$

and hence:

$$\begin{cases} J(v_h) = \dfrac{1}{8} \displaystyle\sum_{T \in \mathcal{T}_h} \frac{1}{m(T)} \left[\left(\sum_{i=1}^{3} \alpha_{iT} v_{iT} \right)^2 + \left(\sum_{i=1}^{3} \beta_{iT} v_{iT} \right)^2 \right] - \\ - \displaystyle\sum_{M \in \mathring{\Sigma}_h} v_M \int_\Omega f(x, y)\, w_M(x, y)\, dx\, dy , \end{cases} \tag{4.26}$$

and

$$v_h \in K_0 \cap V_h \Leftrightarrow \frac{1}{4(m(T))^2} \left[\left(\sum_{i=1}^{3} \alpha_{iT} v_{iT} \right)^2 + \left(\sum_{i=1}^{3} \beta_{iT} v_{iT} \right)^2 \right] \leqslant 1 \quad \forall T \in \mathcal{T}_h , \tag{4.27}$$

which allows the approximate problem (P_{0h}) to be expressed in terms of the v_M. ■

Remark 4.3. In (4.26) the integral

$$\int_\Omega f(x, y)\, w_M(x, y)\, dx\, dy$$

must actually be evaluated on the support of w_M. ■

Remark 4.4. In the terminology of finite element methods, the symmetric positive-definite matrix related to the quadratic part of $J(v_h)$ is called the *stiffness matrix* associated with the approximation considered; in the approximate problem it will play the role of $-\Delta$ in the continuous problem. ■

4.1.5. *On the use of finite element methods of order greater than* 1.

In Sections 4.1.1, 4.1.2, 4.1.3 an approximation was considered which uses polynomials of maximum degree equal to unity over each $T \in \mathcal{T}_h$; this gives a *constant* gradient over T, $\forall T \in \mathcal{T}_h$. The use of polynomials of maximum degree $k\ (\geqslant 2)$ over T, $\forall T \in \mathcal{T}_h$ would, in general, produce over T, $\forall T \in \mathcal{T}_h$ a gradient which is a polynomial function of x, y of degree $k - 1 \geqslant 1$ (and thus *not constant*).

The approximate formulation of the condition $|\operatorname{grad} v(x, y)| \leqslant 1$ a.e. could then be carried out in *various* ways, in particular (this list is not exhaustive):

$$\text{(4.28)} \qquad \begin{cases} |\operatorname{grad} v_h(G_T)| \leqslant 1 \quad \forall T \in \mathcal{T}_h, \\ G_T = \text{centre of gravity of the triangle } T, \end{cases}$$

$$\text{(4.29)} \qquad \frac{1}{m(T)} \int_T |\operatorname{grad} v_h(x, y)|^2 \, dx \, dy \leqslant 1 \quad \forall T \in \mathcal{T}_h,$$

$$\text{(4.30)} \qquad |\operatorname{grad} v_h(x, y)| \leqslant 1 \text{ a.e. in } \Omega_h,$$

$$\text{(4.31)} \qquad \frac{1}{m(T)} \int_T |\operatorname{grad} v_h(x, y)| \, dx \, dy \leqslant 1 \quad \forall T \in \mathcal{T}_h.$$

From a numerical point of view the formulations (4.28) and (4.29) can, without great difficulty, be made explicit in terms of the coefficients of the polynomial approximation considered on T; the satisfaction of the condition (4.30) is only made explicit easily for $k = 2$; in fact, in this particular case, the mapping $g_T : T \to R$, defined by:

$$\text{(4.32)} \qquad g_T(x, y) = |\operatorname{grad} v_h(x, y)|^2 \quad \forall (x, y) \in T,$$

is a polynomial of maximum degree $2(k - 1) = 2$ and its Hessian matrix (ie. that of the second derivatives) is constant and (at least) positive semi-definite; g_T therefore attains its maximum at one of the vertices of the triangle T, so that there are *three constraints* to be satisfied for each triangle $T \in \mathcal{T}_h$. The intermediate numerical calculations necessary to make explicit (and also to satisfy) (4.30) makes this formulation difficult to use for $k \geqslant 3$; this is also the case for (4.31) (for $k \geqslant 2$) since the satisfaction of this relation requires the calculation of the integral on the left-hand side by an *approximate integration* method, the exact calculation being in general impossible. ■

Let V_h be a subspace of $H_0^1(\Omega)$ such that $v_h \in V_h \Rightarrow v_h$ is a polynomial on T, of maximum degree k; the approximation considered is therefore interior in the sense of Chapter 1, Section 4. We note that, except for $k = 1$[1], the fact that v_h satisfies (4.28), (4.29), (4.31) does not in general imply that

$$v_h \in K_0 = [v \mid v \in H_0^1(\Omega),\ |\operatorname{grad} v| \leqslant 1 \text{ a.e}],$$

---

[1] In this case there is an equivalence between (4.28), (4.29), (4.30), (4.31).

the latter only being true for (4.30).

If we use (4.30), the approximate problem, $\underset{v_h \in K_0 \cap V_h}{\text{Min}} J(v_h)$, is (for $k = 2$) a finite-dimensional optimisation problem with $3 \text{ Card } (\mathcal{T}_h) = 3N'$ quadratic constraints compared with N' if $k = 1$ (and with the *same* triangulation)[1]. ■

4.2. <u>An interior approximation method using the eigenfunctions of the operator $-\Delta$ in $H_0^1(\Omega)$.</u>

<u>Synopsis.</u>

We assume that the eigenfunctions in $H_0^1(\Omega)$ of the operator $-\Delta$ are known though this is rarely the case in practical applications; the purpose of this approach is to demonstrate (and this will become apparent in Section 9) the difficulties involved in using high-accuracy methods in one-sided problems which have constraints of the *point* type, as is the case with K_0. ■

4.2.1. *Approximation of* $H_0^1(\Omega)$

Since the domain Ω is bounded, the spectrum of $-\Delta$ in $H_0^1(\Omega)$ is *countable*; we shall use the notation:

$$(4.33) \qquad (\beta_m)_{m \in N} = \text{spectrum of } -\Delta \text{ in } H_0^1(\Omega) \quad (\beta_m > 0,\ \forall m),$$

with w_m the corresponding eigenfunctions; hence

$$(4.34) \qquad \begin{cases} -\Delta w_m = \beta_m w_m & \forall m \in N, \\ w_m \in H_0^1(\Omega) & \forall m \in N. \end{cases}$$

We recall that:

$$(4.35) \qquad \int_\Omega w_m(x)\, w_n(x)\, dx = 0 \quad \text{if} \quad m \neq n;$$

we shall assume that:

$$(4.36) \qquad \int_\Omega |w_m(x)|^2\, dx = 1 \quad \forall m \in N.$$

In particular since the family $(w_m)_{m \in N}$ constitutes a Galerkin basis for $H_0^1(\Omega)$, we approximate $H_0^1(\Omega)$ by the family $(V_M)_{M \in N}$ defined by:

---

(1) The principal motivation behind going from $k = 1$ to $k = 2$ is simply to reduce the number of triangles rather than to increase the accuracy.

(4.37) V_M = subspace of $H_0^1(\Omega)$ generated by $(w_m)_{1 \leq m \leq M}$.

4.2.2. *Approximation of* K_0

Since in practice it is not possible to use the convex set $K_0 \cap V_M$ we shall be content if $v \in V_M$ to satisfy the condition $|\operatorname{grad} v| \leq 1$ on a *finite* set of points of Ω (or of $\bar{\Omega}$), say Ω_h (resp. $\bar{\Omega}_h$); for K_0 this gives the approximation K_{0h}^M (1) defined by

$$K_{0h}^M = [v \mid v \in V_M ;\ |\operatorname{grad} v(P)| \leq 1 \quad \forall P \in \Omega_h] . \tag{4.38}$$

We get a similar definition if we replace Ω_h by $\bar{\Omega}_h$. ■

Remark 4.5. The method for approximating K_0 appears to be of the *collocation* type; a natural method (which will be employed in Section 9), inspired by J.B. Rosen /3/, consists of defining

$$R_h = [x \mid x = (x_1, x_2) \in \mathbb{R}^2,\ x_1 = ih,\ x_2 = jh,\ i, j \in \mathbb{Z}] , \tag{4.39}$$

$$\Omega_h = \Omega \cap R_h \tag{4.40}$$

where $h > 0$ is given (and intended to approach zero).

Similarly we can define $\bar{\Omega}_h = \Omega_h \cup \Gamma_h$ where Γ_h. is a finite subset of points of Γ for which at least one of the coordinates is an integer multiple of h. ■

4.2.3. *Definition of the approximate problem*

With the functional $J : H_0^1(\Omega) \to \mathbb{R}$ once again of the form:

$$J(v) = \frac{1}{2} a(v, v) - \int_\Omega f v \, dx ,$$

with $f \in L^2(\Omega)$, the approximate problem is defined by:

$$\operatorname*{Min}_{v \in K_{0h}^M} J(v) . \quad ■ \tag{4.41}$$

4.2.4. *Solvability of the approximate problem*

As the functional J and K_{0h}^M have the required properties (see Chapter 1, Section 2.1) we have:

---

(1) This approximation will be justified in Section 6.

Proposition 4.3. *The approximate problem* (4.41) *admits one and only one solution,* u_h^M*, characterised by:*

$$(4.42)\qquad \begin{cases} a(u_h^M, v - u_h^M) \geq \displaystyle\int_\Omega f(v - u_h^M)\,\mathrm{d}x \quad \forall v \in K_{0h}^M, \\ u_h^M \in K_{0h}^M . \end{cases}$$

4.2.5. *Explicit formulation of the approximate problem*

If $v \in V_M$ we use the notation:

$$(4.43)\qquad v = \sum_{i=1}^{M} \mu_i w_i, \quad \mu_i \in \mathbb{R} \quad \forall i = 1, \dots, M,$$

and taking into account the *orthonormality* of the $(w_m)_{m \in N}$ we have:

$$(4.44)\qquad \begin{cases} J(v) = \dfrac{1}{2} \displaystyle\sum_{m=1}^{M} \beta_m \mu_m^2 - \sum_{m=1}^{M} f_m \mu_m \quad \forall v \in V_M \\ \text{with} \quad f_m = (f, w_m)_{L^2(\Omega)} . \end{cases}$$

Moreover:

$$(4.45)\qquad |\operatorname{grad} v(P)|^2 = \left(\sum_{m=1}^{M} \mu_m \frac{\partial w_m}{\partial x_1}(P)\right)^2 + \left(\sum_{m=1}^{M} \mu_m \frac{\partial w_m}{\partial x_2}(P)\right)^2 .$$

Putting $\mu = (\mu_1, \mu_2, \dots, \mu_M)$ and taking account of (4.44), (4.45), the approximate problem (4.41) can be written explicitly as

$$(4.46)\qquad \begin{cases} \underset{\mu}{\operatorname{Min}} \left[\dfrac{1}{2} \displaystyle\sum_{m=1}^{M} \beta_m \mu_m^2 - \sum_{m=1}^{M} f_m \mu_m\right] \\ \mu \in \left[\mu \mid \mu \in R^M, \left(\displaystyle\sum_{m=1}^{M} \mu_m \frac{\partial w_m}{\partial x_1}(P)\right)^2 + \left(\sum_{m=1}^{M} \mu_m \frac{\partial w_m}{\partial x_2}(P)\right)^2 \leq \right. \\ \qquad \left. \leq 1, \ \forall P \in \Omega_h (\text{ or } \overline{\Omega}_h)\right]. \ \blacksquare \end{cases}$$

Remark 4.6. In view of (4.44) the matrix associated with the quadratic part of J, considered as a function of μ, is *diagonal*; therefore, from this point of view the situation is optimal in comparison with any other type of approximation; on the other hand the constraints $|\operatorname{grad} v(P)| \leq 1$, $\forall P \in \Omega_h$ (or $\overline{\Omega}_h$) *explicitly* involve *all* the coordinates of μ (this is not the case for the interior approximation considered in Section 4.1 or in the exterior approximations studied in Section 5) and this fact will considerably limit the applicability of this type of approximation. ■

5. EXTERIOR APPROXIMATIONS OF (P_0)

5.1. Approximations of J

5.1.1. *A first exterior approximation of* J

We retain the definitions and the formalism of Section 3.2.1; hence if $f \in L^2(\Omega)$, the approximation of $J(v) = \frac{1}{2} a(v, v) - \int_\Omega fv \, dx$ by $J_h^1 : V_h \to \mathbb{R}$ is defined by:

$$J_h^1(v_h) = \frac{1}{2} \sum_{k=1}^{2} \int_\Omega | \delta_k q_h v_h |^2 \, dx - \int_\Omega f q_h v_h \, dx , \tag{5.1}$$

the *explicit* form of (5.1) being given by:

$$\begin{cases} J_h^1(v_h) = \dfrac{h^2}{4} \displaystyle\sum_{M_{ij} \in \bar{\Omega}_h} \left[\left| \dfrac{v_{i+1j} - v_{ij}}{h} \right|^2 + \left| \dfrac{v_{i-1j} - v_{ij}}{h} \right|^2 + \right. \\ \left. + \left| \dfrac{v_{ij+1} - v_{ij}}{h} \right|^2 + \left| \dfrac{v_{ij-1} - v_{ij}}{h} \right|^2 \right] - h^2 \displaystyle\sum_{M_{ij} \in \Omega_h} f_{ij} v_{ij} , \end{cases} \tag{5.2}$$

with

$$f_{ij} = \frac{1}{h^2} \int_{\varpi_h^0(M_{ij})} f(x) \, dx \quad \forall M_{ij} \in \Omega_h ; \tag{5.3}$$

the sets $\Omega_h, \bar{\Omega}_h$ are as defined in Section 3.2.1 and Remark 3.1 also holds for (5.2). ■

5.1.2. *A second exterior approximation of* J

In the following we shall also use the approximation of J defined by

$$\begin{cases} J_h^2(v_h) = \dfrac{1}{2} \displaystyle\int_{\mathbb{R}^2} \left[\left(\dfrac{\delta_1 q_h v_h\left(x_1, x_2 + \frac{h}{2}\right) + \delta_1 q_h v_h\left(x_1, x_2 - \frac{h}{2}\right)}{2} \right)^2 + \right. \\ \left. + \left(\dfrac{\delta_2 q_h v_h\left(x_1 + \frac{h}{2}, x_2\right) + \delta_2 q_h v_h\left(x_1 - \frac{h}{2}, x_2\right)}{2} \right)^2 \right] \times \\ \times \, dx - \displaystyle\int_\Omega f q_h v_h \, dx . \end{cases} \tag{5.4}$$

Before making (5.4) explicit we define the grid Q_h as follows:

$$Q_h = \left\{ M \mid M \in R^2, \ x_1 = (2i + 1)\frac{h}{2}, \ x_2 = (2j + 1)\frac{h}{2}, \quad i, j \in \mathbb{Z} \right\}, \tag{5.5}$$

and we shall use the notation (see Figure 5.1):

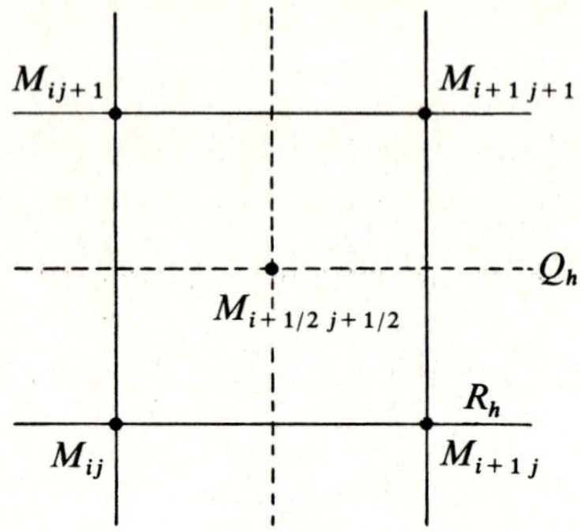

Fig. 5.1.

$$(5.6)\qquad M_{i+1/2\,j+1/2} = \left\{(2i+1)\frac{h}{2},\ (2j+1)\frac{h}{2}\right\}.$$

We define Σ_h by:

$$(5.7)\qquad \Sigma_h = \{ M \mid M \in Q_h,\ \text{at least one of the 4 vertices of } \varpi_h^0(M) \text{ belonging to } \Omega_h \}.$$

We can now express (5.4) explicitly in the form:

$$(5.8)\qquad J_h^2(v_h) = \frac{h^2}{4} \sum_{M_{i+1/2\,j+1/2} \in \Sigma_h} \times$$
$$\times \left[\left(\frac{v_{i+1\,j+1} - v_{ij}}{h}\right)^2 + \left(\frac{v_{i+1\,j} - v_{ij+1}}{h}\right)^2\right] - h^2 \sum_{M_{ij} \in \Omega_h} f_{ij}\, v_{ij},$$

the f_{ij} still being defined by (5.3) and with $v_{pq} = 0$ if $M_{pq} \notin \Omega_h$. ■

5.2. Exterior approximations of K_0

It seems reasonable to approximate K_0 by

$$(5.9)\qquad K_{0h}^1 = \{ v_h \mid v_h \in V_h,\ |\delta_1 q_h v_h|^2 + |\delta_2 q_h v_h|^2 \leqslant 1 \text{ a.e. } \},$$

the function $|\delta_1 q_h v_h|^2 + |\delta_2 q_h v_h|^2$ being constant over square no. 1 (resp. 2, 3, 4) of side $h/2$, in Figure 5.2, and having the value

$$\left|\frac{v_{i+1\,j} - v_{ij}}{h}\right|^2 + \left|\frac{v_{ij+1} - v_{ij}}{h}\right|^2;$$

the explicit form of (5.9) is given by

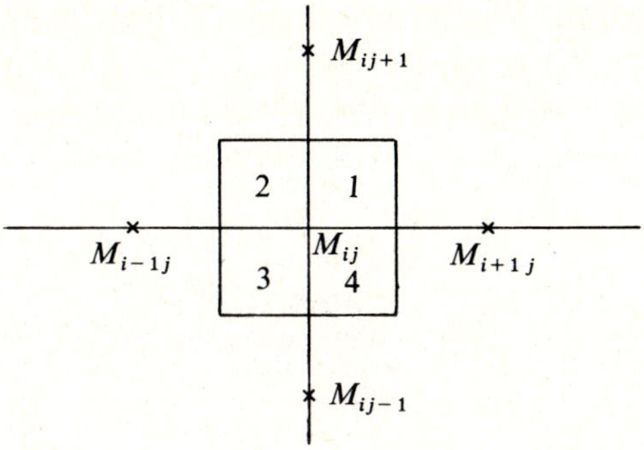

Fig. 5.2.

(5.10)
$$\begin{cases} K_{0h}^1 = \Big\{ v_h \mid v_h \in V_h, \left|\dfrac{v_{i+1j} - v_{ij}}{h}\right|^2 + \left|\dfrac{v_{ij+1} - v_{ij}}{h}\right|^2 \leqslant 1\,, \\ \left|\dfrac{v_{i-1j} - v_{ij}}{h}\right|^2 + \left|\dfrac{v_{ij+1} - v_{ij}}{h}\right|^2 \leqslant 1\,, \\ \left|\dfrac{v_{i-1j} - v_{ij}}{h}\right|^2 + \left|\dfrac{v_{ij-1} - v_{ij}}{h}\right|^2 \leqslant 1 \\ \left|\dfrac{v_{i+ij} - v_{ij}}{h}\right|^2 + \left|\dfrac{v_{ij-1} - v_{ij}}{h}\right|^2 \leqslant 1, \ \forall M_{ij} \in \overline{\Omega}_h \Big\}\,, \end{cases}$$

with, in (5.10), $v_{pq} = 0$ if $M_{pq} \notin \Omega_h$. ■

It results from (5.10) that the use of K_{0h}^1 leads to approximately $4\,\mathrm{Card}\,(\Omega_h)$ quadratic constraints. In order to reduce the number of constraints to around $\mathrm{Card}\,(\Omega_h)$ we can take the mean of the constraints (5.10) at suitably chosen points of R_h or Q_h, which gives:

(5.11)
$$\begin{cases} K_{0h}^2 = \Big\{ v_h \mid v_h \in V_h, \dfrac{1}{2}\left[\left|\dfrac{v_{i+1j} - v_{ij}}{h}\right|^2 + \left|\dfrac{v_{i+1j+1} - v_{ij+1}}{h}\right|^2 + \right. \\ \left. + \left|\dfrac{v_{i+1j+1} - v_{i+1j}}{h}\right|^2 + \left|\dfrac{v_{ij+1} - v_{ij}}{h}\right|^2\right] \leqslant 1 \qquad \forall M_{i+1/2\,j+1/2} \in \Sigma_h \Big\}\,, \end{cases}$$

where the relation defining K_{0h}^2 can be interpreted as an approximation at $M_{i+1/2\,j+1/2}$ of the constraint $|\,\mathrm{grad}\, v\,|^2 \leqslant 1$; K_{0h}^2 can also be defined as:

(5.12)
$$\begin{cases} K_{0h}^2 = \Big\{ v_h \mid v_h \in V_h, \dfrac{1}{2}\left[\left|\delta_1 q_h\, v_h\left(x_1, x_2 + \dfrac{h}{2}\right)\right|^2 + \right. \\ + \left|\delta_1 q_h\, v_h\left(x_1, x_2 - \dfrac{h}{2}\right)\right|^2 + \left|\delta_2 q_h\, v_h\left(x_1 + \dfrac{h}{2}, x_2\right)\right|^2 + \\ \left. + \left|\delta_2 q_h\, v_h\left(x_1 - \dfrac{h}{2}, x_2\right)\right|^2\right] \leqslant 1 \text{ a.e.} \Big\}\,. \end{cases}$$ ■

Before defining a third exterior approximation of K_0 we introduce the set 0_h defined by:

$$0_h = \bigcup_{M_{i+1/2\ j+1/2} \in \Sigma_h} \varpi_h^0(M_{i+1/2\ j+1/2}) . \tag{5.13}$$

We now consider the exterior approximation

$$K_{0h}^3 = \left\{ v_h \mid v_h \in V_h, \frac{\int_{\varpi_h^0(M_{ij})} [| \delta_1 q_h v_h |^2 + | \delta_2 q_h v_h |^2] \, dx}{\text{measure}[\varpi_h^0(M_{ij}) \cap 0_h]} \leqslant 1, \ \forall M_{ij} \in \overline{\Omega}_h \right\} \tag{5.14}$$

The explicit formulation of the membership of K_{0h}^3 leads us to distinguish between three cases:

If $M_{ij} \in \Omega_h$ we have:

$$\frac{1}{2}\left[\left|\frac{v_{i+1j}-v_{ij}}{h}\right|^2+\left|\frac{v_{i-1j}-v_{ij}}{h}\right|^2+\left|\frac{v_{ij+1}-v_{ij}}{h}\right|^2+\left|\frac{v_{ij-1}-v_{ij}}{h}\right|^2\right] \leqslant 1 \tag{5.15}$$

with $v_{pq} = 0$ if $M_{pq} \notin \Omega_h$.

If $M_{ij} \in \Gamma_h$ with a single neighbour in Ω_h (if this neighbour is for example M_{i+1j}, see Figure 5.3) we have

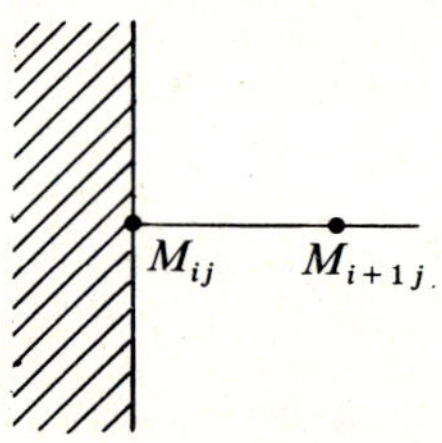

Fig. 5.3. The exterior of O_h is shown hatched.

$$\frac{1}{h} | v_{i+1j} - v_{ij} | \leqslant 1 , \tag{5.16}$$

and, in fact, as $v_{ij} = 0$, we have

$$\frac{1}{h} | v_{i+1j} | \leqslant 1 . \tag{5.17}$$

If $M_{ij} \in \Gamma_h$ with two neighbours in Ω_h (if the two neighbours are, for example M_{i+1j}, M_{ij-1}, see Figure 5.4) we have

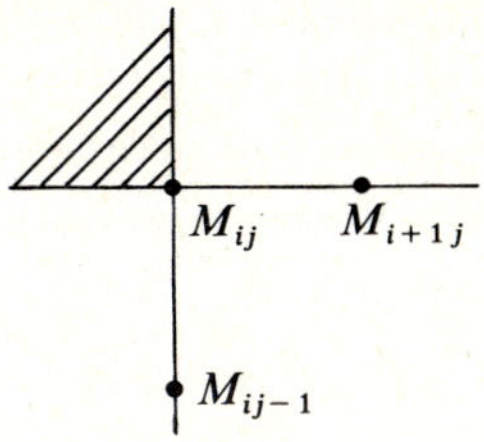

Fig. 5.4. The exterior of O_h is shown hatched.

$$(5.18)\qquad \frac{2}{3}\left[\left|\frac{v_{i+1j}-v_{ij}}{h}\right|^2+\left|\frac{v_{ij-1}-v_{ij}}{h}\right|^2\right]\leqslant 1\,,$$

and as $v_{ij}=0$, we have

$$(5.19)\qquad \frac{2}{3h^2}(v_{i+1j}^2+v_{ij-1}^2)\leqslant 1\,.$$

The relations defining K_{0h}^3 can be interpreted as an approximation at $M_{ij}\in\overline{\Omega}_h$ of the constraint $|\,\mathrm{grad}\,v\,|^2\leqslant 1$. ■

We shall also use the exterior approximation of K_0 defined by:

$$(5.20)\qquad \begin{cases} K_{0h}^4=\{v_h\mid v_h\in V_h, \\ \left|\dfrac{\delta_1 q_h v_h\left(x_1,x_2+\frac{h}{2}\right)+\delta_1 q_h v_h\left(x_1,x_2-\frac{h}{2}\right)}{2}\right|^2+ \\ +\left|\dfrac{\delta_2 q_h v_h\left(x_1+\frac{h}{2},x_2\right)+\delta_2 q_h v_h\left(x_1-\frac{h}{2},x_2\right)}{2}\right|^2 \leqslant 1 \text{ a.e}\}\,, \end{cases}$$

which is expressed explicitly by:

$$(5.21)\qquad K_{0h}^4=\left\{v_h\mid v_h\in V_h, \left|\frac{v_{i+1j+1}-v_{ij}}{h}\right|^2+\left|\frac{v_{i+1j}-v_{ij+1}}{h}\right|^2\leqslant 2 \;\; \forall M_{i+1/2\,j+1/2}\in\Sigma_h\right\}$$

with $v_{pq}=0$ if $M_{pq}\notin\Omega_h$. ■

Remark 5.1. Obviously there are other possible exterior approximations for K_0; for example, in order to approximate $|\,\mathrm{grad}\,v\,|^2\leqslant 1$ at M_{ij} we can use

$$(5.22)\qquad \left|\frac{v_{i+1j}-v_{i-1j}}{2h}\right|^2+\left|\frac{v_{ij+1}-v_{ij-1}}{2h}\right|^2\leqslant 1\,;$$

however, without going into the details, we can state that (5.22) presents serious problems from the point of view of numerical stability. ■

5.3. Formulation of the approximate problem.

With the functionals J_h^l $(l = 1, 2)$ and the convex sets K_{0h}^m $(m = 1, 2, 3, 4)$ defined as in Sections 5.1 and 5.2, we shall take as approximate problems:

$$(5.23) \qquad (P_{0h})_{lm} : \min_{v_h \in K_{0h}^m} J_h^l(v_h) . \quad ■$$

5.4. Solvability of the approximate problem.

The reasoning of Chapter 2, Section 5.3.2 (or Chapter 3, Section 3.2.2) remains valid if we note that the mappings:

$$(5.24) \qquad v_h \to \left(\int_\Omega (|\delta_1 q_h v_h|^2 + |\delta_2 q_h v_h|^2) \, dx \right)^{1/2} ,$$

$$(5.25) \qquad \begin{cases} v_h \to \left(\displaystyle\int_\Omega \left(\left| \dfrac{\delta_1 q_h v_h \left(x_1, x_2 + \frac{h}{2}\right) + \delta_1 q_h v_h \left(x_1, x_2 - \frac{h}{2}\right)}{2} \right|^2 + \right. \right. \\ \left. \left. + \left| \dfrac{\delta_2 q_h v_h \left(x_1 + \frac{h}{2}, x_2\right) + \delta_2 q_h v_h \left(x_1 - \frac{h}{2}, x_2\right)}{2} \right|^2 \right) dx \right)^{1/2} \end{cases}$$

define norms over V_h and that the convex sets K_{0h}^m $(m = 1, 2, 3, 4)$ are closed in V_h; we therefore have both the existence and uniqueness of an optimal solution for $(P_{0h})_{lm}$. ■

The convergence, as $h \to 0$, of the solution of the approximate problem to that of the continuous problem will be investigated in Section 6.

6. CONVERGENCE OF THE INTERIOR AND EXTERIOR APPROXIMATIONS

Synopsis.

In this section we shall demonstrate the convergence of the approximations considered in Sections 4 and 5; we shall refer to the methods of Chapter 1, Section 4 for the interior approximations and of Chapter 1, Section 5 for the exterior approximations.

The proof of the convergence of the various approximations considered is based essentially on a density result. ■

6.1. A lemma concerning density

We shall now prove:

Lemma 6.1. *We have*

$$\overline{\mathcal{D}(\Omega) \cap K_0} = K_0 . \tag{6.1}$$

Proof. We proceed (as in Section 3.3.2, Remark 3.3) using *truncation and regularisation*; we thus have the following two stages:

1) Let $v \in K_0$; we define $v_\varepsilon = \tau_\varepsilon(v)$ (as in (3.34)) by:

$$\tau_\varepsilon(v) = v_\varepsilon = (v - \varepsilon)^+ - (v + \varepsilon)^- , \tag{6.2}$$

where $\varepsilon > 0$ is given; we thus have $\tau_\varepsilon(v) \in H^1(\Omega)$ with $|\operatorname{grad} \tau_\varepsilon(v)| \leqslant 1$ a.e. on Ω, and since:

$$K_0 \subset K_1 = \{ v \mid v \in H_0^1(\Omega),\ |v(x)| \leqslant \delta(x, \Gamma) \quad \text{a.e.} \}$$

we have (as in (3.35)):

$$v_\varepsilon(x) = \begin{cases} 0 & \text{if } \delta(x, \Gamma) \leqslant \varepsilon \\ \leqslant \delta(x, \Gamma) - \varepsilon & \text{otherwise} \end{cases} \tag{6.3}$$

and hence $\tau_\varepsilon(v) \in K_0, \forall \varepsilon > 0$.
However

$$\tau_\varepsilon v \to v \quad \text{in } H^1(\Omega) \text{ as } \varepsilon \to 0, \tag{6.4}$$

so that:

$$\begin{cases} \text{it is sufficient to approximate a function } v \in K_0 \\ \text{which is identically zero in a neighbourhood of } \Gamma . \end{cases} \tag{6.5}$$

2) We extend v $(v \in K_0)$ to $\tilde{v}$ in $\mathbb{R}^2$ by 0 outside Ω; we have $\tilde{v} \in H^1(\mathbb{R}^2)$ and $|\operatorname{grad} \tilde{v}| \leqslant 1$ a.e. on $\mathbb{R}^2$.
If $(\rho_n)_n$ is a *regularising* sequence, we define $\tilde{v}_n \in \mathcal{D}(\mathbb{R}^2)$ by:

$$\tilde{v}_n = \tilde{v} * \rho_n \qquad \left(\tilde{v}_n(x) = \int_{\mathbb{R}^2} \tilde{v}(y)\, \rho_n(x - y)\, \mathrm{d}y \right); \tag{6.6}$$

we have:

$$\tilde{v}_n \to \tilde{v} \quad \text{strongly in } H^1(\mathbb{R}^2) \tag{6.7}$$

We use the notation:

$$v_n = \text{restriction of } \tilde{v}_n \text{ to } \Omega ; \tag{6.8}$$

thus if v satisfies (6.5) and provided n is sufficiently large,

we have:

(6.9) $$\text{support}\,(\tilde{v}_n) \subset \Omega \Rightarrow \begin{cases} v_n \in \mathscr{D}(\Omega) \\ v_n \to v \quad \text{strongly in } H_0^1(\Omega) \text{ as } n \to +\infty . \end{cases}$$

We shall now show that $v_n \in K_0$; in fact

(6.10) $$\text{grad}\,\tilde{v}_n = \rho_n * \text{grad}\,\tilde{v} ,$$

and hence

(6.11) $$|\,\text{grad}\,\tilde{v}_n(x)\,| \leqslant \int_{\mathbb{R}^2} \rho_n(x-\eta)\,|\,\text{grad}\,\tilde{v}(\eta)\,|\,\mathrm{d}\eta \leqslant \int_{\mathbb{R}^2} \rho_n(x-\eta)\,\mathrm{d}\eta = 1 . \quad \blacksquare$$

Remark 6.1. Since the functions of K_0 are equicontinuous on $\bar{\Omega}$ (indeed equi-Lipschitz with constant 1) and since K_0 is bounded in $L^\infty(\Omega)$, K_0 is *strongly compact* in $L^\infty(\Omega)$ (from Ascoli's theorem and hence if $v_n \in K_0$ $\forall n$ with $v_n \to v$ in $\mathscr{D}'(\Omega)$, then $v \in K_0$ and $v_n \to v$ *strongly* in $L^\infty(\Omega)$.(ie. uniformly). ■

6.2. Convergence of the interior approximations.

6.2.1. *Convergence of the finite element method*

Using the results of Chapter 1, Sections 4.3 and 4.4, we shall now investigate the convergence of the interior approximation of Section 4.1; we recall that h was defined by:

(6.14) $$\varphi(h) = h = \underset{T \in \mathscr{T}_h}{\text{Max}} \ (\text{measure of } T).$$

Moreover let us state that $\lim_{h\to 0} (\Omega - \Omega_h) = 0$ if for all $K \subset \Omega, K$ compact, we have $\Omega_h \supset K$ provided h is sufficiently small. We then have:

Theorem 6.1. *If $\Omega - \Omega_h \to 0$ with $\theta \geqslant \theta_0 > 0$ $\forall h$, for any angle θ of T, $\forall T \in \mathscr{T}_h$, we have $u_h \to u$ strongly in $H_0^1(\Omega) \cap L^\infty(\Omega)$ where u_h is the solution of the approximate problem (P_{0h}) defined in Section* 4.1. *relation* (4.13), *and u is the solution of (P_0)*.

Proof. We shall apply Theorem 4.2 of Section 4.4, Chapter 1 For this it is necessary to show that

(i) $\forall v \in H_0^1(\Omega)$, there exists $v_h \in V_h$ such that $v_h \to v$ strongly in $H_0^1(\Omega)$.

(ii) $\forall v \in K_0$, there exists $v_h \in K_0 \cap V_h$ such that $v_h \to v$ strongly in $H_0^1(\Omega)$; for each of these points it is sufficient to take v in a *dense subset* which is in $H_0^1(\Omega)$ for (i), and in K_0 for (ii) (see (4.2) and Remark 4.1, Chapter 1).

In a general manner we take $v \in \mathscr{D}(\Omega)$ and introduce

$$\forall T = \{ \text{triangle } M_1 \, M_2 \, M_3 \} \in \mathscr{T}_h ,$$

the affine linear function Φ_T on T, such that

$$\Phi_T(M_i) = v(M_i) , \qquad i = 1, 2, 3 \tag{6.15}$$

and define v_h by

$$v_h = \Phi_T \quad \text{in } T . \tag{6.16}$$

It can readily be shown, using Taylor series expansions of order 2 [1] that

$$| \Phi_T(x) - v(x) | \leqslant c_1(v) \, h \quad \forall x \in T , \tag{6.17}$$

$$| \operatorname{grad} \Phi_T(x) - \operatorname{grad} v(x) | \leqslant c_2(v) \sqrt{h} \quad \forall x \in T , \tag{6.18}$$

where the constants $c_1(v)$, $c_2(v)$ depend on v .

From these estimates and from the definition (6.16) we deduce straight away that

$$v_h \to v \quad \text{in } H_0^1(\Omega) , \tag{6.19}$$

which gives (i) (on $\mathscr{D}(\Omega)$ dense in $H_0^1(\Omega)$).

In order to show (ii), we use Lemma 6.1, which allows us to take v in $K_0 \cap \mathscr{D}(\Omega)$; since

$$v = \lim (\lambda v) \text{ in } H_0^1(\Omega) , \quad \lambda < 1 , \quad \lambda \to 1 , \tag{6.20}$$

we can confine our attention to v with

$$v \in \mathscr{D}(\Omega) , \qquad | \operatorname{grad} v(x) | \leqslant \lambda < 1 . \tag{6.21}$$

Then, in all $T \in \mathscr{T}_h$ we have:

$$| \operatorname{grad} v_h(x) | = | \operatorname{grad} \Phi_T(x) | \leqslant | \operatorname{grad} v(x) | + c_2(v) \sqrt{h} \leqslant \lambda + c_2(v) \sqrt{h} \tag{6.22}$$

and hence

$$v_h \in K_0 \text{ for sufficiently small } h \text{ (such that } \lambda + c_2(v) \sqrt{h} \leqslant 1 \text{)} \tag{6.23}$$

which, in conjunction with (6.19), gives (ii).

Finally, we have

$$u_h \to u \quad \text{strongly in } L^\infty(\Omega) \tag{6.24}$$

---

[1]) For a systematic study see Ciarlet-Wagschal /1/.

from Remark 6.1. ■

Remark 6.2. Falk /1/ gives an estimate of the approximation error as a function of h; we should point out that the conditions of validity (relating to Ω and $\mathcal{T}_h$) of this estimate are more restrictive than the assumptions of Section 4.1.1 and of Theorem 6.1. ■

6.2.2. *Convergence of the interior approximation method using the eigenfunctions of* $-\Delta$ *in* $H_0^1(\Omega)$

We shall now study the convergence of the interior approximation u_h^M defined in Section (4.2) and using the same notation. We assume that the constraint $|\mathbf{grad}\, v| \leqslant 1$ is satisfied on the set Ω_h defined in Section 4.2.2, Remark 4.5, by relations (4.39), (4.40). We also assume that the spectrum $(\beta_m)_{m \in N}$ of $-\Delta$ has been arranged in order of increasing eigenvalue, i.e.:

$$\beta_1 \leqslant \beta_2 \leqslant \cdots \leqslant \beta_n \leqslant \cdots . \tag{6.25}$$

Then, with u denoting the solution of (P_0), we have:

Theorem 6.2. *As* $M \to +\infty$ *and* $h \to 0$, *with*

$$h\beta_M^r \to 0, \quad \textit{real } r > 1 \tag{6.26}$$

we have:

$$u_h^M \to u \ \textit{strongly in } H_0^1(\Omega). \tag{6.27}$$

Proof. We apply Theorem 4.2 of Chapter 1, Section 4.4, takin[g] account of Remark 4.1 of Chapter 1, Section 4.3; the only point which are not immediate are that we must verify that under the stated conditions for M and h we have:

$$\forall v \in \chi \subset K_0, \ \bar{\chi} = K_0, \ \text{there exists } v_h^M \in K_{0h}^M \text{ such that } v_h^M \to v \text{ strongly in } H_0^1(\Omega) \tag{6.28}$$

and

$$\text{if } v_h^M \in K_{0h}^M, \ v_h^M \to v \ \text{weakly in } H_0^1(\Omega), \text{ then } v \in K_0. \tag{6.29}$$

Verification of (6.28)

We proceed as in the proof of Theorem 6.1, defining χ by:

$$\chi = \bigcup_{0<\lambda<1} \lambda \{ K_0 \cap \mathcal{D}(\Omega) \}. \tag{6.30}$$

Let $v \in \chi$; we define v^M as the sum of the first M terms of the Fourier series expansion of v with respect to the eigenfunctions $(w_m)_{m \in N}$ and hence, since the w_m are orthonormal:

$$(6.31) \qquad v^M = \sum_{m=1}^{M} (w_m, v)_{L^2(\Omega)} w_m .$$

We have $v \in \chi \subset \mathscr{D}(\Omega)$ and hence:

$$(6.32) \qquad (-\Delta)^q v \in H_0^1(\Omega) \quad \forall q \geqslant 0 ,$$

from which it may be deduced (see Lions-Magenes /1/) that, as $M \to +\infty$:

$$(6.33) \qquad v^M \to v \text{ strongly in } H^s(\Omega), \ \forall s \geqslant 0 \Rightarrow v^M \to v \text{ strongly in } C^k(\overline{\Omega}), \quad \forall k \ (^1)$$

and in particular:

$$(6.34) \qquad \operatorname{grad} v^M \to \operatorname{grad} v \text{ strongly in } C^0(\overline{\Omega}) , \quad \forall v \in \chi ;$$

therefore there exists $\eta(M) > 0$ such that $\eta(M) \to 0$, uniformly for $x \in \overline{\Omega}$, as $M \to +\infty$ with :

$$(6.35) \qquad |\operatorname{grad} v^M(x)| \leqslant |\operatorname{grad} v(x)| + \eta(M) \leqslant \lambda + \eta(M) < 1 \quad \forall x \in \overline{\Omega} ,$$

provided M is sufficiently large, and in particular $|\operatorname{grad} v^M(x)| < 1$ on Ω_h, and hence $v^M \in K_{0h}^M$; it will therefore be sufficient to put $v_h^M = v^M$ in order to verify (6.28). ■

Verification of (6.29).

We shall proceed in two stages:

1) We define a family of open sets $\mathscr{O}_h \subset \Omega$ as follows: let

$$(6.36) \qquad \overset{\circ}{\Omega}_h = \{ P \mid P \in \Omega_h, \overline{\omega}_h^0(P) \subset \Omega \}$$

where we recall that

$$\overline{\omega}_h^0(P) = \left] x_1 - \frac{h}{2}, x_1 + \frac{h}{2} \right[\times \left] x_2 - \frac{h}{2}, x_2 + \frac{h}{2} \right[$$

if $P = (x_1, x_2)$; we then define $\mathscr{O}_h$ by:

$$(6.37) \qquad \mathscr{O}_h = \overline{\bigcup_{P \in \overset{\circ}{\Omega}_h} \overset{\circ}{\overline{\omega}_h^0(P)}} ,$$

and we have $\lim_{h \to 0} (\Omega - \mathscr{O}_h) = 0$ in the following sense: $\forall K$ compact inside Ω we have $K \subset \mathscr{O}_h$ provided h is sufficiently small.

(1) This assumes Γ to be sufficiently regular.

Consider then $v_h^M \in K_{0h}^M$ such that $v_h^M \to v$ weakly in $H_0^1(\Omega)$; if $P \in \mathcal{O}_h$, there exists $P^* \in \Omega_h$ with $| PP^* | < h/\sqrt{2}$ and the mean value theorem of Differential Calculus, applied to $\mathbf{grad}\, v_h^M$ gives:

$$\text{(6.38)} \quad \begin{cases} \operatorname{grad} v_h^M(P) = \operatorname{grad} v_h^M(P^*) + \begin{pmatrix} \dfrac{\partial^2 v_h^M}{\partial x_1^2}(Q) & \dfrac{\partial^2 v_h^M}{\partial x_1\,\partial x_2}(Q) \\ \dfrac{\partial^2 v_h^M}{\partial x_1\,\partial x_2}(Q) & \dfrac{\partial^2 v_h^M}{\partial x_2^2}(Q) \end{pmatrix} . \overline{PP}^* \\ \text{with } Q \in [P, P^*], \end{cases}$$

from which we deduce:

$$\text{(6.39)} \quad | \operatorname{grad} v_h^M(P) | \leqslant | \operatorname{grad} v_h^M(P^*) | + \\ + \frac{h}{\sqrt{2}} \sqrt{\left(\frac{\partial^2 v_h^M}{\partial x_1^2}(Q)\right)^2 + 2\left(\frac{\partial^2 v_h^M}{\partial x_1\,\partial x_2}(Q)\right)^2 + \left(\frac{\partial^2 v_h^M}{\partial x_2^2}(Q)\right)^2}\,.$$

In the following C will denote various constants. Since the function $v_h^M \in K_h^M$ and since $P^* \in \Omega_h$ we have, by definition of K_h^M, $| \operatorname{grad} v_h^M(P^*) | \leqslant 1$; we thereby deduce, taking account of (6.39) and the fact that v_h^M belongs to $C^\infty(\overline{\Omega})$:

$$\text{(6.40)} \quad | \operatorname{grad} v_h^M(P) | \leqslant 1 + Ch \, \| v_h^M \|_{C^2(\overline{\Omega})} \quad \forall P \in \mathcal{O}_h .$$

From Sobolev's theorem [1] we have $H^s(\Omega) \subset C^2(\overline{\Omega})$ if $s > 3$, the injection being continuous, and hence:

$$\text{(6.41)} \quad | \operatorname{grad} v_h^M(P) | \leqslant 1 + Ch \, \| v_h^M \|_{H^s(\Omega)} \quad (s > 3) \quad \forall P \in \mathcal{O}_h .$$

Moreover if $v_h^M = \sum_{m=1}^{M} \mu_m w_m$ we have:

$$\text{(6.42)} \quad \| v_h^M \|_{H^s(\Omega)} \leqslant C\left(\sum_{m=1}^{M} \beta_m^s \mu_m^2\right)^{1/2} \leqslant C\beta_M^{(s-1)/2}\left(\sum_{m=1}^{M} \beta_m \mu_m^2\right)^{1/2} = C\beta_M^{(s-1)/2} \, \| v_h^M \|_{H_0^1}$$

We put $r = (s-1)/2$, and hence $s > 3 \Rightarrow r > 1$; moreover, since the sequence v_h^M is weakly convergent in $H_0^1(\Omega)$ it is bounded in $H_0^1(\Omega)$ and hence from (6.41) and (6.42) we finally obtain:

$$\text{(6.43)} \quad | \operatorname{grad} v_h^M(P) | \leqslant 1 + Ch\beta_M^r \quad (r > 1), \quad \forall P \in \mathcal{O}_h . \quad \blacksquare$$

---

(1) More precisely if $\Omega \subset \mathbb{R}^N$ and Γ is sufficiently regular we ha

$$H^s(\Omega) \subset C^k(\overline{\Omega})$$

with continuous injection if $s > \frac{N}{2} + k$ (see Lions-Magenes /1

2) We now show that under the condition $h\beta_M^r \to 0$ as $M \to +\infty$, the weak limit v of v_h^M in $H_0^1(\Omega)$ is in K_0. Let $\varphi \in \mathscr{D}(\Omega)$, $\varphi \geqslant 0$; the functional:

$$w \to \int_\Omega \varphi(|\operatorname{grad} w| - 1)\,\mathrm{d}x \tag{6.44}$$

is convex and strongly continuous in $H_0^1(\Omega)$. From (6.43) and the fact that support $(\varphi) \subset \mathcal{O}_h$ provided h is sufficiently small, we first deduce that

$$\int_\Omega \varphi(|\operatorname{grad} v_h^M| - 1)\,\mathrm{d}x \leqslant Ch\beta_M^r \int_\Omega \varphi\,\mathrm{d}x\,, \tag{6.45}$$

and then, under the assumption that $h\beta_M^r \to 0$ as $M \to +\infty$, we have, in the limit, by weak lower semi-continuity:

$$\begin{cases} \displaystyle\int_\Omega \varphi(|\operatorname{grad} v| - 1)\,\mathrm{d}x \leqslant \liminf \int_\Omega \varphi(|\operatorname{grad} v_h^M| - 1)\,\mathrm{d}x \leqslant \\ \qquad \leqslant C\left(\displaystyle\int_\Omega \varphi\,\mathrm{d}x\right) \lim h\beta_M^r = 0 \quad \forall \varphi \in \mathscr{D}(\Omega)\,, \quad \varphi \geqslant 0\,, \end{cases} \tag{6.46}$$

which implies:

$$|\operatorname{grad} v| \leqslant 1 \quad \text{a.e.} \quad \Leftrightarrow v \in K_0\,, \tag{6.47}$$

and proves (6.29). ■

Taking account of (6.28) we thus have the strong convergence of $u_h^M \to u$ in $H_0^1(\Omega)$. ■

Remark 6.3. Provided Γ is sufficiently regular we can deduce from (6.43) (by using Ascoli's theorem) that we also have $u_h^M \to u$ *strongly* in $L^\infty(\Omega)$, the proof of this property being immediate when Ω is *convex*. ■

6.3. Convergence of the exterior approximations

In this section we shall investigate the convergence of the exterior approximations of (P_0) defined in Section 5, using the results of Chapter 1, Section 5 and a number of results from Section 3.3 of the present chapter. We shall follow the same procedure as that used in Section 3.3 for the problem (P_1); the notation is as in Section 5.

6.3.1. *Reduction of* $(P_{0h})_{lm}$ *to a variational inequality*

We recall that solving (P_0) is equivalent to finding u which is a solution of the variational inequality:

$$(6.48)\qquad \begin{cases} \int_\Omega \operatorname{grad} u \operatorname{grad}(v-u)\,dx \geq \int_\Omega f(v-u)\,dx\,, & \forall v \in K_0\,, \\ u \in K_0\,, \end{cases}$$

for which the approximate analogue is:

$$(6.49)\qquad \begin{cases} a_h^l(u_h, v_h - u_h) \geq \int_\Omega f q_h(v_h - u_h)\,dx\,, & \forall v_h \in K_{0h}^m\,, \\ u_h \in K_{0h}^m\,, \end{cases}$$

where u_h is the solution of $(P_{0h})_{lm}$, $l = 1, 2$, $m = 1, 2, 3, 4$ (see (5.23)), and where the symmetric bilinear forms on V_h, a_h^l ($l = 1, 2$) are defined by:

$$(6.50)\qquad a_h^1(u_h, v_h) = \sum_{k=1}^{2} \int_\Omega \delta_k q_h u_h\, \delta_k q_h v_h\, dx$$

$$(6.51)\qquad \begin{cases} a_h^2(u_h, v_h) = \int_{\mathbb{R}^2} \left(\dfrac{\delta_1 q_h u_h\left(x_1, x_2 + \frac{h}{2}\right) + \delta_1 q_h u_h\left(x_1, x_2 - \frac{h}{2}\right)}{2} \right) \cdot \\ \quad \cdot \left(\dfrac{\delta_1 q_h v_h\left(x_1, x_2 + \frac{h}{2}\right) + \delta_1 q_h v_h\left(x_1, x_2 - \frac{h}{2}\right)}{2} \right) dx + \\ \quad + \int_{\mathbb{R}^2} \left(\dfrac{\delta_2 q_h u_h\left(x_1 + \frac{h}{2}, x_2\right) + \delta_2 q_h u_h\left(x_1 - \frac{h}{2}, x_2\right)}{2} \right) \cdot \\ \quad \cdot \left(\dfrac{\delta_2 q_h v_h\left(x_1 + \frac{h}{2}, x_2\right) + \delta_2 q_h v\left(x_1 - \frac{h}{2}, x_2\right)}{2} \right) dx\,. \end{cases}$$

We recall that here $a(u, v) = \int_\Omega \operatorname{grad} u . \operatorname{grad} v\, dx\,.$ ■

6.3.2. *Exterior approximations for* $H_0^1(\Omega)$, a, K_0, L

We shall now verify for each point in turn (as in Section 3.3.2) that the properties required in Chapter 1, Section 5.2, for Theorem 5.2 of Chapter 1, Section 5.3, to be applicable, are indeed satisfied:

Approximations of $H_0^1(\Omega)$.

We shall consider in particular two cases which correspond to the two approximations of a defined in (6.50) and (6.51), respectively.

1) *A first exterior approximation of* $H_0^1(\Omega)$:

Apart from a slight modification of the notation, this reduces to the approximation of $H_0^1(\Omega)$ studied in Section 3.3.2; we thus have:

(6.52) $$V = H_0^1(\Omega),$$

(6.53) $$F_1 = (L^2(\Omega))^3,$$

and then $\sigma_1 : V \to F_1$, $p_{1h} : V_h \to F_1$ and the norm $\| \ \|_{1h}$ on V_h, defined by:

(6.54) $$\sigma_1 v = \left\{ v, \frac{\partial v}{\partial x_1}, \frac{\partial v}{\partial x_2} \right\},$$

(6.55) $$p_{1h} v_h = \{ q_h v_h, \delta_1 q_h v_h, \delta_2 q_h v_h \},$$

(6.56) $$\| v_{1h} \|_{1h} = [a_{1h}(v_h, v_h)]^{1/2}.$$

We can then apply Proposition 3.1 of Section 3.3.2: under conditions (6.52), (6.53), ... (6.56), the family V_h constitutes an exterior approximation of $H_0^1(\Omega)$. ■

2) *A second exterior approximation of* $H_0^1(\Omega)$.

We once again take $V = H_0^1(\Omega)$, and then:

(6.57) $$F_2 = (L^2(\mathbb{R}^2))^3,$$

and we define $\sigma_2 : V \to F_2$, $p_{2h} : V_h \to F_2$ by:

(6.58) $$\begin{cases} \sigma_2 \tilde{v} = \left\{ \tilde{v}, \dfrac{\partial \tilde{v}}{\partial x_1}, \dfrac{\partial \tilde{v}}{\partial x_2} \right\} \\ \text{with } \tilde{v} = \text{the element of } H^1(\mathbb{R}^2) \text{ obtained by extending} \\ v \text{ by 0 outside } \Omega\,(^1) \end{cases}$$

(6.59) $$p_{2h} v \doteq \{ q_h v_h, \tfrac{1}{2}(\tau_{2(h/2)} \delta_1 q_h v_h + \tau_{2(-h/2)} \delta_1 q_h v_h), \tfrac{1}{2}(\tau_{1(h/2)} \delta_2 q_h v_h + \tau_{1(-h/2)} \delta_2 q_h v_h) \},$$

where, η being a scalar, $\tau_{1\eta}$, $\tau_{2\eta} : L^2(\mathbb{R}^2) \to L^2(\mathbb{R}^2)$ are defined by:

(6.60) $$\begin{cases} (\tau_{1\eta} g)(x_1, x_2) = g(x_1 + \eta, x_2) \text{ a.e.} \\ (\tau_{2\eta} g)(x_1, x_2) = g(x_1, x_2 + \eta) \text{ a.e.} \end{cases}$$

Finally we equip V_h with the norm defined by:

(6.61) $$\| v_h \|_{2h} = [a_h^2(v_h, v_h)]^{1/2}.$$

We have:

<u>Proposition 6.1</u>. *Under conditions* (6.57), (6.58),..., (6.61)

---

(1) We have $\left\| \dfrac{\partial v}{\partial x_k} \right\|_{L^2(\Omega)} = \left\| \dfrac{\partial \tilde{v}}{\partial x_k} \right\|_{L^2(\mathbb{R}^2)}$, $k = 1, 2.$

the family $(V_h)_h$ *constitutes an exterior approximation of* $H_0^1(\Omega)$.

Proof. We must verify (see Chapter 1, Section 5.2):

(6.62) $$\| p_{2h} \|_{\mathscr{L}(V_h,F_2)} \leqslant c \quad \forall h$$

and

(6.63) $$\forall v \in H_0^1(\Omega)\,, \quad \exists v_h \in V_h \quad \text{such that } p_{2h}\, v_h \to \sigma_2\, v \textit{ strongly in } F_2$$

with

(6.64) $$\| v_h \|_{2h} \leqslant c \quad \forall h\,.$$

We shall assume for the moment:

Lemma 6.2. *We have*

(6.65) $$\begin{cases} \dfrac{1}{\gamma} \| q_h\, v_h \|^2_{L^2(\mathbb{R}^2)} \leqslant a_h^2(v_h, v_h) \leqslant a_h^1(v_h, v_h) \quad \forall v_h \in V_h\,, \\ \gamma > 0 \textit{ independent of } h \textit{ provided } h \textit{ is sufficiently small.} \end{cases}$$

From (6.51), (6.59), (6.60), (6.61) we have

(6.66) $$\begin{aligned} \| p_{2h}\, v_h \|_{F_2} &= [\| q_h\, v_h \|^2_{L^2(R^2)} + a_h^2(v_h, v_h)]^{1/2} \\ &= [\| q_h\, v_h \|^2_{L^2(R^2)} + \| v_h \|^2_{2h}]^{1/2}\,, \end{aligned}$$

and hence, from (6.65),

(6.67) $$\| p_{2h}\, v_h \|_{F_2} \leqslant \| v_h \|_{2h} \sqrt{1 + \gamma}\,,$$

which implies (6.62) (with $c = \sqrt{1 + \gamma}$)

In order to verify (6.63), (6.64), as in Section 3.3.2 we define $r_h : H_0^1(\Omega) \to V_h$ by (3.23) and for $v \in H_0^1(\Omega)$, we take

(6.68) $$v_h = r_h\, v\,.$$

We have (see Chapter 1, Section 5.1)

(6.69) $$p_{1h}\, v_h \to \left\{ v, \frac{\partial v}{\partial x_1}, \frac{\partial v}{\partial x_2} \right\} \quad \text{strongly in } (L^2(\Omega))^3$$

which gives

(6.70) $$\| v_h \|_{1h} \leqslant c\,.$$

Still denoting by $\tilde{v}$ the element of $H^1(\mathbb{R}^2)$ obtained by extending v by 0 outside Ω, we deduce from (6.69):

(6.71) $$\begin{cases} q_h\, v_h \to \tilde{v} \quad \text{strongly in } L^2(\mathbb{R}^2) \\ \delta_k q_h\, v_h \to \dfrac{\partial \tilde{v}}{\partial x_k} \quad \text{strongly in } L^2(\mathbb{R}^2) \ (k = 1,2). \end{cases}$$

In order to prove (6.63) from (6.71), we shall use the following result.

Given the two sequences $(g_n)_n$, $g_n \in L^2(\mathbb{R}^2)$ and $(h_n)_n$, $h_n \in R$, such that for $n \to +\infty$

$$(6.72) \qquad \begin{cases} g_n \to g \text{ strongly (resp. weakly) in } L^2(\mathbb{R}^2) \\ h_n \to 0 , \end{cases}$$

it is immediate that

$$(6.73) \qquad \begin{cases} \tau_{1h_n} g_n \to g \quad \text{strongly (resp. weakly) in } L^2(\mathbb{R}^2), \\ \tau_{2h_n} g_n \to g \quad \text{strongly (resp. weakly) in } L^2(\mathbb{R}^2). \end{cases}$$

Taking account of (6.71), (6.72), (6.73) we then have

$$(6.74) \qquad \begin{cases} \dfrac{\tau_{2(h/2)}\, \delta_1 q_h\, v_h + \tau_{2(-h/2)}\, \delta_1 q_h\, v_h}{2} \to \dfrac{\partial \tilde{v}}{\partial x_1} \quad \text{strongly in } L^2(\mathbb{R}^2), \\ \dfrac{\tau_{1(h/2)}\, \delta_2 q_h\, v_h + \tau_{1(-h/2)}\, \delta_2 q_h\, v_h}{2} \to \dfrac{\partial \tilde{v}}{\partial x_2} \quad \text{strongly in } L^2(\mathbb{R}^2) \end{cases}$$

which with (6.59) and the first relation of (6.71) proves (6.63). As regards (6.64), this results from (6.51), (6.61), (6.74) or more simply from (6.56), (6.65), (6.70). Hence we have proved Proposition 6.1, subject to the:

Proof of Lemma 6.2

1) First, we shall prove the inequality on the right; we have

$$(6.75) \qquad \begin{cases} a_h^2(v_h, v_h) = \displaystyle\int_{\mathbb{R}^2} \left[\left(\frac{\tau_{2(h/2)}\, \delta_1 q_h\, v_h + \tau_{2(-h/2)}\, \delta_1 q_h\, v_h}{2}\right)^2 + \right. \\ \qquad \left. + \left(\dfrac{\tau_{1(h/2)}\, \delta_2 q_h\, v_h + \tau_{1(-h/2)}\, \delta_2 q_h\, v_h}{2}\right)^2\right] \mathrm{d}x \\ \leqslant \dfrac{1}{2} \displaystyle\int_{\mathbb{R}^2} [|\, \tau_{2(h/2)}\, \delta_1 q_h\, v_h\,|^2 + |\, \tau_{2(-h/2)}\, \delta_1 q_h\, v_h\,|^2 + \\ + |\, \tau_{1(h/2)}\, \delta_2 q_h\, v_h\,|^2 + |\, \tau_{1(-h/2)}\, \delta_2 q_h\, v_h\,|^2]\, \mathrm{d}x = \\ = \displaystyle\sum_{k=1}^{2} \int_{\mathbb{R}^2} |\, \delta_k q_h\, v_h\,|^2\, \mathrm{d}x\, , \end{cases}$$

with **support** $(\delta_k q_h\, v_h) \subset \Omega$ $(k = 1, 2)$ (V_h and q_h being defined in Section 3.2.1 by (3.6) and (3.7), respectively); from (6.75) we deduce

$$(6.76) \qquad a_h^2(v_h, v_h) \leqslant \sum_{k=1}^{2} \int_{\mathbb{R}^2} |\, \delta_k q_h\, v_h\,|^2\, \mathrm{d}x = \sum_{k=1}^{2} \int_{\Omega} |\, \delta_k q_h\, v_h\,|^2\, \mathrm{d}x = $$
$$= a_h^1(v_h, v_h)\, . \quad \blacksquare$$

2) We shall now prove the first inequality of (6.65) [1] noting that, if we first extend Ω_h into R_h (see (4.39)) and carry out a translation of the origin of coordinates, we can always assume that V_h is generated from Ω_h defined by:

(6.77) $$\Omega_h = \{ M_{ij} \mid M_{ij} \in R_h,\ M_{ij} = (ih, jh),\ 1 \leqslant i \leqslant N-1, \quad 1 \leqslant j \leqslant N-1 \},$$

the associated set $\overline{\Omega}_h$ then being defined by:

(6.78) $$\overline{\Omega}_h = \{ M_{ij} \mid M_{ij} \in R_h,\ M_{ij} = (ih, jh),\ 0 \leqslant i \leqslant N, \quad 1 \leqslant j \leqslant N \},$$

moreover

(6.79) $$\lim_{h \to 0} Nh = L < +\infty$$

the optimal choice of L being defined by

(6.80) $$\begin{cases} L^2 = \text{Minimum area of the squares with sides parallel} \\ \qquad \text{to the coordinate axes and containing } \Omega . \end{cases}$$

We clearly have:

(6.81) $$\dim V_h = (N-1)^2 .$$

We define the symmetric, positive-definite linear operator $-\Delta_h^2$ (an "approximation" of $-\Delta$) by:

(6.82) $$a_h^2(v_h, v_h) = -h^2(\Delta_h^2 v_h, v_h) \quad \forall v_h \in V_h ,$$

where, in (6.82), $(.,.)$ denotes the standard Euclidean inner product in $\mathbb{R}^{(N-1)^2}$; in a more explicit form we have

(6.83) $$-(\Delta_h^2 v_h)_{ij} = \frac{4 v_{ij} - (v_{i+1j+1} + v_{i-1j+1} + v_{i+1j-1} + v_{i-1j-1})}{2h^2},$$

with, in (6.83), $v_{pq} = 0$ if $M_{pq} \notin \Omega_h$.

We shall now verify, using (6.83), that the *eigenvectors* of $-\Delta_h^2$ in V_h are given by:

(6.84) $$\begin{cases} (v_h)_{\mu\nu} = \left\{ v_{ij} \mid v_{ij} = \sin \mu(i-1)\dfrac{\pi}{N} \sin \nu(j-1)\dfrac{\pi}{N}; \right. \\ \qquad \left. i, j = 2, 3, \ldots, N \right\} \\ \text{with } \mu, \nu = 1, 2, \ldots, N-1 , \end{cases}$$

and the corresponding *eigenvalues* by:

(6.85) $$\lambda_{\mu\nu}^2 = \frac{2}{h^2}\left(1 - \cos \mu \frac{\pi}{N} \cos \nu \frac{\pi}{N}\right).$$

---

(1) This is a "discrete Poincaré" type inequality.

The smallest eigenvalue is therefore obtained for $\mu = \nu = 1$ (and $\mu = \nu = N - 1$) and is equal to:

$$(6.86) \qquad \lambda_{11}^2 = \frac{2}{h^2} \sin^2 \frac{\pi}{N} .$$

In view of (6.82) and the orthogonality of the eigenvectors, we therefore have:

$$(6.87) \qquad \begin{cases} a_h^2(v_h, v_h) = - h^2(\Delta_h^2 v_h, v_h) \geqslant \dfrac{2}{h^2} \sin^2 \dfrac{\pi}{N} h^2 \displaystyle\sum_{M_{ij} \in \Omega_h} v_{ij}^2 = \\ \qquad = \dfrac{2}{h^2} \sin^2 \dfrac{\pi}{N} \| q_h v_h \|_{L^2(\mathbb{R}^2)}^2 , \end{cases}$$

and as $h \to 0, N \to +\infty$ with (6.79) we have:

$$(6.88) \qquad \lim \frac{2}{h^2} \sin^2 \frac{\pi}{N} = \frac{2\pi^2}{L^2} > 0 ,$$

which with (6.87) proves Lemma 6.2 ■

Approximation of a :
We have:

Proposition 6.2. *The families* $(a_h^1)_h$ *and* $(a_h^2)_h$ *constitute exterior approximations of* a.

Proof. For the bilinear form a_h^1 as defined in (6.50), and under the conditions (6.52), (6.53), ..., (6.56), this has already been proved (using different notation) in Section 3.3.2, Proposition 3.2; it therefore only remains to prove it for a_h^2 under the conditions (6.57), (6.58), (6.59), (6.61). We clearly have:

$$(6.89) \qquad | a_h^2(u_h, v_h) | \leqslant \| u_h \|_{2h} \| v_h \|_{2h} \quad \forall u_h, v_h \in V_h \quad \text{and} \quad \forall h ,$$

$$(6.90) \qquad a_h^2(v_h, v_h) = \| v_h \|_{2h}^2 \quad \forall v_h \in V_h \quad \text{and} \quad \forall h .$$

It now remains to verify (see Chapter 1, Sections 5.2 and 5.3):

$$(6.91) \qquad \begin{cases} \text{as } p_{2h} v_h \to \sigma_2 v \text{ weakly in } F_2 , p_{2h} w_h \to \sigma_2 w \text{ strongly in } F_2 \\ \text{we have } a_h^2(v_h, w_h) \to a(v, w) , \end{cases}$$

$$(6.92) \qquad \text{if } p_{2h} v_h \to \sigma_2 v \text{ weakly in } F_2 \text{ we have } \liminf a_h^2(v_h, v_h) \geqslant a(v, v) .$$

Since the form

$$(6.93) \qquad \{ (f_1, f_2), (g_1, g_2) \} \to \int_{\mathbb{R}^2} (f_1 g_1 + f_2 g_2) \, dx$$

is bilinear continuous over $(L^2(\mathbb{R}^2))^4$ we deduce, from (6.51), (6.58), (6.59), that under the assumptions of condition (6.91) we have:

$$(6.94) \qquad \lim_{h\to 0} a_h^2(v_h, w_h) = \sum_{k=1}^{2} \int_{\mathbb{R}^2} \frac{\partial \tilde{v}}{\partial x_k} \frac{\partial \tilde{w}}{\partial x_k} \, dx =$$

$$= \sum_{k=1}^{2} \int_{\Omega} \frac{\partial v}{\partial x_k} \frac{\partial w}{\partial x_k} \, dx = a(v, w),$$

which proves (6.91). Regarding (6.92), this comes from (6.51), (6.58), (6.59) and the *weak lower semi-continuity* of the convex and *strongly continuous* functional on $(L^2(\mathbb{R}^2))^2$ defined by

$$(6.95) \qquad (f_1, f_2) \to \int_{\mathbb{R}^2} (f_1^2 + f_2^2) \, dx . \quad \blacksquare$$

Exterior approximations of K_0.

We shall distinguish two cases depending on whether we use the norm $\| \|_{1h}$ or the norm $\| \|_{2h}$ over V_h, the approximation of $H_0^1(\Omega)$. We shall start with the former case which is slightly the simpler of the two; in this case we have

Proposition 6.3. *Under conditions* (6.52), (6.53), ... , (6.56), *the convex sets* K_{0h}^m (m = 1, 2, 3, 4) *defined in Section* 5.2 *constitute exterior approximations of* K_0.

Proof. It is necessary to prove (see Chapter 1, Section 5.2) for m = 1, 2, 3, 4:

(6.96) if $v_h \in K_{0h}^m$ with $\lim_{h\to 0} p_{1h} v_h = \xi$ weakly in F_1, then $\xi \in \sigma_1 K_0$,

and

(6.97) $\forall v \in K_0$, $\exists v_h \in K_{0h}^m$ such that $p_{1h} v_h \to \sigma_1 v$ strongly in F_1 and $\| v_h \|_{1h} \leqslant c$.

• *Verification of* (6.96).

We put $\xi = (\xi_0, \xi_1, \xi_2)$; it then results from Chapter 1, Section 5.1, that:

$$(6.98) \qquad \begin{cases} \xi_0 \in H_0^1(\Omega), \\ \xi_k = \dfrac{\partial \xi_0}{\partial x_k} \quad (k = 1, 2), \end{cases}$$

and hence $\xi \in \sigma_1 H_0^1(\Omega)$.

It now remains to prove that $\xi \in \sigma_1 K_0$; this is immediate for $m = 1$; in fact:

$$v_h \in K_{0h}^1 \Leftrightarrow \sum_{k=1}^{2} | \delta_k q_h v_h |^2 \leqslant 1 \text{ a.e.} \tag{6.99}$$

and from (6.96), (6.98):

$$\delta_k q_h v_h \to \frac{\partial \xi_0}{\partial x_k} \quad \text{weakly in } L^2(\Omega) \quad (k = 1, 2), \tag{6.100}$$

which, in the limit, since the convex set

$$[(f_1, f_2) \mid f_1, f_2 \in L^2(\Omega),\ f_1^2 + f_2^2 \leqslant 1 \text{ a.e}]$$

is closed in $(L^2(\Omega))^2$, gives:

$$\sum_{k=1}^{2} \left| \frac{\partial \xi_0}{\partial x_k} \right|^2 \leqslant 1 \text{ a.e.} \tag{6.101}$$

and hence, with (6.98), $\xi \in \sigma_1 K_0$. ∎

We shall now treat simultaneously the cases where $m = 2, 3, 4$, noting that from (5.14), Section 5.2:

$$\begin{cases} v_h \in K_{0h}^3 \Rightarrow \dfrac{1}{2}\Bigg[\left| \delta_1 q_h v_h\left(x_1 + \dfrac{h}{2}, x_2\right) \right|^2 + \\ \quad + \left| \delta_1 q_h v_h\left(x_1 - \dfrac{h}{2}, x_2\right) \right|^2 + \left| \delta_2 q_h v_h\left(x_1, x_2 + \dfrac{h}{2}\right) \right|^2 + \\ \quad + \left| \delta_2 q_h v_h\left(x_1, x_2 - \dfrac{h}{2}\right) \right|^2 \Bigg] \leqslant 1 \text{ a.e.} \end{cases} \tag{6.102}$$

If the function $\tilde{\xi}_0$ denotes the element of $H^1(\mathbb{R}^2)$ obtained by extending ξ_0 by 0 outside Ω and $q_h v_h$, $\delta_k q_h v_h$ $(k = 1, 2)$ are considered to be elements of $L^2(\mathbb{R}^2)$ (1) we have from (6.96), (6.98),

$$\begin{cases} q_h v_h \to \tilde{\xi}_0 & \text{weakly in } L^2(\mathbb{R}^2) \\ \delta_k q_h v_h \to \dfrac{\partial \tilde{\xi}_0}{\partial x_k} & \text{weakly in } L^2(\mathbb{R}^2) \quad (k = 1, 2). \end{cases} \tag{6.103}$$

We can thus apply (6.72), (6.73), which gives:

$$\lim_{h \to 0} \tau_{k_1(\pm h/2)} \delta_{k_2} q_h v_h = \frac{\partial \tilde{\xi}_0}{\partial x_{k_2}} \quad \text{weakly in } L^2(\mathbb{R}^2). \tag{6.104}$$

Since the sets $[(f_1, f_2) \mid f_1, f_2 \in L^2(\mathbb{R}^2),\ f_1^2 + f_2^2 \leqslant 1 \text{ a.e}]$ resp.

$$\left[(f_i)_{i=1,2,3,4} \mid f_i \in L^2(\mathbb{R}^2),\ i = 1, 2, 3, 4,\ \sum_{i=1}^{4} f_i^2 \leqslant 2 \text{ a.e.} \right]$$

are convex and closed in $(L^2(\mathbb{R}^2))^2$ (resp. $(L^2(\mathbb{R}^2))^4$) , we have in the limit, taking account of (5.12), (6.102), (5.20), (6.60), (6.104):

---

(1) We recall that since the supports of $q_h v_h$, $\delta_k q_h v_h$ $(k = 1, 2)$ are in Ω, these functions are considered to be elements of $L^2(\Omega)$ or $L^2(\mathbb{R}^2)$, as the case may be.

(6.105) $$\sum_{k=1}^{2} \left| \frac{\partial \tilde{\xi}_0}{\partial x_k} \right|^2 \leqslant 1 \text{ a.e. in } \mathbb{R}^2 \Rightarrow \sum_{k=1}^{2} \left| \frac{\partial \xi_0}{\partial x_k} \right|^2 \leqslant 1 \text{ a.e. in } \Omega,$$

and hence, with (6.98), $\xi \in \sigma_1 K_0$. ■

• *Verification of* (6.97).

It is sufficient (see Chapter 1, Section 5.3, Remark 5.4) to verify this for

(6.106) $$\bigcup_{0 \leqslant \lambda < 1} \lambda\{ K_0 \cap \mathscr{D}(\Omega) \}$$

dense in K_0, from Lemma 6.1 [1].

If v belongs to the subset (6.106), we define v_h by:

(6.107) $$v_h = \{ v_{ij} \mid v_{ij} = v(M_{ij}),\ M_{ij} \in \Omega_h \},$$

and since v is in $\mathscr{D}(\Omega)$ we have [2]:

(6.108) $$\begin{cases} q_h v_h \to v & \text{strongly in } L^2(\Omega) \\ \delta_k q_h v_h \to \dfrac{\partial v}{\partial x_k} & \text{strongly in } L^2(\Omega) \qquad (k = 1, 2), \end{cases}$$

that is $p_{1h} v_h \to \sigma_1 v$ strongly in F_1; we clearly have $\| v_h \|_{1h} \leqslant c$ from (6.50), (6.56) and (6.108).

Moreover, a Taylor series expansion of order 2 would show that the relation $| \operatorname{grad} v | \leqslant \lambda < 1$ in Ω implies, using (5.10), (5.11), (5.15), (5.21), that $v_h \in K_{0h}^m$ for m = 1, 2, 3, 4 provided h is sufficiently small. ■

Concerning the second exterior approximation considered we have:

<u>Proposition 6.4</u> *Under conditions* (6.57), (6.58), ... , (6.61) *the convex sets* K_{0h}^m (m = 1, 2, 3, 4) *defined in Section* 5.2 *constitute exterior approximations of* K_0.

Proof. It is necessary to prove (see Chapter 1, Section 5.2), for m = 1, 2, 3, 4, that:

(6.109) If $v_h \in K_{0h}^m$ with $\lim_{h \to 0} p_{2h} v_h = \xi$ weakly in F_2, then $\xi \in \sigma_2 K_0$,

and

(6.110) $\forall v \in K_0$, $\exists v_h \in K_{0h}^m$ such that $p_{2h} v_h \to \sigma_2 v$ strongly in F_2 and $\| v_h \|_{2h} < c$.

---

(1) $v \in C^1(\overline{\Omega})$ and with compact support in Ω would be sufficient.

(2) In fact there is strong convergence in $L^\infty(\Omega)$

• *Verification of* (6.109)

We put $\xi = (\xi_0, \xi_2, \xi_2)$; we then have, from (6.59);

$$(6.111) \quad \begin{cases} q_h v_h \to \xi_0 \text{ weakly in } L^2(\mathbb{R}^2) \text{ and for } k_1, k_2 = 1, 2, k_1 \neq k_2 \\ \frac{1}{2}(\tau_{k_1(h/2)} \delta_{k_2} q_h v_h + \tau_{k_1(-h/2)} \delta_{k_2} q_h v_h) \to \xi_{k_2} \text{ weakly in } L^2(\mathbb{R}^2) \end{cases}$$

We shall treat the case $m = 4$ separately from the three others:

• *Verification of* (6.109) *for* m = 1, 2, 3.

Since the support of $\delta_k q_h v_h$ $(k = 1, 2)$ is contained in Ω, it can easily be shown that the membership conditions for K_{0h}^m (see (5.9) for m = 1, (5.12) for m = 2, (5.14) and (6.102) for m = 3) imply:

$$(6.112) \quad \sum_{k=1}^{2} \int_\Omega |\delta_k q_h v_h|^2 \, dx \, (= a_h^1(v_h, v_h) = \| v_h \|_{1h}^2) \leqslant \text{measure of } \Omega,$$

and hence, from Lemma 6.2 (or from Section 3.3.2, relation (3.22))

$$(6.113) \quad \| p_{1h} v_h \| \leqslant C \quad \text{in } F_1 \, (= L^2(\Omega))^3) \quad \forall h .$$

Therefore there exists an extracted subsequence, also denoted by v_h and $\xi^* \in F_1$, such that if $\xi^* = (\xi_0^*, \xi_1^*, \xi_2^*)$, we have:

$$(6.114) \quad \begin{cases} q_h v_h \to \xi_0^* \text{ weakly in } L^2(\Omega) \\ \delta_k q_h v_h \to \xi_k^* \text{ weakly in } L^2(\Omega). \end{cases}$$

From Proposition 6.3, (6.114) and $v_h \in K_{0h}^m$ imply:

$$(6.115) \quad \begin{cases} \xi_0^* \in K_0 \\ \xi_k^* = \dfrac{\partial \xi_0^*}{\partial x_k} \quad (k = 1, 2) . \end{cases}$$

Applying the translation results (6.72), (6.73) we deduce from (6.111), (6.114), (6.115), that:

$$(6.116) \quad \begin{cases} \xi_0 = \tilde{\xi}_0^* \\ \xi_k = \dfrac{\partial \tilde{\xi}_0^*}{\partial x_k} \quad (k = 1, 2) , \end{cases}$$

where $\tilde{\xi}_0^*$ denotes the element of $H^1(\mathbb{R}^2)$ obtained by extending ξ_0^* by zero outside Ω; from (6.116) and the first relation of (6.115) we thus have $\xi \in \sigma_2 K_0$. ■

• *Verification of* (6.109) *for* m = 4.

The first relation of (6.111) implies (a fortiori):

$$(6.117) \quad q_h v_h \to \xi_0 \quad \text{in } \mathscr{D}'(\mathbb{R}^2)$$

and hence as may be verified immediately:

$$(6.118)\qquad \frac{1}{2}(\tau_{k_1(h/2)}\,\delta_{k_2}q_h\,v_h + \tau_{k_1(-h/2)}\,\delta_{k_2}q_h\,v_h) \to \frac{\partial \xi_0}{\partial x_{k_2}} \quad \text{in } \mathscr{D}'(\mathbb{R}^2)$$
$$(k_1, k_2 = 1, 2)\,,$$

which, together with the second relation of (6.111), implies:

$$(6.119)\qquad \xi_k = \frac{\partial \xi_0}{\partial x_k} \qquad (k = 1, 2)\,.$$

We therefore have:

$$(6.120)\qquad \xi_0 \in H^1(\mathbb{R}^2)\,.$$

Moreover, the convex set $[(f_1, f_2) \mid f_1, f_2 \in L^2(\mathbb{R}^2),\ f_1^2 + f_2^2 \leqslant 1 \text{ a.e.}]$ being closed in $[L^2(\mathbb{R}^2)]^2$, the second relation of (6.111), and the relation expressing the membership of v_h to K_{0h}^4 (see (5.20)) and (6.119) imply that in the limit we have:

$$(6.121)\qquad \sum_{k=1}^{2}\left|\frac{\partial \xi_0}{\partial x_k}\right|^2 \leqslant 1 \quad \text{a.e. in } \mathbb{R}^2\,.$$

In order to demonstrate that $\xi \in \sigma_2 K_0$ it is sufficient, in view of (6.120) and (6.121), to show that ξ_0 is zero outside Ω; the proof of this is immediate.

Indeed, since $q_h v_h$ has its support in Ω, it belongs to the *closed convex subset of* $L^2(\mathbb{R}^2)$ defined by:

$$(6.122)\qquad [g \mid g \in L^2(\mathbb{R}^2),\ g = 0 \ \text{ a.e. in } \mathbb{R}^2 - \bar{\Omega}]$$

the first relation of (6.111) therefore implies that, in the limit, ξ_0 belongs to the convex set (6.122). ■

• *Verification of* (6.110).

As in the proof of the second part of Proposition 6.3, we define v_h using the set (6.106); hence, provided h is sufficiently small, $v_h \in K_{0h}^m \ \ \forall m = 1, 2, 3, 4$. Furthermore (6.108) implies, from (6.72), (6.73) that:

$$(6.123)\qquad \begin{cases} q_h\,v_h \to \tilde{v} \quad \text{strongly in } L^2(\mathbb{R}^2) \\ \frac{1}{2}(\tau_{k_1(h/2)}\,\delta_{k_2}q_h\,v_h + \tau_{k_1(-h/2)}\,\delta_{k_2}q_h\,v_h) \to \dfrac{\partial \tilde{v}}{\partial x_{k_2}} \\ \qquad \text{strongly in } L^2(\mathbb{R}^2) \qquad (k_1, k_2 = 1, 2)\,, \end{cases}$$

and hence

$$(6.124)\qquad \begin{cases} p_{2h}\,v_h \to \sigma_2\,v \quad \text{strongly in } F_2 \\ \|\,v_h\,\|_{2h} \leqslant c\,. \end{cases}$$ ■

Approximation of L.

With the functional $L : H_0^1(\Omega) \to \mathbb{R}$ defined by $L(v) = \int_\Omega fv\,dx$, where $f \in L^2(\Omega)$, and $L_h : V_h \to \mathbb{R}$ by $L_h(v_h) = \int_\Omega fq_h\,v_h\,dx$, the verification of (5.37), (5.38) of Chapter 1, Section 5.2 is immediate for the two types of approximation of H_0^1 considered. ■

6.3.3 *Strong convergence results in* F_1, F_2.

In view of the results proved in Section 6.3.2 for the exterior approximations of $H_0^1(\Omega), a, K_0, L$, we can apply Theorem 5.2 of Chapter 1, Section 5.4 to the problems $(P_{0h})_{lm}$ $(l = 1, 2\ ;\ m = 1, 2, 3, 4)$ and more precisely to the equivalent formulations (6.49); this gives:

Theorem 6.3 *Let* u_{lh}^m *be the solutions of* $(P_{0h})_{lm}$, *and* u *the solution of* (P_0) ; *as* $h \to 0$ *we have*:

1) *Under conditions* (6.52), (6.53), ... , (6.56):

$$(6.125)\qquad \begin{cases} p_{1h}\,u_{1h}^m \to \sigma_1\,u & \text{strongly in } F_1 \\ J_h^1(u_{1h}^m) \to J(u)\,, & \\ \forall m = 1, 2, 3, 4\,; & \end{cases}$$

2) *Under conditions* (6.57), (6.58), ... , (6.61):

$$(6.126)\qquad \begin{cases} p_{2h}\,u_{2h}^m \to \sigma_2\,u & \text{strongly in } F_2 \\ J_h^2(u_{2h}^m) \to J(u)\,, & \\ \forall m = 1, 2, 3, 4\,. & \end{cases}$$ ■

Remark 6.4. In view of (6.72), (6.73), the convergence result (6.125) implies:

$$(6.127)\qquad \begin{cases} p_{2h}\,u_{1h}^m \to \sigma_2\,u & \text{strongly in } F_2 \\ J_h^2(u_{1h}^m) \to J(u)\,, & \\ \forall m = 1, 2, 3, 4\,. & \end{cases}$$ ■

6.3.4 *Uniform convergence results*

In this section we shall prove *the uniform convergence of* $q_h\,u_{lh}^m$ *to* u, for $l =$ 1, 2 and m = 1, 2, 3, 4; the proofs are equally valid for $l =$ 1, 2; in contrast, however, we shall treat the case m = 4 separately from the other three cases. ■

As in the proof of the second part of Lemma 6.2 we can assume that V_h is generated from Ω_h defined by (6.77) and satisfies (6.78), (6.79), (6.80); thus, with $\mathcal{O}_h$ defined by (5.13), we have in this particular case:

(6.128) $\mathcal{O}_h =]0, Nh[\times]0, Nh[\,.$

We consider on $\mathcal{O}_h$ the triangulation $\mathcal{T}_h$ shown in Figure 6.1:

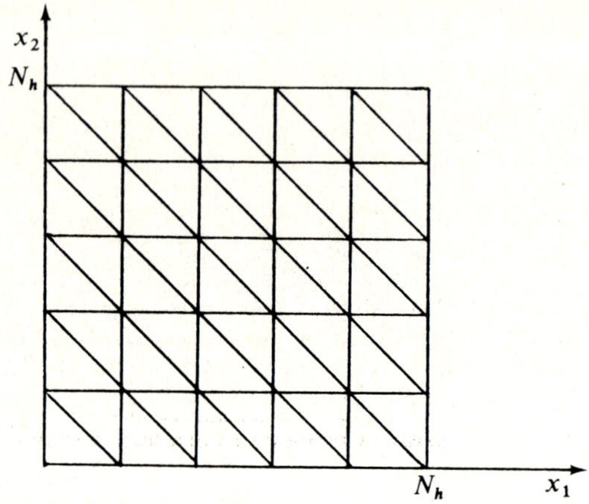

Fig. 6.1.

and we define the mapping $\Pi_h : V_h \to H_0^1(\mathcal{O}_h) \cap C^0(\overline{\mathcal{O}}_h)$ by:

(6.129) $$\begin{cases} \Pi_h v_h \in H_0^1(\mathcal{O}_h) \cap C^0(\overline{\mathcal{O}}_h)\,, \\ \Pi_h v_h \text{ affine linear over the triangles of } \mathcal{T}_h\,, \\ (\Pi_h v_h)(M_{ij}) = v_{ij} \quad \forall M_{ij} \in \overline{\Omega}_h\,. \end{cases}$$

(We recall that $v_{ij} = 0$ if $M_{ij} \notin \Omega_h$). We then have:

Lemma 6.3. *For* m = 1, 2, 3 *the set* $\Pi_h K_{0h}^m$ *is made up of functions which are bounded and Lipschitz on* $\mathcal{O}_h$, *uniformly in* h, *and:*

(6.130) $$\| q_h v_h - \Pi_h v_h \|_{L^\infty(\mathcal{O}_h)} \leqslant Ch\,, \quad C \textit{ independent of } h.$$

Proof. The gradient of $\Pi_h v_h$ is constant over each triangle of $\mathcal{T}_h$ and the conditions expressing the membership of v_h to K_{0h}^m (see Section 5.2) imply:

(6.131) $$\begin{cases} |\operatorname{grad} \Pi_h v_h| \leqslant 1 \text{ a.e. if } m = 1 \\ |\operatorname{grad} \Pi_h v_h| \leqslant \sqrt{2} \text{ a.e. if } m = 2, 3 \end{cases}$$

which with the conditions $\lim\limits_{\substack{h \to 0 \\ N \to +\infty}} Nh = L < +\infty$ and $\Pi_h v_h = 0$ on $\partial\mathcal{O}_h$ implies that the functions $\Pi_h v_h$ belonging to $\Pi_h K_{0h}^m$ $(m = 1, 2, 3)$ are *bounded* and *Lipschitz, uniformly in h.*

Since the function $q_h v_h$ is piecewise constant and equal to v_{ij} on the panel $\varpi_h^0(M_{ij})$ it can be shown, from (6.131), that we have:

$$(6.132)\quad \begin{cases} \| q_h v_h - \Pi_h v_h \|_{L^\infty(\mathcal{O}_h)} \leqslant \dfrac{h}{\sqrt{2}} \text{ if } m = 1, \\ \| q_h v_h - \Pi_h v_h \|_{L^\infty(\mathcal{O}_h)} \leqslant h \text{ if } m = 2, 3, \end{cases}$$

which more precisely defines, and proves, (6.130). ■

From Lemma 6.3, we deduce:

Theorem 6.4. *Let* u_{lh}^m *be the solution of* $(P_{0h})_{lm}$, *and* u *the solution of* (P_0); *as* $h \to 0$ *we have*:

$$(6.133)\quad \begin{cases} q_h u_{lh}^m \to u \quad \textit{strongly in } L^\infty(\Omega) \\ \forall l = 1, 2 \textit{ and } m = 1, 2, 3. \end{cases}$$

Proof. We assume that $\overline{\Omega} \subset \mathcal{O}_h$ (a situation which can always be achieved by extension of Ω_h); from Lemma 6.3, we can apply Ascoli's Theorem to the restriction of $\Pi_h u_{lh}^m$ to $\overline{\Omega}$, and hence as $h \to 0$ and in view of Theorem 6.3 and (6.132) we get:

$$(6.134)\quad \begin{cases} \text{the restriction of } \Pi_h u_{lh}^m \text{ on } \overline{\Omega} \text{ converges strongly to } u \text{ in } L^\infty(\Omega) \\ q_h u_{lh}^m \to u \quad \text{strongly in } L^\infty(\Omega) \\ \forall l = 1, 2 \quad \text{and} \quad m = 1, 2, 3. \quad ■ \end{cases}$$

We shall now prove an analogous result for u_{lh}^4.

We introduce the two disjoint subsets of R_h:

$$(6.135)\quad \begin{cases} R_{1h} = \{ M_{ij} \mid M_{ij} \in R_h, \text{ parity of } i \neq \text{ parity of } j \}, \\ R_{2h} = \{ M_{ij} \mid M_{ij} \in R_h, \text{ parity of } i = \text{ parity of } j \}. \end{cases}$$

We thus have

$$R_h = R_{1h} + R_{2h}.$$

We next introduce

$$(6.136)\quad \begin{cases} \sigma_h(M_{ij}) = \text{ the open panel with centre } M_{ij} \text{ and sides of} \\ \quad \text{length } h\sqrt{2} \text{ parallel to the bisectors of } x_1 O x_2, \end{cases}$$

$$(6.137)\quad \chi_h^{ij} = \text{ characteristic function of } \sigma_h(M_{ij}).$$

It may be noted that

$$(6.138)\quad \overline{\bigcup_{M_{ij} \in R_{rh}} \sigma_h(M_{ij})} = \mathbb{R}^2 \qquad (r = 1, 2)$$

$$(6.139)\quad \varpi_h^0(M_{ij}) \subset \sigma_h(M_{ij}) \subset \varpi_h^1(M_{ij})$$ (1) (see Figure 6.2)

(1) $\varpi_h^1(M_{ij})$ was defined in Chapter 1, Section 5.1, relation (5.7').

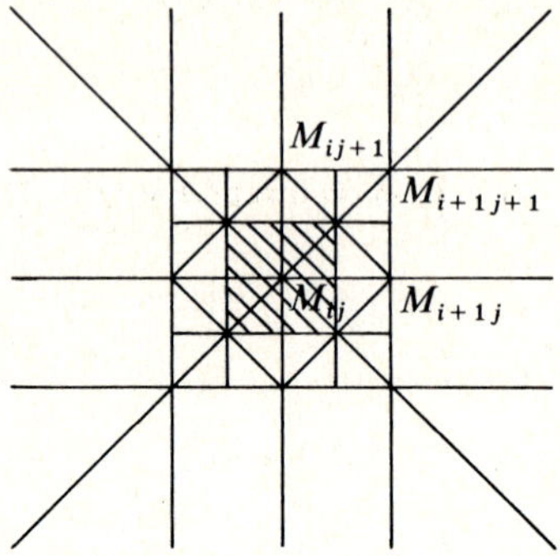

Fig. 6.2. Representation of $\varpi_h^0(M_{ij})$, $\sigma_h(M_{ij})$, $\varpi_h^1(M_{ij})$ ($\varpi_h^0(M_{ij})$ is shown hatched)

We define a mapping $I_{rh} \in \mathscr{L}(L^2(\mathbb{R}^2); L^2(\mathbb{R}^2))$ $(r = 1, 2)$ by:

$$(6.140)\quad \begin{cases} I_{rh}(g) = \sum_{M_{ij} \in R_{rh}} g_{ij}\, \chi_h^{ij}\,, \\ \text{with} \\ g_{ij} = \dfrac{1}{h^2} \displaystyle\int_{\varpi_h^0(M_{ij})} g(x)\, \mathrm{d}x\,. \end{cases}$$

We then have:

<u>Lemma 6.4.</u> *We have* , $\forall r = 1, 2$,

$$(6.141)\qquad \| I_{rh}(g) \|_{L^2(\mathbb{R}^2)} \leqslant \sqrt{2}\, \| g \|_{L^2(\mathbb{R}^2)} \quad \forall g \in L^2(\mathbb{R}^2)$$

(6.142) $I_{rh}(g) \to g$ *strongly in* $L^2(\mathbb{R}^2)$ *as* $h \to 0$.

Proof.

1) *Verification of* (6.141).
We have $(\forall r = 1, 2)$:

$$\| I_{rh}(g) \|_{L^2(\mathbb{R}^2)}^2 = 2\, h^2 \sum_{M_{ij} \in R_{rh}} g_{ij}^2 =$$

$$= 2\, h^2 \sum_{M_{ij} \in R_{rh}} \frac{1}{h^4} \left(\int_{\varpi_h^0(M_{ij})} g(x)\, \mathrm{d}x \right)^2,$$

from which, using the Schwarz inequality, we get:

$$\| I_{rh}(g) \|^2_{L^2(\ ^2)} \leqslant 2 \sum_{M_{ij} \in R_{rh}} \int_{\varpi_h^0(M_{ij})} | g(x) |^2 \,\mathrm{d}x$$

$$\leqslant 2 \sum_{M_{ij} \in R_{rh}} \int_{\sigma_h(M_{ij})} | g(x) |^2 \,\mathrm{d}x$$

$$= 2 \int_{\ ^2} | g(x) |^2 \,\mathrm{d}x = 2 \| g \|^2_{L^2(\mathbb{R}^2)} . \quad \blacksquare$$

2) *Verification of* (6.142).

For $g \in \mathscr{D}(\mathbb{R}^2)$ relation (6.142) is immediate, using the argument of uniform continuity on the support of g; since $\mathscr{D}(\mathbb{R}^2)$ is dense in $L^2(\mathbb{R}^2)$, and taking account of (6.141), (6.142) is true in $L^2(\mathbb{R}^2)$. $\blacksquare$

From this we deduce:

Corollary 6.1. *Consider* $(g_m)_m$ *such that* $g_m \to g$ *strongly in* $L^2(\mathbb{R}^2)$ *as* $m \to +\infty$; *then as* $h \to 0, m \to \infty$ *we have* $\forall q = 1, 2$:

$$I_{qh}(g_m) \to g \quad \textit{strongly in} \quad L^2(\mathbb{R}^2) \qquad \forall q = 1, 2 . \quad \blacksquare$$

Before proving the uniform convergence of $q_h u^4_{lh}$ to u $(l = 1, 2)$ we shall introduce (by analogy with the cases m = 1, 2, 3):

$$\Omega_{rh} = \Omega_h \cap R_{rh} \qquad (r = 1, 2) , \tag{6.143}$$

$$q_{rh} v_h = \sum_{M_{ij} \in \Omega_{rh}} v_{ij} \chi_h^{ij} . \tag{6.144}$$

It may be noted that, from the definition of Ω_h, we have $\varpi_h^1(M_{ij}) \subset \Omega$ if $M_{ij} \in \Omega_h$; in view of (6.139), we therefore have:

$$\text{support } (q_{rh} v_h) \subset \Omega \quad \forall v_h \in V_h ; \qquad r = 1, 2 . \tag{6.145}$$

We also define the triangulation $\mathscr{T}_{rh}$ $(r = 1, 2)$ as shown in Figure 6.3 overleaf, and

$$(6.146) \quad \begin{cases} \Pi_{rh} v_h = \text{the affine linear continuous function over the} \\ \text{triangles of } \mathscr{T}_{rh} \text{, such that:} \\ (\Pi_{rh} v_h)(M_{ij}) = v_{ij} \quad \forall M_{ij} \in R_{rh} \ (1) \quad (r = 1, 2) . \end{cases}$$

We then have:

Lemma 6.5. *The set* $\Pi_{rh} K^4_{0h}$ *is made up of functions which are bounded and Lipschitz on* $\mathbb{R}^2$ *, uniformly in* h*, and:*

---

(1) $v_{ij} = 0 \quad \forall M_{ij} \in R_{rh} - \Omega_{rh} \qquad (r = 1, 2).$

(6.147) $\quad \| q_{rh} v_h - \Pi_{rh} v_h \|_{L^\infty(\mathbb{R}^2)} \leqslant h\sqrt{2} \quad \forall v_h \in K_{0h}^4 ; \quad r = 1, 2 .$

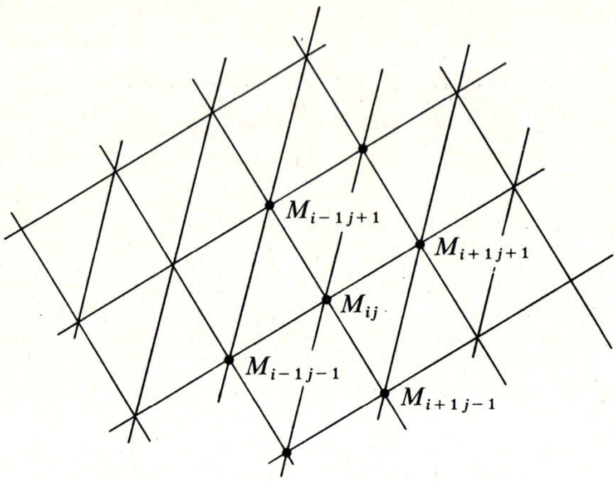

Fig. 6.3. Definition of the triangulation $\mathcal{T}_{rh}$.

Proof. We go back to Lemma 6.3 (by rotating through $\pi/4$ and by replacing h by $h\sqrt{2}$), noting that the membership conditions for K_{0h}^4 (see Section 5.2) imply

(6.148) $\quad | \operatorname{grad} \Pi_{rh} v_h | \leqslant \sqrt{2} \quad$ a.e. $\quad \forall v_h \in K_{0h}^4 ; \quad r = 1, 2 . \quad \blacksquare$

From Lemma 6.5, we deduce:

Theorem 6.5. *Let u_{lh}^4 be the solution of $(P_{0h})_{l4}$, and u the solution of (P_0) ; as $h \to 0$ we have:*

(6.149) $\quad \begin{cases} q_h u_{lh}^4 \to u & \text{strongly in } L^\infty(\Omega) \\ \forall l = 1, 2 . \end{cases}$

Proof.

1) From Theorem 6.3, $q_h u_{lh}^4 \to u$ strongly in $L^2(\Omega)$ as $h \to 0$; by applying Corollary 6.1 of Lemma 6.4, and taking account of (6.145

(6.150) $\quad \begin{cases} q_{rh} u_{lh}^4 \to u & \text{strongly in } L^2(\Omega) \text{ as } h \to 0 \quad (r = 1, 2) , \\ \forall l = 1, 2 . \end{cases}$

2) From Lemma 6.5, we can apply Ascoli's Theorem to the restriction to Ω of $\Pi_{rh} u_{lh}^4$; therefore, from (6.147), (6.150), we have:

$$(6.151)\quad \begin{cases} q_{rh}\, u_{lh}^4 \to u \text{ strongly in } L^\infty(\Omega) \quad (r = 1, 2), \\ \forall l = 1, 2, \end{cases}$$

from which, by *restriction* to $\bigcup_{M_{ij} \in \Omega_{rh}} \varpi_h^0(M_{ij})$ $(r = 1, 2)$ and *reassembling*, we obtain the result (6.149). ■

7. NUMERICAL SOLUTION OF THE APPROXIMATE PROBLEMS RELATING TO (P_0), BY RELAXATION AND POINT OVER-RELAXATION WITH SEQUENTIAL PROJECTION

7.1 Synopsis

In this section we shall study the application of the *relaxation* and *over-relaxation* methods of Chapter 2, Section 1, to the numerical solution of the finite-dimensional problems obtained by applying to (P_0) the approximations of Sections 4.1 and 5 (1).

We should point out from the start that only in the special case of Section 7.3.2 has convergence been proved for the algorithm which will be defined in Section 7.2, since the convex sets approximating K_0 are not "local" in the sense of the definition given in Chapter 2, Section 1.2 (on this subject, see Counter-Example 1.2 of Chapter 2, Section 1.2). Nonetheless, we think it proper to describe this method, inasmuch as in the approximate solution of (P_0), the results obtained coincide (2) with those obtained by other methods for which convergence has been established (3). ■

7.2 Description of the method

We use the procedure described in Chapter 2, Section 1.2, when the convex set K is not local.

7.2.1 *Case of the interior approximation (by finite elements) of Section 4.1.*

We use the notation of Section 4.1. We recall that the

(1) As far as the approximation in Section 4.2 is concerned, the following method is not applicable in view of the "global" character of the approximation of K_0 (see 4.2.5, Remark 4.6).
(2) At least in the examples which we have considered.
(3) For the rate of convergence, see Remark 7.14.

approximate problem defined in Section 4.1 reduces to a *quadratic programming* problem in $\mathbb{R}^N$ ($N = \text{Card}(\mathring{\Sigma}_h)$) with $N' = \text{Card}(\mathscr{T}_h)$ *quadratic constraints*; the variables are the values v_M taken by the functions v_h ($v_h \in V_h \subset H_0^1(\Omega)$) at the nodes $M \in \mathring{\Sigma}_h$.

In the following, v_h will have its usual meaning or will denote the vector [1] $(v_M)_{M \in \mathring{\Sigma}_h}$; by way of simplification we shall index the $M \in \mathring{\Sigma}_h$ by $i = 1, \ldots, N$ and put $v_i = v_{M_i} = v_h(M_i)$.

The algorithm of relaxation with projection is defined by

(7.1) $\quad u_h^0 \in K_0 \cap V_h$ chosen arbitrarily ($u_h^0 = 0$ for example),

and assuming u_h^n to be known, u_i^{n+1} is determined progressively, for $i = 1, \ldots, N$, by:

(7.2)
$$J(u_1^{n+1}, \ldots, u_{i-1}^{n+1}, u_i^{n+1}, u_{i+1}^n, \ldots, u_N^n) \leqslant$$
$$\leqslant J(u_1^{n+1}, \ldots, u_{i-1}^{n+1}, v_i, u_{i+1}^n, \ldots, u_N^n)$$

for v_i such that:

(7.3)
$$\{ u_1^{n+1}, \ldots, u_{i-1}^{n+1}, v_i, u_{i+1}^n, \ldots, u_N^n \} \in K_0 \cap V_h$$

with

(7.4)
$$\{ u_1^{n+1}, \ldots, u_i^{n+1}, u_{i+1}^n, \ldots, u_N^n \} \in K_0 \cap V_h . \quad \blacksquare$$

From a practical point of view we shall make the following remarks with regard to algorithm (7.1), ..., (7.4):

Remark 7.1. The number of constraints to be taken into account in the determination of u_i^{n+1} is equal to the *number of triangles* with M_i as vertex (see Figure 7.1)[2]

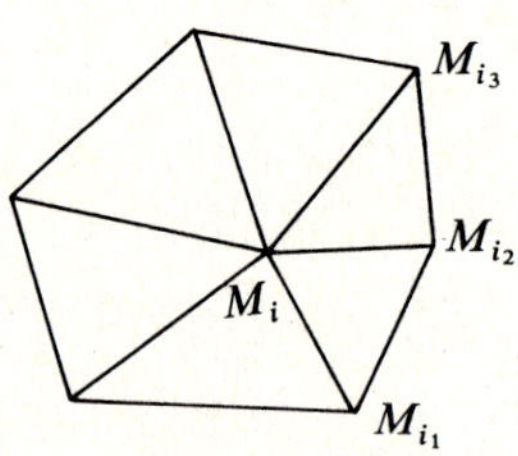

Fig. 7.1.

(1) Rigorously, we have $(v_M)_{M \in \mathring{\Sigma}_h} = r_h v_h$ where r_h is an isomorphism from $v_h \to \mathbb{R}^N$.

(2) Moreover, their union is the support of w_{M_i} (see Section 4.1.1 relation (4.9) for the definition of w_M).

We shall use the following notation:

$$\mathcal{T}_{ih} = \{ T \mid T \in \mathcal{T}_h, M_i \text{ is a vertex of } T \} \tag{7.5}$$

$$N_i = \text{Card}(\mathcal{T}_{ih}). \tag{7.6}$$

Remark 7.2. The conditions (7.3) and (7.4) imply that v_i and u_i^{n+1} belong to a *closed interval* of $\mathbb{R}$ which varies with n, say K_i^n; in practice, K_i^n is determined by noting that

$$K_i^n = \bigcap_{T \in \mathcal{T}_{ih}} I_T^n \tag{7.7}$$

where in (7.7) I_T^n is the *closed interval* of $\mathbb{R}$ defined by:

$$I_T^n = \{ v_i \mid |\text{grad}\, {}^i v_n^{n+1}| \leqslant 1 \text{ on } T \} \tag{7.8}$$

where

$${}^i v_h^{n+1} = \{ u_1^{n+1}, \ldots, u_{i-1}^{n+1}, v_i, u_{i+1}^n, \ldots, u_N^n \}. \tag{7.9}$$

In (7.8), the relation $|\text{grad}\, {}^i v_h^{n+1}|^2 \leqslant 1$ is made explicit, as a function of the coordinates of ${}^i v_h^{n+1}$, by means of relation (4.27) of Section 4.1.4.

It may be noted that the condition $u_h^0 \in K_0 \cap V_h$ leads, progressively, to $K_i^n \neq \varnothing$ since $u_i^n \in K_i^n$. ■

Remark 7.3. We calculate u_i^{n+1} as in Chapter 2, Section 1.2; thus we first define $u_i^{n+1/2}$ as being the solution (unique by strict monotonicity) of:

$$\frac{\partial J}{\partial u_i}(u_1^{n+1}, \ldots, u_{i-1}^{n+1}, u_i^{n+1/2}, u_{i+1}^n, \ldots, u_N^n) = 0 \qquad (i = 1, \ldots, N), \tag{7.10}$$

then:

$$u_i^{n+1} = P_{K_i^n}(u_i^{n+1/2}) \tag{7.11}$$

with:

$$P_{K_i^n}(x) = \max(a_i^n, \min(b_i^n, x)) \quad \forall x \in \mathbb{R} \tag{7.12}$$

if $K_i^n = [a_i^n, b_i^n]$. ■

Remark 7.4. We shall also use the following variant of the algorithm (7.1), ... , (7.4): the relations (7.1), (7.4) remaining the same, we determine $u_i^{n+1/2}$ using (7.10) and u_i^{n+1} using:

$$u_i^{n+1} = P_{K_i^n}((1 - \omega)u_i^n + \omega u_i^{n+1/2}) \tag{7.13}$$

which is of the type over-relaxation (or under-relaxation) with projection (on K_i^n). ■

7.2.2 *The case of the exterior approximations of Section 5*

We use the notation of Section 5; the problems $(P_{0h})_{lm}$ $(l=1,2;$ $m = 1, 2, 3, 4)$ are *quadratic programming* problems in $\mathbb{R}^N$ (N = Card (Ω_h)); the *constraints* are also *quadratic* and there are approximately $4N$ constraints for $m = 1$ and N for $m = 2, 3, 4$. The algorithm of relaxation with projection (a variant of the algorithm (7.1), ... , (7.4)) associated with $(P_{0h})_{lm}$ is then defined by:

$$u_h^0 \in K_{0h}^m, \tag{7.14}$$

chosen arbitrarily (for example $u_h^0 = 0$), then assuming u_h^n to be known, u_{ij}^{n+1} is determined progressively, for $M_{ij} \in \Omega_h$, by:

$$\begin{cases} J_h^l(u_{11}^{n+1}, \dots, u_{i-1j}^{n+1}, u_{ij-1}^{n+1}, u_{ij}^{n+1}, u_{i+1j}^n, u_{ij+1}^n, \dots) \\ \qquad \leqslant J_h^l(u_{11}^{n+1}, \dots, u_{i-1j}^{n+1}, u_{ij-1}^{n+1}, v_{ij}, u_{i+1j}^n, u_{ij+1}^n, \dots) \end{cases} \tag{7.15}$$

for all v_{ij} such that

$$\{ u_{11}^{n+1}, \dots, u_{i-1j}^{n+1}, u_{ij-1}^{n+1}, v_{ij}, u_{i+1j}^n, u_{ij+1}^n, \dots \} \in K_{0h}^m \tag{7.16}$$

with

$$\{ u_{11}^{n+1}, \dots, u_{i-1j}^{n+1}, u_{ij-1}^{n+1}, u_{ij}^{n+1}, u_{i+1j}^n, u_{ij+1}^n, \dots \} \in K_{0h}^m, \tag{7.17}$$

assuming that the iterations are performed with i,j increasing. ■

The following remarks may be made with regard to the algorithm (7.14), ... , (7.17):

Remark 7.5. From the definition of K_{0h}^m $(m = 1, 2, 3, 4)$ (see Section 5.2) the number of constraints to be taken into account in the determination of u_{ij}^{n+1} is equal to N_m $(m = 1, 2, 3, 4)$ with:

$$\begin{cases} N_1 = 12 \\ N_2 = 4 \\ N_3 = 5 \\ N_4 = 4. \end{cases} \tag{7.18}$$

Remark 7.6. Conditions (7.16), (7.17), imply that v_{ij} and u_{ij}^{n+1} belong to a closed interval of $\mathbb{R}$ which varies with n, say K_{ij}^n; in practice, K_{ij}^n is determined by taking the intersection of the N_m intervals of $\mathbb{R}$ determined from the N_m constraints which *explicitly* involve v_{ij}; for these calculations we shall use the formulas (5.10) for $m = 1$, (5.11) for $m = 2$, (5.15), (5.16), (5.18) for $m = 3$ and (5.21) for $m = 4$.

We note that the condition $u_h^0 \in K_{0h}^m$ leads progressively to

$K_{ij}^n \neq \varnothing$, since $u_{ij}^n \in K_{ij}^n$. ■

Remark 7.7. Proceeding as in Section 7.2.1, u_{ij}^{n+1} is calculated by first determining $u_{ij}^{n+1/2}$, the solution of

$$\frac{\partial J_h^l}{\partial v_{ij}}(u_{11}^{n+1}, \dots u_{i-1j}^{n+1}, u_{ij-1}^{n+1}, u_{ij}^{n+1/2}, u_{i+1j}^n, u_{ij+1}^n, \dots) = 0 \quad \forall M_{ij} \in \Omega_h , \tag{7.19}$$

giving:

$$u_{ij}^{n+1/2} = \tfrac{1}{4}(u_{i-1j}^{n+1} + u_{ij-1}^{n+1} + u_{ij+1}^n + u_{i+1j}^n + h^2 f_{ij}) \quad \text{if} \quad l = 1 , \tag{7.20$_1$}$$

$$u_{ij}^{n+1/2} = \tfrac{1}{4}(u_{i-1j-1}^{n+1} + u_{i+1j-1}^{n+1} + u_{i-1j+1}^n + u_{i+1j+1}^n + 2\,h^2 f_{ij}) \quad \text{if} \quad l = 2 , \tag{7.20$_2$}$$

it being assumed in relation $(7.20)_2$ that the traverse is first carried out with increasing j, and then with j fixed and increasing i; in $(7.20)_1$, $(7.20)_2$ $v_{pq} = 0$ if $M_{pq} \notin \Omega_h$. We then determine:

$$u_{ij}^{n+1} = P_{K_{ij}^n}(u_{ij}^{n+1/2}) . \tag{7.21}$$

In the case in which an over-relaxation (or under-relaxation) parameter ω is used, (7.21) is replaced by:

$$u_{ij}^{n+1} = P_{K_{ij}^n}((1 - \omega)\, u_{ij}^n + \omega u_{ij}^{n+1/2}) . \tag{7.22}$$

Remark 7.8. It follows from (7.18) that the use of K_{0h}^1 leads to at least twice as many constraints as the use of K_{0h}^m (m = 2, 3, 4); the use of these last three convex sets is thus preferred for solving (P_0), in an approximate manner, by relaxation with projection. ■

7.3 Convergence results

As was pointed out in Section 7.1, the problem of the convergence of algorithms (7.1), ... , (7.4) and (7.14), ... , (7.17) to the solutions of the corresponding approximate problems seems to be open for the time being. A result concerning the behaviour of the sequence u_h^n when $f \geqslant 0$ and $u_h^0 = 0$ will be given in Section 7.3.1; using the same assumptions for f and u_h^0 the convergence of the algorithm, when Ω is one-dimensional, will be proved in Section 7.3.2.

7.3.1 *A monotonic convergence result*

Definition 7.1. Given two vectors $\boldsymbol{u}$ and $v \in \mathbb{R}^N$ we say that

$u \leqslant v$ if $u_i \leqslant v_i \quad \forall i = 1, N$. ■

We have:

Proposition 7.1. *If* $f \geqslant 0$ *and* $u_h^0 = 0$, *then when* $n \to +\infty$ *the sequence* u_h^n *defined by the algorithms* (7.1)-(7.4) and (7.14)-(7.17) *converges monotonically to* u_h^* *from below, in the sense of Definition* 7.1 (1).

Proof. We have:

$$\| v_h \| \leqslant C \quad \forall v_h \in K_0 \cap V_h \text{ or } K_{0h}^m ; \tag{7.23}$$

it is thus sufficient to prove that

$$u_h^n \leqslant u_h^{n+1} \quad \forall n \geqslant 0 . \tag{7.24}$$

This will first be proved for algorithm (7.14), ... , (7.17) relating to the problems $(P_{0h})_{lm}$:

1) *Proof of* (7.24) *for algorithm* (7.14), ... , (7.17).

By definition, we have

$$f_{ij} = \frac{1}{h^2} \int_{\varpi\mathfrak{R}(M_{ij})} f(x)\, \mathrm{d}x ,$$

thus $\quad f \geqslant 0 \Rightarrow f_{ij} \geqslant 0 \quad \forall M_{ij} \in \Omega_h$.

With the notation

$$u_h^{n+1/2} = (u_{11}^{n+1/2}, \ldots, u_{ij}^{n+1/2}, \ldots) \tag{7.25}$$

we shall prove that we also have:

$$u_h^{n+1/2} \geqslant u_h^{n+1} \quad \forall n \geqslant 0 . \tag{7.26}$$

We argue by recurrence, showing that (7.24), (7.26) being true up to order $n - 1$ implies that they are true for order n.

This property is itself proved by recurrence by assuming:

$$\begin{cases} u_{pq}^n \leqslant u_{pq}^{n+1} & \text{for the coordinates of } u_h^{n+1} \\ \text{calculated } \textit{before} \ u_{ij}^{n+1} . \end{cases} \tag{7.27}$$

We have:

(1) In the case of algorithm (7.1) - (7.4) it is also necessary to assume that the angles of the triangulation $\mathcal{T}_h$ are $\leqslant \pi/2$.

$$(7.28)_1 \qquad u_{ij}^{n+1/2} = \tfrac{1}{4}[u_{i-1j}^{n+1} + u_{ij-1}^{n+1} + u_{i+1j}^{n} + u_{ij+1}^{n} + h^2 f_{ij}] \quad \text{if } l = 1$$

$$(7.28)_2 \qquad u_{ij}^{n+1/2} = \tfrac{1}{4}[u_{i-1j-1}^{n+1} + u_{i+1j-1}^{n+1} + u_{i-1j+1}^{n} + u_{i+1j+1}^{n} + 2h^2 f_{ij}]$$

if $l = 2$

so that, taking account of the above two recurrence hypotheses

$$(7.29) \qquad u_{ij}^{n+1/2} \geqslant u_{ij}^{n-1/2} \geqslant u_{ij}^{n}.$$

Moreover:

$$\left\{\begin{array}{c} u_{ij}^n \in K_{ij}^n \\ u_{ij}^{n+1} = P_{K_{ij}^n}(u_{ij}^{n+1/2}) \end{array}\right\} \Rightarrow (u_{ij}^n - u_{ij}^{n+1})(u_{ij}^{n+1/2} - u_{ij}^n) \leqslant 0$$

so that with (7.29):

$$(7.30) \qquad u_{ij}^{n+1} \geqslant u_{ij}^{n}.$$

It may easily be shown that (7.29), (7.30) are true for the *first* calculated coordinate of $u_h^{n+1/2}$ and u_h^{n+1} and hence (7.24), (7.26), (7.27) follow.

It thus remains to be proved that (7.24), (7.26) are true for $n = 0$: we have $u_{11}^{1/2} = h^2 f_{11}$ if $l = 1$, $u_{11}^{1/2} = 2h^2 f_{11}$ if $l = 2$, so that with $0 = u_{11}^0 \in K_{11}^1$ and $f_{11} \geqslant 0$,

$$(7.31) \qquad u_{11}^{1/2} \geqslant u_{11}^{1} \geqslant u_{11}^{0} = 0.$$

Using a step-by-step argument starting from (7.31), and bearing in mind that $f_{ij} \geqslant 0$, $\forall M_{ij} \in \Omega_h$, and that $0 = u_{ij}^0 \in K_{ij}^1$ we deduce from the formulas $(7.28)_l$ ($l = 1, 2$) that:

$$(7.32) \qquad u_h^{1/2} \geqslant u_h^{1} \geqslant u_h^{0} = 0,$$

which completes the proof of (7.24). ■

2) *Proof of* (7.24) for algorithm (7.1), ... , (7.4).

We have

$$(7.33) \qquad \frac{\partial J}{\partial v_i} = \alpha_i v_i - \sum_{k=1}^{N_i} \beta_{ik} v_{ik} - \gamma_i f_i,$$

where in (7.33) the v_{ik} relate to the N_i vertices neighbouring M_i, i.e. the vertices of the triangles which have M_i as common vertex (see Figure 7.1).

It may easily be shown (if the angles of $\mathcal{T}_h$ are $\leqslant \pi/2$) that:

$$(7.34) \qquad \alpha_i > 0, \quad \beta_{ik} \geqslant 0 \quad \forall k = 1, \dots, N_i, \quad \gamma_i > 0,$$

(7.35) $$\alpha_i = \sum_{k=1}^{N_i} \beta_{ik}$$

(7.36) $$f \geqslant 0 \Rightarrow f_i \geqslant 0 .$$

Taking account of (7.33), (7.34), (7.36), the proof of (7.24) for algorithm (7.1), ... , (7.4) may be achieved by proceeding as above for algorithm (7.14), ... , (7.17). ■

Remark 7.9. The vector u_h^* appearing in the preceding proposition is a "*blockage* point" in the sense of the Counter-Examples 1.1 and 1.2 of Chapter 2, Section 1, for the algorithm considered.

Remark 7.10. The result stated in Proposition 7.1 still holds if we use the variant (7.1), (7.10), (7.13) (resp. (7.14), (7.19), (7.22)) of algorithm (7.1), ... , (7.4) (resp. (7.14),...,(7.17)), with $0 < \omega \leqslant 1$. ■

7.3.2 *A particular case in which the algorithm of relaxation with projection converges to the exact solution of the approximate problem*

Statement of the problem and approximation.

In the particular case $\Omega =]0, 1[$ we define $J : H_0^1(0, 1) \to \mathbb{R}$ by:

(7.37) $$J(v) = \frac{1}{2} \int_0^1 v'^2 \, dx - \int_0^1 fv \, dx ,$$

and we take $f \in L^1(0, 1)$ in (7.37); we thus have for (P_0) the formulation:

(7.38) $$\begin{cases} \text{Min}_{v \in K_0} J(v) , \\ K_0 = [v \mid v \in H_0^1(0, 1), \mid v' \mid \leqslant 1 \text{ a.e. on } (0, 1)] . \end{cases}$$

With N a positive integer (which will tend towards $+\infty$), we put $h = 1/N$ and define the *interior approximations* of $H_0^1(\Omega)$ and K_0 by:

(7.39) $$V_h = [v_h \mid v_h \in H_0^1(0, 1) , \; v_h \text{ linear affine on } [(i-1)h, ih], i = 1, \dots, N] ,$$

(7.40) $$K_{0h} = [v_h \mid v_h \in V_h , \mid v_h' \mid \leqslant 1 \text{ a.e.}] .$$

Writing $v_i = v_h((i-1)h)$ $(i = 1, \dots, N+1)$ we have:

$$(7.41) \qquad v_h \in K_{0h} \Leftrightarrow \begin{cases} \left| \dfrac{v_{i+1} - v_i}{h} \right| \leqslant 1 & (i = 1, ..., N), \\ v_1 = v_{N+1} = 0 . \end{cases}$$

We approximate J by $J_h : V_h \to \mathbb{R}$ defined by:

$$(7.42) \qquad J_h(v_h) = \frac{1}{2} \int_0^1 v_h'^2 \, dx - h \sum_{i=2}^{N} f_i v_i ,$$

with (for example):

$$(7.43) \qquad \begin{cases} f_i = \dfrac{1}{h} \displaystyle\int_{(i-\frac{3}{2})h}^{(i-\frac{1}{2})h} f(x) \, dx , \\ \text{or} \quad f_i = f((i-1)h) \quad \text{if} \quad f \in C^0[0, 1] . \end{cases}$$

The explicit form of (7.42) is given by

$$(7.42') \qquad J_h(v_h) = \frac{h}{2} \sum_{i=1}^{N} \left(\frac{v_{i+1} - v_i}{h} \right)^2 - h \sum_{i=2}^{N} f_i v_i ;$$

we define the approximate problem $(P_0)_h$ by

$$(7.44) \qquad \operatorname*{Min}_{v_h \in K_{0h}} J_h(v_h) .$$

By applying the results of the preceding sections to this elementary particular case, it may be shown that (7.44) admits one and only one solution, u_h , such that, when $h \to 0$, $u_h \to u$ *strongly* in $H^1(0, 1) \cap L^\infty(0, 1)$ where u is the (unique) solution of (7.38). ■

Solution of (7.44) *by point relaxation with projection.*

In the case of problem (7.44), and if the iterations are performed with increasing i, the algorithm described in Section 7.2 takes the form:

$$(7.45) \qquad \begin{cases} u_h^0 \in K_{0h} \quad \text{chosen arbitrarily} \quad (u_h^0 = 0 \text{ for example}), \\ u_1^n = u_{N+1}^n = 0 \\ u_i^{n+1/2} = \frac{1}{2}(h^2 f_i + u_{i-1}^{n+1} + u_{i+1}^n) \qquad (2 \leqslant i \leqslant N), \\ u_i^{n+1} = P_{K_i^n}(u_i^{n+1/2}) \qquad (2 \leqslant i \leqslant N). \end{cases}$$

The interval K_i^n is defined explicitly by:

$$(7.46) \qquad K_i^n = [v_i \mid v_i \in \mathbb{R}, \mid u_{i+1}^n - v_i \mid \leqslant h, \mid u_{i-1}^{n+1} - v_i \mid \leqslant h, i=2, ..., N] ,$$

and is never empty (see Section 7.2) if $u_h^0 \in K_{0h}$. ■

Convergence of algorithm (7.45) *if* $f \geq 0$ *and* $u_h^0 = 0$.

It will be assumed in the following that $f \geq 0$ and $u_h^0 = 0$, so that $f_i \geq 0$ from (7.43). The result of Proposition 7.1 (which also holds for algorithm (7.45)) will be supplemented by showing that $u_h^* = u_h$; the proof of this point will result from the following propositions.

Proposition 7.2. *We have*:

(7.47) *the sequences* u_h^n *and* $u_h^{n+1/2}$ *are increasing*,

(7.48) $u_h^{n+1/2} \geq u_h^{n+1} \quad \forall n \geq 0$,

(7.49) $u_h^n \in K_{0h} \qquad \forall n \geq 0$.

Proof. See that of Proposition 7.1. ■

Proposition 7.3. *If* u_h *is the (unique) solution of* (7.44) *we have*:

(7.50) $u_h \geq 0$

(7.51) $\frac{1}{2}(u_{i+1} + u_{i-1}) \leq u_i \leq \frac{1}{2}(h^2 f_i + u_{i+1} + u_{i-1}), \qquad (2 \leq i \leq N)$.

Proof.

1) *Proof of* (7.50)

Let u_h be the solution of (7.44); we define $\tilde{u}_h \in V_h$, $\tilde{u}_h \geq 0$ by

(7.52) $\tilde{u}_h((i-1)h) = |u_i|, \quad i = 1, ..., N+1$.

From $f_i \geq 0$ and $||u_i| - |u_{i-1}|| \leq |u_i - u_{i-1}|$ we deduce

(7.53) $u_h \in K_{0h} \Rightarrow \tilde{u}_h \in K_{0h}$,

(7.54) $$\sum_{i=1}^{N} \left(\frac{|u_{i+1}| - |u_i|}{h}\right)^2 \leq \sum_{i=1}^{N} \left(\frac{u_{i+1} - u_i}{h}\right)^2,$$

(7.55) $$\sum_{i=1}^{N} f_i u_i \leq \sum_{i=2}^{N} f_i |u_i|,$$

so that

(7.56) $J_h(u_h) \geq J_h(\tilde{u}_h), \qquad \tilde{u}_h \in K_{0h}$,

which implies $\tilde{u}_h = u_h$. ■

2) *Proof of* (7.51).

Let $\bar{u}_i = \frac{1}{2}(h^2 f_i + u_{i+1} + u_{i-1})$ and K_i be the nonempty interval of $\mathbb{R}$ defined by:

$$K_i = [v_i \mid v_i \in \mathbb{R}, \mid v_i - u_{i+1} \mid \leqslant h, \mid v_i - u_{i-1} \mid \leqslant h] . \tag{7.57}$$

Let $w_i = \frac{1}{2}(u_{i+1} + u_{i-1})$, then $f_i \geqslant 0 \Rightarrow \bar{u}_i \geqslant w_i$ and since w_i minimises the function $v_i \to (v_i - u_{i+1})^2 + (v_i - u_{i-1})^2$ over $\mathbb{R}$, we have:

$$\begin{cases} \mid w_i - u_{i+1} \mid = \mid w_i - u_{i-1} \mid , \\ (w_i - u_{i+1})^2 + (w_i - u_{i-1})^2 \leqslant (u_i - u_{i+1})^2 + (u_i - u_{i-1})^2 , \end{cases} \tag{7.58}$$

from which we deduce

$$\mid w_i - u_{i+1} \mid = \mid w_i - u_{i-1} \mid \leqslant \max(\mid u_i - u_{i+1} \mid, \mid u_i - u_{i-1})) \leqslant h \Rightarrow w_i \in K_i \tag{7.59}$$

which, with $u_i = P_{K_i}(\bar{u}_i)$ and $\bar{u}_i \geqslant w_i$ leads to:

$$(w_i - u_i)(\bar{u}_i - u_i) \leqslant 0 \Rightarrow w_i \leqslant u_i \leqslant \bar{u}_i \tag{7.60}$$

hence (7.51). ■

Proposition 7.4. *If for an index* i_0 *we have* $u_{i_0} - u_{i_0-1} = h$ (*resp.* $u_{i_0} - u_{i_0+1} = h$) *then we have* $u_i - u_{i-1} = h \quad \forall i, i = 2, \ldots, i_0$ (*resp.* $u_i - u_{i+1} = h \quad \forall i, i = i_0, \ldots, N$).

Proof. Let i_0 be such that $u_{i_0} - u_{i_0-1} = h$; then taking account of Proposition 7.3 we have $u_{i_0-1} \geqslant \frac{1}{2}(u_{i_0} + u_{i_0-2})$ so that $u_{i_0-1} - u_{i_0-2} \geqslant u_{i_0} - u_{i_0-1} = h$; now $u_h \in K_{0h}$ so that $u_{i_0-1} - u_{i_0-2} = h$, and so on until $i = 2$. A similar argument is used for the other family of relations. ■

Proposition 7.5. *Let* i_M $(2 \leqslant i_M \leqslant N)$ *be such that* $u_{i_M} = \max\limits_{1 \leqslant i \leqslant N+1} u_i$; *then for h sufficiently small we have*:

$$\begin{cases} 0 \leqslant u_{i_M} - u_{i_M-1} < h , \\ 0 \leqslant u_{i_M} - u_{i_M+1} < h , \\ u_{i_M} = \frac{1}{2}(h^2 f_{i_M} + u_{i_M+1} + u_{i_M-1}) . \end{cases} \tag{7.61}$$

Proof. With the notation of Proposition 7.3 we have:

$$\bar{u}_{i_M} = \frac{1}{2}(h^2 f_{i_M} + u_{i_M-1} + u_{i_M+1}) , \tag{7.62}$$

or:

$$\left(\frac{\bar{u}_{i_M} - u_{i_M-1}}{h}\right) + \left(\frac{\bar{u}_{i_M} - u_{i_M+1}}{h}\right) = hf_{i_M} = \int_{e_{i_M}} f(x)\,dx , \tag{7.63}$$

putting

$$e_{i_M} =](i_M - \tfrac{3}{2})\, h, (i_M - \tfrac{1}{2})\, h[\,.$$

From Proposition 7.3 we have $\bar{u}_{i_M} \geqslant u_{i_M}$, so that, with (7.36),

$$(7.64)\qquad \begin{cases} 0 \leqslant \dfrac{\bar{u}_{i_M} - u_{i_M - 1}}{h} \leqslant \displaystyle\int_{e_{i_M}} f(x)\,\mathrm{d}x\,, \\[2ex] 0 \leqslant \dfrac{\bar{u}_{i_M} - u_{i_M + 1}}{h} \leqslant \displaystyle\int_{e_{i_M}} f(x)\,\mathrm{d}x\,. \end{cases}$$

When $h \to 0$ we have $\int_{e_{i_M}} f(x)\,\mathrm{d}x \to 0$, so that for h sufficiently small, $\int_{e_{i_M}} f(x)\,\mathrm{d}x < 1$ so that $\bar{u}_{i_M} \in K_{i_M} \Rightarrow u_{i_M} = \bar{u}_{i_M}$ which together with (7.62), (7.63) proves (7.61).

Proposition 7.6. *The sequence* u_h^n *defined by* (7.45) *converges to* $u_h^* \in K_{0h}$ *from below such that* $u_h^* \leqslant u_h$.

Proof. We shall first prove that $u_h^n \leqslant u_h \;\; \forall n$. We shall argue by recurrence and assume that the property is true up to order n and true for u_h^{n+1} up to the index $i - 1$:

From Propositions 7.2 and 7.3 we have:

$$(7.65)\qquad \begin{cases} \bar{u}_i \;\;\;\;\;\; = \tfrac{1}{2}(h^2 f_i + u_{i-1} + u_{i+1}) \geqslant u_i\,, \\ u_i^{n+1/2} = \tfrac{1}{2}(h^2 f_i + u_{i-1}^{n+1} + u_{i+1}^n) \geqslant u_i^{n+1}\,. \end{cases}$$

If $u_i = \bar{u}_i$ we have, from the recurrence assumptions, $u_i \geqslant u_i^{n+1/2} \geqslant u_i^{n+1}$. If $\bar{u}_i > u_i$ we have from Proposition 7.4,

$$(7.66)\qquad u_i = \min\,(u_{i+1}, u_{i-1}) + h\,,$$

and likewise we always have

$$(7.67)\qquad u_i^{n+1} \leqslant \min\,(u_{i-1}^{n+1}, u_{i+1}^n) + h\,.$$

Now (recurrence assumptions) $u_{i-1}^{n+1} \leqslant u_{i-1}$ and $u_{i+1}^n \leqslant u_{i+1}$, so that:

$$(7.68)\qquad u_i^{n+1} \leqslant \min\,(u_{i-1}^{n+1}, u_{i+1}^n) + h \leqslant \min\,(u_{i-1}, u_{i+1}) + h = u_i\,.$$

Thus if the property is assumed to be true to the order n, since it is true to the order $n + 1$ for $u_1^{n+1} = 0$ it is true for u_h^{n+i} ; now the property is true for $u_h^0 = 0$ and is thus true for all n, that is:

$$(7.69)\qquad u_h^n \leqslant u_h \quad \forall n \geqslant 0\,.$$

Since the sequence u_h^n is increasing and bounded by u_h, it converges from below to $u_h^* \in K_{0h}$ with $u_h^* \leqslant u_h$. ■

Proposition 7.7. *The function* u_h^* *satisfies the properties of* u_h, *the subjects of Propositions* 7.3, 7.4, 7.5.

Proof. From $u_h^0 = 0$ and Proposition 7.2 we deduce:

$$u_h^n \geqslant 0 \quad \forall n \geqslant 0\,, \tag{7.70}$$

then by applying the same method as that used at point 2 of the proof of Proposition 7.3:

$$\tfrac{1}{2}(u_{i+1}^n + u_{i-1}^{n+1}) \leqslant u_i^{n+1} \leqslant \tfrac{1}{2}(h^2 f_i + u_{i+1}^n + u_{i-1}^{n+1})\,, \tag{7.71}$$

so that, proceeding to the limit in (7.70), (7.71):

$$\begin{cases} u_h^* \geqslant 0\,, \\ \tfrac{1}{2}(u_{i+1}^* + u_{i-1}^*) \leqslant u_i^* \leqslant \tfrac{1}{2}(h^2 f_i + u_{i+1}^* + u_{i-1}^*)\,. \end{cases} \tag{7.72}$$

By starting from (7.72) and arguing as in the proof of Proposition 7.4 it may be demonstrated that the conclusions of the latter apply equally to u_h^* ; the same applies for the conclusions of Proposition 7.5. ■

Proposition 7.8. *We have* $u_h^* = u_h$.

Proof.

1) Assume that there exists $i_0,\ 2 \leqslant i_0 \leqslant N$, such that $u_{i_0}^* - u_{i_0-1}^* = h$ (resp. $u_{i_0}^* - u_{i_0+1}^* = h$) ; then $u_{i_0}^* = u_{i_0}$. In fact, from Proposition 7.7, we have:

$$u_{i_0}^* = (i_0 - 1)\,h \ (\text{resp. } u_{i_0}^* = (N + 1 - i_0)\,h)\ ;$$

now $|u_{i+1} - u_i| \leqslant h \quad \forall i = 1, \dots, N$ so that $u_{i_0} \leqslant (i_0 - 1)\,h,\ u_{i_0} \leqslant (N+1-i_0)\,h$ hence $u_{i_0} \leqslant u_{i_0}^*$; thus taking account of $u_h^* \leqslant u_h,\ u_{i_0} = u_{i_0}^*$ and likewise

$$u_i = u_i^* \quad \forall 1 \leqslant i \leqslant i_0 \quad (\text{resp. } \forall i,\ i_0 \leqslant i \leqslant N)\,.$$

2) Moreover, we have

$$J_h(u_h) = J_h(u_h^*) + (J_h'(u_h^*), u_h - u_h^*) + \tfrac{1}{2}(J_h''(u_h^*).u_h - u_h^*, u_h - u_h^*) \tag{7.73}$$

with

$$(J_h'(u_h^*), u_h - u_h^*) = h \sum_{i=2}^{N} \left(\frac{-u_{i-1}^* - u_{i+1}^* + 2\,u_i^*}{h^2} - f_i \right)(u_i - u_i^*)\,, \tag{7.74}$$

$$(J_h''(u_h^*).u_h - u_h^*, u_h - u_h^*) = h \sum_1^N \left(\frac{w_{i+1} - w_i}{h} \right)^2 , \tag{7.75}$$

where $w_h = u_h - u_h^*$.

3) From Proposition 7.7 we have:

$$\frac{-u_{i-1}^* - u_{i+1}^* + 2u_i^*}{h^2} - f_i \leqslant 0 \quad \forall i = 2, \dots, N . \tag{7.76}$$

If the inequality is strict, we have $|u_{i+1}^* - u_i^*| = h$ and/or $|u_i^* - u_{i-1}^*| = h$, so that $u_i = u_i^*$ from point 1. We thus have:

$$\left(\frac{-u_{i-1}^* - u_{i+1}^* + 2u_i^*}{h^2} - f_i \right) (u_i - u_i^*) = 0 \quad \forall i, 2 \leqslant i \leqslant N , \tag{7.77}$$

so that

$$J_h(u_h) - J_h(u_h^*) = \frac{h}{2} \sum_{i=1}^N \left(\frac{w_{i+1} - w_i}{h} \right)^2 \geqslant 0 , \tag{7.78}$$

and since $u_h^* \in K_{0h}$ we then deduce $u_h^* = u_h$. ■

We have thus proved:

<u>Theorem 7.1</u>. *If $f \geqslant 0$ and $u_h^0 = 0$ the sequence u_h^n defined by algorithm* (7.45) *converges to the solution u_h of problem* (7.44) with

$$0 \leqslant u_h^1 \leqslant \cdots \leqslant u_h^n \leqslant u_h^{n+1} \leqslant \cdots \leqslant u_h . \quad ■ \tag{7.79}$$

Remark 7.11. The result stated in Theorem 7.1 still holds if we use the variant of algorithm (7.45) defined by:

$$\begin{cases} u_h^0 = 0 , \\ u_1^n = u_{N+1}^n = 0 , \\ u_i^{n+1/2} = \frac{1}{2}(h^2 f_i + u_{i-1}^{n+1} + u_{i+1}^n) , \\ u_i^{N+1} = P_{K_i^n}(u_i^n + \omega(u_i^{n+1/2} - u_i^n)) , \end{cases} \tag{7.80}$$

with $0 < \omega \leqslant 1$. ■

7.4 <u>Applications. Example 1</u>

We take $\Omega = [x \mid x \in \mathbb{R}^2, r = \sqrt{x_1^2 + x_2^2} < \frac{1}{2}]$ and $f = 10$; taking account of Section 2.3.2 the solution of (P_0) is given by

$$\begin{cases} u(x) = \frac{1}{2} - r & \text{if} \quad 0.2 \leqslant r \leqslant \frac{1}{2} \quad (\textit{plastic zone}) \\ u(x) = 0.4 - 2.5\, r^2 & \text{if} \quad 0 \leqslant r \leqslant 0.2 \quad (\textit{elastic zone}) \end{cases} \tag{7.81}$$

Following Bourgat /1/, we approximate (P_0) using the finite element method of Section 4.1, with the two triangulations $\mathcal{T}_1$, $\mathcal{T}_2$ shown (for 1/12 of the disc) on Figures 7.3 and 7.4; the approximate problems are solved using the algorithm (7.1), (7.10), (7.13) defined in Section 7.2.1, with $u_h^0 = 0$, the termination criterion being chosen as:

$$\frac{\sum_{M_i \in \mathring{\Sigma}_j} |u_i^{n+1} - u_i^n|}{\sum_{M_i \in \mathring{\Sigma}_j} |u_i^{n+1}|} \leqslant 10^{-4} . \tag{7.82}$$

The main results relating to the approximate solution of (P_0) by the above techniques are shown in Table 7.1.

Triangulation	$\mathcal{T}_1$	$\mathcal{T}_2$
Number of vertices	121	433
$N = \text{Card}(\mathring{\Sigma}_j)$ $j = 1, 2$	73	337
ω	1.6	1.6
Number of iterations	7	20
ε_a	1.3×10^{-2}	1.4×10^{-3}
Computation time (CII 10070)	10 s	50 s

Table 7.1.

In Table 7.1, ε_a represents the maximum absolute value of the differences (evaluated at the vertices of the triangulations considered) between the exact solution and the approximate solution, calculated from the approximation and using the above algorithm for the termination criterion (7.82).

The details of the triangulations $\mathcal{T}_1$ and $\mathcal{T}_2$ are shown in Figures 7.2 and 7.3 for 1/12 of the disc (the total triangulation is obtained by rotations of $k\pi/6, k = 1, \ldots, 11$); these figures also show the approximations of the *elastic* zone and the *plastic* zone (hatched region) as well as the interface between the exact elastic and plastic zones (i.e. the circle of radius 0.2 depicted as a dashed line). The reader may consult Bourgat /1/ for further details of the approximate solution of Example 1. ■

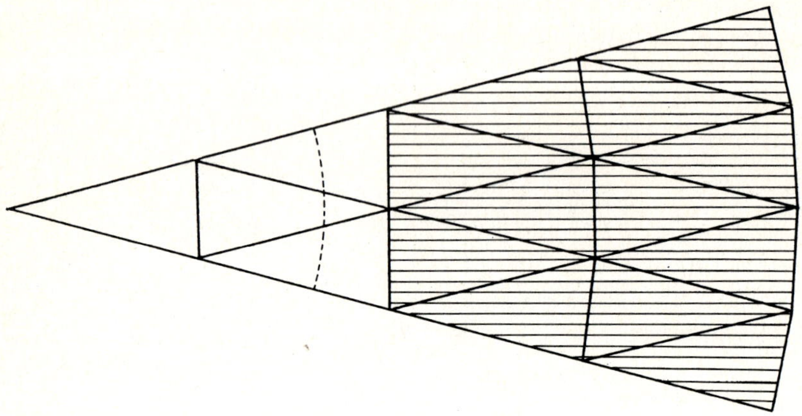

Fig. 7.2 (f = 10). Detail of the triangulation $\mathcal{T}_1$ and representation of the approximated elastic and plastic zones (the plastic zone is shown hatched).

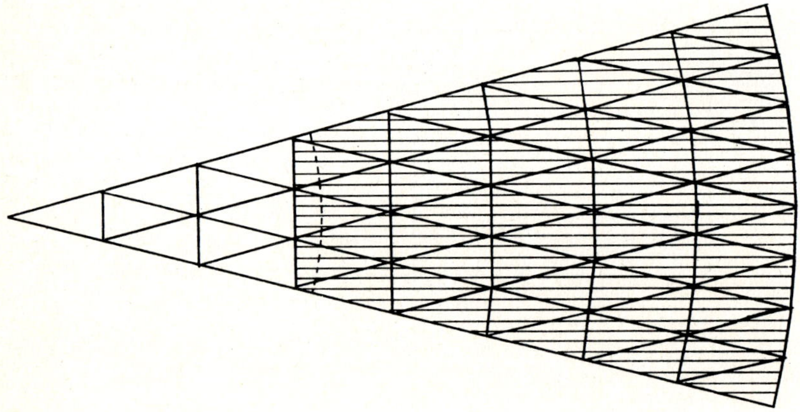

Fig. 7.3 (f = 10). Detail of the triangulation $\mathcal{T}_2$ and representation of the approximated elastic and plastic zones (the plastic zone is shown hatched).

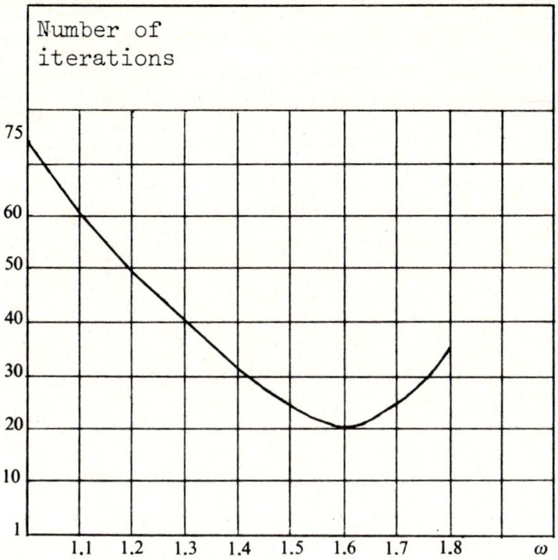

Fig. 7.4. Representation of the number of iterations required for convergence, as a function of ω (Example No. 1).

Remark 7.12. A detailed analysis (see Bourgat /1/) of the behaviour of the successive iterates in algorithm (7.1), (7.10), (7.13) shows that the plastic zone and the plastic component of the solution are rapidly attained, and that the majority of the iterations are spent on the solution of $-\Delta u = f$ in the elastic zone, the convergence being all the more rapid (for a given ω) as the size of the plastic zone increases (i.e. as the number of active constraints increases); this brings together Remarks 5.7, Chapter 2, Section 5.7.2 and 3.7, Chapter 3, Section 3.5.2.

We note (see Figure 7.4) that once again the use of an ω significantly greater than unity allows the convergence to be considerably accelerated. ■

Remark 7.13. The solution u of the problem (P_0) of Example 1 satisfies

(7.83) $$u \in W_0^{1,\infty}(\Omega) \cap W^{2,\infty}(\Omega) .$$

From this it may be deduced (G. Strang, private communication) that under the assumptions of Theorem 6.1, Section 6.2.1, we have:

(7.84) $$\| u - u_h \|_{H_0^1(\Omega)} \leqslant C h^{1/4} , \quad C \text{ independent of } h,$$

with, we recall, $h = \max_{T \in \mathcal{T}_h} \text{area}\,(T)$. ■

7.5 Applications. Example 2.

We consider again the example of Section 3.5 (i.e. $\Omega =]0,1[\times]0,1[$ and $f = 10$) solved previously by replacing the convex set K_0 by K_1 (using the equivalence result of Brézis-Sibony /1/ - see Section 2.2).

Following Goursat /1/ we use, for the approximate solution of (P_0), a method of *exterior approximation* in the sense of Section 5; more precisely, we approximate J by J_h^1 (see Section 5.1.1, relations (5.1), (5.2), (5.3)) and K_0 by K_{0h}^3 (see Section 5.2, relation (5.14)).

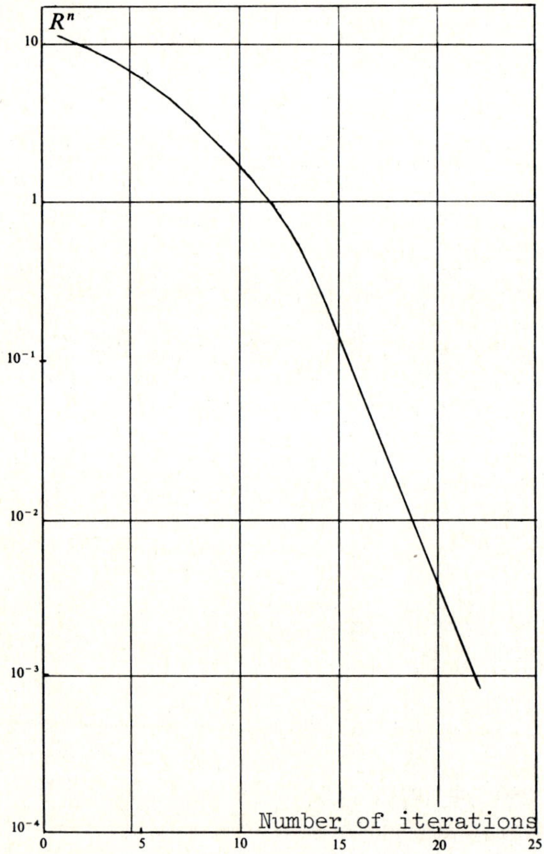

Fig. 7.5. Decrease of R^n as a function of the number of iterations (Example No. 2).

7.5.1 *Numerical values of the parameters*

Mesh size: $h = 1/20$.

Initialisation of algorithm (7.14), (7.19), (7.22): $u_h^0 = 0$.

Termination criterion for algorithm (7.14), (7.19), (7.22):

$$\sum_{M_{ij} \in \Omega_h} | u_{ij}^{n+1} - u_{ij}^n | < 10^{-3} .$$

7.5.2 *Analysis of the numerical results*

For $0 < \omega < 2$, u_h^n converges to a limit which coincides to within a relative value of about 10^{-4} with the solution calculated by solving the corresponding problem (P_{1h}) (see Section 3.5) by the method of Section 3.4, for $h = 1/20$. The optimal value of ω (determined empirically) is about 1.5, which is the optimal value of ω for the corresponding problem (P_{1h}) ; Remark 3.7 (see Section 3.5.2) therefore still holds.

For $\omega = 1.5$ convergence of algorithm (7.14), (7.19), (7.22) is reached in 22 iterations, corresponding to 1.38 sec on an IBM 360/91 or 22 sec. on a CII 10070; the decrease of

$$R^n = \sum_{M_{ij} \in \Omega_h} | u_{ij}^{n+1} - u_{ij}^n |$$

is shown in Figure 7.5 as a function of the number of iterations, for $\omega = 1.5$. It may be said, schematically, that the curvilinear part of the graph is characteristic of the *nonlinearity* of the problem in the course of the solution until the approximate plastic zones and plastic solutions have been determined; the remainder of the graph, which is rectilinear, corresponds to the solution of a *linear* problem with $-\Delta u = f$ (approximately) in the elastic zone; the above applies equally well to Figure 3.1 of Section 3.5.2. ■

7.6 <u>Applications</u>. <u>Example 3</u>.

We take $\Omega =]0, 1[\times]0, 1[$ and f defined by

$$f(x) = \begin{cases} 10 \text{ if } (x_1, x_2) \in]0, \tfrac{1}{2}[\times]\tfrac{1}{2}, 1[, \\ -10 \text{ if } (x_1, x_2) \in]\tfrac{1}{2}, 1[\times]0, \tfrac{1}{2}[, \\ 0 \text{ if } (x_1, x_2) \in]0, \tfrac{1}{2}[\times]0, \tfrac{1}{2}[\cup]\tfrac{1}{2}, 1[\times]\tfrac{1}{2}, 1[. \end{cases} \tag{7.85}$$

With the corresponding problem (P_0) approximated by the *exterior approximations* of the preceding example (see Section 7.5) with $h = 1/20$, convergence of the algorithm (7.14), (7.19), (7.22) for the initialisation $u_h^0 = 0$ and the termination criterion

$$\sum_{M_{ij} \in \Omega_h} | u_{ij}^{n+1} - u_{ij}^n | < 10^{-4} ,$$

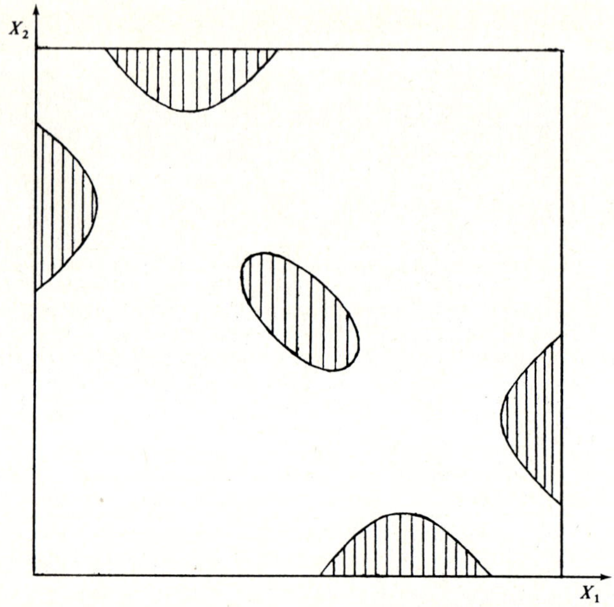

Fig. 7.6. Representation of the zones in which $|\mathbf{grad}\, u| < 1$ and $|\mathbf{grad}\, u| = 1$ (the latter are shown hatched). (Example No. 3).

is reached in 57 iterations (corresponding to a computation time of 1 min. 13 sec. on a CII 10070), if $\omega = 1.7$; the limit thus obtained coincides with the solution of the corresponding problem (P_0), calculated by means of other methods for which convergence can be proved. From the result of the above calculation, the blank region in Figure 7.6 represents the zone in which $|\mathbf{grad}\, u| < 1$ (and $-\Delta u = f$) and the hatched regions represent the zones in which $|\mathbf{grad}\, u| = 1$. ■

7.7 Various remarks

Remark 7.14. The algorithm (7.14), (7.19), (7.22) applied to the solution of (P_0), using the approximations of J and K_0 employed in Examples 2 and 3, is more rapid [1] *in terms of the num-*

[1] For a given ω and with the same initial conditions and termination criteria.

ber of iterations than the algorithm of over-relaxation with projection defined in Section 3.4, when the latter is applied to the solution of the problem (P_1) [1] using the approximation of J defined in Section 3.2.1, relations (3.10), (3.12), [2] and the approximation of K_1 defined in Section 3.2.1, relation (3.11). This behaviour may be explained by noting that the constraints defining K_0 (resp. K_{0h}^3) are more *restrictive* than those defining K_1 (resp. K_{1h}) since $K_0 \subset K_1$ strictly (the analogous property for the approximate convex sets is somewhat more difficult to express).

As far as the *computation time* is concerned, the second algorithm has in fact proved to be somewhat more rapid than the first (all other things being equal, and for the examples which we have considered); this is easily explained by the fact that the projection on the interval K_{ij}^n, in the case of the first algorithm, is more complicated than that on $[-\delta_{ij}, \delta_{ij}]$ in the case of the second, the number of relations defining these two intervals being greater in the case of the first algorithm.

It should be added that of all the methods tried for solving (P_0) in an approximate manner (without making recourse to the equivalence result of Section 2.2), the method considered in this Section 7 is the one which has demonstrated the best performance both in terms of computation time *and* storage requirement. ■

Remark 7.15. In connection with the convergence of the algorithms (7.1), ... , (7.4) and (7.14), ... , (7.17) (and variants with under- or over-relaxation parameters) it may be noted (though this does not constitute a proof) that when $h \to 0$ (or $N \to +\infty$) the approximations of K_0 defined in Section 4.1 (finite elements) and in Section 5.2 (finite differences) become *progressively more local* in the following sense:

the maximum number of constraints which explicitly involve a variable is bounded above by $N_0 = 2\pi/\theta_0$ in the case of the finite element method and under the assumptions of the convergence theorem 6.1, and by the integers N_m $(m = 1, 2, 3, 4)$ given by (7.18) in the case of the exterior approximations of Section 5.2.

We have, of course, $\lim_{h \to 0} (N_m/N) = 0$ $(m = 0, 1, 2, 3, 4)$, where N is the number of variables on which the approximate problem depends. ■

Remark 7.16. In order to solve (P_0) in an approximate manner, it is possible to use the following method of approximation and *decomposition*, the principle of which is due to J. Cea; this method of application is particularly simple for *exterior approximations* $(P_{0h})_{lm}$ $(l = 1, 2; m = 2, 4)$.

---

(1) When, of course, the equivalence result of Section 2.2 holds.

(2) The approximation of J thus obtained therefore coincides with that used for Example 2 in Section 7.5.

The principle is as follows:

With the functional J approximated by J_h^l $(l = 1, 2)$ (see 5.1.1, 5.1.2), the relations expressing membership of K_{0h}^m $(m = 2, 4)$ (see Section 5.2) are imposed at the points of the mesh Q_h (see Section 5.1.2 for the definition of Q_h) which belong to a mesh of interval $2h$ constructed on Q_h (see Figure 7.7) and hence in practice at approximately one point in four of Q_h.

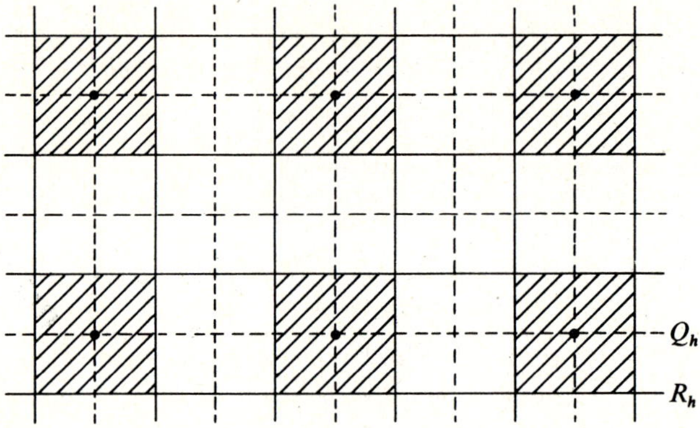

Fig. 7.7.

This method of treating the constraints implies that the relations expressing membership of the approximate convex set become *uncoupled* from each other, since they do not explicitly involve any common variable; this therefore leads to the variables being grouped in blocks of 4, and hence to the approximate problem being solved by a method of constrained under- or over-relaxation by blocks (of 4 variables), which is convergent for $0 < \omega < 2$, as a consequence of the aforementioned uncoupling of the constraints (see Chapter 2, Section 1.4, Remark 1.7 and Glowinski /4/, Section 7). ■

8. SOLUTION OF THE APPROXIMATE PROBLEMS RELATING TO (P_0) BY PENALTY METHODS

8.1 Synopsis

In this section we shall study the application of the *penalty* methods of Chapter 1, Section 3.2 and Chapter 2, Section 3, to the solution of (P_0) and of the finite-dimensional problems obtained by applying to (P_0) the *finite-element* approximation of Section 4.1 and the *exterior* approximations of Section 5.

We shall consider two methods based on penalisation, founded respectively on the following ideas:

(i) We *suppress*, by penalisation, the constraint $|\mathrm{grad}\, v| \leqslant 1$ a.e. or, at least, replace it by constraints which are *simpler* from a numerical point of view (see Section 8.2 below).

(ii) We decompose the problem (P_0) by *adding supplementary functions* and "*artificial constraints*" which are then *suppressed* by penalisation (see Section 8.3 below).

We consider it appropriate, before applying the above methods to the approximate problems, to demonstrate them on the continuous problem, which is simpler in formal terms. We shall then present a number of examples illustrating the results obtained by these methods. ■

8.2 A first penalty method

8.2.1 *Principle of the method*

We associate with J the *penalised functional* $J_\varepsilon : W_0^{1,4}(\Omega) \to \mathbb{R}$ defined by [1]:

$$(8.1) \qquad J_\varepsilon(v) = J(v) + \frac{1}{4\varepsilon}\int_\Omega \{(1 - |\mathrm{grad}\, v|^2)^-\}^2\, \mathrm{d}x\,.$$

We shall assume the following results (for the proofs, see for example, Lions /1/, Chapter 3, Section 5):

Proposition 8.1. *We have the following properties for* J_ε :

(i) J_ε *is continuous and strictly convex on* $W_0^{1,4}(\Omega)$,

(ii) $\lim_{\|v\|_{W_0^{1,4}(\Omega)} \to +\infty} J_\varepsilon(v) = +\infty$,

(iii) J_ε *is Gateaux differentiable over* $W_0^{1,4}(\Omega)$, *its gradient* J'_ε *being defined by*:

$$(8.2) \qquad (J'_\varepsilon(v), w) = a(v, w) - \int_\Omega fw\, \mathrm{d}x + \frac{1}{\varepsilon}\int_\Omega (1 - |\mathrm{grad}\, v|^2)^- \times$$
$$\times\, \mathrm{grad}\, v . \mathrm{grad}\, w\, \mathrm{d}x\,, \quad \forall w \in W_0^{1,4}(\Omega)\,. \quad ■$$

In (8.2), the inner product of $J'_\varepsilon(v)$ with w is that of the duality between $W^{-1,4/3}(\Omega)$ (i.e. the dual of $W_0^{1,4}(\Omega)$), and $W_0^{1,4}(\Omega)$.

---

(1) We recall that

$$v \in K_0 \Rightarrow v \in W_0^{1,\infty}(\Omega) = \left[v \mid v, \frac{\partial v}{\partial x_i} \in L^\infty(\Omega),\ v|_\Gamma = 0\right],$$
$$W_0^{1,\infty}(\Omega) \subset W_0^{1,4}(\Omega)\,.$$

Proposition 8.2. *The penalised problem* (P_ε) *associated with* (P_0) :

$$(8.3) \qquad (P_\varepsilon) \quad \operatorname*{Min}_{v \in W_0^{1,4}(\Omega)} J_\varepsilon(v) \,,$$

admits one and only one solution, u_ε *, characterised by:*

$$(8.4) \qquad \begin{cases} a(u_\varepsilon, v) + \dfrac{1}{\varepsilon}\displaystyle\int_\Omega (1 - |\operatorname{grad} u_\varepsilon|^2)^- \operatorname{grad} u_\varepsilon . \operatorname{grad} v \, dx = \\ \qquad = \displaystyle\int_\Omega fv \, dx \quad \forall v \in W_0^{1,4}(\Omega) \,, \\ u_\varepsilon \in W_0^{1,4}(\Omega) \,. \quad \blacksquare \end{cases}$$

Remark 8.1. We deduce from (8.4) that u_ε is the unique solution, in $W_0^{1,4}(\Omega)$, of the monotone nonlinear elliptic problem:

$$(8.5) \qquad \begin{cases} - \Delta u_\varepsilon - \dfrac{1}{\varepsilon} \displaystyle\sum_{k=1}^{2} \frac{\partial}{\partial x_k} \left((1 - |\operatorname{grad} u_\varepsilon|^2)^- \frac{\partial u_\varepsilon}{\partial x_k} \right) = f \,, \\ u_\varepsilon|_\Gamma = 0 \,. \quad \blacksquare \end{cases}$$

Remark 8.2. It may be shown that the nonlinear operator β :

$$W_0^{1,4}(\Omega) \to W^{-1,4/3}(\Omega)$$

defined by:

$$(8.6) \qquad (\beta(v), w) = \int_\Omega (1 - |\operatorname{grad} v|^2)^- \operatorname{grad} v . \operatorname{grad} w \, dx$$

$$\forall v, w \in W_0^{1,4}(\Omega)$$

is monotone (being the gradient of a convex function), locally Lipschitz, and $\beta(v) = 0 \ \forall v \in K_0$; β is thus a *penalisation operator* associated with K_0, in the sense of Chapter 1, Section 3.2. $\blacksquare$

In connection with the behaviour of u_ε as $\varepsilon \to 0$, we have:

Theorem 8.1. *When* $\varepsilon \to 0$ *we have*:

$$(8.7) \qquad u_\varepsilon \to u \qquad \text{strongly in} \quad H_0^1(\Omega)$$

where u *is the solution of* (P_0) .

Proof. This is a variant of Theorem 3.1, Chapter 1, Section 3.2:

For u a solution of (P_0) , we have $(1 - |\operatorname{grad} u|^2)^- = 0$ so that:

$$(8.8) \qquad J_\varepsilon(u_\varepsilon) \leqslant J_\varepsilon(u) \leqslant J(u) \quad \forall \varepsilon > 0 \,.$$

We thus have:

$$J(u_\varepsilon) \leqslant J(u) \quad \forall \varepsilon > 0\,, \tag{8.9}$$

so that:

$$\| u_\varepsilon \|_{H_0^1(\Omega)} \leqslant C \quad \forall \varepsilon > 0\,. \tag{8.10}$$

There therefore exists $u^* \in H_0^1(\Omega)$ and a subsequence, also denoted by u_ε, such that:

$$\lim_{\varepsilon \to 0} u_\varepsilon = u^* \quad \text{weakly in } H_0^1(\Omega)\,. \tag{8.11}$$

Relation (8.8) implies that:

$$\int_\Omega \{ (1 - |\operatorname{grad} u_\varepsilon|^2)^- \}^2 \, dx \leqslant 4\,\varepsilon(J(u) - J(u_\varepsilon)) \tag{8.12}$$

so that, taking account of (8.10),

$$\| u_\varepsilon \|_{W_0^{1,4}(\Omega)} \leqslant C\,, \tag{8.13}$$

$$\lim_{\varepsilon \to 0} \int_\Omega \{ (1 - |\operatorname{grad} u_\varepsilon|^2)^- \}^2 \, dx = 0\,. \tag{8.14}$$

For the above subsequence we thus have:

$$\lim_{\varepsilon \to 0} u_\varepsilon = u^* \quad \text{weakly in } W_0^{1,4}(\Omega)\,. \tag{8.15}$$

From (8.9), (8.11), (8.14), (8.15) we deduce by weak lower semi-continuity:

$$J(u^*) \leqslant J(u)\,, \tag{8.16}$$

$$\int_\Omega \{ (1 - |\operatorname{grad} u^*|^2)^- \}^2 \, dx = 0\,. \tag{8.17}$$

Relation (8.17) implies that $u^* \in K_0$; we thus have, from (8.16) and the uniqueness of u, $u^* = u$.

We deduce from (8.9), (8.11) that for the above subsequence we have:

$$J(u) \leqslant \liminf J(u_\varepsilon) \leqslant \limsup J(u_\varepsilon) \leqslant J(u)\,,$$

so that:

$$\lim_{\varepsilon \to 0} J(u_\varepsilon) = J(u)\,, \tag{8.18}$$

which implies:

(8.19) $$\lim_{\varepsilon \to 0} \int_\Omega |\operatorname{grad} u_\varepsilon|^2 \, dx = \int_\Omega |\operatorname{grad} u|^2 \, dx .$$

There is thus convergence of the norm, and hence strong convergence since we already have weak convergence. In view of the uniqueness of u, we have $\lim_{\varepsilon \to 0} u_\varepsilon = u$ without restricting ourselves to a subsequence. ■

8.2.2 *A variant*

The penalty method of Section 8.2.1 introduces a *truncation* which has the effect of limiting the differentiability of the functional J_ε to be minimised; in order to get round this difficulty we introduce a slack function $\geqslant 0$ (see Remark 3.6, Chapter 2, Section 3.3); in fact, there is equivalence between:

(8.20) $$|\operatorname{grad} v| \leqslant 1 \quad \text{a.e.}$$

and

(8.21) $$|\operatorname{grad} v|^2 + q = 1 \quad \text{a.e.} \qquad q \geqslant 0 \quad \text{a.e.}$$

and hence we obtain the problem (Π_ε), a variant of (P_ε), by penalising relation (8.21)

(8.22) $$(\Pi_\varepsilon) \quad \min_{(v,q) \in W_0^{1,4}(\Omega) \times L_+^2(\Omega)} \left[J(v) + \frac{1}{4\varepsilon} \int_\Omega (|\operatorname{grad} v|^2 + q - 1)^2 \, dx \right].$$

It may be noted that the penalised functional appearing in (8.22) *is not convex* with respect to the pair (v, q). In view of this we have:

Proposition 8.3. *The problem* (Π_ε) *admits a unique solution* $(u_\varepsilon, p_\varepsilon)$ *where* u_ε *is the solution of* (P_ε) *and where* p_ε *is defined by*:

(8.23) $$p_\varepsilon = \sup(0, 1 - |\operatorname{grad} u_\varepsilon|^2) = (1 - |\operatorname{grad} u_\varepsilon|^2)^+ .$$

Proof. We write

$$I_\varepsilon(v, q) = J(v) + \frac{1}{4\varepsilon} \int_\Omega (|\operatorname{grad} v|^2 + q - 1)^2 \, dx$$

and, given v, we define q_v by:

(8.24) $$q_v = \sup(0, 1 - |\operatorname{grad} v|^2) = (1 - |\operatorname{grad} v|^2)^+ .$$

We then have

(8.25) $$I_\varepsilon(v, q) \geqslant I_\varepsilon(v, q_v) = J_\varepsilon(v) \geqslant J_\varepsilon(u_\varepsilon) \quad \forall (v, q) \in W_0^{1,4}(\Omega) \times L_+^2(\Omega),$$

so that with (8.23)

$$(8.26) \qquad I_\varepsilon(v, q) \geqslant J_\varepsilon(u_\varepsilon) = I_\varepsilon(u_\varepsilon, p_\varepsilon) \quad \forall (v, q) \in W_0^{1,4}(\Omega) \times L_+^2(\Omega),$$

i.e. the result which was to be proved. ■

From Remark 8.1 and Proposition 8.3, it is easy to deduce:

Proposition 8.4. *The pair* $(u_\varepsilon, p_\varepsilon)$, *the solution of* (Π_ε), *is characterised by*:

$$(8.27) \qquad \begin{cases} -\Delta u_\varepsilon - \dfrac{1}{\varepsilon} \displaystyle\sum_{k=1}^{2} \dfrac{\partial}{\partial x_k} (|\operatorname{grad} u_\varepsilon|^2 + p_\varepsilon - 1) \dfrac{\partial u_\varepsilon}{\partial x_k} = f, \\ u_\varepsilon|_\Gamma = 0, \\ p_\varepsilon = \sup(0, 1 - |\operatorname{grad} u_\varepsilon|^2). \end{cases}$$

8.2.3 *Application to the solution of the approximate problems (I). Formulation of the penalised approximate problem.*

Preliminary remark. There is (in principle) no difficulty in applying the penalty method of Sections 8.2.1, 8.2.2 to the various approximations of (P_0) defined in Sections 4 and 5. We shall confine our attention to the approximation $(P_{0h})_{13}$ of (P_0) (see Sections 5.1, 5.2, 5.3), this being the only case which has given rise to numerical applications; the method used is the finite-dimensional analogue of that in Section 8.2.2. ■

Penalisation of the problem $(P_{0h})_{13}$.

We define the space L_h and the convex set L_{h+} by:

$$(8.28) \qquad L_h = \{ q_h \mid q_h = \{ q_{ij} \}_{M_{ij} \in \bar{\Omega}_h}, \quad q_{ij} \in \mathbb{R} \},$$

$$(8.29) \qquad L_{h+} = \{ q_h \mid q_h \in L_h, \quad q_{ij} \geqslant 0 \quad \forall M_{ij} \in \bar{\Omega}_h \},$$

and the functional $I_{\varepsilon h} : V_h \times L_h \to \mathbb{R}$ associated with $(P_{0h})_{13}$ by

$$(8.30)\,(^1) \qquad I_{\varepsilon h}(v_h, q_h) = J_h^1(v_h) + \frac{h^2}{4\varepsilon} \sum_{M_{ij} \in \Omega_h} \alpha_{ij} \, Q_{ij}^2(v_h, q_h),$$

with:

$$(8.31) \qquad \begin{cases} Q_{ij}(v_h, q_h) = \dfrac{1}{2} \beta_{ij} \left(\left| \dfrac{v_{i+1j} - v_{ij}}{h} \right|^2 + \left| \dfrac{v_{i-1j} - v_{ij}}{h} \right|^2 + \left| \dfrac{v_{ij+1} - v_{ij}}{h} \right|^2 + \right. \\ \qquad \left. + \left| \dfrac{v_{ij-1} - v_{ij}}{h} \right|^2 \right) + q_{ij} - 1 \end{cases}$$

---

(1) See Section 5.1.1 for the definition of J_h^1.

where (see Section 5.2, Figures 5.2, 5.3, 5.4):

$$(8.32)\quad \begin{cases} \beta_{ij} = 1 \text{ if } M_{ij} \in \Omega_h , \\ \beta_{ij} = 2 \text{ if } M_{ij} \in \Gamma_h \text{ with a single neighbour in } \Omega_h , \\ \beta_{ij} = \frac{4}{3} \text{ if } M_{ij} \in \Gamma_h \text{ with two neighbours in } \Omega_h , \end{cases}$$

(it is appropriate to take $v_{pq} = 0$ in (8.31) if $M_{pq} \notin \Omega_h$).

Whilst it is possible to take all the α_{ij} equal to unity, a more natural choice is:

$$(8.33)\quad \alpha_{ij} = \frac{1}{\beta_{ij}} \quad \forall M_{ij} \in \bar{\Omega}_h .$$

We then define the problem $(\Pi_{\varepsilon h})$, the discrete analogue of (Π_ε), by:

$$(8.34)\quad (\Pi_{\varepsilon h}) \quad \underset{(v_h,\, q_h) \in V_h \times L_{h+}}{\text{Min}} I_{\varepsilon h}(v_h, q_h) . \quad \blacksquare$$

Writing:

$$(8.35)\quad \begin{cases} G_{ij}(v_h) = \left\{ \frac{1}{2} \left(\left| \frac{v_{i+1j} - v_{ij}}{h} \right|^2 + \left| \frac{v_{i-1j} - v_{ij}}{h} \right|^2 + \left| \frac{v_{ij+1} - v_{ij}}{h} \right|^2 + \right.\right. \\ \qquad \left.\left. + \left| \frac{v_{ij-1} - v_{ij}}{h} \right|^2 \right) \right\}^{1/2} \end{cases}$$

and proceeding as in Sections 8.2.1, 8.2.2, it is easy to prove:

<u>Proposition 8.5</u>. *The problem* $(\Pi_{\varepsilon h})$ *admits one and only one solution*, $(u_{h\varepsilon}, p_{h\varepsilon})$, *in* $V_h \times L_{h+}$ *and when* $\varepsilon \to 0$, $u_{h\varepsilon} \to u_h$, *where* u_h *is the solution of* $(P_{0h})_{13}$. ■

<u>Proposition 8.6</u>. *The pair* $(u_{h\varepsilon}, p_{h\varepsilon})$ *which is the unique solution of* $(\Pi_{\varepsilon h})$ *is characterised by*:

$$(8.36)\quad \begin{cases} \dfrac{\partial I_{\varepsilon h}}{\partial v_{ij}} (u_{h\varepsilon}, p_{h\varepsilon}) = 0 \quad \forall M_{ij} \in \Omega_h , \\ p_{\varepsilon ij} = (1 - \beta_{ij}\, G_{ij}^2(u_h))^+ = \text{Max}\,(0, 1 - \beta_{ij}\, G_{ij}^2(u_{h\varepsilon})) \quad \forall M_{ij} \in \bar{\Omega}_h, \end{cases}$$

where $p_{h\varepsilon} = \{ p_{\varepsilon ij} \}_{M_{ij} \in \bar{\Omega}_h}$. ■

Remark 8.3. The results of Proposition 8.5 are still valid if for the α_{ij} we take *strictly positive* scalars, which may be chosen arbitrarily. ■

Remark 8.4. System (8.36) is the discrete analogue of the system (8.27). ■

8.2.4 *Application to the solution of the approximate problem (II). Solution of the penalised approximate problem.*

In view of Proposition 8.6 it is sufficient, in order to solve $(\Pi_{\varepsilon h})$, to solve the (nonlinear) system (8.36); for this, we shall use a variant of the method of *point over-relaxation with projection.*

In the following (by way of simplification) we shall write $u_{h\varepsilon} = (u_{ij})_{M_{ij} \in \Omega_h}$, $p_{h\varepsilon} = (p_{ij})_{M_{ij} \in \bar{\Omega}_h}$ and since $G_{ij}(v_h)$ explicitly involves, in addition to M_{ij}, only the four immediate neighbours of M_{ij}, we shall write:

$$G_{ij}(v_h) = G_{ij}(v_{i-1j}, v_{ij-1}, v_{ij}, v_{i+1j}, v_{ij+1}).$$

It is then possible to put in explicit form the relation $\frac{\partial I_{\varepsilon h}}{\partial v_{ij}}(u_{h\varepsilon}, p_{h\varepsilon}) = 0$; assuming the α_{ij} to be given by (8.33), we have:

$$(8.37)\quad \begin{cases} -\varepsilon\left(\dfrac{u_{i+1j} + u_{i-1j} + u_{ij+1} + u_{ij-1} - 4u_{ij}}{h^2} - f_{ij}\right) + \\ \quad + \dfrac{1}{2}[\beta_{ij}\, G_{ij}^2(u_{i-1j}, u_{ij-1}, u_{ij}, u_{i+1j}, u_{ij+1}) + p_{ij} - 1] \times \\ \quad \times \left(\dfrac{4u_{ij} - u_{i+1j} - u_{i-1j} - u_{ij+1} - u_{ij-1}}{h^2}\right) + \\ \quad + \dfrac{1}{2}[\beta_{i-1j}\, G_{i-1j}^2(u_{i-2j}, u_{i-1j-1}, u_{i-1j}, u_{ij}, u_{i-1j+1}) + \\ \quad + p_{i-1j} - 1]\left(\dfrac{u_{ij} - u_{i-1j}}{h^2}\right) + \\ \quad + \dfrac{1}{2}[\beta_{ij-1}\, G_{ij-1}^2(u_{i-1j-1}, u_{ij-2}, u_{ij-1}, u_{i+1j-1}, u_{ij}) + \\ \quad + p_{ij-1} - 1]\left(\dfrac{u_{ij} - u_{ij-1}}{h^2}\right) + \\ \quad + \dfrac{1}{2}[\beta_{i+1j}\, G_{i+1j}^2(u_{ij}, u_{i+1j-1}, u_{i+1j}, u_{i+2j}, u_{i+1j+1}) + \\ \quad + p_{i+1j} - 1]\left(\dfrac{u_{ij} - u_{i+1j}}{h^2}\right) + \\ \quad + \dfrac{1}{2}[\beta_{ij+1}\, G_{ij+1}^2(u_{i-1j+1}, u_{ij}, u_{ij+1}, u_{i+1j+1}, u_{ij+2}) + \\ \quad + p_{ij+1} - 1]\left(\dfrac{u_{ij} - u_{ij+1}}{h^2}\right) = 0, \end{cases}$$

with $u_{pq} = 0$ if $M_{pq} \notin \Omega_h$. We denote the left-hand side of (8.37) by

$$F_{\varepsilon ij}(u_{i-2j}, u_{i-1j-1}, u_{i-1j}, u_{i-1j+1}, u_{ij-2}, u_{ij-1}, u_{ij}, u_{ij+1}, u_{ij+2},$$
$$, u_{i+1j-1}, u_{i+1j}, u_{i+1j+1}, u_{i+2j}; p_{ij}, p_{i-1j}, p_{ij-1}, p_{i+1j}, p_{ij+1}),$$

and it may be seen that (8.37) corresponds to the discretisation, at $M_{ij} \in \Omega_h$, of the first relation in (8.27) by means of a 13-point scheme, as shown in Figure 8.1:

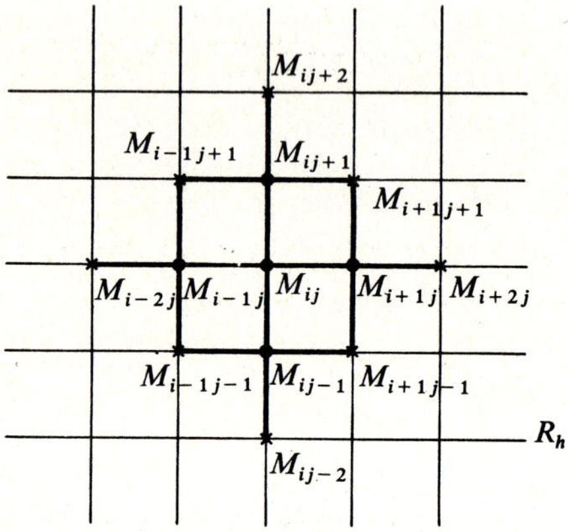

Fig. 8.1.

Description of the over-relaxation algorithm

In the following, it is assumed that the iterations are performed with increasing i, and then with i fixed and increasing j. We thus use the algorithm:

(8.38) $\quad u_h^0, p_h^0$ given arbitrarily,

then, u_h^n, p_h^n being assumed known, the coordinates u_{ij}^{n+1} of u_h^{n+1} are determined using:

$$(8.39)\quad \begin{cases} F_{\varepsilon ij}(u_{i-2j}^{n+1}, u_{i-1j-1}^{n+1}, u_{i-1j}^{n+1}, u_{i-1j+1}^{n+1}, u_{ij-2}^{n+1}, u_{ij-1}^{n+1}, \boxed{u_{ij}^{n+1}}, \\ \quad u_{ij+1}^n, u_{ij+2}^n, u_{i+1j-1}^n, u_{i+1j}^n, u_{i+1j+1}^n, u_{i+2j}^n; \\ \quad p_{ij}^n, p_{i+1j}^n, p_{i-1j}^n, p_{ij-1}^n, p_{ij+1}^n) = \\ = (1-\omega) F_{\varepsilon ij}(u_{i-2j}^{n+1}, u_{i-1j-1}^{n+1}, u_{i-1j}^{n+1}, u_{i-1j+1}^{n+1}, u_{ij-2}^{n+1}, u_{ij-1}^{n+1}, \\ \quad \boxed{u_{ij}^n}, u_{ij+1}^n, u_{ij+2}^n, u_{i+1j-1}^n, u_{i+1j}^n, u_{i+1j+1}^n, u_{i+2j}^n; \\ \quad p_{ij}^n, p_{i+1j}^n, p_{i-1j}^n, p_{ij-1}^n, p_{ij+1}^n) \quad \forall M_{ij} \in \Omega_h, \end{cases}$$

and having calculated u_h^{n+1}, p_h^{n+1} is determined from:

$$(8.40)\quad p_{ij}^{n+1} = \mathrm{Max}\,(0, 1 - \beta_{ij}\, G_{ij}^2(u_h^{n+1})) \quad \forall M_{ij} \in \bar{\Omega}_h .$$

It may be noted that in (8.39) u_{ij}^{n+1} is the solution of an equation of the third *degree* (with a single variable). ■

8.2.5 *Application to the solution of the approximate problems (III). Examples.*

Example 1. We shall consider again the example of Section 7.5; we thus have $\Omega =]0,1[\times]0,1[$ and $f = 10$.

Numerical values of the parameters:

Mesh size: $h = 1/20$

Penalisation parameter: $\varepsilon = 0.625 \times 10^{-3}$

Initialisation of algorithm (8.38) - (8.40): $u_h^0 = 0,\ p_h^0 = 0$.

Termination criterion for algorithm (8.38) - (8.40):

$$\sum_{M_{ij} \in \Omega_h} | u_{ij}^{n+1} - u_{ij}^n | \leqslant 10^{-6} .$$

Solution of (8.39):

The 3rd-degree equation determining u_{ij}^{n+1} from (8.39) was solved by the Newton-Raphson method initialised with u_{ij}^n ; this is a more rapid process than solution by Cardan's method.

Analysis of the numerical results:

For $\omega = 1$ (resp. $\omega = 1.7$) convergence is reached in 391 iterations (resp. 186 iterations) corresponding to an execution time of 10 sec. (resp. 6.2 sec.) on an IBM 360/91. ■

Example 2. We again consider Example 3 of Section 7.6; we thus have $\Omega =]0,1[\times]0,1[$ and f is given by:

$$f(x) = \begin{cases} 10 & \text{if } (x_1, x_2) \in]0, \tfrac{1}{2}[\times]\tfrac{1}{2}, 1[, \\ -10 & \text{if } (x_1, x_2) \in]\tfrac{1}{2}, 1[\times]0, \tfrac{1}{2}[, \\ 0 & \text{if } (x_1, x_2) \in]0, \tfrac{1}{2}[\times]0, \tfrac{1}{2}[\cup]\tfrac{1}{2}, 1[\times]\tfrac{1}{2}, 1[. \end{cases} \tag{8.41}$$

Once again, we take $h = 1/20$, $\varepsilon = 0.625 \times 10^{-3}$, $u_h^0 = 0$, $p_h^0 = 0$ and the same termination criterion for algorithm (8.38) - (8.40) as in the preceding example.

For $\omega = 1$, convergence is then reached in 560 iterations, corresponding to an execution time of 14.5 sec. on an IBM 360/91, with u_{ij}^{n+1} calculated, as above, by the Newton-Raphson method. ■

Remark 8.5. The approximate solutions thus calculated coincide in both of the above examples (to within better than 1%) with the solutions of the same approximate problems, calculated by means of the other methods described in this chapter. ■

Remark 8.6. The rate of convergence may be improved by varying ε; a relatively large value is used to start with, and ε is then reduced in the course of the iterations of algorithm (8.38) - (8.40) (see Section 8.3.5 below). ■

Remark 8.7. Variants of $(\Pi_{\varepsilon h})$ may be defined using the approximation J_h^2 of J (see Section 5.1.2) and/or the approximations K_{0h}^m $(m = 1, 2, 4)$ of K_0 (see Section 5.2); the use of K_{0h}^m $(m = 1, 2, 4)$ would lead to a variant of (8.37) explicitly involving the values of u_{pq} corresponding to the 9 points shown on Figure 8.2 and 12 (resp. 4) slack variables for $m = 1$ (resp. $m = 2, 4$).

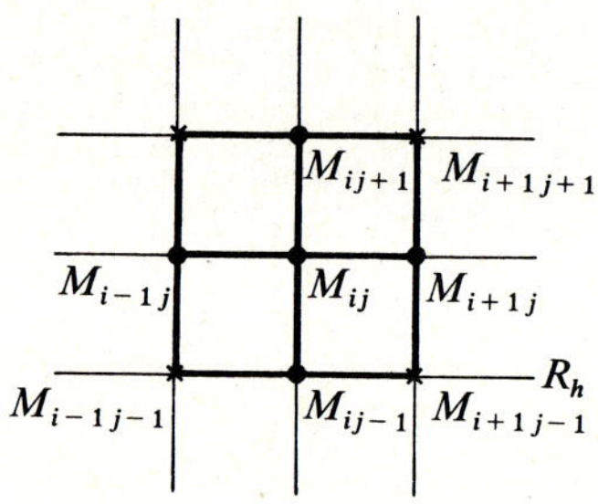

Fig. 8.2.

We recall (see Section 5.2) that for $m = 2, 3, 4$ the number of constraints is approximately equal to the number of variables u_{ij}, so that there should be approximately the same number of slack variables as of variables u_{ij}; conversely for $m = 1$ there would be approximately four times as many slack variables as variables u_{ij}. ■

8.3 A second penalty method

8.3.1 *Principle of the method*

We write:

$$\Lambda = \{ q \mid q \in L^2(\Omega) \times L^2(\Omega),\ q = (q_1, q_2),\ q_1^2(x) + q_2^2(x) \leqslant 1 \text{ a.e.} \}. \tag{8.42}$$

There is then equivalence between:

$$(P_0)\ \operatorname*{Min}_{v \in K_0} J(v),$$

and

$$(\Pi_0)_1 \begin{cases} \operatorname*{Min}_{(v,q) \in \mathscr{K}_0} J(v), \\ \mathscr{K}_0 = \left\{ (v, q) \mid v \in H_0^1(\Omega),\ q \in \Lambda,\ q_1 = \dfrac{\partial v}{\partial x_1},\ q_2 = \dfrac{\partial v}{\partial x_2} \right\}, \end{cases} \tag{8.43}$$

or indeed:

$$(8.44) \qquad (\Pi_0)_2 \quad \underset{(v,q)\in\mathscr{K}_0}{\text{Min}} \left[\frac{1}{2}\int_\Omega (q_1^2 + q_2^2)\,dx - \int_\Omega fv\,dx \right].$$

We shall penalise only the relation:

$$(8.45) \qquad q = \text{grad}\, v\,,$$

so that, for $(\Pi_0)_1$,

$$(8.46) \qquad (\Pi_\varepsilon)_1 \quad \underset{(v,q)\in H_0^1(\Omega)\times\Lambda}{\text{Min}} \left[J(v) + \frac{1}{2\,\varepsilon}\int_\Omega \left| q_1 - \frac{\partial v}{\partial x_1}\right|^2 dx + \frac{1}{2\,\varepsilon}\int_\Omega \left| q_2 - \frac{\partial v}{\partial x_2}\right|^2 dx \right],$$

and, for $(\Pi_0)_2$,

$$(8.47) \qquad (\Pi_\varepsilon)_2 \begin{cases} \underset{(v,q)\in H_0^1(\Omega)\times\Lambda}{\text{Min}} \left[\frac{1}{2}\int_\Omega (q_1^2 + q_2^2)\,dx - \int_\Omega fv\,dx + \frac{1}{2\,\varepsilon} \times \right. \\ \left. \times \int_\Omega \left| q_1 - \frac{\partial v}{\partial x_1}\right|^2 dx + \frac{1}{2\,\varepsilon}\int_\Omega \left| q_2 - \frac{\partial v}{\partial x_2}\right|^2 dx \right]. \end{cases}$$

We denote by $\mathscr{J}_{1\varepsilon}$ (resp. $\mathscr{J}_{2\varepsilon}$) the functional defined by (8.46) (resp. (8.47)) and $\mathscr{V} = H_0^1(\Omega) \times (L^2(\Omega))^2$; the following propositions may then easily be proved:

<u>Proposition 8.6</u>. *We have the following properties for* $\mathscr{J}_{k\varepsilon}$ $(k = 1,\ 2)$:

(i) $\mathscr{J}_{k\varepsilon}$ *is continuous and strictly convex on* $\mathscr{V}$,

(ii) $\lim\limits_{\|(v,q)\|_{\mathscr{V}} \to +\infty} \mathscr{J}_{k\varepsilon}(v, q) = +\infty$,

(iii) $\mathscr{J}_{k\varepsilon}$ *is Gateaux differentiable (and even* C^∞ *in the Frechet sense) on* $\mathscr{V}$. ■

<u>Proposition 8.7</u>. *The penalised problems* $(\Pi_\varepsilon)_k$ $(k = 1,\ 2)$ *each admit one and only one solution on* $H_0^1(\Omega) \times \Lambda$, *say* $(u_\varepsilon^k, p_\varepsilon^k)$, *of which the respective characterisations are:*

for $k = 1$

$$(8.48)\quad \begin{cases} -(1+\varepsilon)\,\Delta u_\varepsilon^1 + \operatorname{div} p_\varepsilon^1 = \varepsilon f \\ u_\varepsilon^1 |_\Gamma = 0 \\ \displaystyle\int_\Omega (p_\varepsilon^1 - \operatorname{grad} u_\varepsilon^1).(q - p_\varepsilon^1)\,dx \geqslant 0 \quad \forall q \in \Lambda \\ p_\varepsilon^1 \in \Lambda, \end{cases}$$

and, for $k = 2$,

$$(8.49)\quad \begin{cases} -\Delta u_\varepsilon^2 + \operatorname{div} p_\varepsilon^2 = \varepsilon f \\ u_\varepsilon^2 |_\Gamma = 0 \\ \displaystyle\int_\Omega \left(p_\varepsilon^2 - \frac{\operatorname{grad} u_\varepsilon^2}{1+\varepsilon}\right).(q - p_\varepsilon^2)\,dx \geqslant 0 \quad \forall q \in \Lambda \\ p_\varepsilon^2 \in \Lambda. \quad \blacksquare \end{cases}$$

Remark 8.8. There is equivalence between the last two relations of (8.48), (8.49) and, respectively:

$$(8.50)\qquad p_\varepsilon^1 = P_\Lambda(\operatorname{grad} u_\varepsilon^1),$$

$$(8.51)\qquad p_\varepsilon^2 = P_\Lambda\left(\frac{\operatorname{grad} u_\varepsilon^2}{1+\varepsilon}\right),$$

where P_Λ denotes the orthogonal projection operator from $L^2(\Omega) \times L^2(\Omega) \to \Lambda$ related to the standard norm of $L^2(\Omega) \times L^2(\Omega)$; explicitly, we have:

$$(8.52)\qquad P_\Lambda(q) = \frac{q}{\sup(1, |q|)} \quad \forall q \in L^2(\Omega) \times L^2(\Omega),$$

with

$$|q| = (q_1^2 + q_2^2)^{1/2}. \quad \blacksquare$$

The use of penalisation is justified by:

Theorem 8.2. *When* $\varepsilon \to 0$ *we have*, $\forall k = 1, 2$:

$$(8.53)\qquad u_\varepsilon^k \to u \quad \textit{strongly in } H_0^1(\Omega)$$

$$(8.54)\qquad p_\varepsilon^k \to \operatorname{grad} u \quad \textit{strongly in } L^2(\Omega) \times L^2(\Omega)$$

where u *is the solution of* (P_0).

Proof This is a variant of the proof of Theorem 3.1, of Chapter 1, Section 3.2; we shall consider only the case $k = 1$, the case $k = 2$ being treated in a similar manner.

Let u be the solution of (P_0) and $p = \text{grad } u$; we have $(u, p) \in H_0^1(\Omega) \times \Lambda$ and:

(8.55) $$J(u_\varepsilon^1) \leqslant \mathscr{I}_{1\varepsilon}(u_\varepsilon^1, p_\varepsilon^1) < \mathscr{I}_{1\varepsilon}(u, p) = J(u) \quad \forall \varepsilon > 0 .$$

We thus have

(8.56) $$J(u_\varepsilon^1) \leqslant J(u) \quad \forall \varepsilon > 0$$

so that

(8.57) $$\| u_\varepsilon^1 \|_{H_0^1(\Omega)} \leqslant C \quad \forall \varepsilon > 0 .$$

Moreover, $p_\varepsilon^1 \in \Lambda \Rightarrow p_\varepsilon^1$ bounded in $L^2(\Omega) \times L^2(\Omega)$, so that there exists a subsequence, also denoted by $(u_\varepsilon^1, p_\varepsilon^1)$, $u^* \in H_0^1(\Omega)$ and $p^* \in L^2(\Omega) \times L^2(\Omega)$ such that when $\varepsilon \to 0$:

(8.58) $$u_\varepsilon^1 \to u^* \quad \textit{weakly} \text{ in } H_0^1(\Omega)$$

(8.59) $$p_\varepsilon^1 \to p^* \quad \textit{weakly} \text{ in } L^2(\Omega) \times L^2(\Omega)$$ [1]

and since Λ is convex and closed in $L^2(\Omega) \times L^2(\Omega)$ (and thus weakly closed) we have:

(8.60) $$p^* \in \Lambda .$$

Relation (8.55) implies that:

(8.61) $$\int_\Omega \left| p_{1\varepsilon}^1 - \frac{\partial u_\varepsilon^1}{\partial x_1} \right|^2 \mathrm{d}x + \int_\Omega \left| p_{2\varepsilon}^1 - \frac{\partial u_\varepsilon^1}{\partial x_2} \right|^2 \mathrm{d}x \leqslant 2\, \varepsilon (J(u) - J(u_\varepsilon^1)) ,$$

so that, taking account of (8.57):

(8.62) $$\lim_{\varepsilon \to 0} \| p_\varepsilon^1 - \text{grad } u_\varepsilon^1 \|_{L^2(\Omega) \times L^2(\Omega)} = 0 .$$

From (8.56), (8.58), (8.59), (8.62) we deduce by (weak) lower semi-continuity

(8.63) $$J(u^*) \leqslant J(u) ,$$

(8.64) $$p^* = \text{grad } u^* ,$$

and (8.60) implies $u^* \in K_0$. We thus have $u^* = u$.

The strong convergence of u_ε^1 and the convergence of the entire $(u_\varepsilon^1)_\varepsilon$ family are proved by proceeding as in the proof of

---

(1) In fact we have $p_\varepsilon^1 \to p^*$ weakly* in $L^\infty(\Omega) \times L^\infty(\Omega)$

Theorem 8.1, the strong convergence of p_ε^1 resulting from that of u_ε^1 and from (8.62). ■

8.3.2 *Application to the solution of approximate problems (I). Formulation of the penalised approximate problem in the case of the finite element method of Section 4.1.*

We shall confine our attention to the finite-dimensional analogue of $(\Pi_\varepsilon)_1$, the other case being treated in a similar manner (by way of simplification, the index 1 will be suppressed in the following); the notation of Section 4 will be retained.

Having defined the space V_h in Section 4.1.1, we introduce the subspace L_h of $L^2(\Omega) \times L^2(\Omega)$:

$$(8.65) \qquad L_h = \left\{ q = (q_1, q_2) \mid q \in L^2(\Omega) \times L^2(\Omega), q_1 = \sum_{T \in \mathcal{T}_h} q_{1T}\,\theta_T, q_2 = \sum_{T \in \mathcal{T}_h} q_{2T}\,\theta_T ;\; q_{1T}, q_{2T} \in \mathbb{R} \right\}.$$

where

$$(8.66) \qquad \theta_T = \text{ characteristic function of } T,$$

then

$$(8.67) \qquad \Lambda_h = L_h \cap \Lambda .$$

It may be noted that:

$$(8.68) \qquad v_h \in V_h \Leftrightarrow \operatorname{grad} v_h \in L_h ,$$

$$(8.69) \qquad v_h \in V_h \cap K_0 \Leftrightarrow \operatorname{grad} v_h \in \Lambda_h ,$$

$$(8.70) \qquad q \in \Lambda_h \Leftrightarrow q_{1T}^2 + q_{2T}^2 \leqslant 1 \quad \forall T \in \mathcal{T}_h .$$

The finite-dimensional analogue of $(\Pi_\varepsilon)_1$ is then defined by:

$$(8.71) \qquad (\Pi_{\varepsilon h}) \quad \operatorname*{Min}_{(v_h, q_h) \in V_h \times \Lambda_h} \mathcal{J}_\varepsilon(q_h, v_h) ,$$

where $\mathcal{J}_\varepsilon$ is defined by (8.46); it is a straightforward matter to prove:

<u>Proposition 8.8</u>. *The problem* $(\Pi_{\varepsilon h})$ *admits one and only one solution,* $(u_{h\varepsilon}, p_{h\varepsilon})$ *, and when* $\varepsilon \to 0$ *we have*

$$(8.72) \qquad u_{h\varepsilon} \to u_h ,$$

(8.73) $$p_{h\varepsilon} \to \text{grad}\, u_h ,$$

where u_h *is the (unique) solution of*

$$\operatorname*{Min}_{v_h \in V_h \cap K_0} J(v_h) . \quad \blacksquare$$

Remark 8.9. The interior approximation method of Section 4.2 (using eigenfunctions of $-\Delta$ in $H_0^1(\Omega)$) is not suitable for the penalisation technique of Section 8.3, for the reasons indicated in Remark 4.6 of Section 4.2.5.

8.3.3 *Application to the solution of the approximate problems (II). Formulation of the penalised approximate problem in the case of the exterior approximations of Section 5.*

Of the exterior approximations of K_0 considered in Section 5.2, K_{0h}^4 is the only one ideally suited to the use of the penalty method described in Section 8.3 (for a discussion of the difficulties connected with the use of the other approximate convex sets of Section 5.2, see Marrocco /1/). Once again, we shall confine our attention to the finite-dimensional analogue of $(\Pi_\varepsilon)_1$ (see (8.46)) and we shall retain the notation of Section 5.

We recall that

$$V_h = \{ v_h \mid v_h = (v_{ij})_{M_{ij} \in \Omega_h} \} ,$$

and we introduce

(8.74) $$L_h = \{ q_h \mid q_h = (q_{1i+1/2\,j+1/2}, q_{2i+1/2\,j+1/2})_{M_{i+1/2j+1/2} \in \Sigma_h} \} ,$$

(8.75) $$\Lambda_h = \{ q_h \mid q_h \in L_h, q_{1i+1/2\,j+1/2}^2 + q_{2i+1/2\,j+1/2}^2 \leqslant 1 \quad \forall M_{i+1/2\,j+1/2} \in \Sigma_h \} .$$

We next define, for $l = 1,2$ the following approximations of $(\Pi_\varepsilon)_1$:

(8.76) $(\Pi_{\varepsilon h}^l)$
$$\begin{cases} \operatorname*{Min}_{(v_h,q_h) \in V_h \times \Lambda_h} \Bigg[J_h^l + \dfrac{h^2}{2\,\varepsilon} \displaystyle\sum_{M_{i+1/2j+1/2} \in \Sigma_h} \times \\ \quad \times \left(q_{1i+1/2\,j+1/2} - \dfrac{v_{i+ij+1} - v_{ij+1} + v_{i+1j} - v_{ij}}{2\,h} \right)^2 + \\ \quad + \dfrac{h^2}{2\,\varepsilon} \displaystyle\sum_{M_{i+1/2j+1/2} \in \Sigma_h} \times \\ \quad \times \left(q_{2i+1/2\,j+1/2} - \dfrac{v_{i+1j+1} - v_{i+1j} + v_{ij+1} - v_{ij}}{h} \right)^2 \Bigg], \end{cases}$$

and it is then easy to prove the following propositions:

Proposition 8.9. *The problem* $(\Pi_{\varepsilon h})$ *admits one and only one solution,* $(u_{h\varepsilon}, p_{h\varepsilon})$ *, and when* $\varepsilon \to 0$ *we have, for* $l = 1$ *or* 2

$$u_{h\varepsilon}^l \to u_h^l \tag{8.77}$$

where u_h^l *is the solution* (see (5.23)) *of*

$$(P_{0h})_{l_4} \quad \operatorname*{Min}_{v_h \in K_{0h}^4} J_h^l(v_h) .$$

Proposition 8.10. *The optimal pair* $(u_{h\varepsilon}, p_{h\varepsilon})$ *is characterised, putting* $u_{h\varepsilon} = (u_{ij})_{M_{ij} \in \Omega_h}$ *and* $p_{h\varepsilon} = (p_{i+1/2\,j+1/2})_{M_{i+1/2\,j+1/2} \in \Sigma_h}$, *by:*

$$(8.78)\quad \left\{ \begin{aligned} & -\varepsilon \frac{u_{i+1j} + u_{i-1j} + u_{ij+1} + u_{ij-1} - 4\,u_{ij}}{h^2} - \\ & - \frac{u_{i+1j+1} + u_{i-1j+1} + u_{i+1j-1} + u_{i-1j-1} - 4\,u_{ij}}{2\,h^2} + \\ & + \frac{p_{1i+1/2\,j+1/2} - p_{1i-1/2\,j+1/2} + p_{1i+1/2\,j-1/2} - p_{1i-1/2\,j-1/2}}{2\,h} + \\ & + \frac{p_{2i+1/2\,j+1/2} - p_{2i+1/2\,j-1/2} + p_{2i-1/2\,j+1/2} - p_{2i-1/2\,j-1/2}}{2\,h} = \\ & = \varepsilon f_{ij} \quad \forall M_{ij} \in \Omega_h \end{aligned} \right.$$

if $l = 1$,

$$(8.79)\quad \left\{ \begin{aligned} & -(1+\varepsilon) \frac{u_{i+1j+1} + u_{i-1j+1} + u_{i+1j-1} + u_{i-ij-1} - 4\,u_{ij}}{2\,h^2} + \\ & + \frac{p_{1i+1/2\,j+1/2} - p_{1i-1/2\,j+1/2} + p_{1i+1/2\,j-1/2} - p_{1i-1/2\,j-1/2}}{2\,h} + \\ & + \frac{p_{2i+1/2\,j+1/2} - p_{2i+1/2\,j-1/2} + p_{2i-1/2\,j+1/2} - p_{2i-1/2\,j-1/2}}{2\,h} = \\ & = \varepsilon f_{ij} \quad \forall M_{ij} \in \Omega_h \end{aligned} \right.$$

if $l = 2$,

$$(8.80)\quad \left\{ \begin{aligned} & \left(p_{1i+1/2\,j+1/2} - \frac{u_{i+1j+1} - u_{ij+1} + u_{i+1j} - u_{ij}}{2\,h} \right) \times \\ & \quad \times (q_{1i+1/2\,j+1/2} - p_{1i+1/2\,j+1/2}) + \\ & \quad + \left(p_{2i+1/2\,j+1/2} - \frac{u_{i+1j+1} - u_{i+1j} + u_{ij+1} - u_{ij}}{2\,h} \right) \times \\ & \quad \times (q_{2i+1/2\,j+1/2} - p_{2i+1/2\,j+1/2}) \geqslant 0 \\ & \quad \forall M_{i+1/2\,j+1/2} \in \Sigma_h , \quad \forall q_h \in \Lambda_h \\ & \quad p_{h\varepsilon}^l \in \Lambda_h \end{aligned} \right.$$

$\forall l = 1, 2.$ ■

Remark 8.9. Relations (8.78), (8.79), (8.80) are the discrete analogues of (8.48). ■

Remark 8.10. There is equivalence between (8.80) and:

$$(8.81) \quad \begin{cases} (p_{1i+1/2\,j+1/2}, p_{2i+1/2\,j+1/2}) = \\ \quad = P_{\overline{D}}\left(\dfrac{u_{i+1j+1} - u_{ij+1} + u_{i+1j} - u_{ij}}{2h}, \right. \\ \quad \left. , \dfrac{u_{i+1j+1} - u_{i+1j} + u_{ij+1} - u_{ij}}{2h}\right) \\ \quad \forall M_{i+1/2\,j+1/2} \in \Sigma_h \end{cases}$$

where

$$(8.82) \qquad \overline{D} = [y \mid y = (y_1, y_2) \in \mathbb{R}^2,\ y_1^2 + y_2^2 \leqslant 1],$$

and where $P_{\overline{D}}$ is the orthogonal projection operator from $\mathbb{R}^2 \to \overline{D}$, related to the Euclidian norm of $\mathbb{R}^2$, or:

$$(8.83) \qquad P_{\overline{D}}(y_1, y_2) = \left\{ \frac{y_1}{\operatorname{Max}(1, \sqrt{y_1^2 + y_2^2})}, \frac{y_2}{\operatorname{Max}(1, \sqrt{y_1^2 + y_2^2})} \right\}$$
$$\forall (y_1, y_2) \in \mathbb{R}^2 . \ ■$$

8.3.4 *Application to the solution of the approximate problems (III). Solution of the penalised approximate problem by over-relaxation with projection.*

We shall confine our attention to the exterior approximations of Section 8.3.3.

Preliminary remark: There is equivalence between:

$$(8.84) \qquad y = P_{\overline{D}}(z),$$

and

$$(8.85) \qquad y = P_{\overline{D}}(y - \rho(y - z)) \quad \forall \rho \geqslant 0 . \ ■$$

Description of the over-relaxation algorithm.

In the following, it will be assumed that the iterations are performed with increasing i, and then with i fixed and increasing j; we thus use the algorithm:

$$(8.86) \qquad u_h^0, p_h^0 \quad \text{given arbitrarily,}$$

then, u_h^n, p_h^n being assumed known, the coordinates u_{ij}^{n+1} of u_h^{n+1} are determined using:

$$(8.87)\quad \begin{cases} -\varepsilon \dfrac{u_{i+1j}^n + u_{i-1j}^{n+1} + u_{ij+1}^n + u_{ij-1}^{n+1} - \boxed{4\,\bar{u}_{ij}^{n+1}}}{h^2} - \\ -\dfrac{u_{i+1j+1}^n + u_{i-1j+1}^{n+1} + u_{i+1j-1}^n + u_{i-1j-1}^{n+1} - \boxed{4\,\bar{u}_{ij}^{n+1}}}{2\,h^2} + \\ +\dfrac{p_{1i+1/2\,j+1/2}^n - p_{1i-1/2\,j+1/2}^n + p_{1i+1/2\,j-1/2}^n - p_{1i-1/2\,j-1/2}^n}{2\,h} + \\ +\dfrac{p_{2i+1/2\,j+1/2}^n - p_{2i+1/2\,j-1/2}^n + p_{2i-1/2\,j+1/2}^n - p_{2i-1/2\,j-1/2}^n}{2\,h} = \\ = \varepsilon f_{ij}\,, \quad \forall M_{ij} \in \Omega_h \end{cases}$$

if $l = 1$,

$$(8.88)\quad \begin{cases} -(1+\varepsilon)\dfrac{u_{i+1j+1}^n + u_{i-1j+1}^{n+1} + u_{i+1j-1}^n + u_{i-1j-1}^{n+1} - \boxed{4\,\bar{u}_{ij}^{n+1}}}{2\,h^2} + \\ +\dfrac{p_{1i+1/2\,j+1/2}^n - p_{1i-1/2\,j+1/2}^n + p_{1i+1/2\,j+1/2}^n - p_{1i-1/2\,j-1/2}^n}{2\,h} + \\ +\dfrac{p_{2i+1/2\,j+1/2}^n - p_{2i+1/2\,j-1/2}^n + p_{2i-1/2\,j+1/2}^n - p_{2i-1/2\,j-1/2}^n}{2\,h} = \\ = \varepsilon f_{ij} \quad \forall M_{ij} \in \Omega_h \end{cases}$$

if $l = 2$, then

$$(8.89)\qquad u_{ij}^{n+1} = u_{ij}^n + \omega_1(\bar{u}_{ij}^{n+1} - u_{ij}^n) \quad \forall M_{ij} \in \Omega_h\,.$$

Having calculated u_h^{n+1} in this way, $p_h^{n+1} \in \Lambda_h$ is determined using:

$$(8.90)\quad \begin{cases} \bar{p}_{1i+1/2\,j+1/2}^{n+1} = (1-\omega_2)\,p_{1i+1/2\,j+1/2}^n + \omega_2 \times \\ \quad \times \left(\dfrac{u_{i+1j+1}^{n+1} - u_{ij+1}^{n+1} + u_{i+1j}^{n+1} - u_{ij}^{n+1}}{2\,h}\right), \\ \bar{p}_{2i+1/2\,j+1/2}^{n+1} = (1-\omega_2)\,p_{2i+1/2j+1/2}^n + \omega_2 \times \\ \quad \times \left(\dfrac{u_{i+1j+1}^{n+1} - u_{i+1j}^{n+1} + u_{ij+1}^{n+1} - u_{ij}^{n+1}}{2\,h}\right), \\ \forall M_{i+1/2\,j+1/2} \in \Sigma_h \end{cases}$$

and

$$(8.91)\qquad (p_{1i+1/2\,j+1/2}^{n+1}, p_{2i+1/2\,j+1/2}^{n+1}) = P_{\bar{D}}(\bar{p}_{1i+1/2\,j+1/2}^{n+1}, \bar{p}_{2i+1/2\,j+1/2}^{n+1}) \quad \forall M_{i+1/2\,j+1/2} \in \Sigma_h\,.$$

In (8.89), (8.91) we take $0 < \omega_1, \omega_2 < 2$. ■

The above algorithm is justified through the characterisation of Proposition 8.10, Remark 8.10 and the preliminary remark made earlier. ■

Convergence of the algorithm

By using a *block* variant of Theorem 1.3 of Chapter 2, Section 1.4 (see Glowinski /4/ Theorem 7.1, Section 7.4) the following proposition may be proved:

Proposition 8.10 *If* $0 < \omega_1 < 2$, $0 < \omega_2 < 2$ *there is convergence of the algorithms* (8.86), (8.87), (8.89), (8.90), (8.91) *and* (8.86), (8.88), (8.89), (8.90), (8.91) *or, when* $n \to +\infty$,

$$u_h^n \to u_{h\varepsilon}^l \quad \forall l = 1, 2 \tag{8.92}$$

$$p_h^n \to p_{h\varepsilon}^l \quad \forall l = 1, 2 . \tag{8.93}$$ ■

8.3.5 *Application to the solution of the approximate problems* (IV). *Examples.*

Example 1. We consider Example 2 of Section 7.5, which was treated earlier (using the other penalty method) in Section 8.2.5, Example 1; we thus have $\Omega =]0, 1[\times]0, 1[$ and $f = 10$.

Approximation of (P_0) : We use the approximation $(P_{0h})_{14}$ (see Section 5.3, Section 8.3.3).

Numerical values of the parameters:

Mesh size: $h = 1/20$

Initialisation of algorithm (8.86) (8.87) (8.89) (8.90) (8.91) : $u_h^0 = 0$, $p_h^0 = 0$.

Termination criterion:

$$\sum_{M_{ij} \in \Omega_h} | u_{ij}^{n+1} - u_{ij}^n | \leqslant \varepsilon_T .$$

Analysis of the numerical results:

With the value of ε *fixed* during the iterative process, the results shown in Table 8.1 were obtained.

The values of $u(0.5, 0.5)$ in Table 8.1 may be compared with the value 0.413 obtained by solving the problem (P_1) (see Section 3) associated with the preceding problem (P_0) using the method of Section 3.4 with $h = 1/40$.

Table 8.1 shows that for small values of ε convergence is very slow; the rate of convergence was improved by reducing ε and ε_T, in parallel, during the iterative process: more precisely, the process starts with $\varepsilon = 10^{-1}$ and $\varepsilon_T = 10^{-2}$ and ε and ε_T are then divided by 10 when

$$\sum_{M_{ij} \in \Omega_h} | u_{ij}^{n+1} - u_{ij}^n | \leqslant \varepsilon_T .$$

The improvement thus brought about, both in terms of the rate of convergence and the precision obtained, is clearly demonstrated by the results given in Table 8.2.

ε	ε_T	ω_1	ω_2	$u(0.5, 0.5)$	Number of iterations	Computation time in secs on CII 10070
10^{-1}	10^{-4}	1.6	1	0.455	183	30 s
10^{-2}	10^{-4}	1.8	1	0.417	848	120 s
10^{-3}	10^{-4}	1.8	1	Value at 3000th iteration: 0.376	Not converged at 3000th iteration	

Table 8.1 (Example 1)

ε	ε_T	ω_1	ω_2	$u(0.5, 0.5)$	Number of iterations	Computation time in secs on CII 10070
10^{-1}	10^{-2}	1.8	1		95	
10^{-2}	10^{-3}	1.8	1		269	
10^{-3}	10^{-4}	1.8	1	0.414	853	120 s

Table 8.2 (Example 1)

The number of iterations indicated in Table 8.2 relates to the overall process (i.e. the iteration "count" is not reset to zero in the course of the execution when convergence of the sub-problems is obtained).

Table 8.3 shows the number of iterations required for convergence, as a function of ω_1 and ε, when $\omega_2 = 1$, $u_h^0 = 0$, $p_h^0 = 0$, $\varepsilon_T = 10^{-4}$.

ω_1 \ ε	10^{-1}	10^{-2}
1.6	183	1 055
1.7	179	943
1.8	180	848
1.9	197	883

Table 8.3 (Example 1)

It may be seen from Table 8.3 that the number of iterations required for convergence is a relatively flat function of ω_1 over quite a wide range either side of the optimal value of ω_1. ■

Example 2. We consider again Example 3 of Section 7.6, which was treated earlier in Section 8.2.5, Example 2; we thus have $\Omega =]0, 1[\times]0, 1[$ and f defined by (8.41).

To solve the above problem numerically we use, as in the preceding example, the approximation $(P_{0h})_{14}$ of (P_0), with $h = 1/20$, the algorithm (8.86), (8.87), (8.89), (8.90), (8.91) being initialised by $u_h^0 = 0, p_h^0 = 0$.

Table 8.4 summarises the results obtained with ε fixed, and Table 8.5 those obtained by decreasing ε and ε_T in the course of the iterations, as in the preceding example:

ε	ε_T	ω_1	ω_2	$u(0.25, 0.75)$	Number of iterations	Computation time in secs on CII 10070
10^{-1}	10^{-4}	1.7	1	0.125	277	48 s
10^{-2}	10^{-4}	1.7	1	0.122	1 895	336 s
10^{-3}	10^{-4}	1.6	1	Value at 3 000th iteration: 0.088	Not converged at 3 000th iteration	

Table 8.4 (Example 2)

ε	ε_T	ω_1	ω_2	$u(0.25, 0.75)$	Number of iterations	Computation time in secs on CII 10070
10^{-1}	10^{-2}	1.8	1		158	
10^{-2}	10^{-3}	1.8	1		472	
10^{-3}	10^{-4}	1.8	1	0.124	732	120 s

Table 8.5 (Example 2)

The variations of the number of iterations required for convergence, as a function of ω_1 and ε, when $\omega_2 = 1$, $u_h^0 = 0$, $p_h^0 = 0$, $\varepsilon_T = 10^{-4}$ are indicated in Table 8.6:

ω_1 \ ε	10^{-1}	10^{-2}
1.6	182	2 473
1.7	277	1 895
1.8	321	2 147
1.9	427	2 835

Table 8.6 (Example 2)

8.4 Comparisons

In view of the preceding results, the two penalty methods considered in Sections 8.2 and 8.3, respectively, appear comparable as far as computation time is concerned; however, the formulation of the approximate problems is relatively complicated in the case of the first method because of the *highly nonlinear* nature of the penalised problem; this difficulty does not exist in the second method, for which the corresponding penalised problem is almost *linear*.

By way of illustration, it is worth pointing out the use by R. Trémolières /1/ of the *method of centres with variable truncation* described in Chapter 2, Section 3.4; this method, the principle of which consists of approximating the solution whilst remaining within the interior of the domain of the constraints, is more costly in computation time since for the problem (P_0), in $\Omega =]0, 1[\times]0, 1[$ with $f = 10$ and with the same characteristics as in Example 1 of Sections 8.2.5, 8.3.5, the solution of the approximate problem:

$$\operatorname*{Min}_{v_h \in K_{0h}^1} J_h^1(v_h)$$

takes 24 sec. on an IBM 360/91 if $h = 1/10$ and 139 sec. if $h = 1/20$, i.e. computation times approximately 20 times longer than those required when the exterior penalty methods of this Section 8 are used.

We refer the reader to Trémolières, *loc. cit.*, for further details of the approximate solution of (P_0) by this method.

9. SOLUTION OF THE APPROXIMATE PROBLEMS RELATING TO (P_0) BY DUALITY METHODS

9.1 Introduction and synopsis

In this section we shall study the application of the *duality* methods of Chapter 1, Sections 3.4, 3.5 and of Chapter 2, Section 4, to the solution of the finite-dimensional problems obtained by applying to (P_0) the approximation methods of Sections 4 and 5.

In order to simplify the presentation, the various approaches used will be demonstrated for the continuous problem.

9.1.1 *A first formulation*

We consider again the formalism of Chapter 2, Section 4.2, Example 4.2; we thus define:

(9.1) $$\Lambda_1 = L_+^2(\Omega)\,,$$

(9.2) $$\begin{cases} \Phi_1 : H_0^1(\Omega) \to L^2(\Omega) \quad \text{by} \\ \Phi_1(v) = |\operatorname{grad} v| - 1\,, \end{cases}$$

and the Lagrangian $\mathcal{L}_1 : H_0^1(\Omega) \times \Lambda_1 \to R$ by

$$\mathcal{L}_1(v, q) = J(v) + \int_\Omega q\Phi_1(v)\,dx\,. \tag{9.3}$$

From the relation

$$\sup_{q \in \Lambda_1} \int_\Omega q\Phi_1(v)\,dx = \begin{cases} 0 & \text{if } |\operatorname{grad} v| \leqslant 1 \\ +\infty & \text{otherwise,} \end{cases} \tag{9.4}$$

we deduce that if $\mathcal{L}_1$ admits a saddle point $\{u, p\}$ on $H_0^1(\Omega) \times \Lambda_1$, then u is the solution of (P_0); hence in this case, for which assumptions (4.10), (4.11) of Chapter 2, Section 4.3 are satisfied, we can solve (P_0) by applying the algorithm (4.12), ... , (4.15) of Chapter 2, Section 4.3; namely:

$$\left\{ \begin{array}{l} \text{with } p^n \text{ known } (\in \Lambda_1), u^n \text{ is defined as being the element} \\ \quad \text{of } H_0^1(\Omega) \text{ which minimises } J(v) + \int_\Omega p^n\,\Phi_1(v)\,dx\,, \end{array} \right. \tag{9.5}$$

$$p^{n+1} = \sup(0, p^n + \rho_n\,\Phi_1(u^n))\,, \tag{9.6}$$

$$\rho_n > 0\,, \quad \text{sufficiently small.} \tag{9.7}$$

The minimisation problem defined by (9.5) (and which admits one and only one solution) is a problem of minimisation of a *non-differentiable* functional, very similar to the flow problem considered in Chapter 1, Sections 1.4, 3.4, and in Chapter 2, Sectio 4.2, Example 4.1; the use of the algorithm (9.5), (9.6), (9.7) for solving (P_0) will be considered in Chapter 5, Section 8. ■

9.1.2 *A second formulation*

We define:

$$\Lambda_2 = L_+^\infty(\Omega)\,, \tag{9.8}$$

$$\left\{ \begin{array}{l} \Phi_2 : H_0^1(\Omega) \to L^1(\Omega) \quad \text{by} \\ \Phi_2(v) = |\operatorname{grad} v|^2 - 1\,, \end{array} \right. \tag{9.9}$$

and the Lagrangian $\mathcal{L}_2 : H_0^1(\Omega) \times \Lambda_2 \to R$ by

$$\mathcal{L}_2(v, q) = J(v) + \frac{1}{2}\int_\Omega q\Phi_2(v)\,dx\,. \tag{9.10}$$

From the relation

$$\sup_{q \in \Lambda_2} \int_\Omega q\Phi_2(v)\,dx = \begin{cases} 0 & \text{if } |\operatorname{grad} v| \leqslant 1 \\ +\infty & \text{otherwise} \end{cases} \tag{9.11}$$

we deduce that if $\mathscr{L}_2$ admits a saddle point $\{u, p\}$ on $H_0^1(\Omega) \times \Lambda_2$, then u is the solution of (P_0).

The considerations of Chapter 2, Section 4, lead - in a totally formal manner, since we are no longer in the Hilbert setting of Chapter 2, Section 3.4 ([1]) - to using the algorithm (4.12), ... , (4.15) of Chapter 2, Section 4.3 in order to solve (P_0); that is:

(9.12) $\begin{cases} \text{with } p^n \text{ known } (\in \Lambda_2),\ u^n \text{ is defined as being the element} \\ \text{of } H_0^1(\Omega) \text{ which minimises } J(v) + \frac{1}{2}\int_\Omega p^n\, \Phi_2(v)\,\mathrm{d}x, \end{cases}$

(9.13) $p^{n+1} = \sup(0, p^n + \rho_n\, \Phi_2(u^n)),$

(9.14) $\rho_n > 0,$ sufficiently small.

It may be noted that, given p^n, u^n is the solution in $H_0^1(\Omega)$ of the elliptic problem

(9.15) $\begin{cases} -\sum_{i=1}^{2} \frac{\partial}{\partial x_i}\left\{(1 + p^n)\frac{\partial u^n}{\partial x_i}\right\} = f, \\ u^n|_\Gamma = 0. \end{cases}$

The use of (9.12), (9.13), (9.14) for solving (P_0) (in fact the interior and exterior approximations of (P_0)) will arise in Sections 9.2, 9.3, 9.4. ■

9.1.3 *A third formulation*

In the two preceding approaches the difficulty of the problem (i.e. the constraint $|\mathbf{grad}\, v| \leqslant 1$) was "eliminated" (at least, formally) by associating with it a Lagrange multiplier (or a Kuhn-Tucker multiplier) satisfying a *non-negativity* condition, since this type of constraint does not present any difficulty *algorithmically* speaking; the new approach is based on the following variant of the considerations of Section 8.3.1.

Once again we write:

(9.16) $\Lambda = \{ q \mid q \in L^2(\Omega) \times L^2(\Omega),\ q = (q_1, q_2),\ q_1^2(x) + q_2^2(x) \leqslant 1 \quad \text{a.e.} \}$

(9.17) $\mathscr{K}_0 = \{ (v, q) \mid v \in H_0^1(\Omega),\ q \in \Lambda,\ q = \mathbf{grad}\, v \};$

([1]) And since, apart from the case $f =$ constant, the problem of the existence of a saddle point for $\mathscr{L}_2$, in $H_0^1(\Omega) \times \Lambda_2$, seems to be open even for $f \in L^2(\Omega)$ (see Remark 9.2 below).

there is then equivalence between (P_0) and

$$\text{(9.18)} \qquad \underset{(v,q)\in\mathscr{K}_0}{\mathrm{Min}} \left[\frac{1}{4}\int_\Omega |\,\mathrm{grad}\, v\,|^2\,\mathrm{d}x + \frac{1}{4}\int_\Omega |\,q\,|^2\,\mathrm{d}x - \int_\Omega fv\,\mathrm{d}x\right].$$

Since the projection $L^2(\Omega) \times L^2(\Omega) \to \Lambda$ does not pose any practical difficulties (see (8.52)), we shall be content to *decouple* v and q, i.e. to eliminate the relation $q = \mathrm{grad}\, v$, by associating with it a Lagrange multiplier (in Section 8.3 we proceeded using *penalisation*) (1)

We thus define:

$$\text{(9.19)} \qquad L = L^2(\Omega) \times L^2(\Omega)$$

$$\text{(9.20)} \qquad \begin{cases} \Phi_3 : H_0^1(\Omega) \times L \to L & \text{by} \\ \Phi_3(v, q) = \mathrm{grad}\, v - q \end{cases}$$

and the Lagrangian $\mathscr{L}_3 : H_0^1(\Omega) \times L \times L \to R$ by

$$\text{(9.21)} \qquad \mathscr{L}_3(v, q\,;\mu) = \frac{1}{4}\int_\Omega |\,\mathrm{grad}\, v\,|^2\,\mathrm{d}x + \frac{1}{4}\int_\Omega |\,q\,|^2\,\mathrm{d}x - \int_\Omega fv\,\mathrm{d}x + {} + \frac{1}{2}\int_\Omega \mu.\Phi(v, q)\,\mathrm{d}x\,.$$

From the relation

$$\text{(9.22)} \qquad \sup_{\mu\in L}\int_\Omega \mu.(\mathrm{grad}\, v - q)\,\mathrm{d}x = \begin{cases} 0 & \text{if } \mathrm{grad}\, v - q = 0 \\ +\infty & \text{otherwise} \end{cases}$$

we deduce that if $\mathscr{L}_3$ admits a saddle point $\{u, p\,;\lambda\}$ on $H_0^1(\Omega) \times \Lambda \times L$ then $\{u, p\}$ is the solution of the problem (9.18) equivalent to (P_0); the algorithm (4.12), ... , (4.15) of Chapter 2, Section 4.3 then takes the form:

$$\text{(9.23)} \qquad \begin{cases} \text{with } \lambda^n \text{ known } (\in L), \text{ we define } \{u^n, p^n\} \text{ as being the} \\ \text{element of } H_0^1(\Omega) \times \Lambda \text{ which minimises } \mathscr{L}_3(v, q\,;\lambda^n)\,, \end{cases}$$

$$\text{(9.24)} \qquad \lambda^{n+1} = \lambda^n + \rho_n\,\Phi_3(u^n, p^n)\,,$$

$$\text{(9.25)} \qquad \rho_n > 0\,, \quad \text{sufficiently small.}$$

---

(1) A method combining both these methods will be found in Section 10.

With the function λ^n given in L, u^n and p^n are, respectively, solutions of

$$(9.26)\qquad \begin{cases} -\Delta u^n = 2f + \operatorname{div} \lambda^n , \\ u^n|_\Gamma = 0 , \end{cases}$$

$$(9.27)\qquad p^n = P_\Lambda(\lambda^n) .$$

The use of (9.23), (9.24), (9.25) for solving (P_0) will be discussed in Section 9.5.

Remark 9.1. The method of Section 9.1.3 is a method of *decomposition-coordination* in the sense of Bensoussan-Lions-Témam /1/, applied to the solution of the problem (9.18) equivalent to (P_0). ■

9.1.4 *Remarks*

Remark 9.2. The problem of the existence of a saddle point for $\mathscr{L}_1$ (resp. $\mathscr{L}_2$, $\mathscr{L}_3$) on $H_0^1(\Omega) \times \Lambda_1$ (resp. $H_0^1(\Omega) \times \Lambda_2$, $H_0^1(\Omega) \times \Lambda \times L$) seems to be open; in this case it is possible, however, by using (for example) Bensoussan-Lions-Témam /1/, Chapter 2, Section 2, subsection 1, to prove the existence of saddle points for

$$(9.28)\qquad \mathscr{L}'_1(v, q) = J(v) + \langle\ |\operatorname{grad} v| - 1, q\ \rangle \qquad \text{in } W_0^{1,\infty}(\Omega) \times (L^\infty(\Omega))'_+ ,$$

$$(9.29)\qquad \mathscr{L}'_2(v, q) = J(v) + \langle\ |\operatorname{grad} v|^2 - 1, q\ \rangle \qquad \text{in } W_0^{1,\infty}(\Omega) \times (L^\infty(\Omega))'_+ ,$$

$$(9.30)\qquad \begin{cases} \mathscr{L}'_3(v, q ; \mu) = \frac{1}{4}\int_\Omega |\operatorname{grad} v|^2 \,dx + \frac{1}{4}\int_\Omega |q|^2 \,dx - \int_\Omega fv \,dx + \\ \quad + \frac{1}{2}\left\langle \frac{\partial v}{\partial x_1} - q_1, \mu_1 \right\rangle + \frac{1}{2}\left\langle \frac{\partial v}{\partial x_2} - q_2, \mu_2 \right\rangle \\ \quad \text{in } W_0^{1,\infty}(\Omega) \times \Lambda \times (L^\infty(\Omega))' \times (L^\infty(\Omega))' \end{cases}$$

where $\langle , \rangle$ denotes the bilinear form of the duality between $L^\infty(\Omega)$ and its dual $(L^\infty(\Omega))'$ with

$$(9.31)\qquad W_0^{1,\infty}(\Omega) = \left\{ v \mid v, \frac{\partial v}{\partial x_i} \in L^\infty(\Omega), \ \ \forall i, \ v|_\Gamma = 0 \right\},$$

$$(9.32)\qquad (L^\infty(\Omega))'_+ = \{ q \mid q \in (L^\infty(\Omega))', \langle v, q \rangle \geqslant 0 \ \ \forall v \in L^\infty_+(\Omega) \} .$$

On the other hand, there is no difficulty in proving the existence of saddle points for the restriction of $\mathscr{L}_1$ (resp. $\mathscr{L}_2$, $\mathscr{L}_3$), to the finite-dimensional convex sets and spaces appearing in the interior approximations of (P_0) defined in Section 4 (see Sections 9.2, 9.3 below); the same applies for the exterior approximations of Section 5 (see Sections 9.4, 9.5 below). ■

Remark 9.3. In the case f = constant (which corresponds to the physical problem of Section 1.2), H. Brézis /5/ has demonstrated the existence of $p \in \Lambda_2 (= L^\infty_+(\Omega))$, with $p = 0$ in the elastic zone, such that $\{u, p\}$, where u is the solution of (P_0), is the *unique* saddle-point of $\mathscr{L}_2$ in $H^1_0(\Omega) \times \Lambda_2$ (for the interpretation of this result in connection with problems in mechanics, see Duvaut-Lions /1/, Chapter 5, Section 6.6). ■

9.2 Application of the duality method of Section 9.1.2 to the solution of the approximate problems. Case of the finite element approximation of Section 4.1.

9.2.1 *On the existence of a saddle point*

We shall retain the notation of Section 4.

Having defined the space V_h in Section 4.1.1 we introduce L_h, a subspace of $L^\infty(\Omega)$:

$$L_h = \left\{ q \mid q = \sum_{T \in \mathscr{T}_h} q_T\, \theta_T,\ q_T \in \mathbb{R} \right\}, \tag{9.33}$$

with

$$\theta_T = \text{ characteristic function of } T, \tag{9.34}$$

and

$$\Lambda_{2h} = \Lambda_2 \cap L_h = \{ q \mid q \in L_h,\ q_T \geqslant 0\ \forall T \in \mathscr{T}_h \}. \tag{9.35}$$

It may be noted that $v_h \in V_h \Rightarrow \operatorname{grad} v_h \in L_h \times L_h$ and that

$$v_h \in K_0 \cap V_h \Rightarrow 1 - |\operatorname{grad} v_h|^2 \in \Lambda_{2h}.$$

With the Lagrangian $\mathscr{L}_2$ defined by (9.9), (9.10) we then have:

Proposition 9.1. *The Lagrangian* $\mathscr{L}_2$ *admits a saddle point* $\{u_h, p_h\}$ *on* $V_h \times \Lambda_{2h}$ *with:*

$$u_h \text{ the solution of } \min_{v_h \in K_0 \cap V_h} J(v_h), \tag{9.36}$$

and

$$p_h(|\operatorname{grad} u_h|^2 - 1) = 0. \tag{9.37}$$

Proof. Since the spaces V_h and L_h are finite-dimensional, this result follows immediately from Rockafellar /3/, Section 28, Theorem 28.2, 28.3. ■

Remark 9.4. Since the pair $\{u_h, p_h\}$ is a saddle-point of $\mathscr{L}_2$ on $V_h \times \Lambda_{2h}$ we have $\mathscr{L}_2(u_h, p_h) \leqslant \mathscr{L}_2(v_h, p_h)\ \forall v_h \in V_h$ so that, explicitly,

$$(9.38)\qquad \frac{1}{2}\int_\Omega (1 + p_h)\,|\operatorname{grad} u_h|^2 - \int_\Omega f u_h\,dx \leqslant$$

$$\leqslant \frac{1}{2}\int_\Omega (1 + p_h)\,|\operatorname{grad} v_h|^2\,dx - \int_\Omega f v_h\,dx\,,$$

and in particular if $v_h = 0$, we have, taking account of (9.37),

$$(9.39)\qquad \int_\Omega p_h\,dx \leqslant 2\int_\Omega f u_h\,dx - \int_\Omega |\operatorname{grad} u_h|^2\,dx\,.$$

Under the conditions for applying Theorem 6.1, Section 6.2.1, when $h \to 0$ we have $u_h \to u$ strongly in $H_0^1(\Omega) \cap L^\infty(\Omega)$, where u is the solution of (P_0); we thus have, since $p_h \geqslant 0$,

$$(9.40)\qquad \|p_h\|_{L^1(\Omega)} \leqslant C \quad \forall h > 0\,,$$

an estimate in $L^1(\Omega)$ for p_h. ■

9.2.2. *Solution of the approximate problem by means of a saddle-point-seeking algorithm* (I). *Description of the algorithm.*

We use the algorithm (4.12), ... , (4.15) of Chapter 2, Section 4.3, which in this particular case takes the form:

$$(9.41)\qquad \begin{cases} \text{with } p_h^n \text{ known } (\in \Lambda_{2h}) \text{ we define } u_h^n \text{ as being the element} \\ \text{of } V_h \text{ which minimises } J(v_h) + \dfrac{1}{2}\displaystyle\int_\Omega p_h^n\,(|\operatorname{grad} v_h|^2 - 1)\,dx\,, \end{cases}$$

$$(9.42)\qquad p_h^{n+1} = \sup\,(0, p_h^n + \rho_n(|\operatorname{grad} u_h^n|^2 - 1))\,,$$

$$(9.43)\qquad \rho_n > 0\,, \quad \text{sufficiently small.}$$

The algorithm (9.41), (9.42), (9.43) is an "approximation" of algorithm (9.12), (9.13), (9.14); it may be noted that, given p_h^n, u_h^n is the *unique* solution of:

$$(9.44)\qquad \begin{cases} \displaystyle\int_\Omega (1 + p_h^n)\operatorname{grad} u_h^n \cdot \operatorname{grad} v_h\,dx = \int_\Omega f v_h\,dx \quad \forall v_h \in V_h\,, \\ u_h^n \in V_h\,, \end{cases}$$

which is a *variational equation* in V_h. ■

Remark 9.5. It is possible to write (9.41) (and/or (9.44)) and (9.42) explicitly in terms of the parameters q_T, $T \in \mathscr{T}_h$, and of the "nodal" parameters

$$v_i = v_h(M_i)\,, \quad u_i^n = u_h^n(M_i)\,, \quad M_i \in \mathring{\Sigma}_h$$

(see Section 4.1.1, relation (4.8)) by using the relations of Section 4.1.4; we refer the reader to Bourgat /1/ for such an explicit formulation. ■

9.2.3. *Solution of the approximate problem by means of a saddle-point-seeking algorithm* (II). *Convergence of the algorithm.*

We have:

Proposition 9.2. *The sequence* $(u_h^n)_n$ *defined by algorithm* (9.41), (9.42), (9.43) *converges to the solution* u_h *of*

$$\operatorname*{Min}_{v_h \in K_h \cap V_h} J(v_h)\,.$$

Proof. Taking account of Proposition 9.1 this follows from Theorem 4.1 of Chapter 2, Section 4.3, as long as the following conditions are satisfied (see relation (4.11'), Chapter 2, Section 4.3):

(i) The mapping $\Phi_{2h} : V_h \to L_h$ defined by:

$$\Phi_{2h}(v_h) = |\operatorname{grad} v_h|^2 - 1 \tag{9.45}$$

is *locally Lipschitz*;

(ii) $\forall q \in \Lambda_{2h}$, the solution of

$$\operatorname*{Min}_{v_h \in V_h} \left\{ J(v_h) + \frac{1}{2} \int_\Omega q \Phi_{2h}(v)\, dx \right\} \tag{9.46}$$

remains within a *bounded region independent of* q.

Point (i) follows immediately since, $\forall v_h \in V_h$, $|\operatorname{grad} v_h|^2$ is quadratic with respect to the "nodal" parameters $v_i = v_h(M_i)$, and thus C^∞; as regards (ii), the solution u_{hq} of (9.46) is the solution in V_h of the variational equation

$$\int_\Omega (1+q) \operatorname{grad} u_{hq} . \operatorname{grad} v_h \, dx = \int_\Omega f v_h \, dx \quad \forall v_h \in V_h\,, \tag{9.47}$$

so that setting $v_h = u_{hq}$ in (9.47) and bearing in mind that $q \geqslant 0$,

$$\left(\int_\Omega |\operatorname{grad} u_{hq}|^2 \, dx \right)^{1/2} \leqslant C \, \| f \|_{L^2(\Omega)} \quad (^1)\,, \tag{9.48}$$

---

(1) C is the Poincaré constant in $H_0^1(\Omega)$ i.e. the reciprocal of th[e] square root of the smallest eigenvalue of $-\Delta$ in $H_0^1(\Omega)$.

which proves (ii) and the proposition. ■

Remark 9.6. By explicitly writing out the calculations in the proof of Theorem 4.1 of Chapter 2, Section 4.3, in the particular case of the algorithm (9.41), (9.42), (9.43), it may be shown that the condition for convergence of the algorithm is given by

$$0 < \alpha_0 \leqslant \rho_n \leqslant \alpha_1 < \frac{4}{1 + \dfrac{C^2 \| f \|^2_{L^2(\Omega)}}{\gamma(\mathcal{T}_h)}}, \tag{9.49}$$

where the constant C is the one appearing in (9.48) and $\gamma(\mathcal{T}_h) = \underset{T \in \mathcal{T}_h}{\text{Min}}$ (Area of T). In practice, the estimate (9.49) is *very pessimistic*, at least for the examples considered below.

9.2.4. *Solution of the approximate problem by a saddle-point-seeking algorithm* (III). *Practical considerations on the use of algorithm* (9.41), (9.42), (9.43).

The work in this section and in the subsequent Section 9.2.5, follows Bourgat /1/. ■

Solution of (9.44).

The determination of u_h^n from p_h^n reduces, using (9.44), to the solution of an (approximate) elliptic problem which is of order 2, linear, and with coefficients which vary in Ω and with n; this problem, which is of Dirichlet type, has been solved by *point over-relaxation* taking as variables the nodal values v_i, $\forall M_i \in \mathring{\Sigma}_h$. ■

Termination criteria: u_h^n is identified to $(u_i^n)_{M_i \in \mathring{\Sigma}_h}$; we then define

$$D_\infty(n) = \underset{M_i \in \mathring{\Sigma}_h}{\text{Max}} | u_i^n - u_i^{n-1} |, \tag{9.50}$$

$$D_1(n) = \frac{\sum_{M_i \in \mathring{\Sigma}_h} | u_i^n - u_i^{n-1} |}{\sum_{M_i \in \mathring{\Sigma}_h} | u_i^n |}, \tag{9.51}$$

and take for algorithm (9.41), (9.42), (9.43), a termination criterion of the form

$$D_1(n) < \varepsilon . \tag{9.52}$$

Moreover, at least at the start of the iterative process, there is no point in solving the Dirichlet sub-problems (9.44) with high precision; for the over-relaxation algorithm applied to problems (9.44), we therefore use a termination criterion of the type (9.52) with ε replaced by ε_n, defined by

$$\varepsilon_n = \min\left(\varepsilon_{n-1}, \frac{D_\infty(n-1)}{10}\right). \quad (9.53)$$ ■

Initialisations: In the following examples, algorithm (9.41), (9.42), (9.43) has been initialised with $p_h^0 = 0$ and, in the determination of u_h^n from p_h^n by solving (9.44), the over-relaxation algorithm has been initialised with u_h^{n-1}. ■

9.2.5 *Solution of the approximate problem using a saddle-point-seeking algorithm* (IV). *Examples*

Example 1. We consider again Example 1 of Section 7.4, that is

$$\Omega = \{ x \mid x \in R^2, x_1^2 + x_2^2 < 1 \} \quad \text{and} \quad f = 10,$$

the exact solution being given explicitly by (7.81); the notation and triangulations used are the same as those in Section 7.4 (see Figures 7.3 and 7.4).

Using the initialisations indicated in Section 9.2.4, we take $\varepsilon = 10^{-3}$ in (9.52), $\varepsilon_1 = 10^{-3}$ in (9.53), $\rho_n = \rho = 1$ in (9.42) and $\omega = 1.5$, where ω is the over-relaxation parameter in the solution of (9.44) (see Section 9.2.4).

The main results concerning the approximate solution of (P_0) by means of the above techniques are shown in Table 9.1 (the notation is the same as that in Table 7.1).

Table 9.2 shows, for the case of triangulation $\mathscr{T}_2$, the number of over-relaxation iterations required to solve the nth problem in (9.44), using the initialisations and termination criterion indicated in Section 9.2.4.

Remark 9.7. The results given in this section and in Table 7.1 of Section 7.4 would appear to indicate, at least, for the particular case considered [1], that the method of over-relaxation with projection described in Section 7 is more rapid than the duality method given above; moreover the method of Section 7 offers the advantages of a simpler implementation (and hence easier programming) and of a smaller storage requirement (for example, there is no need to store the coordinates of p_h^n). It should be pointed out, however, that in the physical problem of elastoplastic torsion (i.e. f = constant) the function p is of great physical interest (see Duvaut-Lions /1/, Chapter 5) and this may justify using algorithm (9.41), (9.42), (9.43) since

[1] This remark in fact is completely general (see below).

this gives, as well as an approximation of the solution u of (P_0), an approximation of the function p. ■

Triangulation	$\mathcal{T}_1$	$\mathcal{T}_2$
Number of vertices	121	433
$N = \text{Card}(\mathring{\Sigma}_j)$ $j = 1, 2$	73	337
$N' = \text{Card}(\mathcal{T}_j)$ $j = 1, 2$	192	768
Number of iterations	8	10
ε_a	1.7×10^{-2}	4×10^{-3}
Computation time IBM 360/91	2 s	8 s
Computation time C II 10070	25 s	90 s

Table 9.1

n	2	4	6	8	10
Number of over-relaxation iterations	28	24	22	7	6

Table 9.2

Further examples. Still keeping $f = 10$, we have also treated the following cases;

Example 2 $\Omega =]0, 1[\times]0, 1[$.

Example 3 Ω is the equilateral triangle with unit side.

Example 4 Ω is the T-shaped domain shown in Figure 9.1.

For the triangulations shown in Figures 9.2, 9.3, 9.4, respectively, Table 9.3 summarises the main results for the solution of the above examples; we refer the reader to Bourgat /1/ for

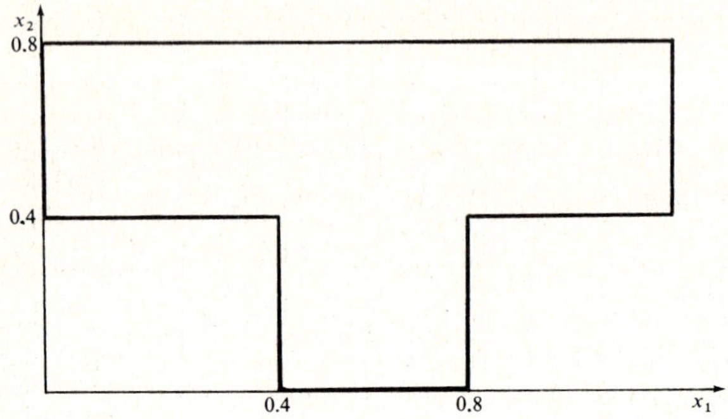

Fig. 9.1.

Ω	Square	equilateral triangle	T-shaped domain
Number of vertices	289	217	239
$N = \text{Card}(\mathring{\Sigma}_h)$	225	169	195
$N' = \text{Card}(\mathcal{T}_h)$	512	384	432
$\rho_n = \rho$	1	1	1
ω	1.5	1.5	1.5
ε	5×10^{-4}	5×10^{-4}	10^{-4}
ε_1	10^{-3}	10^{-3}	10^{-3}
Number of iterations (n)	19	8	16
Comp. time C II 10070	90 s	40 s	60 s

Table 9.3

further details. In addition to the triangulations used, Figures 9.2, 9.3 and 9.4 also show the approximated elastic and plastic zones, a triangle T being taken to lie in a plastic zone when $|\text{grad } u_h| \geq 0.995$ on T.

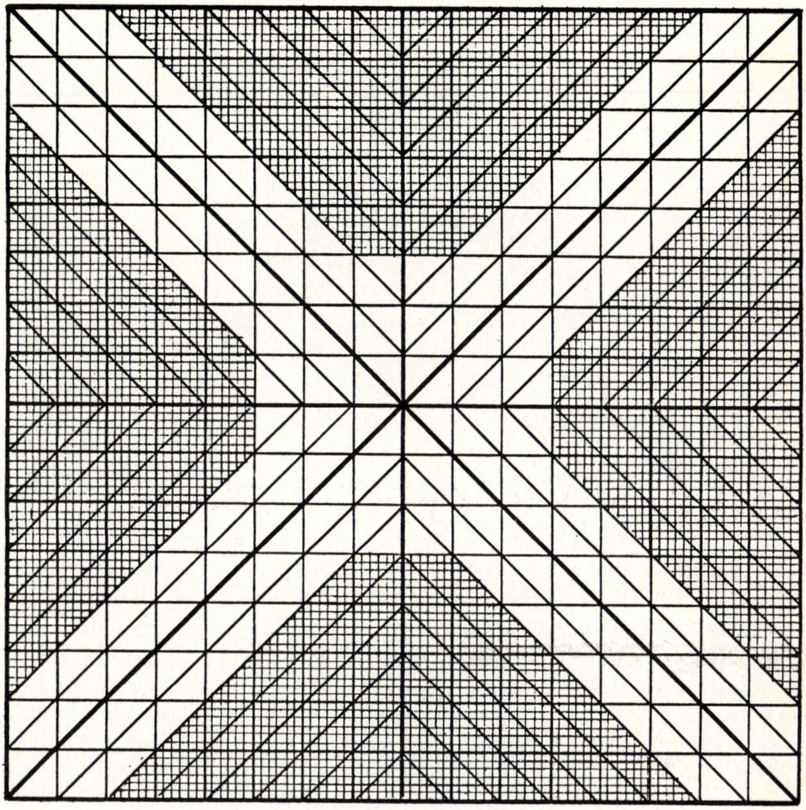

Fig. 9.2. *Square cross-section*: Finite elements (512 triangles), $C = 10$.
Plastic zones: ⊞
Initial triangulation: ▬

9.3 Application of the duality method of Section 9.1.2 to the solution of the approximate problems. Case of the interior approximation by eigenfunctions of $-\Delta$ in $H_0^1(\Omega)$

Preliminary remark. In general the eigenfunctions of $-\Delta$ in $H_0^1(\Omega)$ are not known *explicitly*, and this seriously restricts the use of the approximation described in Section 4.2; we have therefore confined our attention to a single example, corresponding to $\Omega =]0, 1[\times]0$ and $f = 10$, for which the above eigenfunctions and the corresponding eigenvalues are known explicitly and for which approximate solutions for problem (P_0) have been

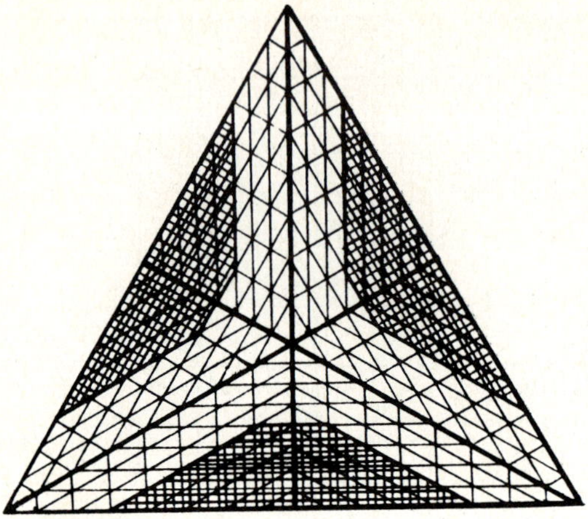

Fig. 9.3. Finite-element calculation of the elasto-plastic torsion of a cylindrical bar with triangular cross-section. Number of finite elements: 384.

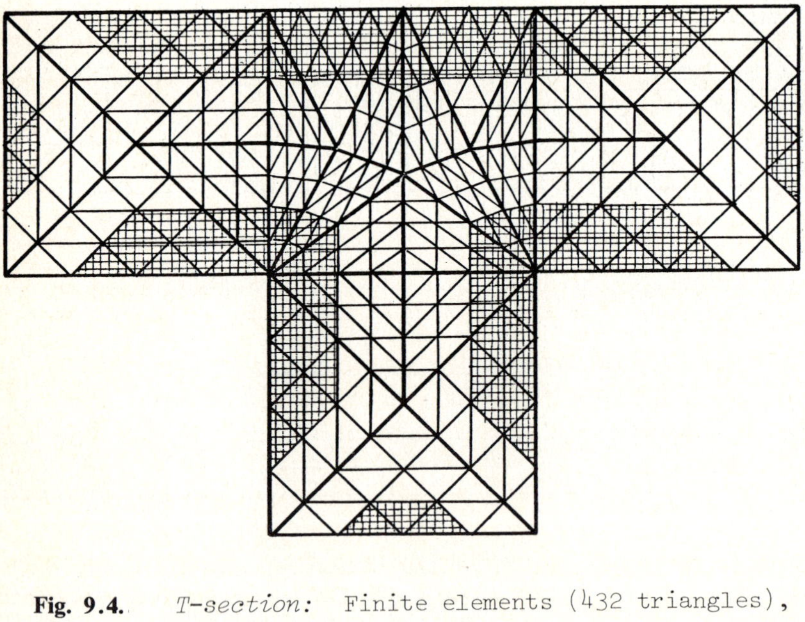

Fig. 9.4. *T-section:* Finite elements (432 triangles), $C = 10$
Plastic zones: ⊞
Initial triangulation: ——

obtained, by other methods earlier in the present volume. ■

9.3.1 *Statement of the continuous problem and spectral decomposition of* $-\Delta$ *in* $H_0^1(\Omega)$

We consider the problem (P_0) relating to $\Omega =]0,1[\times]0,1[$ and $f = 10$, which was treated earlier in Section 3.5 (making use of the equivalence result of Section 2.2) and in Sections 7.5, 8.2.5, 8.3.5 and 9.2.5, by various methods (see also Section 9.4 below).

For the above domain Ω, the *eigenfunctions* of $-\Delta$ in $H_0^1(\Omega)$ are given by:

$$w_{pq}(x_1, x_2) = 2 \sin p\pi x_1 \sin q\pi x_2 , \qquad p, q \in \mathbb{N} , \tag{9.54}$$

and the corresponding *eigenvalues* by:

$$\beta_{pq} = \pi^2(p^2 + q^2) . \tag{9.55}$$

The eigenfunctions (9.54) are orthogonal in $H_0^1(\Omega)$ and orthonormal in $L^2(\Omega)$. ■

9.3.2 *Explicit formulation of the approximate problem*

For a slightly greater degree of generality, it is not absolutely necessary to assume f = constant. With $M \in \mathbb{N}$, we denote by V_M the subspace of $H_0^1(\Omega)$ generated by the w_{pq} for $1 \leqslant p, q \leqslant M$, that is:

$$V_M = \left\{ v \mid v \in H_0^1(\Omega),\ v = \sum_{i \leqslant p,q \leqslant M} \mu_{pq} w_{pq}, \mu_{pq} \in \mathbb{R} \right\}. \tag{9.56}$$

Using the notation (9.56), we have for $v \in V_M$:

$$J(v) = \frac{\pi^2}{2} \sum_{1 \leqslant p,q \leqslant M} (p^2 + q^2)\,\mu_{pq}^2 - \sum_{1 \leqslant p,q \leqslant M} f_{pq}\,\mu_{pq} , \tag{9.57}$$

with (assuming $f \in L^2(\Omega)$):

$$f_{pq} = \int_\Omega w_{pq}(x)\, f(x)\, \mathrm{d}x . \tag{9.58}$$

In the particular case in which f = constant = C, we have:

$$f_{pq} = \begin{cases} \dfrac{8\,C}{\pi^2\,pq} & \text{if } p \text{ and } q \text{ are odd} \\ 0 & \text{otherwise} \end{cases} \tag{9.58'}$$

so that if $v \in V_M$,

$$J(v) = \frac{\pi^2}{2} \sum_{1 \leqslant p,q \leqslant M} (p^2 + q^2)\,\mu_{pq}^2 - \frac{8\,C}{\pi^2} \sum_{\substack{1 \leqslant p,q \leqslant M \\ p,q \text{ odd}}} \frac{\mu_{pq}}{pq} . \tag{9.57'}$$ ■

Remark 9.8. In view of the symmetries of the particular problem under consideration, it is possible to take in (9.57') only μ_{pq} with p and q odd, the other coefficients being zero; however, we have deliberately chosen not to take advantage of this simplification in the following. ■

For the approximation of K_0, we proceed as in Section 4.2.2, Remark 4.5: given M' integer > 0 (which will tend towards infinity) we put $h = 1/M'$ and define:

$$R_h = \{ P_{ij} \mid P_{ij} = (ih, jh), i, j \in \mathbb{Z} \} . \tag{9.59}$$

In the particular case under consideration, the set $\Omega_h = R_h \cap \Omega$ may be written explicitly in the form

$$\Omega_h = \{ P_{ij} \mid P_{ij} = (ih, jh), 1 < i, j < M' - 1 \} , \tag{9.60}$$

giving, for K_0, the approximation

$$K_{0h}^M = \{ v \mid v \in V_M, \mid \text{grad } v(P_{ij}) \mid \leqslant 1 \quad \forall P_{ij} \in \Omega_h \} , \tag{9.61}$$

the condition $|\text{grad } v(P_{ij})| \leqslant 1$ being expressed (writing $x_{1i} = ih, x_{2j} = jh$) in the explicit form

$$2\pi \left[\left(\sum_{1 \leqslant p,q \leqslant M} p\mu_{pq} \cos p\,\pi x_{1i} \sin q\,\pi x_{2j} \right)^2 + \left(\sum_{1 \leqslant p,q \leqslant M} q\mu_{pq} \sin p\,\pi x_{1i} \cos q\,\pi x_{2j} \right)^2 \right]^{1/2} \leqslant 1 . \quad ■ \tag{9.62}$$

Relations (9.57), (9.58), (9.61), (9.62) allow the approximate problem to be expressed explicitly, in terms of $\mu_{pq} (1 \leqslant p, q \leqslant M)$, very simply: $\underset{v \in K_{0h}^M}{\text{Min}} J(v)$. ■

9.3.3 *On the existence of a saddle-point*

We denote by G_{ij} the mapping from $V_M \to \mathbb{R}$ defined by the left-hand side of (9.62) and

$$\Lambda_{2h} = \{ q_h \mid q_h = (q_{ij})_{P_{ij} \in \Omega_h}, \; q_{ij} \geqslant 0 \quad \forall P_{ij} \in \Omega_h \} . \tag{9.63}$$

It is natural to associate with the approximate problem the approximation of $\mathscr{L}_2$ (see Section 9.1.2) defined by:

$$\mathscr{L}_{2h}^M(v_M, q_h) = J(v_M) + \frac{h^2}{2} \sum_{P_{ij} \in \Omega_h} q_{ij}(G_{ij}^2(v_M) - 1) , \tag{9.64}$$

and as the proof of Proposition 9.1 can be adapted without difficulty, $\mathscr{L}_{2h}^M$ admits a saddle-point, $\{ u_M, p_h \}$, on $V_M \times \Lambda_{2h}$ with:

$$u_M \quad \text{the solution of} \quad \underset{v \in K_{0h}^M}{\text{Min}} J(v) , \tag{9.65}$$

(9.66) $\quad p_{ij}(G_{ij}^2(u_M) - 1) = 0 \quad \forall P_{ij} \in \Omega_h$. ■

9.3.4 *Description of the algorithm and convergence*

We use a finite-dimensional variant of algorithm (9.12), (9.13), (9.14), namely:

(9.67) $\left\{ \begin{array}{l} \text{with } p_h^n \text{ known } (\in \Lambda_{2h}), \text{ we define } u_M^n \text{ to be the element} \\ \text{of } V_M \text{ which minimises } \mathscr{L}_{2h}^M(v_M, p_h^n) , \end{array} \right.$

(9.68) $\quad p_{ij}^{n+1} = \max(0, p_{ij}^n + \rho_n(G_{ij}^2(u_M^n) - 1)) \quad \forall P_{ij} \in \Omega_h$,

(9.69) $\quad \rho_n > 0$, sufficiently small.

The proof of Proposition 9.2 can be adapted without difficulty, and it may thus be shown that the sequence $(u_M^n)_n$ defined by (9.67), (9.68), (9.69), converges to the solution of $\underset{\Lambda \in K_{0h}^M}{\text{Min}} J(v)$. ■

9.3.5 *Practical considerations on the use of algorithm* (9.67), (9.68), (9.69)

The terms $\cos p\pi x_{1i}$ (resp. $\cos q\pi x_{2j}$) and $\sin p\pi x_{1i}$ (resp. $\sin q\pi x_{2j}$) which appear in algorithm (9.67), (9.68), (9.69) through the intermediary of the G_{ij} (see (9.62), (9.68)) are obviously calculated once and for all. Moreover, putting $u_M^n = \sum_{1 \leqslant p,q \leqslant M} \lambda_{pq}^n w_{pq}$ in the calculation of u_M^n (with p_h^n known), using (9.67), the vector $(\lambda_{pq}^n)_{1 \leqslant p,q \leqslant M}$ is the solution of a linear system of equations whose matrix (the coefficients of which vary with n) is *symmetric* and *positive-definite* (1); this system has been solved by the *Cholesky* method.

For the initialisation of (9.67), (9.68), (9.69), $p_h^0 = 0$ was always used. ■

9.3.6 *Numerical results*

Algorithm (9.67), (9.68), (9.69) was applied to the particular case corresponding to $f = 10$, and, since the exact solution is not known, the reference solution was taken as the approximate solution calculated by the method of Section 3 (and thus using the equivalence result of Section 2.2) with a mesh size of 1/40.

Results for $h = 1/20$:

- For $M = 10$ (i.e. 100 eigenfunctions w_{pq}), a time of 100 sec was required on an IBM 360/91 to perform 10 iterations; taking

---

(1) And has the disadvantage of being a full matrix.

ρ_n independent of n, i.e. $\rho_n = \rho$, the optimal value of this parameter is very close to unity; for this value the computed solution differed from the reference solution by only about 10^{-3} (in absolute value) by the 5th iteration, and to within 10^{-4} by the 7th iteration. For $0.5 \leqslant \rho \leqslant 1.5$ the reference solution is reproduced in less than 10 iterations, to within 10^{-3} at worst. ∎

• For $M = 5$ (i.e. 25 eigenfunctions w_{pq}) a time of 16 sec. was required on an IBM 360/91 to perform 20 iterations; once again taking $\rho_n = \rho$ the optimal value remains close to unity, and a limit is reached, in practice, which differs from the reference solution by about 10^{-2} in absolute value, in less than 10 iterations; for example, at the point (0.5, 0.5) (resp. 1/20, 1/20) of Ω a value of 0.427 (resp. 0.019) was obtained instead of the 0.413 (resp. 0.029) of the reference solution. ∎

• For $M = 15$, the method is practically unusable, requiring a computation time of several tens of seconds on an IBM 360/91, for *a single iteration*, and with a very large storage requirement. ∎

Results for $h = 1/10$:

For $M = 10$, a time of 50 sec. on an IBM 360/91 was required to carry out 20 iterations and, with $\rho_n = \rho$ again having its optimal value of about unity, convergence was in effect reached in 6 iterations, the computed solution differing by an absolute value of about 10^{-3} from the reference solution. ∎

9.3.7 *Conclusions*

The results given above would appear to indicate that the method of solving the problem (P_0) based on the combined use of approximation by eigenfunctions of $-\Delta$ in $H_0^1(\Omega)$ (see Section 4.2) and of a finite-dimensional variant of algorithm (9.41), (9.42), (9.43) (see Section 9.3.4) is very costly in terms of both computation time and storage requirement. (It should also be remembered that in general the eigenfunctions of $-\Delta$ in $H_0^1(\Omega)$ are not known explicitly). The above drawbacks are linked to the fact that the approximation considered is *global* in nature (the eigenfunctions admit $\bar{\Omega}$ as support) whereas the constraint $|\mathbf{grad}\, v| \leqslant 1$ is *local*; this leads to full matrices of large dimension (varying with n) in the determination of u_M^n in (9.67).

The above difficulties would also be encountered, though admittedly to a lesser degree, if it were required to obtain an approximate solution of (P_0), using an interior approximation of $H_0^1(\Omega)$ by finite elements of order $\geqslant 2$. (see Section 4.1.5). ∎

9.4 Application of the duality method of Section 9.1.2 to the solution of the approximate problems. Case of the exterior approximations of Section 5.

Synopsis: We shall confine our attention to the approximation $(P_{0h})_{13}$ (see Section 5); the other cases may be treated in a very similar manner and, apart from the case $m = 1$ [1], have almost identical storage requirements. ■

9.4.1 *Existence of a saddle-point. Description and convergence of the duality algorithm*

Existence of a saddle-point.

We shall retain the notation of Sections 5 and 8.2.3; thus we define

(9.70) $$L_h = \{ q_h \mid q_h = \{ q_{ij} \}_{M_{ij} \in \overline{\Omega}_h}, \; q_{ij} \in \mathbb{R} \},$$

(9.71) $$\Lambda_{2h} = L_{h+} = \{ q_h \mid q_h \in L_h, \; q_{ij} \geqslant 0 \quad \forall M_{ij} \in \overline{\Omega}_h \},$$

and associate with the approximate problem $(P_{0h})_{13}$ the approximation $\mathscr{L}_{2h}$ of $\mathscr{L}_2$ defined by

(9.72) $$\mathscr{L}_{2h}(v_h, q_h) = J_h^1(v_h) + \frac{h^2}{2} \sum_{M_{ij} \in \overline{\Omega}_h} \alpha_{ij} \, q_{ij}(\beta_{ij} \, G_{ij}^2(v_h) - 1),$$

with α_{ij}, β_{ij}, G_{ij} defined in Section 8.2.3, by (8.33), (8.32), (8.35), respectively.

The proof of Proposition 9.1 can be adapted without difficulty, so that $\mathscr{L}_{2h}$ admits a saddle-point, $\{ u_h, p_h \}$, on $V_h \times \Lambda_{2h}$, with:

(9.73) $$\begin{cases} u_h \text{ the solution of } (P_{0h})_{13}, \\ p_{ij}(\beta_{ij} \, G_{ij}^2(u_h) - 1) = 0 \quad \forall M_{ij} \in \overline{\Omega}_h. \end{cases}$$ ■

Description of the algorithm and convergence.

We use a finite-dimensional variant of algorithm (9.12), (9.13), (9.14), namely:

(9.74) $$\begin{cases} \text{with } p_h^n \text{ known } (\in \Lambda_{2h}), \text{ we define } u_h^n \text{ to be the element} \\ \text{of } V_h \text{ which minimises } \mathscr{L}_{2h}(v_h, p_h^n) \end{cases}$$

(9.75) $$p_{ij}^{n+1} = \max(0, p_{ij}^n + \rho_n(\beta_{ij} \, G_{ij}^2(u_h^n) - 1)) \quad \forall M_{ij} \in \overline{\Omega}_h,$$

(9.76) $$\rho_n > 0, \text{ sufficiently small.}$$

---

(1) See Section 5.2.

The proof of Proposition 9.2 can be adapted without difficulty; it may thus be shown that the sequence $(u_h^n)_n$ defined by (9.74), (9.75), (9.76) converges to the solution of $(P_{0h})_{13}$. ■

9.4.2 *Application*

We shall confine our attention to the case $\Omega =]0, 1[\times]0, 1[$ and $f = 10$ which has been treated in the preceding sections by means of other methods (for further examples, and a more detailed analysis of the numerical results, see Bourgat /1/).

Mesh size: $h = 1/20$.

Solution of (9.74): By point over-relaxation, with increasing i, j.

Initialisations: By $p_h^0 = 0$ for algorithm (9.74), (9.75), (9.76) and by u_h^{n-1} (with $u_h^{-1} = 0$) in the solution of (9.74) by over-relaxation.

Termination criteria: Putting

(9.77) $$D_1(n) = \sum_{M_{ij} \in \Omega_h} | u_{ij}^n - u_{ij}^{n+1} | ,$$

and

(9.78) $$D_\infty(n) = \operatorname*{Max}_{M_{ij} \in \Omega_h} | u_{ij}^n - u_{ij}^{n+1} | ,$$

we take for algorithm (9.74), (9.75), (9.76) the termination criterion:

(9.79) $$D_1(n) \leqslant 10^{-3} .$$

For the solution of the sub-problems (9.74) by over-relaxation, we use a termination criterion of which the severity increases with n, and which has the form:

(9.80) $$D_\infty(p) \leqslant \varepsilon_n ,$$

with

(9.81) $$\varepsilon_n = \min(\varepsilon_{n-1}, 0.2\, D_\infty(n-1)) , \quad \varepsilon_0 = 10^{-3} ,$$

where the integer p in (9.80) represents the number of over-relaxation iterations.

Numerical results.

Table 9.4 summarises the main results for the convergence of algorithm (9.74), (9.75), (9.76):

$\rho_n = \rho$	1
ω	1.7
Computation time C II 10070	30 s
Computation time IBM 360/91	2 s

Table 9.4

The solution thus calculated coincides to within an absolute value of about 10^{-5} with that obtained for the same approximate problem by the method of over-relaxation with projection described in Section 7.5, Example 2. Figure 9.5 shows the restriction to $]0, \frac{1}{2}[\times]0, \frac{1}{2}[$ of the equipotentials of the function p defined in Section 9.1.2, obtained using the approximation p_h provided by algorithm (9.74)-(9.76). Note the continuity of the function p; this is in agreement with results obtained by H. Brézis /5/, Ω being convex. Figure 9.6 shows the variations of the number of iterations required for convergence, as a function of ρ for $\omega = 1.5$, and Figure 9.7 shows the number of over-relaxation iterations required to solve the same sub-problem (9.74) using the termination criterion (9.80), (9.81) and with $\rho_n = \rho = 1, \omega = 1.7$. ■

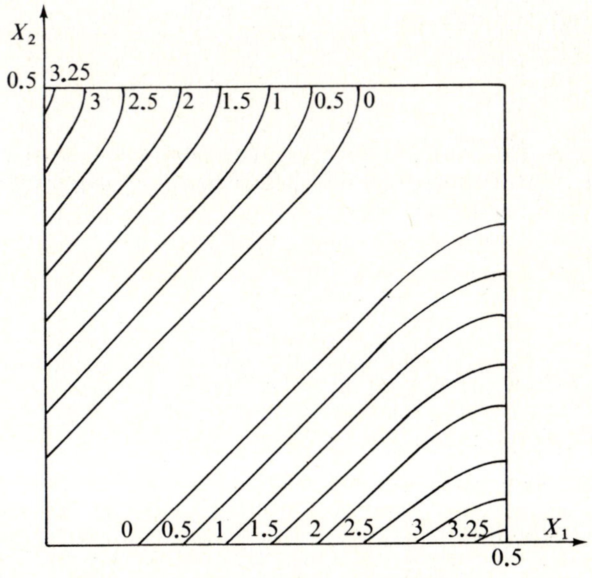

Fig. 9.5. Equipotentials of the dual function p.

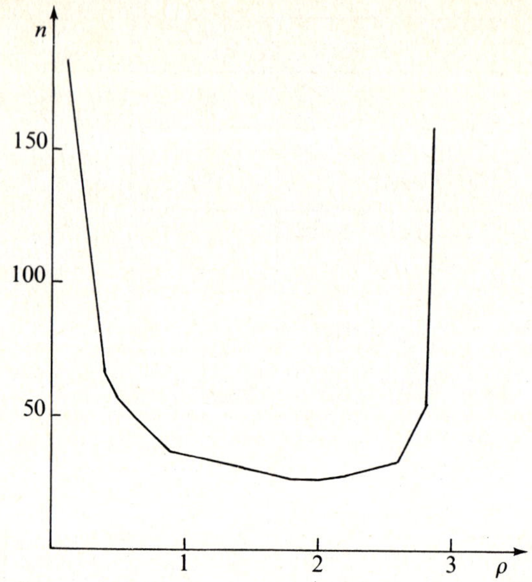

Fig. 9.6. Variation of no. of iterations as a function of ρ ($\omega = 1.5$).

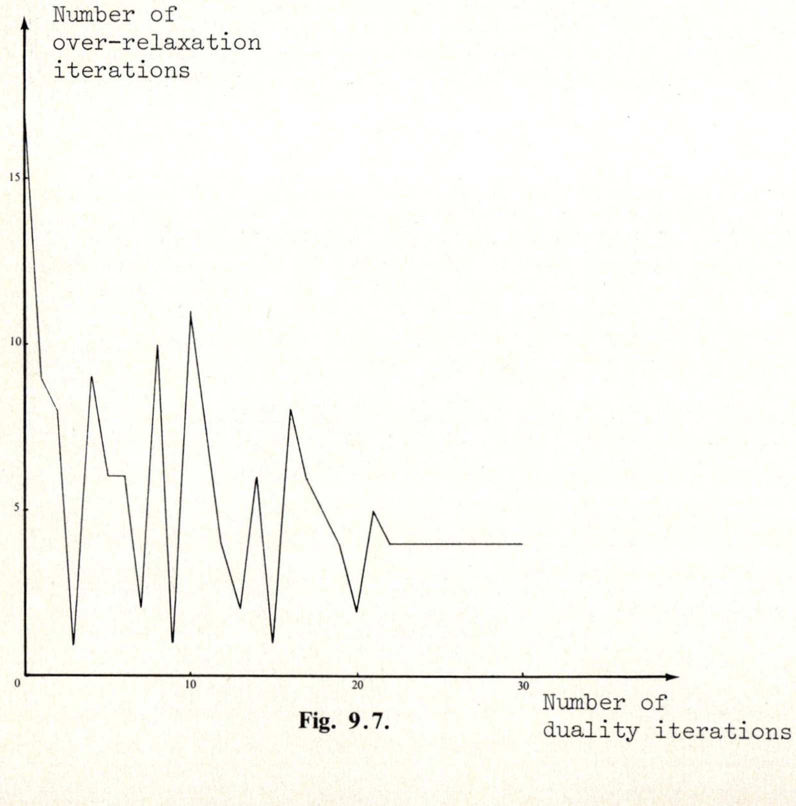

Fig. 9.7.

9.5 Application of the duality method of Section 9.1.3 to the solution of the approximate problems

We shall confine our attention to the finite element approximation of Section 1 and to the exterior approximations of (P_0) using the approximation K_{0h}^4 of K_0 (see Section 5.2).

9.5.1 *Case of the finite element approximation of Section 4.1. (I). Formulation of the approximate problem and existence of a saddle-point.*

We shall retain the notation of Sections 4.1 and 8.3.2; putting:

$$\mathscr{K}_{0h} = \{ (v, q) \mid (v, q) \in \mathscr{K}_0,\ v \in V_h,\ q \in \Lambda_h \}, \tag{9.82}$$

there is then *equivalence* between the problem $\underset{v \in V_h \cap K_0}{\text{Min}} J(v)$ and

$$\underset{(v,q) \in \mathscr{K}_{0h}}{\text{Min}} \left[\frac{1}{4} \int_\Omega |\operatorname{grad} v|^2 \, dx + \frac{1}{4} \int_\Omega |q|^2 \, dx - \int_\Omega fv \, dx \right]. \tag{9.83}$$

With the Lagrangian $\mathscr{L}_3$ defined by (9.20), (9.21), we have:

Proposition 9.3. *The Lagrangian* $\mathscr{L}_3$ *admits a saddle-point,* $\{ u_h, p_h ; \lambda_h \}$ *on* $(V_h \times \Lambda_h) \times L_h$ *with:*

$$u_h \text{ the solution of } \underset{v_h \in K_0 \cap V_h}{\text{Min}} J(v_h) \tag{9.84}$$

$$p_h = \operatorname{grad} u_h \tag{9.85}$$

$$p_h = P_\Lambda(\lambda_h) . \tag{9.86}$$

Proof. Since the problem is finite-dimensional, we may apply Rockafellar /3/, Section 2.8, Theorem 28.2, 28.3. ■

9.5.2 *Case of the finite element approximation of Section 4.1. (II). Description of the algorithm and convergence*

We use algorithm (4.12), ... , (4.15) of Chapter 2, Section 4.3, which in this particular case takes the form:

$$
(9.87) \quad \begin{cases} \text{with } \lambda_h^n \text{ known } (\in L_h), \text{ define } u_h^n \in V_h \text{ and } p_h^n \in \Lambda_h \text{ by} \\ \int_\Omega \operatorname{grad} u_h^n \cdot \operatorname{grad} v_h \, dx = 2 \int_\Omega f v_h \, dx - \int_\Omega \lambda_h^n \cdot \operatorname{grad} v_h \, dx \quad \forall v_h \in V_h, \\ \int_\Omega p_h^n \cdot q_h \, dx = \int_\Omega \lambda_h^n \cdot q_h \, dx \quad \forall q_h \in \Lambda_h \Leftrightarrow p_h^n = p_\Lambda(\lambda_h^n), \end{cases}
$$

$$
(9.88) \qquad \lambda_h^{n+1} = \lambda_h^n + \rho_n(\operatorname{grad} u_h^n - p_h^n),
$$

$$
(9.89) \qquad \rho_n > 0, \quad \text{sufficiently small.}
$$

The algorithm (9.87), (9.88), (9.89) is an "approximation" of algorithm (9.23), (9.24), (9.25). ■

Proposition 9.4. *If* $0 < \alpha \leqslant \rho_n \leqslant \beta < 1$ *the sequence* $\{u_h^n, p_h^n\}$ *defined by* (9.87), (9.88), (9.89) *converges to* $\{u_h, p_h\}$ *with*

$$
(9.90) \qquad u_h \text{ the solution of } \operatorname*{Min}_{v_h \in K_0 \cap V_h} J(v_h)
$$

$$
(9.91) \qquad p_h = \operatorname{grad} u_h. \ ■
$$

9.5.3 *Case of the exterior approximation of Section 5. (I). Formulation of the approximate problem and existence of a saddle-point.*

The preliminary remark of Section 8.3.3 still holds, in the sense that the approximation K_{0h}^4 of K_0 is the only one which is ideally suited to the use of the duality method of Section 9.1.3. We retain the notation of Sections 5 and 8.3.3, that is:

$$
(9.92) \qquad V_h = \{ v_h \mid v_h = (v_{ij})_{M_{ij} \in \Omega_h} \},
$$

$$
(9.93) \qquad L_h = \{ q_h \mid q_h = (q_{1i+1/2\,j+1/2}, q_{2i+1/2\,j+1/2})_{M_{i+1/2\,j+1/2} \in \Sigma_h} \}
$$

$$
(9.94) \qquad \Lambda_h = \{ q_h \mid q_h \in L_h, q_{1i+1/2\,j+1/2}^2 + q_{2i+1/2\,j+1/2}^2 \leqslant 1 \quad \forall M_{i+1/2\,j+1/2} \in \Sigma_h \}.
$$

We introduce the approximation of $\mathscr{K}_0$ defined by

$$
(9.95) \quad \begin{cases} \mathscr{K}_{0h} = \Big\{ (v_h, q_h) \mid v_h \in V_h, q_h \in \Lambda_h, q_{1i+1/2\,j+1/2} = \\ \qquad = \dfrac{v_{i+1j+1} - v_{ij+1} + v_{i+1j} - v_{ij}}{2h}, \\ q_{2i+1/2\,j+1/2} = \dfrac{v_{i+1j+1} - v_{i+1j} + v_{ij+1} - v_{ij}}{2h} \quad \forall M_{i+1/2\,j+1/2} \in \Sigma_h \Big\}. \end{cases}
$$

There is then equivalence between:

$$(9.96) \qquad \operatorname*{Min}_{v_h \in K_{0h}^4} \frac{1}{2}\left(J_h^1(v_h) + J_h^2(v_h)\right)$$

and

$$(9.97) \quad \begin{cases} \displaystyle\operatorname*{Min}_{(v_h,q_h) \in \mathscr{K}_{0h}} \left\{ \frac{h^2}{8} \sum_{M_{ij} \in \bar{\Omega}_h} \left[\left(\frac{v_{i+1j} - v_{ij}}{h}\right)^2 + \left(\frac{v_{i-1j} - v_{ij}}{h}\right)^2 + \right.\right. \\ \displaystyle\quad \left. + \left(\frac{v_{ij+1} - v_{ij}}{h}\right)^2 + \left(\frac{v_{ij-1} - v_{ij}}{h}\right)^2 \right] + \frac{h^2}{4} \times \\ \displaystyle\quad \times \sum_{M_{i+1/2\, j+1/2} \in \Sigma_h} \left(q_{1i+1/2\, j+1/2}^2 + q_{2i+1/2\, j+1/2}^2\right) - \\ \displaystyle\quad \left. - h^2 \sum_{M_{ij} \in \Omega_h} f_{ij}\, v_{ij} \right\} \end{cases}$$

the functional $\frac{1}{2}(J_h^1(v_h) + J_h^2(v_h))$ having the following explicit form (with $v_{pq} = 0$ if $M_{pq} \notin \Omega_h$):

$$(9.98) \quad \begin{aligned} \frac{1}{2}\left(J_h^1(v_h) + J_h^2(v_h)\right) = \frac{h^2}{8} \sum_{M_{ij} \in \bar{\Omega}_h} & \left[\left(\frac{v_{i+1j} - v_{ij}}{h}\right)^2 + \left(\frac{v_{i-1j} - v_{ij}}{h}\right)^2 + \right. \\ & \left. + \left(\frac{v_{ij+1} - v_{ij}}{h}\right)^2 + \left(\frac{v_{ij-1} - v_{ij}}{h}\right)^2 \right] + \\ & + \frac{h^2}{8} \sum_{M_{ij} \in \bar{\Omega}_h} \left[\left(\frac{v_{i+1j+1} - v_{ij}}{h\sqrt{2}}\right)^2 + \left(\frac{v_{i+1j-1} - v_{ij}}{h\sqrt{2}}\right)^2 \right. \\ & \left. + \left(\frac{v_{i-1j+1} - v_{ij}}{h\sqrt{2}}\right)^2 + \left(\frac{v_{i-1j-1} - v_{ij}}{h\sqrt{2}}\right)^2 \right] - h^2 \sum_{M_{ij} \in \Omega_h} f_{ij}\, v_{ij} \,. \end{aligned}$$

Similarly there is equivalence between:

$$(9.99) \qquad \operatorname*{Min}_{v_h \in K_{0h}^4} J_h^2(v_h)$$

and

$$(9.100) \quad \begin{cases} \displaystyle\operatorname*{Min}_{(v_h,q_h) \in \mathscr{K}_{0h}} \left\{ \frac{h^2}{8} \sum_{M_{ij} \in \bar{\Omega}_h} \left[\left(\frac{v_{i+1j+1} - v_{ij}}{h\sqrt{2}}\right)^2 + \left(\frac{v_{i+1j-1} - v_{ij}}{h\sqrt{2}}\right)^2 + \right.\right. \\ \displaystyle\quad \left. + \left(\frac{v_{i-1j+1} - v_{ij}}{h\sqrt{2}}\right)^2 + \left(\frac{v_{i-1j-1} - v_{ij}}{h\sqrt{2}}\right)^2 \right] + \\ \displaystyle\quad + \frac{h^2}{4} \sum_{M_{i+1/2\, j+1/2} \in \Sigma_h} \left(q_{1i+1/2\, j+1/2}^2 + q_{2i+1/2\, j+1/2}^2\right) - \\ \displaystyle\quad \left. - h^2 \sum_{M_{ij} \in \Omega_h} f_{ij}\, v_{ij} \right\}. \quad \blacksquare \end{cases}$$

Remark 9.9. *The approximation of* (P_0) defined by (9.96) was not considered in Sections 5 and 6; it may easily be shown that the results (of convergence, in particular) demonstrated for the other approximations apply equally well to problem (9.96). ■

We denote by $\mathscr{I}_h^l$ the functional from $V_h \times L_h \to \mathbb{R}$ defined by (9.97) if $l = 1$, and by (9.100) if $l = 2$; we associate with problems (9.98), (9.100) the functionals, from $(V_h \times L_h) \times L_h \to \mathbb{R}$, $\mathscr{L}_{3h}^1$ and $\mathscr{L}_{3h}^2$, which are approximations of $\mathscr{L}_3$ and are defined by:

$$(9.101)\quad \begin{cases} \mathscr{L}_{3h}^l(v_h, q_h ; \mu_h) = \mathscr{I}_h^l(v_h, q_h) + \dfrac{h^2}{2} \displaystyle\sum_{M_{i+1/2\, j+1/2} \in \Sigma_h} \Bigg[\mu_{1i+1/2\, j+1/2} \times \\ \quad \times \left(\dfrac{v_{i+1j+1} - v_{ij+1} + v_{i+1j} - v_{ij}}{2\,h} - q_{1i+1/2\, j+1/2}\right) + \\ \quad + \mu_{2i+1/2\, j+1/2}\left(\dfrac{v_{i+1j+1} - v_{i+1j} + v_{ij+1} - v_{ij}}{2\,h} - q_{2i+1/2 j+1/2}\right)\Bigg], \\ \qquad\qquad l = 1, 2 . \end{cases}$$

Using a variant of Proposition 9.3 it may be shown that $\mathscr{L}_{3h}^l$ $(l = 1, 2)$ admits a saddle-point, $\{u_h^l, p_h^l, \lambda_h^l\}$ on $(V_h \times \Lambda_h) \times L_h$ with:

(9.102) u_h^1 the solution of (9.96), u_h^2 the solution of (9.99),

$$(9.103)\quad \begin{Bmatrix} p_{1i+1/2\, j+1/2}^l = \dfrac{u_{i+1j+1}^l - u_{ij+1}^l + u_{i+1j}^l - u_{ij}^l}{2\,h} \\ p_{2i+1/2\, j+1/2}^l = \dfrac{u_{i+1j+1}^l - u_{i+1j}^l + u_{ij+1}^l - u_{ij}^l}{2\,h} \end{Bmatrix} \quad \forall M_{i+1/2\, j+1/2} \in \Sigma_h ,$$

$$(9.104)\quad (p_{1i+1/2\, j+1/2}^l, p_{2i+1/2\, j+1/2}^l) = P_{\bar{D}}(\lambda_{1i+1/2\, j+1/2}^l, \lambda_{2i+1/2\, j+1/2}^l) \quad \forall M_{i+1/2\, j+1/2} \in \Sigma_h \quad (l = 1, 2) ,$$

where, in (9.104) $P_{\bar{D}}$ is defined in Section 8.3.3 by (8.82) (8.83)

9.5.4 *Case of the exterior approximations of Section 5. (II). Description of the algorithm and convergence*

We use algorithm (4.12)-(4.15) of Chapter 2, Section 4.3, which in this particular case is expressed (omitting the upper index l)

1) for $l = 1$, by

$$(9.105)_1\quad \begin{cases} \text{with } \lambda_h^n \text{ known } (\in L_h) \text{ we define } u_h^n \in V_h , \text{ then } p_h^n \in \Lambda_h \text{ by:} \\ -\dfrac{u_{i+1j}^n + u_{i-1j}^n + u_{ij+1}^n + u_{ij-1}^n - 4_{ij}^n}{h^2} = 2\, f_{ij} + \\ \quad + \dfrac{\lambda_{1i+1/2\, j+1/2}^n - \lambda_{1i-1/2\, j+1/2}^n + \lambda_{1i+1/2\, j-1/2}^n - \lambda_{1i-1/2\, j-1/2}^n}{2\,h} + \\ \quad + \dfrac{\lambda_{2i+1/2\, j+1/2}^n - \lambda_{2i+1/2\, j-1/2}^n + \lambda_{2i-1/2\, j+1/2}^n - \lambda_{2i-1/2\, j-1/2}^n}{2\,h} \\ \forall M_{ij} \in \Omega_h ; \end{cases}$$

2) for $l = 2$, by

$$(9.105)_2 \quad \begin{cases} -\dfrac{u^n_{i+1j+1} + u^n_{i-1j+1} + u^n_{i+1j-1} + u^n_{i-1j-1} - 4u^n_{ij}}{2h^2} = 2f_{ij} + \\ + \dfrac{\lambda^n_{1i+1/2\,j+1/2} - \lambda^n_{1i-1/2\,j+1/2} + \lambda^n_{1i+1/2\,j-1/2} - \lambda^n_{1i-1/2\,j-1/2}}{2h} + \\ + \dfrac{\lambda^n_{2i+1/2\,j+1/2} - \lambda^n_{2i+1/2\,j-1/2} + \lambda^n_{2i-1/2\,j+1/2} - \lambda^n_{2i-1/2\,j-1/2}}{2h} \\ \forall M_{ij} \in \Omega_h \end{cases}$$

with, for $l = 1$ or 2

$$(9.106) \quad (p^n_{1i+1/2\,j+1/2}, p^n_{2i+1/2\,j+1/2}) = P_{\bar{D}}(\lambda^n_{1i+1/2\,j+1/2}, \lambda^n_{2i+1/2\,j+1/2}) \quad \forall M_{i+1/2\,j+1/2} \in \Sigma_h\,,$$

$$(9.107) \quad \lambda^{n+1}_{1i+1/2\,j+1/2} = \lambda^n_{1i+1/2\,j+1/2} + \rho_n\left(\frac{u^n_{i+1j+1} - u^n_{ij+1} + u^n_{i+1j} - u^n_{ij}}{2h} - p^n_{1i+1/2\,j+1/2}\right) \quad \forall M_{i+1/2\,j+1/2} \in \Sigma_h\,,$$

$$(9.108) \quad \lambda^{n+1}_{2i+1/2\,j+1/2} = \lambda^n_{2i+1/2\,j+1/2} + \rho_n\left(\frac{u^n_{i+1j+1} - u^n_{i+1j} + u^n_{ij+1} - u^n_{ij}}{2h} - p^n_{2i+1/2\,j+1/2}\right) \quad \forall M_{i+1/2\,j+1/2} \in \Sigma_h\,,$$

$$(9.109) \quad \rho_n > 0\,, \quad \text{sufficiently small.}$$

The above algorithm is an "approximation" of algorithm (9.23), (9.24), (9.25) and relations $(9.105)_l$ $(l = 1,2)$ are the discrete analogues of (9.26). ■

It follows from Chapter 2, Section 4.3, Theorem 4.1 that the sequence $\{u^n_h, p^n_h\}$ defined by $(9.105)_l$ $(l = 1,2)$, (9.106), (9.107), (9.108) is convergent, for ρ_n sufficiently small, to $\{u^l_n, p^l_n\}$. ■

9.5.5 *Case of the exterior approximations of Section 5. (III). A numerical example.*

We consider the problem (P_0) relating to $\Omega =]0,1[\times]0,1[$ and $f = 10$, which was treated earlier; we confine our attention to the approximation defined by (9.96); the algorithm used is thus defined by $(9.105)_1$, ... , (9.109):

Mesh size: $h = 1/20$.

Initialisation of algorithm $(9.105)_1, \ldots, (9.109)$: $\lambda_h^0 = 0$.

Determination of u_h^n *from* λ_h^n : by point over-relaxation, the algorithm being initialised by u_h^{n-1}. With m denoting the iteration count for the over-relaxation, we take as termination criterion:

$$\sum_{M_{ij} \in \Omega_h} | u_{ij}^{m+1} - u_{ij}^m | \leqslant \varepsilon_1 .$$

Termination criterion for algorithm $(9.105)_1, \ldots, (9.109)$

$$\sum_{M_{ij} \in \Omega_h} | u_{ij}^{n+1} - u_{ij}^n | \leqslant \varepsilon .$$

Analysis of the numerical results.

Table 9.5 shows the results relating to the convergence of $(9.105)_1, \ldots, (9.109)$, for an over-relaxation parameter of 1.75 and with $\rho_n = \rho = 0.75$,

ε	5×10^{-3}	5×10^{-3}	5×10^{-4}
ε_1	10^{-3}	10^{-4}	10^{-4}
No. of iterations required for convergence	120	80	90
Approximate value of $u(0.5, 0.5)$	0.415	0.415	0.414
Computation time in seconds on C II 10070	/	/	40 s

Table 9.5

As far as the precision obtained is concerned, the approximate solution thus calculated coincides to within an absolute value of about 10^{-3} with those obtained by the other methods of this chapter. ■

10. DISCUSSION

Problems in elasto-plasticity have given rise to a very large number of works in the field of numerical analysis.

We refer the reader to Argyris /1/, /2/, /3/, Argyris-Scharpf-Spooner /1/, /2/, Argyris-Scharpf /1/, Argyris-Balmer-Doltsinis-Willam /1/, Nayak-Zienkiewicz /1/, Zienkiewicz-Valliapan-King /1/, Zienkiewicz-Valliapan /1/, Zienkiewicz-Cormeau /1/ and to their respective bibliographies.

Likewise the problem of elasto-plastic torsion has given rise to numerous theoretical and numerical investigations; in the numerical field, we may cite Hodge-Herakovich-Stout /1/, Herakovich-Hodge /1/, Stout-Hodge /1/, Tchernousko-Banitchouk /1/, Bourgat /1/, Cea-Glowinski-Nedelec /2/, Goursat /1/, Marrocco /1/, Brézis-Sibony /1/, Haugazeau /2/, and their respective bibliographies.

In the case of the torsion problem, it would appear that the first-order finite elements described in Section 4.1 are the best adapted for practical calculations; these elements, in fact, allow a better approximation of domains with complicated shape than do the finite difference methods of Section 5; moreover, the low regularity of the solutions, together with the drawbacks pointed out in Sections 4 and 6, would appear to limit the scope of finite elements of order 2 or higher.

As far as methods for solving the approximate problems are concerned, we would like to emphasize the advantage offered by methods of the *duality* type based on the use of Uzawa's algorithm (see Section 9), and to point out the following variant which combines the techniques of Sections 8 and 9; the notation used is the same as that used in Section 8.3, and we write:

(10.1) $$L = L^2(\Omega) \times L^2(\Omega)$$

(10.2) $$\Lambda = \{ q \mid q \in L, q = (q_1, q_2), | q |^2 = q_1^2(x) + q_2^2(x) \leqslant 1 \text{ a.e.} \}$$

(10.3) $$j(v, q) = \frac{1}{2} \int_\Omega | q |^2 \, dx - \int_\Omega fv \, dx ;$$

there is then equivalence between (P_0) and [1]:

(10.4) $$\operatorname*{Min}_{(v,q)} j(v, q) , \qquad (v, q) \in H_0^1(\Omega) \times \Lambda , \qquad \nabla v - q = 0 .$$

Let u be the solution of (P_0); then $(u, \nabla u)$ is the unique solution of (10.4). It is natural to associate with (10.4) the *Lagrangian* $\mathscr{L}$, obtained by dualisation of $\nabla v - q = 0$, that is:

$$\mathscr{L}(v, q ; \mu) = j(v, q) + \int_\Omega \mu.(\nabla v - q) \, dx .$$

With μ and q fixed, $\mathscr{L}$ is linear in v and hence *non-coercive*,

---

[1] In the following, we write : $\mathbf{grad}\, v = \nabla v$.

and this is a disadvantage if there is a need to calculate a possible saddle-point of $\mathcal{L}$ on $H_0^1(\Omega) \times \Lambda \times L$ using Uzawa's method; to remedy this situation we apply a principle due to Hestenes /2/ and penalise $\nabla v - q = 0$, giving

$$\mathcal{L}_\varepsilon(v, q ; \mu) = \mathcal{L}(v, q ; \mu) + \frac{1}{2\varepsilon} \int_\Omega |\nabla v - q|^2 \, dx .$$

It may be shown that any saddle-point of $\mathcal{L}_\varepsilon$ on $H_0^1(\Omega) \times \Lambda \times L$ is of the form $(u, \nabla u ; \lambda)$ with u the solution of (P_0).

We apply Uzawa's algorithm to $\mathcal{L}_\varepsilon$, i.e.

(10.5) $\quad \lambda^0 \in L$ given,

with λ^n known, u^n, p^n, λ^{n+1} are determined successively by:

$$\text{(10.6)} \qquad \begin{cases} \mathcal{L}_\varepsilon(u^n, p^n ; \mu^n) \leqslant \mathcal{L}_\varepsilon(v, q ; \mu^n) & \forall (v, q) \in H_0^1(\Omega) \times \Lambda \\ (u^n, p^n) \in H_0^1(\Omega) \times \Lambda \end{cases}$$

$$\text{(10.7)} \qquad \lambda^{n+1} = \lambda^n + \rho_n(\nabla u^n - p^n), \qquad \rho_n > 0 .$$

By assuming for $\mathcal{L}_\varepsilon$ the existence of a saddle point $(u, \nabla u ; \lambda)$ on $H_0^1(\Omega) \times \Lambda \times L$, it may be shown that under the condition

$$\text{(10.8)} \qquad 0 < r_0 \leqslant \rho_n \leqslant r_1 < \frac{2}{\varepsilon}$$

we have:

$$\lim_{n \to +\infty} (u^n, p^n) = (u, \nabla u) \ \textit{strongly} \text{ in } H_0^1(\Omega) \times L$$

It may be noted that the determination of (u^n, p^n), using (10.6) is a finite-dimensional *quadratic programming* problem which can be solved easily by means of the method of *block over-relaxation with projection* described and studied in Cea-Glowinski /2/. The problem of the existence of a saddle-point for $\mathcal{L}_\varepsilon$ on $H_0^1(\Omega) \times \Lambda \times L$ seems to be open in the general case; on the other hand, there is no difficulty in proving such a result for the restriction of $\mathcal{L}_\varepsilon$ to the finite-dimensional approximations of $H_0^1(\Omega) \times \Lambda \times L$ defined by means of a finite element method such as, for example the first-order finite element method described in Section 4. We refer to some recent work by Fortin-Glowinski-Marrocco for a more detailed study of the numerical implementation of (10.5) - (10.7); a study of other nonlinear problems treated by means of variants of (10.5) - (10.7) may also be found in this same article.

Chapter 4

THERMAL CONTROL PROBLEMS, BOUNDARY UNILATERAL PROBLEMS, AND ELLIPTIC VARIATIONAL INEQUALITIES OF ORDER 4

SYNOPSIS

In this chapter we shall consider various problems relating to *numerical* aspects of the methods and technicalities studied in Chapters 1 and 2.

In Sections 1, 2 and 3 we shall study various unilateral elliptic problems of order 2, the unilateral behaviour being related to conditions at the boundary; in Section 4 we shall study the use of elliptic variational inequalities of order 4 for modelling unilateral problems for the case of thin plates. ■

1. PROBLEMS OF THERMAL CONTROL AND OF THE DIFFUSION OF FLUIDS THROUGH SEMI-PERMEABLE WALLS

1.1 Formulation of the problem

A certain number of model problems concerning the thermal control of enclosures Ω, either by means of air-conditioning applied across the boundary, or by diffusion, lead to the determination of a function $u(x)$, $x \in \Omega$ which satisfies

(1.1) $$\mu u - \Delta u = f \quad \text{in } \Omega$$

(1.2) $$\frac{\partial u}{\partial n} + \varphi(u) = 0 \quad \text{on } \Gamma_d$$

(1.3) $$u = 0 \quad \text{on } \Gamma - \Gamma_d$$

or

(1.3)′ $$\frac{\partial u}{\partial n} = 0 \quad \text{on } \Gamma - \Gamma_d .$$

In (1.1)-(1.3)', μ is a constant $\geqslant 0$, Γ_d a part of the boundary Γ of Ω and $\varphi : \mathbb{R} \to \mathbb{R}$ a given *increasing* function.

Following Duvaut-Lions /1/, Chapter 1, the solution of problems (1.1), (1.2), (1.3) and (1.1), (1.2), (1.3)' reduces to the solution of the minimisation problem:

(1.4) $$\begin{cases} J(u) \leqslant J(v) & \forall v \in V \\ u \in V \end{cases}$$

with, assuming that $f \in L^2(\Omega)$,

$$J(v) = \frac{1}{2}\int_\Omega (\mu v^2 + |\operatorname{grad} v|^2)\,dx - \int_\Omega fv\,dx + \int_{\Gamma_d} \Psi(v)\,d\Gamma\ ; \tag{1.5}$$

in (1.4), we take

$$V = \{ v \mid v \in H^1(\Omega),\ v|_{\Gamma-\Gamma_d} = 0 \} \tag{1.6}$$

in the case of problem (1.1), (1.2), (1.3) and

$$V = H^1(\Omega) \tag{1.7}$$

in the case of problem (1.1), (1.2), (1.3)'; in (1.5) the function Ψ is the primitive of φ, which vanishes at the origin, i.e.

$$\Psi(\xi) = \int_0^\xi \varphi(\tau)\,d\tau\,. \tag{1.8}$$

Remark 1.1. In order to help simplify the description, it will be assumed in the following that $\mu > 0$; the reader may consult Duvaut-Lions /1/, Chapter 1, Section 7.3, for a theoretical study of the case $\mu = 0$. ■

Remark 1.2. The function φ considered in this section will be of the following type:

$$\varphi(\xi) = \begin{cases} g & \text{if} \quad \xi \geqslant kg \\ \xi/k & \text{if} \quad -kg < \xi < kg \\ -g & \text{if} \quad \xi \leqslant -kg \end{cases} \tag{1.9}$$

where g and k in (1.9) are *positive* constants. From (1.8), we thus have

$$\Psi(\xi) = \begin{cases} g\xi - \frac{1}{2}kg^2 & \text{if} \quad \xi \geqslant kg \\ \dfrac{\xi^2}{2k} & \text{if} \quad |\xi| \leqslant kg \\ -g\xi - \frac{1}{2}kg^2 & \text{if} \quad \xi \leqslant -kg\,. \end{cases} \tag{1.10}$$

The functions φ and Ψ, respectively, are shown in Figures 1.1 and 1.2.

Some examples relating to functions φ more general than that defined by (1.9), and which may be discontinuous, are given in Duvaut-Lions /1/, Chapter 1. (See also Section 2 of the present chapter for $\Psi(v) = g\,|v|$).

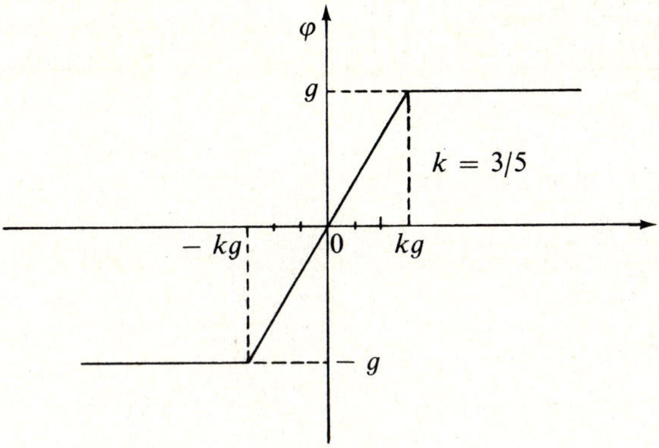

Fig. 1.1. Graph of $\varphi(v)$.

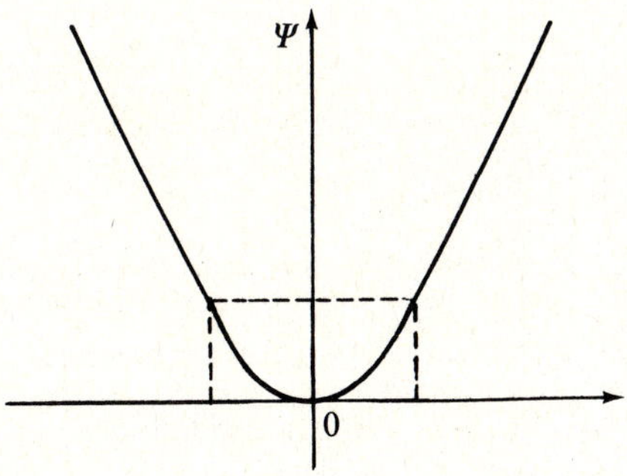

Fig. 1.2. Graph of $\psi(v)$.

1.2 Existence and uniqueness results for problem (1.4), (1.5).

In the following, $\| v \|$ will denote the standard norm of the ace $H^1(\Omega)$, that is

$$\| v \| = \left(\int_\Omega (v^2 + | \operatorname{grad} v |^2 \, dx \right)^{1/2} .$$

We have:

Theorem 1.1 *Assuming Ω to be bounded and with Lipschitz boundary Γ, then the problem* (1.4), (1.5) *admits one and only one solution.*

Proof. The functional J is strictly convex with

$$\lim_{\|v\| \to +\infty} J(v) = +\infty ;$$

since the continuity in $H^1(\Omega)$ of

$$v \to \frac{1}{2} \int_\Omega (|\operatorname{grad} v|^2 + \mu |v|^2) \, dx - \int_\Omega f v \, dx$$

is obvious, in order to apply Theorem 2.1, of Chapter 1, Section 2.1, it is sufficient to prove the continuity of $j : H^1(\Omega) \to \mathbb{R}$ defined by

$$j(v) = \int_{\Gamma_d} \Psi(v) \, d\Gamma .$$

We have:

$$|\Psi(\eta) - \Psi(\xi)| \leqslant g |\eta - \xi| \quad \forall \xi, \eta \in \mathbb{R} \tag{1.11}$$

so that

$$|j(v_2) - j(v_1)| \leqslant g \int_{\Gamma_d} |v_2 - v_1| \, d\Gamma \quad \forall v_2, v_1 \in L^2(\Gamma) ; \tag{1.12}$$

now, with γ_0 denoting the trace mapping, we have

$$\int_{\Gamma_d} |v| \, d\Gamma \leqslant \int_\Gamma |v| \, d\Gamma \leqslant \left(\int_\Gamma d\Gamma \right)^{1/2} \| \gamma_0 v \|_{L^2(\Gamma)} \tag{1.13}$$

and (continuity of the trace mapping)

$$\| \gamma_0 v \|_{L^2(\Gamma)} \leqslant C(\Omega) \| v \|_{H^1(\Omega)} \quad \forall v \in H^1(\Omega) , \tag{1.14}$$

so that the result follows from (1.12) - (1.14). ∎

Remark 1.3. The above result still holds if we assume $k = 0$ in (1.9), in which case

$$j(v) = g \int_{\Gamma_d} |v| \, d\Gamma ;$$

we shall return to the above point in Section 2. ∎

1.3 Approximation by a finite element method

In this section we shall study the approximation of problem (1.4), (1.5), using Lagrangian finite elements of order 1 and 2. To simplify the description it will be assumed in the following that Ω is a *polygonal* domain (and hence Lipschitz).

1.3.1 *Triangulation of* Ω. *Definition of* V_h

We proceed as in Chapter 1, Section 4, Example 4.2 and Chapter 3, Section 4.1; we thus suppose that $\mathcal{T}_h$ is a *finite triangulation* of Ω such that

$$T \subset \overline{\Omega} \quad \forall T \in \mathcal{T}_h, \quad \bigcup_{T \in \mathcal{T}_h} T = \overline{\Omega} \tag{1.15}$$

$$\begin{cases} T \text{ and } T' \in \mathcal{T}_h \Rightarrow T \cap T' = \phi, \text{ or } T \text{ and } T' \text{ have a common side,} \\ \text{or } T \text{ and } T' \text{ and have a common vertex.} \end{cases} \tag{1.16}$$

We put:

$$h = \text{length of the longest side of the } T \in \mathcal{T}_h \tag{1.17}$$

and

$$\begin{cases} \Sigma_h = \text{set of the vertices of } \mathcal{T}_h \\ \Sigma'_h = \text{set of the midpoints of the sides of the triangles of } \mathcal{T}_h. \end{cases} \tag{1.18}$$

In view of (1.15) *all* the vertices of Γ belong to Γ_h; we shall assume, moreover, that Γ_d is the union of sides, *the ends of which lie on* Γ, of elements of $\mathcal{T}_h$.

Classically, the space $H^1(\Omega)$ is approximated by

$$\mathscr{V}_h = \{ v_h \mid v_h \in H^1(\Omega) \cap C^0(\overline{\Omega}),\ v_h|_T \in P_q \ \forall T \in \mathcal{T}_h \} \tag{1.19}$$

in which $v_h|_T$ denotes the restriction of v_h to T and P_q the space of polynomials of degree $\leqslant q$; we shall confine our attention to $q = 1$ and 2.

The space V_h associated with the problem (1.1), (1.2), (1.3) will thus be defined by

$$V_h = \{ v_h \mid v_h \in \mathscr{V}_h,\ v_h|_{\Gamma - \Gamma_d = 0} \}; \tag{1.20}$$

in the case of problem (1.1), (1.2), (1.3)' we shall take

$$V_h = \mathscr{V}_h. \tag{1.21}$$

The natural degrees of freedom for the above approximations are the values taken by v_h on Σ_h (resp. $\Sigma_h \cup \Sigma_h'$) if $q = 1$ (resp. $q = 2$). ■

1.3.2 *Approximation of j*

The functional

$$j(v) = \int_{\Gamma_d} \Psi(v)\, d\Gamma$$

is defined on the above spaces V_h but is not very practical to use; we shall therefore approximate it by j_h which is obtained from j by *numerical integration*, or, which amounts to the same thing, by analytical integration of a suitably chosen *interpolant* of the function $\Psi(v_h)$: we shall denote by W_h the space of the traces on Γ_d of the functions on V_h ; given $v_h \in V_h$, we "approximate" $\Psi(v_h)$ by $\Psi_h(v_h)$ defined by

$$(1.22) \quad \begin{cases} \Psi_h(v_h) \in W_h\,, \ (\Psi_h(v_h))(P) = \Psi(v_h(P)) \\ \forall P \in \Gamma_d \cap \Sigma_h \ \text{ if } q = 1\,, \ \forall P \in \Gamma_d \cap (\Sigma_h \cup \Sigma_h') \quad \text{if} \quad q = 2\,; \end{cases}$$

we then approximate $j(v)$ by

$$(1.23) \qquad j_h(v_h) = \int_{\Gamma_d} \Psi_h(v_h)\, d\Gamma\,;$$

the *convexity* and the *continuity of* j_h are obvious.

With the notation shown in Figure 1.3 for $q = 1$ and in Figure 1.4 for $q = 2$, and summing over the sides of the triangles of $\mathcal{T}_h$

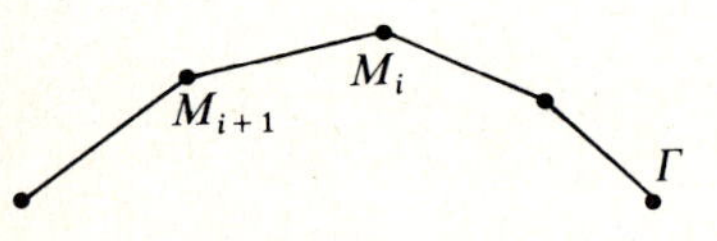

Fig. 1.3.

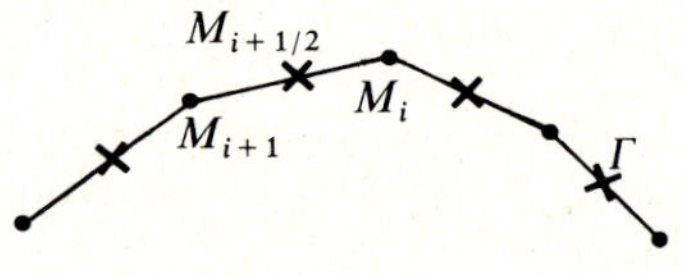

Fig. 1.4.

whose union constitutes Γ_d , we have

$$(1.24) \qquad j_h(v_h) = \tfrac{1}{2} \sum | M_i M_{i+1} | (\Psi(v_h(M_i)) + \Psi(v_h(M_{i+1})))\,, \quad \text{if} \quad q = 1$$

$$(1.25) \qquad j_h(v_h) = \tfrac{1}{6} \sum | M_i M_{i+1} | (\Psi(v_h(M_i)) + 4\, \Psi(v_h(M_{i+1/2})) + \Psi(v_h(M_{i+1})))\,, \quad \text{if} \quad q = 2\,.$$

The formula (1.24) (resp. (1.25)) is clearly related to the *trapezoidal* rule (resp. *Simpson's* rule). ■

1.3.3 *Definition and solvability of the approximate problems*

Using the notation defined in Sections 1.3.1, 1.3.2, and with $a : V \times V \to \mathbb{R}$ denoting the symmetric bilinear form

$$a(u, v) = \int_\Omega (\text{grad } u . \text{grad } v + \mu uv) \, dx \tag{1.26}$$

we approximate problem (1.4), (1.5) by

$$\begin{cases} \frac{1}{2} a(u_h, u_h) + j_h(u_h) - \int_\Omega f u_h \, dx \leqslant \frac{1}{2} a(v_h, v_h) + j_h(v_h) - \\ \qquad - \int_\Omega f v_h \, dx \quad \forall v_h \in V_h \quad u_h \in V_h . \end{cases} \tag{1.27}$$

The minimisation problem (1.27) is equivalent to the variational inequality:

$$\begin{cases} a(u_h, v_h - u_h) + j_h(v_h) - j_h(u_h) \geqslant \int_\Omega f(v_h - u_h) \, dx \quad \forall v_h \in V_h \\ u_h \in V_h . \quad ■ \end{cases} \tag{1.28}$$

The convexity and the continuity of j_h allow the approximate analogue of Theorem 1.1 to be proved, giving:

<u>Theorem 1.2</u>. *The problem* (1.27) *admits one and only one solution.* ■

In order to solve the various approximate problems in an effective manner, it is appropriate to express (1.24) - or (1.25) - explicitly in terms of the values taken by u_h and v_h on Σ_h if $q = 1$, or on $\Sigma_h \cup \Sigma_h'$ if $q = 2$; we shall return to the above point in Section 1.6. ■

To help clarify the study of the convergence of u_h when $h \to 0$, we shall distinguish between the cases $q = 1$ (see Section 1.4) and $q = 2$ (see Section 1.5)

1.4 Convergence of the approximate solutions (I). The case $q = 1$

Throughout this section, C will be used to denote various constants.

1.4.1 *Two lemmas*

For the study of the convergence, when $h \to 0$, we shall make use of the following two lemmas.

Lemma 1.1 *We have*

(1.29) $$j_h(v_h) \geqslant j(v_h) \quad \forall v_h \in V_h\,, \quad \forall h\,.$$

Proof. We use the same notation as that in Section 1.3.2. With $M \in \widehat{M_iM}_{i+1}$, we have $\overrightarrow{M_iM} = \theta\overrightarrow{M_iM}_{i+1}$ where

$$\theta = | M_i\, M |/| M_i\, M_{i+1} |,$$ so that

(1.30) $$0 \leqslant \theta \leqslant 1 \quad \forall M \in \widehat{M_iM}_{i+1}\,;$$

with the function v_h *affine* and Ψ *convex*, we have successively (writing $v_i = v_h(M_i)$) :

(1.31) $$v_h(M) = (1 - \theta)\, v_i + \theta v_{i+1}$$

(1.32) $$\Psi(v_h(M)) \leqslant (1 - \theta)\, \Psi(v_i) + \theta\Psi(v_{i+1})\,.$$

Furthermore, we have

(1.33) $$j(v_h) = \sum \int_{M_i}^{M_{i+1}} \Psi(v_h)\, \mathrm{d}\Gamma$$

and from (1.32), (1.33), we deduce by integration that:

$$j(v_h) \leqslant \sum | M_i\, M_{i+1} | \left[\Psi(v_i) + (\Psi(v_{i+1}) - \Psi(v_i)) \int_0^1 \theta\, \mathrm{d}\theta \right]$$

so that from (1.24)

(1.34) $$j(v_h) \leqslant \tfrac{1}{2} \sum | M_i\, M_{i+1} | \,(\Psi(v_i) + \Psi(v_{i+1})) = j_h(v_h)$$

which proves (1.29). ■

Lemma 1.2 *Let* $s \in [0, 1]$; *we have*

(1.35) $$| j_h(v_h) - j(v_h) | \leqslant C_0\, g h^s \, \| \gamma_0\, v_h \|_{H^s(\Gamma)}$$

with, in (1.35), C_0 *independent of* k, g, h, s, v_h.

Proof. Let $v_h \in V_h$; using the notation shown in Figure 1.5 we define $q_h\, \gamma_0\, v_h \in L^2(\Gamma)$ by:

(1.36) $$q_h\, \gamma_0\, v_h = \sum_{M_i \in \Gamma \cap \Sigma_h} v_h(M_i)\, \varpi_i$$

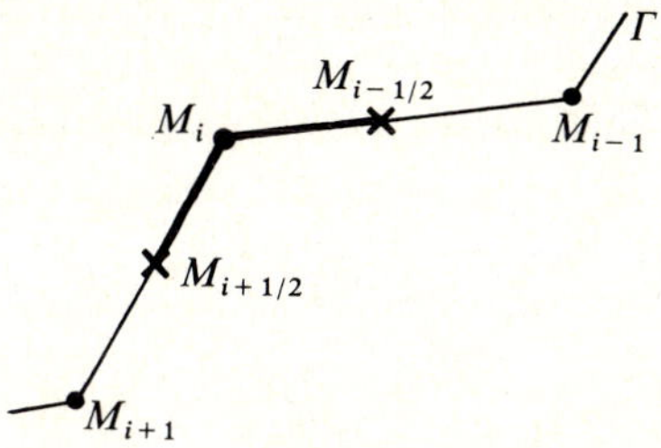

Fig. 1.5.

with

$$\varpi_i = \text{characteristic function of the "arc" } \widehat{M_{i-1/2}M_iM_{i+1/2}}\,. \tag{1.37}$$

For the function $\Psi_h(v_h)$ defined by (1.22), it may be shown that

$$\int_{\Gamma_d} \Psi_h(v_h)\,\mathrm{d}\Gamma = \int_{\Gamma_d} \Psi(q_h\,\gamma_0\,v_h)\,\mathrm{d}\Gamma \tag{1.38}$$

so that

$$j_h(v_h) = \int_{\Gamma_d} \Psi(q_h\,\gamma_0\,v_h)\,\mathrm{d}\Gamma\,. \tag{1.39}$$

From (1.11) and (1.39), we have

$$\begin{aligned} | j_h(v_h) - j(v_h) | &\leqslant g \int_\Gamma | v_h - q_h\,\gamma_0\,v_h |\,\mathrm{d}\Gamma \leqslant \\ &\leqslant g\left(\int_\Gamma \mathrm{d}\Gamma\right)^{1/2} \| \gamma_0\,v_h - q_h\,\gamma_0\,v_h \|_{L^2(\Gamma)}\,. \end{aligned} \tag{1.40}$$

With the notation used in the proof of Lemma 1.1 we have:

$$\begin{aligned} \| \gamma_0\,v_h \|^2_{L^2(\Gamma)} = \int_\Gamma | \gamma_0\,v_h |^2\,\mathrm{d}\Gamma = \frac{1}{6}\sum_i | M_i\,M_{i+1} | \times \\ \times (v_i^2 + (v_i + v_{i+1})^2 + v_{i+1}^2) \end{aligned}$$

and

$$\| q_h \gamma_0 v_h \|^2_{L^2(\Gamma)} = \int_\Gamma | q_h \gamma_0 v_h |^2 \mathrm{d}\Gamma = \frac{1}{2} \sum_i | M_i M_{i+1} | (v_i^2 + v_{i+1}^2) ,$$

Now $\forall \xi$ and $\eta \in \mathbb{R}$ we have

$$\xi^2 + \eta^2 \leqslant \xi^2 + (\xi + \eta)^2 + \eta^2$$

so that

$$(1.41) \qquad \| q_h \gamma_0 v_h \|_{L^2(\Gamma)} \leqslant 2 \| \gamma_0 v_h \|_{L^2(\Gamma)} \quad \forall v_h \in V_h$$

and from (1.41) we deduce

$$(1.42) \qquad \| \gamma_0 v_h - q_h \gamma_0 v_h \|_{L^2(\Gamma)} \leqslant 3 \| \gamma_0 v_h \|_{L^2(\Gamma)} \quad \forall v_h \in V_h . \quad \blacksquare$$

Moreover, we have:

$$\gamma_0 v_h(M) - q_h \gamma_0 v_h(M) = v_h(M) - v_h(M_i) \quad \forall M \in M_i M_{i+1/2}$$

so that, v_h being *affine* over $\widehat{M_i M}_{i+1}$,

$$(1.43) \qquad | \gamma_0 v_h(M) - q_h \gamma_0 v_h(M) | = | M M_i | \frac{| v_{i+1} - v_i |}{| M_i M_{i+1} |} \quad \forall M \in \widehat{M_i M}_{i+1/2} ;$$

and likewise

$$(1.44) \qquad | \gamma_0 v_h(M) - q_h \gamma_0 v_h(M) | = | M M_{i+1} | \frac{| v_{i+1} - v_i |}{| M_i M_{i+1} |} \quad \forall M \in \widehat{M_{i+1/2} M}_{i+1} .$$

By integration over $\widehat{M_i M}_{i+1}$ we deduce from (1.43), (1.44) that

$$(1.45) \qquad \int_{\widehat{M_i M}_{i+1}} | \gamma_0 v_h - q_h \gamma_0 v_h |^2 \mathrm{d}\Gamma = \frac{| M_i M_{i+1} |^3}{12} \frac{| v_{i+1} - v_i |^2}{| M_i M_{i+1} |^2} ,$$

but since $\gamma_0 v_h$ is affine over $\widehat{M_i M}_{i+1}$ we have

$$\left| \frac{\mathrm{d}(\gamma_0 v_h)}{\mathrm{d}\Gamma} \right| = \frac{| v_{i+1} - v_i |}{| M_i M_{i+1} |} \quad \text{over} \quad \widehat{M_i M}_{i+1}$$

so that

$$\int_{\widehat{M_iM_{i+1}}} |\gamma_0 v_h - q_h \gamma_0 v_h|^2 \, d\Gamma = \frac{|M_i M_{i+1}|^2}{12} \times$$

$$\times \int_{\widehat{M_iM_{i+1}}} \left| \frac{d(\gamma_0 v_h)}{d\Gamma} \right|^2 d\Gamma$$

and by summation

$$\int_\Gamma |\gamma_0 v_h - q_h \gamma_0 v_h|^2 \, d\Gamma = \frac{1}{12} \sum_i |M_i M_{i+1}|^2 \times \tag{1.46}$$

$$\times \int_{\widehat{M_iM_{i+1}}} \left| \frac{d(\gamma_0 v_h)}{d\Gamma} \right|^2 d\Gamma .$$

From the definition of h (see (1.17)) we deduce from (1.46) that

$$\| \gamma_0 v_h - q_h \gamma_0 v_h \|^2_{L^2(\Gamma)} \leqslant \frac{h^2}{12} \int_\Gamma \left| \frac{d(\gamma_0 v_h)}{d\Gamma} \right|^2 d\Gamma \leqslant$$

$$\leqslant \frac{h^2}{12} \| \gamma_0 v_h \|^2_{H^1(\Gamma)} ,$$

which may be written:

$$\| \gamma_0 v_h - q_h \gamma_0 v_h \|_{L^2(\Gamma)} \leqslant \frac{h}{2\sqrt{3}} \| \gamma_0 v_h \|_{H^1(\Gamma)} . \tag{1.47}$$

By *interpolation* between $L^2(\Gamma)$ and $H^1(\Gamma)$ we deduce from (1.42), (1.47) that

$$\| \gamma_0 v_h - q_h \gamma_0 v_h \|_{L^2(\Gamma)} \leqslant C h^s \| \gamma_0 v_h \|_{H^s(\Gamma)} \tag{1.48}$$

with C independent of h, s, v_h. Relation (1.48) in conjunction with (1.40) proves (1.35) and hence the lemma. ■

1.4.2 *A convergence theorem*

The following considerations also apply if we assume $k = 0$ in (1.9). We have:

Theorem 1.3. *If, as $h \to 0$, the angles of $\mathcal{T}_h$ are bounded below by a constant $\theta_0 > 0$, we have*

(1.49) $\lim_{h\to 0} u_h = u$ *strongly in* $H^1(\Omega)$

where u *and* u_h *are respectively solutions of the problems* (1.4), (1.5) *and* (1.27).

Proof. This is a variant of the proof of Theorem 4.2 of Chapter 1, Section 4.4. We have:

$$a(u_h, v_h - u_h) + j_h(v_h) - j_h(u_h) \geqslant \int_\Omega f(v_h - u_h)\,\mathrm{d}x \quad \forall v_h \in V_h, \tag{1.50}$$

$$a(u, u_h - u) + j(u_h) - j(u) \geqslant \int_\Omega f(u_h - u)\,\mathrm{d}x. \tag{1.51}$$

By adding (1.50) and (1.51) it may be deduced that $\forall v_h \in V_h$ we have

$$a(u_h - u, u_h - u) + j_h(u_h) - j(u_h) \leqslant \int_\Omega f(u - v_h)\,\mathrm{d}x + $$
$$+ j_h(v_h) - j(u) + a(u_h, v_h - u);$$

from Lemma 1.1, we have $j_h(u_h) - j(u_h) \geqslant 0$, so that

$$a(u_h - u, u_h - u) \leqslant \int_\Omega f(u - v_h)\,\mathrm{d}x + j_h(v_h) - j(u) + a(u_h, v_h - u) \quad \forall v_h \in V_h. \tag{1.52}$$

With $v \in V$, we denote by $r_h v$ the projection of v on V_h relative to the norm

$$v \to \sqrt{a(v, v)}$$

(equivalent to the standard norm of $H^1(\Omega)$); we thus have

$$\begin{cases} a(r_h v, v_h) = a(v, v_h) & \forall v_h \in V_h \\ r_h v \in V_h. \end{cases} \tag{1.53}$$

It is well known (see, for example, Fix-Strang /1/, Ciarlet-Raviart /1/) that under the assumptions adopted for $\mathcal{T}_h$ we have

(1.54) $\lim_{h\to 0} r_h v = v$ *strongly in* V;

putting $v_h = r_h u$ in (1.50) we then deduce from (1.53), (1.54)

$$a(u_h, r_h u - u) = 0 \tag{1.55}$$

(1.56) $$\lim_{h\to 0} \int_\Omega f(u - r_h u)\,dx = 0\,.$$

Moreover

$$| j_h(r_h u) - j(u) | \leqslant | j_h(r_h u) - j(r_h u) | + | j(r_h u) - j(u) | \leqslant$$
$$\leqslant | j_h(r_h u) - j(r_h u) | + Cg \| r_h u - u \|_{L^2(\Gamma)}$$

and taking account of Lemma 1.2 we have

(1.57) $$| j_h(r_h u) - j(u) | \leqslant Cg(\sqrt{h} \| r_h u \|_{H^{1/2}(\Gamma)} + \| r_h u - u \|_{L^2(\Gamma)})\,.$$

Property (1.54) implies that

(1.58) $$\begin{cases} \lim_{h\to 0} \gamma_0 r_h v = \gamma_0 v \quad \text{in } H^s(\Gamma) (\textit{strong}) \\ \forall s \in [0, \tfrac{1}{2}] \end{cases}$$

and (1.57), (1.58) lead to

(1.59) $$\lim_{h\to 0} (j_h(r_h u) - j(u)) = 0\,.$$

Taking account of (1.52) with $v_h = r_h u$, from (1.55), (1.56), (1.59), we have

$$\lim_{h\to 0} a(u_h - u, u_h - u) = 0$$

which proves the strong convergence. ■

Remark 1.4. Theorem 1.3 also holds if in the functional J defined by (1.5) we replace the term $\int_\Omega fv\,dx$ by $L(v)$ where L is an element of the dual V' of the space V; this is important in certain applications. ■

Remark 1.5. If we assume that $f \in L^2(\Omega)$, we have for the solution u of problem (1.4), (1.5), a regularity property of the type

(1.60) $$u \in V \cap H^{1+\alpha}(\Omega)$$

where α is a suitable real number belonging to $]0, 1]$. In view of (1.60), we may expect to obtain estimates of the approximation error as a function of h; this will form the subject of Section 1.4.3 below. ■

1.4.3 *Estimates of the approximation error as a function of h*

In this section we shall give some estimates of the approximation error which are independent of the parameter k in relation (1.9); these estimates are also valid if $k = 0$, in which case we have $j(v) = g \int_{\Gamma_d} |v| \, d\Gamma$. Estimates which are dependent on k will not be considered here since the process for obtaining them comes more under the category of techniques used for solving equations than those used for inequalities. The results of this section are in no way intended to be optimal. In a first reading, the reader may proceed directly to the statement of Theorem 1.4, where the results proved in this section are collected together. ■

In the proof of Theorem 1.3 it was established that the functions u and u_h , which are respectively solutions of problems (1.4), (1.5) and (1.27), satisfy (see (1.52)):

$$(1.61) \qquad a(u_h - u, u_h - u) \leqslant a(u_h, v_h - u) + j_h(v_h) - j(u) + \int_\Omega f(u - v_h) \, dx \qquad \forall v_h \in V_h .$$

We define the *interpolation* operator [1] $\pi_h : C^0(\overline{\Omega}) \to \mathscr{V}_h$ by:

$$(1.62) \qquad \begin{cases} \pi_h v \in \mathscr{V}_h \\ \pi_h v(P) = v(P) \quad \forall P \in \Sigma_h ; \end{cases}$$

suppose $v \in C^0(\overline{\Omega}) \cap V$; then, taking account of the condition imposed on Γ_d , namely that it is the union of sides of triangles of $\mathscr{T}_h$, we in fact have $\pi_h v \in V_h$ whether V_h be defined by (1.20) or (1.21).

We shall take $v_h = \pi_h u$ in (1.61); this is justified since (see Remark 1.5) we have $u \in V \cap H^{1+\alpha}(\Omega)$ $(0 < \alpha \leqslant 1)$ and since - as Ω is a *two-dimensional* open domain with *Lipschitz* boundary - we have (if $\alpha > 0$) $H^{1+\alpha}(\Omega) \subset C^0(\overline{\Omega})$ from a theorem due to Sobolev (see for example, Nečas /1/). We therefore obtain

$$a(u_h - u, u_h - u) \leqslant a(u_h, \pi_h u - u) + j_h(\pi_h u) - j(u) + \int_\Omega f(u - \pi_h u) \, dx ,$$

or

$$(1.63) \qquad a(u_h - u, u_h - u) \leqslant a(u_h - u, \pi_h u - u) + a(u, \pi_h u - u) + \\ + j_h(\pi_h u) - j(u) + \int_\Omega f(u - \pi_h u) \, dx .$$

Since the form a is symmetric we deduce from (1.63) that:

---

(1) $\mathscr{V}_h$ was defined in (1.19).

$$
(1.64)\quad \begin{cases} \frac{1}{2} a(u_h - u, u_h - u) \leqslant \frac{1}{2} a(\pi_h u - u, \pi_h u - u) + \\ \quad + a(u, \pi_h u - u) + j_h(\pi_h u) - j(\pi_h u) + j(\pi_h u) - j(u) + \\ \quad + \int_\Omega f(u - \pi_h u)\,\mathrm{d}x\,. \end{cases}
$$

We shall evaluate the term $a(u, \pi_h u - u)$ by noting that from (1.1) - (1.3)', u is a solution of the equation:

$$
(1.65)\quad \begin{cases} a(u, v) + \int_{\Gamma_a} \varphi(u)\, v\,\mathrm{d}\Gamma = \int_\Omega f v\,\mathrm{d}x \quad \forall v \in V \\ u \in V\,. \end{cases}
$$

Taking $v = \pi_h u - u$ in (1.65) we deduce from (1.64) that:

$$
(1.66)\quad \begin{cases} \frac{1}{2} a(u_h - u, u_h - u) \leqslant \frac{1}{2} a(\pi_h u - u, \pi_h u - u) + j_h(\pi_h u) - \\ \quad - j(\pi_h u) + j(\pi_h u) - j(u) - \int_{\Gamma_a} \varphi(u)\,(\pi u - u)\,\mathrm{d}\Gamma\,. \end{cases}
$$

We have

$$
(1.67)\quad a(\pi_h u - u, \pi_h u - u) \leqslant \operatorname{Max}(1, \mu)\, \| \pi_h u - u \|^2_{H^1(\Omega)}
$$

then from Lemma 1.2

$$
(1.68)\quad \begin{cases} | j_h(\pi_h u) - j(\pi_h u) | \leqslant C_0\, g h^s \| \gamma_0 \pi_h u \|_{H^s(\Gamma)} \\ \text{if } s \in [0, 1] \end{cases}
$$

and finally, from (1.9) and (1.12), (1.13),

$$
(1.69)\quad \begin{cases} | j(\pi_h u) - j(u) - \int_{\Gamma_a} \varphi(u)\,(\pi_h u - u)\,\mathrm{d}\Gamma | \leqslant \\ \quad \leqslant C g \| \gamma_0(\pi_h u - u) \|_{L^2(\Gamma)}, \quad C \text{ independent of } g, k, h, u. \quad \blacksquare \end{cases}
$$

To estimate $a(u_h - u, u_h - u)$ it is thus necessary to estimate

$$
\| \pi_h u - u \|_{H^1(\Omega)}, \quad h^s \| \gamma_0 \pi_h u \|_{H^s(\Gamma)}, \quad \| \gamma_0 \pi_h u - \gamma_0 u \|_{L^2(\Gamma)}
$$

using the regularity property mentioned in Remark 1.5, that is

$u \in V \cap H^{1+\alpha}(\Omega)$, $0 < \alpha \leqslant 1$; we shall distinguish between the two cases (i) and (ii) as follows:

(i) *The case* $V = \{ v \mid v \in H^1(\Omega),\ v\,|_{\Gamma_d - \Gamma} = 0 \}$.

The function u is a solution of the *mixed* (*nonlinear*) problem

$$(1.70)\qquad \begin{cases} -\Delta u + \mu u = f & \text{in } \Omega \\ \dfrac{\partial u}{\partial n} + \varphi(u) = 0 & \text{on } \Gamma_d \\ u = 0 \quad \text{on} \quad \Gamma - \Gamma_d ; \end{cases}$$

we deduce from (1.70), from $|\varphi(u)| \leqslant g$ on Γ_d and from the results concerning the regularity of solutions of mixed problems (see Magènes-Stampacchia /1/), that:

$$(1.71)\qquad \begin{cases} u \in H^{3/2-\varepsilon}(\Omega), \quad \text{so that} \quad \gamma_0 u \in H^{1-\varepsilon}(\Gamma), \\ \varepsilon > 0 \quad \text{arbitrarily small} \end{cases}$$

with

$$(1.72)\qquad \left.\begin{matrix} \| u \|_{H^{3/2-\varepsilon}} \\ \| \gamma_0 u \|_{H^{1-\varepsilon}(\Gamma)} \end{matrix}\right\} \leqslant C(\varepsilon)\,(\| f \|_{L^2(\Omega)} + g), \quad C(\varepsilon) \text{ independent of } f \text{ and } g.$$

To estimate the right-hand sides of the inequalities (1.67) - (1.69) we shall use, once again, *interpolation* in Sobolev spaces.

Estimation of $\| \pi_h u - u \|_{H^1(\Omega)}$.

Following Fix-Strang /1/, Ciarlet-Raviart /1/ etc..., it may be shown that under the assumptions adopted for $\mathcal{T}_h$ in the statement of Theorem 1.3 we have

$$(1.73)\qquad \begin{cases} \| \pi_h v - v \|_{H^1(\Omega)} \leqslant Ch \| v \|_{H^1(\Omega)} \quad \forall v \in H^2(\Omega) \\ C \text{ independent of } h \text{ and } v. \end{cases}$$

Likewise if $\eta > 0$ we have $H^{1+\eta}(\Omega) \subset C^0(\overline{\Omega})$ and

$$(1.74)\qquad \begin{cases} \| \pi_h v - v \|_{H^1(\Omega)} \leqslant C(\eta) \| v \|_{H^{1+\eta}(\Omega)} \quad \forall v \in H^{1+\eta}(\Omega) \\ C(\eta) \text{ independent of } h \text{ and } v. \end{cases}$$

By *interpolation* between $H^{1+\eta}(\Omega)$ and $H^2(\Omega)$ we deduce from (1.71) (1.73), (1.74) that:

$$(1.75)\qquad \begin{cases} \| \pi_h u - u \|_{H^1(\Omega)} \leqslant C(\eta)\, h^{\theta_1} \| u \|_{H^{3/2-\varepsilon}(\Omega)} \\ \text{with } \theta_1 = \dfrac{1}{2} \dfrac{1 - 2(\varepsilon + \eta)}{1 - \eta} \end{cases}$$

where $C(\eta)$ is a constant different from that appearing in (1.74);

from (1.72), (1.75) we deduce

(1.76) $$\| \pi_h u - u \|_{H^1(\Omega)} \leqslant C(\varepsilon, \eta)\, h^{\theta_1} (\| f \|_{L^2(\Omega)} + g) . \quad \blacksquare$$

Estimation of $\| \gamma_0(\pi_h u - u) \|_{L^2(\Gamma)}$.

Since the boundary Γ is *Lipschitz* and *one-dimensional*, we have

$$H^{1/2+\eta}(\Gamma) \subset C^0(\Gamma)$$

if $\eta > 0$. Suppose $v \in C^0(\bar{\Omega})$; the trace of the linear interpolate $\pi_h v$ is the same as the interpolate of the trace $\gamma_0 v$; we have, moreover;

(1.77) $$\| \gamma_0(\pi_h v - v) \|_{L^2(\Gamma)} \leqslant Ch \| \gamma_0 v \|_{H^1(\Gamma)} \leqslant Ch \| v \|_{H^{3/2}(\Omega)} \quad \forall v \in H^{3/2}(\Omega) .$$

Likewise if $\eta > 0$ it may be shown that

(1.78) $$\| \gamma_0(\pi_h v - v) \|_{L^2(\Gamma)} \leqslant C(\eta) \| \gamma_0 v \|_{H^{1/2+\eta}(\Gamma)} \leqslant C(\eta) \| v \|_{H^{1+\eta}(\Omega)} \quad \forall v \in H^{1+\eta}(\Omega) .$$

By interpolation between $H^{1+\eta}(\Omega)$ and $H^{3/2}(\Omega)$ we deduce from (1.71), (1.77), (1.78) that:

(1.79) $$\begin{cases} \| \gamma_0(\pi_h u - u) \|_{L^2(\Gamma)} \leqslant C(\eta)\, h^{\theta_2} \| \gamma_0 u \|_{H^{1-\varepsilon}(\Gamma)} \\ \text{with } \theta_2 = \dfrac{1 - 2(\varepsilon + \eta)}{1 - 2\eta} \end{cases}$$

or, alternatively, using (1.72)

(1.80) $$\| \gamma_0(\pi_h u - u) \|_{L^2(\Gamma)} \leqslant C(\varepsilon, \eta)\, h^{\theta_2} (\| f \|_{L^2(\Omega)} + g) .$$

In the following we shall use (by way of simplification) the same η in (1.75) and (1.79). ■

Estimation of $h^s \| \gamma_0 \pi_h u \|_{H^s(\Gamma)}$.

Following (1.71), we shall take $s = 1 - \varepsilon$; moreover, since the trace of the interpolate $\pi_h v$ is equal to the interpolate of the trace it may be shown that

$$\| \gamma_0 \pi_h v \|_{H^{1-\varepsilon}(\Gamma)} \leqslant C \| \gamma_0 v \|_{H^{1-\varepsilon}(\Gamma)}$$

so that, in view of (1.72) we have:

(1.81) $$\| \gamma_0 \pi_h u \|_{H^{1-\varepsilon}(\Gamma)} \leqslant C(\varepsilon) (\| f \|_{L^2(\Omega)} + g) . \quad \blacksquare$$

Estimation of $\| u_h - u \|_{H^1(\Omega)}$.

We deduce from (1.66)-(1.69), (1.76), (1.80), (1.81):

$$(1.82)\quad \begin{cases} \| u_h - u \|^2_{H^1(\Omega)} \leqslant C(\varepsilon)\, g(\| f \|_{L^2(\Omega)} + g)\, h^{1-\varepsilon} + \\ \quad + C(\varepsilon, \eta)\,(\| f \|_{L^2(\Omega)} + g)^2\, h^{\frac{1-2(\varepsilon+\eta)}{1-\eta}} \times \\ \quad \times \left(1 + \dfrac{g}{\| f \|_{L^2(\Omega)} + g}\, h^{\eta\frac{1-2(\varepsilon+\eta)}{(1-\eta)(1-2\eta)}}\right) \end{cases}$$

with, in (1.82), $C(\varepsilon)$ and $C(\varepsilon,\eta)$ independent of h,k,g,f and ε,η arbitrarily small positive quantities. ∎

(ii) *The case* $V = H^1(\Omega)$.

The function u is a solution of the problem

$$(1.83)\quad \begin{cases} -\Delta u + \mu u = f \quad \text{in } \Omega \\ \dfrac{\partial u}{\partial n} + \varphi(u) = 0 \quad \text{on } \Gamma_d \\ \dfrac{\partial u}{\partial n} = 0 \quad \text{on } \Gamma - \Gamma_d \end{cases}$$

If $\Gamma = \Gamma_d$ it follows (see H. Brézis /2/; Chapter 1), that

$$(1.84)\quad u \in H^{2-\varepsilon}(\Omega)\,,\quad \varepsilon > 0$$

arbitrarily small, and if Ω is convex we can put $\varepsilon = 0$ in (1.84); moreover:

$$(1.85)\quad \| u \|_{H^{2-\varepsilon}(\Omega)} \leqslant C(\varepsilon)\, \| f \|_{L^2(\Omega)}$$

with $C(\varepsilon)$ independent of k,g,f (1). ∎

If $\Gamma_d \neq \Gamma$ we have $\left\| \dfrac{\partial u}{\partial n} \right\|_{L^2(\Gamma)} \leqslant Cg$ so that, from (1.83)

$$(1.86)\quad u \in H^{3/2}(\Omega)$$

with

$$(1.87)\quad \| u \|_{H^{3/2}(\Omega)} \leqslant C(\| f \|_{L^2(\Omega)} + g)\,;$$

we deduce from (1.86) that $\gamma_0 u \in H^1(\Gamma)$ and, since the function φ is Lipschitz (with constant $1/k$), this implies (see, for example,

(1) Since estimate (1.85) is independent of k it still holds if $k = 0$, in which case $j(v) = g \int_\Gamma |v|\, d\Gamma$ (see Section 2 of the present chapter).

H. Brézis /2/, Chapter 1) that $\varphi(\gamma_0 u) \in H^1(\Gamma_d)$ so that

$$\left|\frac{\partial u}{\partial n}\right|_{\Gamma_d} \in H^1(\Gamma_d). \tag{1.88}$$

Since any function on $H^1(\Gamma_d)$, extended by zero in $\Gamma - \Gamma_d$, belongs to $H^{1/2-\varepsilon}(\Gamma)$, $\varepsilon > 0$ arbitrarily small, (1.83) and (1.88) imply that $u \in H^{2-\varepsilon}(\Omega)$. It would be interesting to know whether this regularity property still holds in the limit when $k = 0$ $\left(\text{so that } j(v) = g\int_{\Gamma_d} |v| \,\mathrm{d}\Gamma\right)$, but this problem would appear to be open. ■

In view of (1.84)-(1.87) it would appear that if $V = H^1(\Omega)$ there exists α such that

$$\begin{cases} \frac{1}{2} \leqslant \alpha \leqslant 1 \\ u \in H^{1+\alpha}(\Omega) \end{cases} \tag{1.89}$$

with

$$\begin{cases} \| u \|_{H^{1-\alpha}(\Omega)} \leqslant C(\| f \|_{L^2(\Omega)} + g) \\ C \text{ independent of } k, g, f. \end{cases} \tag{1.90}$$

Using (1.89), (1.90), it may be proved by proceeding as for point (i), that

$$\begin{cases} \| u_h - u \|^2_{H^1(\Omega)} \leqslant Cg(\| f \|_{L^2(\Omega)} + g)\, h + Cg(\| f \|_{L^2(\Omega)} + g) \times \\ \qquad \times h^{1/2+\alpha} + C(\eta)\,(\| f \|_{L^2(\Omega)} + g)^2\, h^{2(\alpha-\eta)/(1-\eta)} \end{cases} \tag{1.91}$$

with in (1.91), C and $C(\eta)$ independent of h, k, g, f, $\eta > 0$, arbitrarily small ($\eta = 0$ if $\alpha = 1$). ■

We summarise the preceding results in:

<u>Theorem 1.4.</u> *Let u be the solution of problem (1.4), (1.5) and u_h the solution of the corresponding approximate problem (1.27); if when $h \to 0$ the angles of $\mathcal{T}_h$ are bounded below by a constant $\theta_0 > 0$, we have the following estimates for the approximation error $\| u_h - u \|_{H^1(\Omega)}$*

(i) *If $V = \{ v \mid v \in H^1(\Omega),\ v|_{\Gamma-\Gamma_d} = 0 \}$ we have:*

$$\begin{cases} \| u_h - u \|^2_{H^1(\Omega)} \leqslant C(\varepsilon)\, g(\| f \|_{L^2(\Omega)} + g)\, h^{1-\varepsilon} + \\ \qquad + C(\varepsilon, \eta)\,(\| f \|_{L^2(\Omega)} + g)^2\, h^{\frac{1-2(\varepsilon+\eta)}{1-\eta}} \times \\ \qquad \times \left(1 + \dfrac{g}{\| f \|_{L^2(\Omega)} + g}\, h^{\eta\, \frac{1-2(\varepsilon-\eta)}{(1-\eta)(1-2\eta)}}\right) \end{cases} \tag{1.92}$$

where in (1.92), $C(\varepsilon)$ and $C(\varepsilon, \eta)$ are independent of h, k, g, f and

ε, η are arbitrarily small positive quantities. ■

(ii) *If $V = H^1(\Omega)$ there exists $\alpha, \frac{1}{2} \leqslant \alpha \leqslant 1$ such that $u \in H^{1+\alpha}(\Omega)$ and we have*

$$(1.93) \quad \begin{cases} \| u_h - u \|^2_{H^1(\Omega)} \leqslant Cg(\| f \|_{L^2(\Omega)} + g)\, h + Cg(\| f \|_{L^2(\Omega)} + g) \times \\ \quad \times\, h^{1/2+\alpha} + C(\eta)\, (\| f \|_{L^2(\Omega)} + g)^2\, h^{2(\alpha-\eta)/(1-\eta)} \end{cases}$$

where in (1.93), C and $C(\eta)$ are independent of h, k, g, f and $\eta > 0$ is arbitrarily small ($\eta = 0$ if $\alpha = 1$). ■

Remark 1.6. If, in the case where $V = \{ v \mid v \in H^1(\Omega),\ v |_{\Gamma - \Gamma_d} = 0 \}$, we have $u \in H^{1+\alpha}(\Omega)$ with $\frac{1}{2} \leqslant \alpha \leqslant 1$ (which is not true in general) we may replace error estimate (1.92) by estimate (1.93). ■

Remark 1.7. If, instead of (1.27), we had taken as the approximate problem:

$$(1.94) \quad \begin{cases} a(u_h, v_h - u_h) + j(v_h) - j(u_h) \geqslant \displaystyle\int_\Omega f(v_h - u_h)\, \mathrm{d}x \quad \forall v_h \in V_h \\ u_h \in V_h \end{cases}$$

which is not very practical from a numerical viewpoint, then instead of (1.92), (1.93) we would have obtained, respectively:

$$(1.95) \quad \| u_h - u \|^2_{H^1(\Omega)} \leqslant C(\varepsilon, \eta)\, (\| f \|_{L^2(\Omega)} + g)^2\, h^{\frac{1-2(\varepsilon+\eta)}{1-\eta}} \times \left(1 + \frac{g h^{\eta \frac{1-2(\varepsilon+\eta)}{(1-\eta)(1-2\eta)}}}{\| f \|_{L^2(\Omega)} + g} \right)$$

$$(1.96) \quad \| u_h - u \|^2_{H^1(\Omega)} \leqslant Cg(\| f \|_{L^2(\Omega)} + g)\, h^{1/2+\alpha} + C(\eta)\, (\| f \|_{L^2(\Omega)} + g)\, h^{2(\alpha-\eta)/(1-\eta)}\,.$$

In view of (1.92), (1.93), (1.95), (1.96) it would appear that the additional error introduced by replacing j by j_h is "practically" of $0(\sqrt{h})$; we may therefore deduce that if, from the point of view of regularity, we have no better than $u \in H^{3/2}(\Omega)$ then the above error is of the same order of magnitude as the approximation error associated with the approximate problem (1.94). On the other hand if we have $u \in H^{1+\alpha}(\Omega)$ with $\alpha > \frac{1}{2}$ it would appear, in view of (1.93), (1.96), that the principal part of the approximation error $\| u_h - u \|_{H^1(\Omega)}$ is the error introduced by replacing j by j_h. ■

1.5 Convergence of the approximate solutions. (II). The case $q = 2$.

Throughout this section, C will be used to denote various constants.

1.5.1 *A lemma*

We shall use the following variant of Lemma 1.2 for the study of the convergence when $h \to 0$:

Lemma 1.3. *Suppose* $s \in [0, 1]$; *we have*

$$|j_h(v_h) - j(v_h)| \leqslant Cgh^s \| \gamma_0 v_h \|_{H^s(\Gamma)} \tag{1.97}$$

where, in (1.97), C *is independent of* k, g, h, s, v_h.

Proof. Suppose $v_h \in V_h$, the notation being that shown in Figure 1.6, and with $|M_{i-1/2} M_{i-5/6}| = |M_{i-1/2} M_{i-1/6}| = \frac{1}{3} |M_{i-1} M_i|$; we define

$$q_h \gamma_0 v_h = \sum_{M_i \in \Gamma \cap \Sigma_h} v_h(M_i) \varpi_i + \sum_{M_{j-1/2} \in \Gamma \cap \Sigma_h} v_h(M_{j-1/2}) \varpi_{j-1/2} \tag{1.98}$$

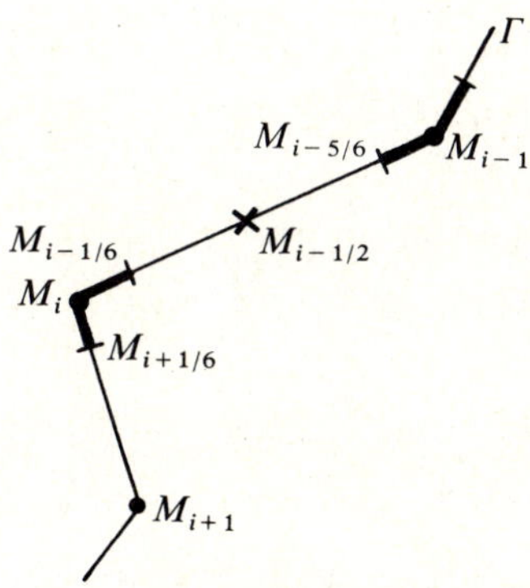

Fig. 1.6.

with

$\varpi_i =$ characteristic function of $M_{i-1/6} M_i M_{i+1/6}$

$\varpi_{j-1/2} =$ characteristic function of $M_{j-5/6} M_{j-1/2} M_{j-1/6}$.

It may be shown that for the function $\psi_h(v_h)$ defined by (1.22) that

$$\int_{\Gamma_d} \psi_h(v_h) \, d\Gamma = \int_{\Gamma_d} \psi(q_h \gamma_0 v_h) \, d\Gamma \tag{1.99}$$

so that

$$j_h(v_h) = \int_{\Gamma_d} \psi(q_h \gamma_0 v_h) \, d\Gamma . \tag{1.100}$$

From (1.11) and (1.100) we have

$$\begin{cases} | j_h(v_h) - j(v_h) | \leqslant g \displaystyle\int_\Gamma | v_h - q_h \gamma_0 v_h | \, d\Gamma \leqslant \\ \leqslant g \left(\displaystyle\int_\Gamma d\Gamma \right)^{1/2} \| \gamma_0 v_h - q_h \gamma_0 v_h \|_{L^2(\Gamma)} \end{cases} \tag{1.101}$$

with

$$\| \gamma_0 v_h \|^2_{L^2(\Gamma)} = \frac{1}{15} \sum_i | M_i M_{i+1} | (2(v_i^2 + v_{i+1}^2) + 8 v_{i+1/2}^2 - - v_i v_{i+1} + 2 v_{i+1/2}(v_i + v_{i+1})) \tag{1.102}$$

and

$$\| q_h \gamma_0 v_h \|^2_{L^2(\Gamma)} = \frac{1}{6} \sum_i | M_i M_{i+1} | (v_i^2 + 4 v_{i+1/2}^2 + v_{i+1}^2) ; \tag{1.103}$$

since

$$\xi^2 + 4 \zeta^2 + \eta^2 \leqslant 2(\xi^2 + \eta^2) + 8 \zeta^2 - \xi\eta + 2 \zeta(\xi + \eta) \tag{1.104}$$

we deduce from (1.102), ... , (1.104) that

$$\| q_h \gamma_0 v_h \|_{L^2(\Gamma)} \leqslant 3 \| \gamma_0 v_h \|_{L^2(\Gamma)} \tag{1.105}$$

so that

$$\| \gamma_0 v_h - q_h \gamma_0 v_h \|_{L^2(\Gamma)} \leqslant 4 \| \gamma_0 v_h \|_{L^2(\Gamma)} \quad \forall v_h \in V_h . \quad \blacksquare \tag{1.106}$$

Before giving an estimate of $\| \gamma_0 v_h - q_h \gamma_0 v_h \|_{L^2(\Gamma)}$ which involves $\| \gamma_0 v_h \|_{H^1(\Gamma)}$, it may be noted that if $\varphi \in P_2$ we have

$$\varphi(\xi) - \varphi(\alpha) = (\xi - \alpha) \, \varphi'\left(\frac{\alpha + \xi}{2} \right), \tag{1.107}$$

so that

$$\int_\alpha^\beta | \varphi(\xi) - \varphi(\alpha) |^2 \, d\xi = \int_\alpha^\beta | \xi - \alpha |^2 \left| \varphi'\left(\frac{\alpha + \xi}{2} \right) \right|^2 d\xi ,$$

or alternatively, writing $\tau = (\alpha + \xi)/2$,

$$(1.108)\qquad \int_\alpha^\beta |\varphi(\xi)-\varphi(\alpha)|^2\,\mathrm{d}\xi = 8\int_\alpha^{(\alpha+\beta)/2} |\tau-\alpha|^2\,|\varphi'(\tau)|^2\,\mathrm{d}\tau \leqslant$$

$$\leqslant 2(\beta-\alpha)^2\int_\alpha^\beta |\varphi'(\tau)|^2\,\mathrm{d}\tau\,.$$

We have

$$(1.109)\qquad \gamma_0 v_h(M) - q_h\gamma_0 v_h(M) = v_h(M) - v_h(M_i)\quad \forall M\in \widehat{M_iM_{i+1/6}}$$

$$(1.110)\qquad \gamma_0 v_h(M) - q_h\gamma_0 v_h(M) = v_h(M) - v_h(M_{i+1/2})\quad \forall M\in \widehat{M_{i+1/6}M_{i+1/2}M_{i+5/6}}$$

$$(1.111)\qquad \gamma_0 v_h(M) - q_h\gamma_0 v_h(M) = v_h(M) - v_h(M_{i+1})\quad \forall M\in \widehat{M_{i+5/6}M_{i+1}}\,,$$

so that, from (1.108)

$$(1.112)\qquad \int_{\widehat{M_iM_{i+1}}} |\gamma_0 v_h - q_h\gamma_0 v_h|^2\,\mathrm{d}\Gamma \leqslant \frac{2}{9}|M_i M_{i+1}|^2 \times$$

$$\times \int_{\widehat{M_iM_{i+1}}} \left|\frac{\mathrm{d}\gamma_0 v_h}{\mathrm{d}\Gamma}\right|^2 \mathrm{d}\Gamma\,.$$

From (1.112) and from the definition of h we deduce by addition that

$$(1.113)\qquad \|\gamma_0 v_h - q_h\gamma_0 v_h\|_{L^2(\Omega)} \leqslant (\sqrt{2}/3)\,h\,\|\gamma_0 v_h\|_{H^1(\Gamma)}\,.$$

By interpolation between $L^2(\Gamma)$ and $H^1(\Gamma)$ we deduce from (1.106), (1.113):

$$(1.114)\qquad \|\gamma_0 v_h - q_h\gamma_0 v_h\|_{L^2(\Gamma)} \leqslant Ch^s\,\|\gamma_0 v_h\|_{H^s(\Gamma)}$$

with C independent of h, s, v_h. Relation (1.114) in combination with (1.101) proves (1.97) and hence the lemma. ■

1.5.2 *A convergence theorem*

The following result also holds if we assume $k = 0$ in (1.9). We have:

Theorem 1.5. *If, when $h \to 0$, the angles of $\mathcal{T}_h$ are bounded below by a constant $\theta_0 > 0$, we have*

$$(1.115) \qquad \lim_{h \to 0} u_h = u \quad \textit{strongly in } H^1(\Omega)$$

where u and u_h are respectively solutions of the problems (1.4), (1.5) *and* (1.27).

Proof. This is a variant of the proof of Theorem 1.3; we thus have $\forall v_h \in V_h$:

$$(1.116) \qquad a(u_h - u, u_h - u) \leqslant a(u_h, v_h - u) - \int_\Omega f(v_h - u)\,dx + $$
$$+ j_h(v_h) - j(u) + j(u_h) - j_h(u_h)\,.$$

With $v \in V$, we denote by $r_h v$ the projection of v on V_h relative to the norm $v \to \sqrt{a(v, v)}$ (equivalent to the standard norm of $H^1(\Omega)$) ; we thus have:

$$(1.117) \qquad \begin{cases} a(r_h v, v_h) = a(v, v_h) & \forall v_h \in V_h \\ r_h v \in V_h\,. \end{cases}$$

Under the assumptions adopted for $\mathcal{T}_h$ we have

$$(1.118) \qquad \lim_{h \to 0} r_h v = v \quad \textit{strongly in } V\,.$$

Putting $v_h = r_h u$ in (1.116) we deduce from (1.117), (1.118) that:

$$(1.119) \qquad a(u_h, r_h u - u) = 0$$

$$(1.120) \qquad \lim_{h \to 0} \int_\Omega f(r_h u - u)\,dx = 0\,.$$

Moreover

$$| j_h(r_h u) - j(u) | \leqslant | j_h(r_h u) - j(r_h u) | + | j(r_h u) - j(u) | \leqslant$$
$$\leqslant | j_h(r_h u) - j(r_h u) | + Cg \, \| r_h u - u \|_{L^2(\Gamma)}$$

and taking account of Lemma 1.3, we have

$$(1.121) \qquad | j_h(r_h u) - j(u) | \leqslant Cg(\sqrt{h} \, \| r_h u \|_{H^{1/2}(\Gamma)} + \| r_h u - u \|_{L^2(\Gamma)})\,.$$

Property (1.118) implies that

$$(1.122) \qquad \begin{cases} \lim_{h \to 0} \gamma_0 r_h v = \gamma_0 v & \textit{strongly in } H^s(\Gamma) \\ \forall s \in [0, 1/2] \end{cases}$$

and (1.121), (1.122) lead to:

$$(1.123)\qquad \lim_{h \to 0} (j_h(r_h u) - j(u)) = 0 .$$

It therefore remains to show that

$$(1.124)\qquad \lim_{h \to 0} (j_h(u_h) - j(u_h)) = 0 ;$$

putting $v_h = 0$ in (1.28) we deduce that

$$\| u_h \|_{H^1(\Omega)} \leqslant C \| f \|_{L^2(\Omega)}$$

so that

$$(1.125)\qquad \| \gamma_0 u_h \|_{H^{1/2}(\Gamma)} \leqslant C \| f \|_{L^2(\Omega)} .$$

Taking account of Lemma 1.3 - with $s = \frac{1}{2}$ in (1.97) - we deduce from (1.125) that:

$$| j_h(u_h) - j(u_h) | \leqslant Cg \sqrt{h} \| f \|_{L^2(\Omega)}$$

which gives (1.124); it follows from (1.116) with $v_h = r_h u$ and from (1.118) - (1.120), (1.123), (1.124) that

$$\lim_{h \to 0} a(u_h - u, u_h - u) = 0$$

from which the theorem then follows. ■

Remark 1.7. Theorem 1.5 still holds if in the functional J defined by (1.5) we replace the term $\int_\Omega fv \, dx$ by $L(v)$ where L is an element of the dual V' of the space V. ■

1.6 Numerical solution of the approximate problems

Synopsis

In this section we shall study the numerical solution of the approximate problems of Section 1.3.3, namely

$$(1.126)\qquad \begin{cases} \operatorname*{Min}_{v_h \in V_h} [J_0(v_h) + j_h(v_h)] , \\ J_0(v_h) = \dfrac{1}{2} \displaystyle\int_\Omega | \operatorname{grad} v_h |^2 \, dx + \dfrac{\mu}{2} \int_\Omega | v_h |^2 \, dx - \int_\Omega f v_h \, dx \end{cases}$$

by means of some of the optimisation methods discussed in Chapter

2. First, it is appropriate to express the functionals $J_0(v_h)$, $j_h(v_h)$ explicitly in terms of the values taken by v_h on Σ_h if $q = 1$, or on $\Sigma_h \cup \Sigma_h'$ if $q = 2$; this will form the subject of Sections 1.6.1, 1.6.2.

1.6.1 *Formulation of the approximation problem.* (I), $q = 1$.

In (1.24) the functional j_h was expressed explicitly in terms of the values taken by v_h on $\Sigma_h \cap \Gamma$, that is:

$$j_h(v_h) = \tfrac{1}{2} \sum | M_i M_{i+1} | (\psi(v_i) + \psi(v_{i+1})) . \tag{1.127}$$

In order to express $\int_\Omega f v_h \, dx$ explicitly, it is convenient to introduce the *basis functions* of the space $\mathscr{V}_h$ (defined by (1.19)) of the same type as those in Chapter 1, Section 4.1, Example 4.2, that is:

$$\begin{cases} (w_i)_{M_i \in \Sigma_h} \\ w_i \in \mathscr{V}_h , \qquad w_i(M_j) = \delta_{ij} \quad \forall M_j \in \Sigma_h ; \end{cases} \tag{1.128}$$

denoting by Ω_i the support of w_i we then deduce

$$\int_\Omega f v_h \, dx = \sum_{M_i \in \Sigma_h} v_i \int_{\Omega_i} f w_i \, dx . \tag{1.129}$$

Using the notation of Figure 1.7, denoting by $A(T)$ the measure of the triangle T, we have

$$\int_T | v_h |^2 \, dx = \frac{A(T)}{12} [(v_{1T} + v_{2T})^2 + (v_{2T} + v_{3T})^2 + (v_{3T} + v_{1T})^2]$$

so that

$$\int_\Omega | v_h |^2 \, dx = \frac{1}{12} \sum_{T \in \mathscr{T}_h} A(T) [(v_{1T} + v_{2T})^2 + (v_{2T} + v_{3T})^2 + (v_{3T} + v_{1T})^2] . \tag{1.130}$$

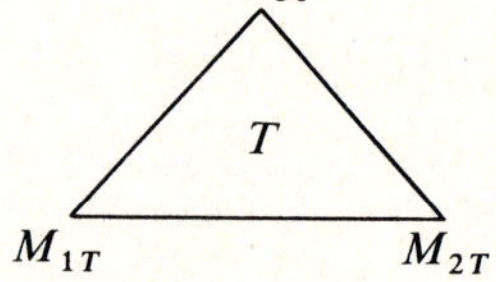

Fig. 1.7.

Finally, by using relations (4.17)-(4.19) of Chapter 3, Section 4.1.4, we may prove

$$\int_T |\operatorname{grad} v_h|^2 \, dx = \frac{1}{4\,A(T)} \times$$
$$\times \left| v_{1T}\,\overrightarrow{M_{2T}M_{3T}} + v_{2T}\,\overrightarrow{M_{3T}M_{1T}} + v_{3T}\,\overrightarrow{M_{1T}M_{2T}} \right|^2$$

so that

$$\int_\Omega |\operatorname{grad} v_h|^2 \, dx = \frac{1}{4} \sum_{T \in \mathcal{T}_h} \frac{1}{A(T)} \times \tag{1.131}$$
$$\times \left| v_{1T}\,\overrightarrow{M_{2T}M_{3T}} + v_{2T}\,\overrightarrow{M_{3T}M_{1T}} + v_{3T}\,\overrightarrow{M_{1T}M_{2T}} \right|^2 .$$

Relations (1.127), (1.129)-(1.131) allow the approximate problem (1.126) to be written explicitly as a finite-dimensional optimisation problem. ■

1.6.2 *Formulation of the approximate problem.* (II), $q = 2$.

In (1.25) the functional j_h was expressed explicitly in terms of the values taken by v_h on $(\Sigma_h \cup \Sigma_h') \cap \Gamma$, that is

$$j_h(v_h) = \tfrac{1}{6} \sum |M_i\,M_{i+1}|\,(\psi(v_i) + 4\,\psi(v_{i+1/2}) + \psi(v_{i+1})) . \tag{1.132}$$

In order to write $\int_\Omega f v_h \, dx$ explicitly, we introduce, as in Section 1.6.1, the basis functions of the space $\mathcal{V}_h$ defined by (1.19), that is

$$\begin{cases} (w_i)_{M_i \in (\Sigma_h \cup \Sigma_h')} \\ w_i \in \mathcal{V}_h , \quad w_i(M_j) = \delta_{ij} \quad \forall M_j \in \Sigma_h \cup \Sigma_h' ; \end{cases} \tag{1.133}$$

denoting by Ω_i the support of w_i , we then deduce that

$$\int_\Omega f v_h \, dx = \sum_{M_i \in (\Sigma_h \cup \Sigma_h')} v_i \int_{\Omega_i} f w_i \, dx . \tag{1.134}$$

Using the notation of Figure 1.8, with $A(T)$ = measure of T, we have

$$(1.135)\quad \left\{ \begin{aligned} &\int_T |v_h|^2\,dx = \frac{A(T)}{3}\left(\frac{1}{10}(v_1^2+v_2^2+v_3^2)+\frac{8}{15}(v_{12}^2+v_{23}^2+v_{31}^2)-\right.\\ &\quad -\frac{1}{30}(v_1 v_2+v_2 v_3+v_3 v_1)+\frac{8}{15}(v_{12} v_{23}+v_{23} v_{31}+v_{31} v_{12})-\\ &\quad \left.-\frac{2}{15}(v_1 v_{23}+v_2 v_{31}+v_3 v_{12})\right) \end{aligned} \right.$$

which allows $\int_\Omega |v_h|^2\,dx$ to be written explicitly since

$$\int_\Omega |v_h|^2\,dx = \sum_{T\in\mathcal{T}_h}\int_\Omega |v_h|^2\,dx\,.$$

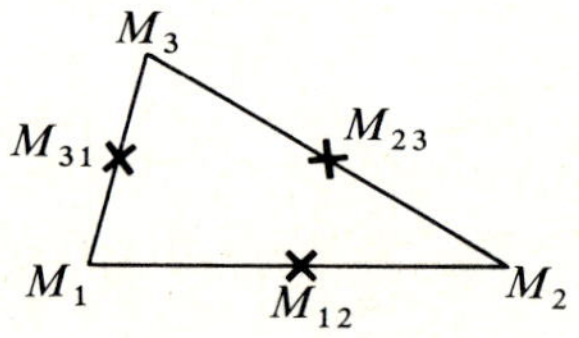

Fig. 1.8.

Moreover, it may be proved that

$$(1.136)\quad \left\{ \begin{aligned} &\int_T |\operatorname{grad} v_h|^2\,dx = \frac{1}{12\,A(T)}\{\,|\,v_1\,\overrightarrow{M_2M_3} - v_2\,\overrightarrow{M_3M_1} +\\ &\quad + v_3\,\overrightarrow{M_1M_2} + 2(v_{12}+v_{23}-v_{31})\,\overrightarrow{M_3M_1}\,|^2 +\\ &\quad + |\,v_1\,\overrightarrow{M_2M_3} + v_2\,\overrightarrow{M_3M_1} - v_3\,\overrightarrow{M_1M_2} +\\ &\quad + 2(v_{23}+v_{31}-v_{12})\,\overrightarrow{M_1M_2}\,|^2 +\\ &\quad + |-v_1\,\overrightarrow{M_2M_3} + v_2\,\overrightarrow{M_3M_1} + v_3\,\overrightarrow{M_1M_2} +\\ &\quad + 2(v_{31}+v_{12}-v_{23})\,\overrightarrow{M_2M_3}\,|^2\,\} \end{aligned} \right.$$

which allows $\int_\Omega |\operatorname{grad} v_h|^2\,dx$ to be written explicitly since

$$\int_\Omega |\operatorname{grad} v_h|^2\,dx = \sum_{T\in\mathcal{T}_h}\int_\Omega |\operatorname{grad} v_h|^2\,dx\,.$$

Using relations (1.132), (1.134)-(1.136) the approximate problem (1.126) may be put in the form of a finite-dimensional optimisation problem. ■

1.6.3 *Solution of the approximate problem by relaxation*

Suppose $N_h = \dim V_h$; hereinafter we shall denote by v_h a vector of $\mathbb{R}^{N_h}$, i.e. $v_h = (v_1, v_2, \ldots, v_{N_h})$. It was seen in Sections 1.6.1, 1.6.2 that the approximate problem (1.126) in fact reduces to minimising in $\mathbb{R}^{N_h}$ a functional denoted by J_h which satisfies the following conditions:

(i) J_h is *strictly convex*

(ii) $$\lim_{\|v_h\| \to +\infty} J_h(v_h) = +\infty$$

(iii) $J_h \in C^1(\mathbb{R}^{N_h})$ since $\psi \in C^1(\mathbb{R})$.

In view of these properties, it follows from Theorem 1.1 of Chapter 2, Section 1.1, that the convergence of the relaxation algorithm applied to the minimisation of J_h in $\mathbb{R}^{N_h}$ is assured. This algorithm is defined as follows:

(1.137) $u_h^0 = (u_1^0, \ldots, u_{N_h}^0)$ given arbitrarily in $\mathbb{R}^{N_h}$;

with u_h^n known, u_h^{n+1} is determined coordinate by coordinate, using

(1.138) $$\frac{\partial J_h}{\partial v_i}(u_1^{n+1}, \ldots, u_{i-1}^{n+1}, u_i^{n+1}, u_{i+1}^n, \ldots, u_{N_h}^n) = 0, \quad 0 \leqslant i \leqslant N_h . \quad \blacksquare$$

Remark 1.8. If the variable v_i relates to a vertex of Σ_h $(q = 1)$, or of $\Sigma_h \cup \Sigma_h'$ $(q = 2)$ which does not belong to Γ_d, equation (1.138) is *linear*. ■

Remark 1.9 If the constant k in (1.9) takes very small values, algorithm (1.137), (1.138) is guaranteed a priori to perform well. In fact, in the limit when $k \to 0$, we have $\psi(\xi) = g|\xi|$ so that

(1.139) $$j_h(v_h) = \frac{g}{2} \sum |M_i M_{i+1}| (|v_i| + |v_{i+1}|), \qquad \text{if} \quad q = 1$$

(1.140) $$j_h(v_h) = \frac{g}{6} \sum |M_i M_{i+1}| (|v_i| + 4|v_{i+1/2}| + |v_{i+1}|), \qquad \text{if} \quad q = 2;$$

we are thus within the conditions for applying Theorem 1.4 of Chapter 2, Section 1.5, and this assures the convergence of the relaxation algorithm - provided, of course, (1.138) (which is not always meaningful) is replaced by:

(1.141) $$J_h(u_1^{n+1}, \ldots, u_{i-1}^{n+1}, u_i^{n+1}, u_{i+1}^n, \ldots) \leqslant$$
$$\leqslant J_h(u_1^{n+1}, \ldots, u_{i-1}^{n+1}, v_i, u_{i+1}^n, \ldots) \quad \forall v_i \in \mathbb{R} .$$

We shall return to the above in Section 2. ■

Remark 1.10. In order to improve the rate of convergence, it can be useful to introduce an *under-* or *over-relaxation* parameter ω; we then obtain the following variant of algorithm (1.137), (1.138):

(1.142) u_h^0 given

(1.143) $$\frac{\partial J_h}{\partial v_i}(u_1^{n+1}, u_2^{n+1}, \ldots, u_{i-1}^{n+1}, u_i^{n+1/2}, u_{i+1}^n, \ldots) = 0$$

(1.144) $$u_i^{n+1} = u_i^n + \omega(u_i^{n+1/2} - u_i^n) .$$ ■

1.6.4 *Solution of the approximate problem by a conjugate gradient method.*

In view of the properties of J_h, it is also possible to use in the solution of the approximate problem (1.126) the variant [1] (2.26) - (2.29) of Chapter 2, Section 2.3, of the *conjugate gradient* algorithm. ■

1.7 Examples

Preliminary remark. In the following examples the approximate problems are all solved by the over-relaxation method (1.142) - (1.144) described in Section 1.6.3; although numerical experiments performed with the conjugate gradient method have given some very satisfactory results, the programming of this method is much lengthier and trickier (for comparable computation times) than that of the over-relaxation method; the corresponding results will therefore not be presented here. ■

1.7.1 *Example* 1

The problem (1.4) considered is defined by:

Geometric data: $\Omega =]0, 1[\times]0, 1[, \Gamma_d = \Gamma$.

Space V : $\Gamma_d = \Gamma$ implies $V = H^1(\Omega)$.

Function f: $f = 1$.

Parameter values: $\mu = 1$, $k = 1, 0.1, 0.01$, $g = 1$ and 0.2.

---

[1] Adapted to non-quadratic cases.

Initialisation of the over-relaxation algorithm: $u_h^0 = 0$.

Termination criterion for the over-relaxation algorithm:

$$\frac{\sum_{i=1}^{N_h} |u_i^{n+1} - u_i^n|}{\sum_{i=1}^{N_h} |u_i^{n+1}|} \leqslant \varepsilon_r \tag{1.145}$$

with $\varepsilon_r = 10^{-5}$ if $q = 1$ (finite elements of order 1), $\varepsilon_r = 10^{-4}$ if $q = 2$ (finite elements of order 2).

Triangulation $\mathcal{T}_h$: Figures 1.9 and 1.10 show the triangulations employed, relating respectively to $q = 1$ and $q = 2$.

Fig. 1.9. ($q = 1$).

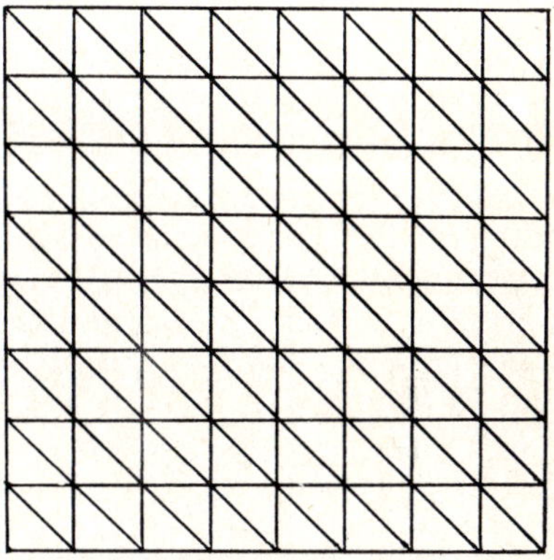

Fig. 1.10. ($q = 2$).

For $q = 1$ (resp. $q = 2$) we have 512 (resp. 128) triangles and 289 nodes, and hence $N_h = 289$.

Analysis of the numerical results.

Table 1.1 gives certain information concerning the solution of the approximate problems and the results obtained; we have denoted by $\mu_1(u_h)$ and $\mu_2(u_h)$ the following two quantities:

$$\mu_1(u_h) = \text{Max}\, u_h(M),\ M \in \Sigma_h \text{ if } q = 1,\quad M \in \Sigma_h \cup \Sigma_h' \text{ if } q = 2\,,$$

$$\mu_2(u_h) = \text{Max}\, u_h(M),\ M \in \Sigma_h \cap \Gamma \text{ if } q = 1,$$
$$M \in (\Sigma_h \cup \Sigma_h') \cap \Gamma \text{ if } q = 2\,.$$

	k	overrelaxation parameter ω		$\mu_1(u_h)$		$\mu_2(u_h)$		Number of iterations	
		$q=1$	$q=2$	$q=1$	$q=2$	$q=1$	$q=2$	$q=1$	$q=2$
$g = 1$	1	1.7	1.7	0.256	0.255	0.206	0.206	50	149
	0.1	1.7	1.7	0.095	0.094	0.029	0.029	56	53
	0.01	1.7	1.7	0.073	0.072	0.003	0.003	56	30
$g = 0.2$	1	1.9	1.9	0.259	0.258	0.209	0.210	99	76
	0.1	1.9	1.9	0.233	0.232	0.183	0.183	116	185
	0.01	1.9	1.9	0.233	0.232	0.183	0.183	125	185

Table 1.1

All the calculations were performed on a CII 10070 with a mean computation time *per iteration* of 0.26 sec. with $q = 1$, and 0.42 sec. with $q = 2$.

In view of these results, it would appear that, for this type of problem at least, approximation by elements of order 1 has the advantage over that by elements of order 2, since for the same precision and the same total number of degrees of freedom, the computation times are approximately half as long.

Figures 1.11 and 1.12 show the equipotentials of the approximate solutions for $g = 0.2$ with $k = 1$ and 0.1.

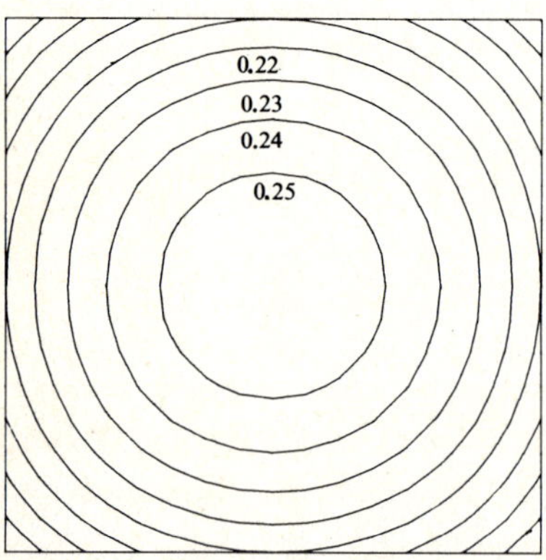

Fig. 1.11.

(Example 1 : $g = 0.2, k = 1$).

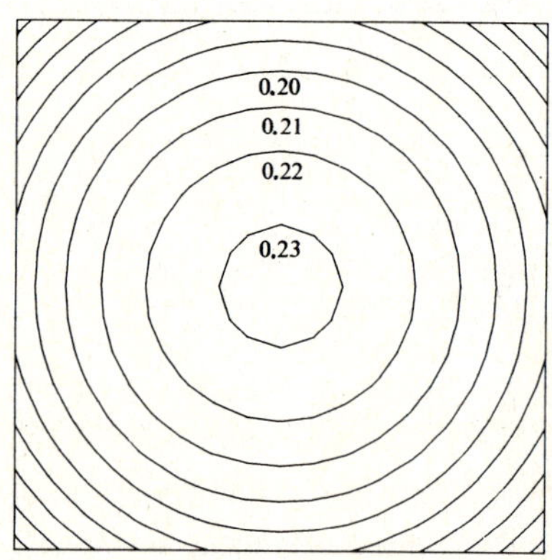

Fig. 1.12.

(Example 2 : $g = 0.2, k = 0.1$).

Figure 1.13 shows the graphs of the functions $u_h(x_1, 0)$ for $g = 0.2$ and with $k = 1, 0.1$.

It may be seen in Figure 1.13 that the fact that the function $u_h(x_1, 0)$ passes beyond the critical value kg does not appear to have any adverse repercussions on the regularity of the trace of the solution. ■

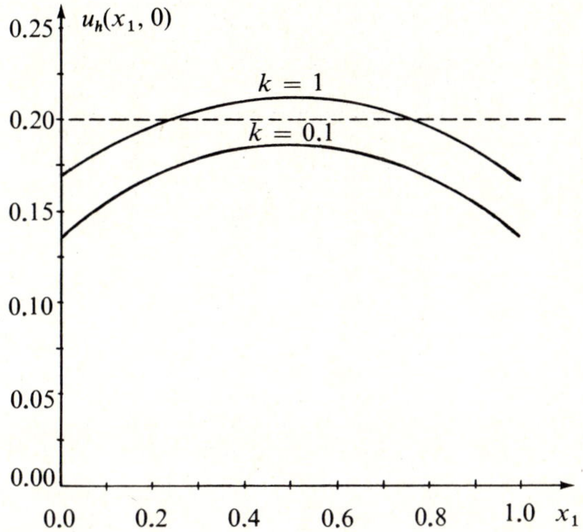

Fig. 1.13. (Example 1: $g = 0.2$). Representation of $u_h(x_1, 0)$.

1.7.2 *Example* 2.

The problem (1.4) is defined by:

Geometric data:

$$\Omega =]0,1[\times]0,1[\,, \quad \Gamma_d = \{(x_1, x_2) \mid 0.25 \leqslant x_1 \leqslant 0.75, x_2 = 0\}.$$

Space V: $V = \{ v \mid v \in H^1(\Omega), v = 0 \quad \text{on} \quad \Gamma - \Gamma_d \}$.

Function f: We have taken f equal to 10 times the character-istic function of the square $]0.375, 0.625[\times]0.375, 0.625[$ i.e. $f(x) = 10$ if x belongs to this square, $f(x) = 0$ otherwise.

Parameter values: We have taken $\mu = 0$, which is permissible since

$$\left(\int_\Omega |\operatorname{grad} v|^2 \, dx \right)^{1/2}$$

efines on V a norm equivalent to the norm induced by $H^1(\Omega)$, and $= 1, 0.1, 0.01$, $g = 1, 0.2$.

Initialisation and termination criterion for the over-relaxation algorithm: see Section 1.7.1.

Triangulation $\mathcal{T}_h$: See Section 1.7.1; the number N_h of degrees of freedom is in this case equal to 234.

Figure 1.14 shows Ω, Γ_d and the support of f.

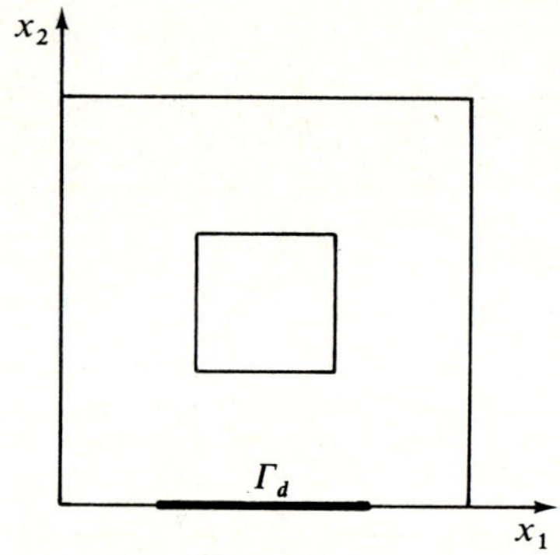

Fig. 1.14. (Example 2)

Analysis of the numerical results

Table 1.2 shows some details relating to the numerical results obtained.

The mean computation times per iteration on a CII 10070 are essentially the same as those in Example 1; since the values obtained are practically the same, this example once again confirms the advantage of working, for this type of problem, with finite elements of order 1.

	k	overrelaxation parameter ω $q=1$	overrelaxation parameter ω $q=2$	Number of iterations $q=1$	Number of iterations $q=2$
$g = 1$	1	1.67	1.7	48	45
	0.1	1.67	1.7	42	39
	0.01	1.67	1.7	40	37
$g = 0.2$	1	1.67	1.7	48	45
	0.1	1.67	1.7	42	39
	0.01	1.67	1.7	48	44

Table 1.2

Figures 1.15, 1.16 and 1.17 show the equipotentials of the approximate solutions for $g = 1$ and $k = 1,\ 0.1,\ 0.01$.

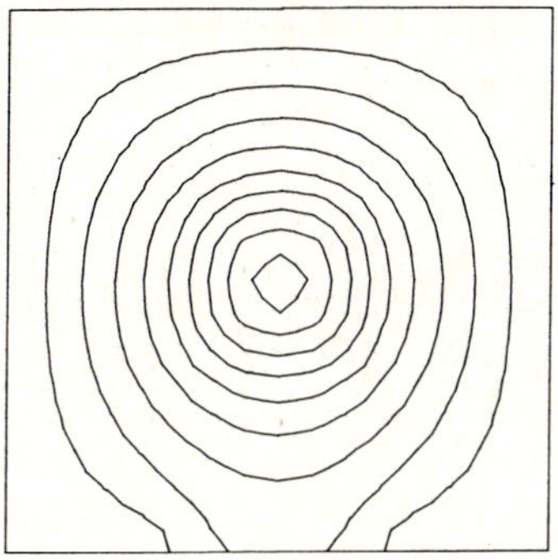

Fig. 1.15. (Example 2: $g = 1,\ k = 1$).

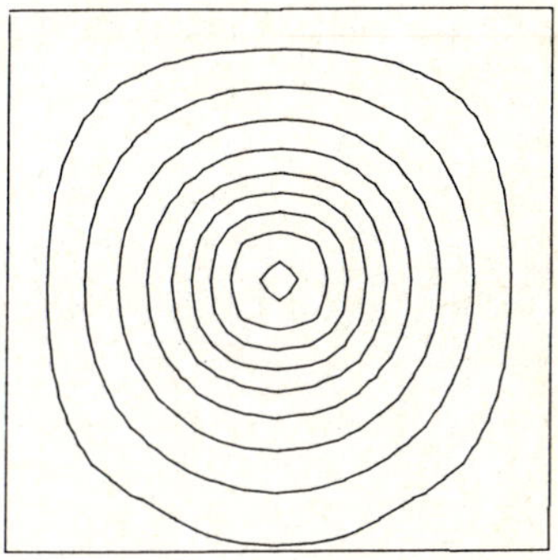

Fig. 1.16. (Example 2: $g = 1,\ k = 0.1$).

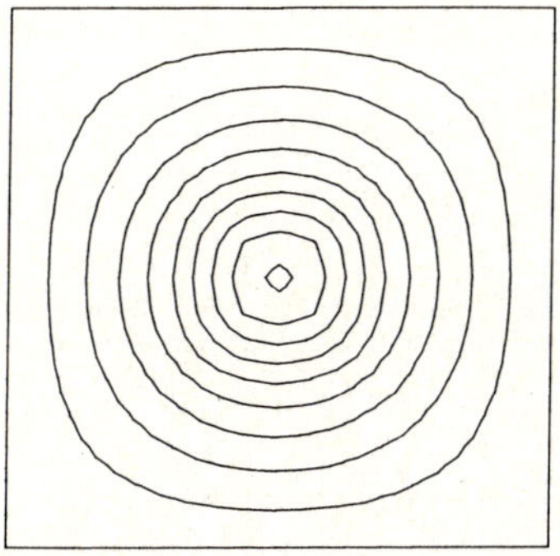

Fig. 1.17. (Example 2: $g = 1,\ k = 0.01$).

1.7.3 *Example* 3

Geometric data: $\Omega = \{ (x_1, x_2) \mid x_1^2 + x_2^2 < \frac{1}{4} \}$, $\Gamma_d =$ the lower semi-ircle (see Figure 1.18).

Space V: $V = \{ v \mid v \in H^1(\Omega), v|_{\Gamma - \Gamma_d} = 0 \}$.

Function f: f is equal to 18 times the characteristic function ˸ the disc of radius 0.25.

Parameter values: $\mu = 0$, $k = 1$, $k = 0.1,\ 0.01$, $g = 1,\ 0.2$.

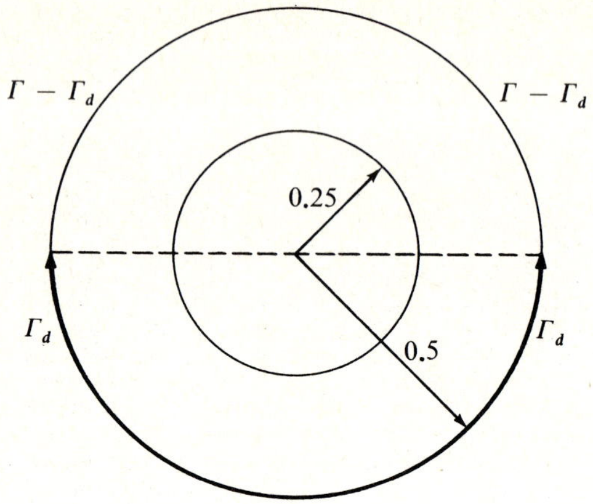

Fig. 1.18.

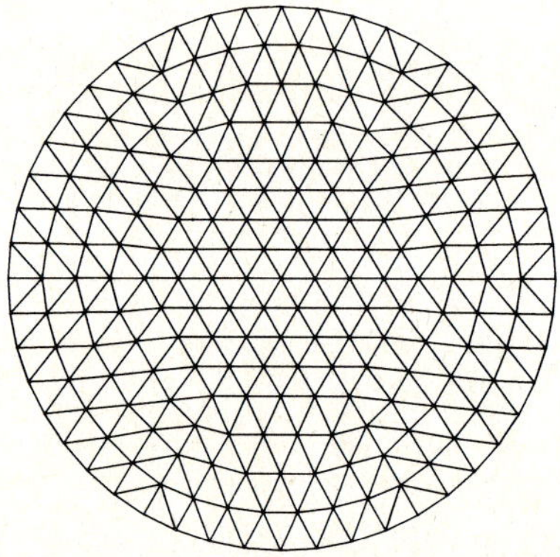

Fig. 1.19.

Initialisation and termination criterion for the relaxation algorithm: see Section 1.7.1.

Triangulation $\mathcal{T}_h$: that of Figure 1.19, i.e. 384 triangles, 217 nodes with $N_h = 193$.

Analysis of the numerical results: We have used only approximation by finite elements of order 1 ($q = 1$); the results obtained are shown in Table 1.3.

	k	ω	Number of iterations
$g = 1$	1	1.76	42
	0.1		35
	0.01		35
$g = 0.2$	1	1.76	42
	0.1		48
	0.01		56

Table 1.3

The mean computation time per iteration is about 0.33 sec. on a CII 10070; Figures 1.20, 1.21, 1.22 show the equipotentials of the approximate solutions for $g = 1$ and with $k = 1, 0.1, 0.01$.

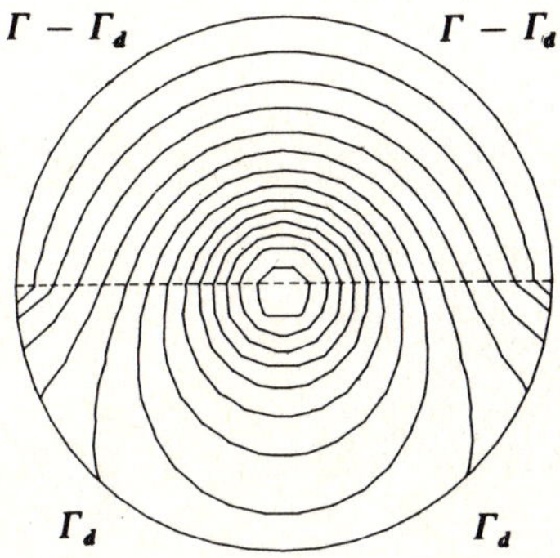

Fig. 1.20. (Example 3: $g = 1$, $k = 1$).

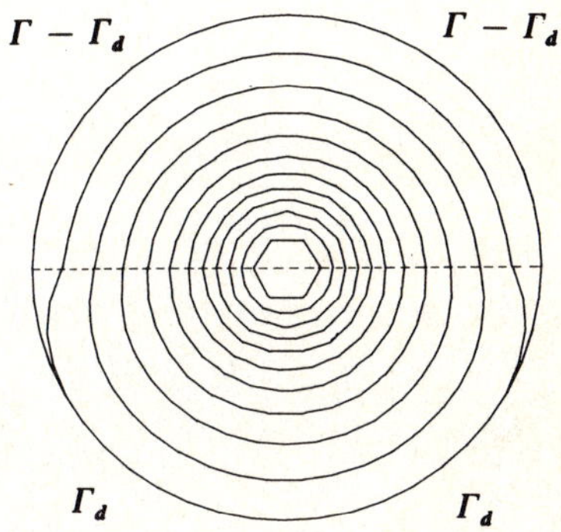

Fig. 1.21. (Example 3: $g = 1$, $k = 0.1$).

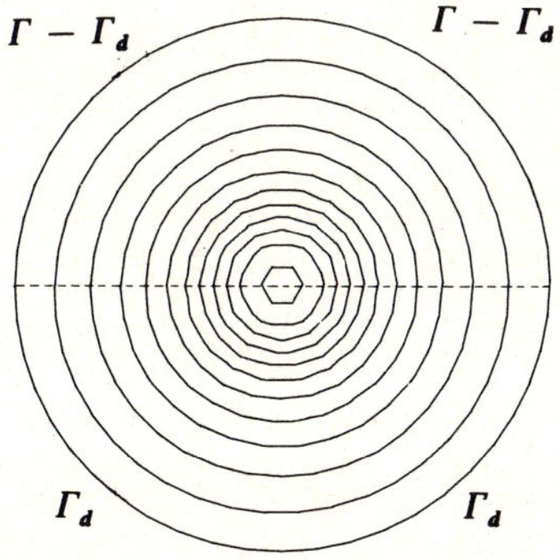

Fig. 1.22. (Example 3: $g = 1$, $k = 0.01$).

Figure 1.23 shows, in *polar coordinates* and for $g = 0.2$, $k = 1$, 0.1, 0.01, the graphs of the functions $u_h(\frac{1}{2}, \theta)$ for $\theta \in [-\pi, 0]$.

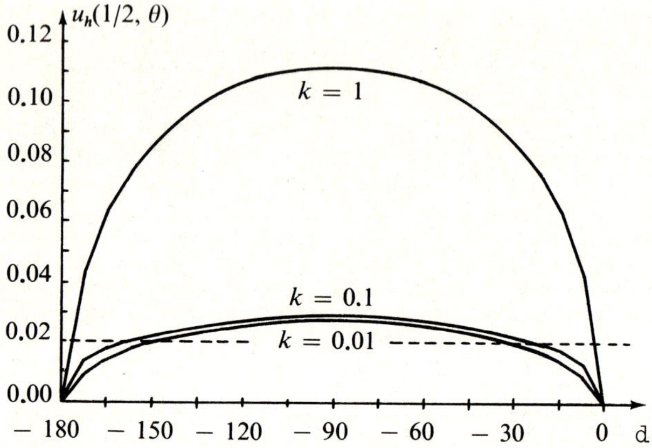

Fig. 1.23. (Example 3: $g = 0.2$).

1.7.4 *Example* 4.

Geometric data: see Section 1.7.3.

Space V: $V = H^1(\Omega)$.

Function f: f is equal to 10 times the characteristic function of the disc of radius 0.25.

Parameter values: $\mu = 1$, $k = 1$, 0.1, 0.01, $g = 1$, 0.2.

Initialisation and termination criterion for the over-relaxation algorithm: see Section 1.7.1.

Triangulation $\mathcal{T}_h$: see Section 1.7.3; in the present case we have $N_h = 217$.

Analysis of the numerical results.

The results obtained are shown in Table 1.4; the mean computation time per iteration was 0.33 sec. on a CII 10070.

Figures 1.24, 1.25, 1.26 show the equipotentials of the approximate solutions for $g = 1$ with $k = 1$, 0.1, 0.01.

	k	ω	Number of iterations
$g = 1$	1	1.91	91
	0.1		96
	0.01		97
$g = 0.2$	1	1.91	91
	0.1		111
	0.01		97

Table 1.4

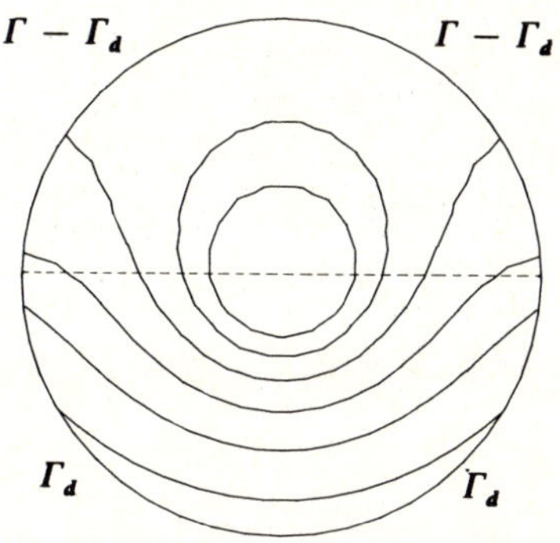

Fig. 1.24. (Example 4: $g = 1$, $k = 1$).

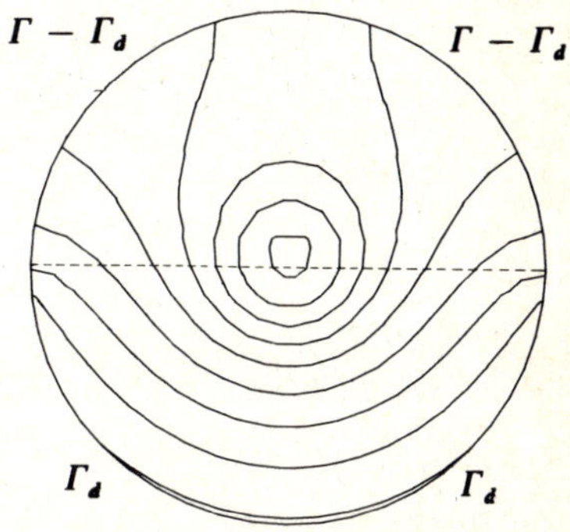

Fig. 1.25. (Example 4: $g = 1$, $k = 0.1$).

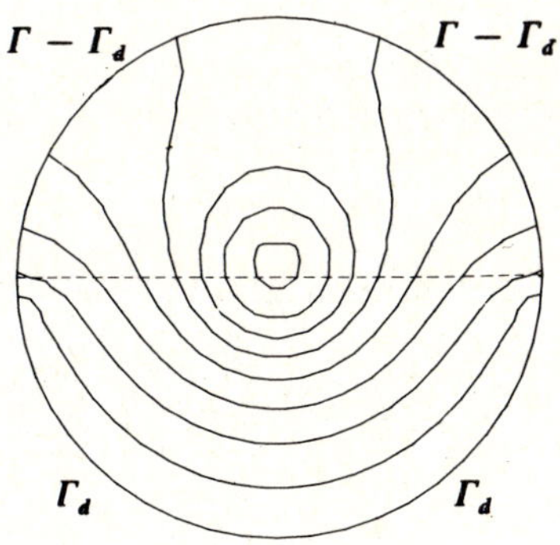

Fig. 1.26. (Example 4: $g = 1$, $k = 0.01$).

2. FRICTION PROBLEMS

2.1 Formulation of the problems

A certain number of model problems [1] relating to the displacement of a solid body Ω, with *friction* on the boundary Γ of Ω, or over a part Γ_d of this boundary, lead to determining a function $u(x)$, $x \in \Omega$, the solution of the *variational* problem

$$(2.1) \qquad \begin{cases} J(u) \leqslant J(v) & \forall v \in V \\ u \in V \end{cases}$$

with

$$(2.2) \qquad J(v) = \frac{1}{2} \int_\Omega (\mu v^2 + |\operatorname{grad} v|^2) \, dx - \int_\Omega f v \, dx + g \int_{\Gamma_d} |v| \, d\Gamma .$$

In (2.2), we take $f \in L^2(\Omega)$, $\mu \geqslant 0$, $g > 0$; as regards V, we shall confine our attention to the cases

$$(2.3) \qquad V = \{ v \mid v \in H^1(\Omega), v = 0 \text{ on } \Gamma - \Gamma_d \}$$

or

$$(2.4) \qquad V = H^1(\Omega) . \quad \blacksquare$$

Remark 2.1. To help simplify the description, we shall assume in the following that $\mu > 0$; we refer the reader to Duvaut-Lions, *loc. cit*, for the study of the case $\mu = 0$. ■

2.2 Existence and uniqueness results for problem (2.1), (2.2)

As for Theorem 1.1 of Section 1.2, we may prove:

Theorem 2.1. *Assuming Ω to be bounded and with Lipschitz boundary Γ, problem* (2.1), (2.2) *admits one and only one solution*

2.3 Relationship with the problems of Section 1

We shall show that problem (2.1), (2.2) may be considered as a limiting case of problem (1.4), (1.5) of Section 1.1, when the parameter k in (1.9) tends to zero; this will result from:

Proposition 2.1. *Let u be the solution of problem* (2.1), (2.2)

[1] See Duvaut-Lions /1/, Chapter 3, for the modelling and study of friction problems which are much more complicated and realistic than those considered in this chapter.

and u_k *that of problem* (1.4), (1.5) *of Section* 1.1; *then*

$$\lim_{k \to 0} u_k = u \quad \textit{strongly in } V .$$

Proof: We write:

$$a(u, v) = \int_\Omega (\mu uv + \operatorname{grad} u . \operatorname{grad} v) \, dx \tag{2.5}$$

$$j(v) = g \int_{\Gamma_d} | v | \, d\Gamma \tag{2.6}$$

$$j_k(v) = \int_{\Gamma_d} \psi_k(v) \, d\Gamma \tag{2.7}$$

where, in (2.7), ψ_k is the function defined by (1.10) and which was denoted by ψ in Section 1.

The solutions u and u_k are characterised respectively by

$$a(u, v - u) + j(v) - j(u) \geqslant \int_\Omega f(v - u) \, dx \tag{2.8}$$

$$a(u_k, v - u_k) + j_k(v) - j_k(u_k) \geqslant \int_\Omega f(v - u_k) \, dx \tag{2.9}$$

so that, taking $v = u_k$ in (2.8), $v = u$ in (2.9), and adding, we obtain:

$$a(u_k - u, u_k - u) \leqslant j_k(u) - j(u) + j(u_k) - j_k(u_k) . \tag{2.10}$$

From (1.10) we deduce:

$$0 \leqslant g \, | \xi | - \psi_k(\xi) \leqslant \tfrac{1}{2} k g^2 \quad \forall \xi \in \mathbb{R} \tag{2.11}$$

so that from (2.6), (2.7), we get:

$$0 \leqslant j(v) - j_k(v) \leqslant \frac{g^2}{2} \left(\int_\Gamma d\Gamma \right) k \quad \forall v \in V . \tag{2.12}$$

From (2.10), (2.12) we deduce

$$a(u_k - u, u_k - u) \leqslant \frac{g^2}{2} \left(\int_\Gamma d\Gamma \right) k$$

so that

$$\lim_{k \to 0} a(u_k - u, u_k - u) = 0 ; \tag{2.13}$$

since $v \to \sqrt{a(v, v)}$ defines a norm on V equivalent to that induced by $H^1(\Omega)$, (2.13) implies the strong convergence of u_k. ■

Remark 2.2. The above result is fundamental for what follows since — being also true for the approximate problems (*for fixed h*) — it implies that the approximation error estimates, *independent of k*, obtained in Section 1, are also valid for problem (2.1), (2.2).

Remark 2.3. Problem (1.4), (1.5) may be considered as being obtained from problem (2.1), (2.2) by *regularisation* of the *non-differentiable* term $j(v) = g \int_{\Gamma_d} |v| \, d\Gamma$. ■

2.4 Dependence of the solution on g

We shall, for the moment, denote by u_g the solution of (2.1), (2.2); relating to the properties of the mapping $g \to u_g$, we have:

Proposition 2.2. *We have*:

(i) *The mapping* $g \to u_g$ *is Lipschitz from* $\mathbb{R}_+ \to V$,

(ii) *The mapping* $g \to \int_{\Gamma_d} |u_g| \, d\Gamma$ *is Lipschitz decreasing.*

(iii) *The mapping* $g \to a(u_g, u_g)$ *is convex decreasing.*

Proof. These properties follow from the more general results of Lemma 2.1 below (this would also apply to Chapter 5, Section 2.2, but, in this case, we have preferred to argue directly). ■

We use, in part, the same formalism as that employed in the preceding sections.

Lemma 2.1. *Suppose we have*:

— V *a real Hilbert space*,

— $a : V \times V \to \mathbb{R}$ *a coercive, symmetric, continuous, bilinear form*

— $L : V \to \mathbb{R}$, *linear continuous*

— $j : V \to \mathbb{R}$ *a seminorm on* V, *continuous on* V,

— g *a scalar* $\geqslant 0$;

we shall denote by J *the functional from* $V \to \mathbb{R}$ *defined by*

$$J(v) = \tfrac{1}{2} a(v, v) + g j(v) - L(v).$$

The problem

$$\begin{cases} J(u_g) \leqslant J(v) & \forall v \in V \\ u_g \in V \end{cases} \tag{2.14}$$

then admits one and only one solution and the mapping $g \to u_g$ *possesses the following properties*:

(i) *the mapping* $g \to u_g$ *is Lipschitz from* $\mathbb{R}_+ \to V$,

(ii) *the mapping* $g \to j(u_g)$ *is Lipschitz decreasing*,

(iii) *the mapping* $g \to a(u_g, u_g)$ *is convex decreasing*.

Proof. In view of the properties of the form $\boldsymbol{a}$, we may always assume that the Hilbert structure of V is defined by the inner product $(u, v) = a(u, v)$, so that for the associated norm, we have $\| v \| = \sqrt{a(v, v)}$; moreover there exists (Riesz's Theorem) $f \in V$, *unique*, such that $L(v) = (f, v) \quad \forall v \in V$, which gives

$$J(v) = \tfrac{1}{2} \| v \|^2 + g j(v) - (f, v) .$$

Finally the continuity of j implies

$$j(v) \leqslant |j| \, \| v \| \quad \forall v \in V \tag{2.15}$$

with

$$|j| = \sup_{\substack{v \in V \\ v \neq 0}} \frac{j(v)}{\| v \|} .$$

Proof of (i). Let $g_1, g_2 \in \mathbb{R}_+$, $u_1 = u_{g_1}$, $u_2 = u_{g_2}$ be the corresponding solutions of (2.14); we have

$$(u_1, v - u_1) + g_1(j(v) - j(u_1)) \geqslant (f, v - u_1) \tag{2.16}$$

$$(u_2, v - u_2) + g_2(j(v) - j(u_2)) \geqslant (f, v - u_2) \tag{2.17}$$

so that, putting $v = u_2$ in (2.16), $v = u_1$ in (2.17), and adding, we obtain:

$$\| u_2 - u_1 \|^2 \leqslant (g_1 - g_2)\,(j(u_2) - j(u_1)) . \tag{2.18}$$

In view of (2.15) we then have

$$\| u_2 - u_1 \|^2 \leqslant | g_2 - g_1 | \, j(u_2 - u_1) \leqslant | g_2 - g_1 | \, |j| \, \| u_2 - u_1 \|$$

so that

$$\| u_2 - u_1 \| \leqslant |j| \, | g_2 - g_1 | . \tag{2.19}$$

and hence (i). ■

Proof of (ii). We have, since j is a seminorm, and from (2.15), (2.19),

$$| j(u_2) - j(u_1) | \leqslant j(u_2 - u_1) \leqslant |j| \, \| u_2 - u_1 \| \leqslant |j|^2 \, | g_2 - g_1 |$$

and $g \to j(u_g)$ is therefore Lipschitz; it is decreasing, as a consequence of (2.18). ■

Proof of (iii). We shall first prove the convexity; u_g satisfies

$$(2.20)\qquad \begin{cases} (u_g, v - u_g) + gj(v) - gj(u_g) \geqslant (f, v - u_g) & \forall v \in V \\ u_g \in V, \end{cases}$$

so that by successively putting $v = 0$ and $v = 2u_g$ in (2.20) we get

$$(2.21)\qquad \| u_g \|^2 + gj(u_g) = (f, u_g).$$

From (2.21) it follows that

$$(2.22)\qquad J(u_g) = -\tfrac{1}{2}\| u_g \|^2.$$

Now

$$J(u_g) = \min_{v \in V}\left[\tfrac{1}{2}\| v \|^2 + gj(v) - (f, v)\right],$$

so that the function $g \to J(u_g)$ thus appears as the lower envelope of a family of affine functions (which are thus *concave*), and is therefore itself *concave*; this result in combination with (2.22) means that the function $g \to \| u_g \|^2$ is *convex*. ■

In order to prove that $g \to \| u_g \|^2$ is decreasing, we shall use a procedure which in fact will also demonstrate a number of other properties of the solution u_g. With

$$V_1 = \mathrm{Ker}\,(j) = \{ v \mid v \in V, j(v) = 0 \},$$

V_1 is a *closed subspace* of V; we shall denote by V_2 the subspace $V_1^\perp$, so that

$$(2.23)\qquad V = V_1^\perp \oplus V_2$$

and the corresponding decompositions of v, f, u_g are

$$(2.24)\qquad v = v_1 + v_2, \qquad f = f_1 + f_2, \qquad u_g = u_{g1} + u_{g2}.$$

It may be shown that

$$(2.25)\qquad j(v) = j(v_2) \quad \forall v \in V$$

and that j is a *norm* on V_2.

From (2.23)-(2.25), we then have

$$(2.26)\qquad \| v \|^2 = \| v_1 \|^2 + \| v_2 \|^2 \quad \forall v \in V$$

$$(2.27)\qquad J(v) = J(v_1, v_2) = \tfrac{1}{2}(\| v_1 \|^2 + \| v_2 \|^2) + gj(v_2) - (f_1, v_1) - (f_2, v_2) \quad \forall v \in V$$

and from (2.26), (2.27) we deduce that u_{g1} is the solution of

$$(2.28) \qquad \operatorname*{Min}_{v \in V_1} \{ \tfrac{1}{2} \| v \|^2 - (f_1, v) \} ,$$

so that

$$(2.29) \qquad u_{g1} = f_1 \quad \forall g .$$

Similarly, u_{g2} is the solution of the problem

$$(2.29)' \qquad \operatorname*{Min}_{v \in V_2} \{ \tfrac{1}{2} \| v \|^2 + gj(v) - (f_2, v) \}$$

and we thus have (cf. (2.21)):

$$(2.30) \qquad \| u_{g2} \|^2 + gj(u_{g2}) = (f_2, u_{g2}) ;$$

Now (2.30) implies

$$(2.31) \qquad \| u_{g2} \| \leqslant \| f_2 \|$$

so that

$$(2.32) \qquad j(u_{g2}) \leqslant \frac{2}{g} \| f_2 \|^2 .$$

From (2.31), (2.32), we then deduce

$$(2.33) \qquad \lim_{g \to \infty} u_{g2} = 0 \quad \textit{weakly in } V$$

and (2.30), (2.33) imply

$$(2.34) \qquad \begin{cases} \lim\limits_{g \to \infty} gj(u_{g2}) = 0 \\ \lim\limits_{g \to \infty} u_{g2} = 0 \quad \textit{strongly in } V . \end{cases}$$

We have $\| u_{g2} \|^2 = \| u_g \|^2 - \| f_1 \|^2$, so that the function $g \to \| u_{g2} \|^2$ is therefore *convex*; moreover

$$(2.35) \qquad \lim_{g \to \infty} \| u_{g2} \|^2 = 0 ;$$

taking account of (2.35) and of $\| u_{g2} \|^2 \geqslant 0 \ \ \forall g$, the convexity of the function $g \to \| u_{g2} \|^2$ implies that it is decreasing and hence that $g \to \| u_g \|^2$ is decreasing, since

$$\| u_g \|^2 = \| f_1 \|^2 + \| u_{g2} \|^2 . \quad \blacksquare$$

Proposition 2.2 may be deduced immediately from Lemma 2.1 by noting that the mapping

$$v \to \int_{\Gamma_d} | v | \, d\Gamma$$

satisfies the properties required of j in the statement of the lemma. ■

Let $V_1 = \{ v \mid v \in V, j(v) = 0 \}$ and let $\hat{u}$ be the solution of

$$(2.36) \qquad \begin{cases} J_0(\hat{u}) \leqslant J_0(v) & \forall v \in V_1 \\ \hat{u} \in V_1 \end{cases}$$

with

$$J_0(v) = \tfrac{1}{2} a(v, v) - L(v) ;$$

It may be seen that by proving point (iii) of Lemma 2.1, we have also proved:

Lemma 2.2. *We have:*

$$\lim_{g \to +\infty} u_g = \hat{u} \quad \textit{strongly in } V. \ ■$$

From this we deduce:

Proposition 2.3. *Let* u_g *be the solution of* (2.1), (2.2), *and* $\hat{u}$ *that of*

$$(2.37) \qquad \begin{cases} \operatorname{Min}_{v \in W} \left\{ \dfrac{1}{2} \displaystyle\int_\Omega (\mu v^2 + |\operatorname{grad} v|^2) \, dx - \int_\Omega f v \, dx \right\} \\ W = \{ v \mid v \in V, v|_{\Gamma_d} = 0 \} ; \end{cases}$$

then

$$\lim_{g \to +\infty} u_g = \hat{u} \quad \textit{strongly in } V. \ ■$$

Remark 2.4. The solution $\hat{u}$ of (2.37) is a solution of the Dirichlet problem

$$\begin{cases} -\Delta \hat{u} + \mu \hat{u} = f \\ \hat{u}|_\Gamma = 0 \end{cases}$$

if V is defined by (2.3), and of the mixed problem

$$\begin{cases} -\Delta \hat{u} + \mu \hat{u} = f \\ \hat{u} = 0 \quad \textit{on} \ \Gamma_d \\ \dfrac{\partial \hat{u}}{\partial n} = 0 \quad \textit{on} \ \Gamma - \Gamma_d \end{cases}$$

if $V = H^1(\Omega)$. ■

2.5 Duality properties

We denote by Λ the closed convex subset of $L^2(\Gamma_d)$ defined by

(2.38) $$\Lambda = \{ q \mid q \in L^2(\Gamma_d), \mid q(x) \mid \leqslant 1 \quad \text{a.e. on} \quad \Gamma_d \} ;$$

we then have:

<u>Theorem 2.2</u> *The solution* u *of problem* (2.1) *is characterised by the existence of* $p \in \Lambda$, *unique, such that*

(2.39) $$\begin{cases} \int_\Omega (\text{grad } u . \text{grad } v + \mu uv) \, \mathrm{d}x + g \int_{\Gamma_d} pv \, \mathrm{d}\Gamma = \int_\Omega fv \, \mathrm{d}x \quad \forall v \in V, \\ u \in V \\ pu = \mid u \mid \quad \text{a.e. on} \quad \Gamma_d \\ p \in \Lambda . \end{cases}$$

Proof. Theorem 2.2 may be proved by using a variant of the proof (based on the Hahn-Banach Theorem) of Theorem 3.3 of Chapter 1, Section 3.4; however, we prefer to give a more constructive proof, based (amongst other things) on the *regularisation* property of Proposition 2.1 (1). Using the same formalism as that employed in Section 2.3, we denote by φ_k the derivative of ψ_k, so that (cf. (1.9)):

(2.40) $$\varphi_k(\xi) = \begin{cases} g & \text{if} \quad \xi \geqslant kg \\ \xi/k & \text{if} \quad - kg \leqslant \xi \leqslant kg \\ - g & \text{if} \quad \xi \leqslant - kg ; \end{cases}$$

the solution u_k of problem (1.4), (1.5) of Section 1.1 is thus also the solution of the nonlinear variational equation

(2.41) $$\begin{cases} a(u_k, v) + \int_{\Gamma_d} \varphi_k(u_k) \, v \, \mathrm{d}\Gamma = \int_\Omega fv \, \mathrm{d}x \quad \forall v \in V \\ u_k \in V . \end{cases}$$

We define p_k by $p_k = \varphi_k(u_k)/g$; it then follows from (2.40), (2.41) that:

(2.42) $$p_k \in \Lambda$$

and

(2.43) $$a(u_k, v) + g \int_{\Gamma_d} p_k \, v \, \mathrm{d}\Gamma = \int_\Omega fv \, \mathrm{d}x \quad \forall v \in V .$$

Since the convex set Λ is weakly compact in $L^2(\Gamma_d)$, we may hence deduce the existence of $p \in \Lambda$ and of an extracted subsequence, also denoted by p_k , such that

---

(1) This process will be taken up again in Chapter 5, Section 6.1.4.

(2.44) $\quad p_k \to p$ weakly in $L^2(\Gamma_d)$ [1]

Taking account of Proposition 2.1 and of (2.44), we may then deduce in the limit in (2.43) that

$$a(u, v) + g \int_{\Gamma_d} pv \, d\Gamma = \int_\Omega fv \, dx \quad \forall v \in V. \tag{2.45}$$

It follows from the proof of Lemma 2.1 that

$$a(u, u) + g \int_{\Gamma_d} |u| \, d\Gamma = \int_\Omega fu \, dx. \tag{2.46}$$

Moreover, putting $v = u$ in (2.45), we may then deduce that:

$$a(u, u) + g \int_{\Gamma_d} pu \, d\Gamma = \int_\Omega fu \, dx,$$

so that with (2.46):

$$\int_{\Gamma_d} |u| \, d\Gamma = \int_{\Gamma_d} pu \, d\Gamma$$

which may be written

$$\int_\Gamma (|u| - pu) \, d\Gamma = 0. \tag{2.47}$$

But $p \in \Lambda$ implies

$$pu \leqslant |pu| \leqslant |u|$$

so that

$$|u| - pu \geqslant 0 \quad \text{a.e. on } \Gamma_d. \tag{2.48}$$

It then follows from (2.47), (2.48) that $|u| = pu$ a.e. on Γ_d. We prove the uniqueness by *reductio ad absurdum*; suppose that there exists $p_1, p_2 \in \Lambda$ satisfying (2.39), so that

$$\int_\Gamma (p_2 - p_1) v \, d\Gamma = 0 \quad \forall v \in V \tag{2.49}$$

and, since the space of *traces* of the functions of V on Γ_d is *dense* in $L^2(\Gamma_d)$, (2.49) implies $p_2 = p_1$, thus proving the uniqueness. Conversely if u obeys (2.39), then we can easily prove that u is the solution of problem (2.1). ■

Remark 2.5. It follows from (2.39) that the solution u of (2.1) satisfies

---

(1) In fact we have $p_k \to p$ weakly* in $L^\infty(\Gamma_d)$.

$$\begin{cases} -\Delta u + \mu u = f & \text{in } \Omega \\ \dfrac{\partial u}{\partial n} + p = 0 & \text{on } \Gamma_d \\ u = 0 & \text{on } \Gamma - \Gamma_d \end{cases}$$

if V is defined by (2.3) and

$$\begin{cases} -\Delta u + \mu u = f & \text{in } \Omega \\ \dfrac{\partial u}{\partial n} + p = 0 \text{ on } \Gamma_d \\ \dfrac{\partial u}{\partial n} = 0 \text{ on } \Gamma - \Gamma_d \end{cases}$$

if $V = H^1(\Omega)$. ■

2.6 Approximation by finite elements

2.6.1 *Formulation of the approximate problem*

We assume, as in Section 1.3, that Ω is polygonal; this then allows $\mathcal{T}_h$ and V_h to be defined as in Section 1.3; in particular, we retain the assumption that " Γ_d *is the union of sides, the ends of which lie on* Γ*, of elements of* $\mathcal{T}_h$ ".

As far as the approximation of $j(v) = g \int_{\Gamma_d} |v| \, d\Gamma$ is concerned, we use the limits (1), when $k \to 0$, of the functionals defined in Section 1.3 by (1.24), (1.25), giving:

$$j_h(v_h) = \frac{g}{2} \sum_i |M_i M_{i+1}| (|v_h(M_i)| + |v_h(M_{i+1})|), \qquad \text{if } q = 1 \tag{2.50}$$

$$j_h(v_h) = \frac{g}{6} \sum_i |M_i M_{i+1}| (|v_h(M_i)| + 4|v_h(M_{i+1/2})| + + |v_h(M_{i+1})|), \quad \text{if } q = 2, \tag{2.51}$$

the summations extending over the sides of the triangles of $\mathcal{T}_h$ whose union constitutes Γ_d .

We then approximate problem (2.1) by the finite-dimensional problem defined (in the form of a variational inequality) by

$$\begin{cases} a(u_h, v_h - u_h) + j_h(v_h) - j_h(u_h) \geq \int_\Omega f(v_h - u_h) \, dx \quad \forall v_h \in V_h \\ u_h \in V_h . \end{cases} \tag{2.52}$$

We immediately have:

---

(1) These are given in explicit form in Section 1.6.3, Remark 1.9, relations (1.139), (1.140).

Theorem 2.3. *The approximate problem* (2.52) *admits one and only one solution.* ■

2.6.2 *Convergence of the approximate solutions*

Using proofs identical to that of Theorem 1.3 of Section 1.4.2 ($q = 1$) and of Theorem 1.5 of Section 1.5.2 ($q = 2$) may prove, for $q = 1, 2$:

Theorem 2.4. *If, when* $h \to 0$, *the angles of* $\mathscr{T}_h$ *are bounded below by a constant* $\theta_0 > 0$, *we have*

$$\lim_{h \to 0} u_h = u \quad \textit{strongly in } H^1(\Omega)$$

where u *and* u_h *are respectively solutions of problems* (2.1) and (2.52). ■

2.6.3 *Estimates of the approximation error.*

For $q = 1$, the estimates of the approximation error obtained in Section 1.4.3, which are *independent of* k, also hold for the approximate solutions of problem (2.1) defined by (2.52). ■

2.7 Numerical solution of the approximate problems

Synopsis

We shall study the solution of the approximate problem (2.52) by the relaxation method considered in Section 1.6.3 and by a duality algorithm of Uzawa type (see Chapter 2, Section 4.3).

2.7.1 *Formulation of the approximate problems*

In order to write the approximate problem (2.52) explicitly, we shall use the expression for j_h given in (2.50) (resp. (2.51)) and the formulas of Section 1.6.1 (resp.(1.6.2)) if $q = 1$ (resp. $q = 2$). ■

2.7.2 *Solution of the approximate problem by relaxation*

In view of Remark 1.9 of Section 1.6.3, the considerations of Section 1.6.3 relating to the approximate problem (1.126) also hold for the approximate problem (2.52) which is equivalent to the finite-dimensional optimisation problem:

$$(2.53) \quad \begin{aligned} &\operatorname*{Min}_{v_h \in V_h} \left[J_0(v_h) + j_h(v_h) \right], \\ &J_0(v_h) = \frac{1}{2} \int_\Omega |\operatorname{grad} v_h|^2 \, \mathrm{d}x + \frac{\mu}{2} \int_\Omega |v_h|^2 \, \mathrm{d}x - \int_\Omega f v_h \, \mathrm{d}x . \end{aligned}$$ ■

2.7.3 *Solution of the approximate problem by a duality method*

We shall first study the *continuous* case; the problem (2.1), (2.2) falls into the category of Chapter 2, Section 4 if we take:

$$J_0(v) = \frac{1}{2}\int_\Omega (|\operatorname{grad} v|^2 + \mu v^2)\,\mathrm{d}x - \int_\Omega fv\,\mathrm{d}x,\ M = V,\ L = L^2(\Gamma_d),$$

$$\Phi(v) = v|_{\Gamma_d},\ \Lambda = \{ q \mid q \in L^2(\Gamma_d),\ |q(x)| \leqslant 1 \text{ a.e. on } \Gamma_d \}.$$

In fact, we then have

(2.54) $$\int_{\Gamma_d} |v|\,\mathrm{d}\Gamma = \operatorname*{Max}_{q\in\Lambda} \int_{\Gamma_d} qv\,\mathrm{d}\Gamma \quad \forall v \in V;$$

next, by introducing the Lagrangian

$$\mathscr{L}(v, \mu) = J_0(v) + g\int_{\Gamma_d} \mu v\,\mathrm{d}\Gamma$$

and using relation (2.39) of the statement of Theorem 2.2 we may prove:

Theorem 2.5. *The Lagrangian* $\mathscr{L}$ *admits a unique saddle point* $\{u, p\}$ *on* $V \times \Lambda$, *where* u *is the solution of* (2.1), (2.2) *and* p *is related to* u *by* (2.39). ■

We denote $v|_{\Gamma_d}$ by $\gamma_d v$; in view of Theorem 2.5, problem (2.1), (2.2) may be solved by the algorithm (4.12)-(4.15) of Chapter 2, Section 4.3, which in this particular case takes the form:

(2.55) $p^0 \in \Lambda$ chosen arbitrarily (for example, zero)

(2.56) $\mathscr{L}(u^n, p^n) \leqslant \mathscr{L}(v, p^n) \quad \forall v \in V,\ u^n \in V$

(2.57) $p^{n+1} = P_\Lambda(p^n + \rho_n g\gamma_d u^n),\ \rho_n > 0.$

Writing (2.56) explicitly, we obtain

(2.58) $$\begin{cases} -\Delta u^n + \mu u^n = f & \text{in } \Omega \\ \dfrac{\partial u^n}{\partial n} + gp^n = 0 \text{ on } \Gamma_d \end{cases}$$

with

(2.59) $u^n = 0$ on $\Gamma - \Gamma_d$ if V is given by (2.3),

(2.60) $\dfrac{\partial u^n}{\partial n} = 0$ on $\Gamma - \Gamma_d$ if $V = H^1(\Omega)$;

moreover

(2.61) $P_\Lambda(q) = \sup(-1, \inf(1, q)) \quad \forall q \in L^2(\Gamma_d)$

which allows (2.57) to be written explicitly.

Since γ_d is a linear continuous mapping from $v \to L^2(\Gamma_d)$, we have

(2.62) $$\| \gamma_d v \|_{L^2(\Gamma_d)} \leqslant \| \gamma_d \| \; \| v \|_V \quad \forall v \in V ;$$

it then follows from Theorem 4.1 of Chapter 2, Section 4.3, that:

(2.63) $u^n \to u$ strongly in V

with the condition (1):

(2.64) $$0 < \alpha_0 \leqslant \rho_n \leqslant \alpha_1 < \frac{2}{g^2 \| \gamma_d \|^2} . \quad \blacksquare$$

The approximate problem (2.52) is solved in a similar manner using the discrete analogue of algorithm (2.55)-(2.57):

We write $\mu_h = \{ \mu_i \}_i$ with $M_i \in \Sigma_h \cap \Gamma_d$ if $q = 1$ or with $M_i \in (\Sigma_h \cap \Sigma_h') \cap \Gamma_d$ if $q = 2$, $\mu_i \in \mathbb{R} \;\; \forall i$; we then define the approximation Λ_h of Λ by

$$\Gamma_h = \{ \mu_h \mid \; | \mu_i | \leqslant 1 \quad \forall i \} ,$$

and then the approximation $\mathscr{L}_h$ of $\mathscr{L}$ by

(2.65) $$\mathscr{L}_h(v_h, \mu_h) = J_0(v_h) + \frac{g}{2} \sum_i | M_i M_{i+1} | (\mu_i v_h(M_i) + \mu_{i+1} v_h(M_{i+1})) \text{ if } q = 1 ,$$

(2.66) $$\mathscr{L}_h(v_h, \mu_h) = J_0(v_h) + \frac{g}{6} \sum_i | M_i M_{i+1} | (\mu_i v_h(M_i) + 4 \mu_{i+1/2} v_h(M_{i+1/2}) + \mu_{i+1} v_h(M_{i+1})) \quad \text{if } q = 2 .$$

The Langrangian $\mathscr{L}_h$ admits a unique saddle point $\{ u_h, \lambda_h \}$ on $V_h \times \Lambda_h$, with u_h the solution of the approximate problem (2.52); the discrete analogue of (2.55)-(2.57) is then

(2.67) $\lambda_h^0 \in \Lambda_h$

(2.68) $$\mathscr{L}_h(u_h^n, \lambda_h^n) \leqslant \mathscr{L}_h(v_h, \lambda_h^n) \quad \forall v_h \in V_h, u_h^n \in V_h$$

(2.69) $$\begin{cases} \lambda_i^{n+1} = \max(-1, \min(1, \lambda_i^n + \rho_n g u_h^n(M_i))) \\ \forall M_i \in \Sigma_h \cap \Gamma_d \text{ if } q = 1, \quad \forall M_i \in (\Sigma_h \cup \Sigma_h') \cap \Gamma_d \text{ if } q = 2 . \end{cases}$$

It may be shown that under the condition $0 < \alpha_0 \leqslant \rho_n \leqslant \alpha_1, \alpha_1$

---

(1) By taking $\left(\int_\Omega (\mu v^2 + | \operatorname{grad} v |^2) \, dx \right)^{1/2}$ as norm on V, the norm of $\gamma_d \in \mathscr{L}(V ; L^2(\Gamma_d))$ corresponding to this choice.

sufficiently small, we have $\lim_{n \to +\infty} u_h^n = u_h$, u_h is the solution of (2.52). ■

2.8 An example

We shall now apply the above methods of approximation and solution to the example defined in Section 2.8.1 below.

2.8.1 *The continuous problem*

The problem (2.1), (2.2) considered is defined by:

Geometrical data: $\Omega =]0,1[\times]0,1[$

$$\Gamma_d = \{ (x_1, x_2) \mid 0 \leqslant x_1 \leqslant 1, x_2 = 0 \} \cup \{ (x_1, x_2) \mid 0 \leqslant x_1 \leqslant 1, x_2 = 1 \}.$$

Space V: $V = \{ v \mid v \in H^1(\Omega), \quad v = 0 \quad \text{on} \quad \Gamma - \Gamma_d \}$.

Function f:

$$f(x) = 10 \text{ if } x \in]0, \tfrac{1}{2}[\times]0,1[\ , \ f(x) = -10 \text{ if } x \in]\tfrac{1}{2}, 1[\times]0,1[\ .$$

Parameter values: $\mu = 0$; $g = 0.2, 0.5, 1, 1.5, 2.$

The condition $v = 0$ on $\Gamma - \Gamma_d$ implies that $v \to \left(\iint_\Omega |\operatorname{grad} v|^2 \, dx \right)^{1/2}$ defines on V a norm which is equivalent to that induced by $H^1(\Omega)$, so that it is permissible to take $\mu = 0$ in (2.2). ■

2.8.2 *The approximate problem*

The case $q = 1$ (resp. $q = 2$) corresponds to approximation by finite elements of order *one* (resp. of order *two*); for the approximate problem defined by (2.52) we have used the triangulation $\mathcal{T}_h$ shown in Figure 1.9 (resp. 1.10) if $q = 1$ (resp. $q = 2$), corresponding to 512 triangles (resp. 128 triangles), 289 nodes and 255 degrees of freedom.

For $q = 1$ (resp. $q = 2$) the approximate problem may be written in explicit form using the relations of Section 1.6.1 (resp. 1.6.2) and of (2.50) (resp. (2.51)). ■

2.8.3 *Solution by over-relaxation*

The approximate problem (2.52) is equivalent to the finite-dimensional minimisation problem

$$(2.70) \qquad \begin{cases} J_h(u_h) \leqslant J_h(v_h) & \forall v_h \in V_h \\ u_h \in V_h \end{cases}$$

with

$$J_h(v_h) = \frac{1}{2}\int_\Omega |\operatorname{grad} v_h|^2 \, dx + j_h(v_h) - \int_\Omega f v_h \, dx ;$$

to solve (2.70), we have used (1) the following algorithm:

(2.71) $u_h^0 = (u_1^0, \dots, u_{N_h}^0)$ chosen in $\mathbb{R}^{N_h}$.

With u_h^n known, we determine u_h^{n+1}, coordinate by coordinate, using

(2.72)
$$J_h(u_1^{n+1}, \dots, u_{i-1}^{n+1}, u_i^{n+1/2}, u_{i+1}^n, \dots) \leqslant \leqslant J_h(u_1^{n+1}, \dots, u_{i-1}^{n+1}, v_i, u_{i+1}^n, \dots) \quad \forall v_i \in \mathbb{R}$$

(2.73) $u_i^{n+1} = u_i^n + \omega(u_i^{n+1/2} - u_i^n)$.

In the example considered, the algorithm (2.71)-(2.73) was in all cases initialised by $u_h^0 = 0$, the termination criterion being defined by

$$\frac{\sum_i | u_i^{n+1} - u_i^n |}{\sum_i | u_i^{n+1} |} \leqslant 10^{-5} .$$

Analysis of the numerical results

Table 2.1 shows certain details relating to the solution of the approximate problems and to the results obtained; we have denoted by $\mu_1(u_h)$ and $\mu_2(u_h)$ the quantities defined by

$$\mu_1(u_h) = \operatorname{Max} u_h(M), \quad M \in \Sigma_h \text{ if } q = 1, \quad M \in \Sigma_h \cup \Sigma_h' \text{ if } q = 2,$$

$$\mu_2(u_h) = \operatorname{Max} u_h(M), \quad M \in \Sigma_h \cap \Gamma \text{ if } q = 1, \quad M \in (\Sigma_h \cup \Sigma_h') \cap \Gamma \text{ if } q = 2 .$$

g	overrelaxation parameter ω		$\mu_1(u_h)$		$\mu_2(u_h)$		Number of iterations	
	$q=1$	$q=2$	$q=1$	$q=2$	$q=1$	$q=2$	$q=1$	$q=2$
0.2	1.74	1.75	0.309	0.309	0.276	0.275	26	40
0.5	1.74	1.75	0.304	0.304	0.221	0.219	26	30
1	1.74	1.75	0.295	0.295	0.129	0.126	26	33
1.5	1.74	1.75	2.286	0.287	0.040	0.038	25	35
2	1.74	1.75	0.284	0.284	0	0	27	37

Table 2.1

(1) The notation is the same as that used in Section 1.6.3.

All the computations were performed on a CII 10070, the mean computation time per iteration being 0.2 sec. for $q = 1$, and 0.4 sec. for $q = 2$.

Figure 2.1 (resp. 2.2) shows the function $x_1 \to u_h(x_1, \frac{1}{2})$ (resp. $x_1 \to u_h(x_1, 0)$) calculated for $q = 1$.

Figures 2.3, 2.4 show the equipotentials of the computed solutions (with $q = 1$), corresponding respectively to $g = 0.2$ and $g = 1.5$.

Remark 2.6. If, all other things being equal, we solve the approximate problem (1.27) of Section 1.3.3, the computed solutions are practically coincident with those of problem (2.52) when $k \leqslant 10^{-3}$; this gives an experimental confirmation of the results of Section 2.3. ■

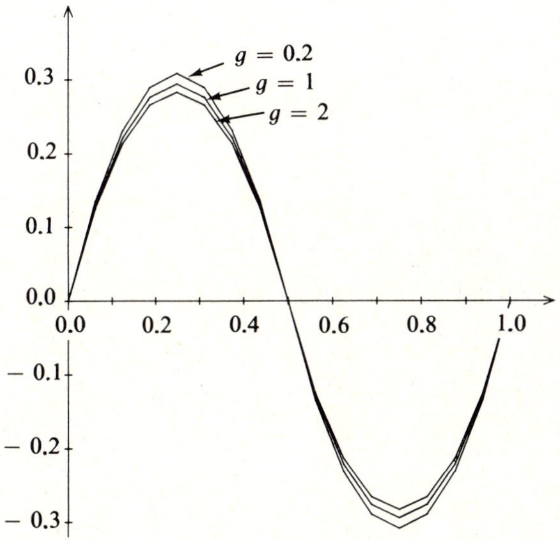

Fig. 2.1. Representation of $u(x_1, 1/2)$ $(q = 1)$.

2.8.4 *Solution by a duality method*

We shall now solve the approximate problem defined in Sections 2.8.1, 2.8.2, using the duality algorithm (2.67)-(2.69) and confining our attention to the case $q = 1$; the conditions used for the computation are as follows:

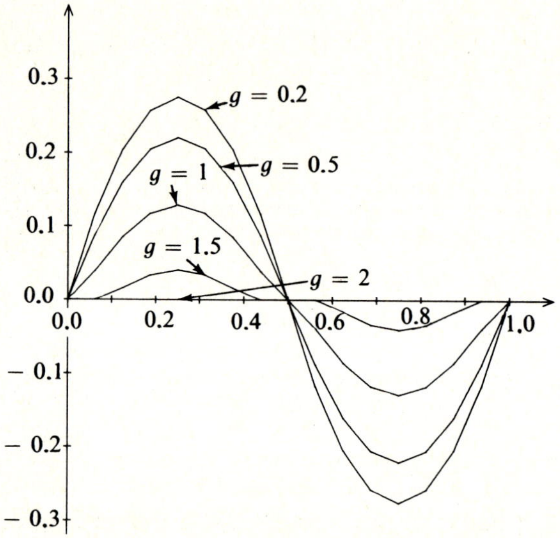

Fig. 2.2. Representation of $u(x_1, 0)$ $(q = 1)$.

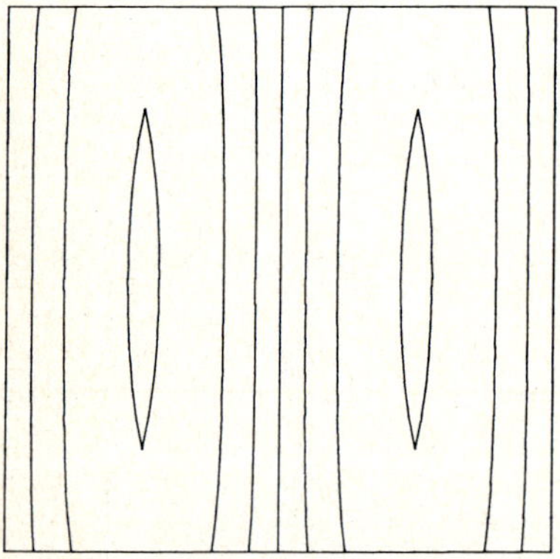

Fig. 2.3. Equipotentials of u $(g = 0.2)$.

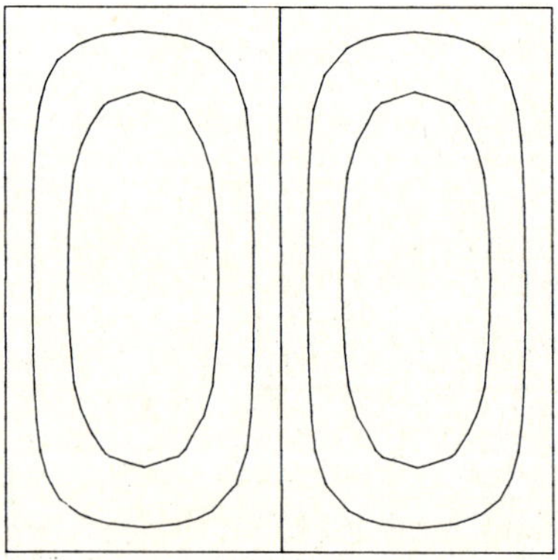

Fig. 2.4. Equipotentials of u $(g = 1.5)$.

Initialisation of (2.67)-(2.69): $\lambda_h^0 = 0$.

Solution of (2.68); by point over-relaxation using an optimal parameter and initialising with u_h^{n-1}.

Choice of ρ_n: this was taken as $\rho_n = \rho$ independent of n

Termination criterion for (2.67)-(2.69):

$$\frac{\sum_i | u_i^{n+1} - u_i^n |}{\sum_i | u_i^n |} \leqslant 10^{-4} . \tag{2.74}$$

The results thus calculated coincide with those obtained using the over-relaxation method described in Section 2.8.3. Table 2.2 shows the number of iterations required for convergence, as a function of the parameter ρ.

Remark 2.7. For $g = 0.2$, 0.5 and 1 it can be seen that $|\lambda_h|$ is everywhere equal to unity; it therefore follows that algorithm (2.67)-(2.69) in fact converges for any positive value of ρ; on the other hand, for $g = 1.5$ and 2, $|\lambda_h|$ takes values strictly less than unity and the algorithm diverges if ρ is too large. ■

Remark 2.8. For the case of the example in Section 2.8.1 we can determine a lower bound for the quantity $\dfrac{2}{g^2 \| \gamma_d \|^2}$ appearing in the conditions for convergence of the duality algorithm (2.55)-(2.57); we recall (see Section 2.7.3, equation (2.64)) that (2.55)-(2.57) converges for

$$0 < \alpha_0 \leqslant \rho_n \leqslant \alpha_1 < \frac{2}{g^2 \| \gamma_d \|^2} .$$

g \ ρ	0.5	1	1.5	2	3	4	5	6	7	8	9
0.2				22	15	12	10	9	8	7	7
0.5		21		12	9	7	6	5	5	5	4
1		17		9	7	6	5	4	3	3	3
1.5		37		21	15	7	9	$+\infty$			
2	25	17	12	$+\infty$							

Table 2.2

We have $V = \{ v \mid v \in H^1(\Omega),\ v = 0 \text{ on } \Gamma - \Gamma_d \}$, so that we can take

$$\| v \| = \left(\int_\Omega | \operatorname{grad} v |^2 \, dx \right)^{1/2}$$

as norm over V.

The eigenfunctions of $-\Delta$ in V, i.e. the solutions of

$$\begin{cases} -\Delta u = \lambda u & \text{in } \Omega \\ u = 0 & \text{on } \Gamma - \Gamma_d \\ \dfrac{\partial u}{\partial n} = 0 & \text{on } \Gamma_d , \end{cases}$$

namely

$$u_{m0} = \sqrt{2} \sin m\pi x_1 ,\ m \geqslant 1$$
$$u_{mn} = 2 \sin m\pi x_1 \cos n\pi x_2 ,\ m, n \geqslant 1 ,$$

form a basis for V (orthonormal in $L^2(\Omega)$), so that $\forall v \in V$:

$$v(x_1, x_2) = \sum_{m \geqslant 1} v_{m0} \sqrt{2} \sin m\pi x_1 + \sum_{m,n \geqslant 1} v_{mn} \, 2 \sin m\pi x_1 \cos n\pi x_2 .$$

From this we deduce that:

$$\| v \|^2 = \int_\Omega | \operatorname{grad} v |^2 \, dx = \pi^2 \left(\sum_{m \geqslant 1} m^2 v_{m0}^2 + \sum_{m,n \geqslant 1} (m^2 + n^2) v_{mn}^2 \right) . \tag{2.75}$$

For the traces on Γ_d, we have

$$v(x_1, 0) = \sum_{m \geqslant 1} \sqrt{2} \sin m\pi x_1 \left(v_{m0} + \sqrt{2} \sum_{n \geqslant 1} v_{mn} \right)$$
$$v(x_1, 1) = \sum_{m \geqslant 1} \sqrt{2} \sin m\pi x_1 \left(v_{m0} + \sqrt{2} \sum_{n \geqslant 1} (-1)^n v_{mn} \right) ,$$

so that

$$\int_0^1 | v(x_1, 0) |^2 \, dx_1 = \sum_{m \geqslant 1} \left(v_{m0} + \sqrt{2} \sum_{n \geqslant 1} v_{mn} \right)^2$$
$$\int_0^1 | v(x_1, 1) |^2 \, dx_1 = \sum_{m \geqslant 1} \left(v_{m0} + \sqrt{2} \sum_{n \geqslant 1} (-1)^n v_{mn} \right)^2$$

which implies

$$\int_{\Gamma_d} v^2 \, d\Gamma \leqslant 2 \sum_{m \geqslant 1} \left(| v_{m0} | + \sqrt{2} \sum_{n \geqslant 1} | v_{mn} | \right)^2 \leqslant$$
$$\leqslant 4 \sum_{m \geqslant 1} \left(| v_{m0} | + \sum_{n \geqslant 1} | v_{mn} | \right)^2 =$$

$$= 4 \sum_{m \geqslant 1} \left(| v_{m0} | + \sum_{n \geqslant 1} n \, | v_{mn} | \frac{1}{n} \right)^2 \leqslant$$

$$\leqslant 4 \sum_{m \geqslant 1} \left(1 + \sum_{n \geqslant 1} \frac{1}{n^2} \right) \left(v_{m0}^2 + \sum_{n \geqslant 1} n^2 \, v_{mn}^2 \right) \leqslant$$

$$\leqslant 4 \left(1 + \frac{\pi^2}{6} \right) \left(\sum_{m \geqslant 1} m^2 \, v_{m0}^2 + \sum_{m,n \geqslant 1} (m^2 + n^2) \, v_{mn}^2 \right) =$$

$$= 4 \left(\frac{1}{\pi^2} + \frac{1}{6} \right) \| v \|^2 .$$

We thus have

$$\| \gamma_d \|^2 \leqslant 4 \left(\frac{1}{\pi^2} + \frac{1}{6} \right)$$

so that the convergence of algorithm (2.55)-(2.57) is assured if

$$0 < \alpha_0 \leqslant \rho_n \leqslant \alpha_1 < \frac{1}{g^2} \frac{3\pi^2}{6 + \pi^2} . \tag{2.76}$$

Noting that $3\pi^2/(6 + \pi^2) \simeq 1.9$, we may then deduce, by looking at Table 2.2, that (2.76) gives a pessimistic estimate of the convergence interval for algorithm (2.67)-(2.69).

2.8.5 *Various remarks*

As pointed out earlier in Remark 5.10 of Chapter 2, Section 5.7, the over-relaxation algorithm is more economical than the duality algorithm [1], both in terms of computation time and storage requirement. However, the duality algorithm has the advantage of also giving the approximation of the 'multiplier' p and allows (2.1), (2.2) to be solved easily, at the cost of trivial modifications if a program for solving elliptic problems of mixed type is available; in fact the essential stage in the implementation of algorithm (2.67)-(2.69) is clearly the solution of problem (2.68). ■

As far as the type of finite elements used is concerned, it would appear that for a given degree of precision the use of elements of order *one* ($q = 1$) is much more economical than that of elements of order *two* ($q = 2$); the advantage of elements of order *one* becomes even greater if the presence of curved boundaries introduces the need for *isoparametric finite elements of order two*. [2]. In fact, in our opinion, the foregoing remark applies for *elliptic problems of order two* whenever the solutions are of *low regularity* (say when the regularity is no better than H^2). ■

---

(1) For solving (2.1), (2.2) at least

(2) See Ciarlet-Raviart /2/ for the definition and theoretical study of curved finite elements of isoparametric type.

3. A PROBLEM WITH UNILATERAL CONSTRAINTS AT THE BOUNDARY

3.1 Synopsis

In this section we shall consider the numerical analysis of the problem

$$(3.1)\qquad \begin{cases} \int_\Omega \operatorname{grad} u.\operatorname{grad}(v-u)\,dx + \mu \int_\Omega u(v-u)\,dx \geqslant \\ \qquad \geqslant \int_\Omega f(v-u)\,dx \quad \forall v \in K \\ u \in K \end{cases}$$

with $f \in L^2(\Omega)$ and

$$(3.2)\qquad K = \{ v \mid v \in H^1(\Omega), v \geqslant 0 \quad \text{a.e. on} \quad \Gamma \}.$$

A variant of this problem (with $\mu = 0$) was considered earlier (from a theoretical viewpoint) in Chapter 1, Sections 1.1 and 2.1, together with an indication of the corresponding physical background. ■

3.2 Existence and uniqueness results for problem (3.1), (3.2)

We shall suppose in the following that $\mu > 0$; it may be noted that there is equivalence between (3.1), (3.2) and

$$(3.3)\qquad \begin{cases} \operatorname*{Min}_{v \in K} J_0(v)\,, \\ J_0(v) = \frac{1}{2}\int_\Omega (|\operatorname{grad} v|^2 + \mu v^2)\,dx - \int_\Omega fv\,dx\,; \end{cases}$$

since the function J_0 is continuous and strictly convex in $H^1(\Omega)$ with

$$\lim_{\|v\| \to +\infty} J_0(v) = +\infty\,,$$

and since K is convex and closed in $H^1(\Omega)$, Theorem 2.1 of Chapter 1, Section 2.1, is applicable, so that the existence and uniqueness of a solution for (3.3) (and, by equivalence, for (3.1),(3.2)), then follow. ■

3.3 Regularity results

If the boundary Γ of Ω is sufficiently regular, it may be shown, (see Lions /1/, Chapter 2, Section 8 and H. Brézis /2/) that the solution u of (3.1), (3.2) belongs to $K \cap H^2(\Omega)$ with

$\| u \|_{H^2(\Omega)} \leqslant C \| f \|_{L^2(\Omega)}$, C independent of u and f. ■

3.4 Duality results

First, we shall prove:

Proposition 3.1. *The solution* u *of* (3.1), (3.2) *is characterised by*

$$(3.4) \qquad \begin{cases} -\Delta u + \mu u = f & \text{in } \Omega \\ u \geqslant 0 \quad \text{a.e. on } \Gamma, & \dfrac{\partial u}{\partial n} \geqslant 0 \text{ a.e. on } \Gamma \\ u \dfrac{\partial u}{\partial n} = 0 \text{ a.e. on } \Gamma . \end{cases}$$

Proof. We write:

$$a(u, v) = \int_\Omega (\operatorname{grad} u . \operatorname{grad} v + \mu u v) \, dx .$$

As K is a convex cone with apex O, the solution u of (3.1), (3.2) is in fact characterised by

$$(3.5) \qquad \begin{cases} a(u, v) \geqslant \displaystyle\int_\Omega f v \, dx \quad \forall v \in K \\ u \in K \\ a(u, u) = \displaystyle\int_\Omega f u \, dx ; \end{cases}$$

we have $\mathscr{D}(\Omega) \subset K$, so that the first relation in (3.5) then implies

$$\pm \left(a(u, v) - \int_\Omega f v \, dx \right) \geqslant 0 \quad \forall v \in \mathscr{D}(\Omega)$$

so that

$$(3.6) \qquad a(u, v) = \int_\Omega f v \, dx \quad \forall v \in \mathscr{D}(\Omega)$$

and (3.6) leads to

$$(3.7) \qquad -\Delta u + \mu u = f \quad \text{in } \Omega .$$

After multiplication by $v \in K$, and integration by parts, we deduce from (3.5), (3.7) that

$$(3.8) \qquad \int_\Gamma v \frac{\partial u}{\partial n} \, d\Gamma = a(u, v) - \int_\Omega f v \, dx \geqslant 0 \quad \forall v \in K ;$$

relation (3.8) then implies $\partial u / \partial n \geqslant 0$ a.e. on Γ, and putting $v = u$ in (3.8), we hence deduce that

$$\int_\Gamma u \frac{\partial u}{\partial n} \, d\Gamma = 0 \, ,$$

now $u \geqslant 0$, $\partial u/\partial n \geqslant 0$ a.e. so that $u(\partial u/\partial n) = 0$ a.e. on Γ, which completes the proof of (3.4). ■

Conversely, starting from (3.4), the characterisation (3.5) may be recovered without difficulty. ■

We denote by $\mathscr{L}$ the Lagrangian defined by

$$\mathscr{L}(v, q) = J_0(v) - \int_\Gamma qv \, d\Gamma$$

and by Λ the positive cone of $L^2(\Gamma)$, that is,

$$\Lambda = \{ q \mid q \in L^2(\Gamma), q \geqslant 0 \text{ a.e. on } \Gamma \} ;$$

noting that $u \in H^2(\Omega)$ implies $\partial u/\partial n \in H^{1/2}(\Gamma) \subset L^2(\Gamma)$ we may then prove ([1]), starting from Proposition 3.1:

Theorem 3.1. *Suppose* u *is the solution of* (3.1), (3.2); *then the Lagrangian* $\mathscr{L}$ *admits* $\{ u, \partial u/\partial n \}$ *as unique saddle point on* $H^1(\Omega) \times \Lambda$. ■

The above results will be used in Section 3.7 in proving the convergence of a duality algorithm which is a variant of that described in Section 2.7.3. ■

3.5 Approximation by finite elements of order one and two

3.5.1 *Formulation of the approximate problem*

We assume, as in Section 1.3, that Ω is polygonal; this allows $\mathscr{T}_h$ to be taken as in Section 1.3.1 and the approximation V_h of the space $H^1(\Omega)$ to be defined by (1.19). We shall approximate K by K_h, a closed convex subset of V_h defined by

$$(3.9)_1 \qquad K_h = K \cap V_h = \{ v_h \mid v_h \in V_h, v_h(P) \geqslant 0 \quad \forall P \in \Sigma_h \cap \Gamma \} \qquad \text{if } q=1$$

$$(3.9)_2 \qquad K_h = \{ v_h \mid v_h \in V_h, v_h(P) \geqslant 0 \quad \forall P \in (\Sigma_h \cup \Sigma_h') \cap \Gamma \} \qquad \text{if } q=2 .$$

It is worth noting that if $q = 2$, we have $K_h \not\subset K$.

The problem (3.1), (3.2), is then approximated by:

$$(3.10) \qquad \begin{cases} a(u_h, v_h - u_h) \geqslant \int_\Omega f(v_h - u_h) \, dx \quad \forall v_h \in K_h \\ u_h \in K_h \end{cases}$$

([1]) The proof of Theorem 3.1 is a direct variant of that of Theorem 5.2 of Chapter 2, Section 5.6.

which admits one and only one solution.

3.5.2 *Convergence of the approximate solutions*

We shall now prove:

Lemma 3.1. *We have*

$$\overline{C^\infty(\bar{\Omega}) \cap K} = K. \tag{3.11}$$

Proof. If $v \in H^1(\Omega)$ then $v = v^+ - v^-$ with $v^+, v^- \in H^1(\Omega)$ and

$$K = \{ v \mid v \in H^1(\Omega), v^- \in H_0^1(\Omega) \}. \tag{3.12}$$

In view of (3.12) and since $\overline{\mathscr{D}(\Omega)} = H_0^1(\Omega)$, it is sufficient, in order to prove (3.11), to show that $\forall v \in H^1(\Omega)$, $v \geqslant 0$ a.e. on Ω, it is possible to find $v_n \in C^\infty(\bar{\Omega})$, $v_n \geqslant 0$ on Ω such that $\lim_{n \to +\infty} v_n = v$ in $H^1(\Omega)$ (strongly). If $v \in H^1(\Omega)$, $v \geqslant 0$ a.e. then v admits an extension $\tilde{v}$ into $H^1(\mathbb{R}^2)$ such that $\tilde{v} \geqslant 0$ a.e. on $\mathbb{R}^2$; in fact if $\tilde{v}$ is an arbitrary extension [1] of v into $H^1(\mathbb{R}^2)$, $v \geqslant 0$ a.e. on Ω implies that $\tilde{v} = |\tilde{v}|$ is also an extension of v into $H^1(\mathbb{R}^2)$. Let $(\rho_n)_n$ be a *regularising sequence* and let $\tilde{v}_n = \tilde{v} * \rho_n$; then $\lim_{n \to +\infty} \tilde{v}_n = \tilde{v}$ in $H^1(\mathbb{R}^2)$ (strongly); moreover, $\tilde{v}_n \in C^\infty(\mathbb{R}^2)$ and

$$\tilde{v}_n(x) = \int_{\mathbb{R}^2} \rho_n(y)\, \tilde{v}(x - y)\, dy \geqslant 0 \quad \forall x \in \mathbb{R}^2$$

since $\rho_n \geqslant 0$ and $\tilde{v} \geqslant 0$ a.e. on $\mathbb{R}^2$.

Let v_n be the restriction of $\tilde{v}_n$ to Ω; it follows from the properties of $\tilde{v}_n$ that $v_n \in C^\infty(\bar{\Omega})$, $v_n \geqslant 0$ on Ω with $\lim_{n \to +\infty} v_n = v$ in $H^1(\Omega)$ (strongly). ■

In view of this lemma we have:

Theorem 3.2. *If, when* $h \to 0$, *the angles of* $\mathscr{T}_h$ *are bounded below by a constant* $\theta_0 > 0$, *we have*

$$\lim_{h \to 0} u_h = u \quad \textit{strongly in} \quad H^1(\Omega)$$

with u *and* u_h *solutions of* (3.1), (3.2) *and* (3.10), *respectively*.

Proof of Theorem 3.2 *for* $q = 1$. We apply Theorem 4.2 of Chapter 1, Section 4.3, taking account of Remark 4.1; it is thus necessary to verify that:

---

1) Since Γ is Lipschitz, such an extension exists (see, for example, Nečas /1/).

$$(3.13)\quad \begin{cases} \text{(i) } \forall v \in \chi,\ \bar{\chi} = K, \text{ we can find } v_h \in K_h \text{ with } v_h \to v \text{ strongly} \\ \qquad \text{in } H^1(\Omega); \\ \text{(ii) if } v_h \in K_h,\ v_h \to v \text{ weakly in } H^1(\Omega), \text{ then } v \in K. \end{cases}$$

In view of the inclusion $K_h \subset K$ point (ii) follows immediately To prove (i) we can take, following Lemma 3.1, $\chi = C^\infty(\bar{\Omega}) \cap K$; let $\pi_h : C^0(\bar{\Omega}) \to V_h$ be the interpolation operator defined by (1.62), Section 1.4.3. If $v \in \chi$ then

$$v(P) \geqslant 0 \quad \forall P \in \Sigma_h \cap \Gamma \Rightarrow \pi_h v(P) \geqslant 0 \quad \forall P \in \Sigma_h \cap \Gamma$$

$$\text{thus } \quad \pi_h v \in K_h \quad \forall v \in \chi\ ;$$

under the condition on the angles of $\mathcal{T}_h$ in the statement of Theorem 3.1, we have, moreover, $\lim_{h \to 0} \pi_h v = v$ in $H^1(\Omega)$ (strongly) so that to prove (i) it is thus sufficient to take

$$\chi = C^\infty(\bar{\Omega}) \cap K \quad \text{and} \quad v_h = \pi_h v\,. \quad \blacksquare$$

Proof of Theorem 3.2 *for* $q = 2$. Once again it is sufficient to verify (3.13); point (i) is treated exactly as for $q = 1$; it thus remains to be shown that (ii) holds. Let $\varphi \in C^0(\Gamma)$; using the same notation as in Section 1.4.1, Figure 1.5, we define the function φ_h, distributed over Γ, by

$$(3.14)\qquad \varphi_h(P) = \sum_i \varphi(M_{i+1/2})\, \chi_{i+1/2}(P)$$

with $\chi_{i+1/2}$ = characteristic function of $\widehat{M_i M_{i+1}}$; we have

$$(3.15)\qquad \lim_{h \to 0} \varphi_h = \varphi \ \textit{uniformly}\ \forall \varphi \in C^0(\Gamma) \text{ and } \varphi_h \geqslant 0 \text{ if } \varphi \geqslant 0\,.$$

Since *Simpson's* rule is *exact* for polynomials of degree $\leqslant 3$, we have

$$\begin{cases} \displaystyle\int_\Gamma \varphi_h v_h \,\mathrm{d}\Gamma = \frac{1}{6} \sum_i |\, M_i M_{i+1} \,| \,[v_h(M_i) + 4\, v_h(M_{i+1/2}) + \\ \qquad + v_h(M_{i+1})]\, \varphi(M_{i+1/2}) \quad \forall v_h \in V_h\,, \quad \forall \varphi \in C^0(\Gamma)\,, \end{cases}$$

so that, from $(3.9)_2$

$$\int_\Gamma \varphi_h v_h \,\mathrm{d}\Gamma \geqslant 0 \quad \forall v_h \in K_h\,, \quad \forall \varphi \in C^0(\Gamma)\,, \quad \varphi \geqslant 0\,.$$

If $v_h \to v$ *weakly* in $H^1(\Omega)$ then $\gamma_0 v_h \to \gamma_0 v$ *strongly* in $L^2(\Gamma)$ so that from (3.15) we have:

$$(3.17)\qquad \lim_{h \to 0} \int_\Gamma \varphi_h v_h \,\mathrm{d}\Gamma = \int_\Gamma \varphi v \,\mathrm{d}\Gamma \quad \forall \varphi \in C^0(\Gamma)\,.$$

Moreover, if $v_h \in K_h$, (3.16), (3.17) imply

$$\int_\Gamma \varphi v \, d\Gamma \geqslant 0 \quad \forall \varphi \in C^0(\Gamma), \quad \varphi \geqslant 0 . \tag{3.18}$$

From (3.18) we have $v \geqslant 0$ on Γ, which completes the proof. ■

3.6 Numerical solution of the approximate problems

Synopsis.

We shall now study the solution of the approximate problem (3.10) by means of the method of point over-relaxation with projection discussed in Chapter 2, Section 1.4, and by means of a duality algorithm of Uzawa type, (see Chapter 2, Section 4.3).

3.6.1 *Formulation of the approximate problems.*

To write the approximate problem (3.10) in explicit form, we shall use $(3.9)_1$ (resp. $(3.9)_2$) and the formulas of Section 1.6.1 (resp. 1.6.2) if $q = 1$ (resp. $q = 2$). ■

3.6.2 *Solution of the approximate problem by over-relaxation with projection*

The approximate problem (3.10) is equivalent to the finite-dimensional problem

$$\begin{cases} \displaystyle\min_{v_h \in K_h} J_0(v_h) \\ J_0(v_h) = \dfrac{1}{2}\displaystyle\int_\Omega (|\operatorname{grad} v_h|^2 + \mu v_h^2) \, dx - \int_\Omega f v_h \, dx . \end{cases} \tag{3.19}$$

In view of the nature of K_h (see $(3.9)_1$ if $q = 1$, $(3.9)_2$ if $q = 2$) we may apply the method of point over-relaxation with projection, discussed in Chapter 2, Section 1.4, to the solution of (3.10); the convergence of this method is assured $\forall \omega \in]0, 2[$. ■

3.6.3 *Solution of the approximate problem by the duality method*

First, we shall consider the *continuous* case; problem (3.1), (3.2) falls into the category of the problems discussed in Chapter 2, Section 4, if we take

$$J_0(v) = \frac{1}{2}\int_\Omega (|\operatorname{grad} v|^2 + \mu v^2) \, dx - \int_\Omega f v \, dx , \quad \mathscr{L}(v, q) = J_0(v) - \int_\Gamma q v \, d\Gamma ,$$

$$M = V = H^1(\Omega), \quad L = L^2(\Gamma), \quad \Phi(v) = -\gamma_0 v , \quad \Lambda = L^2_+(\Gamma).$$

In view of Theorem 3.1 (see Section 3.4) we may apply algorith (4.12)-(4.15) of Chapter 2, Section 4.3 to the solution of (3.1), (3.2); in this particular case this takes the form:

(3.20) $p^0 \in \Lambda$ chosen arbitrarily (for example, zero)

(3.21) $\mathscr{L}(u^n, p^n) \leqslant \mathscr{L}(v, p^n) \quad \forall v \in H^1(\Omega),\ u^n \in V$

(3.22) $p^{n+1} = P_\Lambda(p^n - \rho_n \gamma_0 u^n)\,, \quad \rho_n > 0\,.$

Writing (3.21) explicitly, we obtain the Neumann problem

(3.23) $$\begin{cases} -\Delta u^n + \mu u^n = f & \text{in } \Omega \\ \dfrac{\partial u^n}{\partial n} = \lambda^n & \text{on } \Gamma\,; \end{cases}$$

moreover

(3.24) $(P_\Lambda(q)) = q^+ \quad \forall q \in L^2(\Gamma)$

which allows (3.22) to be put in explicit form.

Since γ_0 is a linear continuous mapping from $H^1(\Omega) \to L^2(\Gamma)$ we have

(3.25) $$\| \gamma_0 v \|_{L^2(\Gamma)} \leqslant \| \gamma_0 \| \left(\int_\Omega (|\operatorname{grad} v|^2 + \mu v^2)\,\mathrm{d}x \right)^{1/2}$$

$$\forall v \in H^1(\Omega)\,;$$

from the theorem in Chapter 2, Section 4.3, we may then deduce that

(3.26) $u^n \to u$ *strongly* in $H^1(\Omega)$

under the condition

(3.27) $0 < \alpha_0 \leqslant \rho_n \leqslant \alpha_1 < \dfrac{2}{\| \gamma_0 \|^2}\,.$ ■

The approximate problem (3.10) is solved in a similar manner using the discrete analogue of algorithm (3.20)-(3.22):

We write $\mu_h = \{ \mu_i \}_i$ with $M_i \in \Sigma_h$ if $q = 1$,
$M_i \in \Sigma_h \cup \Sigma_h'$ if $q = 2$;

We define the approximation Λ_h of Λ by

$$\Lambda_h = \{ \mu_h \mid \mu_i \geqslant 0 \ \forall i \}\,,$$

and then the approximation $\mathscr{L}_h$ of $\mathscr{L}$ by

(3.28) $\mathscr{L}_h(v_h, \mu_h) = J_0(v_h) - \frac{1}{2} \sum_i | M_i M_{i+1} | (\mu_i v_h(M_i) + \mu_{i+1} v_h(M_{i+1}))$ if $q = 1$

(3.29) $$\mathscr{L}_h(v_h, \mu_h) = J_0(v_h) - \frac{1}{6}\sum_i | M_i M_{i+1} | (\mu_i v_h(M_i) + 4 \mu_{i+1/2} \times$$

$$\times v_h(M_{i+1/2})) + \mu_{i+1} v_h(M_{i+1}) \text{ if } q = 2 .$$

The Lagrangian $\mathscr{L}_h$ admits a unique saddle point (1) $\{u_h, \lambda_h\}$ on $V_h \times \Lambda_h$ where u_h is the solution of (3.10); the discrete analogue of (3.20)-(3.22) is then

(3.30) $$\lambda_h^0 \in \Lambda_h$$

(3.31) $$\mathscr{L}_h(u_h^n, \lambda_h^n) \leqslant \mathscr{L}_h(v_h, \lambda_h^n) \quad \forall v_h \in V_h , \ u_h^n \in V_h$$

(3.32) $$\lambda_i^{n+1} = [\lambda_i^n - \rho_n u_h^n(M_i)]^+ \quad \forall M_i \in \Sigma_h \text{ if } q = 1 ,$$

$$\forall M_i \in \Sigma_h \cup \Sigma_h' \text{ if } q = 2 .$$

It may be shown that under the condition $0 < \alpha_0 \leqslant \rho_n \leqslant \alpha_1$, α_1 sufficiently small, we have $\lim_{n \to +\infty} u_h^n = u_h$, u_h being the solution of (3.10). ■

3.7 An example

3.7.1 *Formulation of the continuous problem*

Let $\Omega =]0, 1[\times]0, 1[$ and Γ_d, V, f be as in Section 2.8.1; we have applied the above methods of approximation and solution to the following variant (2) of problem (3.1), (3.2):

(3.33) $$\begin{cases} \operatorname{Min}_{v \in K} J_0(v) \\ J_0(v) = \frac{1}{2}\int_\Omega |\operatorname{grad} v|^2 \, dx - \int_\Omega fv \, dx \end{cases}$$

(3.34) $$K = \{ v \mid v \in V, v \geqslant 0 \text{ a.e. on } \Gamma_d \} .$$

3.7.2 *The approximate problem*

We proceed as in Section 2.8.2, and use the same triangulations.

3.7.3 *Solution by over-relaxation*

The algorithm of point over-relaxation with projection discussed in Chapter 2, Section 1.4, was used under the following conditions (for q = 1 and 2):

Initialisation: $u_h^0 = 0$.

---

1) λ_h is a *Kuhn-Tucker* vector for (3.19)

2) We thus have $\mu = 0$.

Termination criterion:

$$\frac{\sum_i |u_i^{n+1} - u_i^n|}{\sum_i |u_i^{n+1}|} \leqslant 10^{-5}.$$

Value of $\omega : \omega = 1.74.$ ∎

Analysis of the numerical results: For $q = 1$, convergence is attained in 27 iterations and for $q = 2$ in 42 iterations.

The mean time per iteration (on a CII 10070) is 0.2 sec. for $q = 1$, and 0.4 sec. for $q = 2$.

Figure 3.1 shows the graphs of the functions $x_1 \to u_h(x_1, 0)$ and $x_1 \to u_h(x_1, \frac{1}{2})$ and Figure 3.2 the equipotentials of the approximate solution u_h.

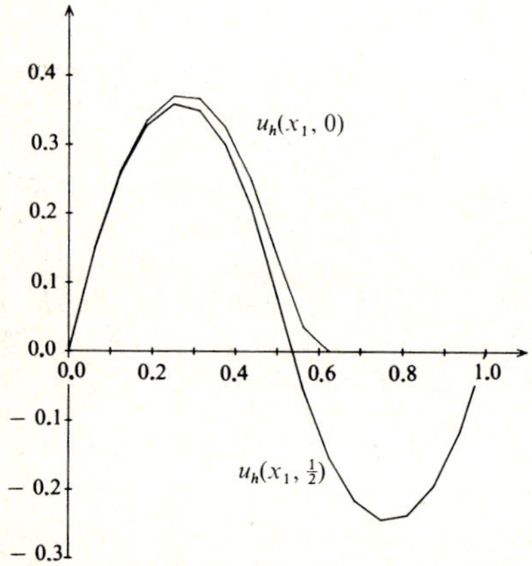

Fig. 3.1.

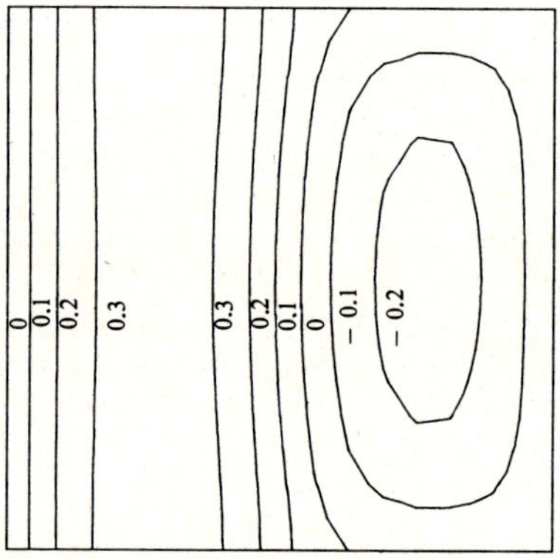

Fig. 3.2.

3.7.4 *Solution by a duality method*

We shall confine our attention to $q = 1$; from a practical point of view, algorithm (3.30)-(3.32) was applied using only the $M_i \in \Sigma_h \cap \Gamma_d$ in (3.32). It may be seen that, under the same operating conditions as those in Section 2.8.4, the computed results coincide with those obtained in Section 3.7.3 using over-relaxation with projection; Table 3.1 shows the number of iterations required for convergence, as a function of the parameter ρ.

ρ	2	4	6	8	10
Number of iterations	27	16	12	17	$+\infty$

Table 3.1

The remarks in Section 2.8.5 regarding the relative merits of the various methods of approximation and numerical solution also hold for the problem considered in this Section 3. ■

4. NUMERICAL ANALYSIS OF VARIATIONAL INEQUALITIES OF ORDER 4

4.1 Synopsis

In Section 4 we shall describe a certain number of specific numerical procedures for variational problems of order 4. The size and complexity of the subject would justify considerably more extensive discussions than the relatively brief outline given here; we therefore refer the reader firstly to Duvaut-Lions /1/ Chapter 4, and /4/, /5/ for the theoretical study and the physical background to variational inequalities of order 4, and secondly [1] to Bégis-Glowinski /1/, Ciarlet-Glowinski /1/, /2/, Ciarlet-Raviart /3/, Glowinski /6/ for further details regarding the finite element methods and iterative techniques described briefly in this section. ■

4.2 An iterative method for solving certain variational problems of order 4

In this section we shall show that certain variational problems of order 4 may be reduced, by means of an iterative process, to the solution of a sequence of second-order problems.

4.2.1 *Reduction of the Dirichlet problem for* Δ^2 *to a sequence of second-order problems*

Let $\Omega \subset \mathbb{R}^N$ be a bounded open domain with a regular boundary Γ; the (homogeneous) Dirichlet problem for Δ^2 is defined by

$$(4.1)\qquad \begin{cases} \Delta^2 u = f & \text{in } \Omega \\ u\,|_\Gamma = \dfrac{\partial u}{\partial n}\Big|_\Gamma = 0\,. \end{cases}$$

---

[1] And to their respective bibliographies.

Functional framework and notation

Let

$$V = H^2(\Omega) \cap H_0^1(\Omega), \quad H_0^2(\Omega) = \overline{\mathcal{D}(\Omega)}^{H^2(\Omega)} =$$
$$= \left\{ v \mid v \in H^2(\Omega), v|_\Gamma = \frac{\partial v}{\partial n}\bigg|_\Gamma = 0 \right\};$$

since the domain Ω is bounded and Γ is sufficiently regular the mapping

$$v \to \| \Delta v \|_{L^2(\Omega)}$$

defines on V a norm which is equivalent to that induced by $H^2(\Omega)$. We shall denote by γ_0 and γ_1 the two trace mappings defined by $\gamma_0 v = v|_\Gamma$, $\gamma_1 v = \partial v/\partial n|_\Gamma$. ■

Variatiational formulation of problem (4.1). *Existence, uniqueness and regularity results.*

Let us assume (for the moment) that $f \in H^{-2}(\Omega)$ ($H^{-2}(\Omega)$: topological dual of $H_0^2(\Omega)$); it is well known that (4.1) admits one and only one solution in $H_0^2(\Omega)$; this solution is also the unique solution of the variational equation:

$$(4.2) \qquad \begin{cases} \displaystyle\int_\Omega \Delta u \, \Delta v \, dx = \langle f, v \rangle & \forall v \in H_0^2(\Omega) \\ u \in H_0^2(\Omega) \end{cases}$$

as well as of the optimisation problem

$$(4.3) \qquad \underset{v \in H^2(\Omega)}{\text{Min}} \left[\frac{1}{2} \int_\Omega |\Delta v|^2 \, dx - \langle f, v \rangle \right].$$

In (4.2) and (4.3), $\langle , \rangle$ represents the bilinear form of the duality between $H^{-2}(\Omega)$ and $H_0^2(\Omega)$.

As far as regularity is concerned, we shall confine our attenion to the following result (for which we refer the reader to Lions-Magènes /1/):

if $f \in L^2(\Omega)$ (1) we have $u \in H^4(\Omega) \cap H_0^2(\Omega)$. ■

Description of the iterative method.

We assume in the following that $f \in L^2(\Omega)$ and we consider the algorithm defined by

---

(1) We thus have $\langle f, v \rangle = \displaystyle\int_\Omega fv \, dx$.

(4.4) λ^0 chosen arbitrarily in $L^2(\Gamma)$ ($\lambda^0 = 0$ for example),

then, having calculated λ^n, we successively determine $u^n \in V$ and $\lambda^{n+1} \in L^2(\Gamma)$ by means of

$$\text{(4.5)} \qquad \begin{cases} \Delta^2 u^n = f \\ \gamma_0 u^n = 0 \\ \gamma_0 \Delta u^n = -\lambda^n \end{cases}$$

$$\text{(4.6)} \qquad \lambda^{n+1} = \lambda^n + \rho_n \gamma_1 u^n, \quad \rho_n > 0. \quad \blacksquare$$

Convergence of algorithm (4.4)-(4.6)

We have:

Theorem 4.1. *The sequence* $(u^n)_n$ *defined by* (4.4)-(4.6) *converges strongly in* V *to the solution* u *of* (4.1), *under the condition*

$$\text{(4.7)} \qquad 0 < r_0 \leqslant \rho_n \leqslant r_1 < 2\sigma_0^2$$

with

$$\text{(4.8)} \qquad \sigma_0 = \inf_{v \in H^2 \cap H_0^1} \frac{\| \Delta v \|_{L^2(\Omega)}}{\left\| \dfrac{\partial v}{\partial n} \right\|_{L^2(\Gamma)}}.$$

Proof. We follow Glowinski /6/; let u be the solution of (4.1); we have $u \in H^4(\Omega) \cap H_0^2(\Omega)$ so that $\gamma_0 \Delta u \in H^{3/2}(\Gamma) \subset L^2(\Gamma)$; we put $\lambda = -\gamma_0 \Delta u$ and $\bar{u}^n = u^n - u, \bar{\lambda}^n = \lambda^n - \lambda$. We have (since $\gamma_1 u = 0$)

$$\text{(4.9)} \qquad \begin{cases} \Delta^2 u = f \\ \gamma_0 u = 0 \\ \gamma_0 \Delta u = -\lambda \end{cases}$$

and

$$\text{(4.10)} \qquad \lambda = \lambda + \rho_n \gamma_1 u$$

so that by subtracting these equations from (4.5), (4.6), we have

$$\text{(4.11)} \qquad \begin{cases} \Delta^2 \bar{u}^n = 0 \\ \gamma_0 \bar{u}^n = 0 \\ \gamma_0 \Delta \bar{u}^n = -\bar{\lambda}^n \end{cases}$$

$$\text{(4.12)} \qquad \bar{\lambda}^{n+1} = \bar{\lambda}^n + \rho_n \gamma_1 \bar{u}^n .$$

With $\mu \in L^2(\Gamma)$, we write $|\mu| = \| \mu \|_{L^2(\Gamma)}$; we may then deduce from (4.12) that

$$(4.13)\qquad |\bar\lambda^n|^2 - |\bar\lambda^{n+1}|^2 = -2\rho_n \int_\Gamma \bar\lambda^n \gamma_1 \bar u^n \, d\Gamma - \rho_n^2 |\gamma_1 \bar u^n|^2 .$$

Using Green's formula

$$(4.14)\qquad \int_\Omega (v\,\Delta w - w\,\Delta v)\,dx = \int_\Gamma \left(v\frac{\partial w}{\partial n} - w\frac{\partial v}{\partial n}\right) d\Gamma ,$$

and (4.11), (4.13), we have

$$0 = \int_\Omega \Delta^2 \bar u^n \bar u^n \, dx = \int_\Omega |\Delta \bar u^n|^2 \, dx - \int_\Gamma \gamma_0 \Delta \bar u^n \gamma_1 \bar u^n \, d\Gamma$$

and then

$$|\bar\lambda^n|^2 - |\bar\lambda^{n+1}|^2 = 2\rho_n \|\Delta \bar u^n\|^2_{L^2(\Omega)} - \rho_n^2 |\gamma_1 \bar u^n|^2_{L^2(\Gamma)} ,$$

so that from the definition of σ_0 we have:

$$(4.15)\qquad |\bar\lambda^n|^2 - |\bar\lambda^{n+1}|^2 \geqslant \rho_n\left(2 - \frac{\rho_n}{\sigma_0^2}\right) \|\Delta \bar u^n\|^2_{L^2(\Omega)} .$$

Under condition (4.7), we deduce from (4.15) that the sequence $|\bar\lambda^n|$ is decreasing and convergent, so that we have $|\bar\lambda^n|^2 - |\bar\lambda^{n+1}|^2 \to 0$ when $n \to +\infty$ and hence $\Delta \bar u^n \to 0$ *strongly* in $L^2(\Omega)$ and this proves the theorem. ■

Remark 4.1. Under condition (4.7), we deduce from (4.15) that λ^n is bounded in $L^2(\Gamma)$, so that, taking account of (4.5), u^n is bounded in $H^{5/2}(\Omega)$, which together with Theorem 4.1 leads to $u^n \to u$ *strongly* in $H^s(\Omega)$ $\forall s < \frac{5}{2}$. ■

Remark 4.2. In the case of the *inhomogeneous problem*, namely

$$\begin{cases} \Delta^2 u = f \quad \text{in } \Omega \\ \gamma_0 u = g_1 , \quad \gamma_1 u = g_2 \end{cases}$$

we can, if g_1 and g_2 are sufficiently regular, use the following generalisation of algorithm (4.4)-(4.6):

$$(4.16)\qquad \lambda^0 \text{ chosen arbitrarily in } L^2(\Gamma)$$

$$(4.17)\qquad \begin{cases} \Delta^2 u^n = f \\ \gamma_0 u^n = g_1 \\ \gamma_0 \Delta u^n = -\lambda^n \end{cases}$$

$$(4.18)\qquad \lambda^{n+1} = \lambda^n + \rho_n(\gamma_1 u^n - g_2) , \qquad \rho_n > 0 ,$$

where the convergence of this algorithm is proved in exactly the same way, and the condition on ρ_n is still (4.7). ■

Remark 4.3. Let $x_0 \in \Omega$ and let $\delta(x_0)$ be the Dirac measure of x_0; if $N = 2$ and $s > 1$ we have $H^s(\Omega) \subset C^0(\overline{\Omega})$ with *continuous injection*, so that $\delta(x_0)$ is thus linear continuous over $H^s(\Omega)$ if $s > 1$. It then follows (see Lions-Magènes /1/) that if $f=\delta(x_0)$ in (4.1) (1), we have $u \in H_0^2(\Omega) \cap H^s(\Omega)$, $\forall s < 3$, so that $\gamma_0\, \Delta u \in L^2(\Gamma)$, which allows - since Theorem 4.1 still holds - algorithm (4.4)-(4.6) to be applied to the solution of (4.1). ■

Remark 4.4. From (4.5) (and this also holds for (4.16)-(4.18)) the determination of u^n from λ^n *factorises* into two Dirichlet problems for $-\Delta$, that is:

$$(4.19)\qquad \begin{cases} -\Delta p^n = f & \text{in } \Omega \\ \gamma_0\, p^n = \lambda^n \end{cases} \qquad \begin{cases} -\Delta u^n = p^n & \text{in } \Omega \\ \gamma_0\, u^n = 0\,. \end{cases}$$

This remark is fundamental since it shows that the solution of (4.1) may be reduced to that of a sequence of Dirichlet problems for $-\Delta$.

It is thus natural to approximate u^n and p^n in $H^1(\Omega)$, and this leads to approximations for u^n which are *exterior*; we shall return to this point in Sections 4.3, 4.4 and 4.5. ■

Duality results. Interpretation of algorithm (4.4)-(4.6).

We shall retain the assumption $f \in L^2(\Omega)$ and the notation $\lambda = -\gamma_0\, \Delta u$; we define $\mathscr{L} : V \times L^2(\Gamma) \to \mathbb{R}$, the *Lagrangian* associated with problem (4.3), as follows:

$$(4.20)\qquad \mathscr{L}(v, \mu) = \frac{1}{2}\int_\Omega |\Delta v|^2\, \mathrm{d}x - \int_\Omega fv\, \mathrm{d}x + \int_\Gamma \mu\gamma_1\, v\, \mathrm{d}\Gamma\,.$$

We may then prove (see Glowinski, *loc. cit.*, Section 1.4):

Theorem 4.2. *The pair* (u, λ) *is the unique saddle point of* $\mathscr{L}$ *on* $V \times L^2(\Gamma)$. ■

In view of Theorem 4.2, we may apply to $\mathscr{L}$ the algorithm (4.12)-(4.13) of Chapter 2, Section 4.3, with $V = H^2(\Omega) \cap H_0^1(\Omega)$, $M = V$, $L = L^2(\Gamma)$ $\Lambda = L$, $\Phi = \gamma_1$; the algorithm thus obtained turns out to be the same algorithm as (4.4)-(4.6). ■

A variant of algorithm (4.4)-(4.6).

If $v \in H^2(\Omega)$, we have $\partial v/\partial n \in H^{1/2}(\Gamma)$, so that it is natural to

(1) This is important in certain applications.

extend to $V \times H^{-1/2}(\Gamma)$ the Lagrangian $\mathscr{L}$ defined by (4.20) so that

$$(4.21) \qquad \tilde{\mathscr{L}}(v, \mu) = \frac{1}{2}\int_\Omega |\Delta v|^2 \, dx - \int_\Omega fv \, dx + \langle \mu, \gamma_1 v \rangle :$$

in (4.21), $\langle , \rangle$ denotes the bilinear form of the duality between $H^{-1/2}(\Gamma)$ and $H^{1/2}(\Gamma)$ (1).

With the operator $S : H^{1/2}(\Gamma) \to H^{-1/2}(\Gamma)$ a duality operator (2), we consider the following variant of algorithm (4.4)-(4.6):

(4.22) $\lambda^0 \in H^{-1/2}(\Gamma)$ chosen arbitrarily ($\lambda^0 = 0$ for example)

then, having calculated λ^n, we successively determine $u^n \in V$ and $\lambda^{n+1} \in H^{-1/2}(\Gamma)$ by means of:

$$(4.23) \qquad \begin{cases} \Delta^2 u^n = f \\ \gamma_0 u^n = 0 \\ \gamma_0 \Delta u^n = -\lambda^n \end{cases}$$

$$(4.24) \qquad \lambda^{n+1} = \lambda^n + \rho_n S\gamma_1 u^n, \quad \rho_n > 0 .$$

Using a variant of the proof of Theorem 4.1 we may prove:

<u>Theorem 4.3</u>. *The sequence* $(u^n, \lambda^n)_n$ *defined by* (4.22)-(4.24) *converges strongly in* $V \times H^{-1/2}(\Gamma)$ *to* $(u, -\gamma_0 \Delta u)$, *where* u *is the solution of* (4.1), *under the condition*:

$$(4.25) \qquad 0 < r_0 \leqslant \rho_n \leqslant r_1 < 2\sigma_1^2$$

with

$$(4.26) \qquad \sigma_1 = \inf_{v \in H^2 \cap H_0^1} \frac{\| \Delta v \|_{L^2(\Omega)}}{\left\| \frac{\partial v}{\partial n} \right\|_{H^{1/2}(\Gamma)}} . \quad \blacksquare$$

Remark 4.5. We may prove that under condition (4.25), we have:

$$(4.27) \qquad \| u^{n+1} - u \|_V \leqslant K \| u^n - u \|_V$$

---

(1) If $\mu \in L^2(\Gamma)$, then $\langle \mu, \gamma_1 v \rangle = \int_\Gamma \mu\gamma_1 v \, d\Gamma$.

(2) We thus have

$$\langle S\mu, \mu \rangle = \| \mu \|^2_{H^{1/2}(\Gamma)} \quad \forall \mu \in H^{1/2}(\Gamma),$$
$$\langle \mu, S^{-1}\mu \rangle = \| \mu \|^2_{H^{-1/2}(\Gamma)} \quad \forall \mu \in H^{-1/2}(\Gamma).$$

(4.28) $$\| \lambda^{n+1} - \lambda \|_{H^{-1/2}(\Gamma)} \leqslant K \| \lambda^n - \lambda \|_{H^{-1/2}(\Gamma)}$$

with [1] $0 < K < 1$; the convergence of (4.22)-(4.24) is thus of first order. ■

Remark 4.6. To prove the convergence of (4.22)-(4.24), it is sufficient to suppose that $\gamma_0 \Delta u \in H^{-1/2}(\Gamma)$; we may then weaken quite considerably the condition $f \in L^2(\Omega)$, and Remark 4.3 holds a fortiori. ■

Remark 4.7. The practical utility of (4.22)-(4.24) is restricted by the difficult numerical use of $H^{-1/2}(\Gamma)$, $H^{1/2}(\Gamma)$, S (however, see Glowinski /7/, /8/. ■

4.2.2 *Generalisation to variational inequalities*

We shall use three simple examples to demonstrate the possibility of generalising algorithm (4.4)-(4.6) to the solution of variational inequalities of order 4. In the following, Ω is a *bounded* open domain of $\mathbb{R}^N$, with regular boundary Γ.

Example 1. *Statement of the problem.*

Let K be the *closed* convex subset of $V = H^2(\Omega) \cap H_0^1(\Omega)$, defined by

$$K = \{ v \mid v \in V,\ \gamma_1 v \geqslant 0 \text{ a.e. } \}$$

and let $f \in L^2(\Omega)$. The variational inequality (of order 4)

(4.29) $$\begin{cases} \displaystyle\int_\Omega \Delta u\, \Delta(v - u)\, \mathrm{d}x \geqslant \int_\Omega f(v - u)\, \mathrm{d}x & \forall v \in K \\ u \in K \end{cases}$$

admits one and only one solution, which is also the solution of the minimisation problem

(4.30) $$\begin{cases} J_0(u) \leqslant J_0(v) & \forall v \in K \\ u \in K \end{cases}$$

with

$$J_0(v) = \frac{1}{2} \int_\Omega | \Delta v |^2\, \mathrm{d}x - \int_\Omega fv\, \mathrm{d}x . \quad ■$$

Regularity and duality results

Following the ideas of Lions /1/, Chapter 2, Section 8.7.2, it may be proved that under the condition $f \in L^2(\Omega)$, we have

---

(1) K depends, amongst other things, on r_0 and r_1.

$u \in H^3(\Omega) \cap H_0^1(\Omega)$, which implies

$$\gamma_0 \Delta u \in H^{1/2}(\Gamma) \subset L^2(\Omega) .$$

From this regularity result, we deduce that u is characterised by

$$(4.31) \qquad \begin{cases} \Delta^2 u = f \text{ in } \Omega , & u \in H^2(\Omega) , \ u = 0 \text{ on } \Gamma \\ \gamma_0(\Delta u) \geqslant 0 , \ \gamma_1 u \geqslant 0 , & \gamma_0(\Delta u)\, \gamma_1 u = 0 \text{ on } \Gamma . \end{cases}$$

Let $\mathscr{L} : V \times L^2(\Gamma) \to \mathbb{R}$ be defined by

$$(4.32) \qquad \mathscr{L}(v, \mu) = J_0(v) - \int_\Gamma \mu \gamma_1 v \, d\Gamma$$

and

$$\Lambda = L_+^2(\Gamma) = \{ \mu \mid \mu \in L^2(\Gamma), \ \mu \geqslant 0 \text{ a.e.} \} ;$$

from the characterisation (4.31), we may deduce:

<u>Theorem 4.4.</u> *Suppose that u is the solution of (4.29), then $\mathscr{L}$ admits $(u, \gamma_0 \Delta u)$ as a unique saddle point on $V \times \Lambda$.* ■

Solution of (4.29) by a duality algorithm.

In view of Theorem 4.4, we may apply to $\mathscr{L}$ the duality algorithm (4.12)-(4.13) of Chapter 2, Section 4.3; we then obtain the following variant of algorithm (4.4)-(4.6):

$$(4.33) \qquad \lambda^0 \text{ chosen arbitrarily in } \Lambda \text{ (for example, zero)}$$

$$(4.34) \qquad \begin{cases} \Delta^2 u^n = f \\ \gamma_0 u^n = 0 \\ \gamma_0 \Delta u^n = \lambda^n , \end{cases}$$

$$(4.35) \qquad \lambda^{n+1} = P_\Lambda(\lambda^n - \rho_n \gamma_1 u^n) = (\lambda^n - \rho_n \gamma_1 u^n)^+ , \quad \rho_n > 0 .$$

The conditions for convergence of (4.33)-(4.35) are exactly those of (4.4)-(4.6), i.e. conditions (4.7); Remarks 4.1, 4.3 and 4.4 still hold. ■

Example 2

Statement of the problem

Let $V = H^2(\Omega) \cap H_0^1(\Omega)$, g constant > 0 and $f \in L^2(\Omega)$; the variational inequality

$$
(4.36)\quad \begin{cases} \displaystyle\int_\Omega \Delta u\, \Delta(v-u)\,\mathrm{d}x + g\int_\Gamma \left|\frac{\partial v}{\partial n}\right| \mathrm{d}\Gamma - g\int_\Gamma \left|\frac{\partial u}{\partial n}\right| \mathrm{d}\Gamma \geqslant \\ \qquad \geqslant \displaystyle\int_\Omega f(v-u)\,\mathrm{d}x \quad \forall v \in V \quad u \in V \end{cases}
$$

admits one and only one solution, which is also the solution of the minimisation problem:

$$
(4.37)\quad \begin{cases} J(u) \leqslant J(v) \quad \forall v \in V \\ u \in V \end{cases}
$$

with

$$
J(v) = \frac{1}{2}\int_\Omega |\Delta v|^2\,\mathrm{d}x + g\int_\Gamma |\gamma_1 v|\,\mathrm{d}\Gamma - \int_\Omega fv\,\mathrm{d}x . \quad \blacksquare
$$

Duality and regularity results.

Let Λ be the *closed* convex subset of $L^2(\Gamma)$ defined by:

$$
(4.38)\quad \Lambda \doteq \{\mu \mid \mu \in L^2(\Gamma),\ |\mu(x)| \leqslant 1 \quad \text{a.e. on } \Gamma\} ;
$$

proceeding as in Chapter 1, Section 1.3 (or alternatively by regularisation as in Chapter 4, Section 2.3 and Chapter 5, Section 6.1.4) we may prove the existence of $\lambda \in \Lambda$ such that the solution u of (4.36) is characterised by

$$
(4.39)\quad \begin{cases} \Delta^2 u = f \quad \text{in } \Omega \\ \gamma_0 u = 0 \\ \gamma_0 \Delta u = -g\lambda \\ \lambda\gamma_1 u = |\gamma_1 u| \text{ a.e.} \end{cases}
$$

With $\mathscr{L} : V \times L^2(\Gamma) \to \mathbb{R}$ defined by

$$
(4.40)\quad \mathscr{L}(v,\mu) = \frac{1}{2}\int_\Omega |\Delta v|^2\,\mathrm{d}x - \int_\Omega fv\,\mathrm{d}x + g\int_\Gamma \mu\gamma_1 v\,\mathrm{d}\Gamma ,
$$

it follows from (4.39) that $\left(u, -\frac{1}{g}\gamma_0 \Delta u\right)$ is the *unique saddle point* of $\mathscr{L}$ on $V \times \Lambda$. $\blacksquare$

As far as the *regularity* of u is concerned, we may deduce from (4.39) that $u \in V \cap H^{5/2}(\Omega)$; in fact, by analogy with the results of Brézis /2/ for problems of order 2, it is reasonable to conjecture that $u \in V \cap H^3(\Omega)$. ■

Solution of (4.36) *by a duality algorithm.*

In view of the above saddle-point result, we may apply to $\mathscr{L}$ the duality algorithm (4.12)-(4.13) of Chapter 2, Section 4.3; we then obtain the following variant of algorithm (4.4)-(4.6):

(4.41) $\lambda^0 \in \Lambda$ chosen arbitrarily (for example, zero)

$$\begin{cases} \Delta^2 u^n = f \\ \gamma_0 u^n = 0 \\ \gamma_0 \Delta u^n = -g\lambda^n \end{cases} \tag{4.42}$$

$$\lambda^{n+1} = P_\Lambda(\lambda^n + \rho_n g \gamma_1 u^n), \quad \rho_n > 0 \tag{4.43}$$

with

$$P_\Lambda(\mu) = \sup(\inf(1, \mu), -1) \quad \forall \mu \in L^2(\Gamma).$$

The conditions for convergence can be defined precisely by

$$0 < r_0 \leqslant \rho_n \leqslant r_1 < (2/g^2)\, \sigma_0^2, \tag{4.44}$$

and Remarks 4.1, 4.3, 4.4 still hold. ■

Example 3

Statement of the problem.

Suppose $\psi \in L^\infty(\Omega)$ and $f \in L^2(\Omega)$; the variational inequality

$$\begin{cases} \displaystyle\int_\Omega \Delta u\, \Delta(v - u)\, dx \geqslant \int_\Omega f(v - u)\, dx \quad \forall v \in K \\ u \in K = \{ v \mid v \in H_0^2(\Omega),\ v \geqslant \psi \text{ a.e.} \} \end{cases} \tag{4.45}$$

admits one and only one solution (1) which is also the solution of the minimisation problem

$$\begin{cases} J_0(u) \leqslant J_0(v) \quad \forall v \in K \\ u \in K \end{cases} \tag{4.46}$$

with

$$J_0(v) = \frac{1}{2}\int_\Omega |\Delta v|^2\, dx - \int_\Omega fv\, dx\,. \quad ■$$

---

(1) Provided we assume that $K \neq \varnothing$, which we do in the following.

Remark 4.8. If $N = 2$, the inequality (4.45) models the small vertical deformations of a horizontal, elastic thin plate, which is *rigidly clamped* around its perimeter Γ, under the influence of a vertical force field of density f, the plate being constrained to remain above an obstacle whose height is given by ψ. ■

Duality results.

If $N = 2$, as will be assumed here, we have $H^2(\Omega) \subset C^0(\bar{\Omega})$ with *continuous injection*, so that if $v \in H_0^2(\Omega)$ we have $v - \psi \in L^\infty(\Omega)$; let $(L^\infty(\Omega))'$ be the topological dual of $L^\infty(\Omega)$, $\ll, \gg$ the bilinear form of the duality between $(L^\infty(\Omega))'$ and $L^\infty(\Omega)$, and Λ the *positive cone* of $(L^\infty(\Omega))'$ i.e.

$$\Lambda = \{ \mu \mid \mu \in (L^\infty(\Omega))', \ll \mu, v \gg \geqslant 0 \ \forall v \in L^\infty(\Omega), v \geqslant 0 \ \text{ a.e. } \}. \tag{4.47}$$

We then define $\mathscr{L} : H_0^2(\Omega) \times (L^\infty(\Omega))' \to \mathbb{R}$ by

$$\mathscr{L}(v, \mu) = J_0(v) + \ll \mu, \psi - v \gg ; \tag{4.48}$$

we assume (by way of simplification) that ψ is negative everywhere in a neighbourhood of Γ, so that we have $K \neq \varnothing$ and there exists $v_0 \in K$ such that $v_0 - \psi \geqslant \delta$, δ being a *strictly positive* constant; it then follows from Ekeland-Témam /1/, Chapter 3, Section 5, that $\mathscr{L}$ admits a saddle point on $H_0^2(\Omega) \times \Lambda$, namely (u, λ), with u the solution of (4.45). ■

Remark 4.9. The above saddle-point result also holds if $f \in H^{-2}(\Omega)$. ■

Solution of (4.45) *by means of duality algorithms.*

In the infinite-dimensional case, Example 3 lies outside the *Hilbert* framework in which the duality algorithms allowing the saddle-points of a certain class of Lagrangians to be calculated were studied in Chapter 2, Section 4. This is not the case for the *finite-dimensional* problems approximating (4.45); to solve these problems, we shall use an approximate form of the following algorithm, which *in general is not meaningful in the infinite-dimensional case*:

$$\lambda^0 \geqslant 0 \quad \text{chosen arbitrarily;} \tag{4.49}$$

having calculated λ^n, we calculate u^n and λ^{n+1} by means of

$$\begin{cases} \Delta^2 u^n = f + \lambda^n & \text{in } \Omega \\ \gamma_0 u^n = \gamma_1 u^n = 0 \end{cases} \tag{4.50}$$

$$\lambda^{n+1} = [\lambda^n + \rho_n(\psi - u^n)]^+ , \quad \rho_n > 0 . \tag{4.51}$$

The biharmonic problem (4.50) may itself by solved by means of the duality method studied in Section 4.2.1. We shall meet in Secion 4.7 a finite-dimensional variant of (4.49)-(4.51) applied to the solution of a problem of the type (4.45). ■

4.3 A new variational formulation of the Dirichlet problem for Δ^2.

Synopsis.

The algorithm (4.4)-(4.6) *explicitly* involves $\gamma_1 u = \partial u/\partial n$ which can present numerical difficulties when applied to the approximate problems; moreover (see Remark 4.4) it would appear desirable to be able to use approximations which are *interior* for problems of order *two* but *exterior* for those of order *four*. To this end we shall give for problem (4.2) an equivalent variational formulation which involves only the spaces $L^2(\Omega)$, $H^1(\Omega)$ and $H_0^1(\Omega)$, and in which the condition $\partial u/\partial n = 0$ is *implicit*; the finite element approximations associated with this new formulation will be discussed in Section 4.5. ■

Although we shall assume that $f \in L^2(\Omega)$, the development which follows would still hold without any modification for $f \in H^{-1}(\Omega)$ (1). Let W be the subspace of $H_0^1(\Omega) \times L^2(\Omega)$ defined by

$$W = \left\{ (v, q) \mid (v, q) \in H_0^1(\Omega) \times L^2(\Omega), \right. \left. \int_\Omega \text{grad } v.\text{grad } \mu \, dx = \int_\Omega q\mu \, dx \quad \forall \mu \in H^1(\Omega) \right\}; \tag{4.52}$$

we then have:

Theorem 4.5. *There is equivalence between problems* (4.2), (4.3) *and the problem*

$$\operatorname*{Min}_{(v,q) \in W} \left[\frac{1}{2} \int_\Omega q^2 \, dx - \int_\Omega fv \, dx \right]. \tag{4.53}$$

Proof. Any solution (u, p) of (4.53) is characterized by

$$\int_\Omega pq \, dx = \int_\Omega fv \, dx \quad \forall (v, q) \in W, \quad (u, p) \in H_0^1(\Omega) \times L^2(\Omega), \tag{4.54}$$

---

(1) The case $f \in H^{-s}(\Omega)$, $s > 1$ would present additional technical difficulties and therefore will not be considered in the rather brief outline given in this section.

(4.55) $$\int_\Omega \operatorname{grad} u . \operatorname{grad} \mu \, dx = \int_\Omega p\mu \, dx \quad \forall \mu \in H^1(\Omega) .$$

Taking $\mu \in H_0^1(\Omega)$ in (4.55), we then deduce

(4.56) $$- \Delta u = p \quad \text{in } \Omega$$

so that, if Γ is sufficiently regular, $u \in H^2(\Omega) \cap H_0^1(\Omega)$.
From (4.55), (4.56) we have

$$\int_\Omega \operatorname{grad} u . \operatorname{grad} \mu \, dx = - \int_\Omega \Delta u \mu \, dx \quad \forall \mu \in H^1(\Omega)$$

so that by applying Green's formula

$$- \int_\Omega \Delta u \mu \, dx + \int_\Gamma \frac{\partial u}{\partial n} \mu \, d\Gamma = - \int_\Omega \Delta u \mu \, dx \quad \forall \mu \in H^1(\Omega) ;$$

we thus obtain

(4.57) $$\int_\Gamma \frac{\partial u}{\partial n} \mu \, d\Gamma = 0 \quad \forall \mu \in H^1(\Omega) \Leftrightarrow \frac{\partial u}{\partial n} = 0$$

which, with $u \in H^2(\Omega) \cap H_0^1(\Omega)$, leads to $u \in H_0^2(\Omega)$.
With $v \in H_0^2(\Omega)$, it may be shown that $(v, - \Delta v) \in W$; it then follows from (4.54) and from the properties of u defined above that

$$\begin{cases} \int_\Omega \Delta u \, \Delta v \, dx = \int_\Omega f v \, dx & \forall v \in H_0^2(\Omega) \\ u \in H_0^2(\Omega) , \end{cases}$$

and u is thus the solution of (4.2), (4.3). ■
Conversely, if u is the solution of (4.2), (4.3) it may be shown that $(u, - \Delta u)$ satisfies the necessary and sufficient conditions (4.54), (4.55). ■

Remark 4.10. In the case of the problem

(4.58) $$\begin{cases} \Delta^2 u = f & \text{in } \Omega \\ \gamma_0 u = g_1 , \quad \gamma_1 u = g_2 ; & g_1, g_2 \text{ sufficiently regular} \end{cases}$$

it is necessary to replace the space W, see (4.52), by

$$W_g = \left\{(v, q)/(v, q) \in H^1(\Omega) \times L^2(\Omega), \int_\Omega \operatorname{grad} v.\operatorname{grad} \mu \, dx = \right.$$
$$\left. = \int_\Omega q\mu \, dx + \int_\Gamma g_2 \, \gamma_0 \, \mu \, d\Gamma \;\; \forall \mu \in H^1(\Omega), \; \gamma_0 \, v = g_1 \right\}$$

where g denotes the pair $\{g_1, g_2\}$.

There is then *equivalence* between (4.58) and

$$\text{(4.59)} \qquad \operatorname*{Min}_{(v,q) \in W_g} \left[\frac{1}{2}\int_\Omega q^2 \, dx - \int_\Omega fv \, dx\right],$$

and an *equivalent variational formulation* of (4.59) is given by

$$\text{(4.60)} \qquad \int_\Omega pq \, dx = \int_\Omega fv \, dx \quad \forall (v, q) \in W, \quad (u, p) \in W_g. \quad \blacksquare$$

4.4 An iterative method for the solution of the Dirichlet problem for Δ^2 based on the variational formulation given in Section 4.3.

We shall give in explicit form a variant of algorithm (4.4)-(4.6) associated with the equivalent variational formulation (4.53 of problem (4.2)-(4.3); it is convenient, and reasonable enough, to interpret (4.53) as an *optimal control problem* in order to demonstrate the following algorithm.

Let $\mathcal{M}$ be *a* complement of $H_0^1(\Omega)$ in $H^1(\Omega)$; we thus have

$$H^1(\Omega) = H_0^1(\Omega) \oplus \mathcal{M}$$

and there is equivalence between (4.53) and the following optimal control problem with *state constraints*:

$$\text{(4.61)} \qquad \operatorname*{Min}_{q \in L^2(\Omega)} \left[\frac{1}{2}\int_\Omega q^2 \, dx - \int_\Omega fv \, dx\right]$$

the *state* variable v being a function of the *control* variable q through the intermediary of the *state equation*

$$\text{(4.62)} \qquad \begin{cases} \displaystyle\int_\Omega \operatorname{grad} v. \operatorname{grad} \mu \, dx = \int_\Omega q\mu \, dx \quad \forall \mu \in H_0^1(\Omega) \\ v \in H_0^1(\Omega) ; \end{cases}$$

q and v must also satisfy the *state constraints*

$$(4.63)\qquad \int_\Omega \text{grad}\, v.\text{grad}\, \mu \, dx = \int_\Omega q\mu \, dx \quad \forall \mu \in \mathcal{M}.$$

Let u be the solution of (4.2), (4.3); problem (4.61)-(4.63) is equivalent to (4.53) and thus to (4.2), (4.3) and therefore admits $p = -\Delta u$ as its *unique* solution, u being the corresponding state function.

A Lagrangian associated with (4.61)-(4.63) is $\mathcal{L} : L^2(\Omega) \times \mathcal{M} \to \mathbb{R}$ (in which v is a function of q through the intermediary of (4.62)) defined by

$$(4.64)\qquad \mathcal{L}(q, \mu) = \frac{1}{2}\int_\Omega q^2 \, dx - \int_\Omega fv \, dx + \int_\Omega \text{grad}\, v.\text{grad}\, \mu \, dx - \int_\Omega q\mu \, dx.$$

We shall now prove:

Theorem 4.6. *Let u be the solution of* (4.2), (4.3) *and* $p = -\Delta u$; *we assume that $u \in H^3(\Omega) \cap H_0^2(\Omega)$*(1); *then $\mathcal{L}$ admits $\{p, p_2\}$ as unique saddle point on $L^2(\Omega) \times \mathcal{M}$, with p_2 defined by*

$$p = p_1 + p_2, \quad p_1 \in H_0^1(\Omega), \quad p_2 \in \mathcal{M}. \quad \blacksquare$$

It follows from Theorem 4.6 that we may apply to $\mathcal{L}$ the algorithm (4.12), (4.13) of Chapter 2, Section 4.3, that is:

$$(4.65)\qquad \lambda^0 \in \mathcal{M} \quad \text{chosen arbitrarily}$$

$$(4.66)\qquad \mathcal{L}(p^n, \lambda^n) \leqslant \mathcal{L}(q, \lambda^n) \quad \forall q \in L^2(\Omega), \quad p^n \in L^2(\Omega)$$

$$(4.67)\qquad \begin{cases} b(\lambda^{n+1}, \mu) = b(\lambda^n, \mu) + \rho_n \left\{ \int_\Omega \text{grad}\, u^n.\text{grad}\, \mu \, dx - \int_\Omega p^n \mu \, dx \right\} \\ \qquad \forall \mu \in \mathcal{M}, \\ \lambda^{n+1} \in \mathcal{M}, \quad \rho_n > 0 ; \end{cases}$$

in (4.67), b is a bilinear form which is continuous, symmetric and coercive on $\mathcal{M} \times \mathcal{M}$.

The strong convergence of p^n to p is assured if $0 < \alpha_0 \leqslant \rho_n \leqslant \alpha_1$, α_1 sufficiently small.

We shall now write problem (4.66) explicitly; in view of (4.64), (4.66) has the form of an *unconstrained optimal control* problem, which admits (see Lions /2/, Chap. 2) a unique solution p^n characterised by the existence of an adjoint state $y^n \in H_0^1(\Omega)$ such that

---

(1) If Γ is regular, then this holds for $f \in H^{-1}(\Omega)$.

$$\begin{cases} \int_\Omega \text{grad}\, u^n . \text{grad}\, \mu \, dx = \int_\Omega p^n \mu \, dx \quad \forall \mu \in H_0^1(\Omega) \\ u^n \in H_0^1(\Omega) \end{cases}$$

(4.68)
$$\begin{cases} \int_\Omega \text{grad}\, y^n . \text{grad}\, \mu \, dx = \int_\Omega f\mu \, dx - \int_\Omega \text{grad}\, \lambda^n . \text{grad}\, \mu \, dx \\ \qquad \forall \mu \in H_0^1(\Omega) \\ y^n \in H_0^1(\Omega) \end{cases}$$

(4.69) $$p^n = y^n + \lambda^n .$$

In view of (4.69) we may replace (4.68), (4.69) by

$$\begin{cases} \int_\Omega \text{grad}\, p^n . \text{grad}\, \mu \, dx = \int_\Omega f\mu \, dx \quad \forall \mu \in H_0^1(\Omega) \\ p^n - \lambda^n \in H_0^1(\Omega) , \end{cases}$$

which gives the explicit form of (4.66):

(4.70)
$$\begin{cases} \int_\Omega \text{grad}\, p^n . \text{grad}\, v \, dx = \int_\Omega fv \, dx \quad \forall v \in H_0^1(\Omega) \\ p^n - \lambda^n \in H_0^1(\Omega) \end{cases}$$

(4.71)
$$\begin{cases} \int_\Omega \text{grad}\, u^n . \text{grad}\, v \, dx = \int_\Omega p^n v \, dx \quad \forall v \in H_0^1(\Omega) \\ \quad u^n \in H_0^1(\Omega) ; \end{cases}$$

the determination of u^n thus reduces to the solution of two Dirichlet problems for $-\Delta$, and it may be noted that in view of (4.71) the strong convergence, in $L^2(\Omega)$, of p^n to $-\Delta u$ implies the strong convergence of u^n to u in $H^2(\Omega)$. ■

4.5 An approximation of problem (4.2), (4.3) by mixed finite elements

4.5.1 *Synopsis*

Following Ciarlet-Raviart /3/, we shall now approximate problem (4.2), (4.3), *via the equivalent formulation* (4.53), by a finite element method which is *interior* in $H^1(\Omega)$ and *exterior* in $H^2(\Omega)$ (see Remark 4.4). ■

4.5.2 *Approximation of* $L^2(\Omega)$, $H^1(\Omega)$, $H_0^1(\Omega)$, $\mathcal{M}$

By way of simplification, we shall assume that Ω is a polygonal domain of $\mathbb{R}^2$; with $\mathcal{T}_h$ a triangulation of Ω which satisfies the assumptions of Section 1.3 of this chapter (the notation of which is retained here) and with P_k denoting the space of polynomials with two variables of degree $\leqslant k$, we approximate $L^2(\Omega)$ and $H^1(\Omega)$ by

$$V_h = \{ v_h \mid v_h \in C^0(\overline{\Omega}),\ v_h|_T \in P_k\ \forall T \in \mathcal{T}_h \}$$

then $H_0^1(\Omega)$ by

$$V_{0h} = \{ v_h \mid v_h \in V_h, v_h = 0 \text{ on } \Gamma \}.$$

For $\mathcal{M}_h$, we may take any arbitrary complement of V_{0h} in V_h, but the simplest is that defined by

$$\mathcal{M}_h = \{ v_h \mid v_h \in V_h, v_h|_T = 0 \quad \forall T \in \mathcal{T}_h, \partial T \cap \Gamma = \varnothing \}$$

and this is the only one which will be considered in the following.

4.5.3 *Approximation of* (4.53)

Let W_h be the approximation of W (see (4.52)) defined by

$$W_h = \left\{ (v_h, q_h) \mid (v_h, q_h) \in V_{0h} \times V_h, \int_\Omega \mathbf{grad}\, v_h . \mathbf{grad}\, \mu_h \, dx = \right.$$
$$\left. = \int_\Omega q_h \mu_h \, dx \quad \forall \mu_h \in V_h \right\}.$$

We approximate (4.53) by

$$\text{(4.72)} \qquad \underset{(v_h, q_h) \in W_h}{\text{Min}} \left[\frac{1}{2} \int_\Omega |q_h|^2 \, dx - \int_\Omega f v_h \, dx \right].$$

Finite element approximations of this type are said to be *mixed*, see Ciarlet-Raviart /3/. ■

4.5.4 *Convergence results* $(k \geqslant 2)$

We suppose that the angles of $\mathcal{T}_h$ are bounded below, *uniformly in h*, by $\theta_0 > 0$, and also that $\mathcal{T}_h$ satisfies the following assumption (the so-called *inverse hypothesis*):

$$\text{(4.73)} \qquad \frac{\underset{T \in \mathcal{T}_h}{\text{Max}}\, h(T)}{\underset{T \in \mathcal{T}_h}{\text{Min}}\, h(T)} \leqslant \tau \quad \forall \mathcal{T}_h, \tau \text{ independent of } h,$$

with, in (4.73), $h(T)$ = length of the longest side of T. It is shown in Ciarlet-Raviart, *loc. cit.*, that using the above assumptions we have

$$\text{(4.74)} \qquad \| u_h - u \|_{H^1(\Omega)} + \| (-\Delta u) - p_h \|_{L^2(\Omega)} \leqslant C \| u \|_{H^{k+2}(\Omega)} h^{k-1}$$

with, in (4.74), C independent of h and u. ■

Remark 4.11. When $u \notin H^{k+2}(\Omega)$ we have, when $h \to 0$, and using the above assumptions on $\mathcal{T}_h$, *strong convergence* of u_h to u in $H^1(\Omega)$ and of p_h to $-\Delta u$ in $L^2(\Omega)$; however, we do not have any estimate of the approximation errors in these spaces. ■

4.5.5 *The case* $k = 1$

If $k = 1$, we do not *in general* know (3) whether $\{u_h, p_h\}$ converges to $\{u, -\Delta u\}$. A *sufficient* condition for having $\{u_h, p_h\} \to \{u, -\Delta u\}$ strongly in $H^1(\Omega) \times L^2(\Omega)$ is to assume that $\mathcal{T}_h$ is generated by 3 *families of parallel straight lines* (except possibly in a neighbourhood of Γ the measure of which tends to zero).

Moreover if, for $k = 1$, we replace in the definition of W_h the condition

(4.75) $$\int_\Omega \text{grad}\, v_h . \text{grad}\, \mu_h \, \mathrm{d}x = \int_\Omega q_h \, \mu_h \, \mathrm{d}x \quad \forall \mu_h \in V_h$$

by (1) (see Figure 4.1)

(4.76) $$\int_\Omega \text{grad}\, v_h . \text{grad}\, \mu_h \, \mathrm{d}x = \frac{1}{3} \sum_{T \in \mathcal{T}_h} A(T) \sum_{i=1}^{3} q_h(M_{iT}) \, \mu_h(M_{iT}) \,,$$

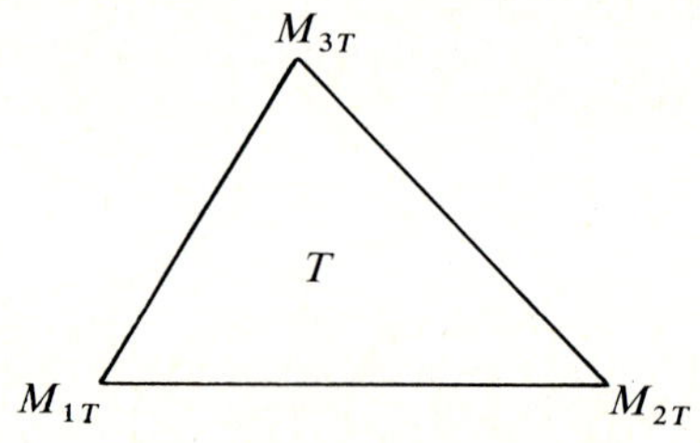

Fig. 4.1.

which is obtained by *numerical integration* of the right-hand side of (4.75), we get, if Ω is a rectangle (2) and for the triangulation shown in Figure 4.2, the classical 13-point scheme for the discretisation of the Dirichlet problem for Δ^2. ■

---

(1) $A(T)$ = area of T.

(2) This also applies if Γ is the union of segments parallel to the coordinate axes; incidentally, the same numerical integration formula is needed to evaluate $\frac{1}{2}\int_\Omega |q_h|^2 \, \mathrm{d}x$ in (4.72).

(3) In fact it has been shown, subsequent to the publication of the original French version of the present volume, that convergence can be proved even in the case $k = 1$ (see Appendix 4).

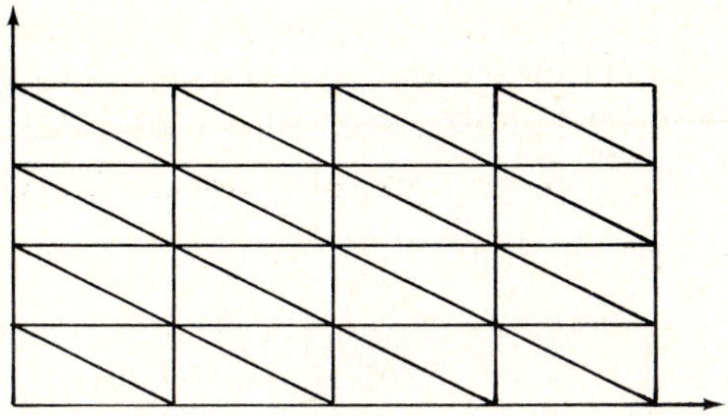

Fig. 4.2.

4.6 An iterative method for the solution of the approximate problem (4.72). Generalisation to inequalities of order 4.

4.6.1 *Description of the algorithm*

The approach used in Section 4.4, based on optimal control, also holds for the finite-dimensional problem (4.72); in order to solve the latter problem we may therefore use the following algorithm [1]:

(4.77) $\lambda_h^0 \in \mathscr{M}_h$ chosen arbitrarily,

then, having calculated $\lambda_h^n \in \mathscr{M}_h$, we calculate $p_h^n, u_h^n, \lambda_h^{n+1}$ by

$$\text{(4.78)} \quad \begin{cases} \displaystyle\int_\Omega \operatorname{grad} p_h^n . \operatorname{grad} v_h \, dx = \int_\Omega f v_h \, dx \quad \forall v_h \in V_{0h} \\ p_h^n - \lambda_h^n \in V_{0h} \end{cases}$$

$$\text{(4.79)} \quad \begin{cases} \displaystyle\int_\Omega \operatorname{grad} u_h^n . \operatorname{grad} v_h \, dx = \int_\Omega p_h^n v_h \, dx \quad \forall v_h \in V_{0h} \\ u_h^n \in V_{0h} \end{cases}$$

$$\text{(4.80)} \quad \begin{cases} \beta_h(\lambda_h^{n+1}, \mu_h) = \beta_h(\lambda_h^n, \mu_h) + \rho_n \left\{ \displaystyle\int_\Omega \operatorname{grad} u_h^n . \operatorname{grad} \mu_h \, dx - \right. \\ \left. \quad - \displaystyle\int_\Omega p_h^n \mu_h \, dx \right\} \quad \forall \mu_h \in \mathscr{M}_h \, , \\ \lambda_h^{n+1} \in \mathscr{M}_h, \rho_n > 0 \, , \end{cases}$$

---

[1] The finite-dimensional variant of (4.65),(4.70),(4.71).(4.67).

where, in (4.80), $\beta_h : \mathcal{M}_h \times \mathcal{M}_h \to \mathbb{R}$ is bilinear, continuous, symmetric and coercive. ■

Remark 4.12. If $k = 1$ we may, of course, replace the right-hand side of (4.79), (4.80) by the term obtained if (4.76) is used instead of (4.75). ■

4.6.2 *Convergence of algorithm* (4.77)-(4.80)

It may be shown, see Ciarlet-Glowinski /1/, that under the condition

$$0 < \alpha_0 \leqslant \rho_n \leqslant \alpha_1 ,$$

α_1 sufficiently small, we have

$$\lim_{n \to +\infty} \{ u_h^n, p_h^n, \lambda_h^n \} = \{ u_h, p_h, \lambda_h \}$$

where $\{ u_h, p_h \}$ is the solution of (4.72) and $\lambda_h \in \mathcal{M}_h$ is a component of p_h in the decomposition $V_h = V_{0h} \oplus \mathcal{M}_h$. ■

In the following, we shall denote by $2\sigma_h^2$ the upper bound of the α_1 values which give convergence of algorithm (4.77)-(4.80).

4.6.3 *Remarks on the choice of* β_h

The bilinear forms β_h defined below possess the properties required in Section 4.6.1:

$$\text{(4.81)} \qquad \beta_h^1(\lambda_h, \mu_h) = \int_\Omega \operatorname{grad} \lambda_h . \operatorname{grad} \mu_h \, dx$$

$$\text{(4.82)} \qquad \beta_h^2(\lambda_h, \mu_h) = \int_\Gamma \lambda_h \, \mu_h \, d\Gamma .$$

These forms have the disadvantage that, in the calculation of λ_h^{n+1} by means of (4.80), a linear system (1) must be solved at each iteration of algorithm (4.77)-(4.80). To remedy this situation we may profitably use a *numerical integration* method to obtain a form β_h whose matrix is *diagonal* with respect to the degrees of freedom defining the finite-element approximation considered; in particular, this will be the case if β_h^2 is approximated by

$$\text{(4.83)} \qquad \beta_h^3(\lambda_h, \mu_h) = \frac{1}{2} \sum_i | M_i M_{i+1} | (\lambda_i \mu_i + \lambda_{i+1} \mu_{i+1}) \quad \text{if } k = 1 ,$$

---

(1) It must be noted, however, that the matrix of this system is symmetric, positive-definite, and of relatively small bandwidth; consequently it is easy to employ a Cholesky factorisation which is performed once and for all.

(4.84) $$\beta_h^4(\lambda_h, \mu_h) = \frac{1}{6}\sum_i |M_i M_{i+1}| (\lambda_i \mu_i + 4 \lambda_{i+1/2} \mu_{i+1/2} + \lambda_{i+1} \mu_{i+1})$$
if $k = 2$;

the notation used in (4.83), (4.84) is the same as that employed in Section 1.3.2 of this chapter.

4.6.4 *Behaviour of* σ_h *when* $h \to 0$

Let σ_0 be the constant defined by (4.8); we suppose that the form β_h used is that defined by (4.82). Using the assumptions made in Section 4.5.4 for $\mathscr{T}_h$ it may then be shown, see Ciarlet-Glowinski /1/, that $\forall k \geqslant 1$, we have

(4.85) $$\lim_{h \to 0} \sigma_h = \sigma_0 .$$

The result (4.85) also holds if we approximate β_h^2 by (4.83) if $k = 1$, and by (4.84) if $k = 2$.

The result (4.85) allows a practical estimate of the quantity σ_h to be obtained when Ω is an open domain for which it is possible to determine, or at least to estimate, σ_0 (see J. Smith /1/). ■

4.6.5 *Application to the solution of variational inequalities of order 4.*

We shall now consider the inequalities of Examples 1 and 2 of Section 4.2.2; we assume in the following, confining our attention to $k = 1$ and 2, that the forms β_h used are those defined by (4.83), (4.84).

Furthermore, the linear mapping $\mathscr{M}_h \to \mathbb{R}$ defined by

$$\mu_h \to \int_\Omega \operatorname{grad} v_h . \operatorname{grad} \mu_h \, dx - \int_\Omega q_h \mu_h \, dx ,$$

may always be put in the following form (using the same notation as Section 1.3.2) [1]:

(4.86) $$\int_\Omega \operatorname{grad} v_h . \operatorname{grad} \mu_h \, dx - \int_\Omega q_h \mu_h \, dx = \sum_{M_i \in \Sigma_h \cap \Gamma} \mu_i G_i(v_h, q_h) ,$$
if $k = 1$

(4.87) $$\begin{cases} \int_\Omega \operatorname{grad} v_h . \operatorname{grad} \mu_h \, dx - \int_\Omega q_h \mu_h \, dx = \sum_{M_i \in \Sigma_h \cap \Gamma} \mu_i G_i(v_h, q_h) + \\ \quad + \sum_{M_{i+1/2} \in \Sigma_h' \cap \Gamma} \mu_{i+1/2} G_{i+1/2}(v_h, q_h) , \text{ if } k = 2 . \end{cases}$$

---

(1) Indeed, this also holds when $k = 1$, if we use (4.76) instead of (4.75).

Approximate solution of inequality (4.29)

We "approximate" $\Lambda = L^2_+(\Gamma)$ by

$$\Lambda_h = \{ \mu_h \mid \mu_h \in \mathscr{M}_h, \mu_h(P) \geqslant 0 \quad \forall P \in \Sigma_h \cap \Gamma \text{ if } k = 1, \\ \forall P \in (\Sigma_h \cup \Sigma'_h) \cap \Gamma \text{ if } k = 2 \} ;$$

in order to solve inequality (4.29) in an approximate manner, we may then use the following algorithm (a finite-dimensional variant of algorithm (4.33)-(4.35)):

(4.88) $\lambda_h^0 \in \Lambda_h$, chosen arbitrarily

(4.89) having calculated $\lambda_h^n \in \Lambda_h$ we determine p_h^n, u_h^n by (4.78), (4.79),

then λ_h^{n+1} by

$$(4.90)_1 \quad \lambda_i^{n+1} = \left[\lambda_i^n + \rho_n \frac{2}{| M_{i-1} M_i | + | M_i M_{i+1} |} G_i(u_h^n, p_h^n) \right]^+ \\ \forall M_i \in \Sigma_h \cap \Gamma, \text{ if } k = 1.$$

$$(4.90)_2 \quad \begin{cases} \lambda_i^{n+1} = \left[\lambda_i^n + \rho_n \dfrac{6}{| M_{i-1} M_i | + | M_i M_{i+1} |} G_i(u_h^n, p_h^n) \right]^+ \\ \forall M_i \in \Sigma_h \cap \Gamma, \\ \lambda_{i+1/2}^{n+1} = \left[\lambda_{i+1/2}^n + \rho_n \dfrac{3}{2 | M_i M_{i+1} |} G_{i+1/2}(u_h^n, p_h^n) \right]^+ \\ \forall M_{i+1/2} \in \Sigma'_h \cap \Gamma, \text{ if } k = 2, \end{cases}$$

with $\rho_n > 0$. ■

Approximate solution of inequality (4.36).

We "approximate" Λ (see 4.38)) by

$$\Lambda_h = \{ \mu_h \mid \mu_h \in \mathscr{M}_h, | \mu_h(P) | \leqslant 1 \quad \forall P \in \Sigma_h \cap \Gamma \text{ if } k = 1, \\ \forall P \in (\Sigma_h \cup \Sigma'_h) \cap \Gamma \text{ if } k = 2 \} ;$$

and then, to solve (4.36) in an approximate manner, we use the following finite-dimensional variant of algorithm (4.41)-(4.43):

(4.91) $\lambda_h^0 \in \Lambda_h$, chosen arbitrarily

(4.92) having calculated $\lambda_h^n \in \Lambda_h$ we determine p_h^n, u^n by (4.78) (4.79), then λ_h^{n+1} by

$$(4.93)_1 \qquad \lambda_i^{n+1} = P_{[-1,+1]}\left(\lambda_i^n + \rho_n g \frac{2}{|M_{i-1} M_i| + |M_i M_{i+1}|} G_i(u_h^n, p_h^n)\right) \quad \forall M_i \in \Sigma_h \cap \Gamma \text{ if } k = 1,$$

$$(4.93)_2 \qquad \begin{cases} \lambda_i^{n+1} = P_{[-1,+1]}\left(\lambda_i^n + \rho_n g \dfrac{6}{|M_{i-1} M_i| + |M_i M_{i+1}|} G_i(u_h^n, p_h^n)\right) \\ \qquad \forall M_i \in \Sigma_h \cap \Gamma, \\ \lambda_{i+1/2}^{n+1} = P_{[-1,+1]}\left(\lambda_{i+1/2}^n + \rho_n g \dfrac{3}{2\,|M_i M_{i+1}|} G_{i+1/2}(u_h^n, p_h^n)\right) \\ \qquad \forall M_{i+1/2} \in \Sigma_h' \cap \Gamma \text{ if } k = 2, \end{cases}$$

with $\rho_n > 0$ and $P_{[-1,+1]}(a) = \max(-1, \min(a, 1)) \quad \forall a \in \mathbb{R}$. ■

Convergence of algorithms (4.88)-(4.90) *and* (4.91)-(4.93).

The constant σ_h being the one appearing in Section 4.6.2, we may show that under the conditions

$$(4.94) \qquad 0 < \alpha_0 \leqslant \rho_n \leqslant \alpha_1 < 2\,\sigma_h^2 \text{ for problem (4.29)},$$

$$(4.95) \qquad 0 < \alpha_0 \leqslant \rho_n \leqslant \alpha_1 < \frac{2}{g^2}\sigma_h^2 \text{ for problem (4.36)},$$

there is convergence of the above algorithms to a pair $\{u_h, p_h\}$ which converges strongly in $H^1(\Omega) \times L^2(\Omega)$, when $h \to 0$ (the hypotheses on $\mathscr{T}_h$ being those of Section 4.5.4) to $\{u, -\Delta u\}$, where u is the solution of the corresponding continuous problem.

Remark 4.13. The numerical solution of problem (4.45) in Example 3, Section 4.2.2, does not present any serious practical difficulties; in fact the solution of problem (4.45), suitably approximated, may be reduced to that of a sequence of problems of the type (4.72), which may then be solved by means of algorithm (4.77)-(4.80) (see Section 4.7 below). ■

4.7 Example and numerical application

Synopsis

In order to illustrate the approximation and solution methods of Section 4, we have chosen a problem which falls into the category of Example 3, Section 4.2.2; for further fourth-order problems treated by analogous methods, see Bourgat /2/. ■

4.7.1 *Formulation of the problem*

Let Ω and E be two open bounded domains of $\mathbb{R}^2$, such that $\bar{E} \subset \Omega$;

we denote by K the closed convex subset of $H_0^2(\Omega)$ defined by

$$K = \{ v \mid v \in H_0^2(\Omega), v(x) \geqslant C \quad \forall x \in E \} .$$

Under these conditions the problem

$$\text{(4.96)} \qquad \underset{v \in K}{\text{Min}} \int_\Omega | \Delta v |^2 \, dx$$

admits one and only one solution, and there is equivalence between (4.96) and the following fourth-order inequality:

$$\text{(4.97)} \qquad \begin{cases} \displaystyle\int_\Omega \Delta u \, \Delta(v - u) \, dx \geqslant 0 \quad \forall v \in K \\ u \in K . \quad \blacksquare \end{cases}$$

4.7.2 *Approximation of problem* (4.96), (4.97)

We assume in the following that Ω is polygonal; to approximate (4.96), (4.97) we shall use the finite element approximation of Section 4.5, with finite elements of order 1; this is permissible if (see Section 4.5.5) the triangulation $\mathcal{T}_h$ used is generated (except possibly in the neighbourhood of Γ) by three families of parallel straight lines.

With the space W_h defined by

$$W_h = \{ (v_h, q_h) \mid (v_h, q_h) \in V_{0h} \times V_h , (v_h, q_h) \text{ satisfies } (4.76) \} ,$$

we denote by Σ_h the set of the vertices of $\mathcal{T}_h$ and we approximate, respectively, K and (4.96), (4.97) by

$$\text{(4.98)} \qquad K_h = \{ (v_h, q_h) \mid (v_h, q_h) \in W_h, v_h(P) \geqslant C \quad \forall P \in \overline{E} \cap \Sigma_h \}$$

$$\text{(4.99)} \qquad \underset{(v_h, q_h) \in K_h}{\text{Min}} \frac{1}{3} \sum_{T \in \mathcal{T}_h} A(T) \sum_{i=1}^{3} q_h^2(M_{iT}) .$$

The approximate problem (4.99) admits one and only one solution, (u_h, p_h). $\blacksquare$

4.7.3 *Iterative methods of solution*

A first method of solution.

This is a discretised version of algorithm (4.49)-(4.51); with M_h denoting the number of elements of $\Sigma_h \cap \overline{E}$, we define $\mathcal{L}_h : \quad W_h \times \mathbb{R}^{M_h} \to \mathbb{R}$, the Lagrangian associated with (4.99), by

$$\text{(4.100)} \qquad \mathcal{L}_h(v_h, q_h ; z_h) = \frac{1}{6} \sum_{T \in \mathcal{T}_h} A(T) \sum_{i=1}^{3} q_h^2(M_{iT}) + \frac{1}{3} \sum_{P \in \overline{E} \cap \Sigma_h} \overline{A}(P) \, (C - v_h(P)) \, z_h(P)$$

where, in (4.100), $\bar{A}(P)$ = sum of the areas of the triangles on $\mathcal{T}_h$ which admit P as common vertex. Since the Lagrangian $\mathcal{L}_h$ admits a saddle-point on $W_h \times \mathbb{R}_+^{M_h}$, namely (u_h, p_h, y_h) where (u_h, p_h) is the solution of (4.99), then in order to solve this latter problem we may use the following "approximation" of algorithm (4.49)-(4.51):

(4.101) $\quad y_h^0 \in \mathbb{R}_+^{M_h}$, chosen arbitrarily

then having calculated y_h^n, we determine (u_h^n, p_h^n) and y_h^{n+1} by

$$(4.102) \qquad \begin{cases} \mathcal{L}_h(u_h^n, p_h^n ; y_h^n) \leqslant \mathcal{L}_h(v_h, q_h ; y_h^n) \quad \forall (v_h, q_h) \in W_h \\ (u_h^n, p_h^n) \in W_h \end{cases}$$

$$(4.103) \qquad y_h^{n+1}(P) = [y_h^n(P) + \rho_n(C - u_h^n(P))]^+ \quad \forall P \in \bar{E} \cap \Sigma_h, \rho_n > 0 .$$

Under the condition $0 < \alpha_0 \leqslant \rho_n \leqslant \alpha_1$, α_1 sufficiently small, we have convergence of the sequence (u_h^n, p_h^n) to (u_h, p_h). It may be noted that problem (4.102) is of the type considered in Section 4.6.1 and hence may be solved using algorithm (4.77)-(4.80). ■

A second method of solution.

By combining algorithms (4.101)-(4.103) and (4.77)-(4.80) we obtain (using the notation of Sections 4.5.5 and 4.6.5) the following algorithm:

(4.104) $\quad (\lambda_h^0, y_h^0) \in \mathcal{M}_h \times \mathbb{R}_+^{M_h}$, chosen arbitrarily

then having calculated (λ_h^n, y_h^n) we determine (u_h^n, p_h^n) and $(\lambda_h^{n+1}, y_h^{n+1})$ by

$$(4.105) \qquad \begin{cases} \displaystyle\int_\Omega \text{grad}\, p_h^n . \text{grad}\, v_h \, dx = \int_\Omega f v_h \, dx + \frac{1}{3} \sum_{P \in \bar{E} \cap \Sigma_h} \bar{A}(P)\, y_h^n(P)\, v_h(P) \\ \qquad \forall v_h \in V_{0h} , \\ p_h^n - \lambda_h^n \in V_{0h} , \end{cases}$$

$$(4.106) \qquad \begin{cases} \displaystyle\int_\Omega \text{grad}\, u_h^n . \text{grad}\, v_h \, dx = \frac{1}{3} \sum_{T \in \mathcal{T}_h} A(T) \sum_{i=1}^{3} p_h^n(M_{iT})\, v_h(M_{iT}) \\ \qquad \forall v_h \in V_{0h} , \\ u_h^n \in V_{0h} , \end{cases}$$

$$(4.107) \qquad \lambda_i^{n+1} = \lambda_i^n + \rho_n^1 \frac{2}{|M_{i-1} M_i| + |M_i M_{i+1}|} G_i(u_h^n, p_h^n) \\ \qquad \forall M_i \in \Sigma_h \cap \Gamma, \rho_n^1 > 0 ,$$

$$(4.108) \qquad y_h^{n+1}(P) = [y_h^n(P) + \rho_n^2(C - u_h^n(P))]^+ \quad \forall P \in \bar{E} \cap \Sigma_h, \rho_n^2 > 0 .$$

It may be shown that under the conditions $0 < \alpha_0^i \leqslant \rho_n^i \leqslant \alpha_1^i$, $i = 1, 2$, with α_1^i sufficiently small, we have convergence of

(u_h^n, p_h^n) to (u_h, p_h) ; the problem of the determination of the sequence $(\rho_n^1, \rho_n^2)_n$ which gives (4.104)-(4.108) the optimal rate of convergence requires further attention. ■

4.7.4 *An application*

Geometric data:

$\Omega =]0, 1[\times]0, 1[$, E: disc with centre (0.5, 0.5) and radius 0.25, $C = 0.062\,5$.

Triangulation of Ω:

The triangulation used is that of Figure 1.9 in Section 1.7.1 of this chapter, comprising 800 triangles, 441 nodes and 361 interior nodes. It may be noted that $\mathcal{T}_h$ is indeed generated by three families of parallel straight lines. It may also be noted that

$$\bar{A}(P) = \frac{3}{400} \quad \forall P .$$

Solution algorithm:

We use (4.104)-(4.108).

Initialisation of algorithm (4.104)-(4.108):

$$\lambda_h^0 = 0, \; y_h^0(P) = 100 \quad \forall P \in \bar{E} \cap \Sigma_h .$$

Solution of the Dirichlet problems (4.105), (4.106):

By the Cholesky method.

Parameters ρ_n^1, ρ_n^2:

These were taken as $\rho_n^1 = \rho^1 = 1.25$, $\rho_n^2 = \rho^2 = 750$; it may be noted that the orders of magnitude of these two parameters are very different.

Analysis of the results.

Table 4.1 shows the variations of

$$R_{1h}^n = \underset{P \in \Sigma_h}{\operatorname{Max}} \left| u_h^{n+1}(P) - u_h^n(P) \right|$$

and of

$$R_{2h}^n = \sum_{M_i \in \Sigma_h \cap \Gamma} \left| G_i(u_h^n, p_h^n) \right|$$

as a function of the number of iterations; the significance of R_{2h}^n is that this quantity is an approximation of $\int_\Gamma \left|\frac{\partial u}{\partial n}\right| d\Gamma$, and this therefore allows us to ascertain whether the condition $\partial u/\partial n = 0$ has been properly satisfied (approximately at least).

Number of iterations	R_{1h}^n	R_{2h}^n
50	9×10^{-5}	1.5×10^{-2}
100	1.3×10^{-4}	7.3×10^{-3}
200	1.7×10^{-5}	1.2×10^{-3}
300	1.3×10^{-6}	1.5×10^{-4}
400	1.2×10^{-6}	1.07×10^{-4}
500	9.7×10^{-7}	8.6×10^{-5}
600	7.7×10^{-7}	6.9×10^{-5}
700	6.25×10^{-7}	5.6×10^{-5}
750	5.6×10^{-7}	5×10^{-5}

Table 4.1

The computation time is 2 min. 40 sec. for 100 iterations on a CII 10070, the calculations being performed in double precision.

Figure 4 shows the equipotentials of the computed solution u_h, and Figure 4.4 represents the function $x_1 \to u_h(x_1, 0.5)$.

Remark 4.14. With the domain Ω supporting an *elastic thin plate*, it appears - and this could be predicted theoretically - that the *zone of contact* between the plate and the *obstacle* (i.e. the cylinder of base E and of height C), is limited to a circle of radius 0.25 and height C. It then follows that the multiplier y_h has no limit in $L^2(\Omega)$ when $h \to 0$; this is thus typical of a situation in which the algorithm (4.49)-(4.51) is not meaningful. It may also be noted that the multiplier y_h is zero except at the points of $\Sigma_h \cap \bar{E}$ neighbouring the circle ∂E, the boundary of E; at these points, y_h takes values which become progressively larger as the value of h decreases (of the order of 1000 for the triangulation $\mathcal{T}_h$ used). ■

Remark 4.15. It is evident that for the shape of Ω and the triangulation $\mathcal{T}_h$ used, the approximate problem (4.99) relating to the example in Section 4.7.4 could have been described in terms

---

(1) In fact the limit of y_h is not a function.

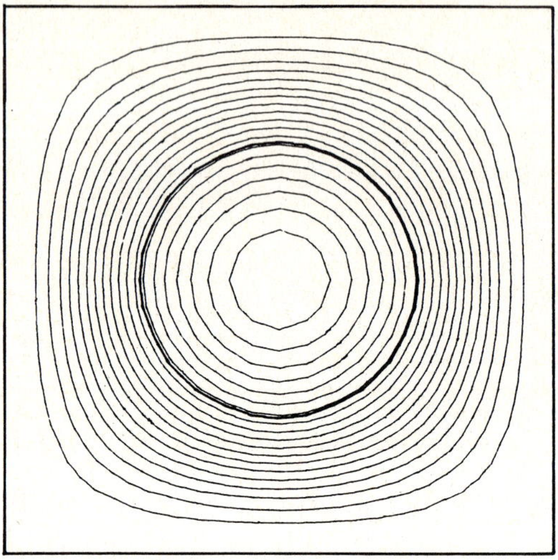

Fig. 4.3. Equipotentials of the function u_h.

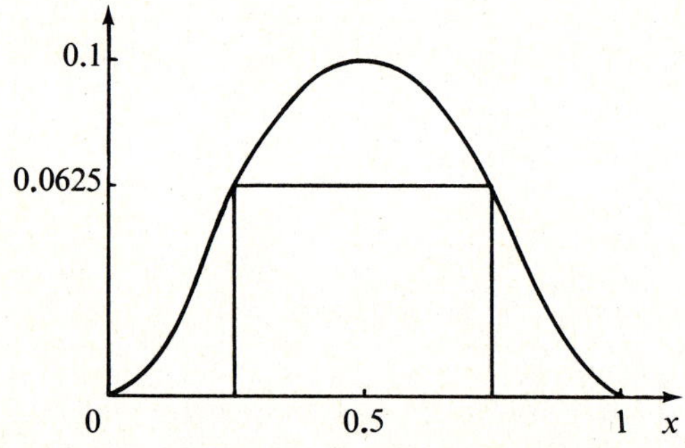

Fig. 4.4. Graph of the function $u_h(x_1, 0.5)$.

of a *finite difference* approximation; we have not pursued this viewpoint, however, since the application of finite difference methods for problems of order 4 is *in practice* limited to domains Ω whose boundaries consist of *segments parallel to the coordinate axes*; the finite element approximations studied in Section 4 do not suffer from this limitation. ■

5. DISCUSSION

Sections 1, 2, 3 are given for reasons of methodology; in fact the techniques employed are of a general interest and are the ones used in a number of interesting applied problems. *For example*, the reader may consult Frémond /2/.

The methods of Section 4 seem to be well suited to the solution of certain problems in Fluid Mechanics. In fact, using the notation of Section 4.3, the functions u and p ($p = -\Delta u$) appear respectively as the *stream function* ([1]) and the *vorticity* ([2]) when, for example, Stokes and Navier-Stokes flows are being modelled by partial differential equations of order 4; the finite-element method of Section 4.5 - which follows Ciarlet-Raviart /3/ - possesses, amongst other advantages, that of giving a continuous approximation of the stream function *and* of the vorticity. Moreover, it results from Section 4.6 - which follows Ciarlet-Glowinski /1/, /2/ - that in order to solve the Dirichlet problem for Δ^2, it is sufficient to have available a finite-element program for the Dirichlet problem for $-\Delta$ (in fact, the approximation described in Section 4.5 was constructed to this end). Some studies of iterative algorithms of the type considered in Section 4 may be found, for example ([3]), in Smith /1/, /2/, /3/, Bossavit /1/, Ehrlich /1/; however, these methods are suitable only for *finite difference* approximations, and for domains whose boundaries are formed by the union of segments parallel to the coordinate axes. The application of these methods, in conjunction with some approximations by *finite elements*, appears in Glowinski /6/, where an interpretation is also given of the algorithms mentioned above as being algorithms for the calculation of saddle-points; this latter point of view is used systematically in Ciarlet-Glowinski /1/, /2/ and in Section 4 of the present chapter, where the possibility of combining these iterative methods with practically *any* finite element approximation (except elements of order 1; see Section 4.5.5) ([4]) is demonstrated.

([1]) Usually denoted by Ψ.

([2]) Usually denoted by ω.

([3]) We refer to Glowinski /6/ and Ciarlet-Glowinski /1/, /2/ for a detailed bibliography on this subject.

([4]) Subsequent to the publication of the original French version of the present volume, it has been shown that even this limitation does not apply (see Appendix 4).

Chapter 5

NUMERICAL ANALYSIS OF THE STEADY FLOW OF A BINGHAM FLUID IN A CYLINDRICAL DUCT

INTRODUCTION

In this chapter we shall investigate the application of the results and methods of Chapters 1 and 2 to the numerical analysis of the flow problem considered in Chapter 1, Section 1.4. ■

1. STATEMENT OF THE CONTINUOUS PROBLEM. PHYSICAL MOTIVATION. SYNOPSIS

1.1 Statement of the continuous problem

The notation is that of Chapter 1, Section 1.4; Ω is an open *bounded* domain of $\mathbb{R}^2$ with boundary Γ. On $H_0^1(\Omega)$ we define:

(1.1) $$a(u, v) = \mu \int_\Omega \left(\frac{\partial u}{\partial x_1} \frac{\partial v}{\partial x_1} + \frac{\partial u}{\partial x_2} \frac{\partial v}{\partial x_2} \right) \mathrm{d}x , \quad \mu \text{ constant } > 0 ,$$

(1.2) $$J_0(v) = \frac{1}{2} a(v, v) - \int_\Omega fv \, \mathrm{d}x \quad (f \in L^2(\Omega)) ,$$

(1.3) $$j(v) = \int_\Omega |\operatorname{grad} v| \, \mathrm{d}x ,$$

(1.4) $$J_g(v) = J_0(v) + gj(v) , \quad g \text{ constant } \geqslant 0 ,$$

and we consider the variational problem:

(1.5) $$(P_0) : \min_{v \in H_0^1(\Omega)} J_g(v) .$$

Since the functional J_g is continuous and strictly convex on $H_0^1(\Omega)$, with

$$\lim_{\|v\|_{H_0^1(\Omega)} \to +\infty} J_g(v) = +\infty ,$$

Theorem 2.1 of Chapter 1, Section 2.1 applies, giving the existence and uniqueness of a solution of (P_0). We shall denote this solution by u_g ; u_g is also the unique solution of the variational inequality:

(1.6) $$\begin{cases} a(u, v - u) + gj(v) - gj(u) \geqslant \int_\Omega f(v - u) \, \mathrm{d}x \quad \forall v \in H_0^1(\Omega) , \\ u \in H_0^1(\Omega) , \end{cases}$$

equivalent to (P_0). ■

1.2. Physical motivation

In Duvaut-Lions /1/, Chapter 6, following the definition of a Bingham fluid, it is shown that in the special case of steady laminar flow in a cylindrical duct of cross-section Ω, the flow velocity is the solution of (1.5), (1.6), where f is the linear pressure drop (constant in practice), g is a positive constant representing a stress threshold below which the fluid behaves as a rigid material and above which it behaves as an incompressible fluid with viscosity μ; hereafter we shall take $\mu = 1$.

For the numerical analysis of more general Bingham flows the reader is referred to Bégis /1/, Fortin /1/ and Chapter 6 of the present volume. ■

1.3. Synopsis

In Section 2 we shall give some of the properties of the solution of problem (P_0); from the Numerical Analysis point of view the major difficulty of the problem considered is the *non-differentiability* of the functional J_g; in Sections 6 and 7 we shall describe techniques of *regularisation* (cf. Chapter 1, Section 3.3) and *duality* (cf. Chapter 1, Section 3.4 and Chapter 2, Section 4) which allow us to solve (P_0) (1), or, more precisely, approximations of (P_0); these approximations are investigated in Sections 3, 4 and 5.

In Section 8 we shall apply the results of the preceding sections to the solution of the elasto-plastic torsion problem investigated in Chapter 4, using the dual formulation of Chapter 3, Section 9.1.1. ■

2. SOME PROPERTIES OF THE SOLUTION OF THE CONTINUOUS PROBLEM

2.1. Regularity results

If Ω is a bounded open domain with a sufficiently regular boundary Γ then it is shown in H. Brézis /6/ that $f \in L^2(\Omega)$ implies, for the solution u_g of (P_0):

$$u_g \in H^2(\Omega) \cap H_0^1(\Omega)\,. \tag{2.1}$$

Remark 2.1. We shall see in Section 2.3 that the above property is optimal, at least in Sobolev spaces of whole number order. ■

2.2. Dependence of the solution on g

It is shown in Duvaut-Lions /1/, Chapter 6, Section 8.3 that:

---

(1) See also Section 9 for an algorithm of the augmented Lagrangian type (cf. Chapter 2, Section 6).

Theorem 2.1. *The function* $g \to u_g$ *is Lipschitz continuous from* $\mathbb{R}^+ \to H_0^1(\Omega)$. *Moreover:*

(2.2) $u_g = 0$ *for* $g \geqslant g_0 =$ *critical threshold depending on* Ω & f,

and

(2.3) $\left\{\begin{array}{l}\textit{the function } g \to a(u_g, u_g) \textit{ (resp. } g \to j(u_g)\textit{) is continuous} \\ \textit{decreasing (zero for } g \geqslant g_0\textit{).}\end{array}\right.$ ■

In the special case f = const. (which is of interest in practical applications), (2.2) can be rendered precise by showing (see Glowinski /2/) that:

(2.4) $g_0 = \gamma \mid f \mid$,

with

(2.5) $\frac{1}{\gamma} = \inf_{\substack{v \in H_0^1(\Omega) \\ \|v\|_{L^1(\Omega)} \geqslant 1}} j(v)$. ■

We supplement (2.3) by:

Theorem 2.2 *The mapping* $g \to a(u_g, u_g)$ *is convex from* $\mathbb{R}^+ \to \mathbb{R}^+$.

Proof. The function u_g is the solution of (1.6); by taking $v = 0$ and $v = 2\,u_g$ in (1.6), we then deduce:

(2.6) $a(u_g, u_g) + gj(u_g) = \int_\Omega f u_g \, dx$

whence

(2.7) $J_g(u_g) = -\frac{1}{2} a(u_g, u_g)$.

From the definition of u_g we have:

(2.8) $J_g(u_g) = \operatorname{Inf}_{v \in H_0^1(\Omega)} \left(\frac{1}{2} a(v, v) + gj(v) - \int_\Omega fv \, dx \right)$;

the mapping $g \to J_g(u_g)$, the *lower envelope* of a family of affine linear functions of g, is thus concave, so that $g \to a(u_g, u_g)$ is *convex* from (2.7)[1].

Remark 2.2. Other properties of the solution of (P_0) will be found in Duvaut-Lions, *loc. cit.*; in particular if $f \geqslant 0$ on Ω we have $u_g \geqslant 0$ on Ω. ■

Remark 2.3. We shall apply the duality theory due to

---

[1] This proof *does not use* the special structure of $a(u, v)$ and $j(v)$.

Rockafellar /1/ to the present problem (we also refer to the account by Ekeland-Témam /1/)

We put [1]:

$$(2.9)\qquad F(v) = g\int_\Omega |\operatorname{grad} v|\,dx - (f, v)\,, \quad v \in H_0^1(\Omega) = V$$

$$(2.10)\qquad G(q) = \frac{1}{2}\int_\Omega |q|^2\,dx\,, \qquad q \in (L^2(\Omega))^n = Y\,,$$

$$(2.11)\qquad \Lambda v = \operatorname{grad} v \;\;(^2)\;\; \Lambda \in \mathscr{L}(V; Y)\,.$$

The problem is equivalent to

$$(2.12)\qquad \inf_v\,[F(v) + G(\Lambda v)]\;;$$

we then have

$$(2.13)\qquad \inf_v\,[F(v) + G(\Lambda v)] = \sup_q\,[-F^*(\Lambda^* q) - G^*(-q)]\,,$$

where F^* and G^* are the conjugate functions of F and G and Λ^* is the adjoint of Λ.

Thus

$$(2.14)\qquad \begin{cases} G^*(q) = \sup_\varphi\left[\int_\Omega q.\varphi\,dx - G(\varphi)\right] = \frac{1}{2}\int_\Omega |q|^2\,dx\,, \\ \Lambda^* q = -\operatorname{div} q \end{cases}$$

and

$$(2.15)\qquad F^*(\Lambda^* q) = \sup_v\left[-\int_\Omega v \operatorname{div} q\,dx - g\int_\Omega |\operatorname{grad} v|\,dx + (f, v)\right].$$

If w is the solution of

$$(2.16)\qquad -\Delta w = f\,, \qquad w|_\Gamma = 0\,,$$

then

$$(f, v) = (\operatorname{grad} w, \operatorname{grad} v)$$

and thus, if we put:

$$p = q + \operatorname{grad} w\,,$$

---

(1) This also holds for $n > 2$.

(2) This Λ should not be confused with the convex set denoted later on by the same term.

we have:

$$F^*(\Lambda^* q) = \sup_v \left[\int_\Omega p.\text{grad}\, v\, dx - g \int_\Omega |\text{grad}\, v|\, dx \right].$$

From this we deduce that

$$F^*(\Lambda^* q) = \begin{cases} 0 \text{ if } \left| \int_\Omega p.\text{grad}\, v\, dx \right| \leqslant g \int_\Omega |\text{grad}\, v|\, dx \quad \forall v, \\ +\infty \text{ otherwise} \end{cases}$$

Let

$$K = \left\{ p \mid p \in (L^\infty(\Omega))^n, \left| \int_\Omega p.\text{grad}\, v\, dx \right| \leqslant \int_\Omega |\text{grad}\, v|\, dx \right.$$

$$\text{(2.17)} \qquad \left. \forall v \in H_0^1(\Omega) \right\}.$$

Then

$$F^*(\Lambda^* q) = \begin{cases} 0 \text{ if } p \in gK \\ +\infty \text{ otherwise.} \end{cases}$$

If the dimension is 1 we have:

$$K = \{ p, \ |p(x)| \leqslant 1 \}.$$

The dual problem is then

$$\sup_{p \in gK} -\tfrac{1}{2} |p - \text{grad}\, w|^2, \qquad |\quad| = \text{norm}\,(L^2(\Omega))^n.$$

Consequently

$$\text{(2.18)} \qquad a(u_g, u_g) = \inf_p |p - \text{grad}\, w|^2, \quad p \in gK.$$

If $n = 1$ and if $K = \{ p \mid p(x) | \leqslant 1 \}$ we deduce *explicitly*

$$a(u_g, u_g) = \int_{|h(x)| \geqslant g} (g - f(x))^2\, dx$$

where

$h = \dfrac{dw}{dx}$ belongs to $H^1(\Omega) \subset C^0(\overline{\Omega})$, from (2.16), if $f \in L^2(\Omega)$.

We then deduce that

$$\frac{d}{dx} a(u_g, u_g) = 2 \int_{|h(x)| \geqslant g} (g - |h(x)|)\, dx \leqslant 0,$$

$$\frac{d^2}{dg^2} a(u_g, u_g) = 2 \int_{|h(x)| \geqslant g} dx.$$

If the set on which $|h(x)| = c$ has measure zero $\forall c$, we then have

$$\frac{d^2}{dg^2} a(u_g, u_g) = 0 \text{ if } g = g_c .$$

We deduce $a(u_g, u_g) \sim -m(g - g_c)^3$ in the neighbourhood of $g_c (m \geqslant 0)$ and

$$\frac{d^3}{dg^3} a(u_g, u_g) = -2 \lim \frac{1}{\varepsilon} \text{ measure} \{ x \mid g \leqslant |h| \leqslant g + \varepsilon \} \leqslant 0.$$

In the general case (2.18),

$$a(u_g, u_g) = g^2 \inf_{k \in K} \left| k - \frac{h^2}{g} \right| .$$

Let P_K = orthogonal projection on K in $(L^2(\Omega))^n$. We have:

$$a(u_g, u_g) = g^2 \, | P_K(h/g) - h/g |^2 . \tag{2.19}$$

Duality relation.

Let q_0 be the solution of the dual problem, so that

$$q_0 = p_0 - \text{grad } w , \qquad p_0 = g P_K \left(\frac{\text{grad } w}{g} \right).$$

We have:

$$G(\Lambda u) + G^*(-q_0) = \langle \Lambda u, -q_0 \rangle$$

i.e.

$$\tfrac{1}{2} | \Lambda u |^2 + \tfrac{1}{2} | q_0 |^2 = -(\Lambda u, q_0)$$

i.e.

$$\Lambda u + q_0 = 0 .$$

Thus

$$\text{grad } u = \text{grad } w - g P_K \left(\frac{\text{grad } w}{g} \right). \tag{2.20}$$ ■

2.3. Some particular cases for which the solution is known

2.3.1 *A first example*

We take $\Omega =]0, 1[$ and $f = c > 0$; this corresponds to a Bingham flow between two planes unit distance apart.

Problem (P_0) is thus defined by:

$$\min_{v \in H_0^1(\Omega)} \left[\frac{1}{2} \int_0^1 v'^2 \, dx + g \int_0^1 | v' | \, dx - c \int_0^1 v \, dx \right], \tag{2.21}$$

with $v' = dv/dx$.

The solution of (2.21) is given by:

$$u_g = 0 \quad \text{if} \quad g \geqslant g_0 = c/2\,, \tag{2.22}$$

and if $g < c/2$ by:

$$\begin{cases} u_g(x) = \frac{c}{2}x(1-x) - gx & \text{if } 0 \leqslant x \leqslant \frac{1}{2} - \frac{g}{c}\,, \\ u_g(x) = \frac{c}{2}\left(\frac{1}{2} - \frac{g}{c}\right)^2 & \text{if } \frac{1}{2} - \frac{g}{c} \leqslant x \leqslant \frac{1}{2} + \frac{g}{c}\,, \\ u_g(x) = \frac{c}{2}x(1-x) - g(1-x) & \text{if } \frac{1}{2} + \frac{g}{c} \leqslant x \leqslant 1\,. \end{cases} \tag{2.23}$$

If can be verified that for $g < c/2$ we have $u_g \in H_0^1(0, 1) \cap H^2(0, 1)$ but that $u_g \notin H^3(0, 1)$. ■

Figure 2.1 shows the solution of (2.21) corresponding to $g = 1$ and $c = 10$ (we have $\frac{1}{2} - g/c = 0.4$).

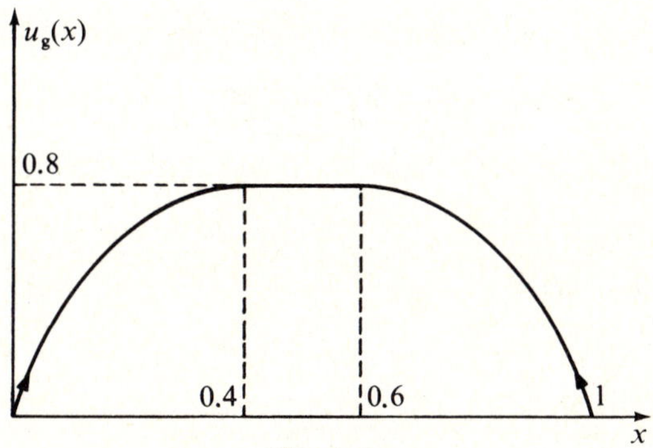

Fig. 2.1.

2.3.2 *A second example*

We take $\Omega = \{ x \mid x = (x_1, x_2),\ x_1^2 + x_2^2 < R^2 \}, f = c > 0$; the solution of (P_0) is given by:

$$u_g = 0 \quad \text{if } g \geqslant g_0 = cR/2\,, \tag{2.24}$$

and if $g < cR/2$ by:

$$\begin{cases} u_g(x) = \dfrac{(R-r)}{2}\left[\dfrac{c}{2}(R+r) - 2g\right] & \text{if } R' \leqslant r \leqslant R\,, \\ u_g(x) = \dfrac{(R-R')}{2}\left[\dfrac{c}{2}(R+R') - 2g\right] & \text{if } 0 \leqslant r \leqslant R'\,, \end{cases} \tag{2.25}$$

with

$$(2.26) \qquad r = \sqrt{x_1^2 + x_2^2}\,, \qquad R' = 2\,g/c\,.$$

Remark on the non-uniqueness of the solution of the dual problem.

It can easily be shown that in the case of the example of Section 2.3.1 the dual problem

$$(2.27) \qquad \begin{cases} \max\limits_{\mu \in \Lambda} \; \min\limits_{v \in H_0^1(0,1)} \left[\dfrac{1}{2} \displaystyle\int_0^1 v'^2 \, \mathrm{d}x + g \int_0^1 \mu v' \, \mathrm{d}x - c \int_0^1 v \, \mathrm{d}x \right], \\ \Lambda = \{ \mu \mid \mu \in L^\infty(0,1), \; |\mu(x)| \leqslant 1 \text{ a.e.} \} \end{cases}$$

admits one and only one solution, the reason for this being obvious!

In the case of the example of Section 2.3.2, we shall see that the dual problem admits an infinite number of solutions:

Consider a system of *polar coordinates* in which we use the notation $\mu = \{ \mu_r, \mu_\theta \}$; we then have:

$$(2.28) \qquad \operatorname{div} \mu = \frac{1}{r} \frac{\partial}{\partial_r} (r \mu_r) + \frac{1}{r} \frac{\partial \mu_\theta}{\partial \theta}\,.$$

Using (2.28) and the characterisation:

$$(2.29) \qquad \begin{cases} -\Delta u - g \operatorname{div} \lambda = c\,, \\ u|_\Gamma = 0\,, \\ \lambda \in \Lambda\,, \\ \lambda . \operatorname{grad} u = |\operatorname{grad} u|\,, \end{cases}$$

where u (resp. λ) is the solution of (P_0) (or the dual problem), it can be shown that λ, defined by:

$$(2.30) \qquad \lambda_r = -1\,, \qquad \lambda_\theta = 0 \text{ if } R' < r < R\,,$$

$$(2.31) \qquad \lambda_r = -r/R'\,, \qquad \lambda_\theta = \beta(r) \text{ if } 0 < r < R'\,,$$

with:

$$(2.32) \qquad \beta \in L^\infty(0, R')\,, \qquad |\beta(r)| \leqslant \sqrt{1 - (r/R')^2} \quad \text{a.e. on } (0, R')\,,$$

is a solution of the dual problem which thus admits an infinite number of solutions, but only one radially symmetric solution, obtained by taking $\beta = 0$ in (2.31). ■

Remark 2.4. Other exact solutions may be found in Mosolov-Miasnikov /1/, in particular for the case where Ω is a circular ring.

Remark 2.5. We can now complete Remark 2.1; in the two examples above, in which the data is very regular, we have, for sufficiently small g,

$$u_g \in C^1(\bar{\Omega}) \cap H^2(\Omega) \cap H_0^1(\Omega) \text{ but } u_g \notin C^2(\bar{\Omega}),\ u_g \notin H^3(\Omega). \quad \blacksquare$$

3. INTERIOR APPROXIMATION OF (P_0) BY A FINITE ELEMENT METHOD

In the following sections, unless stated otherwise, the solution of (P_0) will simply by denoted by u.

Synopsis.

In this section we shall define a method of approximating (P_0) using first-order finite elements, i.e. using continuous functions on $\bar{\Omega}$ which are piecewise linear and affine on a triangulation of Ω; the convergence of this approximation will be investigated in Section 5 and numerical methods for solving the approximate problem will be given in Sections 6 and 7. ■

3.1. Triangulation of Ω. Definition of V_h.

We proceed as in Chapter 3, Section 4.1.1, whose notation we retain; we can thus construct an *interior* approximation of $H_0^1(\Omega)$ — in the sense of Chapter 1, Section 4 — using a family V_h of subspaces of $H_0^1(\Omega)$. ■

3.2. Definition of the approximate problem

We define the approximate problem (P_{0h}) by:

$$(3.1) \qquad (P_{0h}) \min_{v_h \in V_h} J_g(v_h) . \quad \blacksquare$$

3.3. Solvability of the approximate problem

Since the functional J_g is continuous and strictly convex on V_h, with $\lim_{\|v_h\| \to +\infty} J_g(v_h) = +\infty$, it results from Chapter 1, Section 2.1, that (P_{0h}) admits one and only one solution, u_h say. The approximate solution u_h is obviously characterised by:

$$(3.2) \qquad \begin{cases} a(u_h, v_h - u_h) + gj(v_h) - gj(u_h) \geq \int_\Omega f(v_h - u_h)\,dx & \forall v_h \in V_h , \\ u_h \in V_h . \end{cases}$$

3.4. Explicit formulation of the approximate problem

The formulas of Chapter 3, Section 4.1.4 allow us to express $J_g(v_h)$ as a function of the nodal values $v_M = v_h(M),\ M \in \Sigma_h$; the approximate problem (3.1) then comes down to the minimisation of a nondifferentiable functional on $\mathbb{R}^N$ $(N = \dim V_h = \text{card}\,(\mathring{\Sigma}_h))$. ■

3.5. On the use of finite elements of order greater than 1

In the case of an approximation by first-order finite elements, for which the gradient of v_h is constant over each triangle of $\mathcal{T}_h$, there is no difficulty in expressing $j(v_h) = \int_\Omega |\text{grad}\, v_h|\,\mathrm{d}x$ as an explicit function of the nodal values v_M; on the other hand, such a calculation is impossible if we use an approximation by finite elements of order k, $k \geqslant 2$; it is then necessary to use an approximation of this integral; of all the possible choices, we cite the following in which $A(T)$= area of T, G_T= centre of gravity of T, in which the notation is as shown in Figure 3.1:

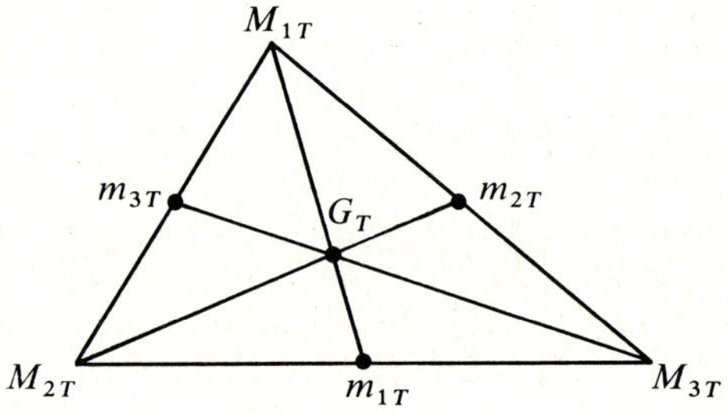

Fig. 3.1.

$$\sum_{T \in \mathcal{T}_h} A(T)\,|\text{grad}\, v_h(G_T)| \tag{3.3}$$

$$\frac{1}{3} \sum_{T \in \mathcal{T}_h} A(T) \sum_{i=1}^{3} |\text{grad}\, v_h\,(M_{iT})| \tag{3.4}$$

$$\frac{1}{3} \sum_{T \in \mathcal{T}_h} A(T) \sum_{i=1}^{3} |\text{grad}\, v_h(m_{iT})|\,. \tag{3.5}$$

In Section 7 we shall meet an example taken from Fortin /1/, in which the solution of (P_0) is approximated by a finite element method of order 2, the integral $\int_\Omega |\text{grad}\, v|\,\mathrm{d}x$ being evaluated using (3.3) and (3.5). ■

Incidentally, it may be noted that because of the relatively low regularity of the solution, the advantage of using finite elements of order greater than one, for an *equal number* of degree of freedom, is not clearly established. ■

4. EXTERIOR APPROXIMATIONS OF (P_0)

We retain the definitions and the formalism of Chapter 3, Sections 3.2.1 and 5.

4.1. Approximation of J_0

We approximate

$$J_0(v) = \frac{1}{2} a(v, v) - \int_\Omega fv \, dx \text{ by } J_{0h} : V_h \to \mathbb{R}$$

defined by:

$$J_{0h}(v_h) = \frac{1}{2} \sum_{k=1}^{2} \int_\Omega | \delta_k q_h v_h |^2 \, dx - \int_\Omega f q_h v_h \, dx , \tag{4.1}$$

the explicit form of (4.1) being given by:

$$\begin{aligned} J_{0h}(v_h) = \frac{h^2}{4} \sum_{M_{ij} \in \overline{\Omega}_h} \Bigg[\left(\frac{v_{i+1j} - v_{ij}}{h} \right)^2 + \left(\frac{v_{i-1j} - v_{ij}}{h} \right)^2 + \\ + \left(\frac{v_{ij+1} - v_{ij}}{h} \right)^2 + \left(\frac{v_{ij-1} - v_{ij}}{h} \right)^2 \Bigg] - h^2 \times \\ \times \sum_{M_{ij} \in \Omega_h} f_{ij} v_{ij} ; \quad v_{pq} = 0 \text{ if } M_{pq} \notin \Omega_h , \end{aligned} \tag{4.2}$$

with:

$$f_{ij} = \frac{1}{h^2} \int_{\varpi_h^0(M_{ij})} f(x) \, dx \quad \forall M_{ij} \in \Omega_h . \tag{4.3}$$

The sets Ω_h, $\overline{\Omega}_h$ were defined in Chapter 3, Section 3.2.1. ■

Remark 4.1. We could also use the exterior approximation of J_0 given in Chapter 3, Section 5.1.2, equation (5.4), but since this approximation has not been used in any numerical calculation we restrict our attention in the following to (4.2). ■

4.2. Exterior approximations of j

We shall give below four possible exterior approximations for j; the most natural approximation is clearly:

$$j_h^1(v_h) = \int_\Omega \sqrt{| \delta_1 q_h v_h |^2 + | \delta_2 q_h v_h |^2} \, dx , \tag{4.4}$$

which can be expressed in terms of the v_{ij} by:

$$
(4.5)\quad \begin{cases} j_h^1(v_h) = \dfrac{h^2}{4} \displaystyle\sum_{M_{ij} \in \bar{\Omega}_h} \Bigg[\left(\left| \dfrac{v_{i+1j} - v_{ij}}{h} \right|^2 + \left| \dfrac{v_{ij+1} - v_{ij}}{h} \right|^2 \right)^{1/2} + \\ \quad + \left(\left| \dfrac{v_{i-1j} - v_{ij}}{h} \right|^2 + \left| \dfrac{v_{ij+1} - v_{ij}}{h} \right|^2 \right)^{1/2} + \\ \quad + \left(\left| \dfrac{v_{i-1j} - v_{ij}}{h} \right|^2 + \left| \dfrac{v_{ij-1} - v_{ij}}{h} \right|^2 \right)^{1/2} + \\ \quad + \left(\left| \dfrac{v_{i+1j} - v_{ij}}{h} \right|^2 + \left| \dfrac{v_{ij-1} - v_{ij}}{h} \right|^2 \right)^{1/2} \Bigg], \\ v_{pq} = 0 \quad \text{if} \quad M_{pq} \notin \Omega_h . \end{cases}
$$

It results from (4.5) that $j_h^1(v_h)$ is approximately the sum of $4\,\mathrm{card}\,(\Omega_h)$ nondifferentiable terms; to reduce this quantity to about $\mathrm{card}\,(\Omega_h)$, we can use the following approximations:

$$
(4.6)\quad \begin{cases} j_h^2(v_h) = \\ = \displaystyle\int_{\mathbb{R}^2} \sqrt{\dfrac{1}{2} \left| \delta_1 q_h\, v_h \left(x_1, x_2 + \dfrac{h}{2} \right) \right|^2 + \dfrac{1}{2} \left| \delta_1 q_h\, v_h \left(x_1, x_2 - \dfrac{h}{2} \right) \right|^2 + \dfrac{1}{2} \left| \delta_2 q_h\, v_h \left(x_1 + \dfrac{h}{2}, x_1 \right) \right|^2 + \dfrac{1}{2} \left| \delta_2 q_h\, v_h \left(x_1 - \dfrac{h}{2}, x_2 \right) \right|^2 }\, \mathrm{d}x \end{cases}
$$

$$
(4.7)\quad \begin{cases} j_h^3(v_h) = \\ = \displaystyle\int_{\mathbb{R}^2} \sqrt{\dfrac{1}{2} \left| \delta_1 q_h\, v_h \left(x_1 + \dfrac{h}{2}, x_2 \right) \right|^2 + \dfrac{1}{2} \left| \delta_1 q_h\, v_h \left(x_1 - \dfrac{h}{2}, x_2 \right) \right|^2 + \dfrac{1}{2} \left| \delta_2 q_h\, v_h \left(x_1, x_2 + \dfrac{h}{2} \right) \right|^2 + \dfrac{1}{2} \left| \delta_2 q_h\, v_h \left(x_1, x_2 - \dfrac{h}{2} \right) \right|^2 }\, \mathrm{d}x \end{cases}
$$

$$
(4.8)\quad \begin{cases} j_h^4(v_h) = \displaystyle\int_{\mathbb{R}^2} \sqrt{ \left| \dfrac{\delta_1 q_h\, v_h \left(x_1, x_2 + \frac{h}{2} \right) + \delta_1 q_h\, v_h \left(x_1, x_2 - \frac{h}{2} \right)}{2} \right|^2 + \left| \dfrac{\delta_2 q_h\, v_h \left(x_1 + \frac{h}{2}, x_2 \right) + \delta_2 q_h\, v_h \left(x_1 - \frac{h}{2}, x_2 \right)}{2} \right|^2 }\, \mathrm{d}x . \end{cases}
$$

(4.6), (4.7) and (4.8) can readily be expressed in terms of the v_{ij}, giving:

$$
(4.9)\quad \begin{cases} j_h^2(v_h) = h^2 \displaystyle\sum_{M_{i+1/2\, j+1/2} \in \Sigma_h} \left(\dfrac{1}{2} \left| \dfrac{v_{i+1j+1} - v_{ij+1}}{h} \right|^2 + \dfrac{1}{2} \left| \dfrac{v_{i+1j} - v_{ij}}{h} \right|^2 + \right. \\ \quad \left. + \dfrac{1}{2} \left| \dfrac{v_{i+1j+1} - v_{i+1j}}{h} \right|^2 + \dfrac{1}{2} \left| \dfrac{v_{ij+1} - v_{ij}}{h} \right|^2 \right)^{1/2}, \\ v_{pq} = 0 \ \text{if} \ M_{pq} \notin \Omega_h , \end{cases}
$$

$$\text{(4.10)}\quad \begin{cases} j_h^3(v_h) = h^2 \sum_{M_{ij} \in \bar{\Omega}_h} \left(\frac{1}{2}\left|\frac{v_{i+1j} - v_{ij}}{h}\right|^2 + \frac{1}{2}\left|\frac{v_{i-1j} - v_{ij}}{h}\right|^2 + \right. \\ \left. + \frac{1}{2}\left|\frac{v_{ij+1} - v_{ij}}{h}\right|^2 + \frac{1}{2}\left|\frac{v_{ij-1} - v_{ij}}{h}\right|^2\right)^{1/2}, \; v_{pq} = 0 \text{ if } M_{pq} \notin \Omega_h, \end{cases}$$

$$\text{(4.11)}\quad \begin{cases} j_h^4(v_h) = h^2 \sum_{M_{i+1/2\,j+1/2} \in \Sigma_h} \left(\left|\frac{v_{i+1j+1} - v_{ij}}{h\sqrt{2}}\right|^2 + \left|\frac{v_{ij+1} - v_{i+1j}}{h\sqrt{2}}\right|^2\right)^{1/2}, \\ v_{pq} = 0 \text{ if } M_{pq} \notin \Omega_h . \end{cases}$$

Remark 4.2. Approximation (4.10) can be replaced by

$$\text{(4.10')}\quad \begin{cases} j_h^3(v_h) = h^2 \sum_{M_{ij} \in \bar{\Omega}_h} \alpha_{ij} \left(\frac{1}{2}\left|\frac{v_{i+1j} - v_{ij}}{h}\right|^2 + \frac{1}{2}\left|\frac{v_{i-1j} - v_{ij}}{h}\right|^2 + \right. \\ \left. + \frac{1}{2}\left|\frac{v_{ij+1} - v_{ij}}{h}\right|^2 + \frac{1}{2}\left|\frac{v_{ij-1} - v_{ij}}{h}\right|^2\right)^{1/2}, \\ \text{with:} \\ \alpha_{ij} = 1 \text{ if } M_{ij} \in \Omega_h,\ \alpha_{ij} = \sqrt{3}/2 \text{ if } M_{ij} \notin \Omega_h \text{ with two neighbours} \\ \text{in } \Omega_h,\ \alpha_{ij} = \sqrt{2}/2 \text{ if } M_{ij} \notin \Omega_h \text{ with a single neighbour in } \Omega_h . \end{cases}$$

In fact this corresponds to the situation portrayed in Chapter 3, Section 5.2 by Figures 5.3, 5.4. ■

4.3. Formulation of the approximate problem

For the approximate problem we take:

$$\text{(4.12)}\qquad (P_{0h})_l : \min_{v_h \in V_h} \left[J_{0h}(v_h) + g j_h^l(v_h)\right] ; \quad l = 1, 2, 3, 4 .$$

4.4. Solvability of the approximate problem

We can apply the results of Chapter 1, Section 2.1, by noting that the mapping

$$\text{(4.13)}\qquad v_h \to \left(\int_\Omega \left(|\delta_1 q_h v_h|^2 + |\delta_2 q_h v_h|^2\right) \mathrm{d}x\right)^{1/2}$$

defines a norm on V_h; we therefore have — in view of the continuity and convexity of j_h^l $(l = 1, 2, 3, 4)$ — the existence and uniqueness of a solution of $(P_{0h})_l$, $l = 1, 2, 3; 4$. ■

5. CONVERGENCE OF THE INTERIOR AND EXTERIOR APPROXIMATIONS

Synopsis.

In this section we shall prove the convergence of the approximation considered in Sections 3 and 4; we shall make use of the methods of Chapter 1, Section 4 for the interior approximations and Chapter 1, Section 5 for the exterior approximations.

The convergence proofs for the above approximations are based essentially on the density of $\mathscr{D}(\Omega)$ in $H_0^1(\Omega)$. ∎

5.1. Convergence of the finite element method of Section 3.

We shall investigate the convergence of the first-order finite element approximation of Section 3, using the results of Chapter 1, Sections 4.3 and 4.4; the notation is that of Section 3 and Chapter 3, Section 6.2.1.

We have:

Theorem 5.1. *We assume that if θ denotes an angle of T, $T \in \mathscr{T}_h$, there exists $\theta_0 > 0$ such that $\theta \geqslant \theta_0$ $\forall T \in \mathscr{T}_h$, $\forall h$. We also assume that $\Omega - \Omega_h \to 0$ (in the sense that, for all compact sets E of Ω we have : $E \subset \Omega_h$ for h sufficiently small). Let u_h (resp. u) be the solution of (P_{0h}) (resp. (P_0)). Then $u_h \to u$ strongly in $H_0^1(\Omega)$*

Proof.

1) We put $v_h = 0$ in relation (3.2) of Section 3.3, giving:

$$a(u_h, u_h) + gj(u_h) \leqslant \int_\Omega f u_h \, \mathrm{d}x \tag{5.1}$$

from which we deduce, since $j(u_h) \geqslant 0$

$$\| u_h \|_{H_0^1(\Omega)} \leqslant C_0 \| f \|_{L^2(\Omega)} \quad \forall h , \tag{5.2}$$

where, in (5.2), $C_0 = 1/\sqrt{\lambda_0}$, λ_0 being the smallest eigenvalue of $-\Delta$ in $H_0^1(\Omega)$.

2) It results from the proof of Theorem 6.1, Chapter 3, Section 6.2.1, and from the density of $\mathscr{D}(\Omega)$ in $H_0^1(\Omega)$, that $\forall v \in H_0^1(\Omega)$ we can construct a sequence $r_h v$ such that

$$\begin{cases} r_h v \in V_h \quad \forall h , \\ r_h v \to v \quad \text{strongly in } H_0^1(\Omega). \end{cases} \tag{5.3}$$

If u is the solution of (P_0), we then have (see (1.6), Section 1.1):

$$a(u, u_h - u) + gj(u_h) - gj(u) \geqslant \int_\Omega f(u_h - u) \, \mathrm{d}x \quad \forall h . \tag{5.4}$$

In the same way we have (see (3.2), Section 3.3):

$$a(u_h, r_h u - u_h) + gj(r_h u) - gj(u_h) \geqslant \int_\Omega f(r_h u - u_h) \, \mathrm{d}x , \tag{5.5}$$

so that, by addition

(5.6)
$$a(u_h - u, u_h - u) \leqslant a(u_h, r_h u - u) + g(j(r_h u) - j(u)) + \int_\Omega f(u - r_h u)\,dx .$$

In view of (5.2) it can readily be deduced from (5.6) that

(5.7)
$$\begin{cases} \| u_h - u \|^2_{H^1_0(\Omega)} \leqslant \{ C_0 \| f \|_{L^2(\Omega)} + g\sqrt{\text{meas.}(\Omega)} \} \times \\ \quad \times \| r_h u - u \|_{H^1_0(\Omega)} + \| f \|_{L^2(\Omega)} \| r_h u - u \|_{L^2(\Omega)} , \end{cases}$$

and hence $(u_h)_h$ is strongly convergent in $H^1_0(\Omega)$ since, from (5.3), $r_h u \to u$ strongly in $H^1_0(\Omega)$. ■

Remark 5.1. A fairly general definition of the convergence of Ω_h to Ω was given in Chapter 3, Section 6.2.1; under more restrictive conditions - which are almost always satisfied in practical applications - and for which we refer for example to Ciarlet-Raviart /1/, we can construct $r_h : H^1_0(\Omega) \to V_h$ such that (1)

(5.8)
$$\begin{cases} \forall v \in H^1_0(\Omega) \cap H^2(\Omega) \quad \text{we have} \\ \| r_h v - v \|_{H^1_0(\Omega)} \leqslant C_1(v)\, h^{1/2} , \\ \| r_h v - v \|_{L^2(\Omega)} \leqslant C_2(v)\, h . \end{cases}$$

In particular (5.8) is satisfied if we define r_h by:

(5.9)
$$\begin{cases} r_h v = \text{projection of } v \text{ on } V_h \\ \text{for the norm } v \to \sqrt{a(v, v)} \end{cases}$$

for which case it is clear that

(5.10)
$$a(v_h, r_h v - v) = 0 \quad \forall v \in H^1_0(\Omega) , \quad \forall v_h \in V_h .$$

In view of (5.6), (5.8) and (5.10) we then have

(5.11)
$$\| u_h - u \|^2_{H^1_0(\Omega)} \leqslant \alpha g h^{1/2} + \beta h , \quad \alpha \text{ and } \beta \text{ independent of } g \text{ and } h. \quad ■$$

5.2. Convergence of the exterior approximations

In this section we shall investigate the convergence of the exterior approximations defined in Section 4, using the results of Chapter 1, Section 5; the notation is that of Section 4.

5.2.1. *Reduction of* $(P_{0h})_l$ *to a variational inequality*

The approximate problem $(P_{0h})_l$, $l = 1, 2, 3, 4$ is equivalent to the variational inequality:

---

(1) We recall that $h = \max_{T \in \mathcal{T}_h} \text{Area}(T)$.

$$(5.12)\qquad \begin{cases} a_h(u_h, v_h - u_h) + gj_h^l(v_h) - gj_h^l(u_h) \geqslant \int_\Omega fq_h(v_h - u_h)\,\mathrm{d}x \\ \forall v_h \in V_h\,, \quad u_h \in V_h\,; \quad l = 1, 2, 3, 4 \end{cases}$$

with

$$(5.13)\qquad a_h(u_h, v_h) = \sum_{k=1}^{2} \int_\Omega \delta_k q_h\, u_h\, \delta_k q_h\, v_h\, \mathrm{d}x\,. \quad \blacksquare$$

With the function f we associate $L : H_0^1(\Omega) \to \mathbb{R}$ defined by:

$$(5.14)\qquad L(v) = \int_\Omega fv\,\mathrm{d}x\,.$$

5.2.2. *Exterior approximations for* $H_0^1(\Omega)$, a, j, L

1) *Exterior approximation for* $H_0^1(\Omega)$, a, L

Since the bilinear form a_h is identical to that denoted by a_h^1 in Chapter 3, Section 6.3.1, we refer to Chapter 3, Section 6.3.2 where it was proved that

$$V_h,\ a_h,\ v_h \to \int_\Omega fq_h\, v_h\,\mathrm{d}x$$

constitute exterior approximations of $H_0^1(\Omega)$, a, L, respectively, in the sense of Chapter 1, Section 5. $\blacksquare$

2) *Exterior approximations of* j

We have:

Proposition 5.1. *The functionals* j_h^l, $l = 1, 2, 3, 4$, *constitute exterior approximations of* j.

Proof. We have to show (see Chapter 1, Section 5.2) that

$$(5.15)\qquad \begin{cases} \{ q_h v_h, \delta_1 q_h v_h, \delta_2 q_h v_h \} \to \left\{ v, \dfrac{\partial v}{\partial x_1}, \dfrac{\partial v}{\partial x_2} \right\} \\ \text{weakly in } (L^2(\Omega))^3 \quad \Rightarrow \liminf j_h^l(v_h) \geqslant j(v)\,;\ l = 1, 2, 3, 4\,. \end{cases}$$

$$(5.16)\qquad \begin{cases} \{ q_h v_h, \delta_1 q_h v_h, \delta_2 q_h v_h \} \to \left\{ v, \dfrac{\partial v}{\partial x_1}, \dfrac{\partial v}{\partial x_2} \right\} \\ \text{strongly in } (L^2(\Omega))^3 \quad \Rightarrow \lim j_h^l(v_h) = j(v)\,;\ l = 1, 2, 3, 4\,. \end{cases}$$

Verification of (5.15), (5.16) is immediate for j_h^1; moreover, by using the notation of Chapter 3, Section 6.3.2 we have:

$$(5.17)\qquad j_h^2(v_h) = \int_{\mathbb{R}^2} \sqrt{\frac{1}{2}\,|\,\tau_{2(h/2)}\,\delta_1 q_h\, v_h\,|^2 + \frac{1}{2}\,|\,\tau_{2(-h/2)}\,\delta_1 q_h\, v_h\,|^2 + \frac{1}{2}\,|\,\tau_{1(h/2)}\,\delta_2 q_h\, v_h\,|^2 + \frac{1}{2}\,|\,\tau_{1(-h/2)}\,\delta_2 q_h\, v_h\,|^2}\,\mathrm{d}x\,,$$

$$(5.18)\qquad j_h^3(v_h) = \int_{\mathbb{R}^2} \sqrt{\frac{1}{2}\,|\,\tau_{1(h/2)}\,\delta_1 q_h\, v_h\,|^2 + \frac{1}{2}\,|\,\tau_{1(-h/2)}\,\delta_1 q_h\, v_h\,|^2 + \frac{1}{2}\,|\,\tau_{2(h/2)}\,\delta_2 q_h\, v_h\,|^2 + \frac{1}{2}\,|\,\tau_{2(-h/2)}\,\delta_2 q_h\, v_h\,|^2}\,\mathrm{d}x\,,$$

$$(5.19)\qquad j_h^4(v_h) = \int_{\mathbb{R}^2} \sqrt{\left|\frac{\tau_{2(h/2)}\,\delta_1 q_h\, v_h + \tau_{2(-h/2)}\,\delta_1 q_h\, v_h}{2}\right|^2 + \left|\frac{\tau_{1(h/2)}\,\delta_2 q_h\, v_h + \tau_{1(-h/2)}\,\delta_2 q_h\, v_h}{2}\right|^2}\,\mathrm{d}x\,.$$

It is then almost obvious that properties (5.15), (5.16) result from the *translation* properties described in Chapter 3, Section 6.3.2 in the proof of Proposition 6.1 (see (6.72), (6.73)). ■

5.2.3. *A strong convergence result*

In view of Sections 5.2.1 and 5.2.2 we can apply Theorem 5.2 of Chapter 1, Section 5.4 to the problems $(P_{0h})_l$, $l = 1, 2, 3, 4$, and more precisely to the equivalent formulations (5.12). Hence:

Theorem 5.2. *Let* u_h^l *be the solution of* $(P_{0h})_l$, u *the solution of* (P_0); *when* $h \to 0$ *we have*:

$$(5.20)\qquad \begin{cases} \{\, q_h\, u_h^l,\ \delta_1 q_h\, u_h^l,\ \delta_2 q_h\, u_h^l \,\} \to \left\{ u, \dfrac{\partial u}{\partial x_1}, \dfrac{\partial u}{\partial x_1} \right\} & \textit{strongly in } (L^2(\Omega))^3 \\ J_{0h}(u_h^l) \to J_0(u)\,, \\ j_h^l(u_h^l) \to j(u) \\ \forall l = 1, 2, 3, 4\,. \quad ■ \end{cases}$$

6. METHODS OF SOLUTION BY REGULARISING j

Synopsis.

In this section we shall make use of the *regularisation* techniques introduced in Chapter 1, Section 3.3. In Section 6.1 we shall consider the regularisation of the continuous problem (P_0) and investigate the approximation error thus created as a function of the regularisation parameter ε. In Sections 6.2 and 6.3 respectively, we shall investigate the regularisation of the approximate problems defined in Sections 3 and 4, and the numerical solution methods for the regularised approximate problems; these techniques will be applied to several examples in Section 6.4. ■

6.1. Regularisation of the continuous problem (P_0)

6.1.1 *Formulation of the regularised problem*

There are various ways of regularising the nondifferentiable term $\int_\Omega |\operatorname{grad} v| \, dx$, but we shall restrict our attention to the two approximations defined below.

We consider the two numerical functions $\phi_{1\varepsilon}$ and $\phi_{2\varepsilon}$, convex, $\geqslant 0$, and of class C^1 on $\mathbb{R}$, defined by:

$$\phi_{1\varepsilon}(\tau) = \sqrt{\tau^2 + \varepsilon^2} - \varepsilon, \tag{6.1}$$

$$\phi_{2\varepsilon}(\tau) = \begin{cases} \dfrac{2\varepsilon}{\pi}\left(1 - \cos\dfrac{\pi\tau}{2\varepsilon}\right) & \text{if } |\tau| \leqslant \varepsilon, \\ |\tau| - \varepsilon\left(1 - \dfrac{2}{\pi}\right) & \text{if } |\tau| > \varepsilon. \end{cases} \tag{6.2}$$

With these we associate (P_ε^l), approximations which are obtained by regularising (P_0) and defined by:

$$(P_\varepsilon^l) \quad \min_{v \in H_0^1(\Omega)} \left[J_0(v) + g \int_\Omega \phi_{l\varepsilon}(|\operatorname{grad} v|) \, dx \right]; \quad l = 1, 2. \tag{6.3}$$ ■

6.1.2 *Solvability of the regularised problems. Convergence and estimation of the regularisation error.*

We write $j_{l\varepsilon}(v) = \int_\Omega \phi_{l\varepsilon}(|\operatorname{grad} v|) \, dx$, $l = 1, 2$; we then have:

<u>Proposition 6.1</u>. *The approximate problems* (P_ε^l), $l = 1, 2$, *admit one and only one solution,* u_ε^l, *characterised by:*

$$\begin{cases} -\Delta u_\varepsilon^l + g j_{l\varepsilon}'(u_\varepsilon^l) = f & \text{on } \Omega, \\ u_\varepsilon^l = 0 & \text{on } \Gamma, \end{cases} \tag{6.4}$$

(*with* $j_{l\varepsilon}'$ = *gradient of* $j_{l\varepsilon}$) *and by*:

$$\begin{cases} a(u_\varepsilon^l, v - u_\varepsilon^l) + g j_{l\varepsilon}(v) - g j_{l\varepsilon}(u_\varepsilon^l) \geqslant \displaystyle\int_\Omega f(v - u_\varepsilon^l) \, dx \quad \forall v \in H_0^1(\Omega), \\ u_\varepsilon^l \in H_0^1(\Omega). \end{cases} \tag{6.5}$$

Proof. Since the functionals $v \to J_0(v) + g j_{l\varepsilon}(v)$ are strictly convex, continuous and differentiable with

$$\lim_{\|v\| \to +\infty} [J_0(v) + g j_{l\varepsilon}(v)] = +\infty,$$

this results from Chapter 1, Sections 2 and 3.3, and from the general theory of monotone operators (see, e.g. Lions /1/). ■

The regularisation process defined above is justified by:

Theorem 6.1. *If* u_ε^l *is the solution of* (P_ε^l) *and* u *that of* (P_0), *we have, for* $l = 1, 2$:

$$\| u_\varepsilon^l - u \|_{H_0^1(\Omega)} \leqslant C_l \sqrt{\varepsilon}\,;\ C_l \text{ independent of } \varepsilon. \tag{6.6}$$

Proof. We have (see (1.6) and (6.5)):

$$a(u, u_\varepsilon^l - u) + gj(u_\varepsilon^l) - gj(u) \geqslant \int_\Omega f(u_\varepsilon^l - u)\,\mathrm{d}x \tag{6.7}$$

$$a(u_\varepsilon^l, u - u_\varepsilon^l) + gj_{l\varepsilon}(u) - gj_{l\varepsilon}(u_\varepsilon^l) \geqslant \int_\Omega f(u - u_\varepsilon^l)\,\mathrm{d}x\,, \tag{6.8}$$

so that, by addition,

$$\begin{aligned} \| u_\varepsilon^l - u \|^2_{H_0^1(\Omega)} &= a(u_\varepsilon^l - u, u_\varepsilon^l - u) \leqslant \\ &\leqslant g[(j_{l\varepsilon}(u) - j(u)) + (j(u_\varepsilon^l) - j_{l\varepsilon}(u_\varepsilon^l))]\,. \end{aligned} \tag{6.9}$$

It can readily be verified that:

$$\begin{cases} \big|\, |\tau| - \phi_{1\varepsilon}(\tau) \,\big| \leqslant \varepsilon\,, \\ \big|\, |\tau| - \phi_{2\varepsilon}(\tau) \,\big| \leqslant \left(1 - \dfrac{2}{\pi}\right)\varepsilon\,, \end{cases} \tag{6.10}$$

whence, by definition of $j_{\varepsilon l}$,

$$\| u_\varepsilon^l - u \|^2_{H_0^1(\Omega)} \leqslant 2\,g\gamma_l \text{ measure}(\Omega)\,\varepsilon\,, \tag{6.11}$$

with $\gamma_1 = 1$, $\gamma_2 = (1 - 2/\pi)$, which proves (6.6) with

$$\begin{cases} C_1 = \sqrt{2\,g \text{ measure}(\Omega)}\,, \\ C_2 = \sqrt{2\,g(1 - 2/\pi) \text{ measure}(\Omega)}\,. \end{cases} \tag{6.12}$$

■

Remark 6.1. The above proof holds if the continuous linear form $v \to \int_\Omega fv\,\mathrm{d}x$ is replaced by $L \in H^{-1}(\Omega)$. ■

6.1.3. *Estimates of the regularisation error in the one-dimensional case.*

It is reasonable to suppose that the estimate (6.6) of the regularisation error is not optimal. In this section we shall therefore give an estimate which is more precise than (6.6) for the particular one-dimensional case considered earlier in Section 2.3.1; we shall restrict our attention to regularisation by the function $\phi_{1\varepsilon}$.

The problem (P_0) is defined by:

$$(6.13)\qquad (P_0)\quad \min_{v \in H_0^1(0,1)} \left[\frac{1}{2}\int_0^1 |v'|^2\,\mathrm{d}x + g\int_0^1 |v'|\,\mathrm{d}x - c\int_0^1 v\,\mathrm{d}x\right],$$

and the regularised problem (P_ε) by

$$(6.14)\qquad (P_\varepsilon)\quad \min_{v \in H_0^1(0,1)} \left[\frac{1}{2}\int_0^1 |v'|^2\,\mathrm{d}x + g\int_0^1 \sqrt{\varepsilon^2 + v'^2}\,\mathrm{d}x - c\int_0^1 v\,\mathrm{d}x\right].$$

Problems (P_0) and (P_ε) admit unique solutions, denoted in the following by u_0 and u_ε respectively. We shall now prove:

<u>Theorem 6.2.</u> *If* $c \geqslant 2g > 0$ *we have*:

$$(6.15)\qquad \| u_0 - u_\varepsilon \|_{L^\infty(0,1)} \leqslant C_1\,\varepsilon,\ C_1 \textit{ independent of } \varepsilon,$$

$$(6.16)\qquad \| u_0 - u_\varepsilon \|_{H_0^1(0,1)} \leqslant C_2\,\varepsilon\sqrt{-\operatorname{Log}\varepsilon},\ C_2 \textit{ independent of } \varepsilon.$$

Proof. This is an exercise in calculus; by virtue of symmetry we may restrict our attention to the interval $0 \leqslant x \leqslant \frac{1}{2}$. We put $p_0 = u_0'$ and $p_\varepsilon = u_\varepsilon'$; on the one hand we have (see Section 2.3.1).

$$(6.17)\qquad p_0(x) = \begin{cases} c(\frac{1}{2} - x) - g & \text{if } 0 \leqslant x \leqslant \frac{1}{2} - g/c,\\ 0 & \text{if } \frac{1}{2} - g/c \leqslant x \leqslant \frac{1}{2},\end{cases}$$

and on the other hand

$$(6.18)\qquad p_\varepsilon(x)\left(1 + \frac{g}{\sqrt{\varepsilon^2 + p_\varepsilon^2(x)}}\right) = c\left(\frac{1}{2} - x\right) \quad \forall x \in [0, 1].$$

p_0 and p_ε are shown graphically in Figure 6.1.

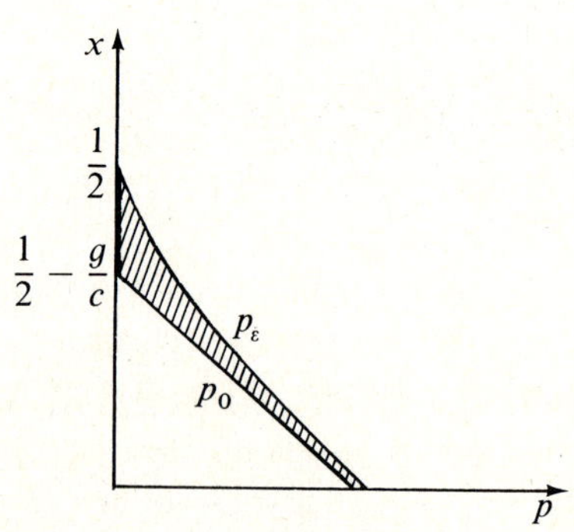

Fig. 6.1.

Proof of (6.15). Twice the area of the hatched region in Figure 6.1 is equal to $\| u_\varepsilon' - u_0' \|_{L^1(0,1)}$. Hence, in view of (6.17), (6.18):

$$\| u_\varepsilon' - u_0' \|_{L^1(0,1)} \leqslant \frac{2g}{c} \int_0^{+\infty} \left(1 - \frac{p}{\sqrt{\varepsilon^2 + p^2}} \right) \mathrm{d}p = \frac{2g}{c} \varepsilon . \tag{6.19}$$

From (6.19) and from $u_\varepsilon(x) - u_0(x) = \int_0^x (u_\varepsilon'(\xi) - u_0'(\xi)) \, \mathrm{d}\xi$ we deduce

$$\| u_\varepsilon - u_0 \|_{L^\infty(0,1)} \leqslant \frac{g}{c} \varepsilon , \tag{6.20}$$

which proves (6.15) and renders it precise. ■

Proof of (6.16). Since the solution u_ε of (6.14) is a solution of:

$$\begin{cases} - u_\varepsilon'' - g \dfrac{\mathrm{d}}{\mathrm{d}x} \left(\dfrac{u_\varepsilon'}{\sqrt{\varepsilon^2 + u_\varepsilon'^2}} \right) = c , \\ u_\varepsilon(0) = u_\varepsilon(1) = 0 , \end{cases} \tag{6.21}$$

it is thus C^∞; we deduce from equation (6.18) that

$$\| p_\varepsilon \|_{L^\infty(0,1)} \leqslant c/2 \quad \forall \varepsilon , \tag{6.22}$$

and

$$p_\varepsilon' \left(1 + \frac{g\varepsilon^2}{(\varepsilon^2 + p_\varepsilon^2)^{3/2}} \right) = - c , \tag{6.23}$$

so that

$$\| p_\varepsilon' \|_{L^\infty(0,1)} \leqslant c \quad \forall \varepsilon . \tag{6.24}$$

The family $(p_\varepsilon)_\varepsilon$ is thus *equi-Lipschitzian*; in view of (6.22) we can apply Ascoli's theorem to $(p_\varepsilon)_\varepsilon$. Hence:

$$p_\varepsilon \to p_0 \text{ uniformly,} \tag{6.25}$$

and in particular

$$p_\varepsilon(\tfrac{1}{2} - g/c) \to 0 . \tag{6.26}$$

In the following we shall write $p_c = p_\varepsilon(\frac{1}{2} - g/c)$; we thus have

$$p_c + g \frac{p_c}{\sqrt{\varepsilon^2 + p_c^2}} = g , \tag{6.27}$$

and hence

$$(6.28)\qquad p_c = \frac{g\varepsilon^2}{(p_c + \sqrt{\varepsilon^2 + p_c^2})\sqrt{\varepsilon^2 + p_c^2}}\,.$$

When $\varepsilon \to 0$, $p_c \to 0$ and $p_c/\sqrt{p_c^2 + \varepsilon^2} \to 1$; we thus have $\varepsilon/p_c \to 0$ and hence

$$(6.29)\qquad p_c \simeq \frac{g\varepsilon^2}{2\,p_c^2} \Rightarrow p_c \simeq \sqrt[3]{g/2}\,\varepsilon^{2/3}\,. \quad \blacksquare$$

We have

$$\| u_\varepsilon - u_0 \|_{H_0^1(0,1)} = \left[2\int_0^{1/2-g/c} | p_\varepsilon(x) - p_0(x) |^2\, dx + \right.$$
$$\left. + 2\int_{1/2-g/c}^{1/2} | p_\varepsilon(x) - p_0(x) |^2\, dx\right]^{1/2};$$

we put

$$I_1 = \int_{1/2-g/c}^{1/2} | p_\varepsilon(x) - p_0(x) |^2\, dx \quad \text{and} \quad I_2 = \int_0^{1/2-g/c} | p_\varepsilon(x) - p_0(x) |^2\, dx\,.$$

Estimation of I_1.

On $[\frac{1}{2} - g/c, \frac{1}{2}]$ we have $p_0(x) = 0$ and hence

$$(6.30)\qquad I_1 = \int_{1/2-g/c}^{1/2} p_\varepsilon^2(x)\, dx\,.$$

We put

$$(6.31)\qquad p_\varepsilon = \varepsilon \operatorname{sh} \phi\,,$$

which gives

$$(6.32)\qquad \varepsilon \operatorname{sh} \phi + g \operatorname{th} \phi = c(\tfrac{1}{2} - x)\,,$$

and

$$(6.33)\qquad I_1 = \frac{\varepsilon^2}{c}\int_0^{\phi_c} \operatorname{sh}^2 \phi \left(\varepsilon \operatorname{ch} \phi + \frac{g}{\operatorname{ch}^2 \phi}\right) d\phi,\ \phi_c = \operatorname{Arg\,sh} \frac{p_c}{\varepsilon}\,.$$

An elementary calculation shows that

$$(6.34)\qquad I_1 = \frac{\varepsilon^2}{c}\left(\frac{\varepsilon}{3} \operatorname{sh}^3 \phi_c + g\phi_c - g \operatorname{th} \phi_c\right).$$

We deduce from (6.29), (6.33), (6.34) that

$$(6.35) \qquad I_1 \simeq -(g/3c)(\varepsilon^2 \operatorname{Log} \varepsilon) . \quad \blacksquare$$

Estimation of I_2.

On $[0, \frac{1}{2} - g/c]$ we have $p(x) + g \operatorname{th} \phi = c(\frac{1}{2} - x)$ and $p_0(x) + g = c(\frac{1}{2} - x)$, and hence

$$(6.36) \qquad p(x) - p_0(x) = g(1 - \operatorname{th} \phi) = g(e^{-\phi}/\operatorname{ch} \phi) ,$$

and, in view of (6.32)

$$(6.37) \qquad I_2 \leqslant \frac{g^2}{c} \int_{\phi_c}^{+\infty} \frac{e^{-2\phi}}{\operatorname{ch}^2 \phi} \left(\varepsilon \operatorname{ch} \phi + \frac{g}{\operatorname{ch}^2 \phi} \right) d\phi .$$

Putting $t = e^{\phi}$, and performing the necessary manipulations we obtain:

$$(6.38) \qquad \int_{\phi_c}^{+\infty} \frac{e^{-2\phi}}{\operatorname{ch}^2 \phi} \left(\varepsilon \operatorname{ch} \phi + \frac{g}{\operatorname{ch}^2 \phi} \right) d\phi =$$

$$= 2\varepsilon(e^{-\phi_c} - \operatorname{Arctg} e^{-\phi_c}) + \frac{8g}{3} \frac{1}{(e^{2\phi_c} + 1)^3} ,$$

and using the fact that $e^{\phi_c} \simeq \sqrt[3]{4g}\, \varepsilon^{-1/3}$:

$$(6.39) \qquad \int_{\phi_c}^{+\infty} \frac{e^{-2\phi}}{\operatorname{ch}^2 \phi} \left(\varepsilon \operatorname{ch} \phi + \frac{g}{\operatorname{ch}^2 \phi} \right) d\phi \simeq \frac{\varepsilon^2}{3g} ,$$

which gives

$$(6.40) \qquad I_2 \leqslant \frac{g}{3c} \varepsilon^2 [1 + \alpha(\varepsilon)] \quad \text{with} \quad \lim_{\varepsilon \to 0} \alpha(\varepsilon) = 0 . \quad \blacksquare$$

From (6.35) and (6.40) we immediately deduce that

$$(6.41) \qquad \| u_\varepsilon - u_0 \|_{H_0^1(0,1)} \simeq \sqrt{2g/3c}\, \varepsilon \sqrt{-\operatorname{Log} \varepsilon} ,$$

which proves Theorem 6.2. ■

Remark 6.2. Consider the problem

$$(P_0) \quad \min_{v \in H_0^1(\Omega)} \left[\frac{1}{2} a(v, v) + g j(v) - c \int_\Omega v \, dx \right]$$

regularised by

$$(P_\varepsilon) \quad \min_{v \in H_0^1(\Omega)} \left[\frac{1}{2} a(v, v) + g \int_\Omega \sqrt{\varepsilon^2 + |\operatorname{grad} v|^2} \, dx - c \int_\Omega v \, dx \right] .$$

If Ω is a disc of radius R, it can be proved by proceeding as in the proof of Theorem 6.2 that

(6.42) $\| u_\varepsilon - u_0 \|_{L^\infty(\Omega)} \leqslant C_1\varepsilon$, C_1 independent of ε,

(6.43) $\| u_\varepsilon - u_0 \|_{H_0^1(\Omega)} \leqslant C_2\,\varepsilon\sqrt{-\operatorname{Log}\varepsilon}$, C_2 independent of ε,

where u_0 (resp. u_ε) is the solution of (P_0) (resp. (P_ε)). ■

Remark 6.3. Since the domain Ω is arbitrary in $\mathbb{R}^2$ we consider, with $f \in H^{-1}(\Omega)$:

$$\min_{v \in H_0^1(\Omega)} [\tfrac{1}{2} a(v, v) + gj(v) - \langle f, v \rangle], \tag{6.44}$$

$$\min_{v \in H_0^1(\Omega)} \left[\frac{1}{2} a(v, v) + g \int_\Omega \sqrt{\varepsilon^2 + |\operatorname{grad} v|^2}\, dx - \langle f, v \rangle\right]. \tag{6.45}$$

If u_0 and u_ε are the respective solutions of (6.44) and (6.45), we conjecture on the basis of particular explicit cases that:

(6.46) $\| u_\varepsilon - u_0 \|_{H_0^1(\Omega)} \leqslant C\varepsilon^{2/3}$, C independent of ε. ■

6.1.4. *Application of the regularisation process to obtaining duality results.*

We shall use the above regularisation process to give an elementary demonstration of the ("duality" type) characterisation (3.48) - (3.51) of Chapter 1, Section 3.4; we assume in the following that $f \in H^{-1}(\Omega)$.

We formulate (P_0) in terms of the equivalent variational inequality

$$\begin{cases} a(u, v - u) + gj(v) - gj(u) \geqslant \langle f, v - u \rangle & \forall v \in H_0^1(\Omega), \\ u \in H_0^1(\Omega), \end{cases} \tag{6.47}$$

and the corresponding *regularised* problem by

$$\begin{cases} a(u_\varepsilon, v - u_\varepsilon) + g\left(\displaystyle\int_\Omega \sqrt{\varepsilon^2 + |\operatorname{grad} v|^2}\, dx - \right. \\ \qquad \left. - \displaystyle\int_\Omega \sqrt{\varepsilon^2 + |\operatorname{grad} u_\varepsilon|^2}\, dx\right) \geqslant \langle f, v - u_\varepsilon \rangle \quad \forall v \in H_0^1(\Omega), \\ u_\varepsilon \in H_0^1(\Omega). \end{cases} \tag{6.48}$$

By putting $v = 0$ and $v = 2u$ in (6.47) it can readily be shown that

$$a(u, u) + gj(u) = \langle f, u \rangle ; \tag{6.49}$$

moreover the solution u_ε of (6.48) is characterised by

$$\begin{cases} a(u_\varepsilon, v) + g \displaystyle\int_\Omega \frac{\operatorname{grad} u_\varepsilon . \operatorname{grad} v}{\sqrt{\varepsilon^2 + |\operatorname{grad} u_\varepsilon|^2}}\, dx = \langle f, v \rangle \quad \forall v \in H_0^1(\Omega), \\ u_\varepsilon \in H_0^1(\Omega). \end{cases} \tag{6.50}$$

Let $\Lambda = \{ \mu \mid \mu = (\mu_1, \mu_2) \in L^2(\Omega) \times L^2(\Omega),\ \mu_1^2(x) + \mu_2^2(x) \leqslant 1$ a.e.$\}$; we define $\lambda_\varepsilon \in \Lambda$ by

$$\lambda_\varepsilon = \frac{\text{grad } u_\varepsilon}{\sqrt{\varepsilon^2 + |\text{ grad } u_\varepsilon|^2}}. \tag{6.51}$$

There then exists $\lambda \in \Lambda$ and an extracted subsequence, again denoted by $(\lambda_\varepsilon)_\varepsilon$, such that

$$\lambda_\varepsilon \to \lambda \text{ weakly* in } L^\infty(\Omega) \times L^\infty(\Omega). \tag{6.52}$$

From (6.48) we have

$$a(u_\varepsilon, v) + g \int_\Omega \lambda_\varepsilon.\text{grad } v \, dx = \langle f, v \rangle \quad \forall v \in H_0^1(\Omega), \tag{6.53}$$

$$a(u_\varepsilon, u_\varepsilon) + g \int_\Omega \lambda_\varepsilon.\text{grad } u_\varepsilon \, dx = \langle f, u_\varepsilon \rangle, \tag{6.54}$$

and because of the *strong* convergence of u_ε to u (see Theorem 6.1) we have, in the limit in (6.53), (6.54):

$$a(u, v) + g \int_\Omega \lambda.\text{grad } v \, dx = \langle f, v \rangle, \tag{6.55}$$

$$a(u, u) + g \int_\Omega \lambda.\text{grad } u \, dx = \langle f, u \rangle. \tag{6.56}$$

From (6.49), (6.56), we deduce that:

$$\int_\Omega \lambda.\text{grad } u \, dx = \int_\Omega |\text{ grad } u \,| \, dx. \tag{6.57}$$

We also have the equivalence

$$\left\{ \begin{array}{l} \displaystyle\int_\Omega \lambda.\text{grad } u \, dx = \int_\Omega |\text{grad } u| \, dx \\ \lambda \in \Lambda \end{array} \right\} \Leftrightarrow \left\{ \begin{array}{l} \lambda.\text{grad } u = |\text{grad } u| \\ \lambda \in \Lambda \end{array} \right. \tag{6.58}$$

which, with (6.55), gives

$$\left\{ \begin{array}{l} -\Delta u - g \text{ div } \lambda = f, \\ u|_\Gamma = 0, \end{array} \right. \tag{6.59}$$

$$\left\{ \begin{array}{l} \lambda.\text{grad } u = |\text{ grad } u \,|, \\ \lambda \in \Lambda, \end{array} \right. \tag{6.60}$$

i.e. the characterisation required. ∎

6.2. Regularisation of the approximate problems

6.2.1. *Case of the interior approximation of Section* 3

The notation is that of Sections 3 and 6.1.1, 6.1.2; the

regularised problem corresponding to the approximate problem (3.1) of Section 3.2 is defined by

$$(6.61)\quad \begin{cases} a(u^l_{\varepsilon h}, v_h - u^l_{\varepsilon h}) + g j_{l\varepsilon}(v_h) - g j_{l\varepsilon}(u^l_{\varepsilon h}) \geqslant \int_\Omega f(v_h - u^l_{\varepsilon h})\,\mathrm{d}x \quad \forall v_h \in V_h\,, \\ u^l_{\varepsilon h} \in V_h\,, \end{cases}$$

with $l = 1, 2$. It is simple to prove:

Theorem 6.3. *Problem (6.61) admits one and only one solution, and when* $\varepsilon \to 0$ *we have,* $\forall l = 1, 2$

$$(6.62)\quad u^l_{\varepsilon h} \to u_h\,,$$

where u_h *is the solution of the approximate problem* (P_{0h}) (see (3.1), Section 3.2). ■

Remark 6.4. The following convergence results can be readily proved:

$$(6.63)\quad \begin{cases} \lim\limits_{h\to 0} u^l_{\varepsilon h} = u^l_\varepsilon \ \textit{strongly in}\ H^1_0(\Omega)\,, \\ \text{where } u^l_\varepsilon \text{ is the solution of } (P^l_\varepsilon) \text{ (see (6.3) Section 6.1.1)}, \end{cases}$$

$$(6.64)\quad \lim_{\varepsilon, h\to 0} u^l_{\varepsilon h} = u \ \textit{strongly in}\ H^1_0(\Omega)\,,\ u \text{ being the solution of } (P_0).$$

In fact, under the conditions for which the estimate (5.11) was obtained (see Section 5.1 , Remark 5.1) it can be shown that if u is the solution of (P_0),

$$(6.65)\quad \begin{cases} \| u - u^l_{\varepsilon h} \|^2_{H^1_0(\Omega)} \leqslant \alpha h + g(\beta h^{1/2} + \gamma_l\,\varepsilon)\,, \\ \alpha, \beta, \gamma_l \text{ independent of } h \text{ and } \varepsilon;\ \alpha, \beta \text{ independent of } l. \end{cases}$$ ■

From the point of view of the *numerical solution* of (6.61), we shall in fact use the equivalent formulation, given in *variational* form by

$$(6.66)\quad \begin{cases} a(u^l_{\varepsilon h}, v_h) + g(j'_{l\varepsilon}(u^l_{\varepsilon h}), v_h) = \int_\Omega f v_h\,\mathrm{d}x \quad \forall v_h \in V_h\,, \\ u^l_{\varepsilon h} \in V_h\,, \end{cases}$$

with, for $l = 1$,

$$(6.67)\quad (j'_{l\varepsilon}(u^1_{\varepsilon h}), v_h) = \int_\Omega \frac{\operatorname{grad} u^1_{\varepsilon h} \cdot \operatorname{grad} v_h}{\sqrt{\varepsilon^2 + |\operatorname{grad} u^1_{\varepsilon h}|^2}}\,\mathrm{d}x\,,$$

the case $l = 2$ being a little more complicated to write explicitly.

There is in principle no difficulty in expressing (6.66), using the formulas of Chapter 3, Section 4.1.4 in the form of a system of N[1] nonlinear equations whose unknowns are the values

[1] $N = \dim V_h = \operatorname{card}(\mathring{\Sigma}_h)$.

of $u^l_{\varepsilon h}$ at the nodes $M \in \mathring{\Sigma}_h$ of the triangulation $\mathscr{T}_h$. ■

6.2.2. *Case of the exterior approximation of Section 4*

The notation is that of Sections 4 and 6.1.1, 6.1.2; the approximations are defined by regularising the functionals j^k_h, $k = 1, 2, 3, 4$, of Section 4.2 using, respectively

(6.68)
$$\begin{cases} j^{1l}_{h\varepsilon}(v_h) = \dfrac{h^2}{4} \displaystyle\sum_{M_{ij} \in \bar{\Omega}_h} \left[\phi_{l\varepsilon}\left(\sqrt{\left|\dfrac{v_{i+1j} - v_{ij}}{h}\right|^2 + \left|\dfrac{v_{ij+1} - v_{ij}}{h}\right|^2}\right) + \right. \\ \quad + \phi_{l\varepsilon}\left(\sqrt{\left|\dfrac{v_{i-1j} - v_{ij}}{h}\right|^2 + \left|\dfrac{v_{ij+1} - v_{ij}}{h}\right|^2}\right) + \\ \quad + \phi_{l\varepsilon}\left(\sqrt{\left|\dfrac{v_{i-1j} - v_{ij}}{h}\right|^2 + \left|\dfrac{v_{ij-1} - v_{ij}}{h}\right|^2}\right) + \\ \quad \left. + \phi_{l\varepsilon}\left(\sqrt{\left|\dfrac{v_{i+1j} - v_{ij}}{h}\right|^2 + \left|\dfrac{v_{ij-1} - v_{ij}}{h}\right|^2}\right)\right], \end{cases}$$

(6.69)
$$j^{2l}_{h\varepsilon}(v_h) = h^2 \sum_{M_{i+1/2\, j+1/2} \in \Sigma_h} \phi_{l\varepsilon} \times \left(\sqrt{\frac{1}{2}\left|\frac{v_{i+1j+1} - v_{ij+1}}{h}\right|^2 + \frac{1}{2}\left|\frac{v_{i+1j} - v_{ij}}{h}\right|^2 + \frac{1}{2}\left|\frac{v_{ij+1} - v_{ij}}{h}\right|^2 + \frac{1}{2}\left|\frac{v_{i+1j+1} - v_{i+1j}}{h}\right|^2}\right),$$

(6.70)
$$j^{3l}_{h\varepsilon}(v_h) = h^2 \sum_{M_{ij} \in \bar{\Omega}_h} \alpha^2_{ij}\, \phi_{l\varepsilon} \times \left(\frac{1}{\alpha_{ij}} \sqrt{\frac{1}{2}\left|\frac{v_{i+1j} - v_{ij}}{h}\right|^2 + \frac{1}{2}\left|\frac{v_{ij+1} - v_{ij}}{h}\right|^2 + \frac{1}{2}\left|\frac{v_{i-1j} - v_{ij}}{h}\right|^2 + \frac{1}{2}\left|\frac{v_{ij-1} - v_{ij}}{h}\right|^2}\right),$$

(6.71)
$$j^{4l}_{h\varepsilon}(v_h) = h^2 \sum_{M_{i+1/2\, j+1/2} \in \Sigma_h} \phi_{l\varepsilon} \times \left(\sqrt{\left|\frac{v_{i+1j+1} - v_{ij}}{h\sqrt{2}}\right|^2 + \left|\frac{v_{ij+1} - v_{i+1j}}{h\sqrt{2}}\right|^2}\right),$$

with $v_{pq} = 0$ if $M_{pq} \notin \Omega_h$; in (6.70) we take $\alpha_{ij} = 1$ if j^3_h is given by (4.10), or the values given in (4.10') if j^3_h, given by this latter equation, is regularised.

The regularised approximate problems of (4.12) corresponding to (6.68) ... (6.71) are given by

(6.72)
$$\min_{v_h \in V_h} [J_{0h}(v_h) + g j^{kl}_{h\varepsilon}(v_h)]; \quad k = 1, 2, 3, 4, \quad l = 1, 2.$$

It is simple to prove:

Theorem 6.4. *Problem (6.72) admits one and only one solution,* $u_{h\varepsilon}^{kl}$*, and when* $\varepsilon \to 0$ *we have*, $\forall l = 1, 2$,

$$u_{h\varepsilon}^{kl} \to u_h^k , \tag{6.73}$$

where u_h^k *is the solution of the approximate problem* $(P_{0h})_k$ *defined by (4.12) in Section 4.3.*

The problems (6.72) are equivalent to the system of N [1] non-linear equations in N unknowns given by

$$\begin{cases} h^2\left(-\dfrac{u_{i+1j} + u_{i-1j} + u_{ij+1} + u_{ij-1} - 4\,u_{ij}}{h^2} - f_{ij}\right) + \\ \qquad + g\dfrac{\partial}{\partial v_{ij}} j_{h\varepsilon}^{kl}(u_{h\varepsilon}^{kl}) = 0\,, \quad M_{ij} \in \Omega_h\,, \\ u_{pq} = 0 \ \text{ if } M_{pq} \notin \Omega_h\,, \end{cases} \tag{6.74}$$

with $u_{h\varepsilon}^{kl} = (u_{ij})_{M_{ij} \in \Omega_h}$.

It is clear that the numerical solution of (6.72) will be carried out on the equivalent formulation (6.74).

Remark 6.5. For $k = 1, 2, 4$ (resp. $k = 3$) system (6.74) corresponds to a discretisation of the regularised continuous problem (6.4) by a 9-*point* (resp. 13-*point*) *finite difference* scheme; see Chapter 3, Section 8.2.5, Figure 8.2 (resp. Section 8.2.4, Figure 8.1) for the location of the points in question. ■

6.3. Solution of the regularised approximate problem (I). Method of point over-relaxation

6.3.1. *Description and convergence of the method*

The various regularised approximate problems defined in Section 6.2 are of the following type (we use the notation of Chapter 2, Section 1):

$$\min_{v \in \mathbb{R}^N} J(v)\,, \tag{6.75}$$

with $J : \mathbb{R}^N \to \mathbb{R}$ class C^1, strongly convex [2] and satisfying

$$\lim_{\|v\| \to +\infty} J(v) = +\infty\,;$$

thus for the actual solution of (6.75) we can use the relaxation algorithm described in Chapter 2, Section 1.1, whose convergence follows as a result of Theorem 1.1 of the same section; that is:

$$u^0 = \{ u_1^0, \ldots, u_N^0 \} \text{ given;} \tag{6.76}$$

with u^n known, we calculate u^{n+1}, coordinate by coordinate, using

---

(1) $N = \dim V_h = \operatorname{card}(\Omega_h)$.

(2) i.e. $(J'(v) - J'(u), v - u) \geqslant \alpha \| v - u \|^2 \quad \forall u, v \in \mathbb{R}^N$, $\alpha > 0$.

$$(6.77)\qquad \begin{cases} J(u_1^{n+1}, \ldots, u_{i-1}^{n+1}, u_i^{n+1}, u_{i+1}^n, \ldots) \leqslant J(u_1^{n+1}, \ldots, u_{i-1}^{n+1}, v_i, u_{i+1}^n, \ldots) \\ \quad \forall v_i \in \mathbb{R}, \quad i = 1, \ldots, N. \end{cases}$$

There is a well known equivalence between (6.77) and

$$(6.78)\qquad \frac{\partial J}{\partial v_i}(u_1^{n+1}, \ldots, u_{i-1}^{n+1}, u_i^{n+1}, u_{i+1}^n, \ldots) = 0 .$$

We can seek to improve the speed of convergence of the algorithm by introducing an under- or over-relaxation parameter (for the case of linear equations see Chapter 2, Section 1.4), giving the following two possible variants of (6.76), (6.77):

First algorithm.

$$(6.79)\qquad u^0 = \{ u_1^0, \ldots, u_N^0 \} \text{ given}$$

with u^n known, we calculate u^{n+1}, coordinate by coordinate, using

$$(6.80)\qquad \frac{\partial J}{\partial v_i}(u_1^{n+1}, \ldots, u_{i-1}^{n+1}, \bar{u}_i^{n+1}, u_{i+1}^n, \ldots) = 0 ,$$

$$(6.81)\qquad u_i^{n+1} = u_i^n + \omega(\bar{u}_i^{n+1} - u_i^n) ,$$

with $i = 1, 2, \ldots, N.$ ■

Second algorithm.

With u^n known, we calculate u^{n+1}, coordinate by coordinate, using

$$(6.82)\qquad \begin{cases} \dfrac{\partial J}{\partial v_i}(u_1^{n+1}, \ldots, u_{i-1}^{n+1}, u_i^{n+1}, u_{i+1}^n, \ldots) = \\ \qquad = (1 - \omega) \dfrac{\partial J}{\partial v_i}(u_1^{n+1}, \ldots, u_{i-1}^{n+1}, u_i^n, u_{i+1}^n, \ldots) , \end{cases}$$

with $i = 1, 2, \ldots, N.$ ■

In the examples solved numerically by regularisation and over-relaxation we have used the regularisation of $\int_\Omega |\operatorname{grad} v| \, dx$ by $\int_\Omega \sqrt{\varepsilon^2 + |\operatorname{grad} v|^2} \, dx$. Under these conditions the functional J of the corresponding problem (6.75) is C^∞ and its Hessian matrix at the point v, $J''(v)$ is positive definite $\forall v \in \mathbb{R}^N$; from Schechter /1/, /2/, /3/, these two properties are sufficient to ensure the convergence of (6.79), (6.80), (6.81) under the condition:

$$(6.83)\qquad 0 < \omega < \beta \leqslant 2 ,$$

where in practice the bound β is generally difficult to determine explicitly. To the best of our knowledge, algorithm (6.79), (6.82) does not appear to have been subject to any theoretical investigation, although in numerous circumstances it has proved more "robust" than (6.79), (6.80), (6.81). ■

Remark 6.6. In the examples considered (see Section 6.3.2 below) the nonlinear equations (6.80) and (6.82) were solved by Newton's method; that is, in the case of the calculation of $\bar{u}_i^{n+1}$ from (6.80):

$$\bar{u}_i^{n+1,0} = u_i^n, \tag{6.84}$$

$$\bar{u}_i^{n+1,m+1} = \bar{u}_i^{n+1,m} - \frac{\dfrac{\partial J_h}{\partial v_i}(u_1^{n+1}, \ldots, u_{i-1}^{n+1}, \bar{u}_i^{n+1,m}, u_{i+1}^n, \ldots)}{\dfrac{\partial^2 J_h}{\partial v_i^2}(u_1^{n+1}, \ldots, u_{i-1}^{n+1}, \bar{u}_i^{n+1,m}, u_{i+1}^n, \ldots)}. \tag{6.85}$$

Similar relations would be obtained for the calculation of u_i^{n+1} from (6.82).

On the subject of Newton's method, it should be pointed out that in Schechter, *loc. cit.*, it is shown in connection with algorithm (6.79), (6.80), (6.81) that under the assumptions $J \in C^\infty(\mathbb{R}^N)$, $J''(v)$ positive definite $\forall v \in \mathbb{R}^N$, ω "suitably chosen", only *a single* iteration of Newton's method need be carried out in the calculation of u_i^{n+1}, and thus, using the notation of (6.84), (6.85), we can define u_i^{n+1} by

$$u_i^{n+1} = u_i^n + \omega(\bar{u}_i^{n+1,1} - u_i^n), \tag{6.86}$$

and still have convergence of the algorithm. ■

Remark 6.7. If, in the algorithms (6.79), (6.80), (6.81) and (6.79), (6.82), only one Newton iteration is carried out in the calculation of u_i^{n+1}, then the new algorithms thus obtained coincide. ■

6.3.2. *Applications.*

For Ω we take the disc of radius 4 and f = const = C; the solution of (P_0) is then given by equations (2.19),...(2.21) in Section 2.3.2.

The approximation used is the interior approximation, using first-order finite elements, of Section 3; Figure 6.2 shows the corresponding triangulation $\mathcal{T}_h$: there are 1024 triangles, 545 nodes and 481 interior nodes, thus giving an approximate problem with 481 unknowns.

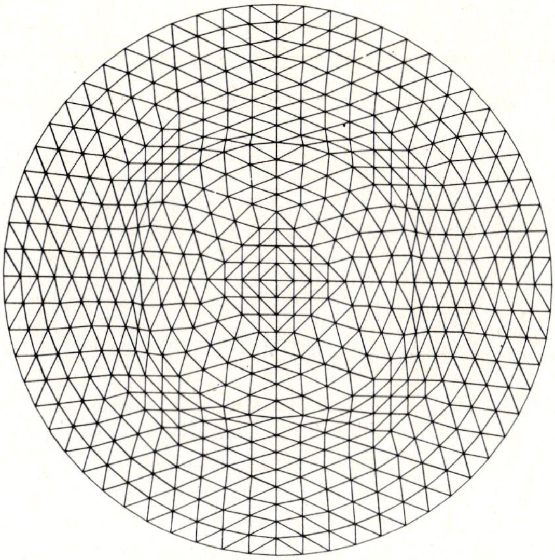

Fig. 6.2.

We restrict our attention to regularisation by $\phi_{l\varepsilon}$, and hence, with the notation of Section 6.2.1, to the regularised approximate problem

$$(6.87)\quad \begin{cases} a(u^1_{\varepsilon h}, v_h - u^1_{\varepsilon h}) + g\int_\Omega \sqrt{\varepsilon^2 + |\operatorname{grad} v_h|^2}\,dx - \\ \quad - g\int_\Omega \sqrt{\varepsilon^2 + |\operatorname{grad} u^1_{\varepsilon h}|^2}\,dx \geqslant \\ \quad \geqslant C\int_\Omega (v_h - u^1_{\varepsilon h})\,dx \quad \forall v_h \in V_h, \\ u^1_{\varepsilon h} \in V_h, \end{cases}$$

equivalent to

$$(6.88)\quad \min_{v_h \in V_h}\left[\frac{1}{2}\int_\Omega |\operatorname{grad} v_h|^2\,dx + g\int_\Omega \sqrt{\varepsilon^2 + |\operatorname{grad} v_h|^2}\,dx - C\int_\Omega v_h\,dx\right].$$

The results of Section 6.3.1 can be applied to problem (6.88) which, in particular, comes within the scope of Remark 6.6; we can thus solve (6.88) by nonlinear over-relaxation, carrying out only a single Newton iteration to determine [1] $u^{n+1}_M (M \in \mathring{\Sigma}_h)$.

The method described above was applied using the following parameter values

(1) The suffix ε has been omitted.

(6.89) $C = 10\,, \quad g = 1 \quad \text{and} \quad 10\,, \quad \varepsilon = 10^{-2}\,.$

We introduce the "residual" $R(n)$ defined by:

(6.90) $$R(n) = \frac{\sum_{M \in \mathring{\Sigma}_h} |u_M^{n+1} - u_M^n|}{\sum_{M \in \mathring{\Sigma}_h} |u_M^{n+1}|}\,.$$

With the nonlinear over-relaxation algorithm - using one Newton iteration - initialised by

(6.91) $u_h^0 = 0\,,$

the following phenomena are observed:

- Each iteration takes about 2 sec on a CII 10070.
- For $g = 1$, it appears that the optimum value of ω is of the order of 1.4 and that for this value of ω the quantity $R(n)$ is of the order of 10^{-5} after 500 iterations; this therefore leads to a computation time which in our opinion is too long, particularly when compared with those obtained with the methods described in subsequent sections; the approximation error relative to the exact solution is acceptable since, for $n = 500$, we have

(6.92) $$\max_{M \in \mathring{\Sigma}_h} \frac{|u_M^n - u(M)|}{|u(M)|} < 10^{-3}\,,$$

where, in (6.92), u denotes the solution of (P_0).
- For $g = 10$, the computation times required to give an acceptable approximation to the exact solution are far too long for the method used to be regarded as a production tool. ■

6.4. Solution of the regularised approximate problems (II). Gradient method with auxiliary operator.

6.4.1. *Description and convergence of the method*

It was shown in Chapter 2, Section 2.1 - referring to Céa /2/ for the proofs - that problems of the type (6.75) can be solved by means of the algorithm:

(6.93) u^0 given,

(6.94) $u^{n+1} = u^n - \rho_n J'(u^n)\,,$

(6.95) $\rho_n > 0$ suitably chosen.

A useful variant of this algorithm is as follows: since $S \in \mathscr{L}(\mathbb{R}^N; \mathbb{R}^N)$ is symmetric and positive definite, the algorithm (6.93), (6.94), (6.95) can be generalised by determining u^{n+1}

using

$$u^{n+1} = u^n - \rho_n S^{-1} J'(u^n)\,, \tag{6.96}$$

the case (6.93), (6.94), (6.95) then corresponds to $S = I$; with a suitable choice of S (on this point, see Section 6.4.2 below) we can considerably accelerate the convergence compared with the case $S = I$, at least when $J'(u)$ is the discretised form of an unbounded operator which may be linear or nonlinear; such is the case for the examples considered in this book.

Regarding the convergence of algorithm (6.93), (6.96), (6.95), we have:

Theorem 6.5. *Under the assumptions*

$$(J'(v_2) - J'(v_1),\, v_2 - v_1) \geqslant \alpha \| v_2 - v_1 \|^2\,, \quad \alpha > 0\,, \tag{6.97}$$

$$\| J'(v_2) - J'(v_1) \| \leqslant \beta \| v_2 - v_1 \|\,, \quad \beta \text{ independent of } v_1\,,\ v_2 \in \mathbb{R}^N\,, \tag{6.98}$$

and under the condition

$$0 < r_1 \leqslant \rho_n \leqslant r_2 < 2\,\alpha\sigma/\beta^2\,, \tag{6.99}$$

where σ is the smallest eigenvalue of S, we have, $\forall u^0 \in \mathbb{R}^N$, convergence to the solution u of (6.75)[1] *of the sequence u^n defined by (6.93), (6.96).*

Proof. This is a variant of the standard proof of the convergence of the gradient method; we put $|||\, v\, ||| = (Sv, v)^{1/2}$. Since the solution u of (6.75) satisfies

$$u = u - \rho_n S^{-1} J'(u) \quad \forall n\,, \tag{6.100}$$

and, by subtracting this from (6.96) (writing $\bar{u}^n = u^n - u$), we have

$$\bar{u}^{n+1} = \bar{u}^n - \rho_n S^{-1}(J'(u^n) - J'(u))\,, \tag{6.101}$$

and hence, by scalar squares,

$$|||\, \bar{u}^{n+1}\, |||^2 = |||\, \bar{u}^n\, |||^2 - 2\,\rho_n (J'(u^n) - J'(u), \bar{u}^n) + \rho_n^2 (S^{-1}(J'(u^n) - J'(u)),\, J'(u^n) - J'(u))\,, \tag{6.102}$$

or, in view of (6.97), (6.98) and using the fact that $\| S^{-1} \| = 1/\sigma$, we have

(1) This "u" has nothing to do (at least for the moment) with the solution of (P_0) (defined by (1.5)).

(6.103) $$|||\,\bar{u}^n\,|||^2 - |||\,\bar{u}^{n+1}\,|||^2 \geqslant \rho_n(2\,\alpha - \rho_n\,\beta^2/\sigma)\,\|\,\bar{u}^n\,\|^2\,,$$

and hence convergence of u^n under the condition (6.99). ∎

Remark 6.8. Using the conventional notation $\rho(S) = \|\,S\,\|$ it is easy to show, still using (6.97), (6.98), (6.103),

(6.104) $$|||\,\bar{u}^{n+1}\,|||^2 \leqslant |||\,\bar{u}^n\,|||^2 \left(1 - 2\,\rho_n \frac{\alpha}{\rho(S)} + \rho_n^2 \frac{\beta^2}{\sigma^2}\right),$$

giving the *geometric* convergence of $|||\,u^n - u\,|||$ to zero under the condition

(6.105) $$0 < r_1 \leqslant \rho_n \leqslant r_2 < \frac{2\,\alpha\sigma}{\beta^2}\left(\frac{\sigma}{\rho(S)}\right)$$

which is more restrictive than (6.99) (except when $S = I$). ∎

Remark 6.9. An investigation of the convergence of algorithms of the type (6.93), (6.96) can be found in Brezis-Sibony /2/ and Sibony /1/, under assumptions which are somewhat weaker than those of Theorem 6.5, with application to the solution of non-linear boundary-value problems. ∎

Remark 6.10. The reader may check that algorithm (3.20) of Chapter 1, Section 3.1, Remark 3.2 is an infinite-dimensional variant (given in variational form) of algorithm (6.93), (6.96). ∎

The gradient of the regularised approximate functionals of Section 6.2 clearly satisfies the assumptions for the application of Theorem 6.5, the Lipschitz constant β being of the form

(6.106) $$\beta = \frac{\beta_0}{\varepsilon}(1 + 0(\varepsilon))\,,\ \lim_{\varepsilon \to 0} 0(\varepsilon) = 0\,.$$ ∎

6.4.2. *An example of application*

We consider the particular (P_0) problems defined by

$$\Omega =]0, 1[\times]0, 1[, \quad \mu = 1, f = 10, g = 1 \text{ and } 1.6.$$

Type of discretisation: By finite differences (exterior approximation) the nondifferentiable term being approximated by the variant (4.10') of j_h^3 and by j_h^4 (see equation (4.11)).

Type of regularisation: By the function $\phi_{2\varepsilon}$ (1) (see relation (6.2)).

Mesh size: $h = 1/20$.

(1) This approximation of $\tau \to |\,\tau\,|$, although more complicated than that using $\tau \to \sqrt{\varepsilon^2 + \tau^2}$, is widely used (at least, its first and second derivatives) for the transient analysis of electrical networks, since it is known how to produce it physically

Regularisation parameter: $\varepsilon = 4\times10^{-4}$.

Implementation of the algorithm (6.93), (6.96):

In view of the exterior approximations used and with the corresponding notation, algorithm (6.93), (6.96) applied to the solution of the regularised approximate problems takes the form

$$(6.107)\qquad \begin{cases} u_h^0 \text{ given} \\ u_h^{n+1} = u_h^n - \rho_n S_h^{-1} F_h(u_n^n), \end{cases}$$

where F_h is the operator from $V_h \to V_h$ defined by the left-hand side of (6.74) (with $l = 2$ and $k = 3, 4$) and S_h the discretised form of a suitably chosen elliptic operator. We initially took $S = S_1 = I$ but with such a choice the convergence proved to be far too slow; we therefore took, as suggested by Godunov-Prokopov /1/ in connection with linear elliptic problems,

$$S = S_2 = \left(I - \frac{\partial^2}{\partial x_1^2}\right)\left(I - \frac{\partial^2}{\partial x_1^2}\right),$$

the boundary conditions being of Dirichlet type, homogeneous with respect to each factor. Another suitable choice is $S_3 = -\Delta$ for homogeneous Dirichlet conditions. For a given termination criterion (see (6.108) below) the two methods are largely equivalent in terms of the number of iterations; however, the possibility, with S_2, of solving the auxiliary problem at each iteration by direct methods of the Gauss or Choleski type (in view of the *tridiagonal* structure of the discretised forms of $I - \partial^2/\partial x_1^2, I - \partial^2/\partial x_2^2$) leads for the examples considered to a method about *ten times as fast*, in terms of computation time, as a method using $S_3 = -\Delta$ with solution of the auxiliary Dirichlet problems by point over-relaxation with optimal parameter.

Numerical results.

With the *termination criterion* for the iteration in the form:

$$(6.108)\qquad R^n = \sum_{M_{ij} \in \Omega_h} | u_{ij}^n - u_{ij}^{n-1} | \leqslant 10^{-3},$$

and with u_h^0 given by the restriction of the function $xy(1-x) \times (1-y)$ to Ω_h, the results relating to the convergence of algorithm (6.107) using $S = S_2$ are summarised in Table 6.1 overleaf.

The IBM 360/91 machine time used is of the order of 3 sec. for each problem.

With regard to the accuracy obtained, we can state that there is coincidence to within a relative accuracy of at least 10^{-3} compared with the results obtained for the same problems using the duality methods of Section 7 below; for further details we refer to Goursat /2/. ∎

g	j_h^k	ρ_n	Number of iterations required for convergence
1	j_h^3	Starting with $\rho_0 = 1$ retaining $\rho_n = 1$ until R^n ceases to decrease, and then taking: $\rho_n = 1/20$	134
	j_h^4		164
1.6	j_h^3		133
	j_h^4		210

Table 6.1

7. DUALITY METHODS

Synopsis.

In this section we shall investigate the application of the duality results and algorithms of Chapter 1, Sections 3.4, 3.5 and Chapter 2, Section 4. First, the principles of the method for the continuous problem (1.5) are reviewed in Section 7.1, and then variants of the continuous case will be employed, in Sections 7.1 and 7.2 respectively, to solve the approximate problems defined in Section 4 (exterior approximations) and Section 3 (finite element approximations). ■

7.1. Application to the solution of the continuous problem

It results from Chapter 1, Section 3.4 that the solution u of (1.5) is characterised by the existence of

$$\lambda \in \Lambda = \{ \mu \mid \mu \in L^2(\Omega) \times L^2(\Omega), \mid \mu(x) \mid \leqslant 1 \text{ a.e.}\}$$

such that

$$\lambda . \operatorname{grad} u = \mid \operatorname{grad} u \mid , \tag{7.1}$$

$$\begin{cases} - \Delta u - g \operatorname{div} \lambda = f & \text{in } \Omega , \\ u = 0 \quad \text{on } \Gamma . \end{cases} \tag{7.2}$$

It follows immediately from (7.1), (7.2) that $\{ u, \lambda \}$ is a *saddle point* on $H_0^1(\Omega) \times \Lambda$ of the Lagrangian $\mathscr{L} : H_0^1(\Omega) \times (L^2(\Omega))^2 \to \mathbb{R}$ defined by

$$\mathscr{L}(v, \mu) \doteq \frac{1}{2} a(v, v) - (f, v) + g \int_\Omega \mu . \operatorname{grad} v \, dx . \tag{7.3}$$

We can use (as has already been mentioned in Chapter 2, Sectio 4.4, Remark 4.9) the duality algorithm (4.12), (4.13) of Chapter 2, Section 4.3 with Λ defined as above and

$$V = H_0^1(\Omega) , \quad M = V , \tag{7.4}$$

$$(7.5) \qquad L = L^2(\Omega) \times L^2(\Omega),$$

$$(7.6) \qquad \phi : V \to L \quad \text{defined by} \quad \phi v = \text{grad } v.$$

With regard to the operator $P_\Lambda : L \to \Lambda$ this is defined by

$$(7.7) \qquad \begin{cases} P_\Lambda(\mu) = \dfrac{\mu}{\sup(1, |\mu|)} \\ |\mu| = \sqrt{\mu_1^2 + \mu_2^2} \text{ if } \mu = \{\mu_1 \, \mu_2\}. \end{cases}$$

The explicit form of the algorithm mentioned above is then given by:

$$(7.8) \qquad \lambda^0 \text{ given (arbitrarily);}$$

with λ^n known ($\in \Lambda$), we define u^n, and then λ^{n+1} by

$$(7.9) \qquad \begin{cases} -\Delta u^n - g \operatorname{div} \lambda^n = f \text{ in } \Omega, \\ u^n = 0 \text{ on } \Gamma, \end{cases}$$

$$(7.10) \qquad \lambda^{n+1} = P_\Lambda(\lambda^n + \rho_n \text{ grad } u^n).$$

The following proposition can be deduced from the proof of Theorem 4.1 of Chapter 2, Section 4.3:

Proposition 7.1. *Under the condition*

$$(7.11) \qquad 0 < \alpha_0 \leqslant \rho_n \leqslant \alpha_1 < 2/g,$$

for the sequence u^n *defined by* (7.8), (7.9), (7.10) *we have*

$$(7.12) \qquad u^n \to u \quad \textit{strongly in } H_0^1(\Omega)$$

where u *is the solution of* (1.5). ■

7.2. Application to the solution of the approximate problems (I). Case of the exterior approximations of Section 4.

7.2.1. *Duality algorithm for the approximate problems of Section 4.*

We shall in fact confine our attention to the approximate problem (4.12) (see Section 4.3) relating to j_h^4 (defined by (4.11)), since – as can be readily verified – of all the approximate problems (4.12) this is the one which possesses the simplest dual problem [1].

With the notation of Section 4 and of Chapter 3, Section 5.1.2, we define the space L_h by

---

[1] See the article by Céa-Glowinski /1/ on this subject.

(7.13) $L_h = \{ \mu_h \mid \mu_h = \{ \mu_h^1, \mu_h^2 \}, \mu_h^k = \{ \mu_{i+1/2\, j+1/2}^k \}_{M_{i+1/2\, j+1/2} \in \Sigma_h}, k = 1, 2 \},$

and the closed convex set Λ_h of L_h by

(7.14) $$\Lambda_h = \{ \mu_h \mid \mu_h \in L_h, | \mu_{i+1/2\, j+1/2}^1 |^2 + + | \mu_{i+1/2\, j+1/2}^2 |^2 \leqslant 1 \quad \forall M_{i+1/2\, j+1/2} \in \Sigma_h \},$$

and, finally, the mapping $G_h : V_h \to L_h$ (the discrete analogue of grad) by

(7.15) $$\begin{cases} G_{i+1/2\, j+1/2}^1(v_h) = \dfrac{v_{i+1j+1} - v_{ij+1} + v_{i+1j} - v_{ij}}{2\, h}, \\ G_{i+1/2\, j+1/2}^2(v_h) = \dfrac{v_{i+1j+1} - v_{i+1j} + v_{ij+1} - v_{ij}}{2\, h}. \end{cases}$$

Under these conditions, it can be shown that the solution [1] u_h of (P_{0h}) is characterised by the existence of $\lambda_h \in \Lambda_h$ such that

(7.16) $$\begin{cases} -\dfrac{u_{i+1j} + u_{i-1j} + u_{ij+1} + u_{ij-1} - 4\, u_{ij}}{h^2} = f_{ij} + \\ + g\left(\dfrac{\lambda_{i+1/2\, j+1/2}^1 - \lambda_{i-1/2\, j+1/2}^1 + \lambda_{i+1/2\, j-1/2}^1 - \lambda_{i-1/2\, j-1/2}^1}{2\, h} \right. \\ \left. + \dfrac{\lambda_{i+1/2\, j+1/2}^2 - \lambda_{i+1/2\, j-1/2}^2 + \lambda_{i-1/2\, j+1/2}^2 - \lambda_{i-1/2\, j-1/2}^2}{2\, h}\right) \\ \forall M_{ij} \in \Omega_h, u_{pq} = 0 \quad \text{if} \quad M_{pq} \notin \Omega_h, \end{cases}$$

(7.17) $$\sqrt{| G_{i+1/2\, j+1/2}^1(u_h) |^2 + | G_{i+1/2\, j+1/2}^2(u_h) |^2} = = \sum_{k=1}^{2} \lambda_{i+1/2\, j+1/2}^k\, G_{i+1/2\, j+1/2}^k(u_h), \quad \forall M_{i+1/2 j+1/2} \in \Sigma_h.$$

which are discrete analogues of (7.1), (7.2); the pair $\{ u_h, \lambda_h \}$ is a saddle point in $V_h \times \Lambda_h$ of the Lagrangian $\mathscr{L}_h$ defined by

(7.18) $$\mathscr{L}_h(v_h, \mu_h) = J_{0h}(v_h) + + gh^2 \sum_{M_{i+1/2\, j+1/2} \in \Sigma_h} \sum_{k=1}^{2} \mu_{i+1/2\, j+1/2}^k\, G_{i+1/2\, j+1/2}^k(v_h).$$

We can thus apply the results of Chapter 2, Section 4 to this finite-dimensional situation, giving the discrete analogue of algorithm (7.8), (7.9), (7.10), that is:

(7.19) λ_h^0 given;

---

(1) Henceforward the superscript 4 will be omitted.

with λ_h^n known $(\in \Lambda_h)$, we define u_h^n and then λ_h^{n+1} by

$$(7.20)\quad \begin{cases} -\dfrac{u_{i+1j}^n + u_{i-1j}^n + u_{ij+1}^n + u_{ij-1}^n - 4u_{ij}^n}{h^2} = f_{ij} + \\ + g\left(\dfrac{\lambda_{i+1/2\,j+1/2}^{1n} - \lambda_{i-1/2\,j+1/2}^{1n} + \lambda_{i+1/2\,j-1/2}^{1n} - \lambda_{i-1/2\,j-1/2}^{1n}}{2h}\right. \\ \left. + \dfrac{\lambda_{i+1/2\,j+1/2}^{2n} - \lambda_{i+1/2\,j-1/2}^{2n} + \lambda_{i-1/2\,j+1/2}^{2n} - \lambda_{i-1/2\,j-1/2}^{2n}}{2h}\right) \\ \forall M_{ij} \in \Omega_h,\ u_{pq}^n = 0 \quad \text{if } M_{pq} \notin \Omega_h, \end{cases}$$

$$(7.21)\quad \begin{cases} (\lambda_{i+1/2\,j+1/2}^{1n+1}, \lambda_{i+1/2\,j+1/2}^{2n+1}) = P_D(\lambda_{i+1/2\,j+1/2}^{1n} + \\ \quad + \rho_n G_{i+1/2\,j+1/2}^1(u_h^n), \lambda_{i+1/2\,j+1/2}^{2n} + \rho_n G_{i+1/2\,j+1/2}^2(u_h^n)) \\ \forall M_{i+1/2\,j+1/2} \in \Sigma_h, \end{cases}$$

using the notation of Chapter 3, Section 8.3.3, Remark 8.10.

Since Theorem 4.1 of Chapter 3, Section 4.3 can also be applied here, we have convergence of the sequence u_h^n, defined by (7.19), (7.20), (7.21), to the solution u_h of $(P_{0h})_4$ for

$$0 < \alpha_{0h} \leqslant \rho_n \leqslant \alpha_{1h}, \quad \alpha_{1h} \text{ sufficiently small.} \quad ■$$

Remark 7.1. Equations (7.16), (7.17), (7.20), (7.21) constitute "natural" discretisations of the corresponding continuous equations (7.2), (7.1), (7.9), (7.10). ■

7.2.2. *Solution of a particular case for which the exact solution is known.*

Construction of the test problem.

The exact solution of (P_0) when f = const = C and Ω is a disc of radius R was given in Section 2.3.2. Let u be the restriction to $\Omega =]0, 1[\times]0, 1[$ of the solution of (P_0) relating to the disc of radius $R = \sqrt{2}/2$ and centre (0.5, 0.5) for $\mu = 1$, $C = 10$, $g = 1.25 = 5/4$ (hence $R' = \frac{1}{4}$, see Figure 7.1).

The essence of what has already been said in the present chapter applies to the (non-homogeneous) variational problem

$$(7.22)\quad \begin{cases} \min\limits_{v \in H} [\frac{1}{2} a(v, v) + gj(v) - (f, v)], \\ H = \{ v \mid v \in H^1(\Omega), v|_\Gamma = u|_\Gamma \}, \end{cases}$$

whose solution is clearly u; in particular, (7.22) can be solved approximately using algorithm (7.19), (7.20), (7.21); it is sufficient to replace in (7.20) the condition $u_{pq} = 0$ if $M_{pq} \notin \Omega_h$ (in which case M_{pq} then belongs to Γ) by the condition $u_{pq}^n = u(M_{pq})$ if $M_{pq} \in \Gamma$. ■

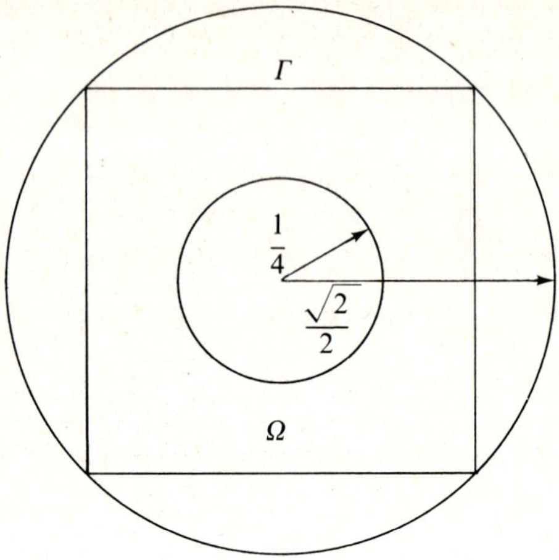

Fig. 7.1.

Solution of the approximate Dirichlet problems (7.20).

This is carried out using point over-relaxation with optimal parameter; if we denote by $u_h^{n,m}$ the sequence of iterates given by the above over-relaxation algorithm, the calculation of u_h^n is initialised by

$$u_h^{n,0} = u_h^{n-1}, \tag{7.23}$$

the termination criterion used being

$$\sum_{M_{ij} \in \Omega_h} | u_{ij}^{n,m} - u_{ij}^{n,m-1} | \leqslant 10^{-4}. \quad \blacksquare \tag{7.24}$$

Initialisation of algorithm (7.19), (7.20), (7.21):

$$\lambda_h^0 = 0. \tag{7.25}$$

Termination criterion for algorithm (7.19), (7.20), (7.21):

$$R^n = \sum_{M_{ij} \in \Omega_h} | u_{ij}^n - u_{ij}^{n-1} | \leqslant 3{\times}10^{-4}. \tag{7.26}$$

Choice of ρ_n : We take $\rho_n = \rho = 1/g = \frac{4}{5}$.

Analysis of the numerical results:

We use the notation

$$E^n = \sum_{M_{ij} \in \Omega_h} | u_{ij}^n - u(M_{ij}) | \tag{7.27}$$

where u is the exact solution of (7.22); results relating to the convergence and accuracy obtained are shown in Table 7.1.

h	$\frac{1}{10}$	$\frac{1}{20}$	$\frac{1}{40}$
Number of iterations N required for convergence	25	25	36
R^N	2.9×10^{-4}	2.6×10^{-4}	2.8×10^{-4}
E^N	5.3×10^{-2}	2.1×10^{-2}	2.7×10^{-2}
Computation time on IBM 360/91	0.35 s	1.53 s	9.62 s

Table 7.1

Remark 7.2. For $h = 1/20, \max_{M_{ij} \in \Omega_h} |u_{ij}^{25} - u(M_{ij})| \simeq 10^{-4}$. ■

A more detailed analysis of the convergence of algorithm (7.19), (7.20), (7.21) and of the numerical results obtained, including the dependence of the solutions on g, may be found in Section 7.2.3. below. ■

7.2.3. *Other examples* (I)

We shall consider examples whose exact solution is not known (1).

Definition of the domain and values of the parameters.

We take $\Omega =]0, 1[\times]0, 1[$, $\mu = 1$, $f = 10$; we shall take several values for g, with the aim of investigating the dependence of the solutions with respect to this parameter.

Mesh size: $h = 1/20$.

Solution of the approximate Dirichlet problems (7.20):

By point over-relaxation with optimal parameter.

(1) Except when g is sufficiently large, in which case the solution is exactly zero.

Initialisation and termination criteria of the algorithms:

Those of the example in Section 7.2.2.

Choice of ρ_n: We take $\rho_n = \rho$.

Analysis of the numerical results:

Figure 7.2 shows the number of iterations required for convergence, as a function of ρ, when $g = 1$.

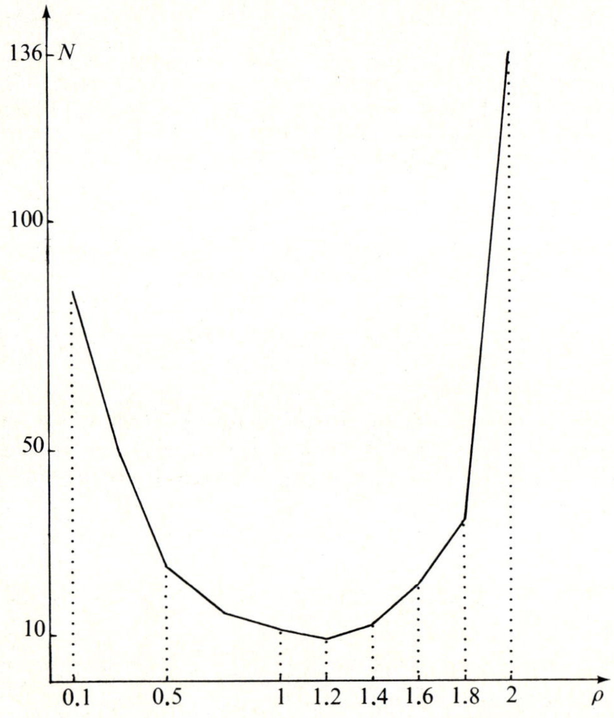

Fig. 7.2. Variation of the number of iterations required for convergence, as a function of ρ, ($g = 1$).

For $\rho = 2/g = 2$ there is still convergence, in 136 iterations, but for $\rho = 2.1$, divergence; the restriction (7.11), which relates to the continuous problem, is thus very realistic for the approximate problem $(P_{0h})_4$. For $\rho = 1.1$ (which roughly corresponds to the optimal value for ρ) the convergence of algorithm (7.19), (7.20), (7.21) is attained in 12 iterations, giving an execution time of 1 sec. on an IBM 360/91.

Figure 7.3 shows the number of over-relaxation iterations required to solve the approximate Dirichlet problems (7.20) with $g = 1$, $\rho = 1$, using the initialisation (7.23) and the termination criterion (7.24).

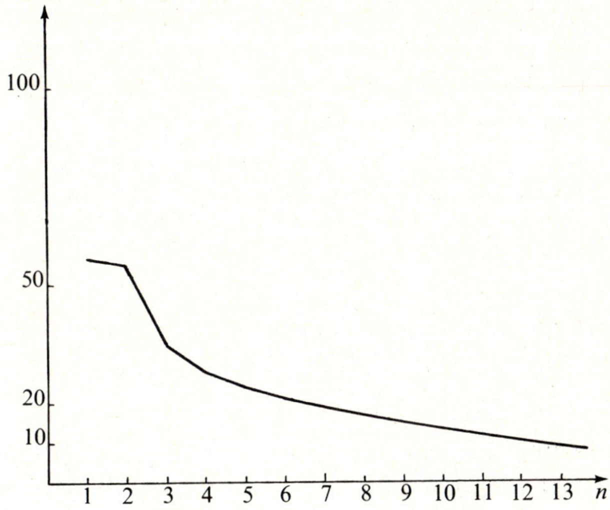

Fig. 7.3. Variation of the number of over-relaxation iterations required to solve the approximate Dirichlet problems (7.20) ($g = 1$, $\rho = 1$).

Figures 7.4, 7.5 show the fluid zones (1) and rigid zones (2) (hatched) corresponding to $g = 1$ and $g = 1.8$ respectively.

Figure 7.6 represents the function $x_1 \to u(x_1, \frac{1}{2})$ for $\mu = 1$, $g = 1$, $f = 10$.

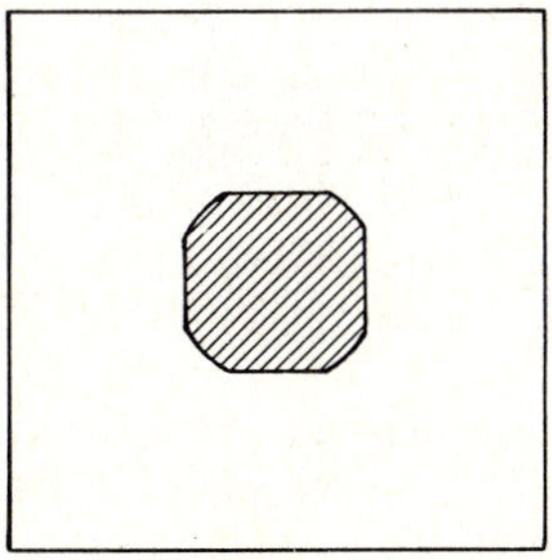

Fig. 7.4. $\mu = 1, g = 1, f = 10$, $u_{\max} = 0.291$.

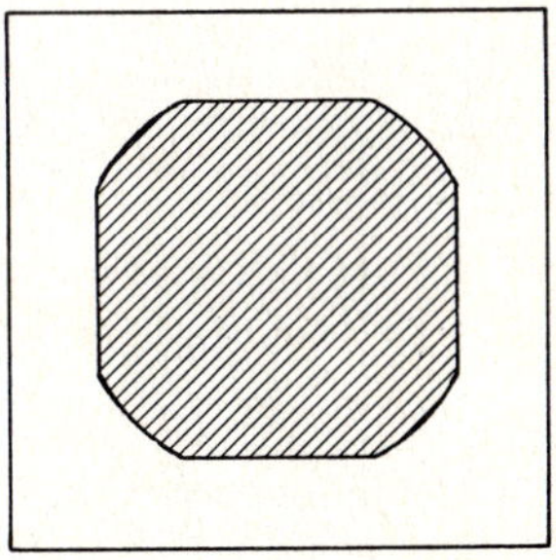

Fig. 7.5. $\mu = 1, g = 1.8, f = 10$, $u_{\max} = 0.080$.

(1) i.e. $|\operatorname{grad} u| > 0$.
(2) i.e. $|\operatorname{grad} u| = 0$.

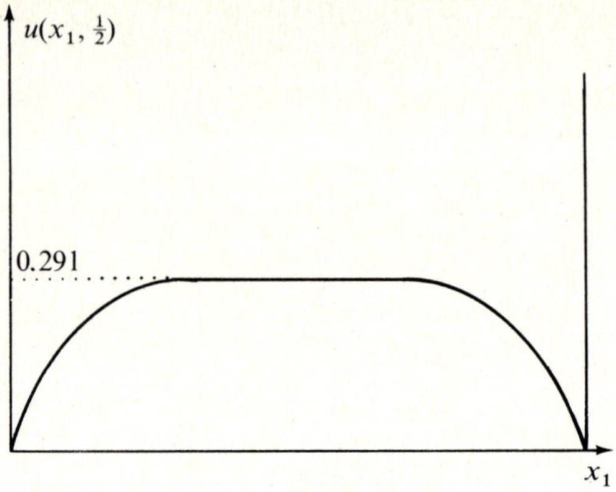

Fig. 7.6. Representation of $u(x_1, 1/2), \mu = 1, g = 1, f = 10.$

We consider it appropriate to show, in Figure 7.7, the decrease of $\| u_g \|_{H_0^1(\Omega)}$ as a function of g; in the case when Ω is a disc and f is a constant it is easy to deduce from equations (2.20) of Section 2.3.2 that $\| u_g \|^2_{H_0^1(\Omega)}$ is proportional to $|g - g_c|^3$; we have therefore plotted $g \to \| u_g \|^{2/3}_{H_0^1(\Omega)}$, using calculations carried out for a fairly large number of values of g; since the graph obtained is a straight line it appears that in the case when Ω is a square, and f is a constant, $\| u_g \|_{H_0^1(\Omega)}$ again varies as $|g - g_c|^{3/2}$ (apart from approximation and calculation errors).

7.2.4. *Other examples* (II).

We once again consider an example for which the exact solution is not known (at least for g sufficiently small).

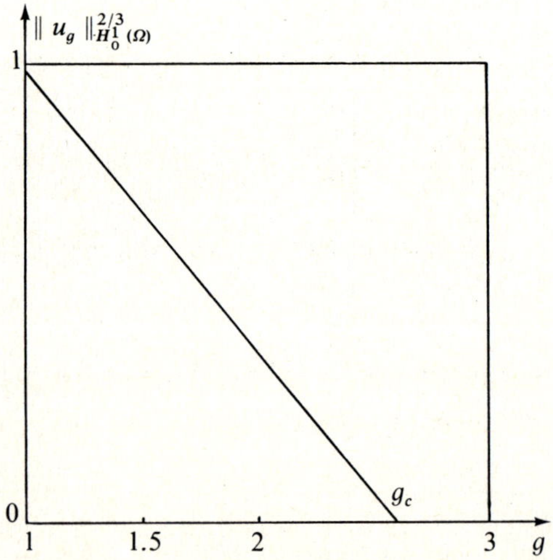

Fig. 7.7. Representation of $g \to \| u_g \|^{2/3}_{H_0(\Omega)}$.

Definition of the domain and values of the parameters:

For Ω we take the half-disc of radius $\frac{1}{2}$, $\mu = 1, f = 10, g = 0.75$.

Mesh size: $h = 1/20$.

Analysis of the numerical results:

We use algorithm (7.19), (7.20), (7.21) as in the previous examples, which gives the results shown in Figure 7.8 for the fluid and rigid (hatched) zones.

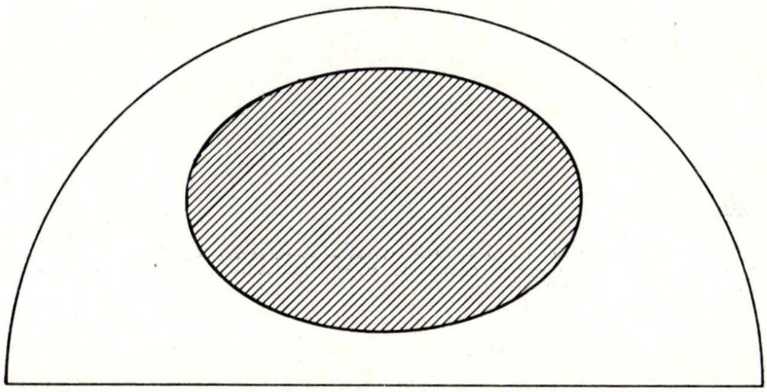

Fig. 7.8. Semi-circular duct ($\mu = 1, f = 10, g = 0.75$).

7.3. Application to the solution of the approximate problems (II). The case of finite element approximations.

7.3.1. *The case of the finite element approximation of Section 3.*

Description and convergence of the duality algorithm.

We use the notation of Section 3 and of Chapter 3, Section 8.3.2; the solution u_h of (3.1) (and of the equivalent problem (3.2)) is characterised by the existence of $\lambda_h \in \Lambda_h$ such that

$$(7.28) \qquad \begin{cases} a(u_h, v_h) + g \int_\Omega \lambda_h \cdot \text{grad } v_h \, dx = \int_\Omega f v_h \, dx \quad \forall v_h \in V_h , \\ u_h \in V_h , \end{cases}$$

$$(7.29) \qquad | \text{grad } u_h | = \lambda_h \cdot \text{grad } u_h ,$$

which are approximate analogues of (7.1), (7.2); the pair $\{ u_h, \lambda_h \}$ is a saddle point in $V_h \times \Lambda_h$ of the Lagrangian $\mathscr{L}_h$ defined by:

(7.30) $$\mathscr{L}_h(v_h, \mu_h) = \frac{1}{2} a(v_h, v_h) + g \int_\Omega \mu_h . \text{grad } v_h \, dx - \int_\Omega f v_h \, dx .$$

We can thus apply to this infinite-dimensional situation the results of Chapter 2, Section 4, which yield the approximate analogue of algorithm (7.8), (7.9), (7.10) given by:

(7.31) λ_h^0 given;

with λ_h^n known $(\in \Lambda_h)$, we define u_h^n and then λ_h^{n+1} by

(7.32) $$\begin{cases} a(u_h^n, v_h) + g \displaystyle\int_\Omega \lambda_h^n . \text{grad } v_h \, dx = \displaystyle\int_\Omega f v_h \, dx \quad \forall v_h \in V_h , \\ u_h^n \in V_h , \end{cases}$$

(7.33) $$\lambda_h^{n+1} = P_{\Lambda_h}(\lambda_h^n + \rho_n \text{ grad } u_h^n) ,$$

with, we recall, $P_{\Lambda_h} : L_h \to \Lambda_h$ defined by

(7.34) $$P_{\Lambda_h}(q_h) = \frac{q_h}{\max(1, |q_h|)} .$$

Since Theorem 4.1 of Chapter 2, Section 4.3 may also be applied here, the convergence of the sequence u_h^n defined by (7.31), (7.32), (7.33) to the solution u_h of problem (3.1) can thus be deduced for

(7.35) $$0 < \alpha_0 \leqslant \rho_n \leqslant \alpha_1 < 2/g ,$$

as in the case of the continuous problem. ■

Remark 7.2. It is easy to express equations (7.28), (7.29) and (7.32), (7.33) in terms of the nodal values $v_M = v_h(M), M \in \mathring{\Sigma}_h$ using the formulas of Chapter 3, Section 4.1.4. ■

7.3.2. *The case of the finite element approximation of Section* 3. (II). *Examples* (1)

Solution of the approximate Dirichlet problems (7.32).

By point over-relaxation with $\omega = 1.5$, for all the examples considered; the calculation of u_h^n is initialised by $u_h^{0,n} = u_h^{n-1}$ the superscript p being used to denote the over-relaxation iterations, with $u_h^{p,n}$ for the pth iterate.

(1) The numerical treatment of the examples below is due to J.F. Bourgat.

Termination criteria:

We define $D_m(n)$ by

$$D_m(n) = \sum_{M \in \mathring{\Sigma}_h} | u_M^n - u_M^{n-1} | \Big/ \sum_{M \in \mathring{\Sigma}_h} | u_M^n | , \tag{7.36}$$

and $D_m^n(p)$ by

$$D_m^n(p) = \sum_{M \in \mathring{\Sigma}_h} | u_M^{p,n} - u_M^{p-1,n} | \Big/ \sum_{M \in \mathring{\Sigma}_h} | u_M^{p,n} | . \tag{7.37}$$

For all the examples treated below we then take as termination criteria

$$D_m(n) \leqslant 10^{-4} \tag{7.38}$$

for algorithm (7.31), (7.32), (7.33) and

$$D_m^n(p) \leqslant \varepsilon_n \tag{7.39}$$

with

$$\varepsilon_n = \min(\varepsilon_{n-1}, D_m(n-1)/5) , \quad \varepsilon_1 = 10^{-3} \tag{7.40}$$

for the over-relaxation algorithm.

Initialisation of algorithm (7.31), (7.32), (7.33) : $\lambda_h^0 = 0$.

Values of ρ_n: We take $\rho_n = \rho = 0.7$ for all the examples considered.

Example 1. We take an example for which the exact solution is known and given by formulas (2.25) of Section 2.3.2 , namely: Ω is the disc of radius $\frac{1}{2}$, $\mu = 1$, $f = 10$, $g = 1.25$, giving $R' = \frac{1}{4}$. With the triangulation shown in Figure 7.9, consisting of 512 triangles and 225 interior nodes, convergence of algorithm (7.31), (7.32), (7.33) is reached in 10 iterations under the above conditions, giving a computation time of 90 sec. on a CII 10070; if u is the exact solution we have

$$\max_{M \in \mathring{\Sigma}_h} | u_h^{10}(M) - u(M) | = 6 \times 10^{-4} .$$

Figure 7.9 shows the fluid and rigid (hatched) zones thus calculated.

Example 2. We take $\Omega =]0, 1[\times]0, 1[$, $\mu = 1$, $f = 10$, $g = 1$. Using the triangulation shown in Figure 7.10, consisting of 512 triangles, convergence is reached in 14 iterations, giving a computation time of 2 min 30 sec on a CII 10070.

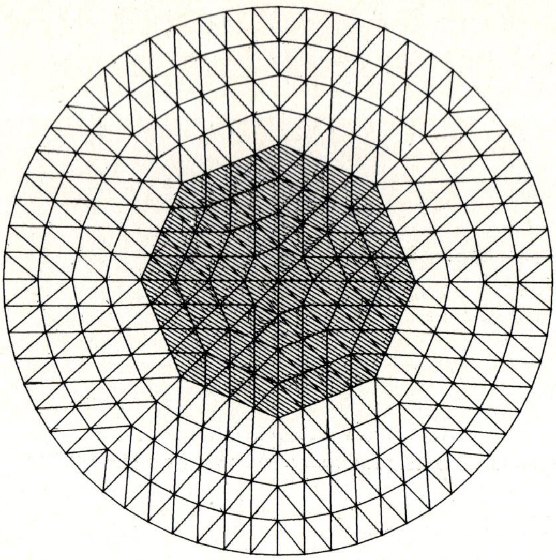

Fig. 7.9. $\mu = 1, \rho = 10, g = 1.25.$

Figure 7.10 also shows the calculated fluid and rigid (hatched) zones.

Example 3. We take Ω to be the half-disc of radius $\frac{1}{2}$, $\mu = 1$, $f = 10$, $g = 0.75$. With the triangulation shown in Figure 7.11, consisting of 256 triangles, convergence is reached in 23 iterations, giving a computation time of 1 min. 10 sec on a CII 10070. Figure 7.11 shows the calculated fluid and rigid (hatched) zones.

7.3.3. *The case of approximation by second-order Lagrangian finite elements.*

General considerations.

The following considerations are taken from Fortin /1/. For an approximation of $H_0^1(\Omega)$ given by a family of subspaces V_h formed from continuous functions, and whose restriction to the triangles T of a triangulation $\mathscr{T}_h$ of Ω is of degree $\leqslant 2$, it has already been pointed out in Section 3.5 that the use of this type of finite element in the solution of problem (1.5) poses certain difficulties connected with the approximation of the nondifferentiable term $g \int_\Omega |\operatorname{grad} v| \, dx$.

The approximation used being of Lagrangian type in which the

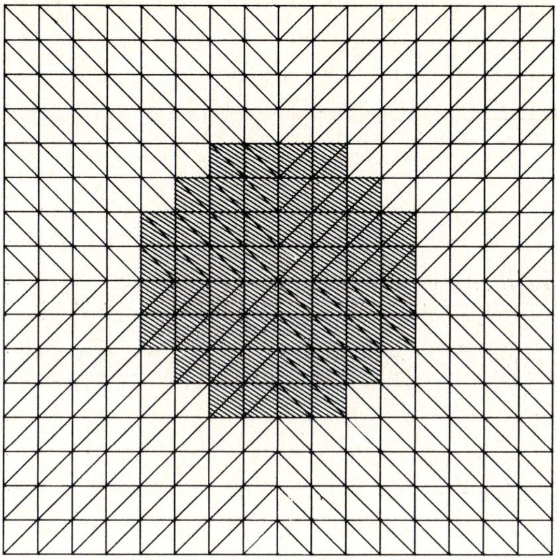

Fig. 7.10. $\mu = 1, f = 10, g = 1.$

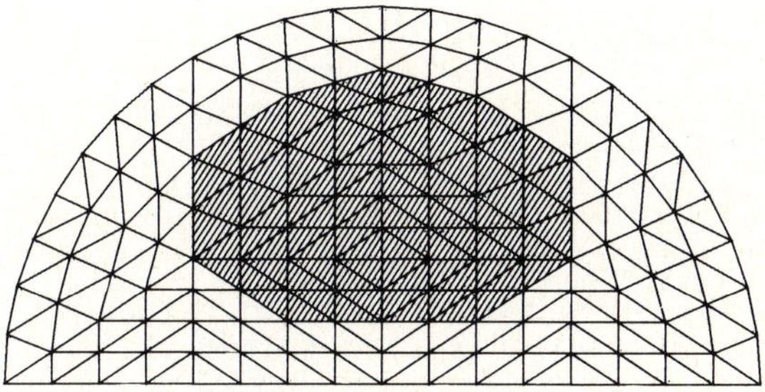

Fig. 7.11. $\mu = 1, f = 10, g = 0.75.$

degrees of freedom are the values taken by v_h at the points M_{iT}, m_{iT} $(i = 1, 2, 3)$ of Figure 3.1 of Section 3.5, we shall use *numerical integration* formulas to approximate the functional $g \int_\Omega |\operatorname{grad} v| \, dx$; we approximate the functional

$$J_g(v) = \frac{1}{2} a(v, v) + gj(v) - \int_\Omega fv \, dx ,$$

first by

$$(7.41) \qquad J_{gh}^1(v_h) = \frac{1}{2} a(v_h, v_h) + g \sum_{T \in \mathcal{T}_h} A(T) \, |\operatorname{grad} v_h(G_T)| - \int_\Omega fv_h \, dx ,$$

and secondly by

(7.42) $$J^2_{gh}(v_h) = \frac{1}{2}\, a(v_h, v_h) + g \sum_{T \in \mathcal{T}_h} \frac{A(T)}{3} \sum_{i=1}^{3} |\operatorname{grad} v_h(m_{iT})| - \int_\Omega f v_h \, dx\,,$$

with G_T = centre of gravity of T and $A(T)$ = area of T.

It can be shown without difficulty that the approximate problems:

(7.43) $$\min_{v_h \in V_h} J^1_{gh}(v_h)\,,$$

(7.44) $$\min_{v_h \in V_h} J^2_{gh}(v_h)\,,$$

admit, respectively, one and only one solution u_h^1, u_h^2 say, and that when $h \to 0$ as in the statement of Theorem 5.1 of Section 5.1, we have

(7.45) $$u_h^i \to u \ \textit{strongly in} \ H^1_0(\Omega) \qquad (i = 1, 2)\,,$$

where u is the solution of (P_0) (1).

Given the Lagrangians $\mathcal{L}_h^1$, $\mathcal{L}_h^2$ defined by

(7.46) $$\mathcal{L}_h^1(v_h, \mu_h) = \frac{1}{2}\, a(v_h, v_h) + g \sum_{T \in \mathcal{T}_h} A(T)\, \mu_T . \operatorname{grad} v_h(G_T) - \int_\Omega f v_h \, dx\,,$$

(7.47) $$\mathcal{L}_h^2(v_h, \mu_h) = \frac{1}{2}\, a(v_h, v_h) + g \sum_{T \in \mathcal{T}_h} \frac{A(T)}{3} \sum_{i=1}^{3} \times$$
$$\times\, \mu_{iT} . \operatorname{grad} v_h(m_{iT}) - \int_\Omega f v_h \, dx\,,$$

with μ_T, μ_{iT} $(i = 1, 2, 3) \in \mathbb{R}^2$, it can be proved that the following finite-dimensional variants of algorithm (7.8), (7.9), (7.10) define sequences $u_h^{i,n}$ $(i = 1, 2)$ which converge, respectively, to the solution u_h^i of (7.43), (7.44):

Case i = 1:

(7.48) $$\{ \lambda_T^0 \}_{T \in \mathcal{T}_h} \ \text{given};$$

with λ_T^n known, we determine u_h^n and λ_T^{n+1} by

(7.49) $$\begin{cases} a(u_h^n, v_h) + g \displaystyle\sum_{T \in \mathcal{T}_h} A(T)\, \lambda_T^n . \operatorname{grad} v_h(G_T) = \int_\Omega f v_h \, dx \quad \forall v_h \in V_h\,, \\ u_h^n \in V_h\,, \end{cases}$$

---

(1) This last point is a relatively immediate consequence of general results on the convergence of finite element approximations (see for example Fix-Strang /1/, Ciarlet-Raviart /1/).

(7.50) $\lambda_T^{n+1} = P_{\bar{D}}(\lambda_T^n + \rho_n A(T) \operatorname{grad} u_h^n(G_T))\,, \quad T \in \mathcal{T}_h\,,$

(7.51) $0 < \alpha_{0h}^1 \leqslant \rho_n \leqslant \alpha_{1h}^1\,,$ α_{1h}^1 sufficiently small. ■

Case $i = 2$:

(7.52) $\{ (\lambda_{1T}^0, \lambda_{2T}^0, \lambda_{3T}^0) \}_{T \in \mathcal{T}_h}$ given;

with $(\lambda_{1T}^n, \lambda_{2T}^n, \lambda_{3T}^n)$ known, we determine u_h^n and $(\lambda_{1T}^{n+1}, \lambda_{2T}^{n+1}, \lambda_{3T}^{n+1})$ by

(7.53) $$\begin{cases} a(u_h^n, v_h) + g \displaystyle\sum_{T \in \mathcal{T}_h} \frac{A(T)}{3} \sum_{i=1}^{3} \lambda_{iT}^n \cdot \operatorname{grad} v_h(m_{iT}) = \int_\Omega f v_h \, dx \\ \forall v_h \in V_h\,, \quad u_h^n \in V_h\,, \end{cases}$$

(7.54) $\lambda_{iT}^{n+1} = P_{\bar{D}}\left(\lambda_{iT}^n + \rho_n \dfrac{A(T)}{3} \operatorname{grad} u_h^n(m_{iT})\right), \quad i = 1, 2, 3,\ T \in \mathcal{T}_h\,,$

(7.55) $0 < \alpha_{0h}^2 \leqslant \rho_n \leqslant \alpha_{1h}^2\,,$ α_{1h}^2 sufficiently small. ■

We recall that if $y = (y_1, y_2) \in \mathbb{R}^2$ we have $P_{\bar{D}}(y) = \dfrac{y}{\max(1, \sqrt{y_1^2 + y_2^2})}$.

Application to an example for which the exact solution is known.

We consider again the (nonhomogeneous) example of Section 7.2.2; we shall not go into the details of the obvious modifications which need to be applied to (7.49), (7.53), nor into those which relate to the use of the duality algorithms given earlier. We shall simply state that with a triangulation $\mathcal{T}_h$ of 128 triangles we obtain approximate solutions u_h^i such that

(7.56) $$\max_P | u_h^1(P) - u(P) | = 6\times 10^{-3}\,, \quad P \in \{ (m_{iT})_{i=1,2,3} \}_{T \in \mathcal{T}_h} \cup \cup \{ (M_{iT})_{i=1,2,3} \}_{T \in \mathcal{T}_h}\,,$$

(7.57) $$\max_P | u_h^2(P) - u(P) | = 8.5\times 10^{-4}\,, \quad P \in \{ (m_{iT})_{i=1,2,3} \}_{T \in \mathcal{T}_h} \cup \cup \{ (M_{iT})_{i=1,2,3} \}_{T \in \mathcal{T}_h}\,,$$

these maximum errors being obtained in the neighbourhood of the interface between the rigid and fluid zones (where derivatives of order $\geqslant 2$ are discontinuous). ■

8. APPLICATION TO THE SOLUTION OF THE ELASTO-PLASTIC TORSION PROBLEM OF CHAPTER 3.

8.1. Synopsis

A duality algorithm was formulated in Chapter 3, Section 9.1.1, equations (9.5), (9.6), (9.7), which allows the solution of the elasto-plastic torsion problem of Chapter 3 to be obtained. The application of this algorithm entails, among other things, the minimisation of a nondifferentiable functional of a type very similar to that considered in Chapter 5, Sections 1 to 7; we shall therefore use a variant of algorithm (7.8), (7.9), (7.10) of Section 7.1 to solve the above nondifferentiable problem. ■

8.2. Reformulation of algorithm (9.5), (9.6), (9.7) of Chapter 3, Section 9.1.1.

For reasons of clarity we shall reformulate algorithm (9.5), (9.6), (9.7) of Chapter 3, Section 9.1.1, using the same formalism; we thus define

$$\Lambda_1 = L^2_+(\Omega)\,, \tag{8.1}$$

$$\begin{cases} \phi_1 : H^1_0(\Omega) \to L^2(\Omega) \quad \text{by} \\ \phi_1(v) = |\operatorname{grad} v| - 1\,, \end{cases} \tag{8.2}$$

and the Lagrangian $\mathscr{L}_1 : H^1_0(\Omega) \times \Lambda_1 \to \mathbb{R}$ by

$$\mathscr{L}_1(v, q) = -\frac{1}{2}\,a(v, v) - \int_\Omega fv\,\mathrm{dx} + \int_\Omega q\phi_1(v)\,\mathrm{dx}\,. \tag{8.3}$$

From the relation

$$\sup_{q \in \Lambda_1} \int_\Omega q\phi_1(v)\,\mathrm{dx} = \begin{cases} 0 \text{ if } |\operatorname{grad} v| \leqslant 1\,, \\ +\infty \text{ otherwise}, \end{cases} \tag{8.4}$$

we deduce that if $\mathscr{L}_1$ admits a saddle point $\{u, p\}$ on $H^1_0(\Omega) \times \Lambda_1$, u is the solution of the problem (P_0) :

$$(P_0) \min_{v \in K_0} \left[\frac{1}{2}\,a(v, v) - \int_\Omega fv\,\mathrm{dx}\right], \tag{8.5}$$

$$K_0 = \{ v \mid v \in H^1_0(\Omega),\ |\operatorname{grad} v| \leqslant 1 \quad \text{a.e.}\}\,. \tag{8.6}$$

Thus, in the case where the assumptions (4.10), (4.11) of Chapter 2, Section 4.3 are satisfied, we can solve (P_0) by applying algorithm (4.12),...,(4.15) of Chapter 2, Section 4.3, that is:

$$\text{(8.7)} \quad \begin{cases} \text{with } p^n \text{ known } (\in \Lambda_1),\ u^n \text{ is defined as being the element} \\ \text{of } H_0^1(\Omega) \text{ which minimises } \frac{1}{2}a(v, v) - \int_\Omega fv\,dx + \int_\Omega p^n \phi_1(v)\,dx\,, \end{cases}$$

$$\text{(8.8)} \quad p^{n+1} = \sup(0, p^n + \rho_n \phi_1(u^n))\,,$$

$$\text{(8.9)} \quad \rho_n > 0 \text{ sufficiently small.} \quad \blacksquare$$

8.3. Approximate implementation of algorithm (8.7), (8.8), (8.9).

We shall restrict our attention to the solution of the approximate problem $(P_{0h})_{14}$ of Chapter 3, Section 5.3; we shall use the notation of Chapter 3, Section 5 and Chapter 5, Section 7.2.

8.3.1. *Existence of a saddle point. Description and convergence of the duality algorithm*

Existence of a saddle point.

We define

$$\text{(8.10)} \quad \Lambda_{1h} = \{ q_h \mid q_h = \{ q_{i+1/2\,j+1/2} \}_{M_{i+1/2\,j+1/2} \in \Sigma_h},\ q_{i+1/2\,j+1/2} \geqslant 0 \quad \forall M_{i+1/2\,j+1/2} \in \Sigma_h \}\,,$$

and we associate with the approximate problem $(P_{0h})_{14}$ the approximation $\mathscr{L}_{1h}$ of $\mathscr{L}_1$ defined by

$$\text{(8.11)} \quad \begin{cases} \mathscr{L}_{1h}(v_h, q_h) = J_{0h}(v_h) + h^2 \sum_{M_{i+1/2\,j+1/2} \in \Sigma_h} q_{i+1/2\,j+1/2} \times \\ \quad \times \left(\sqrt{|G^1_{i+1/2\,j+1/2}(v_h)|^2 + |G^2_{i+1/2\,j+1/2}(v_h)|^2} - 1\right). \end{cases}$$

Since the proof of Proposition 9.1 of Chapter 3, Section 9.2.1 can be adapted without difficulty, $\mathscr{L}_{1h}$ admits a saddle point, $\{u_h, p_h\}$, on $V_h \times \Lambda_{1h}$ with

$$\text{(8.12)} \quad \begin{cases} u_h \text{ the solution of } (P_{0h})_{14}\,, \\ p_{i+1/2\,j+1/2}\left(\sqrt{|G^1_{i+1/2\,j+1/2}(v_h)|^2 + |G^2_{i+1/2\,j+1/2}(v_h)|^2} - 1\right) = 0 \\ \quad \forall M_{i+1/2\,j+1/2} \in \Sigma_h\,. \end{cases}$$

Description of the algorithm and convergence.

We use a finite-dimensional variant of algorithm (8.7), (8.8), (8.9):

$$\text{(8.13)} \quad \begin{cases} \text{with } p_h^n \text{ known } (\in \Lambda_{1h}), \text{ we define } u_h^n \text{ as the element of} \\ V_h \text{ which minimises } \mathscr{L}_{1h}(v_h, p_h^n)\,, \end{cases}$$

(8.14)
$$\begin{cases} p_{i+1/2\,j+1/2}^{n+1} = \max\,(0, p_{i+1/2\,j+1/2}^{n} + \\ \quad + \rho_n(\sqrt{|\,G_{i+1/2\,j+1/2}^{1}(u_h^n)\,|^2 + |\,G_{i+1/2\,j+1/2}^{2}(u_h^n)\,|^2} - 1))\,, \\ \quad \forall M_{i+1/2\,j+1/2} \in \Sigma_h\,, \end{cases}$$

(8.15) $\rho_{1n} > 0$, sufficiently small.

It results from Theorem 4.1 of Chapter 2, Section 4.3 that the sequence $(u_h^n)_n$ defined by (8.13), (8.14), (8.15) converges to the solution of $(P_{0h})_{14}$. ■

8.3.2. *Solution of the nondifferentiable problem* (8.13).

There is no difficulty in adapting the considerations of Section 7.2.1 to the solution of the nondifferentiable problem (8.13), and we then obtain the following variant of algorithm (7.19), (7.20), (7.21) of Section 7.2.1 for calculating u_h^n from p_h^n :

(8.16) $\lambda_h^{n,0}$ given (we take $\lambda_h^{n,0} = \lambda_h^{n-1}$);

with $\lambda_h^{n,m}$ known, we define $u_h^{n,m}$ and then $\lambda_h^{n,m+1}$ by:

(8.17)
$$\begin{cases} -\dfrac{u_{i+1\,j}^{n,m} + u_{i-1\,j}^{n,m} + u_{i\,j+1}^{n,m} + u_{i\,j-1}^{n,m} - 4\,u_{ij}^{n,m}}{h^2} = \\ \quad = f_{ij} + \left(\dfrac{\lambda_{i+1/2\,j+1/2}^{1n,m} - \lambda_{i-1/2\,j+1/2}^{1n,m} + \lambda_{i+1/2\,j-1/2}^{1n,m} - \lambda_{i-1/2\,j-1/2}^{1n,m}}{2\,h} + \right. \\ \quad \left. + \dfrac{\lambda_{i+1/2\,j+1/2}^{2n,m} - \lambda_{i+1/2\,j-1/2}^{2n,m} + \lambda_{i-1/2\,j+1/2}^{2n,m} - \lambda_{i-1/2\,j-1/2}^{2n,m}}{2\,h}\right) \end{cases}$$

$\forall M_{ij} \in \Omega_h$, $u_{pq}^{n,m} = 0$ if $M_{pq} \notin \Omega_h$,

(8.18)
$$\begin{cases} (\lambda_{i+1/2\,j+1/2}^{1n,m+1}, \lambda_{i+1/2\,j+1/2}^{2n,m+1}) = \\ \quad = P_{p_{i+1/2\,j+1/2}^{n}}(\lambda_{i+1/2\,j+1/2}^{1n,m} + \rho_{2m}^{n}\,G_{i+1/2\,j+1/2}^{1}(u_h^{n,m}), \\ \quad \lambda_{i+1/2\,j+1/2}^{2n,m} + \rho_{2m}^{n}\quad G_{i+1/2\,j+1/2}^{2}(u_h^{n,m}))\quad \forall M_{i+1/2\,j+1/2} \in \Sigma_h\,, \end{cases}$$

with, for $y = (y_1, y_2) \in \mathbb{R}^2$ and $q \geq 0$:

(8.19)
$$\begin{cases} P_q(y) = 0 \quad \text{if } q = 0 \quad \forall y \in \mathbb{R}^2\,, \\ P_q(y) = \dfrac{qy}{\max\,(q, \sqrt{y_1^2 + y_2^2})} \quad \text{if} \quad q > 0\,. \end{cases}$$

Since Theorem 4.1 of Chapter 2, Section 4.3 also applies here, the sequence $u_h^{n,m}$ defined by (8.16), (8.17), (8.18) converges to the solution u_h^n of (8.13) for $\rho_{2n}^m > 0$ sufficiently small. ■

8.4. Application to an example [1]

We consider again the basic example of Chapter 3, namely $\Omega =]0, 1[\times]0, 1[, f = 10$.

Mesh size : $h = 1/20$.

Solution of the approximate Dirichlet problems (8.17):

By point over-relaxation with optimal parameter, initialising with $u_h^{n,m-1}$, using a termination criterion of the type (7.24). ■

Initialisation of algorithm (8.16), (8.17), (8.18):

$$\lambda_h^{n,0} = \lambda_h^{n-1} . \tag{8.20}$$

Termination criterion for algorithm (8.16), (8.17), (8.18):

$$\sum_{M_{ij} \in \Omega_h} | u_{ij}^{n,m} - u_{ij}^{n,m-1} | \leqslant 10^{-3} . \tag{8.21}$$

Initialisation of algorithm (8.13), (8.14), (8.15):

$$p_h^0 = 0 . \tag{8.22}$$

Termination criterion for algorithm (8.13), (8.14), (8.15):

$$\sum_{M_{ij} \in \Omega_h} | u_{ij}^n - u_{ij}^{n-1} | \leqslant 10^{-3} . \tag{8.23}$$

Choice of ρ_{1n} *and* ρ_{2m}^n : $\rho_{1n} = \rho_1 = 1, \quad \rho_{2m}^n = \rho_2 = 0.1$.

Numerical results:

Convergence was attained in 12.57 sec on an IBM 360/91, the total number of over-relaxation iterations carried out being 794. ■

8.5. A variant of algorithm (8.13), (8.14), (8.15)

As the computation time for the example in Section 8.4 appears prohibitive, we have also used the following algorithm:

$$(u_h^0, p_h^0, \lambda_h^0) \quad \text{given;} \tag{8.24}$$

with $u_h^n, p_h^n, \lambda_h^n$ known, we calculate $p_h^{n+1}, \lambda_h^{n+1}, u_h^{n+1}$ by

$$\begin{cases} p_{i+1/2\,j+1/2}^{n+1} = \max\,(0, p_{i+1/2\,j+1/2}^n + \\ \quad + \rho_{1n}\,(\sqrt{| G_{i+1/2\,j+1/2}^1(u_h^n) |^2 + | G_{i+1/2\,j+1/2}^2(u_h^n) |^2} - 1))\,, \\ \forall M_{i+1/2\,j+1/2} \in \Sigma_h\,, \end{cases} \tag{8.25}$$

---

(1) This numerical implementation is due to M. Goursat.

$$
(8.26) \qquad \begin{cases} (\lambda^{1n+1}_{i+1/2\,j+1/2}, \lambda^{2n+1}_{i+1/2\,j+1/2}) = \\ \qquad = P_{p^{n+1}_{i+1/2\,j+1/2}}(\lambda^{1n}_{i+1/2\,j+1/2} + \rho_{2n}\, G^1_{i+1/2\,j+1/2}(u^n_h), \\ \qquad \lambda^{2n}_{i+1/2\,j+1/2} + \rho_{2n}\, G^2_{i+1/2\,j+1/2}(u^n_h)), \quad \forall M_{i+1/2\,j+1/2} \in \Sigma_h\,; \end{cases}
$$

and finally determine u_h^{n+1} by using *only one* over-relaxation iteration, starting from u_h^n, in the solution of the linear system

$$
(8.27) \qquad \begin{cases} -\dfrac{v_{i+1j} + v_{ij+1} + v_{i-1j} + v_{ij-1} - 4\,v_{ij}}{h^2} = \\ = f_{ij} + \left(\dfrac{\lambda^{1n+1}_{i+1/2\,j+1/2} - \lambda^{1n+1}_{i-1/2\,j+1/2} + \lambda^{1n+1}_{i+1/2\,j-1/2} - \lambda^{1n+1}_{i-1/2\,j-1/2}}{2\,h} + \right. \\ \left. + \dfrac{\lambda^{2n+1}_{i+1/2\,j+1/2} - \lambda^{2n+1}_{i+1/2\,j-1/2} + \lambda^{2n+1}_{i-1/2\,j+1/2} - \lambda^{2n+1}_{i-1/2\,j-1/2}}{2\,h} \right) \\ \forall M_{ij} \in \Omega_h\,, \quad v_{pq} = 0 \text{ if } M_{pq} \notin \Omega_1\,. \end{cases}
$$

Starting from $u_h^0 = 0, p_h^0 = 0, \lambda_h^0 = 0$ and taking as termination criterion

$$
(8.28) \qquad \sum_{M_{ij} \in \Omega_h} |\, u^n_{ij} - u^{n-1}_{ij} \,| \leqslant 10^{-3}\,,
$$

with $\rho_{1n} = \rho_1 = 0.4, \quad \rho_{2n} = \rho_2 = 0.2$, convergence is attained in 140 iterations or 3.9 sec. execution time on an IBM 360/91. ■

Remark 8.1. The duality method developed in this section with a view to solving the elasto-plastic torsion problem of Chapter 3, appears to have a performance inferior to those of the other duality methods considered in Chapter 3, Section 9; in particular this method has a greater core storage requirement than the method of Section 9.4. The only advantage is that the matrix of the approximate Dirichlet problem to be solved at each iteration is independent of the iteration under consideration; this is not the case with algorithm (9.74), (9.75), (9.76) of Chapter 3, Section 9.4.1 in which the matrix of the approximate Dirichlet problem (9.74) depends on p_h^n. ■

9. DISCUSSION

Sections 1 to 7 of this chapter represent an extension of the paper of Céa-Glowinski /1/ in which the authors restricted their attention to finite difference approximations.

The results of this chapter would appear to indicate that the method best suited to the solution of the variational problem (P_0) defined by (1.5) for domains Ω of general shape, consists of the combined use of the *first order finite element* approximation of Section 3 and the variant (7.31), (7.32), (7.33) (see Section 7.3.1) of the *duality* algorithm (7.8), (7.9), (7.10) of Section 7.1. However, we considered it appropriate to dwell at some length (in Section 6) on *regularisation*, insofar as this type of method seems - at least at present - the most suitable for the numerical solution of certain *optimal control* problems in which the usual *state equation* is replaced by a *variational inequality* involving a nondifferentiable term; indeed regularisation allows the above optimal control problem to be approximated by a problem in which the "state inequality" is replaced by a nonlinear state equation for which it is possible to state explicitly conditions necessary for optimality; on this subject, see Yvon /1/, /2/. With regard to regularisation, we note that in his proof of the H^2 regularity of the solution of (P_0) (mentioned in Section 2.1), Brézis /6/ uses - among other things - a regularisation of the nondifferentiable term $\int_\Omega |\operatorname{grad} v|\,dx$.

With regard to duality methods, we should mention in connection with the solution of problem (1.5) that the use of a finite-dimensional variant of the duality algorithm [1]:

$$\lambda^0 \text{ given } (\in \Lambda),\quad u^0 \text{ given } (\in H_0^1(\Omega)), \tag{9.1}$$

$$u^{n+1} = u^n - \rho_1 S^{-1}(-\Delta u^n - g \operatorname{div} \lambda^n - f),\quad \rho_1 > 0, \tag{9.2}$$

$$\lambda^{n+1} = P_\Lambda(\lambda^n + \rho_2 \operatorname{grad} u^{n+1}),\quad \rho_2 > 0, \tag{9.3}$$

taking S defined by

$$S = \left(I - \frac{\partial^2}{\partial x^2}\right)\left(I - \frac{\partial^2}{\partial y^2}\right), \tag{9.4}$$

with homogeneous Dirichlet boundary conditions on Γ, did not lead [2] to a computation time less than those obtained with the finite-dimensional variants of algorithm (7.8), (7.9), (7.10) of Section 7.1. ■

To conclude this chapter, we would like to mention another algorithm for solving (1.5); this is a variant of algorithm (10.5)-(10.7) of Chapter 3, Section 10, the notation of which we shall use here. We thus write:

$$L = L^2(\Omega) \times L^2(\Omega)$$

$$j(v, q) = \frac{1}{2}\int_\Omega q^2\,dx + g\int_\Omega |q|\,dx - \int_\Omega fv\,dx;$$

there is then equivalence between (P_0) and

---

[1] Of Arrow-Hurwicz type (see Chapter 2, Section 4.4)

[2] For rectangular domains and finite-difference approximations, which are the only practical cases for which it is possible (and worthwhile) to apply algorithm (9.1), (9.2), (9.3) with S defined by (9.4).

(9.5) $\quad \text{Min}\, j(v, q)\,, \quad (v, q) \in H_0^1(\Omega) \times L\,, \quad \nabla v - q = 0\,.$

Let u be the solution of (P_0). Then $(u, \nabla u)$ is the unique solution of (9.5); a *Lagrangian* naturally associated with (9.5) is

$$\mathscr{L}(v, q\,;\mu) = j(v, q) + \int_\Omega \mu.(\nabla v - q)\,\mathrm{d}x\,.$$

With μ and q fixed, since $\mathscr{L}$ is non-coercive in v, we shall again penalise $\nabla v - q = 0$, giving

$$\mathscr{L}_\varepsilon(v, q\,;\mu) = \mathscr{L}(v, q\,;\mu) + \frac{1}{2\,\varepsilon}\int_\Omega |\,\nabla v - q\,|^2\,\mathrm{d}x\,.$$

It can be shown that $\mathscr{L}_\varepsilon$ admits a unique saddle point on $H_0^1(\Omega) \times L \times L$, of the form $(u, \nabla u\,;\lambda)$ where u is the solution of (P_0) and λ the solution of the dual problem of (P_0) described in Chapter 1, Section 3.5.

We shall apply Uzawa's algorithm to $\mathscr{L}_\varepsilon$, giving

(9.6) $\quad \lambda^0 \in L$ given;

with λ^n known, we successively determine u^n, p^n, λ^{n+1} by

(9.7) $$\begin{cases} \mathscr{L}_\varepsilon(u^n, p^n\,;\mu^n) \leqslant \mathscr{L}_\varepsilon(v, q\,;\mu^n) \quad \forall (v, q) \in H_0^1(\Omega) \times L \\ (u^n, p^n) \in H_0^1(\Omega) \times L \end{cases}$$

(9.8) $\quad \lambda^{n+1} = \lambda^n + \rho_n(\nabla u^n - p^n)\,, \quad \rho_n > 0\,.$

In view of the existence of a saddle point of $\mathscr{L}_\varepsilon$ on $H_0^1(\Omega) \times L \times L$, under the condition

(9.9) $\quad 0 < r_0 \leqslant \rho_n \leqslant r_1 < 2/\varepsilon$

we have

$$\lim_{n\to+\infty} (u^n, p^n) = (u, \nabla u) \quad (\textit{strong}) \text{ in } H_0^1(\Omega) \times L$$

It may be noted that the determination of (u^n, p^n) using (9.7) is an infinite-dimensional problem of the minimisation of a *nondifferentiable* functional; however, because of the very special structure of this functional, we can use the relaxation methods described and studied in Céa-Glowinski /2/ to solve (9.7). We refer the reader to Glowinski-Marrocco /3/ for a detailed study of the application of (9.6)-(9.8) to an approximation of (P_0) using first-order finite elements.

Chapter 6

GENERAL METHODS FOR THE APPROXIMATION AND SOLUTION OF TIME-DEPENDENT VARIATIONAL INEQUALITIES

INTRODUCTION

In this chapter we study time-dependent problems. We shall consider the general subject of variational *inequalities*, examining the *approximation* and *solution* of these inequalities in some detail.

Three fundamental types of inequalities will be investigated: parabolic inequalities of type I, parabolic inequalities of type II and time-dependent inequalities of the second order in time. Each of these types of problem will be considered separately, the emphasis being laid on concrete examples which bring out the nature of the general formulation used. This general formulation will enable us to make an in-depth study of the finite difference approximation [1] and then for each type of inequality we shall investigate the numerical solution of several typical problems.

To help identify the nature of the problems in question, let us say that it is natural from both the physical and mathematical viewpoints to extend the general formalism of steady-state inequalities:

$$(Au - f, v - u) + j(v) - j(u) \geqslant 0 \quad \forall v \in K, u \in K,$$

to the case in which the solution evolves in time. This leads to an examination of problems of the following types:

- *Parabolic inequalities of type* I: find a function $t \to u(t)$ for $t \in [0, T]$, with values in $K \subset V$ such that

$$\left(\frac{\mathrm{d}u}{\mathrm{d}t}(t) + Au(t) - f(t), v - u(t)\right) + j(v) - j(u(t)) \geqslant 0$$

$$\forall v \in K, u(t) \in K, u(0) = u_0 \text{ given.}$$

- *Parabolic inequalities of type* II: find a function $t \to u(t)$ for $t \in [0, T]$, with values in V such that

---

(1) All of what follows is valid (and is somewhat simpler) for *interior* approximations of the "finite element" type; moreover, it is an interior method which we shall use in Section 11.

$$\left(\frac{du}{dt}(t) + Au(t) - f(t), v - \frac{du}{dt}(t)\right) + j(v) - j\left(\frac{du}{dt}(t)\right) \geqslant 0$$

$$\forall v \in K, \frac{du}{dt}(t) \in K .$$

- *Hyperbolic inequalities* (or well posed in the Petrowski sense): find a function $t \to u(t)$ for $t \in [0, T]$, with values in V such that

$$\left(\frac{d^2u}{dt^2}(t) + Au(t) - f(t), v - \frac{du}{dt}(t)\right) + j(v) - j\left(\frac{du}{dt}(t)\right) \geqslant 0 \quad \forall v \in K, \frac{du}{dt}(t) \in K .$$

1. BACKGROUND

Following on from the initial concepts outlined in Chapter 1, Section 1.1, we give here a brief review of the principal concepts of functional analysis which will be of use later. The fundamental aspects may be found, for example, in Duvaut-Lions /1/, Chapter 1. More advanced developments are given in Lions-Magènes /1/, Schwartz /1/ and Sobolev /1/.

1.1 Spaces of vector-valued distributions and functions

Given an interval $[0, T] \subset \mathbb{R}$ (in general $T < +\infty$) and a Banach space X with norm $\|\cdot\|_X$, we denote by $L^p(0, T; X)$ the space of (classes of) functions $t \to f(t)$ which are measurable from $[0, T] \to X$ (for the measure dt) such that

$$\| f \|_{L^p(0,T;X)} = \left(\int_0^T \| f(t) \|_X^p \, dt\right)^{1/p} < +\infty \quad (p \neq +\infty),$$

$$\| f \|_{L^\infty(0,T;X)} = \sup_{t \in (0,T)} \text{ess} \, \| f(t) \|_X < +\infty .$$

The spaces $L^p(0, T; X)$ are Banach spaces for the first norm if $p \neq +\infty$, and for the second norm if $p = +\infty$.

If X is a Hilbert space equipped with an inner product $(,)_X$ then the space $L^2(0, T; X)$ is also a Hilbert space for the inner product

$$(f, g)_{L^2(0,T;X)} = \int_0^T (f, g)_X \, dt .$$

We denote by $\mathscr{D}(]0,T[\,;X)$ the space of indefinitely differentiable functions with compact support in $]0,T[$ and values in X, and by $\mathscr{D}'(]0,T[\,;X)$ the space of distributions on $]0,T[$ with values in X, defined by

$$\mathscr{D}'(]0,T[\,;X) = \mathscr{L}(\mathscr{D}(]0,T[)\,;X),$$

where in a general manner $\mathscr{L}(M\,;N)$ denotes the space of linear continuous mappings from $M \to N$.

In particular we can associate with $f \in L^p(0,T\,;X)$ a distribution $\tilde{f} : \mathscr{D}(]0,T[) \to X$ defined by

$$\tilde{f}(\phi) = \int_0^T f(t)\,\phi(t)\,\mathrm{d}t\,, \quad \phi \in \mathscr{D}(]0,T[)\,.$$

The mapping $\phi \to \tilde{f}(\phi)$ is indeed linear continuous from $\mathscr{D}(]0,T[) \to X$. Moreover, $f \to \tilde{f}$ is an injection and it is appropriate to identify f and $\tilde{f}$. Also $f \to 0$ in $L^p(0,T\,;X) \Rightarrow \tilde{f} \to 0$ in $\mathscr{D}'(]0,T[\,;X)$, i.e. $\tilde{f}(\phi) \to 0$, $\forall \phi \in \mathscr{D}(]0,T[)$. We thus have, algebraically and topologically

$$L^p(0,T\,;X) \subset \mathscr{D}'(]0,T[\,;X)\,.$$

For $f \in \mathscr{D}'(]0,T[\,;X)$, we define $\mathrm{d}^k f/\mathrm{d}t^k = f^{(k)}$ by

$$f^{(k)}(\phi) = (-1)^k f(\phi^{(k)}) \quad \forall \phi \in \mathscr{D}(]0,T[)\,,$$

and hence, in particular, $f^{(k)} \in \mathscr{D}'(]0,T[\,;X)$, $\forall k$. ■

1.2 Functional setting

Whereas in *steady-state* problems we have used *one* Hilbert space V (or, in certain cases, one Banach space), in the case of *time-dependent* problems we *must* make use of *two Hilbert spaces* (and in certain cases, two Banach spaces; however, we shall restrict our attention to the case of Hilbert spaces). Thus let V and H be two Hilbert spaces with:

(1.1) $\quad V \subset H$, the injection of V in H is continuous, V dense in H.

We denote by $((,))$ (resp. $(,)$) the inner product in V (resp. H) and by $\|\ \|$ (resp. $|\ |$) the corresponding norm. We thus have

(1.2) $\quad |v| \leqslant c\,\|v\| \quad \forall v \in V\,.$

We identify H *with its dual.* Then, in the identification compatible with the above, we have:

(1.3) $\quad V \subset H \subset V' \qquad (V' = \text{dual of } V)\,.$

We equip V' with the "dual" norm

$$\| w \|_* = \sup_{v \in V} (w, v), \quad \text{with} \quad \| v \| = 1 .$$

We define the space $W(0, T)$ by

$$W(0, T) = \left\{ v \mid v \in L^2(0, T; V), v' = \frac{dv}{dt} \in L^2(0, T; V') \right\}.$$

With the inner product

$$(u, v)_{L^2(0,T;V)} + (u', v')_{L^2(0,T;V')} ,$$

$W(0, T)$ is a Hilbert space. Every $v \in W(0, T)$ is, after possible modification on a set of measure zero, continuous from $[0, T] \to H$.

We can then define a closed affine subspace of $W(0, T)$ by

$$W_0(0, T) = \{ v \mid v \in W(0, T), v(0) = u_0, \quad u_0 \text{ given in } H \}.$$

2. INTRODUCTION TO PARABOLIC TIME-DEPENDENT INEQUALITIES OF TYPE I

In this section we shall introduce by means of two examples the general formalism of parabolic time-dependent inequalities of type I.

2.1 Examples of parabolic inequalities of type I

2.1.1 *Example* I

In this first example we again consider the diffusion problem given as a first example in Chapter 1, Section 1.1. Here, however, we shall consider the general case in which we seek to calculate the *evolution* of the fluid pressure when a pressure difference is applied at the outer surface of the *thin semi-permeable* wall. Thus, instead of attempting to find what the pressure will be when equilibrium is reached, we seek to determine the evolution of the pressure at each point of the enclosure Ω, *starting from the initial instant* at which a pressure difference is established at the boundary Γ of Ω, until a given instant T (if T is sufficiently large, the pressure must evolve towards the steady-state solution)

This problem leads to the following formalism: find a function $u(x, t)$ for $t \in [0, T]$, $x \in \Omega$ such that

$$\text{(2.1)} \qquad \frac{\partial u}{\partial t} - \Delta u = f, \quad \text{for } (x, t) \in Q =]0, T[\times \Omega ,$$

with *the boundary conditions* ($h(x)$ being the external pressure applied at $x \in \Gamma$):

(2.2) $$u(x, t) > h(x) \Rightarrow \frac{\partial u}{\partial n}(x, t) = 0, \quad x \in \Gamma, t \text{ fixed}$$

(with the internal pressure higher than the external pressure, the semi-permeable wall prevents any exchange of fluid),

(2.3) $$u(x, t) \leqslant h(x) \Rightarrow \frac{\partial u}{\partial n}(x, t) \geqslant 0, \quad x \in \Gamma,$$

(with the external pressure higher than the internal pressure, the semi-permeable wall allows a positive flow from the outside towards the inside of Ω; thus by virtue of continuity, the only possible case is $u(x, t) = h(x)$). The *initial condition* for the problem is:

(2.4) $$u(x, 0) = u_0(x) = \text{given}, \quad x \in \Omega \quad (u_0 \geqslant h),$$

(it is clear that the evolution of the pressure is dependent on the pressure at the initial instant).

We shall pose this problem in the form of a *time-dependent variational inequality*. To this end we introduce

(2.5) $$V = H^1(\Omega), \qquad H = L^2(\Omega),$$

(2.6) $$a(u, v) = \int_\Omega \operatorname{grad} u \operatorname{grad} v \, dx,$$

(2.7) $$(f, g) = \int_\Omega fg \, dx,$$

(2.8) $$K = \{ v \mid v \in V, v \geqslant h \text{ on } \Gamma \},$$

and denote by $u(t)$ the function $x \to u(x, t)$; we then consider the problem:

(2.9)
find $u(t) \in K$, t traversing the interval $[0, T]$, such that

$$\left(\frac{\partial u}{\partial t}, v - u\right) + a(u, v - u) - (f, v - u) \geqslant 0, \quad \forall v \in K, \quad \text{with } u(0) = u_0.$$

Applying Green's formula in (2.9) we obtain

(2.10) $$\left(\frac{\partial u}{\partial t} - \Delta u - f, v - u\right) + \int_\Gamma \frac{\partial u}{\partial n}(v - u) \, d\Gamma \geqslant 0, \quad \forall v \in K, u \in K.$$

Using this inequality it may readily be shown that the problems (2.1) - (2.4) and (2.8), (2.9) are equivalent. An equivalent formalism is as follows:

We put:

$$(2.11)\quad \left|\begin{aligned} \mathscr{K} &= \{ v \mid v \in W(0, T), v(t) \in K, \text{ a.e. } t \in [0, T] \}, \\ \mathscr{K}_0 &= \{ v \mid v \in W_0(0, T), v(t) \in K, \text{ a.e. } t \in [0, T] \}, \end{aligned}\right.$$

and consider the problem derived from (2.9) by integration:

$$(2.12)\quad \left|\begin{aligned} &\text{find } u \in \mathscr{K}_0 \text{ such that} \\ &\int_0^T (u', v - u)\,\mathrm{d}t + \int_0^T a(u, v - u)\,\mathrm{d}t - \int_0^T (f, v - u)\,\mathrm{d}t \geqslant 0, \\ &\quad \forall v \in \mathscr{K}. \end{aligned}\right.$$

It can be shown without difficulty that (2.12) implies (2.9).

2.1.2 *Example* II

In the previous example we introduced a time-dependent inequality relating to a *convex set*. In the example which follows we shall introduce a time-dependent inequality which involves a *non-differentiable* term. This inequality will be obtained from the *time-dependent* formulation of a *thermal control* problem (cf. Duvaut-Lions /1/, Chapter 1, 2.3.1).

The notation will be as follows: Ω = domain inside the enclosure, Γ = boundary of Ω, $u(x, t)$ = temperature at $x \in \Omega$ at the instant t. Air-conditioning units are present whose function is to inject heat fluxes across Γ when the temperature $u(x, t)$, $x \in \Gamma$, lies outside an interval $[h_1, h_2]$. When

$$u(x, t) \notin [h_1, h_2], \qquad x \in \Gamma,$$

we assume that we can produce a heat flux whose value is proportional to the difference between $u(x, t)$, $x \in \Gamma$ and the nearer of the two numbers h_1, h_2.

The problem is formulated as follows: find $u(x, t)$ for $t \in [0, T]$, $x \in \Omega$ such that:

$$(2.13)\quad \frac{\partial u}{\partial t} - \Delta u = f \text{ for } (x, t) \in Q = \Omega \times]0, T[,$$

with the *boundary conditions*

$$(2.14)\quad -\frac{\partial u}{\partial n} = \phi(u), \quad x \in \Gamma,$$

where $\phi(v)$ is a function, which is not differentiable everywhere, defined by:

$$(2.15) \qquad \phi(\lambda) = \begin{cases} g_1(\lambda - h_1) & \text{if} \quad \lambda \leqslant h_1\,, \\ 0 & \text{if} \quad h_1 \leqslant \lambda \leqslant h_2\,, \\ g_2(\lambda - h_2) & \text{if} \quad \lambda \geqslant h_2\,, \end{cases}$$

with

$$g_1 < 0 < g_2\,, \quad h_1 \leqslant h_2\,,$$

and *the initial condition*

$$(2.16) \qquad u(x, 0) = u_0(x) = \text{given initial temperature}, \quad x \in \Omega\,.$$

This problem can be written in the form of a *time-dependent variational inequality*. To this end we introduce (2.5), (2.6), (2.7) and

$$(2.17) \qquad j(v) = \int_\Gamma \phi(v)\, \mathrm{d}\Gamma\,,$$

and we consider the problem

$$(2.18) \qquad \left| \begin{array}{l} \text{find } u(t) \in V,\ t \text{ traversing the interval } [0, T], \text{ such that} \\ \left(\dfrac{\partial u}{\partial t}, v - u\right) + a(u, v - u) - (f, v - u) + j(v) - j(u) \geqslant 0\,, \\ \forall v \in V, \text{ with } u(0) = u_0\,. \end{array} \right.$$

Applying Green's formula in (2.18) we obtain

$$(2.19) \qquad \left(\frac{\partial u}{\partial t} - \Delta u - f, v - u\right) + \int_\Gamma \frac{\partial u}{\partial n}(v - u)\, \mathrm{d}\Gamma + \\ + \int_\Gamma [\phi(v) - \phi(u)]\, \mathrm{d}\Gamma \geqslant 0\,.$$

Using this inequality it is possible to show that the problems (2.13)-(2.16) and (2.17), (2.18) are equivalent, these problems themselves being equivalent to the following:

$$(2.20) \qquad \left| \begin{array}{l} \text{find } u \in W_0(0, T) \text{ such that} \\ \displaystyle\int_0^T (u', v - u)\, \mathrm{d}t + \int_0^T a(u, v - u)\, \mathrm{d}t - \int_0^T (f, v - u)\, \mathrm{d}t + \\ \qquad + \displaystyle\int_0^T [j(v) - j(u)]\, \mathrm{d}t \geqslant 0\,, \quad \forall v \in W(0, T)\,. \quad \blacksquare \end{array} \right.$$

2.2 Abstract formulation

Throughout the rest of the discussion of parabolic inequalities of type I, we shall use the following setting and notation:

(2.21) $V, V', H, \| \|, ((,)), | |, (,)$ defined as in (1.1), (1.2), (1.3)

(2.22) K, non-empty convex set of V

(2.23) $u_0 \in K$, given initial condition

(2.24) $A \in \mathscr{L}(V; V')$ with

(2.25) $(Av, v) \geqslant \alpha \| v \|^2$, α constant > 0.

We shall put

(2.26) $a(u, v) = (Au, v), \quad a(u) = (Au, u)$,

(2.27)
$$\begin{cases} j : K \to \mathbb{R}, \neq \pm \infty \text{ on } K, \text{ l.s.c. on } K \text{ and integrable} \\ \text{for all } v \in L^2(0, T; K) \text{ i.e.} \left| \int_0^T j(v) \, dt \right| < +\infty , \end{cases}$$

(2.28) $f \in L^2(0, T; V')$,

(2.29)
$$\left| \begin{array}{l} \mathscr{K} = \{ v \mid v \in W(0, T), v(t) \in K, \quad \text{a.e. } t \in [0, T] \} , \\ \mathscr{K}_0 = \{ v \mid v \in W_0(0, T), v(t) \in K, \quad \text{a.e. } t \in [0, T] \} . \end{array} \right.$$

The general problem investigated is

(2.30)
$$\left| \begin{array}{l} \text{find } u \in \mathscr{K}_0 \text{ such that} \\ \displaystyle\int_0^T (u' + Au - f, v - u) \, dt + \int_0^T [j(v) - j(u)] \, dt \geqslant 0 , \quad \forall v \in \mathscr{K}, \end{array} \right.$$

or in an equivalent local (or instantaneous) form:

(2.31)
$$\left| \begin{array}{l} \text{find } u \in \mathscr{K}_0 \text{ such that} \\ (u' + Au - f, v - u) + j(v) - j(u) \geqslant 0 , \quad \forall v \in K, \quad \text{a.e. } t \in [0, T] . \end{array} \right.$$

A solution u of (2.30) or (2.31), when it exists, is called a *strong* solution.

The following results can be proved:

Theorem 2.1. *Under the conditions* (2.21),...,(2.29) *and under the supplementary assumptions*

(2.32) $f' = \dfrac{\partial f}{\partial t} \in L^2(0, T; V')$,

(2.33) $j(v) \equiv 0$,

(2.34) $f(0) - Au_0 \in H$,

the problem (2.31) (*or* (2.30)) *admits a unique solution* u. *Moreover this solution satisfies*

$$u, u' \in L^2(0, T; V) \cap L^\infty(0, T; H). \quad \blacksquare$$

Remark 2.1. The results of the above theorem are still valid when $j(v) \not\equiv 0$ provided assumption (2.34) is replaced by an assumption of the same type involving the value of $j(u_0)$. See Duvaut-Lions /1/, Chapter 1, Section 5.2 and Brézis /1/, /2/.

Remark 2.2. The numerical methods described below allow us to *prove* this theorem. ■

Remark 2.3. Because of the assumptions necessary for the existence of the solution of (2.30) or (2.31), we are led to consider a "weak" formulation of problems (2.30), (2.31). To this end, we define

$$\mathscr{K}_f = \{ v \mid v \in L^2(0, T; V), v(t) \in K, \text{ a.e. } t \in [0, T] \},$$

and consider the problem

$$(2.35) \quad \left| \begin{array}{l} \text{find } u \in \mathscr{K}_f \text{ such that} \\ \displaystyle\int_0^T (v' + Au - f, v - u)\, dt + \int_0^T [j(v) - j(u)]\, dt \geqslant 0, \quad \forall v \in \mathscr{K}_0 . \end{array} \right.$$

It can be shown without difficulty that any solution of (2.30) is a solution of (2.35). However the converse is *not true*, unless supplementary assumptions are made (for example (2.32)-(2.34)).

A solution u of (2.35) is termed a *weak solution* of the problem.

We have:

Theorem 2.2. *With conditions* (2.21),...,(2.29) *the problem* (2.35) *possesses a unique weak solution.* ■

Remark 2.4. When $j(v)$ is differentiable everywhere - and convex - with gradient $j'(v)$, then in the inequalities (2.30) and (2.35) we can replace the term

$$\int_0^T [j(v) - j(u)]\, dt \quad \text{by} \quad \int_0^T (j'(v), v - u)\, dt,$$

and in (2.31) the term

$$j(v) - j(u) \quad \text{by} \quad (j'(v), v - u). \quad \blacksquare$$

Remark 2.5. We assume here that A, j *do not depend on* $t \in [0, T]$. For problems in which K depends on t, see Brézis /4/, Attouch and Damlamian /1/, J. Moreau /4/ and J.C. Peralba /1/. This assumption is made solely to avoid encumbering the investigation with an excessive number of assumptions. ■

Remark 2.6. When $K = V$ and when $j(v)$ is differentiable and convex, the inequalities *reduce to equations*; for example (2.31) becomes

$$\left| \begin{array}{l} \text{find } u \in W_0(0, T) \text{ such that} \\ u' + Au - f + j'(u) = 0, x \in \Omega, \text{ a.e. } t \in [0, T]. \end{array} \right.$$ ■

3. APPROXIMATION OF PARABOLIC INEQUALITIES OF TYPE I

Synopsis.

In this section we shall introduce formalisms *for the numerical solution* of the inequality (2.30). For this we shall define a family of inequalities "close" to

$$(3.1) \qquad \int_0^T (u' + Au - f, v - u)\,\mathrm{d}t + \int_0^T [j(v) - j(u)]\,\mathrm{d}t \geqslant 0, \quad \forall v \in \mathscr{K}, u \in \mathscr{K}_0,$$

or to the equivalent instantaneous inequality:

$$(3.2) \qquad (u' + Au - f, v - u) + j(v) - j(u) \geqslant 0, \quad \forall v \in K, u \in \mathscr{K}_0, \text{ a.e. } t \in [0, T],$$

and for which it will be possible to calculate the solutions more easily. The solutions of these inequalities will be sought in spaces $L^2(0, T; V_h)$, V_h being *finite-dimensional*. Denoting one of these solutions by u_h we shall show that when the dimension of the spaces V_h tends to infinity we have

$$u_h \to u \text{ weakly in } L^2(0, T; V)$$

where u is the "weak" solution of

$$(3.3) \qquad \int_0^T (v' + Au - f, v - u)\,\mathrm{d}t + \int_0^T [j(v) - j(u)]\,\mathrm{d}t \geqslant 0, \quad \forall v \in \mathscr{K}_0, u \in \mathscr{K}_f,$$

and, in the case when (3.1) possesses a unique *strong solution* u we shall show that

$$u_h \to u \text{ strongly in } L^2(0, T; V)$$

In general, the approximation of time-dependent problems consists of

- a spatial discretisation (discretisation of Ω);

- a time discretisation (discretisation of $[0, T]$);
- an approximation of the space V by a space V_h of functions allowing each element of V to be suitably approximated, and the investigation of the way in which V_h approximates H (1);
- an approximation of the convex set $K \subset V$ by a convex set K_h of V_h such that with each element of K we can associate a sufficiently "close" element taken in K_h;
- an approximation of the inequality (3.1) by an inequality which is compatible with the other characteristics of the approximation.

In the following section we shall examine these points and give the fundamental assumptions leading to the formulation of problems which are *numerically close* to the general problem (3.2). ■

3.1 Fundamental assumptions for the approximation

As in Chapter 1, Section 5, we shall first "axiomatise" the approximation assumptions. For the moment we shall restrict our attention to *exterior approximations* (see Chapter 1, Section 5.2); the theory which follows is valid (and in fact even a little simpler) in the case of *interior approximations* (of the *finite element* type).

3.1.1 *Approximation of V and H*

We consider $V, H, \| \|, | |$ as in Section 1.2 of the present chapter, and then introduce a *simultaneous exterior approximation of V and H*, using the following:

(3.4) Φ, a Hilbert space, with inner product $[,]_\Phi$ and with norm $[\ ]_\Phi$,

(3.5) $F = H \times \Phi$ equipped with the inner product and the product norm (2)

(3.6) $\pi : F \to H$, first canonical projection

(3.7) $\overline{\omega} : F \to \Phi$, second canonical projection

(3.8) $\sigma : V \to F$, extension isomorphism from V onto $\sigma V \subset F$ with:

(3.9) $\pi\sigma =$ identity

(3.10) h, parameter which tends to zero.

---

(1) This is one of the elements which are new compared with the steady-state case.

(2) For simplification we shall not introduce the space $\overline{H}$, the "extension" of H (see Céa /3/, Raviart /1/).

(3.11) V_h, a Hilbert space with inner product $((,))_h$ and norm $\| \|_h = \| \|_{V_h}$ (in general this space will be of finite dimension $N(h)$ with $N(h) \to +\infty$, when $h \to 0$).

(3.12) p_h, a linear operator from $V_h \to F$ called the extension of V_h to F, satisfying a *stability* condition:

(3.13) $\| p_h v_h \|_F \leqslant C \| v_h \|_h$, C a constant independent of h, $\forall v_h \in V_h$,

and a *convergence* condition (see (5.27), Chapter 1):

(3.14) $\forall v \in V$, $\exists v_h \in V_h$ with $p_h v_h \to \sigma v$ strongly in F and $\| v_h \|_h \leqslant C$.

In addition we have

(3.15) q_h, a linear operator from $V_h \to H$, injective such that $q_h = \pi p_h$.

We define a *second inner product* and a second norm on V_h,

$(v_h, w_h)_h$ and $| v_h |_h$,

which "approximate" the inner product $(,)$ and the norm $| |$ on H, by

$$(3.16)\qquad \left| \begin{array}{l} (v_h, w_h)_h = (q_h v_h, q_h w_h), \\ | v_h |_h = | q_h v_h |, \end{array} \right.$$

and we assume that $\forall v_h \in V_h$:

(3.17) $| v_h |_h \leqslant C \| v_h \|_h$ (in the same way that $| | \leqslant C \| \|$);

and that

(3.18) $\| v_h \|_h \leqslant S(h) | v_h |_h$, *with, in general* $S(h) \to \infty$ *if* $h \to 0$.

The definition of V_h and the approximation Ω_h of Ω will be specified at the time of application. ■

Typical example. In the setting of Chapter 1, Section 5.1,

$$V = H_0^1(\Omega),\ H = L^2(\Omega),\ \Phi = (L^2(\Omega))^n,$$

$$\sigma v = \left(v, \frac{\partial v}{\partial x_1}, \ldots, \frac{\partial v}{\partial x_n} \right);$$

if

$$f = \{ f_0, \phi \} \in H \times \Phi : \ \pi f = f_0, \ \overline{\omega} f = \phi;$$

V_h is given by (5.10), Chapter 1:

$$p_h v_h = \{ v_h, \delta_1 v_h, \ldots, \delta_n v_h \}, \quad q_h v_h = v_h .$$

It can then be verified that we have (3.18) with:

$$S(h) = \frac{C}{|h|}. \quad \blacksquare$$

3.1.2 *Discretisation of time and of the differential operator*

We divide $[0, T]$ into $N = N(k)$ intervals of length k, where k tends to zero as $N(k) \to +\infty$. We seek an "approximation" to the solution of the inequality (3.1) in the class of step functions on $[0, T]$ with values in V_h.

Let $v_h^i \in V_h$, $i = 0, \ldots, N$ and let $\chi_k^i(t)$ be the characteristic function of $[ik, (i+1)k[$. We put ([1])

$$v_{h,k} = \sum_{i=0}^{N} v_h^i \chi_k^i.$$

On the space of these functions we approximate the operator $u \to u' = \dfrac{du}{dt}$ by

$$(3.19) \qquad \delta v_{h,k}(t) = \frac{v_{h,k}(t+k) - v_{h,k}(t)}{k}$$

or by

$$(3.20) \qquad \bar{\delta} v_{h,k}(t) = \frac{v_{h,k}(t) - v_{h,k}(t-k)}{k} = \delta v_{h,k}(t-k). \quad \blacksquare$$

Remark 3.1. We could also approximate $v \to v'$ by discretisation operators of second order, third order, etc... To second order we would put:

$$\delta^* v_{h,k}(t) = \frac{1}{2k}\left[3 v_{h,k}(t) - 4 v_{h,k}(t-k) + v_{h,k}(t-2k)\right]. \quad \blacksquare$$

3.1.3 *Approximation of $\mathscr{K}$ and of the initial condition*

We adopt (2.22), (2.29) for the definition of K and $\mathscr{K}$. We take K_h as a closed convex set of V_h.

We shall make use of the following easily-proved results, see for example Trémolières /4/:

Lemma 3.1 (consistency). *We assume that* $\{K_h, K\}$ *is consistent* (Chapter 1, (5.33)).

Let $v_{h,k}$ *with* $h, k \to 0$, *be a sequence of functions such that:*

([1]) If necessary, we may define $v_{h,k}(t)$ over $[0, T+k]$ by summing from 0 to $N+1$.

$$\left|\begin{array}{l} v_{h,k}(t) \in K\,, \quad t \in [0, T]\,, \\ q_h\, v_{h,k} \to v \ \text{ weakly* in } L^\infty(0, T; H)\ \text{(1)} \\ p_h\, v_{h,k} \to \tilde{v} \ \text{ weakly in } L^2(0, T; F) \end{array}\right.$$

then $v \in K$, a.e. $t \in [0, T]$ and $\tilde{v} = \sigma v$.

Lemma 3.2 (convergence). *We assume that* $\{K_h, K\}$ *is convergent* (Chapter 1, (5.34)). *Then*:

$$\left|\begin{array}{l} \forall v \in L^2(0, T; K)\,, \quad \exists v_{h,k} \in L^2(0, T; K_k) \quad \text{(where } h \text{ is not connected with } k\text{) such that} \\ q_h\, v_{h,k} \to v \ \text{ strongly in } L^2(0, T; H) \\ p_h\, v_{h,k} \to \sigma v \ \text{ strongly in } L^2(0, T; F) \\ \text{and if in addition } v' \in L^2(0, T; V') \text{ then} \end{array}\right.$$

$$q_h\, \delta v_{h,k} \to v' \quad \text{strongly in } L^2(0, T; V').$$

Moreover we assume that the given initial condition $u_0 \in K \subset V$ can be approximated by a sequence $u_{0,h} \in K_h (h \to 0)$ such that

(3.21) $\quad p_h\, u_{0,h} \to \sigma u_0 \quad$ strongly in F

(3.22) $\quad \| u_{0,h} \|_h \leqslant C\,, \quad C$ constant, independent of k. ■

3.1.4 *Approximation of* $a(u, v)$

Given $a(u, v) = (Au, v)$ as in (2.24)-(2.26) we take $a_h(u_h, v_h) = (A_h\, u_h, v_h)_h$ satisfying assumptions (5.28)-(5.31) of Chapter 1, and $A_h \in \mathscr{L}(V_h; V_h)$.

The following results can be verified without difficulty:

Lemma 3.3 (consistency). *Given* $v_{h,k}(t) \in K_h$, a.e. $t \in [0, T]$ *such tha*

$$p_h\, v_{h,k} \to \sigma v \quad \text{weakly in } L^2(0, T; F)$$

then if the pair $\{ a_h(,), a(,) \}$ *is consistent* (Chapter 1, (5.31)) *we have*:

$$\liminf \int_0^T a_h(v_{h,k}, v_{h,k})\, \mathrm{d}t \geqslant \int_0^T a(v, v)\, \mathrm{d}t\,.$$

---

(1) i.e. $\displaystyle\int_0^T (q_h\, v_{h,k}, \phi)\, \mathrm{d}t \to \int_0^T (v, \phi)\, \mathrm{d}t\,, \quad \forall \phi \in L^1(0, T; H)\,.$

Lemma 3.4 (convergence). *Given* $v \in L^2(0, T; K)$, $w \in L^2(0, T; K)$ *and*

$v_{h,k}(t) \in K_h$, $w_{h,k}(t) \in K_h$, a.e. $t \in [0, T]$ *such that*

$p_h v_{h,k} \to \sigma v$ *weakly in* $L^2(0, T; F)$

$p_h w_{h,k} \to \sigma w$ *strongly in* $L^2(0, T; F)$

then if the pair $\{ a_h(,), a(,) \}$ *is convergent* (cf. Chapter 1, (5.30)) *we have*

$$\int_0^T a_h(v_{h,k}, w_{h,k}) \, dt \to \int_0^T a(v, w) \, dt ,$$

$$\int_0^T a_h(w_{h,k}, v_{h,k}) \, dt \to \int_0^T a(w, v) \, dt . \quad \blacksquare$$

3.1.5 *Approximation of* $j(v)$

With $j(v)$ defined as in (2.27) we take a family of approximations of j by convex l.s.c. functions j_h which are integrable for all $v_h \in L^2(0, T; K_h)$, and which satisfy (5.35), (5.36) of Chapter 1.

The following results can then be proved without difficulty:

- lower semi-continuity:

(3.23)
$$\left| \begin{array}{l} v_{h,k} \to u_{h,k} \text{ strongly in } L^2(0, T; V_h) \, ((h, k) \text{ fixed}), \text{ with} \\ v_{h,k}(t), \; u_{h,k}(t) \in K_h, \quad \text{implies} \\ \liminf \displaystyle\int_0^T j_h(v_{h,k}) \, dt \geqslant \int_0^T j_h(u_{h,k}) \, dt ; \end{array} \right.$$

- consistency:

(3.24)
$$\left| \begin{array}{l} v_{h,k}(t) \in K_h, \text{ a.e. } t \in [0, T], \\ p_h v_{h,k} \to \sigma v \text{ weakly in } L^2(0, T; F) \\ \Rightarrow \liminf \displaystyle\int_0^T j_{h,k}(v_{h,k}) \, dt \geqslant \int_0^T j(v) \, dt ; \end{array} \right.$$

- convergence:

(3.25)
$$\left| \begin{array}{l} v_{h,k}(t) \in K_h, \text{ a.e. } t \in [0, T], \\ p_h v_{h,k} \to \sigma v \text{ strongly in } L^2(0, T; F) \\ \Rightarrow \displaystyle\int_0^T j_h(v_{h,k}) \, dt \to \int_0^T j(v) \, dt . \quad \blacksquare \end{array} \right.$$

3.1.6 *Approximation of* $f(t)$

We take $f \in L^2(0, T; V')$ and a family of functions $f_{h,k} \in L^2(0, T; V_h)$ such that

$$\text{(3.26)} \quad \begin{cases} t \to f_{h,k}(t) \quad \text{constant in } t \text{ over each interval} \\ [ik, (i+1)k], \quad i = 0, \dots, N, \text{ with} \\ \| f_{h,k}(t) \|_h^* \leqslant C, \text{ constant independent of } h \text{ and } k, \end{cases}$$

and satisfying

$$\text{(3.27)} \quad \begin{cases} v_{h,k}(t) \in K, \text{ a.e. } t \in [0, T], \\ p_h v_{h,k} \to \sigma v \text{ weakly in } L^2(0, T; F) \\ \Rightarrow \int_0^T (f_{h,k}, v_{h,k})_h \, dt \to \int_0^T (f, v) \, dt. \end{cases}$$

For example, if $f \in L^2(0, T; H)$ we can take

$$f_{h,k} = \sum_{i=0}^{N} f_h^i \chi_k^i$$

where χ_k^i is the characteristic function of $[ik, (i+1)k[$ with

$$f_h^i = \frac{1}{k} \int_{ik}^{(i+1)k} f_h(t) \, dt$$

and $f_h(t)$ approximates $f(t)$ as in Chapter 1, Section 5.2; alternatively, if f is continuous and belongs to $\mathscr{C}(0, T; H)$, we can take

$$\text{(3.27')} \quad f_h^i = f_h(ik);$$

we can also take $f_h^i = f((i-1)k)$ or $f((i - \frac{1}{2})k)$; the numerical illustrations which follow will be given with

$$f_h^i = f_h(ik). \quad \blacksquare$$

3.2 Approximation schemes for parabolic inequalities of type I

In this section we shall give the principal ideas behind the approximation of the general parabolic inequality given in (3.2). We recall this inequality:

$$\text{(3.28)} \quad \begin{cases} \text{find } u \in \mathscr{K}_0 \text{ such that, a.e. } t \in [0, T], \\ (u' + Au - f, v - u) + j(v) - j(u) \geqslant 0, \quad \forall v \in K. \end{cases}$$

In view of the approximation assumptions given in the previous section, we are led quite naturally to consider the following problem - *or scheme*:

$$(3.29)\quad \left| \begin{array}{l} \text{find } u_{h,k} = \sum_{i=0}^{N} u_h^i \chi_k^i \text{ such that} \\ u_h^0 = u_{0,h}, \text{ and for all } t \in [k, T] \text{ such that} \\ (\bar{\delta} u_{h,k} + A_h u_{h,k} - f_{h,k}, v_h - u_{h,k})_h + j_h(v_h) - j_h(u_{h,k}) \geqslant 0 \\ \forall v_h \in K_h, \quad u_{h,k}(t) \in K_h, \quad \forall t \in [0, T]. \end{array} \right.$$

This problem may be written in a more "numerical" manner:

$$(3.30)\quad \left| \begin{array}{l} \text{find } u_h^i, \quad i = 0, 1, \dots, N, \quad \text{with } u_h^0 = u_{0,h}, \\ \qquad \text{such that for } i = 0, \dots, N, \\ \left(\dfrac{u_h^{i+1} - u_h^i}{k} + A_h u_h^{i+1} - f_h^{i+1}, v_h - u_h^{i+1} \right)_h + \\ \qquad + j_h(v_h) - j_h(u_h^{i+1}) \geqslant 0, \\ \forall v_h \in K_h, \quad u_h^{i+1} \in K_h. \end{array} \right.$$

This scheme is termed fully "*implicit*" because of the term $A_h u_h^{i+1}$ which appears in the inequality. This term often gives rise to a somewhat intricate numerical solution and for certain problems (as we shall see in the numerical applications) it can be useful to replace it by the term $A_h u_h^i$, in which case we talk of an "*explicit*" scheme (1).

We note that (3.30) can be written:

$$(3.31)\quad \begin{array}{l} ((I/k + A_h) u_h^{i+1} - u_h^i/k - f_h^{i+1}, v_h - u_h^{i+1})_h + \\ \qquad + j_h(v_h) - j_h(u_h^{i+1}) \geqslant 0 \end{array}$$

and, with the reservation that A_h be *symmetric*, assuming u_h^i to be known (which is true for u_h^0), we obtain u_h^{i+1} as a solution of the *minimisation* problem:

$$(3.32)\quad \underset{v_h \in K_h}{\text{minimise}} \ \frac{1}{2}\left(\left(\frac{I}{k} + A_h\right) v_h, v_h\right)_h - \left(\frac{u_h^i}{k} + f_h^{i+1}, v_h\right)_h + j_h(v_h). \ \blacksquare$$

With the aim of simplifying somewhat the investigation of the implicit and explicit schemes we shall assume that $j = 0$, and in order to avoid having to repeat very similar proofs we introduce "*scheme* I":

---

(1) In fact this is a misnomer: the calculation of u_h^{i+1} assumes the solution of a variational inequality with respect to the identity operator in K_h, i.e. the calculation of the projection P_{K_h} on K_h in the sense of $(,)_h$. See also (3.34).

$$(3.33)\quad \left| \begin{array}{l} \left(\dfrac{u_h^{i+1} - u_h^i}{k} + A_h u_h^{i+\theta} - f_h^{i+1}, v_h - u_h^{i+1}\right)_h \geqslant 0 \quad \forall v_h \in K_h, \\ u_h^{i+1} \in K_h, \quad u_h^0 = u_{0h}, \quad \text{with} \\ u_h^{i+\theta} = u_h^i + \theta(u_h^{i+1} - u_h^i), \end{array} \right.$$

where, in (3.33) θ *is fixed* in $[0, 1]$.

For $\theta = 1$ (resp. $\theta = 0$) this is an implicit scheme (resp. explicit).

With $\theta = \frac{1}{2}$ and in the case when $K_h = V_h$, this is the classical so-called Crank-Nicholson scheme. ■

Remark 3.2. Another category of scheme is given by *scheme* II (which is a "decoupled" scheme, as opposed to scheme I which is not decoupled).

$$(3.34)\quad \left| \begin{array}{l} \text{find} : u_{h,k}^* = \sum_{i=0}^{N} u_h^{i+1/2} \chi_k^i \ \text{ and } \ u_{h,k} = \sum_{i=0}^{N} u_h^i \chi_k^i \ \text{ with} \\ u_h^0 = u_{0,h}, \qquad \text{such that} \\ \dfrac{u_h^{i+1/2} - u_h^i}{k} + A_h(\theta^- u_h^i + \theta^* u_h^{i+1/2} + \theta^+ u_h^{i+1}) - f_h^{i+1} = 0, \quad \text{and} \\ u_h^{i+1} = P_{K_h}[u_h^{i+1/2}] \quad \text{(projection on } K_h \text{ in the sense of the} \\ \qquad \text{norm } (,)_h) \text{ with } \theta^-, \theta^*, \theta^+ \geqslant 0, \quad \theta^- + \theta^* + \theta^+ = 1. \end{array} \right.$$

It can be verified that this scheme includes scheme I (non-decoupled) as a special case (put $\theta^* = 0$ and it can be seen that if $u_{h,k}$ satisfies (3.34) then it also satisfies (3.33)). ■

3.3 Convergence of the approximate inequalities

In this section we shall investigate the complete approximation of scheme I (for the case in which $j \not\equiv 0$ and for the investigation of (3.34), we refer the reader to Trémolières /4/).

For convenience we recall the "continuous" inequality and the "discrete" inequality:

$$(3.35)\quad \left| \begin{array}{l} (u' + Au - f, v - u) \geqslant 0, \quad \text{a.e. } t \in [0, T], \\ \forall v \in K, \quad u \in \mathscr{K}_0, \end{array} \right.$$

$$(3.36)\quad \left| \begin{array}{l} \left(\dfrac{u_h^{i+1} - u_h^i}{k} + A_h u_h^{i+\theta} - f_h^{i+1}, v_h - u_h^{i+1}\right)_h \geqslant 0, \\ \forall v_h \in K_h, \quad u_h^{i+1} \in K_h, \quad i = 0, 1, \dots, N, \quad \text{with } u_h^0 = u_{0,h} \text{ and} \\ u_h^{i+\theta} = u_h^i + \theta(u_h^{i+1} - u_h^i), \quad \theta \in [0, 1]. \end{array} \right.$$

We shall show that for suitable values of h and k, the solution

$$(3.37) \qquad u_{h,k} = \sum_{i=0}^{N} u_h^i \chi_k^i$$

of (3.36) can be considered as realising an approximation of the solution u of the inequality (3.35).

The proof takes place in four stages:

- definition of the "weak" discrete inequality,
- investigation of the stability,
- weak convergence,
- strong convergence.

First we note that the existence and uniqueness of u_h^{i+1} are immediate consequences of the coercivity of the operator $(I/k+\theta A_h)$, $\forall\theta \in [0, 1]$ (see Chapter 1, Theorems 2.1 and 2.2). (In the case when $\theta = 0$ the coercivity must be considered with the norm $|\ |_h \geqslant 1/S(h) \|\ \|_h$, $S(h) \in]0, +\infty[$, $\forall h$.)

3.3.1 *Definition of the "weak" discrete inequality*

We have:

<u>Lemma 3.5</u> *Let* $v_{h,k} = \sum_{i=0}^{N} v_h^i \chi_k^i$, $u_{h,k} = \sum_{i=0}^{N} u_h^i \chi_k^i$. *Then*

$$(3.38) \qquad \left(\frac{v_h^{i+1} - v_h^i}{k}, v_h^{i+1} - u_h^{i+1}\right)_h = \left(\frac{u_h^{i+1} - u_h^i}{k}, v_h^{i+1} - u_h^{i+1}\right)_h + \\ + \frac{1}{2k} |v_h^{i+1} - u_h^{i+1}|_h^2 - \frac{1}{2k} |v_h^i - u_h^i|_h^2 + \\ + \frac{1}{2k} |v_h^{i+1} - v_h^i - (u_h^{i+1} - u_h^i)|_h^2,$$

and if we have

$$(3.39) \qquad \int_k^T (\bar\delta u_{h,k} + A_h u_{h,k}^\theta - f_{h,k}, v_{h,k} - u_{h,k})_h \, dt \geqslant 0,$$

then we also have

$$(3.40) \qquad \int_k^T (\bar\delta v_{h,k} + A_h u_{h,k}^\theta - f_{h,k}, v_{h,k} - u_{h,k})_h \, dt \geqslant \\ \geqslant \frac{1}{2} |v_{h,k}(T) - u_{h,k}(T)|_h^2 - \frac{1}{2} |v_{h,k}(0) - u_{h,k}(0)|_h^2 . \quad \blacksquare$$

Proof. To obtain (3.38) we note that

$$(v_h^{i+1} - v_h^i, v_h^{i+1} - u_h^{i+1})_h = (u_h^{i+1} - u_h^i, v_h^{i+1} - u_h^{i+1})_h + \\ + (v_h^{i+1} - v_h^i - (u_h^{i+1} - u_h^i), v_h^{i+1} - u_h^{i+1})_h$$

and use the classical relation

$$(a-b, a) = \tfrac{1}{2}|a|^2 - \tfrac{1}{2}|b|^2 + \tfrac{1}{2}|a-b|^2 . \tag{3.41}$$

By summing (3.38) from $i = 0$ to $N - 1$ we obtain

$$\int_k^T (\bar{\delta} u_{h,k}, v_{h,k} - u_{h,k})_h \, dt \leqslant \int_k^T (\bar{\delta} v_{h,k}, v_{h,k} - u_{h,k}) \, dt - \\ - \frac{1}{2}\sum_{i=1}^{N} |v_h^i - u_h^i|^2 + \frac{1}{2}\sum_{i=0}^{N-1} |v_h^i - u_h^i|^2 ,$$

and hence we deduce (3.40) by substituting this upper bound in (3.39). ■

Remark 3.3. In the case in which $v_{h,k}(0) = u_{h,k}(0)$ we deduce from (3.39) that

$$\int_k^T (\bar{\delta} v_{h,k} + A_h u_{h,k}^{\theta} - f_{h,k}, v_{h,k} - u_{h,k})_h \, dt \geqslant 0 . \quad ■ \tag{3.42}$$

3.3.2 *Stability investigation*

We shall show that the solution $u_{h,k}$ defined by (3.37) as the (unique) solution of (3.36) remains, with h and k possibly subject to stability conditions, within *bounded* sets. (In order to facilitate the proof we shall assume that $0 \in K$). With $v = 0$ in (3.36) and omitting the suffix h for the time being, we obtain:

$$\frac{1}{k}(u^{i+1} - u^i, u^{i+1}) + a(u^{i+\theta}, u^{i+1}) \leqslant (f^{i+1}, u^{i+1}) .$$

By using relation (3.41) we obtain

$$\frac{1}{2k}[|u^{i+1}|^2 - |u^i|^2 + |u^{i+1} - u^i|^2] + a(u^{i+\theta}, u^{i+1}) \leqslant (f^{i+1}, u^{i+1}) . \tag{3.43}$$

However

$$a(u^{i+\theta}, u^{i+1}) = \theta a(u^{i+1}, u^{i+1}) + (1-\theta)\, a(u^i, u^{i+1}) = \\ = \theta a(u^{i+1}, u^{i+1}) + (1-\theta)\, a(u^i, u^i) + (1-\theta)\, a(u^i, u^{i+1} - u^i) ,$$

and

$$(f^{i+1}, u^{i+1}) = (f^{i+1}, u^{i+\theta}) + (f^{i+1}, u^{i+1} - u^{i+\theta}) = \\ = (1-\theta)(f^{i+1}, u^i) + \theta (f^{i+1}, u^{i+1}) + \\ + (1-\theta)(f^{i+1}, u^{i+1} - u^i) ,$$

so that (3.43) gives [1]

$$| u^{i+1} |^2 - | u^i |^2 + | u^{i+1} - u^i |^2 + 2k[\theta a(u^{i+1}) + (1-\theta) a(u^i)] \leqslant$$
$$\leqslant 2k[\theta(f^{i+1}, u^{i+1}) + (1-\theta)(f^{i+1}, u^i) + + (1-\theta)(f^{i+1}, u^{i+1} - u^i)] - 2k(1-\theta) a(u^i, u^{i+1} - u^i).$$

However, from the classical relation

$$(3.43') \qquad 2(a, b) \leqslant \frac{1}{\varepsilon} | a |^2 + \varepsilon | b |^2, \quad \forall \varepsilon > 0,$$

we have the upper bounds

$$(f^{i+1}, u^{i+1}) \leqslant \| f^{i+1} \|_* \| u^{i+1} \| \leqslant \leqslant \frac{1}{2\alpha} \| f^{i+1} \|_*^2 + \frac{1}{2} a(u^{i+1}), \quad (\varepsilon = \alpha),$$

$$(f^{i+1}, u^i) \leqslant \| f^{i+1} \|_* \| u^i \| \leqslant \frac{1}{\alpha} \| f^{i+1} \|_*^2 + \frac{1}{4} a(u^i), \quad \left(\varepsilon = \frac{\alpha}{2}\right),$$

$$(f^{i+1}, u^{i+1} - u^i) \leqslant \| f^{i+1} \|_* \| u^{i+1} - u^i \| \leqslant \leqslant \frac{1}{2\varepsilon} \| f^{i+1} \|_*^2 + \frac{\varepsilon}{2} S(h)^2 | u^{i+1} - u^i |^2, \qquad (\varepsilon > 0 \text{ arbitrary}),$$

$$| a(u^i, u^{i+1} - u^i) | \leqslant C \| u^i \| \| u^{i+1} - u^i \| \leqslant \leqslant \frac{1}{4} a(u^i) + \frac{C^2}{\alpha} S(h)^2 | u^{i+1} - u^i |^2, \quad \left(\varepsilon = \frac{2}{\alpha}\right),$$

and substituting in (3.43) we obtain $\forall \varepsilon > 0$:

$$(3.44) \qquad | u^{i+1} |^2 - | u^i |^2 + \left[1 - \left(\varepsilon + \frac{2}{\alpha} C^2\right) kS(h)^2 (1-\theta)\right] \times \times | u^{i+1} - u^i |^2 + (1-\theta) ka(u^i) + \theta ka(u^{i+1}) \leqslant \leqslant Ck \| f^{i+1} \|_*^2.$$

We then introduce the following assumption, termed the *stability assumption*,

$$(3.45) \qquad 1 - \frac{2C^2}{\alpha} kS(h)^2 (1-\theta) \geqslant \beta > 0, \quad \forall h, k.$$

(The assumption is clearly redundant when $\theta = 1$).
(3.44) then gives (reintroducing the suffix h)

$$| u_h^{i+1} |_h^2 - | u_h^i |_h^2 + \beta | u_h^{i+1} - u_h^i |_h^2 + \alpha k[(1-\theta) \| u_h^i \|_h^2 + + \theta \| u_h^{i+1} \|_h^2] \leqslant Ck \| f_h^{i+1} \|_{*,h}^2.$$

---

[1] We write $a(v) = a(v, v)$.

With assumption (3.26) on $f_{h,k}$ we obtain, by summing from $i = 0$ to $n \leqslant N - 1$

$$| u_h^{n+1} |_h^2 + \beta \sum_{i=0}^{n} | u_h^{i+1} - u_h^i |_h^2 + \\ + \alpha \sum_{i=0}^{n} k[(1 - \theta) \| u_h^i \|_h^2 + \theta \| u_h^{i+1} \|_h^2] \leqslant C .$$

We hence deduce that, $\forall h, k$ satisfying (3.45):

(3.46) $p_h u_{h,k}$ remains in a bounded set of $L^2(0, T; F)$,

(3.47) $q_h u_{h,k}$ remains in a bounded set of $L^\infty(0, T; H)$,

(3.48) $\sum_{i=0}^{N-1} | u_h^{i+1} - u_h^i |_h^2 \leqslant C$, fixed constant. ■

3.3.3 *Weak convergence*

Let $v \in \mathscr{K}$ and $v_{h,k}$ be as in Lemma 3.2. By substituting $v_{h,k}$ into (3.36) we obtain

$$\int_k^T (\bar{\delta} u_{h,k} + A_h u_{h,k}^\theta - f_{h,k}, v_{h,k} - u_{h,k})_h \, dt \geqslant 0 ,$$

and in view of Lemma 3.5 and Remark 3.3, and taking $v_{h,k}(0) = u_{0,h}$, we obtain

$$\int_k^T (\bar{\delta} v_{h,k} + A_h u_{h,k}^\theta - f_{h,k}, v_{h,k})_h \, dt \geqslant \\ \geqslant \int_k^T (\bar{\delta} v_{h,k}, u_{h,k}) \, dt + \int_k^T (A_h u_{h,k}^\theta, u_{h,k}) \, dt - \\ - \int_k^T (f_{h,k}, u_{h,k}) \, dt , \tag{3.49}$$

where we recall that:

$$u_{h,k}^\theta(t) = u_{h,k}(t - k) + \theta[u_{h,k}(t) - u_{h,k}(t - k)] .$$

However, from (3.46), (3.47) and Lemma 3.1 (consistency) we can extract from $u_{h,k}$ a subsequence again denoted by $u_{h,k}$ such that

$$\left| \begin{array}{ll} q_h u_{h,k} \to u & \text{weakly* in } L^\infty(0, T; H) \\ p_h u_{h,k} \to \sigma u & \text{weakly in } L^2(0, T; F) \\ \text{with } u \in K . & \end{array} \right.$$

Proceeding to the limit in (3.49) we obtain:

$$(3.50) \qquad \int_0^T (v' + Au - f, v - u)\,\mathrm{d}t \geqslant \liminf \int_k^T (A_h u_{h,k}^\theta, u_{h,k} - u_{h,k}^\theta)_h\,\mathrm{d}t\,.$$

It remains to show that the right-hand side is $\geqslant 0$.
We have the upper bound:

$$\begin{aligned} Y_{h,k} &= \left| \int_k^T (A_h u_{h,k}^\theta, u_{h,k} - u_{h,k}^\theta)_h\,\mathrm{d}t \right| = \\ &= \left| \sum_{i=0}^N k(A_h u_h^{i+\theta}, u_h^{i+1} - u_h^{i+\theta})_h \right| = \\ &= \sum_{i=0}^N k(1-\theta) \,|\, (A_h u_h^{i+\theta}, u_h^{i+1} - u_h^i)_h \,| \leqslant \\ &\leqslant (1-\theta)\sqrt{k}\, S \sum_{i=0}^N \sqrt{k}\, \| u_h^{i+\theta} \|_h \,|\, u_h^{i+1} - u_h^i \,|_h \leqslant \\ &\leqslant (1-\theta)\, S(h)\sqrt{k} \left(\sum_{i=0}^N k \,\| u_h^{i+\theta} \|_h^2 \right)^{1/2} \times \\ &\times \left(\sum_{i=0}^N | u_h^{i+1} - u_h^i |_h^2 \right)^{1/2} . \end{aligned}$$

By using (3.46), (3.48) and the *convergence assumption* (redundant when $\theta = 1$):

$$(3.51) \qquad (1-\theta)\, kS(h)^2 \to 0\,, \quad \text{if} \quad h, k \to 0\,,$$

we obtain

$$Y_{h,k} \to 0 \quad \text{if} \quad h, k \to 0\,,$$

and hence

$$\int_0^T (v' + Au - f, v - u)\,\mathrm{d}t \geqslant 0\,, \quad \forall v \in \mathscr{K}_0\,, \quad u \in \mathscr{K}_f\,,$$

and u is indeed a "weak" solution [1]. ■

3.3.4 *Strong convergence*

We shall now show that $u_{h,k}$ "approximates" the (strong) solution of the inequality (3.35). To this end we assume first that the existence and uniqueness conditions for this solution are realised. We then know (see Remark 2.2) that the inequalities:

---

(1) We note that, in contrast to the case of equations, *stability does not necessarily imply convergence.*

(3.52) $$\int_0^T (u' + Au - f, v - u)\,\mathrm{d}t \geqslant 0\,, \quad \forall v \in \mathscr{K}\,, \quad u \in \mathscr{K}_0\,,$$

(3.53) $$\int_0^T (v' + Au - f, v - u)\,\mathrm{d}t \geqslant 0\,, \quad \forall v \in \mathscr{K}_0\,, \quad u \in \mathscr{K}_f\,,$$

are equivalent, that is, that the solution $u \in \mathscr{K}_f$ of (3.53) is in fact such that $u \in \mathscr{K}_0$ and u satisfies (3.52). We therefore know that $p_h u_{h,k}$ tends *weakly* towards σu in $L^2(0, T; F)$ with u a *strong solution* of (3.52). In fact it may be shown that $p_h u_{h,k}$ tends *strongly* towards σu in $L^2(0, T; F)$. For this we consider a sequence $\tilde{u}_{h,k}$ which approximates u as in Lemma 3.2 and we shall evaluate

(3.54) $$X_{h,k} = \int_k^T (A_h u_{h,k} - A_h \tilde{u}_{h,k}, u_{h,k} - \tilde{u}_{h,k})_h\,\mathrm{d}t\,,$$

and show that this quantity tends to zero.

From Lemma 3.5, taking account of Remark 3.3 and putting $v_{h,k} = \tilde{u}_{h,k}$ in inequality (3.42), we obtain

(3.55) $$\int_k^T (A_h u^\theta_{h,k}, u_{h,k})_h\,\mathrm{d}t \leqslant \int_k^T (\bar{\delta}\tilde{u}_{h,k}, \tilde{u}_{h,k} - u_{h,k})_h\,\mathrm{d}t + \\ + \int_k^T (A_h u^\theta_{h,k}, \tilde{u}_{h,k})_h\,\mathrm{d}t - \int_k^T (f_k, \tilde{u}_{h,k} - u_{h,k})_h\,\mathrm{d}t\,.$$

However

(3.56) $$X_{h,k} = \int_k^T (A_h u_{h,k}, u_{h,k})_h\,\mathrm{d}t - \int_k^T (A_h u_{h,k}, \tilde{u}_{h,k})_h\,\mathrm{d}t + \\ + \int_k^T (A_h \tilde{u}_{h,k}, \tilde{u}_{h,k} - u_{h,k})_h\,\mathrm{d}t\,.$$

Now

(3.57) $$u^\theta_{h,k}(t) = u_{h,k}(t) + (\theta - 1)\,(u_{h,k}(t) - u_{h,k}(t - k))$$

and hence

(3.58) $$\int_k^T (A_h u_{h,k}, u_{h,k})_h\,\mathrm{d}t = \int_k^T (A_h u^\theta_{h,k}, u_{h,k})_h\,\mathrm{d}t - \\ - (\theta - 1)\int_k^T (A_h u_{h,k}(t) - A_h u_{h,k}(t - k), u_{h,k})_h\,\mathrm{d}t\,,$$

so that (3.58), (3.55) in (3.56) give

$$X_{h,k} \leqslant \int_k^T (\bar{\delta}\tilde{u}_{h,k}, \tilde{u}_{h,k} - u_{h,k})_h \, dt + \int_k^T (A_h u^\theta_{h,k}, \tilde{u}_{h,k})_h \, dt -$$
$$- \int_k^T (f_{h,k}, \tilde{u}_{h,k} - u_{h,k})_h \, dt -$$
$$- (\theta - 1) \int_k^T (A_h u_{h,k}(t) - A_h u_{h,k}(t-k), u_{h,k})_h \, dt -$$
$$- \int_k^T (A_h u_{h,k}, \tilde{u}_{h,k})_h \, dt + \int_k^T (A_h \tilde{u}_{h,k}, \tilde{u}_{h,k} - u_{h,k})_h \, dt ,$$

or alternatively, using (3.57),

$$X_{h,k} \leqslant \int_k^T (\bar{\delta}\tilde{u}_{h,k} + A_h \tilde{u}_{h,k} - f_{h,k}, \tilde{u}_{h,k} - u_{h,k})_h \, dt + \tag{3.59}$$
$$+ (\theta - 1) \int_k^T (A_h u_{h,k}(t) - A_h u_{h,k}(t-k) ,$$
$$\tilde{u}_{h,k}(t) - u_{h,k}(t))_h \, dt .$$

We shall now show that

$$Z_{h,k} = (1 - \theta) \left| \sum_{i=0}^{N-1} k(A_h(u_h^{i+1} - u_h^i), \tilde{u}_h^{i+1} - u_h^{i+1})_h \right|$$

tends towards zero. We obtain

$$Z_{h,k} \leqslant (1 - \theta) \, S(h) \sum_{i=0}^{N-1} \sqrt{k} \, | u_h^{i+1} - u_h^i |_h \sqrt{k} \, \| \tilde{u}_h^{i+1} - u_h^{i+1} \|_h \tag{3.60}$$

and hence

$$Z_{h,k} \leqslant (1 - \theta) \sqrt{k} \, S(h) \left(\sum_{i=0}^{N-1} | u_h^{i+1} - u_h^i |_h^2 \right)^{1/2} \times$$
$$\times \left(\sum_{i=0}^{N-1} k \, \| \tilde{u}_h^{i+1} - u_h^{i+1} \|_h^2 \right)^{1/2} .$$

By using (3.48), together with the fact that $p_h \tilde{u}_{h,k}$ and $p_h u_{h,k}$ remain bounded in $L^2(0, T; F)$, and by imposing the *convergence* condition (3.51), we obtain

$$Z_{h,k} \to 0 .$$

Consequently $X_{h,k} \to 0$ if

$$\int_h^T (\bar{\delta}\tilde{u}_{h,k} + A_h \tilde{u}_{h,k} - f_{h,k}, \tilde{u}_{h,k} - u_{h,k})_h \, dt \to 0 ,$$

which follows immediately by virtue of (3.46), (3.47), the properties given in Lemma 3.2 for $\tilde{u}_{h,k}$, the results of Lemma 3.4 for $A_h \tilde{u}_{h,k}$ and the assumptions (3.26), (3.27) on f_h. Then $X_{h,k} \to 0$ and we can deduce from the uniform coercivity of A_h in h that

$$p_h(u_{h,k} - \tilde{u}_{h,k}) \to 0 ,$$

and thus

$$p_h u_{h,k} \to \sigma u \quad \text{strongly in } L^2(0, T; F) .$$

The set of results obtained enables us to state:

Theorem 3.1. *For* $\theta \in [0, 1], h, k \to 0,$ *and with the supplementary assumption* $(1 - \theta) kS(h)^2 \to 0$ (*redundant if* $\theta = 1$), *the solutions* $u_{h,k} = \sum_{i=0}^{N} u_h^i \chi_k^i$ *of scheme* I (3.33) *satisfy*:

$$q_h u_{h,k} \to u \quad \textit{weakly* in } L^\infty(0, T; H)$$

$$p_h u_{h,k} \to \sigma u \quad \textit{weakly in } L^2(0, T; F)$$

where u *is the unique* (weak) *solution of* (3.53). *When this solution is the unique strong solution of* (3.52) *we have*

$$p_h u_{h,k} \to \sigma u \quad \textit{strongly in } L^2(0, T; F) . \qquad \blacksquare$$

Remark 3.4. When h and k are able to tend to zero without it being necessary that $kS(h)^2 \to 0$ we say that the scheme is *unconditionally* stable and convergent. This is thus the case here for $\theta = 1$. In the other cases $(0 \leqslant \theta < 1)$ the scheme is said to be *conditionally* stable and convergent. ■

Remark 3.5. It is shown in Trémolières /4/ that scheme II given in Remark 3.2 is *conditionally* stable and convergent unless

(3.61) $\quad \theta^+ = 1 , \quad \theta^- = 0 , \quad \theta^* = 0 ,$

and unless

(3.62) $\quad \theta^* \geqslant \frac{1}{2} , \quad \theta^- = 0 , \quad$ and $\quad K_h = V_h , \quad K = V .$

In this latter case the problem to be solved, if j is differentiable, is a variational *equation*. Under assumptions (3.61) or (3.62), scheme II is *unconditionally* stable and convergent. In particular we recover the result that Crank-Nicholson schemes are *unconditionally stable and convergent* (see Raviart /1/, Témam /3/). ■

Remark 3.6. The results of Section 3.3.3 establish the *existence* of a weak solution. ■

4. NUMERICAL SOLUTION OF SOME PARABOLIC INEQUALITIES OF TYPE I

In this section we propose to solve three examples of parabolic inequalities of type I. These examples will allow us to assess the relative merits of *explicit* schemes (scheme I, with $\theta = 0$), schemes which are "*implicit in projection*" (scheme I, with $\theta = 1$), and schemes which are "*implicit in equation*" (scheme II, with $\theta^* = 1$, $\theta^- = \theta^+ = 0$). At the same time we shall investigate the *sensitivity* of the schemes to the quantities h, k and the ratio $k/h^2 \simeq kS(h)^2$.

We shall carry out this investigation on:

- A model problem using the convex sets studied in Chapter 3;
- A friction problem;
- A problem of the deformation of a membrane.

4.1 A model problem

Consider the parabolic inequality

$$(4.1) \qquad (u' - \Delta u - f, v - u) \geqslant 0\,, \quad \forall v \in K\,, \quad u \in \mathscr{K}_0\,, \qquad \text{a.e. } t \in [0, T]\,,$$

with:

$$K = \{ v \mid v \in H_0^1(\Omega),\ | v(x) | \leqslant \mathrm{d}(x, \Gamma), \ \text{ a.e. on } \Omega \}\,,$$

$$\mathscr{K}_0 = \{ v \mid v \in W_0(0, T),\ v(t) \in K, \ \text{ a.e. } t \in [0, T] \}\,,$$

where $\mathrm{d}(x, \Gamma)$ is the Euclidian distance from $x \in \Omega$ to the boundary Γ of Ω. We take $\Omega =]0, 1[$ and the time interval $[0, 1]$. It can be shown that for

$$f(x, t) = \begin{cases} 100 \text{ if } x < t/2\,, \text{ or if } x > 1 - t/2\,, \\ -[1/(t-1)^2](x^2 - x - t(t-2)/4) - 2/(t-1) \\ \quad \text{if } t/2 \leqslant x \leqslant 1 - t/2\,, \end{cases}$$

and the initial condition

$$u_0(x) = x(1 - x)\,,$$

the problem has the exact solution (see Figure 4.1):

$$u(x, t) = \begin{cases} x \text{ if } 0 \leqslant x \leqslant t/2\,, \\ [1/(t-1)]\,[x^2 - x + t^2/4] \text{ if } t/2 < x < 1 - t/2\,, \\ 1 - x \text{ if } 1 - t/2 \leqslant x \leqslant 1\,. \end{cases}$$

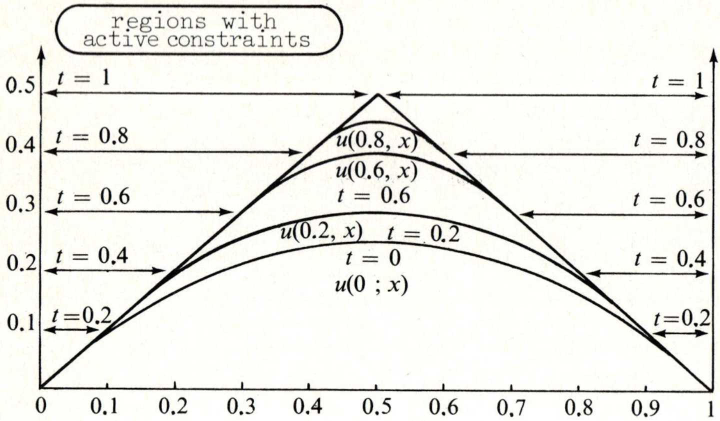

Fig. 4.1. Evolution of the solution of the elasto-plastic problem.

We discretise $\Omega =]0, 1[$ into $M + 1$ intervals of length h and the time interval $[0, T] = 1$ into N intervals of length k.

We shall seek the discrete solution in the following form (θ_h^i being the characteristic function of the element $\omega_h^0(ih)$) :

$$u_{h,k}(t) = \sum_{i=0}^{M+1} u_i^n \, \theta_h^i , \quad t \in [nk, (n + 1) k[,$$

and we shall evaluate the quality of the solution found at the instant nk by means of (*):

$$\text{ECART}_2 = \left(\sum_{i=1}^{M} | u(ih, nk) - u_i^n |^2 \right)^{1/2} .$$

We take

$$\begin{aligned} & u_i^0 = ih(1 - ih) , \quad i = 0, \dots, M + 1 , \\ (4.2) \qquad & f_i^n = f(ih, nk) , \quad i = 0, \dots, M + 1 \quad \text{and} \quad n = 0, \dots, N , \\ & d_i = \min \{ ih, M + 1 - ih \} . \end{aligned}$$

All the results relating to problem (4.1) were obtained on a BULL-GE-265 and the times are quoted in sixths of a second, including compilation time.

(*) *Translator's note*: The original French terminology, retained here, means "deviation".

4.1.1 *Investigation of the explicit scheme*

The scheme can be written, for $n = 0, \ldots, N-1$

$$u_1^{n+1} = \max\left(-d_1, \min\left(u_1^n + \frac{k}{h^2}(-2u_1^n + u_2^n) + kf_1^{n+1}, d_1\right)\right),$$

$$u_i^{n+1} = \max\left(-d_i, \min\left(u_i^n + \frac{k}{h^2}(u_{i-1}^n - 2u_i^n + u_{i+1}^n) + hf_i^{n+1}, d_i\right)\right),$$

$$i = 1, \ldots, M-1,$$

$$u_M^{n+1} = \max\left(-d_M, \min\left(u_M^n + \frac{k}{h^2}(u_{M-1}^n - 2u_M^n) + kf_M^{n+1}, d_M\right)\right),$$

and clearly, with the initial condition (4.2) and, for all n,

$$u_0^n = u_{M+1}^n = 0 .$$

The results are given in Table 4.1.

4.1.2 *Investigation of the "implicit in equation" scheme*

The scheme can be written, for $n = 0, \ldots, N,$

$$(4.3)\quad \left| \begin{array}{l} u_i^{n+1/2} - \dfrac{k}{h^2}(u_{i+1}^{n+1/2} - 2u_i^{n+1/2} + u_{i-1}^{n+1/2}) = u_i^n + kf_i^{n+1}, \\ i = 1, \ldots, M, \end{array}\right.$$

$$(4.4)\quad \left| \begin{array}{l} u_i^{n+1} = \max(-d_i, \min(u_i^{n+1/2}, d_i)), \quad i = 1, \ldots, M, \\ \text{with } u_i^0 \text{ given by (4.2) and, for all } n, u_0^n = u_{M+1}^n = 0 . \end{array}\right.$$

For each $n, n = 0, 1, \ldots$ we have to solve the system of equations (4.3), which gives $u_i^{n+1/2}$, $i = 1, \ldots, M$ and we then obtain u_i^{n+1}, $i = 1, \ldots, M$ using (4.4). System (4.3) was solved by the Gauss-Seidel and conjugate gradient methods. The common termination criterion in the search for the solution $u^{n+1/2}$ at the instant $t = nk$ by each of these two methods is

$$\sqrt{\sum_{i=1}^{M} |J_i'(v^n)|^2} \leqslant 10^{-6},$$

with

$$J_i'(v^n) = \left(1 + 2\frac{k}{h^2}\right) v_i^n - \frac{k}{h^2} v_{i-1}^n - \frac{k}{h^2} v_{i+1}^n - u_i^{n-1} - kf_i^n$$

where u^{n-1} is the solution found at the instant $t = (n-1)k$.

The calculations were carried out on a BULL-GE-265, the computation times being quoted in sixths of a second.

Characteristics		Values of $h^{-\frac{1}{2}} \times \mathrm{ECART}_2$: (exact solution - calculated solution) for various values of $t \in [0, 1]$											Solution time
h	k	$t = k$	0.1	0.2	0.3	0.4	0.5	0.6	0.7	0.8	0.9	1	
1/10	1/1 000	0	0.003 3	0.002 1	0.004 8	0.001 8	0.006 5	0.001 0	0.010 8	0.000 3	0.022 8	0	548
	1/600	0	0.003 5	0.002 3	0.005 1	0.002 0	0.006 9	0.001 1	0.011 0	0.000 4	0.023 2	0	244
	1/500	0	0.003 6	0.002 4	0.005 2	0.002 1	0.007 0	0.001 2	0.011 2	0.000 5	0.023 5	0	207
	1/400	0	0.003 8	0.002 6	0.005 5	0.002 3	0.007 3	0.001 3	0.011 4	0.000 6	0.023 9	0	170
	1/250	0.000 1	0.004 2	0.003 3	0.006 2	0.003 0	0.008 0	0.001 8	0.012 3	0.001 0	0.025 0	0	128
	1/200	0.000 1	0.004 6	0.003 8	0.006 7	0.003 4	0.008 5	0.002 2	0.012 8	0.001 2	0.025 0	0	115
Instability	1/150	0.000 3	0.005 8	0.091 6	28.9	1 004	35 381	10^6	10^7	10^9	10^{10}	10^{12}	101
	1/100	0.000 6	0.112 6	135	84 054	5×10^2	10^{10}	10^{13}	10^{16}	10^{18}	10^{21}	10^{24}	95
1/20	1/1 000	0	0.001 0	0.001 3	0.001 4	0.001 4	0.001 2	0.001 0	0.000 7	0.000 4	0.000 3	0	677
	1/800	0	0.001 1	0.001 5	0.001 6	0.001 5	0.001 4	0.001 1	0.000 8	0.000 4	0.000 3	0	580
Instability	1/700	0	0.563 0	3 500	—	—	—	—	—	—	—	—	—
	1/100	0.000 8	3 632	—	—	—	—	—	—	—	—	—	—

Table 4.1. Explicit scheme.

For the Gauss-Seidel method, the values of the "optimal" parameter ω_{opt} are given in Table 4.2 as a function of h and k [1].
Figure 4.2 shows the variation of ω_{opt} as a function of h and k.

h	k	$\omega_{\text{opt. theo.}}$	Computation time
	1/600	1.014 58	54
	1/400	1.026 54	52
	1/200	1.064 03	53
	1/150	1.087 32	57
1/10	1/100	1.127 86	73
	1/50	1.212 96	53
	1/10	1.404 81	56
	1/5	1.456 76	55
	1/2	1.496 33	58
	1/800	1.069 90	91
1/20	1/100	1.352 60	108
	1/10	1.639 48	112
	1/5	1.679 12	113
1/30	1/100	1.499 55	109
	1/10	1.743 24	137

Table 4.2

The results obtained by the above two methods are recorded in Table 4.3. In this table, the figures in the upper triangles ◸ give the numbers of iterations required by the conjugate gradient method to obtain the solution at the corresponding instant $t = nk$. The figures in the lower triangles ◢ give the numbers of iterations required by the Gauss-Seidel method used with optimum parameter ($\omega_{\text{opt. theo.}}$: see Table 4.2) to obtain the solution at the corresponding instant $t = nk$.
The figures in the last column of the table give, in the upper triangles, the total solution time using the conjugate gradient method, and in the lower triangles, the total solution time using the Gauss-Seidel method.

4.1.3 *Investigation of the "implicit in inequality" scheme.*

The scheme can be written, for $n = 0, \dots, N$,

---

(1) This parameter was obtained by applying the iterative method of Varga /1/ to the unconstrained problem. (See also Trémolières /4/ and Bourgat-Trémolières /1/).

Characteristics		$h^{-\frac{1}{2}} \times \mathrm{ECART}_2$: (exact solution - calculated solution)											Computation time
Values of $t \in [0, T]$		k	0.1	0.2	0.3	0.4	0.5	0.6	0.7	0.8	0.9	1	
$h = 1/10$	$k = 1/1\,000$	2 / 0 / 3	2 / 0.003 0 / 3	5 / 0.008 5 / 5	4 / 0.027 9 / 5	4 / 0.025 2 / 5	4 / 0.026 7 / 5	3 / 0.182 / 5	3 / 0.021 1 / 5	5 / 0.019 1 / 5	3 / 0.250 / 5	3 / 0 / 5	1 302 / 1 103
	$k = 1/600$	2 / 0 / 4	2 / 0.003 0 / 4	6 / 0.013 4 / 6	5 / 0.048 8 / 6	6 / 0.041 7 / 6	5 / 0.047 7 / 6	5 / 0.417 / 5	5 / 0.032 4 / 5	6 / 0.042 7 / 5	4 / 0.250 / 5	4 / 0 / 5	958 / 740
	$k = 1/400$	2 / 0 / 5	2 / 0.003 1 / 5	6 / 0.020 5 / 7	5 / 0.075 2 / 6	4 / 0.052 3 / 6	4 / 0.079 2 / 6	4 / 0.533 / 6	5 / 0.053 1 / 6	4 / 0.050 0 / 6	4 / 0.025 0 / 6	4 / 0 / 6	715 / 567
	$k = 1/150$	3 / 0 / 8	2 / 0.003 1 / 8	4 / 0.038 2 / 10	8 / 0.122 5 / 9	6 / 0.114 8 / 10	6 / 0.112 1 / 9	6 / 0.087 1 / 5	6 / 0.075 9 / 7	4 / 0.050 0 / 7	3 / 0.025 0 / 7	4 / 0 / 7	360 / 302
	$k = 1/100$	3 / 0 / 8	3 / 0.003 1 / 8	3 / 0.084 2 / 10	6 / 0.164 3 / 10	5 / 0.159 2 / 10	6 / 0.140 4 / 10	6 / 0.106 1 / 10	6 / 0.075 9 / 10	8 / 0.050 0 / 10	6 / 0.025 0 / 10	5 / 0 / 9	232 / 242
	$k = 1/50$	3 / 0 / 11	4 / 0.003 2 / 11	4 / 0.148 9 / 13	4 / 0.202 0 / 12	4 / 0.172 7 / 11	4 / 0.140 4 / 11	5 / 0.106 1 / 12	5 / 0.075 9 / 12	5 / 0.050 0 / 12	5 / 0.025 0 / 12	6 / 0 / 12	138 / 169
$h = 1/20$	$k = 1/800$	3 / 0 / 6	4 / 0.008 9 / 9	4 / 0.035 5 / 10	4 / 0.044 2 / 10	3 / 0.045 2 / 10	4 / 0.041 4 / 10	3 / 0.040 9 / 10	4 / 0.038 0 / 10	4 / 0.034 0 / 9	5 / 0.025 0 / 9	5 / 0 / 8	2 346 / 2 789

Conjugate gradient method ◸

Gauss-Seidel method with $\omega_{\mathrm{opt.\,theo.}}$ ◿

Table 4.3. "Implicit in equation" scheme. (See text for explanation of table).

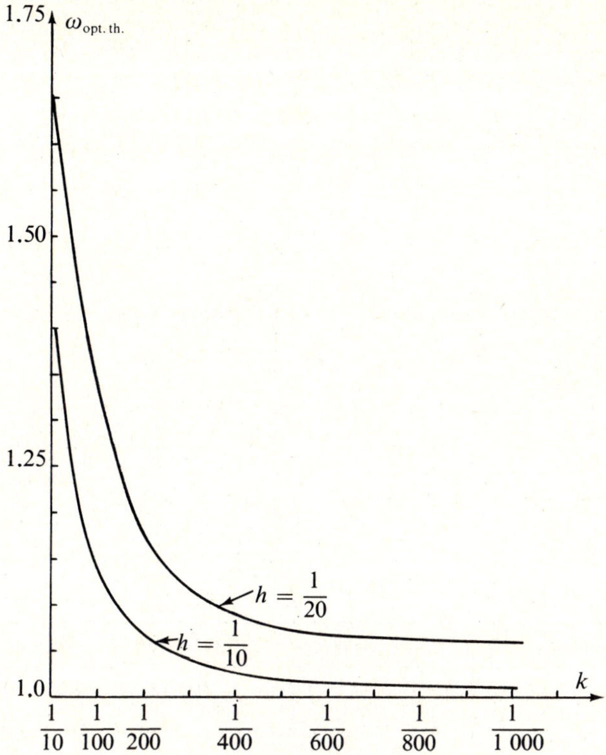

Fig. 4.2.

$$\left|\begin{array}{l} \sum_{i=1}^{M} \left(u_i^{n+1} - u_i^n - \frac{k}{h^2}(u_{i+1}^{n+1} - 2\,u_i^{n+1} + u_{i-1}^{n+1}) - kf_i^{n+1},\ v_i - u_i^{n+1} \right) \geqslant 0\,, \\ \forall v = (v_1, \ldots, v_N) \text{ with } |\,v_i\,| \leqslant d_i\,,\quad i = 1, \ldots, M\,, \\ \text{and } u^{n+1} = (u_1^{n+1}, \ldots, u_N^{n+1}) \text{ such that } |\,u_i^{n+1}\,| \leqslant d_i,\ i = 1, \ldots, M\,, \\ \text{with } u_i^0 \text{ given by (4.2) and, for all } n,\quad u_0^n = u_{M+1}^n = 0\,. \end{array}\right.$$

This scheme can be solved at the instant $t = nk$ by *minimising* over

$$K = \{\, v \mid |\,v_i\,| \leqslant d_i,\ i = 0, \ldots, M + 1 \,\}$$

the function $J(v)$ which has the gradient $J'(v) = \{\, J_1'(v), \ldots, J_M'(v) \,\}$ with:

$$J_i'(v) = v_i - u_i^n - \frac{k}{h^2}(v_{i+1} - 2\,v_i + v_{i-1}) - kf_i^{n+1}\,.$$

This minimisation was carried out using the Gauss-Seidel method in projection, with the "optimal" parameter given by Table 4.2 (1). The termination criterion in the search for the solution at the instant $t = nk$ is given by:

$$\sqrt{\sum_{i=1}^{M} |\tilde{J}_i'(v^n)|^2} \leqslant 10^{-6},$$

with

$$\tilde{J}_i'(v^n) = \begin{cases} J_i'(v^n) & \text{if} \quad |v_i^n| < d_i, \\ J_i'(v^n) & \text{if} \quad |v_i^n| = d_i \quad \text{and} \quad J_i'(v^n)\, v_i^n \geqslant 0, \\ 0 & \text{if} \quad v_i^n = d_i \quad \text{and} \quad J_i'(v^n)\, v_i^n < 0. \end{cases}$$

The results are given in Table 4.4.

		Values of $h^{-\frac{1}{2}} \times ECART_2$ (exact solution - calculated solution) for various values of t.										
h	k	$t=k$	$=0.1$	$=0.2$	$=0.3$	$=0.4$	$=0.5$	$=0.6$	$=0.7$	$=0.8$	$=0.9$	$=1$
1/10	1/100	10^{-5}	0.003 1	0.001 3	0.004 4	0.001 2	0.006 3	0.000 6	0.010 0	0.000 1	0.022	0
	1/50	10^{-4}	0.003 2	0.001 2	0.004 6	0.001 1	0.006 5	0.000 6	0.010 6	0.000 2	0.022	0
	1/10	10^{-3}	0.003 8	0.000 4	0.005 5	0.000 7	0.007 9	0.000 6	0.012 1	0.000 3	0.024	0
1/20	1/100	10^{-5}	0.000 3	0.000 5	0.000 5	0.000 6	0.000 5	0.000 4	0.000 3	0.000 1	10^{-5}	0
	1/10	10^{-4}	0.000 8	0.001 0	0.001 0	0.000 8	0.000 6	0.000 4	0.000 8	0.000 1	10^{-6}	0
1/30	1/100	10^{-5}	0.000 5	0.000 2	0.000 7	0.000 3	0.001 0	0.000 3	0.001 8	0.000 1	0.004 3	0
	1/10	10^{-4}	0.000 5	0.001 1	0.000 5	0.000 9	0.000 9	0.000 3	0.002 1	0.000 1	0.004 6	0

Table 4.4. "Implicit in inequality" scheme.

Remark 4.1. The "implicit in inequality" scheme has also been solved by a special method, adapted from the conjugate gradient method and using truncation. This method (see Trémolières /4/ gives slightly better results than the Gauss-Seidel method;

(1) By "optimal" parameter we mean the theoretical optimum value of the parameter ω for the *unconstrained* case. In fact in the case of inequalities this parameter is in no way optimal Here we should take the ω_{opt} of the Dirichlet problem defined on the domain enclosed by the free boundary, i.e. the zone in which $|u| < d$ (see Chapter 3, Remark 3.7).

however it is more awkward to program. ■

4.1.4 *Conclusions of* 4.1

With regard to the model problem considered here, we have the following conclusions:

1. *Explicit scheme.*

- very simple programming,
- importance of a good choice of k and h : $k/h^2 \leqslant$ const. $\sim \frac{1}{5}$,
- divergence and overflow if k is too large,
- thus, in the absence of the evaluation of the stability *constant*, this method necessitates "numerical trial and error".

2. "*Implicit in equation*" *scheme.*

- fairly simple programming (when the "projection" does not cause difficulty),
- for a given choice of k and h, the scheme appears to be *less accurate and half as fast* as the explicit scheme,
- the danger of divergence appears to be reduced.

3. "*Implicit in inequality*" *scheme.*

- more awkward programming,
- very high accuracy [1], and, for a given choice of h and k, the scheme is faster than that above, and, *for the same precision*, *faster than the explicit scheme.* ■

4.2 A model problem of time-dependent friction

For $t \in [0, T]$, we consider the inequality

$$(u', v - u) + a(u, v - u) + j(v) - j(u) \geqslant (f, v - u), \quad \forall v \in V, \quad u \in \mathscr{V}_0, \tag{4.5}$$

with

---

[1] Observed experimentally.

$$
\left|\begin{array}{l}
\Omega =]0, 1[\,, \quad T = 1\,, \\
V = \{ v \mid v \in H^1(\Omega),\ v\,|_{x=0} = 0 \}\,, \\
\mathscr{V}_0 = \{ v \mid v \in W(0, T),\ v(0) = u_0 \ \text{given} \in V \}\,, \\
a(u, v) = \displaystyle\int_\Omega \operatorname{grad} u \operatorname{grad} v \, dx\,, \\
j(v) = g \mid v\,|_{x=1}\,, \quad g = 10 \mid t(\tfrac{3}{4} - t) \mid\,, \\
f = 20\, x(\tfrac{3}{2} - x)\,(\tfrac{3}{4} - 2\,t) + 40\, t(\tfrac{3}{4} - t)\,.
\end{array}\right.
$$

This problem is also equivalent to

$$
\left|\begin{array}{l}
u' - \Delta u = f \quad \text{in } \Omega\,, \\
\left|\dfrac{\partial u}{\partial n}\right| \leqslant g\,, \ \text{at } x = 1\,, \\
u\dfrac{\partial u}{\partial n} + g \mid u \mid = 0\,, \ \text{at } x = 1\,.
\end{array}\right.
$$

With the above properties and with $u_0 = 0$ the inequality has the exact solution

$$
u = 20\, x(\tfrac{3}{2} - x)\, t\,(\tfrac{3}{4} - t)\,.
$$

This solution is depicted in Figure 4.3 below.

This problem was solved by the explicit scheme and by the "implicit in inequality" scheme. The solution of the implicit scheme was carried out for each time iteration by a *dual method* (see Chapter 2), in view of the nondifferentiable term due to $j(v)$.

For the spatial discretisation over $]0, 1[$ we seek u_h in the form

$$
u_h = \sum_{i=1}^{M} u_i\, \theta_h^i\,, \quad \text{i.e.} \quad u_h \in V_h = \{ v_h \mid v_h = \sum_{i=1}^{M} v_i\, \theta_h^i \}
$$

where θ_h^i is the characteristic function of the interval $[(i - \frac{1}{2})\,h, (i + \frac{1}{2})\,h[$ for $i = 1, \ldots, M - 1$ and θ_h^M is that of $[1 - h/2, 1[$, with M the integer $M = 1/h$.

Before discussing the numerical results, we shall briefly indicate how these schemes can be obtained.

4.2.1 *Obtaining the implicit scheme*

For convenience we shall first consider the case of semi-discrete approximations (discretisation in time but not in space). At the time $t = nk$, ($k = 1/N$, integer N), with $n = 0, 1, \ldots, N$, the inequality can be written:

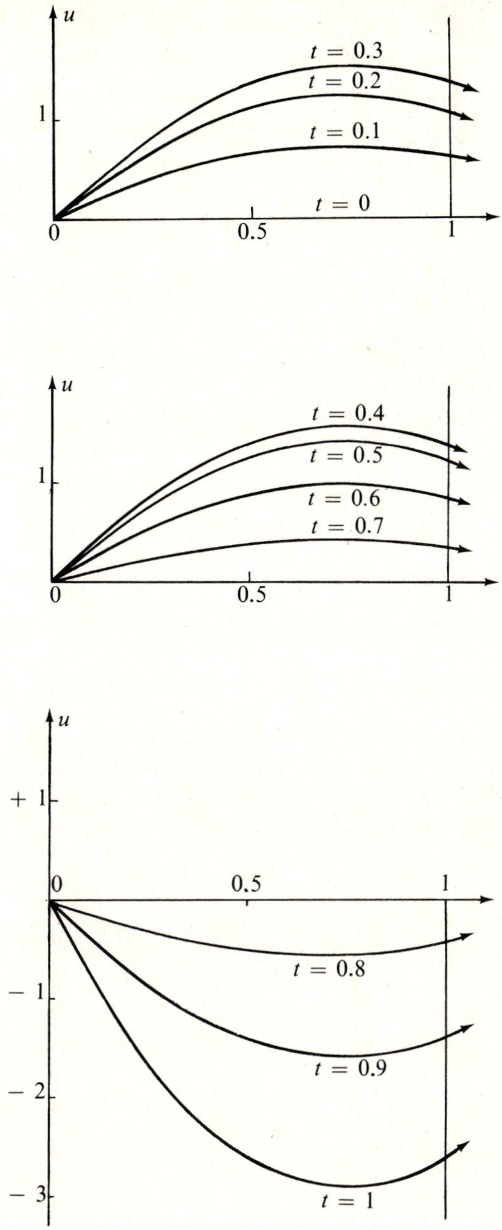

Fig. 4.3. Variation of the exact solution of the time-dependent friction problem.

(4.6)
$$\left| \begin{array}{l} \left(\dfrac{u^{n+1} - u^n}{k}, v - u^{n+1}\right) + \displaystyle\int_\Omega \operatorname{grad} u^{n+1} . \operatorname{grad} (v - u^{n+1}) \, dx - \\ \quad - (f^{n+1}, v - u^{n+1}) + j^{n+1}(v) - j^{n+1}(u^{n+1}) \geqslant 0 , \\ \forall v \in V , \quad u^{n+1} \in V , \quad u^0 = u_0 \quad \text{given.} \end{array} \right.$$

It is easy to see that this inequality is equivalent to the problem

(4.7)
$$\begin{array}{l} \underset{v \in V}{\underline{\text{minimise}}} \ \dfrac{1}{2} \displaystyle\int_\Omega | v |^2 \, dx - (v, u^n) + \dfrac{k}{2} \int_\Omega | \operatorname{grad} v |^2 \, dx - \\ \quad - k(f^{n+1}, v) + k j^{n+1}(v) , \end{array}$$

or, because of the special form of $j(v)$ (see Chapter 4), equivalent to solving

(4.8)
$$\begin{array}{l} \underset{\substack{|p| \leqslant 1 \\ p \in \mathbb{R}}}{\underline{\max}} \ \underset{v \in V}{\underline{\min}} \dfrac{1}{2} \displaystyle\int_\Omega | v |^2 \, dx - (v, u^n) + \dfrac{k}{2} \int_\Omega | \operatorname{grad} v |^2 \, dx - \\ \quad - k(f^{n+1}, v) + k p g^{n+1} v_M . \end{array}$$

We *then discretise in space*, so that (dividing by h, noting that the measure of the last interval is $h/2$, and putting $u_0^{n+1} = 0$):

(4.9)
$$\begin{array}{l} \underset{|p| \leqslant 1}{\underline{\max}} \ \underset{v_h \in V_h}{\underline{\min}} \ L_h^n(v_h, p) = \dfrac{1}{2} \displaystyle\sum_{i=1}^{M-1} | v_i |^2 + \dfrac{1}{4} | v_M |^2 - \\ \quad - \displaystyle\sum_{i=1}^{M-1} v_i u_i^n - \dfrac{1}{2} v_M u_M^n + \dfrac{k}{2 h^2} \sum_{i=1}^{M} | v_i - v_{i-1} |^2 - \\ \quad - \displaystyle\sum_{i=1}^{M-1} k f_i^{n+1} v_i - \dfrac{k}{2} f_M^{n+1} v_M + \dfrac{k}{h} p g^{n+1} v_M . \end{array}$$

We then write the conditions for optimality: find $u_1^{n+1}, \dots, u_M^{n+1}$ and p such that:

(4.10)
$$\begin{array}{l} \dfrac{u_i^{n+1} - u_i^n}{k} - \dfrac{1}{h^2} (u_{i+1}^{n+1} - 2 u_i^{n+1} + u_{i-1}^{n+1}) - f_i^{n+1} = 0 , \\ i = 1, \dots, M - 1 , \end{array}$$

(4.11)
$$\frac{u_M^{n+1} - u_M^n}{k} + \frac{2}{h^2} (u_M^{n+1} - u_{M-1}^{n+1}) - f_M^{n+1} + \frac{2}{h} p g^{n+1} = 0 ,$$

(4.12)
$$\left(\frac{k}{h} g^{n+1} u_M^{n+1}, q - p\right) \geqslant 0 , \quad \forall q , \ | q | \leqslant 1 \ \text{ and } \ | p | \leqslant 1 .$$

It is then possible to refer to a semi-implicit dual method to define the following algorithm which, for u_h^n known (recall that $u_h^0 = 0$ is given), allows us to determine u_h^{n+1} as follows (we denote by $u_h^{n+1(i)}$ the ith iterate in the search for u_h^{n+1}):

- take any arbitrary p^0, for example $p^0 = \text{sign}\,(u_M^n) \times 1$, and a suitable interval $\rho > 0$; then
- solve (4.10), (4.11), for $p = p^0$, which defines $u_M^{n+1(0)}$,
- generate $p^1 = P[p^0 + \rho g^{n+1} u_M^{n+1(0)}]$ where P is the projection on $[0, 1]$,
- solve (4.10), (4.11), for $p = p^1$ (which defines $u_M^{n+1(1)}$),
- generate $p^2 = P[p^1 + \rho g^{n+1} u_M^{n+1(1)}]$, etc.

A multiplicity of variants of these schemes may be defined: for example, in (4.9) we can replace

$$\frac{1}{h^2} \mid u_M^{n+1} - u_{M-1}^{n+1} \mid^2 ,$$

by

$$\left| \frac{\partial}{\partial n_h} (u_h)_{x=1} \right|^2 = \left| \frac{1}{2\,h} (3\,u_M - 4\,u_{M-1} + u_{M-2}) \right|^2 ,$$

this quantity corresponding to the square of the "quadratic" approximation of the derivative at $x = 1$.

4.2.2 *Obtaining the fully explicit scheme*

Although this has not been justified theoretically, we can replace (4.10), (4.11), (4.12) by

$$\frac{u_i^{n+1} - u_i^n}{k} - \frac{1}{h^2}(u_{i+1}^n - 2\,u_i^n + u_{i-1}^n) - f_i^{n+1} = 0\,, \tag{4.13}$$

$$i = 1, \ldots, M - 1\,,$$

$$\frac{u_M^{n+1} - u_M^n}{k} + \frac{2}{h^2}(u_M^n - u_{M-1}^n) - f_M^{n+1} + \frac{2}{h} p^n g^{n+1} = 0\,, \tag{4.14}$$

$$p^n = \text{sign}\,(u_M^n) \times 1\,. \tag{4.15}$$

4.2.3 *Numerical results*

With the aim of assessing the merits of these schemes and of the approximation, we define the following quantities:

- the discrete normal derivative of the exact solution at $x = 1$ and $t = nk$,

$$\nabla_h u^n_{ex}\,|_{x=1} = \frac{u(1, nk) - u(1 - h, nk)}{h},$$

- the discrete normal derivative of the approximate solution at $x = 1$ and at $t = nk$,

$$\nabla_h u^n_h\,|_{x=1} = \frac{u^n_M - u^n_{M-1}}{k},$$

- then

$$\nabla = \max_{n=1,\dots,N=1/k} |\nabla_h u^n - \nabla_h u^n_h\,|_{x=1},$$

$$U = \max_{\substack{n=1,\dots,N\\ i=1,\dots,M}} |u(ih, nk) - u_h(ih, nk)|,$$

- and finally the difference between the exact solution and the calculated solution at $t = nk$,

$$\mathrm{ECART}_2^{t=nk} = \left(\sum_{i=1}^{M} [u(ih, nk) - u^n_i]^2_h\right)^{1/2}.$$

The two schemes were investigated on a BULL-GE-265. The times are quoted in sixths of a second, including compilation time.

Table 4.5 gives the results of the explicit scheme and Table 4.6 those of the implicit scheme. For this scheme equations (4.10), (4.11) were solved by the Gauss-Seidel method with the values $\omega_{\text{opt. exp.}}$ of the over-relaxation parameter obtained experimentally (see Table 4.7) and the parameter ρ fixed at 0.1. Table 4.8 shows U and Δ in terms of h and k for the explicit and implicit schemes.

4.2.4 *Conclusions of* 4.2

With regard to the model problem of time-dependent friction considered here, we have:

1. *Fully explicit scheme*

- danger of instability
- stability appears to correspond to $k/h^2 \leqslant 1/2$.

2. *Implicit scheme*

- high accuracy, practically independent of h and k (but the exact solution is "quadratic"),
- for the same precision, *much slower* than the explicit scheme. ■

h	k	ECART$_2$ $t=k$	0.1	0.2	0.3	0.4	0.5	0.6	0.7	0.8	0.9	1	Time
1/10	1/100	0.004	18	10^6	10^{11}	10^{16}	10^{21}	10^{25}	10^{30}	10^{35}	...	...	—
	1/500	0.000 1	0.001 5	0.001 2	0.000 5	0.001 4	0.003 3	0.005 6	0.008 3	0.011 1	0.014 2	0.017 4	145
	1/1 000	0.000 045	0.000 78	0.000 64	0.000 25	0.000 70	0.001 67	0.002 84	0.004 15	0.005 58	0.007 10	0.008 69	247
	1/2 000	0.000 011	0.000 39	0.000 32	0.000 12	0.000 35	0.000 83	0.001 42	0.002 07	0.002 76	0.003 53	0.004 34	463
1/20	1/100	0.006 5	10^8	10^{20}	10^{32}	...							—
	1/500	0.002 6	10^{13}	10^{30}	...								—
	1/1 000	0.000 066	0.000 77	0.000 64	0.000 24	0.000 68	0.001 64	0.002 80	0.004 10	0.005 52	0.007 02	0.008 59	427
	1/2 000	0.000 016	0.000 38	0.000 32	0.000 12	0.000 34	0.000 82	0.001 40	0.002 05	0.002 76	0.003 51	0.004 25	802
1/30	1/100	...	...	...									—
	1/500	...	...	...									
	1/1 000	0.000 081	10^{36}	...									
	1/2 000	0.000 020	0.000 38	0.000 32	0.000 11	0.000 33	0.000 82	0.001 39	0.002 04	0.002 74	0.003 49	0.004 27	1 548

Table 4.5. Time-dependent friction: explicit scheme

		ECART$_2$											
h	k	$t = k$	0.1	0.2	0.3	0.4	0.5	0.6	0.75 (1)	0.8	0.9	1.0	Time
1/10	1/100	90×10^{-5}	81×10^{-4}	14×10^{-3}	19×10^{-3}	23×10^{-3}	26×10^{-3}	28×10^{-3}	31×10^{-3}	32×10^{-3}	33×10^{-3}	34×10^{-3}	210
	1/500	36×10^{-6}	16×10^{-4}	29×10^{-4}	39×10^{-4}	47×10^{-4}	53×10^{-4}	58×10^{-4}	63×10^{-4}	64×10^{-4}	67×10^{-4}	68×10^{-4}	506
	1/1 000	91×10^{-7}	81×10^{-5}	14×10^{-4}	19×10^{-4}	23×10^{-4}	26×10^{-4}	29×10^{-4}	31×10^{-4}	32×10^{-4}	33×10^{-4}	35×10^{-4}	840
1/20	1/100	88×10^{-5}	79×10^{-4}	14×10^{-3}	19×10^{-3}	22×10^{-3}	25×10^{-3}	28×10^{-3}	30×10^{-3}	31×10^{-3}	32×10^{-3}	33×10^{-3}	586
	1/500	36×10^{-6}	16×10^{-4}	28×10^{-4}	38×10^{-4}	46×10^{-4}	52×10^{-4}	56×10^{-4}	61×10^{-4}	63×10^{-4}	65×10^{-4}	67×10^{-4}	1 349
	1/1 000	90×10^{-7}	80×10^{-5}	14×10^{-4}	19×10^{-4}	23×10^{-4}	26×10^{-4}	28×10^{-4}	30×10^{-4}	31×10^{-4}	32×10^{-4}	33×10^{-4}	2 029
1/30	1/100	82×10^{-5}	78×10^{-4}	14×10^{-3}	18×10^{-3}	22×10^{-3}	25×10^{-3}	28×10^{-3}	30×10^{-3}	31×10^{-3}	32×10^{-3}	33×10^{-3}	1 183
	1/500	36×10^{-6}	15×10^{-4}	28×10^{-4}	38×10^{-4}	45×10^{-4}	51×10^{-4}	56×10^{-4}	61×10^{-4}	62×10^{-4}	64×10^{-4}	66×10^{-4}	2 786
	1/1 000	50×10^{-7}	80×10^{-4}	19×10^{-4}	19×10^{-4}	23×10^{-4}	25×10^{-4}	28×10^{-4}	30×10^{-4}	31×10^{-4}	32×10^{-4}	33×10^{-4}	4 333

Table 4.6. Time-dependent friction: implicit scheme

(1) For this value of t the solution is zero and $j(v)$ is "nondifferentiable".

h	1/10			1/20			1/30		
k	1/100	1/500	1/1 000	1/100	1/500	1/1 000	1/100	1/500	1/1 000
$\omega_{\text{opt.exp.}}$	1.4	1.1	1.1	1.45	1.15	1.1	1.5	1.25	1.12
iterations	11	6	5	20	9	7	28	13	10

Table 4.7.

h	k	Explicit scheme U	Explicit scheme Δ	Implicit scheme U	Implicit scheme Δ
1/10	1/100	Divergence and overflow		8×10^{-3}	2×10^{-4}
	1/500	2×10^{-3}	10^{-2}	2×10^{-3}	4×10^{-5}
	1/1 000	1.5×10^{-3}	6×10^{-3}	10^{-3}	2×10^{-5}
	1/2 000	10^{-3}	5×10^{-3}	not considered	
1/20	1/100	Divergence and overflow		8×10^{-3}	10^{-4}
	1/500			2×10^{-3}	2×10^{-5}
	1/1 000	3×10^{-3}	5×10^{-3}	10^{-3}	10^{-5}
	1/2 000	2×10^{-3}	2×10^{-3}	not considered	
1/30	1/100	Divergence and overflow		8×10^{-3}	5×10^{-5}
	1/500			2×10^{-3}	10^{-5}
	1/1 000			10^{-3}	5×10^{-6}
	1/2 000	10^{-3}	2×10^{-3}	not considered	

Table 4.8. Relative accuracy of the schemes.

4.3 A model problem of the deformation of a membrane

We consider a membrane fixed over a horizontal rectangular frame under a constant tension F and subject to a (positive or negative) load $q(x, t)$. The deformations of this membrane are restricted by a fixed horizontal plane set at a distance a from the frame. We are required to determine the deflection $u(x, t)$ of the membrane.

Let Ω be the domain bounded by the frame and Ω_0 the region in which the membrane is in contact with the horizontal plane. The problem can be expressed in the form: for $t \in [0, T]$ with $u(x, 0) = u_0(x)$ given, find $u(x, t)$ such that

$$\left| \begin{array}{ll} u' - \Delta u = \dfrac{q(x, t)}{F} & \text{in } \Omega - \Omega_0 , \\ u(x, t) = a & \text{on } \Omega_0 . \end{array} \right.$$

With $f = q/F, a = 0$ (1), we arrive at the following problem (2): for t traversing the interval $[0, T]$, solve the inequality:

$$(4.16) \quad \left| \begin{array}{l} (u' - \Delta u - f, v - u) \geqslant 0 , \quad \forall v \in K, u \in \mathscr{K} , \\ \text{with} \\ K = \{ v \mid v \in H_0^1(\Omega), v \geqslant 0 \text{ on } \Omega \} , \\ \mathscr{K} = \{ v \mid v \in W(0, T), v(0) = u_0 \text{ given, } v \in K \} . \end{array} \right.$$

For this application we shall take $\Omega =]0, 1[\times]0, 1[$ and $T = 1$. We put

$$\alpha(t) = \tfrac{1}{2} + \tfrac{1}{4} \sin (2 \pi t) ,$$

and for the exact solution we take (putting $x_1 = x, x_2 = y$):

$$u = \begin{cases} - 100\, x(x - \alpha(t))^2 y(1 - y) , & x \leqslant \alpha(t) , \\ 0 , & x > \alpha(t) , \end{cases}$$

with the initial condition:

$$u_0 = \begin{cases} - 100\, x(x - \frac{1}{2})^2 y(1 - y) , & x \leqslant \frac{1}{2} , \\ 0 , & x > \frac{1}{2} . \end{cases}$$

Corresponding to this solution we have

$$(4.17) \quad f = \begin{cases} u' - \Delta u, & x \leqslant \alpha(t) , \\ \leqslant 0 \ \text{arbitrary} , & x > \alpha(t) . \end{cases}$$

Figure 4.4 shows a cross-section of the exact solution for $y = \frac{1}{2}$, $x \in [0, 1]$ for the instants $t = 0, 0.1, 0.2, ..., 1$. In this figure a double arrow $\leftrightarrow$ represents the zones of Ω where $u = 0$ (i.e. the zones where the constraints are active, for which we say the convex set K is "saturated").

The spatial discretisation is the standard discretisation of $H_0^1(\Omega)$ (see Chapter 3, Section 3)

$$h = h_x = h_y = 1/20 \text{ or } 1/40 ,$$

(thus with $N_h = 1/h - 1$, we have N_h^2 variables),

---

(1) Here we have arbitrarily assumed that the membrane rests on the horizontal plane.

(2) A time-dependent variant of the model problem considered in Chapter 2, Section 5.

$$\left|\begin{array}{l} u_{0,h} = \sum_{M \in \Omega_h} u_0(M)\, \theta_h^M, \text{ where } \theta_h^M \text{ is the characteristic function} \\ \qquad \text{of the panel } \overline{\omega}_h^0(M), \\ V_h = \text{space of step functions on } \Omega_h, \text{ zero at the edge,} \\ \qquad \text{and constant on the panels } \overline{\omega}_h^0(M),\ M \in \Omega_h, \\ K_h = \{ v_h \mid v_h \in V_h,\ v_h \geqslant 0 \text{ on } \Omega_h \}. \end{array}\right.$$

For the time discretisation we take

$$k = 5\times 10^{-4},\quad 10^{-3},\quad 2\times 10^{-3},\quad 5\times 10^{-3}.$$

The approximation of f is, for $t \in [nk, (n+1)k[$,

$$f_{h,k}^n = \sum_{M \in \Omega_h} f(M, (n+1)k)\, \theta_h^M .$$

Moreover, since for $x > \alpha(t)$ the choice of $f \leqslant 0$ is arbitrary, we carried out tests with $f = 0$, $f = -\frac{1}{2}(x - \alpha(t))$, $f = -4(x - \alpha(t))$, $f = -100(x - \alpha(t))$ for $x > \alpha(t)$. For all the schemes used, the numerical results are practically identical (see Trémolières /4/). The numerical results given below relate to the case $f = -\frac{1}{2}(x - \alpha(t))$.

The approximation of Δ is given by

$$\Delta_h u_h(x) = \sum_{i=1,2} \frac{u_h(x + h\vec{e}_i) - 2\, u_h(x) + u_h(x - h\vec{e}_i)}{h^2}.$$

4.3.1 *Approximation schemes*

The schemes investigated are as follows:

S1 - *Explicit scheme*

$$\left|\begin{array}{l} \dfrac{u_h^{n+1/2} - u_h^n}{k} - \Delta_h u_h^n - f_{h,k}^n = 0, \\ u_h^{n+1} = P_{K_h}[u_h^{n+1/2}] \text{ where } P_{K_h} \text{ is the projection in the sense} \\ \qquad \text{of the inner product } (,)_h \\ u_h^0 = u_{0,h}. \end{array}\right.$$

S2 - "*Implicit in equation*" *scheme*

$$\left|\begin{array}{l} \dfrac{u_h^{n+1/2} - u_h^n}{k} - \Delta_h u_h^{n+1/2} - f_{h,k}^n = 0, \\ u_h^{n+1} = P_{K_h}[u_h^{n+1/2}], \\ u_h^0 = u_{0,h}. \end{array}\right.$$

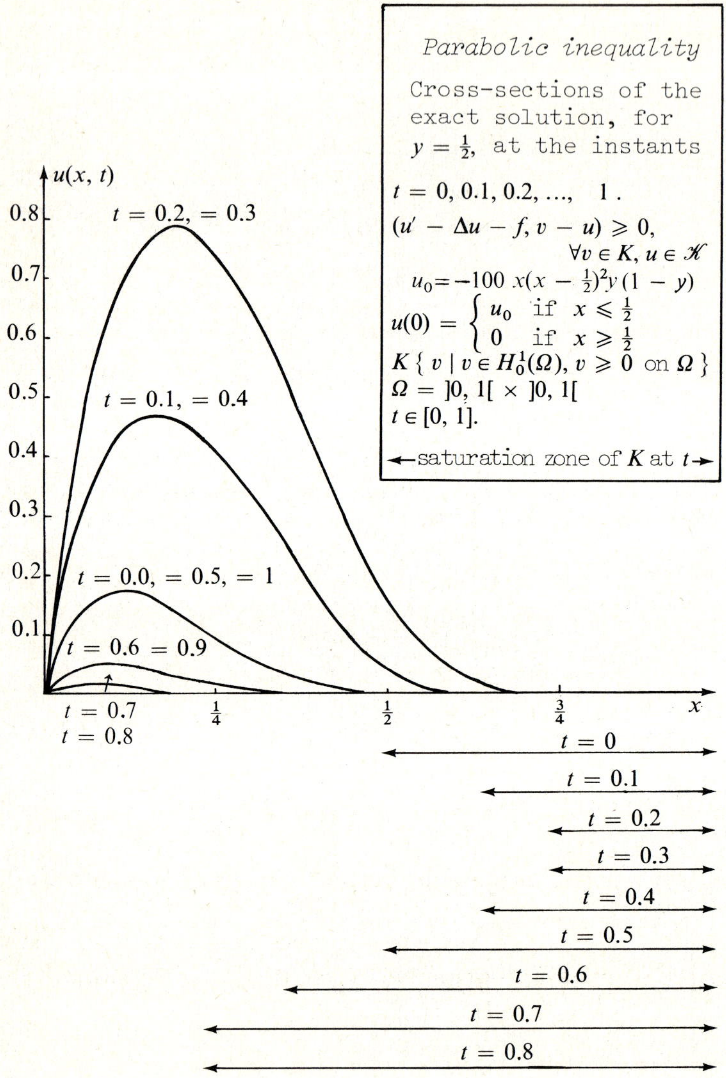

Parabolic inequality
Cross-sections of the exact solution, for $y = \frac{1}{2}$, at the instants
$t = 0, 0.1, 0.2, \ldots, \quad 1$.
$(u' - \Delta u - f, v - u) \geqslant 0, \quad \forall v \in K, u \in \mathcal{K}$
$u_0 = -100\, x(x - \frac{1}{2})^2 y(1 - y)$
$u(0) = \begin{cases} u_0 & \text{if } x \leqslant \frac{1}{2} \\ 0 & \text{if } x \geqslant \frac{1}{2} \end{cases}$
$K \{ v \mid v \in H_0^1(\Omega), v \geqslant 0 \text{ on } \Omega \}$
$\Omega =]0, 1[\times]0, 1[$
$t \in [0, 1]$.
← saturation zone of K at t →
$u(x, t)$
0.8
0.7
0.6
0.5
0.4
0.3
0.2
0.1
$t = 0.2, = 0.3$
$t = 0.1, = 0.4$
$t = 0.0, = 0.5, = 1$
$t = 0.6 = 0.9$
$t = 0.7$
$t = 0.8$
$\frac{1}{4}$
$\frac{1}{2}$
$\frac{3}{4}$
x
$t = 0$
$t = 0.1$
$t = 0.2$
$t = 0.3$
$t = 0.4$
$t = 0.5$
$t = 0.6$
$t = 0.7$
$t = 0.8$

Fig. 4.4.

S3 - *Crank-Nicholson scheme* (1)

$$\left| \begin{array}{l} \left(\dfrac{u_h^{n+1}-u_h^n}{k} - \Delta_h\left(\dfrac{u_h^{n+1}+u_h^n}{2}\right) - f_{h,k}^n, v_h - u_h^{n+1}\right)_h \geq 0\,, \\ \forall v_h \in K_h,\ u_h^{n+1} \in K_h\,, \\ u_h^0 = u_{0,h}\,. \end{array} \right.$$

S4 - *Fully implicit scheme*

$$\left| \begin{array}{l} \left(\dfrac{u_h^{n+1}-u_h^n}{k} - \Delta_h u_h^{n+1} - f_{h,k}^n, v_h - u_h^{n+1}\right)_h \geq 0 \\ \forall v_h \in K_h,\ u_h^{n+1} \in K_h\,, \\ u_h^0 = u_{0,h}\,. \end{array} \right.$$

4.3.2 *Numerical results*

Except in the case of the explicit scheme, the calculation of u_h^{n+1} is carried out by the Gauss-Seidel method with truncation $(u_h^n \geq 0)$, the relaxation parameter ω being chosen in accordance with the formula (see Varga /1/) (2)

$$\omega = \frac{2}{1 + \left(1 - \left[\left(\dfrac{(1-\pi h^2/2)\,4\,h^2}{4 + 2\,h^2/k^2}\right)^2\right]^{1/2}\right.}\,.$$

The termination criterion used in the calculation of u_h^{n+1} by the Gauss-Seidel method with projection is

$$\varepsilon_0 = \sum_{M \in \Omega_h} |u_h^{n+1(m+1)}(M) - u_h^{n+1(m)}(M)| \leq 10^{-6}\,,$$

where $u_h^{n+1(m)}$ is the mth iteration vector and $u_h^{n+1(m+1)}$ the $(m+1)^{\text{th}}$. (One iteration corresponds to a complete sweep over the coordinates).

The relative accuracy of the schemes will be measured at each instant $t = nk$ in terms of the quantity

---

(1) A variant of this scheme is given by:

$$\left| \begin{array}{l} \left(\dfrac{u_h^{n+1}-u_h^n}{k} - \Delta_h\left(\dfrac{u_h^{n+1}+u_h^n}{2}\right) - f_{h,k}^n, v_h - \dfrac{u_h^n+u_h^{n+1}}{2}\right)_h \geq 0 \\ \forall v_h \in K_h\,, \quad \dfrac{u_h^n+u_h^{n+1}}{2} \in K_h\,, \\ u_h^0 = u_{0,h} \end{array} \right.$$

(2) This is the optimal parameter in the absence of constraints.

$$\text{ECART} = \sum_{M \in \Omega_h} | u_h^n(M) - u(M ; nk) |$$

where u_h^n is the solution calculated at $t = nk$ and $u(M ; nk)$ is the exact solution at $t = nk$, at the point M.

With scheme S1 divergence was observed for $h = 1/20$ and $k = 10^{-3}$ during the initial time-iterations (erroneous results and overflow before $t = 100\,k = 0.1$, with $T = 1$). A second test with $h = 1/20$, $k = 5 \times 10^{-4}$ gave accurate results.

For the S2 scheme, several tests showed that with $h = 1/20$, the best time-step appeared to be $k = 10^{-3}$; similarly for S3 and S4.

Table 4.9 shows the best results for these four schemes.

4.3.3 *Conclusions of* 4.3

For the solution of the inequality (4.16) with constraints distributed over Ω we have:

1. *Explicit scheme*

- danger of divergence and overflow,
- an accurate stability estimate is $k/h^2 \leqslant \frac{1}{5}$,
- for the same precision, *twice as fast* as the other schemes.

Scheme	Characteristics		ECART											Time
			$t=0.1$	0.2	0.3	0.4	0.5	0.6	0.7	0.8	0.9	1	1 1	
S1	$h=\frac{1}{20}$	$k=0.5\times10^{-3}$	0.28	0.51	0.78	0.74	0.38	0.12	0.04	0.39	0.08	0.16	0.28	10 min 5
S2	$h=\frac{1}{20}$	$k=10^{-3}$	0.29	0.53	0.81	0.74	0.38	0.12	0.04	0.04	0.08	0.16	0.29	24 min
S3			0.27	0.54	0.82	0.74	0.38	0.13	0.04	0.04	0.07	0.16	0.28	24 min
S4			0.43	0.70	0.78	0.56	0.27	0.10	0.03	0.03	0.09	0.17	0.43	26 min

Table 4.9. Comparison of the various schemes

2. *The other schemes.*

- The "implicit in equation", "semi-implicit in projection" and "fully implicit" schemes appear to be similar in this case. ■

5. INTRODUCTION TO PARABOLIC INEQUALITIES OF TYPE II

To start with we shall give two examples arising in the theory of thermal servo-control.

5.1 Example I

We consider a continuous medium occupying a domain Ω with boundary Γ and whose edge temperature is *servo-controlled* so as not to diminish with time, this being achieved by injecting heat across Γ (the wall thickness being negligible). The problem may be formulated as follows (see Duvaut-Lions /1/):

- heat equation in Ω,

$$\frac{\partial u}{\partial t} - \Delta u = f, \quad x \in \Omega, \quad t \in]0, T[;$$

- given initial temperature,

$$u(x, 0) = u_0(x); \tag{5.1}$$

- boundary conditions on $\Gamma \times]0, T[$ (servo-control),

$$\begin{cases} \dfrac{\partial u}{\partial t} \geqslant 0, \\ \dfrac{\partial u}{\partial n} \geqslant 0, \\ \dfrac{\partial u}{\partial t}\dfrac{\partial u}{\partial n} = 0. \end{cases}$$

By using Green's formula, this problem may be expressed in the form: find $u(x, t)$ satisfying (5.2) with $u \geqslant 0$ on Γ, such that, a.e. $t \in [0, T]$, we have

$$(u' - \Delta u - f, v - u') \geqslant 0, \quad \forall v \geqslant 0 \text{ on } \Gamma. \tag{5.2}$$

In order for the problem to be meaningful we shall seek u such that,

$$u, u' \in L^2(0, T; H^1(\Omega)) \tag{5.3}$$

and such that (5.2) is satisfied $\forall v \in H^1(\Omega)$ with $v \geqslant 0$ on Γ. ■

5.2 Abstract formulation and existence theorem

We again use (1.1)-(1.3) and (2.21)-(2.26) (1), (2.28), (2.29), and we assume, as is essential here (2), that

(5.4) A is *symmetric* and independent of time (3).

We put

$$\mathscr{K} = \{ v \mid v \in W(0, T), v(t) \in K, \quad \text{a.e. } t \in [0, T] \},$$

$$\mathscr{K}_0 = \{ v \mid v \in W_0(0, T), v(t) \in K, \quad \text{a.e. } t \in [0, T] \},$$

$$\mathscr{K}_0^* = \{ u \mid u \in L^2(0, T; V), u' \in \mathscr{K}_0 \}.$$

The general problem whose approximation and solution we shall study is:

(5.5)
$$\left| \begin{array}{l} \text{find } u \in \mathscr{K}_0^* \text{ such that} \\ \displaystyle\int_0^T (u' + Au - f, v - u') \, dt \geq 0, \quad \forall v \in \mathscr{K}. \end{array} \right.$$

We recall (see Duvaut-Lions /1/, Brézis /1/) the following existence result:

Theorem 5.1. *Let f and u_0 be such that:*

(5.6) $f, f' \in L^2(0, T; H), \quad u_0 \in V,$

(5.7) $\exists v_0 \in K$ *which satisfies* $(v_0 + Au_0 - f(0), v - v_0) \geq 0, \quad \forall v \in K.$

There then exists a unique function u such that

(5.8) $u, u' \in L^\infty(0, T; V),$

(5.9) $u(0) = u_0,$

(5.10) $u'(t) \in K, \quad \text{a.e. } t \in [0, T],$

(5.11) $(u' + Au - f, v - u') \geq 0, \quad \forall v \in K, \text{ a.e. } t \in [0, T];$

---

(1) For simplification we shall not consider functionals j which come under the classification of parabolic inequalities of type II. However, the proofs can be extended without difficulty to the case in which $j \neq 0$. See also Remark 5.4.

(2) We require at least that the "principal part" of A be symmetric

(3) This point is purely to simplify the proofs a little.

moreover

(5.12) $u'' \in L^2(0, T; H)$,

(5.13) $u'(0) = v_0$.

Remark 5.1. It can be shown without difficulty that the inequalities (5.5) and (5.11) are equivalent. ■

Remark 5.2. Assumption (5.7) is always satisfied if

(5.14) K is *bounded* in V

or indeed if

(5.15) $-[Au_0 - f(0)] \in K$. ■

Remark 5.3. The following inequality can be established

$$\sup_{t \in [0,T]} \| u'(t) \|^2 \leqslant \frac{1}{\alpha} \left\{ a(v_0, v_0) + \int_0^T | f'(t) |^2 \, dt \right\}.$$

This result enables us to conclude, *when* v_0 *exists* (see Remark 5.2), that u' is *bounded* in K for the norms $L^2(0, T; V)$ and $L^\infty(0, T; V)$. It is then easy to see that if K is not bounded, *it is possible to define a convex set*

$$K_R = \{ v \mid v \in K, \| v \| \leqslant R < +\infty \},$$

R sufficiently large, such that problem (5.11) defined on K_R in fact has the *same solution* as the initial problem. We can then consider, *without loss of generality*, (but still under assumption (5.7)) that

(5.16) K is *bounded* in V.

This remark is useful for the study of the approximation. ■

Remark 5.4. We can likewise solve the following inequality, which is more general than (5.5):

$$(5.17) \quad \int_0^T (u' + Au - f, v - u') \, dt + \int_0^T [j(v) - j(u')] \, dt \geqslant 0, \quad \forall v \in \mathscr{K}, u \in \mathscr{K}_0$$

where j is a functional of the type (2.27). ■

6. APPROXIMATION OF PARABOLIC INEQUALITIES OF TYPE II

6.1 Fundamental assumptions for the approximation

In studying the approximation we shall retain everything said in Section 3.1 with, in addition, the following assumptions:

(6.1) $a_h(u_h, v_h)$ is *symmetric*,

(6.2) K_h is bounded in V_h, *independently* of h, i.e.
$$\| v_h \|_h \leqslant C\,, \quad \forall v_h \in K_h \quad \text{and} \quad \forall h, \quad h \to 0\,.$$

(For this latter assumption, see Remark 5.3).

We shall transpose the results given in Section 3.1.3 to $\mathscr{K}$ and to $\mathscr{K}_0^*$.

In view of (5.6) we have $f \in \mathscr{C}^0(0, T; H)$ and we take

$$f_{h,k} = \sum_{i=0}^{M} f_h^i \chi_k^i\,,$$

as in Section 3.1.6 with f_h^i defined by (3.27'). ■

6.2 Approximation scheme for parabolic inequalities of type II

Taking account of the above assumptions, we shall now study the approximation of the inequality

$$(u' + Au - f, v - u') \geqslant 0\,, \quad \forall v \in \mathscr{K}, u \in \mathscr{K}_0^*, \text{ a.e. } t \in [0, T]\,, \tag{6.3}$$

by

$$\left| \begin{array}{l} (\delta u_{h,k} + A_h u_{h,k}^\theta - f_{h,k}, v_h - \delta u_{h,k})_h \geqslant 0\,, \\ \forall v_h \in K_h, \delta u_{h,k} \in K_h, t \in [0, T], u_{h,k}(0) = u_{0,h}\,, \end{array} \right. \tag{6.4}$$

where

$$u_{h,k}^\theta(t) = u_{h,k}(t) + \theta[u_{h,k}(t + k) - u_{h,k}(t)] = u_{h,k}(t) + k\theta\delta u_{h,k}(t)\,,$$
$$\theta \in [0, 1]\,.$$

Written "pointwise" (6.4) becomes: find $u_h^1, u_h^2, \dots, u_h^N$ such that:

$$\left| \begin{array}{l} \left(\dfrac{u_h^{i+1} - u_h^i}{k} + A_h[u_h^i + \theta(u_h^{i+1} - u_h^i)] - f_h^i, v - \dfrac{u_h^{i+1} - u_h^i}{k} \right)_h \geqslant 0 \\ \forall v \in K, \dfrac{u_h^{i+1} - u_h^i}{k} \in K\,, \text{ with } u_h^0 = u_{0,h}\,. \end{array} \right. \tag{6.5}$$

We note that inequality (6.5) can be written:

(6.6) $$((I_h + \theta k A_h)\,\delta_h^i + A_h u_h^i - f_h^i, v - \delta_h^i)_h \geqslant 0\,, \quad \forall v \in K\,,$$
$$\delta u_h^i \in K\,, \quad u_h^0 = u_{0,h}\,,$$

with

(6.7) $$\delta_h^i = \frac{u_h^{i+1} - u_h^i}{k}.$$

This possesses a unique solution δ_h^i ; thus one by one all the

$$u_h^{i+1}, i = 0, 1, \ldots, (u_h^{i+1} = u_h^i + k\delta_h^i)\,,$$

exist and are uniquely defined.

Moreover we note (since A_h is symmetric) that the u_h^{i+1}, for $i=0, 1, \ldots$ are uniquely defined from the respective solutions of the N minimisation problems $(i = 0, \ldots, N-1)$

(6.8)
minimise $\frac{1}{2}\,|\,\delta_h^i\,|_h^2 + \frac{1}{2}\,\theta k(A_h\,\delta_h^i, \delta_h^i)_h - (A_h\,u_h^i - f_h^i, \delta_h^i)_h$
subject to the constraints
$$\delta_h^i = \frac{u_h^{i+1} - u_h^i}{k} \in K_h$$
where u_h^i, $(i \geqslant 1)$, is given (the solution of the previous problem), and $u_h^0 = u_{h,0}$. We put $u_h^{i+1} = u_h^i + k\delta u_h^i$.

We use the following terminology:

- *Explicit* scheme if $\theta = 0$,
- Crank-Nicholson scheme if $\theta = \frac{1}{2}$,
- *Implicit* scheme if $\theta = 1$.

Remark 6.1. In the same spirit, the solution of the inequality

$$(u' + Au - f, v - u') + j(v) - j(u') \geqslant 0\,, \quad \forall v \in \mathscr{K}, u \in \mathscr{K}_0^*\,,$$

(where $j(v)$ is defined as in (2.27)) leads to the schemes (omitting the suffix h)

$$\left(\frac{u^{i+1} - u^i}{k} + A[u^i + \theta(u^{i+1} - u^i)] - f^i, v - \frac{u^{i+1} - u^i}{k}\right) + \\ + j(v) - j\left(\frac{u^{i+1} - u^i}{k}\right) \geqslant 0\,, \quad \forall v \in K, \frac{u^{i+1} - u^i}{k} \in K,$$

or, since A is symmetric, to the solution of the problems:

minimise $\frac{1}{2}\,|\,\delta_h^i\,|_h^2 + \frac{1}{2}\,\theta k(A_h\,\delta_h^i, \delta_h^i) - (A_h\,u_h^i - f_h^i, \delta_h^i)_h + j(\delta_h^i)$,
subject to the same constraints as in (6.8). ■

Remark 6.2. In the same way as for parabolic inequalities of type I, we can here also define *decoupled schemes*:

$$\left| \begin{array}{l} \dfrac{\tilde{u}_h^{i+1} - u_h^i}{k} + A_h[u_h^i + \theta(\tilde{u}_h^{i+1} - u_h^i)] - f_h^i = 0\,, \\ u^{i+1} = u^i + kP_{K_h}\left[\dfrac{\tilde{u}_h^{i+1} - u_h^i}{k}\right], \end{array} \right.$$

where P_{K_h} is the projection on K_h in the sense of the norm $(,)_h$. ■

6.3 Convergence of the approximate inequalities

We shall now study scheme (6.5) which we rewrite (omitting the subscript h for clarity) as follows

$$(6.9)\qquad \left| \; (\delta^i + Au^{i+\theta} - f^i, v - \delta^i) \geqslant 0\,, \quad \forall v \in K,\ \delta^i \in K,\ i = 0, 1, \ldots, N\,, \right.$$

with

$$\left| \begin{array}{l} u^0 = u_0 \quad \text{given}, \\ \delta^i = \dfrac{u^{i+1} - u^i}{k}, \\ u^{i+\theta} = u^i + \theta[u^{i+1} - u^i]\,. \end{array} \right.$$

In addition we shall use the following notation:

$$\left| \begin{array}{l} \delta^{i+\theta} = \dfrac{u^{i+1+\theta} - u^{i+\theta}}{k} = \delta^i + \theta[\delta^{i+1} - \delta^i]\,, \\ \gamma^i = \dfrac{\delta^{i+1} - \delta^i}{k} = \dfrac{u^{i+2} - 2\,u^{i+1} + u^i}{k^2}, \\ g^i = \dfrac{f^{i+1} - f^i}{k}. \end{array} \right.$$

6.3.1 *Stability*

We shall now establish:

Theorem 6.1. *When* $(h, k) \to 0$

- *in a general fashion if* $\frac{1}{2} \leqslant \theta \leqslant 1$,
- *under the stability condition*

$$(6.10)\qquad CkS(h)^2 \leqslant 1 - \beta \quad \text{if} \quad 0 \leqslant \theta < \tfrac{1}{2}, \quad (\beta \text{ fixed} \in]0, 1[)\,,$$

we have, for all $n = 0, 1, \ldots, N$,

$$(6.11)\qquad \| u_h^n \|_h \leqslant C\,,$$

(6.12) $$\left\| \frac{u_h^n - u_h^{n-1}}{k} \right\|_h \leqslant C ,$$

and

(6.13) $$\sum_{i=1}^{N-1} k \left| \frac{u^{i+1} - 2\, u^i + u^{i+1}}{k^2} \right|_h^2 \leqslant C .$$

Proof. Writing (6.9) for $i + 1$ we have

(6.14) $$(\delta^{i+1} + Au^{i+1+\theta} - f^{i+1}, v - \delta^{i+1}) \geqslant 0 , \quad \forall v \in K, \delta^{i+1} \in K .$$

With $v = \delta^{i+1}$ in (6.9), $v = \delta^i$ in (6.14) and adding, we obtain

$$| \delta^{i+1} - \delta^i |^2 + a(u^{i+1+\theta} - u^{i+\theta}, \delta^{i+1} - \delta^i) \leqslant (f^{i+1} - f^i, \delta^{i+1} - \delta^i) ,$$

or, dividing by k,

$$k \,| \gamma^i |^2 + a(\delta^{i+\theta}, \delta^{i+1} - \delta^i) \leqslant k\left(\frac{f^{i+1} - f^i}{k}, \gamma^i\right) \leqslant$$

$$\leqslant \frac{k}{2} | g^i |^2 + \frac{k}{2} | \gamma^i |^2 ,$$

and hence, multiplying by 2,

$$k \,| \gamma^i |^2 + 2\, a(\delta^i, \delta^{i+1} - \delta^i) + 2\, \theta a(\delta^{i+1} - \delta^i) \leqslant k \,| g^i |^2 ,$$

or, since a is *symmetric*,

(6.15) $$k \,| \gamma^i |^2 + a(\delta^{i+1}) - a(\delta^i) + (2\, \theta - 1)\, a(\delta^{i+1} - \delta^i) \leqslant k \,| g^i |^2 .$$

We then distinguish the following two cases:

1st case : $\frac{1}{2} \leqslant \theta \leqslant 1$.

Since the term in $(2\, \theta - 1)$ is positive we can suppress it in (6.15), and by summing from $i = 0$ to $n \leqslant N - 1$ we obtain

$$\sum_{i=0}^{n} k \,| \gamma^i |^2 + a(\delta^{n+1}) \leqslant \sum_{i=0}^{n} k \,| g^i |^2 + a(\delta^0) .$$

As a consequence of assumption (6.2) we have $a(\delta^0) \leqslant C$. In view of (5.6) and (3.27') we then obtain (6.13) immediately. Using the coercivity of $a(,)$ we deduce (6.12). The upper bound (6.11) is an immediate consequence of (6.12) and of the fact that $a(\delta^0) \leqslant C$ by noting that

$$u_h^{n+1} = u_h^0 + k \sum_{i=0}^{n} \delta^0 .$$

2nd case : $0 \leqslant \theta < \frac{1}{2}$.

We use the upper bound

$$a(\delta^{i+1} - \delta^i) = h^2\, a(\gamma^i) \leqslant Ck^2\, S(h)^2\, |\gamma^i|^2\,,$$

whence

$$(2\,\theta - 1)\, a(\delta^{i+1} - \delta^i) \geqslant (2\,\theta - 1)\, Ck^2\, S(h)^2\, |\gamma^i|^2\,,$$

and (6.15) implies that

$$k[1 + (2\,\theta - 1)\, CkS(h)^2]\, |\gamma^i|^2 + a(\delta^{i+1}) - a(\delta^i) \leqslant k\, |g^i|^2\,. \tag{6.16}$$

Hence if we assume, in the case $2\,\theta - 1 < 0$, that

$$1 - CkS(h)^2 \geqslant \beta > 0,\ \ \beta \text{ fixed } \in]0, 1[\,,$$

it may be deduced from (6.16) that

$$\beta k\, |\gamma^i|^2 + a(\delta^{i+1}) - a(\delta^i) \leqslant k\, |g^i|^2\,,$$

and by summation we obtain (6.11), (6.12), (6.13), as in the 1st case. Thus Theorem 6.1 is proved. ■

Remark 6.3. In the case where $\forall h, k \to 0$, the inequality (6.9) is in fact, for $i = 0$, an *equation* (convex set not "saturated"); we have

$$\delta^0 + Au^{0+\theta} - f^0 = 0\,,$$

and hence

$$\|\,\delta^0\,\| = \|\, A(u^0 + \theta k\delta^0) - f^0\,\| \leqslant \|\, Au^0\,\| + \theta k\, \|\, A\delta^0\,\| + \|\, f^0\,\|\,,$$

from which we deduce, with $\|\, Au^0\,\| \leqslant C$, $\|\, A\delta^0\,\| \leqslant C$ that

$$(1 - \theta kC)\, \|\,\delta^0\,\| \leqslant C\,,$$

so that, for k sufficiently small, δ^0 is *bounded* for the norm V_h *independently* of h and k, and assumption (6.2) is in fact unnecessary. ■

6.3.2. *Weak convergence*

From Theorem 6.1, when $h, k \to 0$ (with (6.10) if $0 \leqslant \theta \leqslant \frac{1}{2}$) there exists a subsequence, again denoted by $u_{h,k}$, and elements

$$u \in L^\infty(0, T; V),\ v \in L^\infty(0, T; V),\ w \in L^2(0, T; H),\ \xi \in V\,, \tag{6.17}$$

such that

$$\begin{cases} p_h\, u_{h,k} \to \sigma u \ \text{ weakly* in } L^\infty(0, T; F) \\ q_h\, u_{h,k} \to \ u \ \text{ weakly* in } L^\infty(0, T; H) \end{cases} \tag{6.18}$$

(6.19) $$\begin{cases} p_h\, \delta u_{h,k} \to \sigma v \text{ weakly * in } L^\infty(0, T; F) \\ q_h\, \delta u_{h,k} \to v \text{ weakly * in } L^\infty(0, T; H) \end{cases}$$

(6.20) $$q_h\, \delta^2 u_{h,k} \to w \quad \text{weakly in} \quad L^2(0, T; H)$$

(6.21) $$\begin{cases} p_h\, u_h^N \to \sigma\xi \text{ weakly in } F, \\ q_h\, u_h^N \to \xi \text{ weakly in } H. \end{cases}$$

In the same spirit as in Lemma 3.1 it may then be shown (see Viaud /1/) that

(6.22) $$v = u', \quad w = u'', \quad \xi = u(T), \quad u' \in K,$$

so that u satisfies (5.8), (5.9), (5.10), (5.12). We shall show that u satisfies the inequality (5.11).

Let $v \in \mathscr{K}$ and $v_{h,k} = \sum_{i=0}^{N} v_h^i \chi_k^i$ be a sequence approximating v as in Lemma 3.2. We take $v = v_h^i$ in (6.9), sum from $i = 0$ to $N - 1$ and multiply by k; we obtain (once more omitting the suffix h for the time being)

(6.23) $$\sum_{n=0}^{N-1} [k(\delta^i, v^i) + ka(u^{i+\theta}, v^i) - k(f^i, v^i - \delta^i)] \geqslant \sum_{i=0}^{N-1} k\, |\, \delta^i\, |^2 + X,$$

with

(6.24) $$X = \sum_{i=0}^{N-1} ka(u^{i+\theta}, \delta^i)\,.$$

Now

$$\begin{aligned} X &= \sum_{i=0}^{N-1} a(u^{i+\theta}, u^{i+1} - u^i) = \\ &= \sum_{i=0}^{N-1} \frac{1}{2} [a(u^{i+1}) - a(u^i) + (2\,\theta - 1)\, a(u^{i+1} - u^i)] = \\ &= \frac{1}{2} a(u^N) - \frac{1}{2} a(u^0) + Y, \end{aligned}$$

with

$$Y = (2\,\theta - 1) \sum_{i=0}^{N-1} a(u^{i+1} - u^i)\,.$$

We shall distinguish the following two cases (reintroducing the suffix h):

1st case : $\frac{1}{2} \leqslant \theta \leqslant 1$.

Then

$$Y \geqslant 0\,, \quad \text{so that } X \geqslant \tfrac{1}{2}\, a_h(u_h^N) - \tfrac{1}{2}\, a_h(u_h^0) \quad \text{and}$$

(6.25) $$\liminf_{h,k \to 0} X \geqslant \tfrac{1}{2}\, a(u(t)) - \tfrac{1}{2}\, a(u^0)\,.$$

Using (6.23) and (6.31) together with the various consistency assumptions on $\mathscr{K}$, $a(,)$, f and the results (6.18), (6.19), we obtain:

$$\int_0^T (u' + Au - f, v - u')\,\mathrm{d}t \geqslant 0$$

which is true $\forall v \in \mathscr{K}$. From this we deduce that u is a solution of problem (5.5) or (6.3).

2nd case : $0 \leqslant \theta < \frac{1}{2}$.

We then have $Y \leqslant 0$, but the above result remains valid if we have $Y \to 0$.

Now

(6.26) $\quad a_h(u_h^{i+1} - u_h^i) \leqslant Ck^2\, a_h(\delta_h^i) \leqslant$ from (6.12) $\leqslant Ck^2$.

Hence

(6.27) $\quad |Y| \leqslant kC \sum_{i=0}^{N} k \leqslant kC,$

so that we indeed have $Y \to 0$.

To conclude, we note that as a consequence of the uniqueness of u, the results (6.24),...,(6.27) are true for the entire sequence.

We can now state:

Theorem 6.2. *When* $h, k \to 0$ *and in the case when* $0 \leqslant \theta < \frac{1}{2}$ *with*

$$CkS(h)^2 \leqslant 1 - \beta, \quad \beta \textit{ fixed } \in]0, 1[,$$

we have

(6.28)
$$\left|\begin{array}{lll} \left.\begin{array}{l} p_h u_{h,k} \to \sigma u \\ p_h \delta u_{h,k} \to \sigma u' \end{array}\right\} & \textit{weakly * in} & L^\infty(0, T; F) \\ \left.\begin{array}{l} q_h u_{h,k} \to u \\ q_h \delta u_{h,k} \to u' \end{array}\right\} & \textit{weakly * in} & L^\infty(0, T; H) \\ q_h \delta^2 u_{h,k} \to u'' & \textit{weakly * in} & L^2(0, T; H) \\ p_h u_h^N \to \sigma u(T) & \textit{weakly in} & F \\ q_h u_h^N \to u(T) & \textit{weakly in} & H. \end{array}\right.$$
■

Remark 6.4. We note that we have thus *proved the existence* of a solution of problem (5.5), (6.3). ■

6.3.3. *Strong convergence*

We shall now prove (cf. Viaud /1/) the following result:

Theorem 6.3. *When* $h, k \to 0$ *and in the case where* $0 \leqslant \theta < \frac{1}{2}$

with

$$(6.29) \qquad CkS(h)^2 \leqslant 1 - \beta\,, \quad \beta \text{ fixed } \in]0, 1[\,,$$

we have:

$$\left.\begin{aligned} &(6.30) \qquad p_h\, u_{h,k} \to \sigma u \\ &(6.31) \qquad p_h\, \delta u_{h,k} \to \sigma u' \end{aligned}\right\} \text{ strongly in } L^\infty(0, T; F)\,,$$

$$\left.\begin{aligned} &(6.32) \qquad q_h\, u_{h,k} \to u \\ &(6.33) \qquad q_h\, \delta u_{h,k} \to u' \end{aligned}\right\} \text{ strongly in } L^\infty(0, T; H)\,. \quad \blacksquare$$

Proof of Theorem 6.3. We assume here that we can "extend" $a(,)$ to F; in other words, that we can define $\tilde{a}(,)$ on $F \times F$ such that

$$\tilde{a}(\sigma u, \sigma v) = a(u, v)\,, \quad \forall u, v \in V\,,$$

and such that $\tilde{a}(,)$ retains on F the same coercivity constant α and continuity constant C as $a(,)$, and stays symmetric (see Remark 6.5 below); in addition, we take $a_h(u_h, v_h) = \tilde{a}(p_h\, u_h, p_h\, v_h)$.

We then put (for simplification omitting subscripts indicating certain dependencies on h and k)

$$(6.34) \qquad X(t) = \nu \int_0^t |\, q_h\, \delta u_{h,k} - u' \,|^2\, \mathrm{d}s + \frac{1}{2}\, \tilde{a}(\sigma u(t) - p_h\, u_{h,k}(t)) \qquad (^1)$$

with

$$\nu = \begin{cases} 1 & \text{if } \theta \geqslant \frac{1}{2}\,, \\ 1 - (\frac{1}{2} - \theta)\, CkS(h)^2 & \text{if } \theta < \frac{1}{2}\,, \end{cases}$$

so that $\nu \to 1$ if $h \to 0$.

We write

$$X(t) = X_1(t) + X_2(t) + X_3(t)\,,$$

with

$$X_1(t) = \nu \int_0^t |\, u' \,|^2\, \mathrm{d}s + \frac{1}{2}\, a(u(t))\,,$$

$$X_2(t) = -\, 2\, \nu \int_0^t (q_h\, \delta u_{h,k}, u')\, \mathrm{d}s - \tilde{a}(\sigma u(t), p_h\, u_h(t))\,,$$

$$X_3(t) = \nu \int_0^t |\, q_h\, \delta u_{h,k} \,|^2\, \mathrm{d}s + \frac{1}{2}\, a_h(u_h(t))\,.$$

We shall now examine these three terms in succession in order to show that $X(t) \to 0$.

1st term : $X_1(t)$.

---

(1) $\tilde{a}(v) = \tilde{a}(v, v),\ v \in F$.

From $u, u' \in L^\infty(0, T; V)$, $u(0) = u_0$ and the fact that $a(u, v)$ is symmetric, we deduce that

$$\frac{1}{2} a(u(t)) = \frac{1}{2} a(u_0) + \int_0^t a(u, u')\, ds ,$$

and hence

$$X_1(t) = \int_0^t \nu[| u' |^2 + a(u, u')]\, ds + \frac{1}{2} a(u_0) .$$

2nd term : $X_2(t)$.

We put $\tilde{u}_k(t) = \sum_{i=0}^{N-1} u((i+1)k)\, \chi_k^i$. It can then readily be verified that for $t \in [nk, (n+1)k[$, we have

$$\tilde{a}(\sigma u(t), p_h u_{h,k}(t)) = \tilde{a}(\sigma u_0, p_h u_h^0) + \int_0^t \tilde{a}(\sigma u', p_h u_{h,k})\, ds + $$

$$+ \int_0^{nk} \tilde{a}(\sigma \tilde{u}_k, p_h \delta_k u_{h,k})\, ds ,$$

and hence

$$X_2(t) = -2\nu \int_0^t (q_h \delta_k u_{h,k}, u')\, ds - \int_0^t \tilde{a}(\sigma u', p_h u_{h,k})\, ds -$$

$$- \int_0^{nk} \tilde{a}(\sigma \tilde{u}_k, p_h \delta_k u_{h,k})\, ds - \tilde{a}(\sigma u_0, p_h u_h^0) .$$

By noting that $\tilde{u}_k \to u$ strongly in $L^2(0, T; V)$, using (6.28) and taking account of (6.29) we have, proceeding to the limit,

$$X_2(t) \to -2 \int_0^t [| u' |^2 + a(u', u)]\, ds - a(u_0) .$$

3rd term : $X_3(t)$.

From Lemma 3.2 we can consider a sequence $v_{h,k} \in L^2(0, T; K)$ which approximates u' in the following sense

$$\text{(6.35)} \quad \left| \begin{array}{ll} q_h v_{h,k} \to u' & \text{strongly in } L^2(0, T; H) \\ p_h v_{h,k} \to \sigma u' & \text{strongly in } L^2(0, T; F) \\ q_h \delta v_{h,k} \to u'' & \text{strongly in } L^2(0, T; V') . \end{array} \right.$$

By substituting $v_{h,k}$ into the discrete inequality (6.4) and integrating, we obtain (omitting the suffix k):

$$\text{(6.36)} \quad \int_0^t [| \delta u_h |_h^2 + a_h(u_h^\theta, \delta u_h)]\, ds \leqslant \int_0^t [(\delta u_h + A_h u_h^\theta, v_h)_h - (f_h, v_h - \delta u_h)_h]\, ds .$$

However for $t \in [nk, (n+1)k[$ it can be shown without difficulty that

$$\int_0^t a_h(u_h^\theta, \delta u_h)\,ds = \frac{1}{2}[a_h(u_h(t)) - a_h(u_h^0)] + + k\left(\theta - \frac{1}{2}\right)\int_0^{nk} a_h(\delta u_h, \delta u_h)\,ds + \int_{nk}^t a_h(u_h^\theta, \delta u_h)\,ds. \tag{6.37}$$

By using, in the case $\theta < \frac{1}{2}$, the following upper bound:

$$0 \geqslant k\left(\theta - \frac{1}{2}\right)\int_0^{nk} a_h(\delta u_h, \delta u_h)\,ds \geqslant \left(\theta - \frac{1}{2}\right) CkS(h)^2 \int_0^t |\delta u_h|_h^2\,ds,$$

we obtain

$$\int_0^t [|\delta u_h|_h^2 + a_h(u_h^\theta, \delta u_h)]\,ds \geqslant \frac{1}{2}[a_h(u_h(t)) - a_h(u_h^0)] + + \nu \int_0^t |\delta u_h|_h^2\,ds + \int_{nk}^t a_h(u_h^\theta, \delta u_h)\,ds. \tag{6.38}$$

Combining (6.36) and (6.38) we obtain:

$$X_3(t) \leqslant \int_0^t [(\delta u_h + A_h u_h^\theta, v_h)_h - (f_h, v_h - \delta u_h)_h]\,ds + + \frac{1}{2} a_h(u_h^0) - \int_{nk}^t a_h(u_h^\theta, \delta u_h)\,ds.$$

Noting that the last integral on the right-hand side tends to zero, and using the convergence assumptions (6.28), (6.35), we obtain, on proceeding to the limit:

$$\limsup_{h,k\to 0} X_3(t) \leqslant \int_0^t [|u'|^2 + a(u, u')]\,ds + \frac{1}{2} a(u^0).$$

Collecting the results obtained for X_1, X_2, X_3, we have

$X(t) \to 0$, when $h, k \to 0$ (with (6.29)).

Hence we deduce that

$$\left| \begin{array}{l} a_h\, \delta u_{h,k} \to u' \text{ strongly in } L^2(0, T; H) \\ p_h\, u_h(t) \to \sigma u(t) \text{ strongly in } F, \text{ a.e. } t \in [0, T]. \end{array} \right.$$

However the fact that $p_h u_{h,k} \to \sigma u$ weakly * in $L^\infty(0, T; F)$ and strongly a.e. in F implies, by Lebesgue's theorem, that

$$p_h u_{h,k} \to \sigma u \text{ strongly in } L^2(0, T; F).$$

Moreover, for $t \in [nk, (n+1)k[$, with

$$\tilde{u}_k(t) = \sum_{i=0}^{N-1} u(ik)\, \chi_k^i ,$$

it can easily be verified that (omitting the suffix k)

$$\| \sigma u(t) - p_h u_h(t) \|_F^2 = \| \sigma u_0 - p_h u_h^0 \|_F^2 + \\ + 2 \int_0^t ((\sigma u', \sigma u - p_h u_h))_F \, ds - k \int_0^{nk} \| \delta u_h \|_h^2 \, ds + \\ + 2 \int_0^{nk} ((p_h \delta u_h(s), p_h u_h(s+k) - \sigma \tilde{u}_k(s+k))) \, ds ,$$

and with

$$\tilde{u}'_k(t) = \sum_{i=0}^{N-1} u'(ik)\, \chi_k^i ,$$

it can similarly be verified that

$$| u'(t) - q_h \delta u_h(t) |^2 = | v_0 - q_h \delta u_h^0 |^2 + \\ + 2 \int_0^t (u'', u' - q_h \delta u_h) \, ds - k \int_0^{nk} | \delta^2 u_h |^2 \, ds + \\ + 2 \int_0^{nk} (q_h \delta^2 u_h(s), q_h \delta u_h(s+h) - \tilde{u}'_k(s+k)) \, ds .$$

It may then be deduced from these last two formulas that:

$$\left| \begin{array}{lll} p_h u_{h,k} \to \sigma u & \text{strongly in} & L^\infty(0, T; F) \\ q_h \delta u_{h,k} \to u' & \text{strongly in} & L^\infty(0, T; H) \end{array} \right.$$

which proves the theorem. ∎

Remark 6.5. In general, with V such that $H_0^1(\Omega) \subset V \subset H^1(\Omega)$, we take $F = L^2(\Omega) \times L^2(\Omega)^N$, where Ω is an open bounded region of $\mathbb{R}^N$, and with

$$a(u, v) = \int_\Omega uv \, dx + \sum_{i=1}^N \int_\Omega \frac{\partial u}{\partial x_i} \frac{\partial v}{\partial x_i} \, dx ,$$

we put

$$\tilde{a}(\tilde{u}, \tilde{v}) = \int_\Omega u_0 v_0 \, dx + \sum_{i=1}^N \int_\Omega u_i v_i \, dx$$

where $u = (u_0, u_1, \dots, u_N) \in F$, $\tilde{v} = (v_0, v_1, \dots, v_N) \in F$ and for $u \in V$, we define

$$\sigma u = \left(u, \frac{\partial u}{\partial x_1}, \dots, \frac{\partial u}{\partial x_N} \right) .$$

With:

$$p_h : v_h \in V_h \to (v_h |_\Omega ; \nabla_{h_1} v_h |_\Omega, \ldots, \nabla_{h_N} v_h |_\Omega)$$

and

$$\| v_h \|_h = \| p_h v_h \|_F$$

it can then be verified without difficulty that $\tilde{a}(,)$ satisfies the conditions required at the start of the proof of Theorem 6.3. ■

7. NUMERICAL SOLUTION OF PARABOLIC INEQUALITIES OF TYPE II

In this section we shall indicate some results concerning the numerical solution of the two examples given in Section 5.

First, we recall that the discretisation of these two examples leads to inequalities of the form:

$$(7.1)\quad \begin{cases} \left(\dfrac{u_h^{i+1} - u_h^i}{k} + A_h[u_h^i + \theta(u_h^{i+1} - u_h^i)] - f_h^i, v_h - \left(\dfrac{u_h^{i+1} - u_h^i}{k}\right)\right)_h \geqslant 0, \\ \forall v_h \in K_h \quad \text{and} \quad \dfrac{u_h^{i+1} - u_h^i}{k} \in K_h, i = 0, 1, \ldots, N = \dfrac{T}{k}, \\ u_h^0 = u_{0,h} \text{ given}, \ \theta \in [0, 1]. \end{cases}$$

This may also be stated in the form: determine, for $i = 0, 1, \ldots$, the solution w_h^{i+1} of the inequality

$$(7.2)\quad \begin{cases} (w_h^{i+1} + A_h[u_h^i + k\theta w_h^{i+1}] - f_h^i, v_h - w_h^{i+1})_h \geqslant 0, \\ \forall v_h \in K_h, w_h^{i+1} \in K_h, \end{cases}$$

and define

$$(7.3)\quad u_h^{i+1} = u_h^i + k w_h^{i+1}.$$

Putting $\tilde{f}_h^i = A_h u_h^i - f_h^i$ it can be seen that (7.2) may be written

$$(7.4)\quad ((I_h + k\theta A_h) w_h^{i+1} - \tilde{f}_h^i, v_h - w_h^{i+1})_h \geqslant 0, \quad \forall v_h \in K_h, w_h^{i+1} \in K_h,$$

or (since A_h is symmetric):

$$(7.5)\quad \underset{w_h \in K_h}{\text{minimise}}\ J_h(w_h) = \tfrac{1}{2}((I_h + k\theta A_h) w_h, w_h)_h - (\tilde{f}_h^i, w_h),$$

the solution of this last problem giving w_h^{i+1}.

For the two examples which follow, problems (7.5) will be solved by the method of relaxation with constraints (see Chapter 2, Section 1.2). Note that the optimal value of the relaxation parameter is directly related to the term $(I_h + \theta k A_h)$, and moreover, depends on the zones of saturation of the convex set K_h (i.e. where the constraints are active), as well as, obviously, on h. In the following numerical applications we have chosen to take values of ω giving results which are not too expensive in terms of computation time. The parameter was taken as indicated (as a function of $\lambda = \theta k$) in Figure 7.1.

To assess the merits of the various schemes and the stability and convergence assumptions, we shall rely on problems for which *exact analytic solutions* are available.

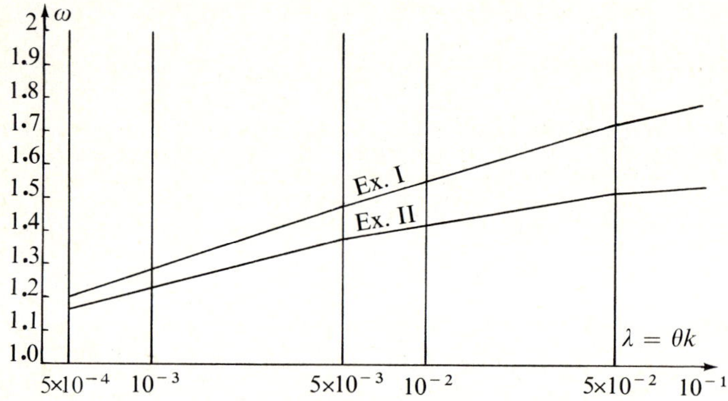

Fig. 7.1. Values of the relaxation parameter ω as a function of $\lambda = \theta k$ for the two examples considered.

For the two examples I and II we shall take:

$$\Omega =]0, 1[\times]0, 1[, T = 0.1 ,$$
$f(x, t)$ and $u(x, t)$ continuous.

Having specified the discretisation of Ω for each example, we shall take

$$f_h^i = \sum_{M \in \Omega_h} f[M, ik]\, \theta_h^M ,$$
$$u_{0,h} = \sum_{M \in \Omega_h} u(M, 0)\, \theta_h^M ,$$

where θ_h^M is the characteristic function of the panel $\overline{\omega}_h^0(M)$.

The investigation will be carried out for

$h = 1/20$,

k between 10^{-1} and 2×10^{-4} ,

$\theta=0$ (explicit scheme)

$=\frac{1}{2}$ (Crank-Nicholson scheme)

$=1$ (implicit scheme). ■

The computations were performed on a CII 10070. ■

7.1. Solution of Example I

The particulars relating to this example are:

$$(7.6)\quad \begin{cases} V = H^1(\Omega), \quad H = L^2(\Omega), \\ K = \{ u \in H^1(\Omega) \mid u(x) \geqslant 0 \quad \text{a.e. } x \in \Gamma \}. \end{cases}$$

On $\Omega =]0,1[\times]0,1[$ we consider the grid

$$R_h = \{ M = (m_1 h, m_2 h), m_1, m_2 = 0, 1, \dots, 20 = 1/h \}$$

and we define

$$\Omega_h = \{ M \in R_h \mid \overline{\omega}_h^1(M) \cap \Omega \neq \phi \},$$
$$\Gamma_h = \{ M \in R_h \mid M \in \Gamma \},$$
$$V_h = \left\{ u_h(x) = \sum_{M \in \Omega_h} u_h^M \, \theta_h^M(x), u_h^M \in \mathbb{R} \right\},$$

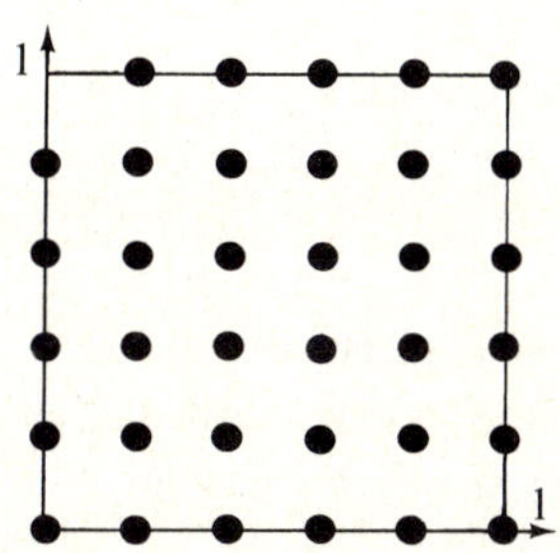

Fig. 7.2. Ω_h for $h = 1/5$.

$$(7.7)\quad K_h = \{ u_h \in V_h \mid u_h(M) \geqslant 0, \quad \forall M \in \Gamma_h \},$$

$$(7.8)\quad (u_h, v_h)_h = \int_\Omega u_h(x)\, v_h(x)\, \mathrm{d}x, \quad u_h \quad \text{and} \quad v_h \in V_h,$$

$$(7.9)\quad (A_h u_h, v_h)_h = \sum_{i=1,2} \int_\Omega \delta_i u_h(x)\, \delta_i v_h(x)\, \mathrm{d}x, \quad \forall u_h \quad \text{and} \quad v_h \in V_h,$$

with

$$(7.10)\quad \delta_i u_h(x) = \frac{u_h\left(x + \frac{h}{2}\vec{e}_i\right) - u_h\left(x - \frac{h}{2}\vec{e}_i\right)}{h}.$$

It can be shown in the present case that

$$S(h) = 2\sqrt{2}/h.$$

As indicated previously, at each time "iteration" ($i = 0, 1, \ldots$ $\ldots, N = T/k$) we calculate w_h^{i+1} — and hence deduce u_h^{i+1} — by solving a problem of the type (7.5) by the relaxation method, using a relaxation parameter as indicated in Figure 7.1.

In the determination of w_h^{i+1}, the relaxation algorithm is terminated when the difference in the L^∞ norm between two successive iterates $w_h^{i+1(n-1)}$ and $w_h^{i+1(n)}$ is less than 10^{-4}, i.e. when

$$\sum_{M \in \Omega_h} | w_h^{i+1(n)}(M) - w_h^{i+1(n-1)}(M) | \leqslant 10^{-4} . \tag{7.11}$$

In order to assess quantitatively the merits of the various schemes, we shall also investigate the relative error $E(u)$ in the L^2 norm, relative to the various exact solutions, as a function of the time:

$$E(u) = \left(\sum_{M \in \Omega_h} | u_h^i(M) - u(M ; ik) |^2 \Big/ \sum_{M \in \Omega_h} | u(M ; ik) |^2 \right)^{1/2} . \tag{7.12}$$

(For the exact solution considered here, the denominator remains > 0).

We consider the following exact solution on $Q =]0, T[\times \Omega =]0, 0.1[\times]0, 1[\times]0, 1[$

$$\begin{aligned} u(x, y, t) = 1 &+ 100(2x + 1)(x - 1)^2 y^2[(\sin 10t - y)^+]^3 + \\ &+ 100(2y + 1)(y - 1)^2 x^2[(\sin 10t - x)^+]^3 - \\ &- 100 x^2(1 - x) y(1 - y)^2 [(y - \sin 10t)^+]^3 - \\ &- 100 y^2(1 - y) x(1 - x)^2[(x - \sin 10t)^+]^3 , \end{aligned} \tag{7.13}$$

with, for a quantity g:

$$g^+ \begin{cases} = g & \text{if } g \geqslant 0 , \\ = 0 & \text{if } g \leqslant 0 . \end{cases}$$

With the data

$$u_0(x, y) = u(x, y, 0) ,$$

and

$$f(x, y, t) = u'(x, y, t) - \Delta u(x, y, t) , \tag{7.14}$$

the problem

$$(u' - \Delta u - f, v - u) \geqslant 0 , \quad \forall v \in K, u \in \mathscr{K}_0^* \tag{7.15}$$

(where K is given by (7.6) and $\mathscr{K}_0^*$ is defined as in Section 5.2) has the exact solution (7.13)[1].

It may be noted, moreover, that an arbitrary constant can be added to u without modifying f (see (7.14)). Since it would therefore be possible to give results with as small a relative error as desired, we shall note that

$$\max_{(x,y,t)\in Q} u(x, y, t) \simeq 3.2 .$$

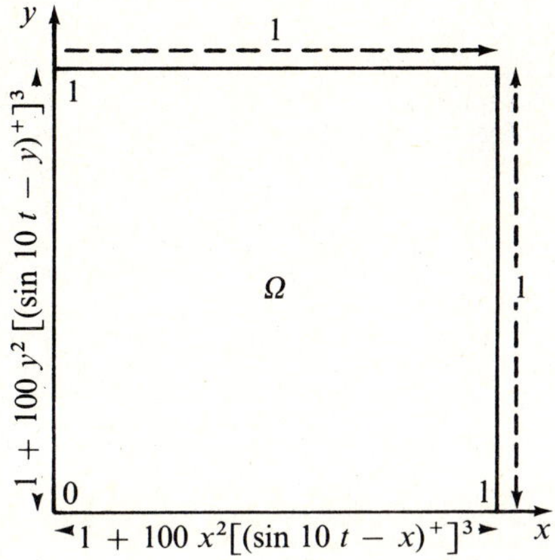

Fig. 7.3. u on Γ.

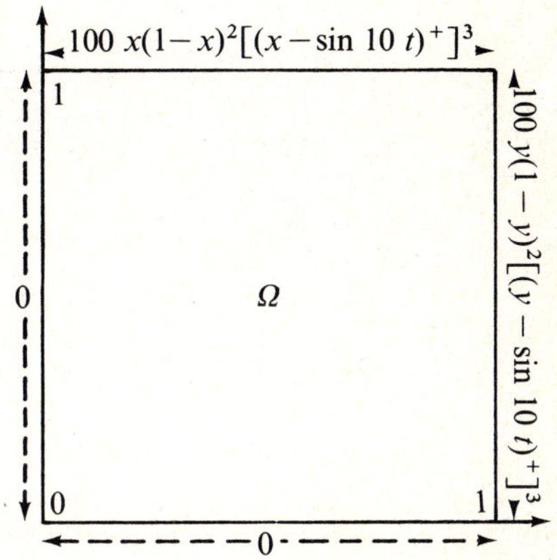

Fig. 7.4. $\partial u/\partial n$ on Γ.

Figures 7.3 and 7.4 define the values of u and $\partial u/\partial n$ on Γ.

Figures 7.5, 7.6, 7.7 show the evolution of the error $E(u)$ for various values of k and for the implicit, Crank-Nicholson and explicit schemes. With $h = 1/20$, the implicit and Crank-Nicholson schemes give a minimum error for $k = 10^{-3}$ and these schemes are almost equivalent for this value. We observe that in the explicit scheme overflow occurs for $k \geqslant 7\times10^{-4}$ and that the value $k = 2\times10^{-4}$ gives results comparable to those obtained using the other schemes with $k=10^{-3}$. The computation times (on a CII 10070), were as follows:

- Implicit and semi-implicit schemes with $h = 1/20$, $k = 10^{-3}$: time $\simeq$ 2.5 min.
- Explicit scheme with $h = 1/20$, $k = 2\times10^{-4}$: time $\simeq$ 6 min. ■

(1) As a consequence of choosing f in accordance with (7.14), we note that the solution (7.13) also corresponds to problem (7.15) with $K = H^1(\Omega)$ in place of (7.6). Thus the inequality (7.15) is here equivalent to an equation.

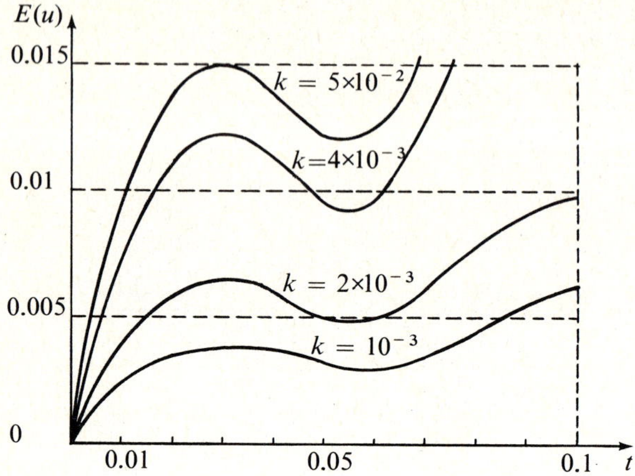

Fig. 7.5. Implicit scheme.

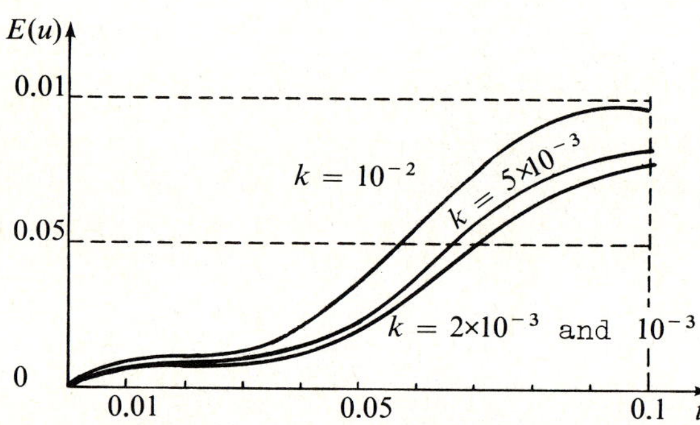

Fig. 7.6. Semi-implicit scheme.

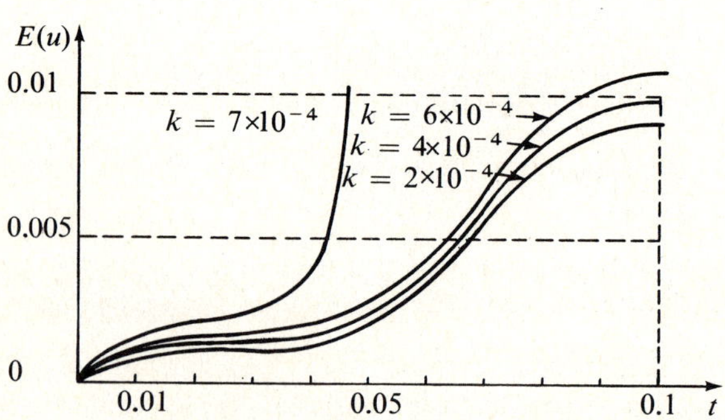

Fig. 7.7. Explicit scheme.

7.2. Solution of Example II

The particulars relating to this example are

$$(7.16)\quad \left| \begin{array}{l} V = H_0^1(\Omega)\,, \quad H = L^2(\Omega)\,, \\ K = \{ u \in H_0^1(\Omega) \mid u(x) \geqslant 0 \quad \text{a.e. } x \in \Omega \}\,. \end{array} \right.$$

Taking $\Omega =]0, 1[\times]0, 1[$ we define, with R_h as in Section 7.1,

$$\left| \begin{array}{l} \Omega_h = \{ M \in R_h \mid \overline{\omega}_h^1(M) \subset \Omega \}\,, \\ V_h = \left\{ u_h(x) = \sum_{M \in \Omega_h} u_h^M \, \theta_h^M(x),\ u_h^M \in \mathbb{R} \right\}, \\ K_h = \{ u_h \in V_h \mid u_h(M) \geqslant 0\,, \quad \forall M \in \Gamma \} \\ (,)_h,\ A_h,\ f_h^i,\ u_{0,h} \quad \text{as in Section 7.1.} \end{array} \right.$$

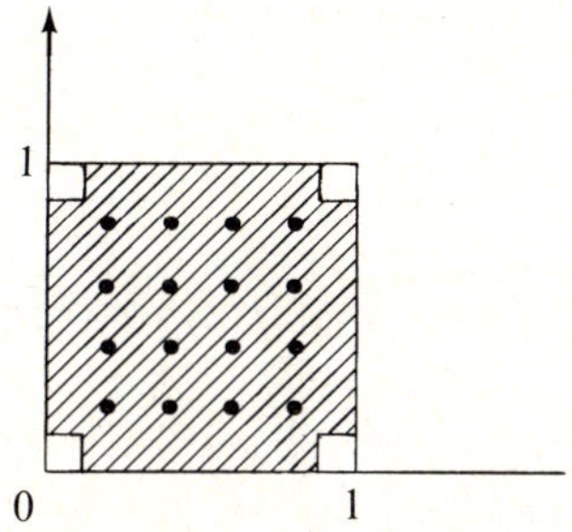

Fig. 7.8. Ω_h for $h = 1/5$.

In this case the problem can again be written in the form (7.5) and is solved by relaxation, with ω taken as in Figure 7.1 and using the termination criterion given by (7.11).

We shall assess the schemes in terms of $E(u)$ given by (7.12), and with

$$E(u^i) = \left(\sum_{M \in \Omega_h} \left| \frac{u_h^{i+1}(M) - u_h^i(M)}{k} - u'(M\,; (i + \theta)\,k) \right|^2 \times \right.$$
$$\left. \times \sum_{M \in \Omega_h} | u'(M\,; (i + \theta)\,k) |^2 \right)^{1/2}.$$

Here it can be shown once again that $S(h) = 2\sqrt{2}/h$.

We shall now investigate the approximate solution of the inequality

$$(7.17)\qquad (u' - \Delta u - f, v - u') \geqslant 0\,, \quad \forall v \in K,\ u \in \mathscr{K}_0^*$$

(where K is given by (7.16) and $\mathscr{K}_0^*$ is defined as in Section 5.2).

We take:

(7.18) $$f(x, y, t) = u' - \Delta u - 100(10 \sin t - x^2 - y^2)^+ ,$$

(7.19) $$u(x, y, 0) = u_0(x, y) ,$$

with the exact solution

(7.20) $$u(x, y, t) = 10 \sin x \sin y(1 - x)(1 - y) - \\ - 10\, xy(1 - x)(1 - y)\left[(x^2 + y^2 - 10 \sin t)^+\right]^3 .$$

Over the domain

$$\Omega_0^t = \{ (x, y) \in \Omega \mid 10 \sin t \geqslant x^2 + y^2 \} ,$$

we have $u' = 0$ (see Figure 7.9).

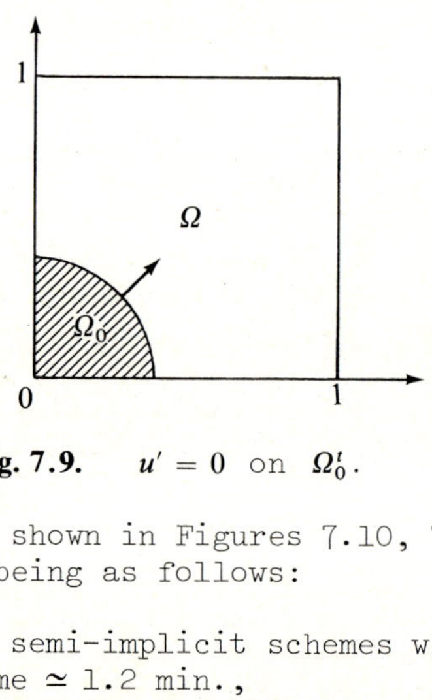

Fig. 7.9. $u' = 0$ on Ω_0^t.

The results are shown in Figures 7.10, 7.11, 7.12, 7.13, the computation times being as follows:

- Implicit and semi-implicit schemes with $h = 1/20$, $k = 10^{-3}$: time $\simeq$ 1.2 min.,
- Explicit scheme with $h = 1/20$, $k = 2 \times 10^{-4}$: time $\simeq$ 3.5 min.

For the explicit scheme, overflow occurred for $k \geqslant 7 \times 10^{-4}$ (see Figure 7.13).

In this case the implicit scheme gives the best results. ■

7.3. Conclusions

In conclusion, for the two examples considered it was observed that for similar accuracy the implicit schemes were two or three times faster than the explicit schemes, and that the latter diverge when the timestep is not small enough. Other examples which corroborate those presented here for inequalities of type II are given in Viaud /1/. ■

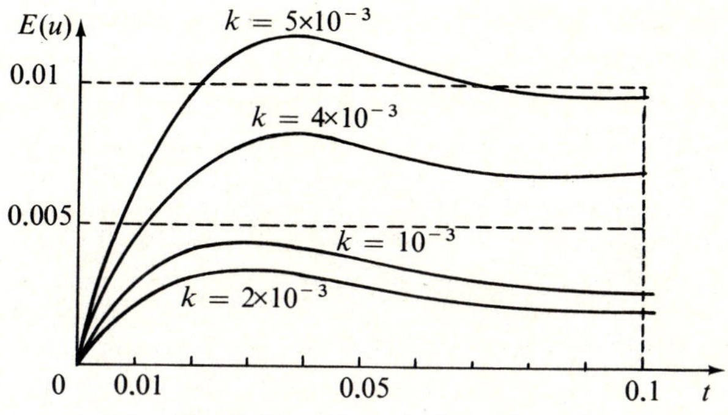

Fig. 7.10. Implicit scheme.

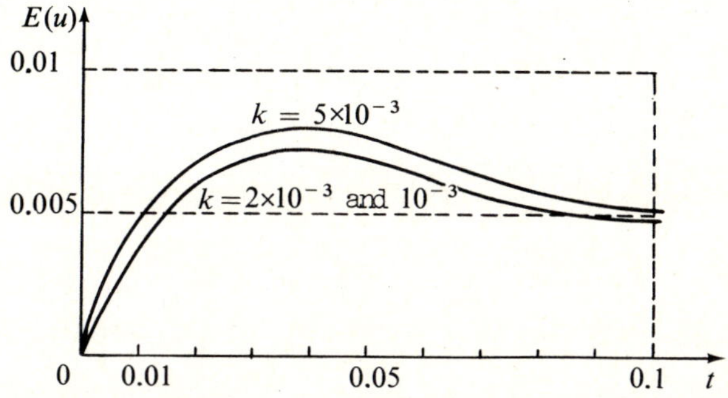

Fig. 7.11. Semi-implicit scheme.

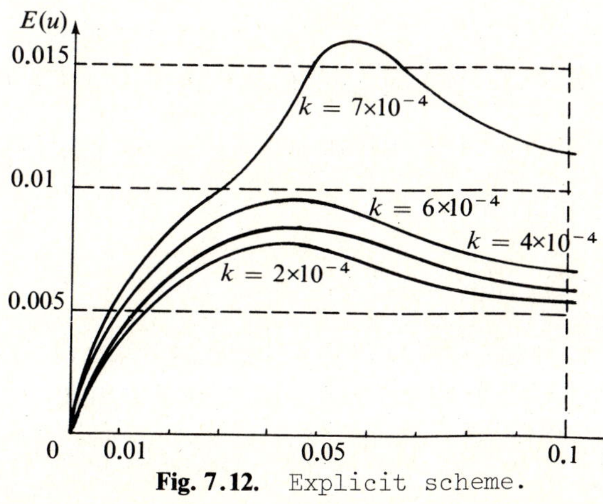

Fig. 7.12. Explicit scheme.

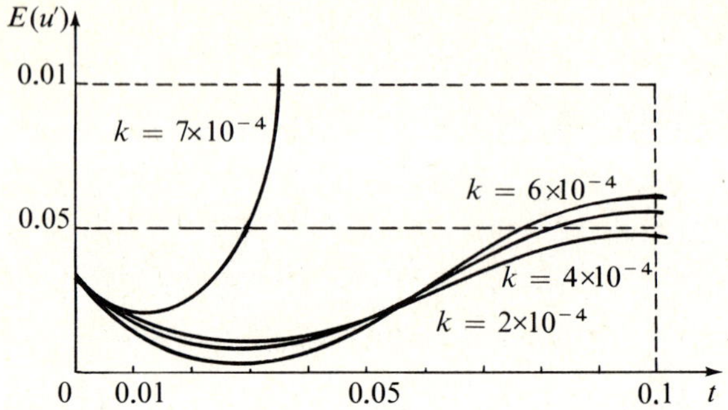

Fig. 7.13. $E(u')$ for the explicit scheme.

8. INTRODUCTION TO TIME-DEPENDENT INEQUALITIES OF THE SECOND ORDER IN t

8.1. Example I

The theory of visco-elastic materials with edge friction leads to problems (see Duvaut-Lions /1/) for which the following is a model example: find a function $u = u(x, t)$ such that

$$\frac{\partial^2 u}{\partial t^2} - \Delta \frac{\partial u}{\partial t} - \Delta u = f \quad \text{in} \quad \Omega \times]0, T[\,,$$

with the initial data

$$u(x, 0) = u_0(x)\,, \quad \frac{\partial u}{\partial t}(x, 0) = u_1(x)\,,$$

and the boundary conditions:

$$u = 0 \ \text{ on one part } \ \Gamma_0 \times]0, T[\ \text{ of the boundary } \ \Gamma \times]0, T[\,,$$

$$\left| \frac{\partial u}{\partial n} \right| \leqslant g \quad \text{on} \quad \Gamma_1 \times]0, T[\,, \quad \Gamma_0 \cup \Gamma_1 = \Gamma\,,$$

$$u \frac{\partial u}{\partial n} + g\,|\,u\,| = 0 \quad \text{on} \quad \Gamma_1 \times]0, T[\,.$$

It is easy to show that the above problem can be formulated as a *variational inequality of the second order in* t:

$$(8.1) \quad \left| \begin{array}{l} (u'', v - u') + a_1(u', v - u') + a_0(u, v - u') + j(v) - j(u') \geqslant \\ \qquad \geqslant (f, v - u') \quad \forall v \in V\,, \quad u' \in V \\ u(0) = u_0\,, \quad u'(0) = u_1\,, \end{array} \right.$$

where

$$V = \{ v \mid v \in H^1(\Omega),\ v = 0 \text{ on } \Gamma_0 \},$$

$$a_0(u, v) = a_1(u, v) = \int_\Omega \operatorname{grad} u \operatorname{grad} v \, dx,$$

$$j(v) = \int_{\Gamma_1} g \mid v \mid d\Gamma . \quad \blacksquare$$

8.2. Example II

A number of physical situations (see Duvaut-Lions /1/) lead to the search for a function u such that

$$\left| \begin{array}{l} u'' - \Delta u - f \geqslant 0, \quad u' \geqslant 0, \\ u'(u'' - \Delta u - f) = 0 \quad \text{in} \quad \Omega \times]0, T[, \end{array} \right.$$

with the initial data as above and with the boundary conditions given, for example, by

$$u = 0 \text{ on } \Gamma \times]0, T[.$$

This problem can also be formulated as a variational inequality of the second order in t, namely

$$\left| \begin{array}{l} (u'', v - u') + a(u, v - u') \geqslant (f, v - u') \quad \forall v \in K, \\ u'(t) \in K, \quad u(0) = u_0, \quad u'(0) = u_1, \end{array} \right. \tag{8.2}$$

where

$$K = \{ v \mid v \in H_0^1(\Omega),\ v \geqslant 0 \quad \text{a.e. in } \Omega \},$$

$$a(u, v) = \int_\Omega \operatorname{grad} u \operatorname{grad} v \, dx . \quad \blacksquare$$

8.3. Abstract formulation

Using the same functional setting as in Section 5.2 ($V \subset H \subset V'$) we introduce, with K a closed convex set of V, the convex sets

$$\mathcal{K} = \{ v \mid v \in W(0, T),\ v(t) \in K, \quad \text{a.e. } t \in [0, T] \},$$

$$\mathcal{K}^* = \{ v \mid v \in L^2(0, T; V),\ v' \in \mathcal{K} \},$$

$$\mathcal{K}_0^* = \{ v \mid v \in \mathcal{K}^*,\ v(0) = u_0,\ v'(0) = u_1 \},$$

where u_0, u_1 are given, and we shall consider inequalities of the type

$$L(u, v) = \int_0^T (u'' - Au - f, v - u') \, dt \geqslant 0, \quad \forall v \in \mathcal{K}, \quad u \in \mathcal{K}_0^*, \tag{8.3}$$

with $u(0) = u_0$, $u'(0) = u_1$ where A and f have the same properties as in

Section 5.2; in particular A is assumed to be *symmetric* [1] (and is clearly coercive and continuous).

The inequality (8.3) is more often written in the form

$$(8.4)\qquad (u'' - Au - f, v - u') \geqslant 0 \quad \forall v \in K\,, \quad u \in \mathscr{K}_0^*\,.$$

The existence of u, the solution of (8.3), is given (see Lions /1/) by:

Theorem 8.1. *With*

$$(8.5)\qquad f, f' \in L^2(0, T; H)\,,$$

$$(8.6)\qquad Au_0 \in H\,,$$

$$(8.7)\qquad u_1 \in K\,,$$

there exists a unique function u *satisfying* (8.3) (*or* (8.4))*such that*

$$(8.8)\qquad u, u' \in L^\infty(0, T; V)\,,$$

$$(8.9)\qquad u'' \in L^\infty(0, T; H)\,. \quad \blacksquare$$

Remark 8.1. By introducing

$$\mathscr{K}_f = \{\, v \mid v \in L^2(0, T; V),\ v' \in L^2(0, T; V),\ v(0) = u_0,\ v'(t) \in K, \quad \text{a.e. } t \in [0, T] \,\}$$

we can investigate the existence of *weak* solutions of the inequality

$$\int_0^T (v'' - Au - f, v' - u')\, \mathrm{d}t \geqslant 0\,, \quad \forall v \in \mathscr{K}_0^*\,, \quad u \in \mathscr{K}_f\,.$$

For this problem we refer the reader to Brézis /2/, Brézis-Lions /1/. ■

Remark 8.2. A formalism somewhat more general than (8.3) is to consider the inequality

$$(8.10)\qquad \int_0^T (u'' + Au - f, v - u')\, \mathrm{d}t + \int_0^T [j(v) - j(u')]\, \mathrm{d}t \geqslant 0\,, \quad \forall v \in \mathscr{K}\,, \quad u \in \mathscr{K}_0^*$$

---

[1] It is sufficient that the *principal part* of A be symmetric. We can also consider problems in which A is dependent on t. The case of (8.1) may be treated by the same methods.

where j is a functional of the type (2.27). ■

Remark 8.3. Result (8.8) allows us to consider that if K is not bounded then it is possible to define a convex set K_R (defined as in Remark 5.3) which is *bounded in* V, and which is such that the problem (8.3) *defined on* K_R (rather than K) has the same solution as the initial problem. We can thus consider that

(8.11) K is *bounded* in V.

In the same way as in the parabolic case of type II, this remark is of use in investigating the approximation. ■

9. APPROXIMATION OF INEQUALITIES OF THE SECOND ORDER IN t

In this section we embark on the investigation of the approximation of hyperbolic inequalities of the form (8.4).

9.1. Assumptions.

We shall work within the setting defined in 6.1. As usual we shall denote the approximate functions by $v_{h,k}$ with

$$v_{h,k}(t) = \sum_{i=0}^{N} v_h^i \chi_k^i(t), \quad v_h^i \in V_h .$$

We shall use the notation

$$\left| \begin{aligned} \delta v_{h,k}(t) &= \frac{v_{h,k}(t+k) - v_{h,k}(t)}{k}, \\ d v_{h,k}(t) &= \frac{v_{h,k}(t+k) - v_{h,k}(t-k)}{2k}, \\ \gamma v_{h,k}(t) &= \frac{\delta v_{h,k}(t) - \delta v_{h,k}(t-k)}{k}. \end{aligned} \right.$$

For the approximation of u_0, u_1, $\mathscr{K}^*$, $a(u, v) = (Au, v)$, $f(t)$ we make the following assumptions:

Approximation of the initial values.

With $u_0 \in V, u_1 \in K$, we assume that there exist

(9.1) $u_{0,h} \in V_h$ such that $p_h u_{0,h} \to \sigma u_0$ strongly in F and $\| u_{0,h} \|_h \leqslant C$,

(9.2) $u_{1,h} \in K_h$ such that $p_h u_{1,k} \to \sigma u_1$ strongly in F and $\| u_{1,h} \|_h \leqslant C$.

Approximation of $\mathcal{K}^*$.

We assume that the approximation is *consistent*; that is, for each family of functions $v_{h,k} \in L^2(0, T; V_h)$, $(h, k \to 0)$, such that

$$
(9.3)\quad \left|\begin{array}{l}
dv_{h,k}(t) \in K_h, \quad t \in [0, T], \\
q_h v_{h,k} \to v \text{ weakly * in } L^\infty(0, T; H) \\
p_h v_{h,k} \to \sigma v_1 \text{ weakly * in } L^\infty(0, T; F) \\
p_h v_{h,k}(T) \to \sigma v_2 \text{ and } p_h v_{h,k}(0) \to \sigma v_2^0 \text{ weakly in } V \\
q_h dv_{h,k} \to v_3 \text{ weakly * in } L^\infty(0, T; H) \\
\qquad (\text{or with } q_h \delta v_{h,k} \to v_3) \\
p_h dv_{h,k} \to \sigma v_4 \text{ weakly * in } L^\infty(0, T; F) \\
\qquad (\text{or with } q_h \delta v_{h,k} \to \sigma v_4) \\
q_h \delta v_{h,k}(T-k) \to v_5 \text{ and } q_h \delta v_{h,k}(0) \to v_5^0 \text{ weakly in } H \\
q_h \gamma v_{h,k} \to v_6 \text{ weakly * in } L^\infty(0, T; F)
\end{array}\right.
$$

we have $v \in \mathcal{K}^*$ and

$$
v_1 = \sigma v, \quad v_2 = \sigma v(T), \quad v_2^0 = \sigma v(0), \quad v_3 = v', \\
v_4 = \sigma v', \quad v_5 = v'(T), \quad v_5^0 = v'(0), \quad v_6 = v''.
$$

We also assume that the approximation is *convergent*, i.e. that for each function $v \in L^2(0, T; V)$ with $v(t) \in K$, a.e. $t \in [0, T]$ there exists a sequence $v_{h,k}$ $(h, k \to 0)$ such that

$$
(9.4)\quad \left|\begin{array}{l}
v_{h,k}(t) \in K_h, \quad t \in [0, T], \\
q_h v_{h,k} \to v \text{ strongly in } L^2(0, T; H) \\
p_h v_{h,k} \to \sigma v \text{ strongly in } L^2(0, T; F).
\end{array}\right.
$$

We shall assume (cf. Remark 8.3 and (8.11)) that

(9.5) $\quad K_h$ is *bounded* in V_h.

For this last assumption, cf. Remark 9.4. ■

Approximation of $a(u, v)$.

We shall assume (cf. (8.6)) that

(9.6) $\quad |A_h u_{0,h}|_h \leqslant C$, constant independent of h.

Otherwise the assumptions are the same as in Section 6.2. In particular, $a_h(u_h, v_h)$ is *symmetric*.

Approximation of f.

The same remarks apply as in Section 6.2. We shall assume here that

$$f, f' \in L^\infty(0, T; H),$$

and therefore

$$f \in \mathscr{C}^0([0, T]; H),$$

and we take

$$f_{h,k}(t) = \sum_{i=0}^{N} f_h^i \chi_k^i(t)$$

such that

(9.7)
$$\left| \begin{array}{l} q_h f_{h,k} \to f \ \text{ in } L^\infty(0, T; H) \ \text{ and } \ \| f_{h,k} \|_h \leqslant C, \\ p_h \delta f_{h,k} \to f' \ \text{ in } L^\infty(0, T; H) \ \text{ and } \ \| \delta f_{h,k} \|_h^* \leqslant C. \end{array} \right.$$

9.2. Approximation schemes

Taking account of the approximation assumptions and with

$$u_{h,k}^\theta(t) = u_{h,k}(t) + \theta[u_{h,k}(t + k) - u_{h,k}(t)], \quad \theta \in [0, 1],$$

we shall investigate the approximation of (8.4) by

(9.8)
$$\left| \begin{array}{l} (\gamma u_{h,k} + A_h u_{h,k}^\theta - f_{h,k}, v_h - \mathrm{d}u_{h,k})_h \geqslant 0, \\ \forall v_h \in K_h, \quad \mathrm{d}u_{h,k} \in K_h, \quad t \in [k, T], \\ \text{with } u_{h,k}(0) = u_{0,h}, \quad u_{h,k}(k) = u_{0,h} + k u_{1,k}. \end{array} \right.$$

In the following we shall use the notation (omitting the suffix h)

$$\left| \begin{array}{l} \delta^i = \dfrac{u^{i+1} - u^i}{k}, \\ \mathrm{d}^i = \dfrac{u^{i+1} - u^{i-1}}{2k} = \dfrac{\delta^i + \delta^{i-1}}{2} = \left(\dfrac{k\gamma^i}{2} + \delta^{i-1}\right), \\ \gamma^i = \dfrac{\delta^i - \delta^{i-1}}{k} = \dfrac{u^{i+1} - 2u^i + u^{i-1}}{k^2} = 2\left(\dfrac{d^i - \delta^{i-1}}{k}\right), \\ u^{i+\theta} = u^i + \theta[u^{i+1} - u^i] = u^i + \theta k \delta^i = u^i + 2\theta k d^i - \theta k \delta^{i-1}, \\ \delta^{i+\theta} = \dfrac{u^{i+1+\theta} - u^{i+\theta}}{k} = \delta^i + \theta[\delta^{i+1} - \delta^i], \\ g^i = \dfrac{f^{i+1} - f^i}{k}. \end{array} \right.$$

Scheme (9.8) can thus be written: find $u^2, u^3, \ldots, u^N$ such that

(9.9)
$$\left| \begin{array}{l} \left(\dfrac{u^{i+1} - 2u^i + u^{i-1}}{k^2} + A u^{i+\theta} - f^i, v - \dfrac{u^{i+1} - u^{i-1}}{2k}\right) \geqslant 0, \\ (i \geqslant 1), \quad \forall v \in K, \quad \dfrac{u^{i+1} - u^{i-1}}{2k} \in K, \end{array} \right.$$

(9.10)
$$u^0 = u_0, \quad u^1 = u_0 + k u_1.$$

From the theory of elliptic inequalities, the inequality (9.9), which by putting

$$\tilde{f}^i = -\delta^{i-1} + \frac{k}{2}[Au^i - \theta k A\delta^{i-1} - f^i],$$

can also be written

$$((I + \theta k^2 A)\, d^i + \tilde{f}^i, v - d^i) \geq 0,$$

possesses a unique solution, d^i.

This enables us one by one to define uniquely all the

$$u^{i+1} = u^{i-1} + kd^i, \quad (i \geq 1),$$

with the data (9.10).

Since A is symmetric, we note that the u^{i+1} $(i \geq 1)$ can be obtained from the solutions d^i of the $N-1$ minimisation problems $(i = 1, \ldots, N-1)$:

minimise $\frac{1}{2}|d^i|^2 + \frac{1}{2}\theta k^2(Ad^i, d^i) + (\tilde{f}^i, d^i)$,

subject to the constraints

$d^i \in K$,

where $\delta^{i-1}, u^{i-1}, u^i = u^{i-1} + k\delta^{i-1}$ are, for $i \geq 2$, given by the solution of the previous problem and, for $i = 1$, $\delta^0 = u_1, u^0 = u_0, u^1 = u_0 + k\delta^1$.

We use the following terminology:

- *explicit* scheme if $\theta = 0$,
- *semi-implicit* scheme if $0 < \theta < 1$,
- *implicit* scheme if $\theta = 1$. ■

Remark 9.1. In the same spirit, the solution of the inequality

$$(u'' + Au - f, v - u') + j(v) - j(u') \geq 0, \quad \forall v \in K, \quad u \in \mathscr{K}_0^*,$$

where $j(v)$ is defined as in (2.27), leads to the scheme

$$\left(\frac{u^{i+1} - 2u^i + u^{i-1}}{k^2} + Au^{i+\theta} - f^i, v - \frac{u^{i+1} - u^{i-1}}{2k}\right) + $$
$$+ j(v) - j\left(\frac{u^{i+1} - u^{i-1}}{2k}\right) \geq 0, \quad \forall v \in K, \quad \frac{u^{i+1} - u^{i-1}}{2k} \in K,$$

or, if A is symmetric, to the solution of the problems

minimise $\frac{1}{2}|d^i|^2 + \frac{1}{2}\theta k^2(Ad^i, d^i) + (\tilde{f}^i, d^i) + j(d^i)$,

subject to the constraints $d^i \in K$. ■

Remark 9.2. It is apparent from the above remark that the

methods of Chapter 2, which were originally designed for solving *elliptic variational inequalities*, now also play a fundamental part in the solution of *time-dependent variational inequalities*. ■

Remark 9.3. A very large number of other schemes is also possible. Some of these are mentioned below (omitting the suffix h):

Scheme I

$$\text{(9.11)}\quad \begin{cases} \left(\dfrac{u^{i+1} - 2u^i + u^{i-1}}{k^2} + A\bar{u}^i - f^i, v - \dfrac{u^{i+1} - u^{i-1}}{2k}\right) \geqslant 0, \\ \forall v \in K, \quad \dfrac{u^{i+1} - u^{i-1}}{2k} \in K, \end{cases}$$

with

$$\text{(9.12)}\quad \begin{cases} \bar{u}^i = \theta^- u^{i-1} + \theta^0 u^i + \theta^+ u^{i+1}, \\ \text{where} \quad \theta^-, \theta^0, \theta^+ \in [0, 1] \quad \text{and} \quad \theta^- + \theta^0 + \theta^+ = 1. \end{cases}$$

Schemes II

$$\text{(9.13)}\quad \begin{cases} \left(\dfrac{u^{i+1} - 2u^i + u^{i-1}}{k^2} + A\bar{u}^i - f^i, v - \dfrac{u^{i+1} - u^i}{k}\right) \geqslant 0, \\ \forall v \in K, \quad \dfrac{u^{i+1} - u^i}{k} \in K, \\ \text{and } \bar{u}^i \text{ as in (9.12).} \end{cases}$$

Schemes III (decoupled schemes)

$$\begin{cases} \dfrac{u^{i+1/2} - 2u^i + u^{i-1}}{k^2} + A\bar{u}^{i+1/2} - f^i = 0, \\ u^{i+1} = u^i + kP_{K_h}\left[\dfrac{u^{i+1/2} - u^i}{k}\right], \\ \text{or} \\ u^{i+1} = u^{i-1} + 2kP_{K_n}\left[\dfrac{u^{i+1/2} - u^{i-1}}{2k}\right], \end{cases}$$

where P_{K_h} is the projection on K_h in the sense of the $(,)_h$ norm. ■

The approximation of schemes I and II is studied in Trémolières /4/. ■

9.3. Convergence of the approximate inequalities

We shall now investigate the stability and, for $h, k \to 0$, the convergence of scheme (9.9). Using the notation of Section 9.2, scheme (9.9) is expressed as follows:

$$(9.14)\quad \left| \begin{array}{l} (\gamma^i, v - d^i) + a(u^{i+\theta}, v - d^i) \geqslant (f^i, v - d^i), \quad (i \geqslant 1), \\ \forall v \in K, \quad d^i \in K. \end{array} \right.$$

Before starting the investigation we recall Gronwall's Lemma:

Lemma 9.1. *If* p_i, $i = 0, 1, ..., n$, *are numbers which satisfy*

$$0 \leqslant p_1 \leqslant C_1, \quad \textit{constant} \geqslant 0,$$

$$0 \leqslant p_{n+1} \leqslant C_1 + C_2 \sum_{i=1}^{n} k_i p_i, \quad n = 1, ..., N,$$

with C_2 *constant* $\geqslant 0$ *and*

$$k_i \geqslant 0, \quad i = 1, ..., N,$$

$$\sum_{i=1}^{N} k_i = T, \textit{ positive constant}$$

then

$$p_n \leqslant C_1 \exp\left(C_2 \sum_{i=1}^{n} k_i\right) \leqslant C_1 \exp(C_2 T),$$

for $n = 2, 3, ..., N$. ■

9.3.1. *Stability*

We shall now prove:

Theorem 9.1. *When* $(h, k) \to 0$

$$(9.15)\quad \left| \begin{array}{l} \text{— in an arbitrary manner if } \theta = 1, \\ \text{— and, if } 0 \leqslant \theta < 1, \text{ with the stability condition,} \\ Ck^2 S(h)^2 \leqslant 1 - \beta, \quad \beta \text{ fixed} \in]0, 1[, \end{array} \right.$$

we have, for all $n = 0, 1, ...$

$$(9.16)\quad \| u_h^n \|_h \leqslant C,$$

$$(9.17)\quad \left\| \frac{u_h^n - u_h^{n-1}}{k} \right\|_h \leqslant C, \qquad \left\| \frac{u_h^{n+1} - u_h^{n-1}}{2k} \right\|_h \leqslant C,$$

$$(9.18)\quad \left| \frac{u_h^{n+1} - 2u_h^n + u_h^{n-1}}{k^2} \right|_h \leqslant C.$$

First, we establish:

Lemma 9.2. *Under assumptions* (9.6) *and* (9.5) *we have*

$$(9.19)\quad |\gamma^1|_h^2 \leqslant C,$$

where C *is a constant independent of* h *and* k.

Proof of Lemma 9.2. *We note that*

$$(\gamma^1 + A(u^1 + \theta k\delta^1) - f^1, v - d^1) \geqslant 0\,, \quad \forall v \in K\,,$$

and hence by taking $v = \delta^0$, *noting that*

$$d^1 - \delta^0 = \frac{k}{2}\gamma^1 \text{ and that } u^1 + \theta k\delta^1 = u^0 + (1+\theta)\,k\delta^0 + \theta k^2\,\gamma^1\,,$$

$$\left(\gamma^1 + A(u^0 + (1 + \theta)\,k\delta^0 + \theta k^2\,\gamma^1) - f^1, \frac{k}{2}\gamma^1\right) \leqslant 0\,,$$

and by multiplying by $2/k$, and suppressing the term $a(\gamma^1, \gamma^1)$, which is positive, we have

$$|\gamma^1|^2 \leqslant |a(u^0, \gamma^1)| + (1 + \theta)\,|a(\delta^0, \delta^1 - \delta^0)|\,,$$

and (9.19) can then be deduced immediately using (9.6) and (9.5).■

Proof of Theorem 9.1. Writing (9.14) for $i + 1$, we have

$$(9.20) \qquad (\gamma^{i+1}, v - d^{i+1}) + a(u^{i+1+\theta}, v - d^{i+1}) \geqslant (f^{i+1}, v - d^{i+1})\,.$$

Taking $v = d^{i+1}$ in (9.14), $v = d^i$ in (9.20), and adding, we obtain

$$(\gamma^{i+1} - \gamma^i, d^{i+1} - d^i) + a(u^{i+1+\theta} - u^{i+\theta}, d^{i+1} - d^i) \leqslant$$
$$\leqslant (f^{i+1} - f^i, d^{i+1} - d^i)\,.$$

Now

$$d^{i+1} - d^i = \frac{\delta^{i+1} + \delta^i}{2} - \frac{\delta^i + \delta^{i-1}}{2} = \frac{\delta^{i+1} - \delta^{i-1}}{2} =$$
$$= \frac{\delta^{i+1} - \delta^i + \delta^i - \delta^{i-1}}{2} = \frac{k}{2}(\gamma^{i+1} + \gamma^i)$$

and hence, after multiplication by $2/k$,

$$(9.21) \qquad (\gamma^{i+1} - \gamma^i, \gamma^{i+1} + \gamma^i) + a(\delta^{i+\theta}, \delta^{i+1} - \delta^{i-1}) \leqslant k(g^i, \gamma^{i+1} + \gamma^i)\,,$$

but, noting that $a(,)$ is symmetric, we have

$$a(\delta^{i+\theta}, \delta^{i+1} - \delta^{i-1}) = (1 - \theta)\cdot a(\delta^i, \delta^{i+1} - \delta^{i-1}) +$$
$$+ \theta a(\delta^{i+1}, \delta^{i+1} - \delta^{i-1}) = (1 - \theta)\,[a(\delta^i, \delta^{i+1}) -$$
$$- a(\delta^i, \delta^{i-1})] + \frac{\theta}{2}[a(\delta^{i+1}) - a(\delta^{i-1}) + a(\delta^{i+1} - \delta^{i-1})]\,,$$

and, for $1 \leqslant n \leqslant N - 1$,

$$\sum_{i=1}^{n} a(\delta^{i+\theta}, \delta^{i+1} - \delta^{i-1}) = (1-\theta)\left[a(\delta^n, \delta^{n+1}) - a(\delta^0, \delta^1)\right] + \frac{\theta}{2}\left[a(\delta^{n+1}) + a(\delta^n) - a(\delta^1) - a(\delta^0) + \sum_{i=1}^{n} a(\delta^{i+1} - \delta^{i-1})\right] = (1-\theta)\left[a(\delta^n, \delta^{n+1} - \delta^n) - a(\delta^0, \delta^1)\right] + \left(1 - \frac{\theta}{2}\right) a(\delta^n) + \frac{\theta}{2}\left[a(\delta^{n+1}) - a(\delta^1) - a(\delta^0) + \sum_{i=1}^{n} a(\delta^{i+1} - \delta^{i-1})\right].$$

Now

$$|a(\delta^0, \delta^1)| \leqslant \tfrac{1}{2} a(\delta^0) + \tfrac{1}{2} a(\delta^1),$$
$$|a(\delta^n, \delta^{n+1} - \delta^n)| \leqslant \tfrac{1}{2} a(\delta^n) + \tfrac{1}{2} k^2 a(\gamma^{n+1}),$$

and hence, suppressing the term $\dfrac{\theta}{2} a(\delta^{n+1})$, which is positive

$$\sum_{i=1}^{n} a(\delta^{i+\theta}, \delta^{i+1} - \delta^{i-1}) \geqslant (1-\theta)\left[-\frac{1}{2} a(\delta^n) - \frac{1}{2} k^2 a(\gamma^{n+1})\right] + \left(1 - \frac{\theta}{2}\right) a(\delta^n) - \frac{1}{2} a(\delta^0) - \frac{1}{2} a(\delta^1) \geqslant \frac{1}{2} a(\delta^n) - \frac{(1-\theta)}{2} k^2 a(\gamma^{n+1}) - \frac{1}{2} a(\delta^0) - \frac{1}{2} a(\delta^1).$$

Moreover

$$\sum_{i=1}^{n} k(g^i, \gamma^{i+1} + \gamma^i) \leqslant \sum_{i=1}^{n} Ck(|\gamma^{i+1}| + |\gamma^i|) \leqslant \sum_{i=1}^{n} 2\, Ck\, |\gamma^i| + Ck\, |\gamma^{n+1}| \leqslant C + C \sum_{i=1}^{n} k\, |\gamma^i|^2 + \frac{1}{2} |\gamma^{n+1}|^2.$$

So that, summing (9.21) from $i = 1$ to $n \leqslant N - 1$, we obtain

$$\frac{1}{2} |\gamma^{n+1}|^2 + \frac{1}{2} a(\delta^n) \leqslant C + C \sum_{i=1}^{n} k\, |\gamma^i|^2 + \frac{1-\theta}{2} k^2 a(\gamma^{n+1}) + \frac{1}{2} a(\delta^0) + \frac{1}{2} a(\delta^1) + |\gamma^1|^2.$$

From Lemma 9.2 $(|\gamma^1| \leqslant C)$, assumption (9.5) $(a(\delta^1) \leqslant C)$ and assumption (9.2) $(a(\delta^0) \leqslant C)$, we obtain, after multiplication by 2,

$$[1 - (1-\theta)\, Ck^2\, S(h)^2]\, |\gamma^{n+1}|^2 + a(\delta^n) \leqslant C + C \sum_{i=1}^{n} k\, |\gamma^i|^2. \tag{9.22}$$

By then imposing (9.15) if $0 \leqslant \theta < 1$, we obtain (9.18) by Gronwall's lemma. The upper bounds (9.17) and (9.16) then follow immediately (see (9.22)). ■

Remark 9.4. In the case in which $\forall h, k \to 0$, the inequality (9.14) is in fact, for $i = 1$, an equation (the convex set is not "saturated"); we have

$$\frac{\delta^1 - \delta^0}{k} + Au^{1+\theta} - f^0 = 0\,,$$

and hence

$$\| \delta^1 \| = \| \delta^0 + kAu^1 + \theta k^2 A\delta^1 - kf^0 \| \leqslant$$
$$\leqslant \| \delta^0 \| + k \| Au^1 \| + \theta k^2 \| A\delta^1 \| + k \| f^0 \|\,,$$

from which it can be deduced that when k is sufficiently small, δ^1 is *bounded* for the V_h norm independently of h and k and assumption (9.5) *is in fact unnecessary*. ■

9.3.2. *Weak convergence*

From assumptions (9.1), (9.2), (9.3) and under the conditions of Theorem 9.1, the sequence

$$u_{h,k} = \sum u_h^i \chi_k^i$$

satisfies, for an extracted subsequence also denoted by $u_{h,k}$,

$$\text{(9.23)}\quad \left|\begin{array}{l} q_h u_{h,k} \to \bar{u} \text{ weakly * in } L^\infty(0, T; H)\,, \quad \bar{u} \in \mathcal{K}^*\,, \\ p_h u_{h,k} \to \sigma\bar{u} \text{ weakly * in } L^\infty(0, T; F) \\ p_h \,\mathrm{d}u_{h,k} \to \sigma\bar{u}' \text{ weakly * in } L^\infty(0, T; F) \\ q_h \gamma u_{h,k} \to u'' \text{ weakly * in } L^\infty(0, T; H) \\ p_h u_{h,k}(T) \to \sigma\bar{u}(T) \text{ and } p_h u_{h,k}(0) \to \sigma\bar{u}(0) \text{ weakly in } V \\ q_h \delta u_{h,k}(T - k) \to u'(T) \text{ and } q_h \delta u_{h,k}(0) \to \sigma\bar{u}'(0) \text{ weakly in } H. \end{array}\right.$$

Now let v be arbitrary with $v \in L^2(0, T; V)$, $v(t) \in K$, a.e. $t \in [0, T]$ and let $v_{h,k}$ be a sequence which converges to v in the sense of (9.4).

Replacing v by v_h^i in (9.14), we obtain

$$\text{(9.24)}\quad \begin{aligned} &\left(\frac{u_h^{i+1} - 2\,u_h^i + u_h^{i-1}}{k^2} + A_h\, u_h^{i+\theta}, v_h^i\right)_h - \left(f_h^i, v_h^i - \frac{u_h^{i+1} - u_h^{i-1}}{2\,k}\right)_h \geqslant \\ &\geqslant D_{h,k}^i = \left(\frac{u_h^{i+1} - 2\,u_h^i + u_h^{i-1}}{k^2} + A_h\, u_h^{i+\theta}, \frac{u_h^{i+1} - u_h^{i-1}}{2\,k}\right)_h . \end{aligned}$$

We shall show that by multiplying by k and summing from $i = 1$ to $N - 1$ we obtain

$$\text{(9.25)}\quad \int_0^T (\bar{u}'' + A\bar{u}, v)\,\mathrm{d}t - \int_0^T (f, v - \bar{u}')\,\mathrm{d}t \geqslant \int_0^T (\bar{u}'' + A\bar{u}, \bar{u}')\,\mathrm{d}t\,,$$

thus establishing that $\bar{u}$ is a solution of the problem. Since the left-hand side of (9.25) is obtained immediately, it remains to examine the right-hand side of (9.24).

Multiplying the right-hand side of (9.24) by $2k$, we obtain

$$2kD_{h,k}^{i} = (\delta^{i} - \delta^{i-1}, \delta^{i} + \delta^{i-1})_{h} + (A_{h} u_{h}^{i+\theta}, u_{h}^{i+1} - u_{h}^{i-1}) = \\ = |\delta^{i}|_{h}^{2} - |\delta^{i-1}|_{h}^{2} + (A_{h} u_{h}^{i+\theta}, u_{h}^{i+1} - u_{h}^{i-1}).$$

However (omitting the suffix h)

$$(9.26)\qquad 2a(u^{i+\theta}, u^{i+1} - u^{i-1}) = a(u^{i+1}) - a(u^{i-1}) + \\ + (1-\theta)\left[a(u^{i} - u^{i-1}) - a(u^{i+1} - u^{i})\right] + \theta a(u^{i+1} - u^{i-1}).$$

Hence, summing from $i = 1$ to $N - 1$, and omitting the suffix h, we obtain

$$(9.27)\qquad 2D_{h,k} = \sum_{i=1}^{N-1} 2kD_{h,k}^{i} = |\delta^{N-1}|^{2} - |\delta^{0}|^{2} + \\ + \frac{1}{2}\Big(a(u^{N}) + a(u^{N-1}) - a(u^{1}) - a(u^{0}) + \\ + (1-\theta)\left[a(u^{1} - u^{0}) - a(u^{N} - u^{N-1})\right] + \\ + \theta \sum_{i=1}^{N-1} a(u^{i+1} - u^{i-1})\Big).$$

Now we have the following results

$\liminf |\delta^{N-1}|^{2} \geqslant |\bar{u}'(T)|^{2}$ (from (9.23)),
$|\delta^{0}| \to |u_{1}|^{2} = |\bar{u}'(0)|^{2}$ (from (9.2) and (9.23)),
$\liminf a(u^{N}) \geqslant a(\bar{u}(T))$, $\liminf a(u^{N-1}) \geqslant a(\bar{u}(T))$ (from the consistency of $a(,)$ and (9.23))

$a(u^{1}) \to a(u_{0}) = a(\bar{u}(0))$, $a(u^{0}) \to a(u_{0}) = a(\bar{u}(0))$ (from the convergence of $a(,)$ and (9.23)),

and hence

$$\liminf 2D_{h,k} \geqslant 2D = |\bar{u}'(T)|^{2} - |\bar{u}'(0)|^{2} + a(\bar{u}(T)) - \\ - a(\bar{u}(0)) + \liminf r,$$

with

$$r = \frac{1}{2}\Big((1-\theta)\left[a(u^{1} - u^{0}) - a(u^{N} - u^{N-1})\right] + \\ + \theta \sum_{i=1}^{N-1} a(u^{i+1} - u^{i-1})\Big).$$

Using Green's formula (Lemma 1.4)(1) and dividing by 2 we have,

$$D = \int_0^T (\bar{u}'', \bar{u}')\,dt + \int_0^T a(\bar{u}, \bar{u}')\,dt + \liminf \frac{r}{2}.$$

By summing and proceeding to the limit in (9.24) we then obtain

$$\int_0^T (\bar{u}'' + A\bar{u} - f, v - \bar{u}')\,dt \geqslant \liminf \frac{r}{2}. \tag{9.28}$$

We note at the outset that if $\theta = 1$ then $\liminf r \geqslant 0$ and $\bar{u}$ is indeed a solution of the initial inequality (8.3). In the other cases $(\theta < 1)$ we have

$$r = \frac{1}{2}\left\{(1 - \theta)\,[k^2\,a(\delta^0) - k^2\,a(\delta^{N-1})] + \theta \sum_{i=1}^{N-1} k^2\,a(d^i)\right\}.$$

Now

$$a(\delta^0) \leqslant C\,\|\,\delta^0\,\|_h^2 \leqslant C\,, \quad \text{(by hypothesis)}$$

$$a(\delta^{N-1}) \leqslant C\,\|\,\delta^{N-1}\,\|_h^2 \leqslant C\,, \quad \text{(from (9.17))},$$

$$\sum_{i=1}^{N-1} k^2\,a(d^i) \leqslant C \sum_{i=1}^{N-1} k^2 \left\|\frac{\delta^i + \delta^{i-1}}{2}\right\|^2 \leqslant C \sum_{i=1}^{N-1} k^2$$

(from (9.17)),

and hence

$$|\,r\,| \leqslant (1 - \theta)\,k^2\,C + \theta k C \sum_{i=1}^{N-1} k \leqslant (1 - \theta)\,k^2\,C + \theta k C\,,$$

and we have $\liminf |\,r\,| = 0$ and hence clearly,

$$\liminf r = 0\,, \tag{9.29}$$

which proves (see (9.28) with (9.29)) that $\bar{u}$ is indeed a solution of the initial inequality. Since the solution $\bar{u}$ is unique, it is not necessary to extract a subsequence, and the entire sequence converges towards u. We can now state:

Theorem 9.2. *When* $(h, k) \to 0$

- *in an arbitrary manner if* $\theta = 1$,
- *and, if* $0 \leqslant \theta < 1$, *with the stability condition*

$$Ck^2\,S(h)^2 \leqslant 1 - \beta\,, \quad \beta \textit{ fixed } \in\,]0, 1[\,,$$

---

(1) Since $A(t)$ is symmetric, for all v such that $v, v' \in L^2(0, T; V)$ we have

$$\int_0^T (A(t)\,v, v')\,dt = \int_0^T \frac{1}{2}\frac{d}{dt}(A(t)\,v, v)\,dt - \frac{1}{2}\int_0^T (A'\,v, v)\,dt\,.$$

we have, with u *a solution of the initial inequality* (8.3),

$$q_h u_{h,k} \to u \quad \text{weakly * in} \quad L^\infty(0, T; H)$$
$$p_h u_{h,k} \to \sigma u \quad \text{weakly * in} \quad L^\infty(0, T; F)$$
$$p_h \delta u_{h,k} \to \sigma u' \quad \text{weakly * in} \quad L^\infty(0, T; F)$$
$$q_h \gamma u_{h,k} \to u'' \quad \text{weakly * in} \quad L^\infty(0, T; H).$$

Remark 9.5. The above proves the existence of $u \in \mathscr{K}_0^*$, the solution of 8.3. ■

9.3.3. *Strong convergence*

In this last section we shall establish:

Theorem 9.3. *When* $(h, k) \to 0$

- *in an arbitrary manner if* $\theta = 1$,
- *and, if* $0 \leqslant \theta < 1$, *with the condition*

$$Ck^2 S(h)^2 \leqslant 1 - \beta, \quad \beta \text{ fixed } \in]0, 1[,$$

we have, with u *a solution of the initial inequality* (8.3),

$$p_h u_{h,k}(t) \to \sigma u(t) \text{ strongly in } V, \ t \in [0, T],$$
$$q_h \delta u_{h,k}(t) \to u'(t) \text{ strongly in } H, \ t \in [0, T]. \quad ■$$

Before commencing the proof of this theorem, it should be pointed out that we take $a(,)$, $\tilde{a}(,)$ as in 6.3.3 and that we shall omit using the operators q_h, p_h, σ in cases where there is no possible ambiguity.

We shall now evaluate the quantity

$$X(t) = | u' - q_h \delta u_{h,k} |^2 + \tilde{a}(u(t) - p_h u_{h,k}(t)).$$

By expanding, we obtain

$$X(t) = X_1(t) - 2 X_2(t) + X_3(t),$$

with

$$X_1(t) = | u' |^2 + \tilde{a}(u),$$
$$X_2(t) = (u', \delta u_{h,k}) + \tilde{a}(u, p_h u_{h,k}),$$
$$X_3(t) = | \delta u_{h,k} |^2 + \tilde{a}(p_h u_{h,k}).$$

Using the assumptions and the lemmas concerning integration-by-parts, the *first term* becomes

$$X_1(t) = 2 \int_0^t (u'', u')\, ds + 2 \int_0^t a(u, u')\, ds + a(u_0) + | u'(0) |^2. \quad ■$$

Let us now examine the *second term*. Let nk be such that

$t \in [nk, (n+1)k[$ and put

$$\tau_k u = \sum_{i=0}^{N-1} u^{i+1} \chi_k^i, \quad \text{with} \quad u^i = u(ik).$$

We recall that $u_{h,k} = \sum_{i=0}^{N} u_h^i \chi_k^i$ and that $a(,)$ is symmetric and does not depend on time. We readily obtain (omitting p_h) that

$$\tilde{a}(u(t), p_h u_{h,k}(t)) = \int_{nk}^{t} \tilde{a}(u', u_h^n) \, ds + \tilde{a}(u^n, u_h^n).$$

However for $i = 1, \ldots, N$, we have:

$$\int_{(i-1)k}^{ik} \tilde{a}(u', u_h^{i-1}) \, ds = \tilde{a}(u^i, u_h^{i-1}) - \tilde{a}(u^{i-1}, u_h^{i-1}),$$

and hence, by adding $\tilde{a}(u^i, u_h^i)$ to each side

$$\tilde{a}(u^i, u_h^i) = \int_{(i-1)k}^{ik} \tilde{a}(u', u_h^{i-1}) \, ds + \tilde{a}(u^i, u_h^i - u^{i-1}) + \tilde{a}(u^{i-1}, u_h^{i-1}),$$

and finally

$$\tilde{a}(u(t), p_h u_{h,k}(t)) = \int_0^t \tilde{a}(u', p_h u_{h,k}) \, ds + \int_0^{nk} \tilde{a}(\tau_k u, p_h \delta u_{h,k}) \, ds + \\ + \tilde{a}(u(0), u_{h,k}(0)).$$

Still for the second term, let us now evaluate $(u', \delta u_{h,k})$. We have

$$(u', \delta u_{h,k}) = \int_{nk}^{t} (u'', \delta^n) \, ds + (u'(nk), \delta^n).$$

But

$$\int_{(i-1)k}^{ik} (u'', \delta^{i-1}) \, ds = (u'(ik), \delta^{i-1}) - (u'((i-1)k), \delta^{i-1}),$$

and, by adding $(u'(ik), \delta^{i-1})$ to each side

$$(u'(ik), \delta^i) = \int_{(i-1)k}^{ik} (u'', \delta^{i-1}) \, ds + \int_{(i-1)k}^{ik} (u'(ik), \gamma^i) \, ds + \\ + (u'((i-1)k), \delta^{i-1}),$$

and hence

$$(u', \delta u_{h,k}) = \int_0^t (u'', q_h \delta u_{h,k}) \, ds + \int_0^{nk} (\tau_k u', q_h \gamma u_{h,k}) \, ds + \\ + (u'(0), q_h \delta^0).$$

We can then proceed to the limit in X_2 and obtain

$$\lim X_2(t) = X_1(t) .$$

Let us now examine the *third term*. We first write the inequality in integrated form. For this purpose we shall use the notation

$$\left|\begin{aligned} u_{h,k}(t) &= \sum_{i=0}^{N} u_h^i\, \chi_k^i(t) ,\\ u_{h,k}^\theta(t) &= \sum_{i=0}^{N-1} (u_h^i + \theta[u_h^{i+1} - u_h^i])\, \chi_k^i(t) ,\\ \delta u_{h,k}(t) &= \sum_{i=1}^{N-1} \left(\frac{u_h^{i+1} - u_h^i}{k}\right) \chi_k^i(t) ,\\ \mathrm{d}u_{h,k}(t) &= \sum_{i=1}^{N-1} \frac{u_h^{i+1} - u_h^{i-1}}{2\,k} \chi_k^i(t) ,\\ \gamma u_{h,k}(t) &= \sum_{i=1}^{N-1} \frac{u_h^{i+1} - 2\,u_h^i + u_h^{i-1}}{k^2} \chi_k^i(t) .\end{aligned}\right.$$

We can then write (9.14) in integrated form for $t \in [k, T-k]$:

$$(9.30) \qquad \int_k^t (\gamma u_{h,k} + A_h\, u_{h,k}^\theta, \mathrm{d}u_{h,k})\, \mathrm{d}s \leqslant \int_k^t (\gamma u_{h,k} + A_h\, u_{h,k}^\theta, v_{h,k})\, \mathrm{d}s + \\ + \int_k^t (f_{h,k}, \mathrm{d}u_{h,k} - v_{h,k})\, \mathrm{d}s ,$$

where we take $v_{h,k} = \sum_{i=1}^{N-1} v_h^i\, \chi_k^i$ such that

$$p_h\, v_{h,k} \to \sigma u' \text{ strongly in } L^2(0, T; F) \text{ and } v_h^i \in K_h .$$

However, for $t \in [nk, (n+1)\,\beta[$, $n \geqslant 1$, we have

$$(9.31) \qquad \int_k^t = \int_k^{(n+1)k} - \int_t^{(n+1)k} = \sum_{i=1}^{n} \int_{ik}^{(i+1)k} - \int_t^{(n+1)k}$$

$$(9.32) \qquad \int_{ik}^{(i+1)k} (\gamma u_{h,k}^\theta + A_h\, u_{h,k}, \mathrm{d}u_{h,k})\, \mathrm{d}s = \\ = \left(\frac{\delta^i - \delta^{i-1}}{k} + Au^{i+\theta}, \frac{\delta^i + \delta^{i-1}}{2}\right) k = \\ = \frac{1}{2}\{\, |\, \delta^i\, |^2 - |\, \delta^{i-1}\, |^2 + (Au^{i+\theta}, \delta^i + \delta^{i-1})\, k\, \} ,$$

and hence, using (9.31), (9.32) and (9.26), with $\varepsilon(k) \to 0$, when $k \to 0$,

$$(9.33)\qquad 2\int_k^t (\gamma u_{h,k} + A_h\, u_{h,k}^\theta, \mathrm{d}u_{h,k})\,\mathrm{d}s = \varepsilon(k) + |\,\delta u_{h,k}(t)\,|^2 - \\ - |\,\delta u_{h,k}(0)\,|^2 + \frac{1}{2}\Big\{[a(u_{h,k}(t+k)) + a(u_{h,k}(t)) - \\ - a(u_{h,k}(k)) - a(u_{h,k}(0))] + (1-\theta)\,[a(u^1 - u^0) - \\ - a(u^{n+1} - u^n)] + \theta \sum_{i=1}^n a(u^{i+1} - u^{i-1})\Big\}.$$

We then have, with $\frac{1}{2}\, a(u_{h,k}(t)) = (1 - \frac{1}{2})\, a(u_{h,k}(t))$, and substituting (9.33) into (9.30):

$$(9.34)\qquad X_3(t) \leqslant |\,\delta u_{h,k}(0)\,|^2 - \frac{1}{2}\Big\{[a(u_{h,k}(t+k)) + a(u_{h,k}(t)) - \\ - a(u_{h,k}(k)) - a(u_{h,k}(0))] + (1-\theta)\,[a(u^1 - u^0) - \\ - a(u^{n+1} - u^n)] + \theta \sum_{i=1}^n a(u^{i+1} - u^{i-1})\Big\} + \\ + 2\int_k^t (\gamma u_{h,k} + A_h\, u_{h,k}^\theta, v_{h,k})\,\mathrm{d}s + \\ + 2\int_k^t (f_{h,k}, \mathrm{d}u_{h,k} - v_{h,k})\,\mathrm{d}s\,.$$

Proceeding to the limit, we obtain

$$\lim X_3(t) = |\,u'(0)\,|^2 + a(u(0)) + 2\int_0^t (u'' + Au, u')\,\mathrm{d}s\,.$$

Assembling the results we have

$$\lim X(t) = \lim X_1(t) - 2\lim X_2(t) + \lim X_3(t) = 0\,,$$

which establishes the theorem. ■

10. NUMERICAL SOLUTION OF INEQUALITIES OF THE SECOND ORDER IN *t*.

In this section we propose to solve, using simplified data, Examples I and II of Section 8. The numerical investigation will enable us to appreciate the relative merits of the implicit and explicit schemes, and to assess the approximations made, in terms of h and k.

10.1. Solution of Example I

Considering Example I, we shall seek to solve the inequality

$$
(10.1)\quad \left|\begin{array}{l} (u'', v - u') + a_1(u', v - u') + a_0(u, v - u') + j(v) - \\ \quad - j(u') \geqslant (f, v - u'), \quad \forall v \in V, \quad u' \in V, \\ u(0) = u_0, \quad u'(0) = u'_0, \end{array}\right.
$$

for $t \in]0, T]$ with

$$
\left|\begin{array}{l} \Omega \text{ an open bounded domain of } \mathbb{R}^2 \text{ with boundary } \Gamma = \Gamma_0 \cup \Gamma_1, \\ V = \{ v \in H^1(\Omega) \mid v = 0 \text{ on } \Gamma_0 \}, \\ j(v) = g \displaystyle\int_{\Gamma_1} |v| \, \mathrm{d}\Gamma. \end{array}\right.
$$

We shall compare the solution u of (10.1) with the solution of the "quasi-static" problem (see Duvaut-Lions /1/),

$$
(10.2)\quad \left|\begin{array}{l} a_1(\tilde{u}', v - \tilde{u}') + a_0(\tilde{u}, v - \tilde{u}') + j(v) - j(\tilde{u}') \geqslant (f, v - \tilde{u}'), \\ \forall v \in V, \quad \tilde{u}' \in V, \quad \tilde{u}(0) = \tilde{u}_0, \quad \tilde{u}'(0) = \tilde{u}'_0. \end{array}\right.
$$

For these two problems we take

$$a_0 = a_1 = a \quad \text{with} \quad a(u, v) = \int_\Omega \operatorname{grad} u \operatorname{grad} v \, \mathrm{d}x,$$

$\Omega =]0, 1[\times]0, 1[$,

$\Gamma = \Gamma_0 \cup \Gamma_1$ (see Figure 10.1),

$f = 40 \sin 2\pi x$,

$g = 0.2$.

The discretisation of Ω is carried out in the usual manner and the discretised V, $a(u, v)$ and f are defined in the normal way.

We take

$$h = 1/20, \qquad k = 10^{-2}.$$

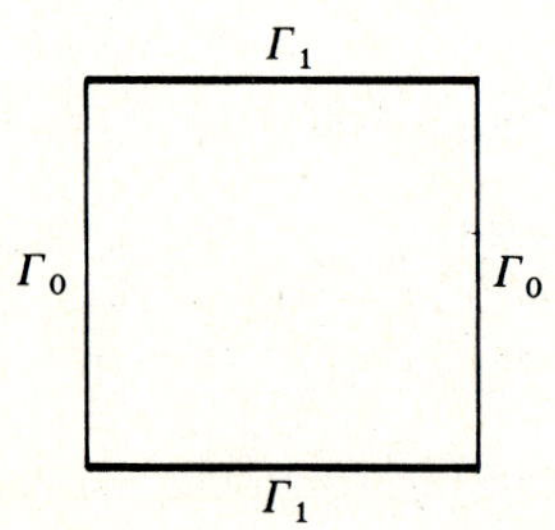

Fig. 10.1.

For the initial data we take

$$
(10.3)\quad \begin{cases} u_0 = 0, & u'_0 = 0, \\ \tilde{u}_0 = 0, & \tilde{u}'_0 = 0. \end{cases}
$$

$$(10.4) \qquad \begin{cases} u_0 = 0\,, & u_0' = 0.2\,, \\ \tilde{u}_0 = 0\,, & \tilde{u}_0' = 0.2\,. \end{cases}$$

We approximate the two inequalities by semi-implicit schemes. Putting

$$\delta^n = \frac{u^{n+1} - u^n}{k}, \qquad \frac{u^n + u^{n+1}}{2} = u^n + \frac{k}{2}\delta^n\,,$$

the discretisation of (10.1) can then be written

$$\left(\frac{\delta^n - \delta^{n-1}}{k}, v - \delta^n\right) + a_1(\delta^n, v - \delta^n) + \\ + a_0\left(u^n + \frac{k}{2}\delta^n, v - \delta^n\right) + j(v) - j(\delta^n) \geqslant (f, v - \delta^n)\,, \\ \forall v \in V\,,$$

and the discretisation of (10.2) can be written

$$a_1(\delta^n, v - \delta^n) + a_0\left(u^n + \frac{k}{2}\delta^n, v - \delta^n\right) + j(v) - j(\delta^n) \geqslant (f, v - \delta^n), v \in V\,.$$

In order to test the convergence $u(t) \to \tilde{u}(t)$, when $t \to \infty$, we put

$$D = \sup_{M \in \Omega_h} |\, u_h(M) - \tilde{u}_h(M)\,|\,.$$

The results shown in Table 10.1 confirm the convergence results.

D \ t	0.5	2	4	8
case(10.3)	2.1×10^{-2}	4×10^{-3}	1.5×10^{-3}	1×10^{-3}
case(10.4)	2.2×10^{-2}	9×10^{-3}	4×10^{-3}	1.5×10^{-3}

Table 10.1.

For the two inequalities (solved by over-relaxation at each time-step) the solution required 5 min to reach $t = 10$ on a CII 10070. Results completely analogous to those of Table 10.1 were obtained by replacing V by

$$K = \{\, v \in H^1(\Omega) \mid v = 0 \text{ on } \Gamma_0, v \geqslant 0 \text{ on } \Gamma_1 \,\}\,,$$

in (10.1) and (10.2) and taking $g = 0$. ■

10.2 Solution of Example II

This involves solving the inequality

$$(10.5) \qquad (u'' - \Delta u - f, v - u') \geqslant 0\,, \quad \forall v \in K, u \in \mathscr{K}_0$$

with

$$\left|\begin{array}{l} V \;= H_0^1(\Omega),\ \Omega \text{ an open bounded domain of } \mathbb{R}^2\,, \\ K \;= \{\, v \mid v \in V,\ v \geqslant 0, \text{ a.e. on } \Omega \,\}\,, \\ \mathcal{K}_0 = \{\, v \mid v, v' \in L^2(0, T; V),\ v'' \in L^2(0, T; V')\ , \\ \qquad v(0) = u_0 \text{ given},\ v'(0) = u_1 \text{ given} \\ \qquad v' \geqslant 0, \text{a.e. on } Q = \left]0, T\right[\times \Omega \,\}\,. \end{array}\right.$$

We take $\Omega = \left]0, 1\right[\times \left]0, 1\right[$ with f, u_0, u_1 defined in such a way that the exact solution is

$$u = \begin{cases} 10 \sin\left(\dfrac{3\pi}{2} t\right) x(1 - x)\, y(1 - y)\,, & 0 \leqslant t \leqslant \dfrac{1}{3}\,, \\ 10\, x(1 - x)\, y(1 - y) & \dfrac{1}{3} < t < \dfrac{2}{3}\,, \\ 10\left(1 - \sin \dfrac{3\pi}{2} t\right) x(1 - x)\, y(1 - y)\,, & \dfrac{2}{3} \leqslant t \leqslant 1 = T\,. \end{cases}$$

The solution (see Figure 10.2) "ascends" from $t = 0$ to $\frac{1}{3}$, stabilises between $\frac{1}{3}$ and $\frac{2}{3}$, and "re-ascends" for $t = \frac{2}{3}$ to 1. In order to represent this problem in the form of an actual inequality (rather than an equation), we take

$$f = \begin{cases} u'' - \Delta u & , \quad 0 \leqslant t \leqslant \frac{1}{3}\,, \\ -\Delta u - (t - \frac{1}{3})(\frac{2}{3} - t)\,, & \quad \frac{1}{3} < t < \frac{2}{3}\,, \\ u'' - \Delta u & , \quad \frac{2}{3} \leqslant t \leqslant 1\,. \end{cases}$$

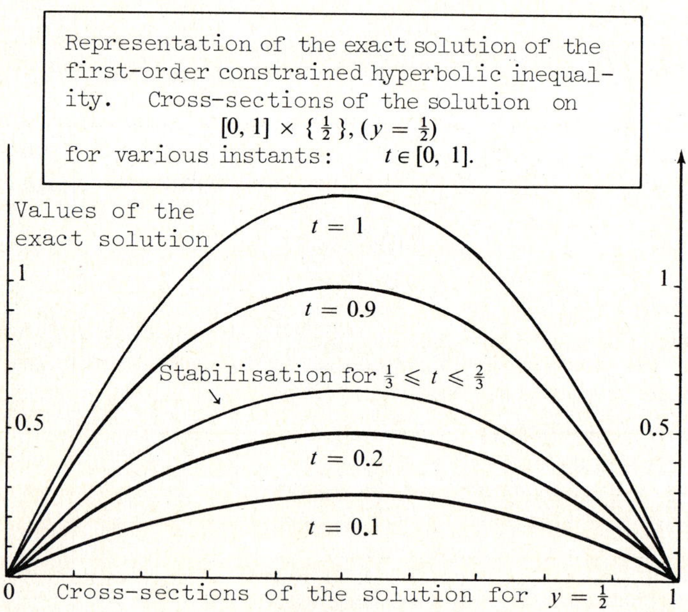

Fig. 10.2.

The standard discretisation was used for V, with $h = 1/20$. ■

10.2.1 *Description of the schemes*

The inequality was approximated by the scheme described below. In each case

$$u_h^0 = u(0) = 0\,,\quad \delta_h^0(M) = u'(0, M) \text{ and } u_h^1 = u_h^0 + k\delta_h^0\,,$$

$$\delta_h^i = (u_h^{i+1} - u_h^i)/k\,.$$

For the schemes (omitting the suffix h) we have, with $i \geqslant 0$,

Explicit scheme

$$\left|\begin{array}{l} \dfrac{\tilde{\delta}^i - \delta^{i-1}}{k} - \Delta u^i - f^i = 0\,, \\ \delta^i = \max\,[\tilde{\delta}^i, 0]\,,\quad u^{i+1} = u^i + k\delta^i\,. \end{array}\right.$$

"Implicit in equation" scheme

$$\left|\begin{array}{l} \dfrac{\tilde{\delta}^i - \delta^{i-1}}{k} - \Delta\tilde{u}^i - f^i = 0\,,\quad \text{with}\quad \tilde{u}^i = u^{i-1} + k\delta^i\,, \\ \delta^i = \max\,[\tilde{\delta}^i, 0]\,,\quad u^{i+1} = u^i + k\delta^i\,. \end{array}\right.$$

"Semi-implicit in inequality" scheme

$$\left|\begin{array}{l} \left(\dfrac{\delta^i - \delta^{i-1}}{k} - \Delta\left(\dfrac{u^i + u^{i+1}}{2}\right) - f^i, v - \delta^i\right) \geqslant 0\,,\quad \forall v \geqslant 0, \delta^i \geqslant 0\,, \\ \text{with } u^{i+1} = u^i + k\delta^i\,. \end{array}\right.$$

"Implicit in inequality" scheme

$$\left|\begin{array}{l} \left(\dfrac{\delta^i - \delta^{i-1}}{k} - \Delta u^{i+1} - f^i, v - \delta^i\right) \geqslant 0\,,\quad \forall v \geqslant 0, \delta^i \geqslant 0\,, \\ \text{with } u^{i+1} = u^i + k\delta^i\,. \end{array}\right.$$

10.2.2 *Results of the various schemes*

The following table (Table 10.2) gives the results for the schemes described in 10.2.1.

For schemes other than the explicit scheme, the solution method used at each timestep is the Gauss-Seidel method, or the Gauss-Seidel method with projection, with the values of ω corresponding to

$$\omega = 2\left[1 + \left(\frac{1 - 2(1 - (\pi^2/2)\,h^2)\,h^2}{2 + h^2/k^2}\right)^{1/2}\right]^{-1}.$$

The termination criterion for the Gauss-Seidel method is

$$\sum_{M \in \Omega_h} \left|\,\delta_h^{i(n+1)}(M) - \delta_h^{i(n)}(M)\,\right| < 10^{-5}$$

where $\delta_h^{i(n)}$ is the solution calculated at the *i*th iteration of the method. The quality of the approximations is measured in terms of

$$\mathrm{ECART}_1 = \sum_{M \in \Omega_h} | u_h^i(M) - u(M\,;ik) | . \quad (*)$$

$h = 1/20$			$1/h\,\mathrm{ECART}_1$ (exact solution - calculated solution)										
Scheme	k	Time	$t=0.1$	0.2	0.3	0.4	0.5	0.6	0.7	0.8	0.9	1	1.1
Explicit	10^{-3}	4min	2.05	7.47	14.14	16.68	16.68	16.68	16.59	14.77	12.19	10.6	—
	5×10^{-4}	16min	1.00	3.63	7.12	9.71	9.71	9.71	9.62	8.02	6.75	4.58	—
	2×10^{-4}	30min	0.39	1.42	2.63	3.00	3.00	3.00	2.89	2.73	2.66	2.12	—
	10^{-4}	57min	0.19	0.70	1.19	1.48	1.48	1.48	1.43	1.38	1.31	1.02	—
Implicit in equation	10^{-3}	16min	0.05	0.19	0.36	0.41	0.41	0.41	0.31	0.01	0.23	0.32	0.24
Semi-implicit in inequality	10^{-3}	16min	0.02	0.10	0.19	0.22	0.22	0.22	0.12	0.16	0.37	0.46	0.40
Implicit in inequality	10^{-3}	16min	0.05	0.19	0.36	0.41	0.41	0.41	0.31	0.01	0.23	0.32	0.24

Table 10.2. Hyperbolic inequality of Example II.

The problems were solved on a CII 10070. ■

10.2.3. *Conclusions of Section* 10

For the problem (10.5) considered here, we have:

Explicit scheme:

- always the danger of divergence, with overflow,
- stability appears to correspond to $k/h \leqslant 2\times10^{-3}$.

The other schemes:

- virtually identical,
- an admissible value of k/h is 2×10^{-2},

(*) *See Translator's Note on page* 432.

- For the same precision, these schemes are three times faster than the explicit scheme. ■

11. NUMERICAL COMPUTATION OF THE FLOW OF BINGHAM FLUIDS

11.1 Notation and statement of the problem

We shall give here only those results which are essential to the understanding of the numerical results, and we refer the reader to Duvaut-Lions /1/ for a more complete discussion of the problem; Ω will denote an open bounded domain of $\mathbb{R}^n$ with boundary Γ. Unless stated otherwise, we shall restrict our attention hereafter to the *two-dimensional* case $(n = 2)$. If $v = \{ v_1, \dots, v_n \}$ is a function with values in $\mathbb{R}^n$, we put

$$(11.1) \qquad D_{ij}(v) = \frac{1}{2}\left(\frac{\partial v_i}{\partial x_j} + \frac{\partial v_j}{\partial x_i}\right)$$

$$(11.2) \qquad D_{\mathrm{II}}(v) = \frac{1}{2} \sum_{i,j=1}^{n} (D_{ij}(v))^2$$

$$(11.3) \qquad a(v, w) = 2 \sum_{i,j=1}^{n} \int_{\Omega} D_{ij}(v)\, D_{ij}(w)\, \mathrm{d}x$$

$$(11.4) \qquad j(v) = 2 \int_{\Omega} (D_{\mathrm{II}}(v))^{1/2}\, \mathrm{d}x$$

$$(11.5) \qquad b(u, v, w) = \sum_{i,j=1}^{n} \int_{\Omega} u_i \frac{\partial v_j}{\partial x_i} w_j\, \mathrm{d}x\,,$$

$$(11.6) \qquad V = \{ v \mid v \in (H_0^1(\Omega))^n \text{ and } \operatorname{div} v = 0 \}$$

$$(11.7) \qquad H = \{ v \mid v \in (L^2(\Omega))^n \text{ and } \operatorname{div} v = 0 \text{ and } v.n\,|_\Gamma = 0 \}\,;$$

H is the completion of V in $(L^2(\Omega))^n$ and we have:

$$(11.8) \qquad V \subset H \subset V'\,;$$

V is equipped with the topology of $(H_0^1(\Omega))^n$ and we shall denote by $\| v \|$ the norm of v in V. Similarly $|v|$ will denote the norm of v in H.

For $v \in V$, it is clear that:

$$(11.9) \qquad a(v, v) \geqslant \alpha \| v \|^2\,; \qquad \alpha > 0\,.$$

We consider the following problem:

Problem 1. Find $u(t)$ which satisfies the following conditions in $]0, T[$:

(11.10) $\quad u \in L^2(0, T; V) \cap L^\infty(0, T; H), \quad u' \in L^2(0, T; V')$

(11.11) $\quad (u'(t), v - u(t)) + \nu a(u(t), v - u(t)) + b(u(t), u(t), v) +$
$\qquad + gj(v) - gj(u(t)) \geqslant (f(t), v - u(t)), \quad \forall v \in V,$ a.e. in t.

(11.12) $\quad u(0) = u_0 \in H, \qquad f \in L^2(0, T; V').$

The variational inequality (11.11), together with (11.10) and (11.12) constitutes a model describing the flow of a rigid visco-plastic Bingham fluid, with viscosity ν and plasticity threshold g (see Duvaut-Lions, *loc. cit*). Note that for $g = 0$, we recover the classical variational formulation of the Navier-Stokes equations for Newtonian fluids. For $g > 0$, we observe zones in which the behaviour is that of a fluid with viscosity ν. The variational formulation automatically takes account of the free boundary separating the two zones; the precise determination of this boundary will be one of the principal difficulties involved in the numerical treatment of the problem.

We recall here simply that in a Bingham fluid the stress tensor σ_{ij} and the rate-of-deformation tensor D_{ij} are related by the constitutive law:

$$\sigma_{ij} = -p\delta_{ij} + g(D_{ij}/(D_{\mathrm{II}})^{1/2}) + 2\mu D_{ij}.$$

This expression is only meaningful if $D_{\mathrm{II}} \neq 0$.

If $D_{\mathrm{II}} = 0$, the stress tensor is indeterminate and we have a rigid zone; if g increases, the rigid zones grow and are liable to block the flow completely for g sufficiently large. The scalar $-p$ represents the spherical part of the stress tensor. We may take p to be the pressure.

Remark 11.1. In practice, it is probable that in the case of Bingham fluids $(g > 0)$ it is permissible to neglect the non-linear terms introduced into (11.11) by the trilinear form $b(u, u, v)$. If we accept this simplification, most of the results which we shall state for the two-dimensional case are also valid to any dimension.

The following results are proved in Duvaut-Lions /1/.

Theorem 11.1 *In two-dimensional space* $(n = 2)$ *problem* I *possesses a unique solution.*

Theorem 11.2 *For* $n = 2$, *if* $u(t)$ *denotes the solution of problem* I, *there exist functions* m_{ij} $(1 \leqslant i, j \leqslant 2)$ *such that*:

(11.13) $\quad m_{ij} \in L^\infty(Q), \quad m_{ij} = m_{ji} (Q = \Omega \times]0, T[)$

(11.14) $\sum_{i,j=1}^{2} m_{ij} m_{ij} \leqslant 1$ *a.e. in* Q.

(11.15) $\sum_{i,j=1}^{2} m_{ij} D_{ij}(u) = \left(\sum_{i,j=1}^{2} D_{ij}(u) D_{ij}(u) \right)^{1/2}$ *a.e. in* Q.

(11.16) $(u', v) + \nu a(u, v) + b(u, u, v) + \sqrt{2} g \int_\Omega \sum_{i,j=1}^{2} m_{ij} D_{ij}(v) \, dx = (f, v)$

$\forall v \in V$, *a.e. in* t.

Conversely if (11.13)-(11.16) *are satisfied, with:*

(11.17) $u \in L^2(0, T; V) \cap L^\infty(0, T; H)$, $u' \in L^2(0, T; V')$, $u(0) = u_0$

then u *is the solution of Problem* I. ■

11.2 Numerical schemes

We shall consider here an extremely simple scheme which has in practice proved effective for approximating Problem I. The problems posed by its practical application will be examined in Section 11.3. As regards the spatial discretisation, we shall assume for simplicity that we are considering an *interior* approximation V_h of V [1]; we shall mention in passing the minor modifications which must be made in the case where an *exterior* approximation $\tilde{V}_h$ is used. It would be possible to transpose the results to any other type of discretisation very easily, without affecting the principle of the scheme. For any integer n, $0 \leqslant n \leqslant N-1$, we put

(11.18) $f_k^{n+\theta} = \frac{1}{k} \int_{nk}^{(n+1)k} f(t) \, dt \in V' \quad (0 \leqslant \theta \leqslant 1)$.

Given a triangulation τ_h of Ω, we assume u_h^0 given in V_h, chosen so that u_h^0 tends to u_0 in H when h tends to zero. We shall assume that $\| u_h^0 \|$ is bounded. We shall define by recurrence a sequence u_h^n of elements of V_h with which we shall then associate "approximating functions". We write:

(11.19) $u_h^{n+\theta} = \theta u_h^{n+1} + (1-\theta) u_h^n \quad (0 \leqslant \theta \leqslant 1)$.

To simplify the notation, we denote by u_h^n a sequence which actually depends on h and k.

---

(1) Some indication of the construction of the spaces V_h, $\tilde{V}_h$ is given at the end of this section.

Assuming that $u_h^0, \ldots, u_h^n$ are known, we define u_h^{n+1} as the solution in V_h of:

$$(11.20)\qquad \frac{1}{k}(u_h^{n+1} - u_h^n, v_h - u_h^{n+\theta}) + \nu a(u_h^{n+\theta}, v_h - u_h^{n+\theta}) + \\ + b(u_h^{n+\theta}, u_h^{n+\theta}, v_h) + gj(v_h) - gj(u_h^{n+\theta}) \geqslant \\ \geqslant (f_k^{n+\theta}, v_h - u_h^{n+\theta}), \quad \forall v_h \in V_h .$$

For $\theta > 0$, we have

$$(11.21)\qquad u_h^{n+1} = \frac{1}{\theta} u_h^{n+\theta} - \frac{(1-\theta)}{\theta} u_h^n$$

and hence

$$(11.22)\qquad \frac{u_h^{n+1} - u_h^n}{k} = \frac{u_h^{n+\theta} - u_h^n}{\theta k}.$$

We thus have to solve:

$$(11.23)\qquad (u_h^{n+\theta}, v_h - u_h^{n+\theta}) + k\theta \nu a(u_h^{n+\theta}, v_h - u_h^{n+\theta}) + k\theta b(u_h^{n+\theta}, u_h^{n+\theta}, v_h) + \\ + k\theta gj(v_h) - k\theta gj(u_h^{n+\theta}) \geqslant (u_h^n, v_h - u_h^{n+\theta}) + k\theta(f_k^{n+\theta}, v_h - u_h^{n+\theta}), \\ \forall v_h \in V_h .$$

It is clear that (11.23) is meaningless for $\theta = 0$. In this case we would have to consider a semi-implicit scheme:

$$(11.24)\qquad \frac{1}{k}(u_h^{n+1} - u_h^n, v_h - u_h^{n+1}) + \nu a(u_h^n, v_h - u_h^{n+1}) + b(u_h^n, u_h^n, v_h - u_h^{n+1}) + \\ + gj(v_h) - gj(u_h^{n+1}) \geqslant (f_k^{n+1}, v_h - u_h^{n+1}), \quad \forall v_h \in V_h .$$

However problem (11.24) is not coercive on V_h, which brings in difficulties regarding the existence (and computation) of u_h^{n+1}. ■

Using the methods of Chapter 4, we may verify:

Proposition 2.1. *In two-dimensional space* ($n = 2$), (11.20), (*and thus also* (11.23)), *with* $\theta > 0$ *and* k *"sufficiently small", define* $u_h^{n+\theta}$ *and* u_h^{n+1} *uniquely.*

By methods analogous to those above, it can be shown (details of the proofs are given in M. Fortin /1/) that the u_h^n and $u_h^{n+\theta}$ defined by (11.20) satisfy, for $\theta \geqslant \frac{1}{2}$:

$$(11.25)\qquad |u_h^n|^2 \leqslant C \qquad (0 \leqslant n \leqslant N)$$

$$(11.26)\qquad k \sum_{n=0}^{N-1} \| u_h^{n+\theta} \|^2 \leqslant C ,$$

$$(11.27)\qquad \sum_{n=0}^{N-1} | u_h^{n+1} - u_h^n |^2 \leqslant C ,$$

C = Const. independent of k and h provided that:

(11.28) $\quad kS^2(h) \leqslant C_0$, an arbitrary constant.

For $\theta > \frac{1}{2}$, condition (11.28) is superfluous and in addition we have:

$$k \sum_{n=0}^{N-1} \| u_h^{n+1} - u_h^n \|^2 \leqslant C_1 \,. \tag{11.29}$$

The above estimates are *sufficient*, in the case of inequalities for *linear operators*, to demonstrate the *convergence*, but they are insufficient in the case of inequalities (and equations) for *non-linear* operators. For these cases we have to obtain a supplementary estimate allowing us to proceed to the limit *by compactness*. To this end we use a method due to L. Tartar and R. Témam /1/ (see also the work of Témam /2/ and the bibliography therein).

We write (11.20) in the form (1)

$$\begin{aligned}\frac{1}{k}(u_h^{n+1} - u_h^n, u_h^{n+\theta} - v_h(t)) &\leqslant va(u_h^{n+\theta}, u_h^{n+\theta} - v_h(t)) - \\ &- b(u_h^{n+\theta}, u_h^{n+\theta}, u_h^{n+\theta} - v_h(t)) + gj(v_h(t)) - gj(u_h^{n+\theta}) + \\ &+ (f_k^{n+\theta}, u_h^{n+\theta} - v_h(t)), \quad \forall v_h \in L^2(0, T; V_h), \\ &\forall t \in [nk, (n+1)k[\,.\end{aligned} \tag{11.30}$$

Consider the left-hand side of (11.30). Defining $u_h(t)$ as the linear function over each interval $[nk, (n+1)k]$ which is equal to u_h^n at nk, we have:

$$\frac{1}{k}(u_h^{n+1} - u_h^n, u_h^{n+\theta} - v_h(t)) = \frac{1}{k}(u_h^{n+1} - u_h^n, u_h^{n+\theta}) - (u_h'(t), v_h(t)) \,. \tag{11.31}$$

Then let $v_h(t) = \frac{1}{\sigma}\int_{t-\sigma}^{t} u_h(s)\,\mathrm{d}s$, by convention taking $u_h(s) = u_h^0$ for $s \leqslant 0$.

Integrating from 0 to T we obtain:

$$\begin{aligned}-\int_0^T (u_h'(t), v_h(t))\,\mathrm{d}t &= \int_0^T (u_h(t), v_h'(t))\,\mathrm{d}t + (u_h(0), v_h(0)) - \\ &- (u_h(T), v_h(T)) = \int_0^T \frac{(u_h(t), u_h(t) - u_h(t-\sigma))}{\sigma}\,\mathrm{d}t + \\ &+ | u_h^0 |^2 - (u_h(T), v_h(T)) \,.\end{aligned} \tag{11.32}$$

---

(1) Note that $b(u, v, w) = -\, b(u, w, v) - \sum_{j=1}^{n} \int_\Omega (\operatorname{div} u)\, v_j\, w_j\,\mathrm{d}x$ (integration by parts) and $b(u, v, v) = 0$, with $u \in V$ and $\forall v \in (H_0^1(\Omega))^n$.

Moreover

$$(11.33)\qquad \int_0^T \frac{(u_h^{n+1} - u_h^n, u_h^{n+\theta})}{k}\,\mathrm{d}t = \sum_{n=0}^{N-1} (u_h^{n+1} - u_h^n, u_h^{n+\theta}) =$$

$$= \frac{(2\theta - 1)}{2} \sum_{n=0}^{N-1} | u_h^{n+1} - u_h^n |^2 + \frac{1}{2} | u_h^N |^2 - \frac{1}{2} | u_h^0 |^2 .$$

Moreover we have:

$$(11.34)\qquad \int_0^T \frac{(u_h(t), u_h(t) - u_h(t-\sigma))}{\sigma}\,\mathrm{d}t = \int_0^T \frac{| u_h(t) |^2}{\sigma}\,\mathrm{d}t -$$

$$- \int_\sigma^N \frac{(u_h(t), u_h(t-\sigma))}{\sigma}\,\mathrm{d}t - \int_0^\sigma \frac{(u_h(t), u_h^0)}{\sigma}\,\mathrm{d}t =$$

$$= \frac{1}{2} \int_\sigma^T \frac{| u_h(t) - u_h(t-\sigma) |^2}{\sigma}\,\mathrm{d}t + \frac{1}{2} \int_0^\sigma \frac{| u_h(t) |^2}{\sigma}\,\mathrm{d}t -$$

$$- \int_0^\sigma \frac{(u_h(t), u_h^0)}{\sigma}\,\mathrm{d}t + \frac{1}{2} \int_{T-\sigma}^T \frac{| u_h(t) |^2}{\sigma}\,\mathrm{d}t .$$

Thus, by (11.30), we have:

$$(11.35)\qquad \frac{1}{2} \int_\sigma^T \frac{| u_h(t) - u_h(t-\sigma) |^2}{\sigma}\,\mathrm{d}t + \frac{(2\theta - 1)}{2} \sum_{n=0}^{N-1} | u_h^{n+1} - u_h^n |^2 +$$

$$+ \frac{1}{2} | u_h^N |^2 + \frac{1}{2} \int_{T-\sigma}^T \frac{| u_h(t) |^2}{\sigma}\,\mathrm{d}t - (u_h^N, v_h(T)) +$$

$$+ \frac{1}{2} | u_h^0 |^2 + \frac{1}{2} \int_0^\sigma \frac{| u_h(t) |^2}{\sigma}\,\mathrm{d}t - \left(u_h^0, \frac{1}{\sigma} \int_0^\sigma u_h(t)\,\mathrm{d}t \right) \leqslant$$

$$\leqslant \int_0^T (f_h(t), \tilde{u}_h(t) - v_h(t))\,\mathrm{d}t + g \int_0^T j(v_h(t))\,\mathrm{d}t ,$$

where $\tilde{u}_h$ denotes the step function equal to $u_h^{n+\theta}$ in $[nk, (n+1)k]$.

However we have:

$$(11.36)\qquad | (u_h^N, v_h(T)) | \leqslant \frac{1}{2} | u_h^N |^2 + \frac{1}{2} | v_h(T) |^2$$

and in addition

$$(11.37)\qquad | v_h(T) |^2 \leqslant \frac{1}{\sigma^2} \left| \int_{T-\sigma}^T u_h(t)\,\mathrm{d}t \right|^2 \leqslant \frac{1}{\sigma} \int_{T-\sigma}^T | u_h(t) |^2\,\mathrm{d}t .$$

Similarly:

$$(11.38)\qquad \left|\left(u_h^0, \frac{1}{\sigma}\int_0^\sigma u_h(t)\,\mathrm{d}t\right)\right| \leqslant \frac{1}{2}\int_0^\sigma \frac{|\,u_h(t)\,|^2}{\sigma}\,\mathrm{d}t + \frac{1}{2}\,|\,u_h^0\,|^2 .$$

Thus

$$(11.39)\qquad \frac{1}{2}\int_\sigma^T \frac{|\,u_h(t) - u_h(t-\sigma)\,|^2}{\sigma}\,\mathrm{d}t \leqslant \int_0^T (f_h(t), \tilde{u}_h(t) - v_h(t))\,\mathrm{d}t + $$

$$+ g\int_0^T j(v_h(t))\,\mathrm{d}t .$$

We shall now majorise the right-hand side. First, consider the linear terms

$$(11.40)\qquad \nu\int_0^T a(\tilde{u}_h(t), \tilde{u}_h(t) - v_h(t))\,\mathrm{d}t \leqslant \frac{\nu}{2}\int_0^T \|\,\tilde{u}_h(t)\,\|^2\,\mathrm{d}t + $$

$$+ \frac{\nu}{2}\int_0^T \|\,\tilde{u}_h(t) - v_h(t)\,\|^2\,\mathrm{d}t \leqslant \frac{\nu}{2}\int_0^T \|\,\tilde{u}_h(t)\,\|^2\,\mathrm{d}t + $$

$$+ \nu\int_0^T \|\,\tilde{u}_h(t)\,\|^2\,\mathrm{d}t + \nu\int_0^T \|\,v_h(t)\,\|^2\,\mathrm{d}t .$$

Now $\displaystyle\int_0^T \|\,v_h(t)\,\|^2\,\mathrm{d}t \leqslant C\int_0^T \|\,u_h(t)\,\|^2\,\mathrm{d}t \leqslant$ const. if $kS^2(h) \leqslant C_0$.

Similarly,

$$(11.41)\qquad \int_0^T (f_k^{n+\theta}, \tilde{u}_h(t) - v_h(t))\,\mathrm{d}t \leqslant C\int_0^T |\,f(t)\,|_*^2\,\mathrm{d}t + \text{const.}$$

For the nonlinear terms we have

$$(11.42)\qquad \int_0^T b(\tilde{u}_h(t), \tilde{u}_h(t), \tilde{u}_h(t) - v_h(t))\,\mathrm{d}t = \int_0^T b(\tilde{u}_h(t), v_h(t), \tilde{u}_h(t))\,\mathrm{d}t$$

and hence by using the estimates of Chapter 6, Section 3 of Duvaut-Lions /1/:

$$(11.43) \quad \int_0^T b(\tilde{u}_h(t), v_h(t), \tilde{u}_h(t))\,dt \leqslant C\int_0^T |\tilde{u}_h(t)|\,\|\tilde{u}_h(t)\|\,\|v_h(t)\|\,dt \leqslant$$

$$\leqslant C'\int_0^T \|\tilde{u}_h(t)\|^2\,dt + C'\int_0^T \|v_h(t)\|^2\,dt \leqslant \text{const.}$$

Finally

$$(11.44) \quad g\int_0^T j(v_h(t))\,dt \leqslant C + g\int_0^T \|v_h(t)\|^2\,dt \leqslant \text{const.}$$

Consequently, we have shown that the *approximate function* $u_h(t)$ *satisfies*:

$$(11.45) \quad \int_0^{T-\sigma} |u_h(t+\sigma) - u_h(t)|^2\,dt \leqslant C\sigma$$

on condition that $kS^2(h) \leqslant C_0$ (*arbitrary constant*); *thus for all* $\varepsilon > 0$, *we have*:

(11.46) $u_h(t)$ resides in a bounded set of $H^{1/2-\varepsilon}(0, T; H)$. ■

The above estimates, combined with the compactness of the injection from $V \to H$, allow us to proceed to the limit. We hav

Theorem 11.3. *We assume that* h *and* $k \to 0$, *in an arbitrary manner if* $\theta > \frac{1}{2}$, *and with condition* (11.28) *if* $\theta = \frac{1}{2}$. *Then* $u_h \to u$ *weakly* in* $L^\infty(0, T; H)$ *and weakly in* $L^2(0, T; V)$. ■

Remark 11.2. We refer to M. Fortin /1/ for details of the proof of Theorem 11.3 and for the *strong* convergence results. ■

Construction of interior and exterior approximations of V.

There are essentially *three* classes of possibilities for appro imating V by spaces V_h :

(i) interior approximation $V_h \subset V$;
(ii) interior approximation of $(H_0^1(\Omega))^n$ $(n = 2$ or $3)$, the condition $\operatorname{div} v = 0$ being satisfied only *approximately*;
(iii) approximation by spaces of functions *which are not* in $(H^1(\Omega))^n$ (non-conforming approximations). ■

An exhaustive study of this topic will not be undertaken here.

We shall merely give some general indications and refer the reader to the bibliography for proofs, and in particular to the work of Crouzeix and Raviart /1/.

We suppose that Ω is a polyhedral domain of $\mathbb{R}^n$; let $\mathcal{T}_h$ be a triangulation into non-degenerate simplexes K; if $h(K)$ (resp. $\rho(K)$) denotes the diameter of K (resp. the diameter of the sphere inscribed inside K) we assume

$$\sup_{K \in \mathcal{T}_h} \frac{h(K)}{\rho(K)} = \sigma < \infty .$$

If $K \in \mathcal{T}_h$, with vertices $a_{i,K}$, $i = 1, 2, \dots, (n+1)$, we denote by $\lambda_{i,K}(x)$ the barycentric coordinates of a point x of $\mathbb{R}^n$ with respect to the points $a_{i,K}$.

Example 1 (Situation (ii)). In two-dimensions (to simplify the description) we denote by W_h the space of functions V_h continuous in $\overline{\Omega}$ such that:

1) $V_h|_K \in$ space of polynomials of degree $\leqslant 2(P_2)$, $\forall K \in \mathcal{T}_h$,
2) $V_h|_\Gamma = 0$.

A function $v_h \in W_h$ is characterised by its values at the vertices $a_{i,K}$ and at the midpoints $a_{ij,K}$ of $[a_{i,K}\, a_{j,K}]$.

We then define V_h as a subspace of $(H_0^1(\Omega))^2$, *but not as a subspace of* V, by : $v_h \in (W_h)^2$ and

$$\int_K \operatorname{div} v_h \, dx = 0 \quad \forall K \in \mathcal{T}_h . \tag{11.47}$$

If u_h (resp. u) denotes the solution of the approximate linear (resp. continuous) steady-state problem, it can be shown (see Crouzeix-Raviart /1/) that

$$\| u_h - u \|_{(H^1(\Omega))^2} \leqslant Ch , \tag{11.48}$$

where h denotes the maximum diameter of $K \in \mathcal{T}_h$. ■

Example 2 (Situation (ii)). The estimate (11.48) is not what might be expected for the order of an approximation using *second degree* polynomials. Clearly this arises from the fact that, formally, the approximation (11.47) of the condition " $\operatorname{div} v = 0$ " is only *first order* accurate.

We can improve upon this without introducing *general* third degree polynomials.

If we write λ_i in place of $\lambda_{i,K}(x)$, we introduce

$$\left| \begin{array}{l} P_K = \text{space generated by the polynomials } \lambda_1^2, \lambda_2^2, \lambda_3^2, \\ \quad \lambda_1\lambda_2, \quad \lambda_2\lambda_3, \quad \lambda_3\lambda_1 \quad \textit{and} \quad \lambda_1\lambda_2\lambda_3 . \end{array} \right. \tag{11.49}$$

We then introduce

$$(11.50)\quad \left| \begin{array}{l} W_h = \{ v_h \mid v_h \in C^0(\overline{\Omega}) ,\quad v_h|_K \in P_k \quad \forall K \in \mathcal{T}_h ,\\ \qquad v_h = 0 \text{ on } \Gamma \} ; \end{array} \right.$$

an element v_h of W_h is characterised by its value at the vertices $a_{i,K}$, at the midpoints $a_{ij,K}$ and at the barycentres of K, $K \in \mathcal{T}_h$.
We then introduce

$$(11.51)\quad V_h = \left\{ v_h \mid v_h \in (W_h)^2_; \int_K q \operatorname{div} v_h \, dx = 0 \ \forall q \in P_1 \text{ (polynomials of degree } \leqslant 1 \text{) and } \forall K \in \mathcal{T}_h \right\} .$$

It can be shown (see Crouzeix-Raviart /1/) that

$$(11.52)\quad \| u_h - u \|_{(H^1(\Omega))^2} \leqslant C h^2 . \quad ■$$

The method of Example 1 was introduced by Fortin /1/, in which the extension of the above example into three dimensions and the investigation of higher-order approximations can also be found. ■

Example 3 (Situation (i)). *In two dimensions*, a classical means of constructing "test functions" which satisfy the conditio " $\operatorname{div} v = 0$ " is to work with the stream function and to make use of the fact that since Ω is a simply-connected bounded open domai the mapping

$$(11.53)\quad \psi \to u = \left\{ \frac{\partial \psi}{\partial x_2} , - \frac{\partial \psi}{\partial x_1} \right\}$$

establishes an isomorphism from $H_0^2(\Omega)$ onto V. We then construct an approximation of $H_0^2(\Omega)$ formed from polynomials of degree 5 on K, whose image is taken using (11.53). See Fortin /1/, Thomasset /1/ and the account by R. Témam /2/. ■

Example 4 (Situation (iii)). We introduce (still assuming for simplicity that $n = 2$):

$$(11.54)\quad \left| \begin{array}{l} W_h = \{ v_h \mid v_h|_K \in P_1 \quad \forall K,\ v_h \text{ continuous at the midpoints } a_{ij,K} \\ \qquad \text{of } [a_{i,K}, a_{j,K}], \text{ and zero at the points } a_{ij,K} \text{ which lie on } \Gamma \} . \end{array} \right.$$

We next introduce

$$(11.55)\quad V_h = \left\{ v_h \mid v_h \in (W_h)^2 , \int_K \operatorname{div} v_h \, dx = 0 \ \forall K \right\} .$$

Then $a(u, v)$ *is not* defined on V_h , but we introduce

(11.56) $a_h(u_h, v_h) = 2 \sum_{K \in \mathcal{T}_h} \int_K D_{ij}(u_h) D_{ij}(v_h) \, dx ,$

which enables us to define the solution u_h of the Stokes problem

(11.57) $a_h(u_h, v_h) = (f, v_h) \quad \forall v \in V_h, u_h \in V_h .$

We then obtain the same estimate as (11.48); see Crouzeix-Raviart /1/. ■

11.3 Numerical results

Having chosen the space V_h, the problem reduces to the numerical solution of (11.23) [1]. If we introduce the nonlinear operator A (in finite dimensions) by:

(11.58) $(Au, v) = (u, v) + k\theta a(u, v) + k\theta b(u, u, v)$

we then have to solve:

(11.59) $$(Au_h^{n+\theta}, v_h - u_h^{n+\theta}) + k\theta g[j(v_h) - j(u_h^{n+\theta})] \geqslant$$
$$\geqslant (u_h^n + k\theta f_k^{n+\theta}, v_h - u_h^{n+\theta}) \quad \forall v_h \in V_h .$$

Using the discrete analogue of Theorem 11.2, it can be seen that (11.59) is *equivalent* to the following (Lagrangian) problem (which expresses the fact that $\{ u_h^{n+\theta}, q_h^{n+\theta} \}$ is a saddle point):

(11.60) $$\left| \begin{array}{l} (Au_h^{n+\theta}, v_h) + k\theta \sqrt{2}\, g(q_h^{n+\theta}, Dv_h) = (u_h^n + k\theta f_k^{n+\theta}, v_h) \\ \forall v_h \in V_h \ \ (^2), \end{array} \right.$$

(11.61) $(Du_h^{n+\theta}, q - q_h^{n+\theta}) \leqslant 0 \quad \forall q \in K$

where

(11.62) $K = \{ q \mid q_{ij} = q_{ji} \in L^\infty(\Omega), \ |q| \leqslant 1 \} .$

Uzawa's algorithm gives the following: to simplify the notation we put:

(11.63) $u_h^{n+\theta} = w , \qquad q_h^{n+\theta} = m .$

Then, starting from arbitrary m^0 we define : $w^1, m^1, w^2, m^2, \ldots$, as follows:

---

1) Hereafter we put : $Dv = \{ D_{ij}(v) \}$.

2) Where, in the case of *non-conforming* elements, the integrals over Ω are replaced by the sum of the integrals over the simplexes of the triangulation.

(11.64)
$$\begin{cases} (Aw^{s+1}, v_h) + k\theta\sqrt{2}\,g(m^s, Dv_h) = (u_h^n + k\theta f_k^{n+\theta}, v_h) & \forall v_h \in V_h \\ m^{s+1/2} = m^s + \rho\theta k\sqrt{2}\,gDw^{s+1}, \\ m^{s+1} = P_K(m^{s+1/2}), \quad P_K \text{ projector on } K \text{ in } (L^2(\Omega))^4. \end{cases}$$

If in (11.64) we choose the parameter $\rho > 0$, such that

(11.65) $\rho < \nu/k\theta g^2$

then it can be shown (Fortin /1/) that the above algorithm converges. ■

Fortin, *loc. cit.*, also contains an investigation of the Arrow-Hurwicz algorithm together with various further remarks on the application of these algorithms. We also refer to Bégis /1/ for an investigation of the various parameters for accelerating the convergence. ■

Naturally, one of the essential features of the numerical experiments carried out is the determination of the rigid zones [1].

In this context we note that if $\sum m_{ij}^2 < 1$ then condition (11.15) together with (11.14) can only be satisfied if $D_{ij}u = 0$.

Thus

(11.66) if $\sum m_{ij}^2(x) < 1$, x lies in the rigid zone.

This information is used in conjunction with $|Du_h^{n+\theta}(x)|$. This allows an accurate practical determination of the rigid zones. ■

Some numerical results.

We present only *some* of the numerical results obtained, notabl[y] by M. Fortin /1/ and Bégis /3/.

The work of Fortin uses the finite element method employed in Example 1. We shall present some results due to Bégis *loc. cit.* who uses finite difference methods, and thus exterior approximati[ons] (the above considerations may be extended to this case, but with additional technical difficulties).

Figure 11.1 shows the domain of rigidity (hatched) at the instant 0.005 for zero right-hand side and non-zero *boundary cond-itions* (cases which come within the general theory).

Figure 11.2 shows the domain of rigidity for a value of g greater than that in Figure 11.1, all other quantities being equa[l].

---

(1) About which there are very few results other than "numerica[l] evidence".

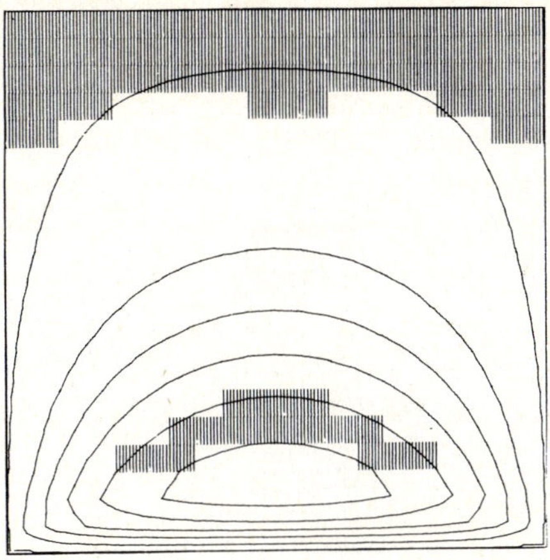

Fig. 11.1. Time 0.005 : $g = 2.5$.

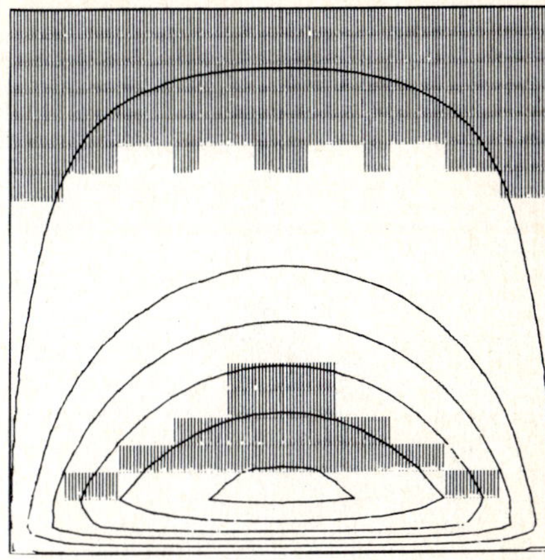

Fig. 11.2. Time 0.005 : $g = 5$.

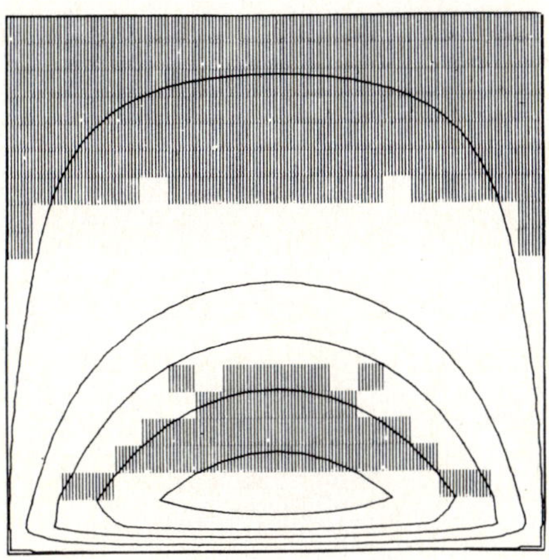

Fig. 11.3. Time 0.005 : $g = 10$.

PARAMETERS

VISCOSITY 1.0
EXTERNAL FORCE F1(X1,X2,T) = 0.0 F2(X1,X2,T) = 0.0

BOUNDARY CONDITIONS

NORMAL COMPONENT OF VELOCITY 0.0
TANGENTIAL COMPONENT OF VELOCITY 0.0 AT X1 = 0, 1 AND X2 = 1

INITIAL CONDITION

INITIAL VELOCITY 0.0

- STREAM FUNCTION CONTOURS -

||||||| RIGID ZONES

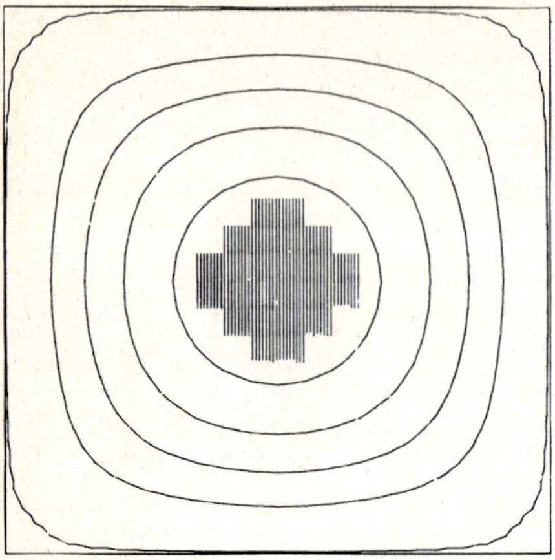

Fig. 11.4. Time 0.005.

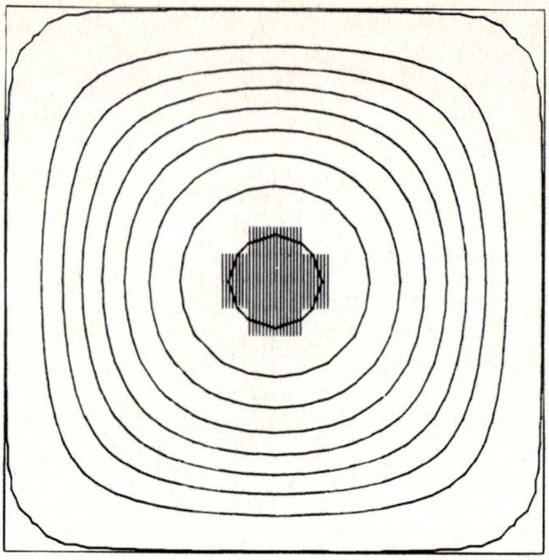

Fig. 11.4′. Time 0.010.

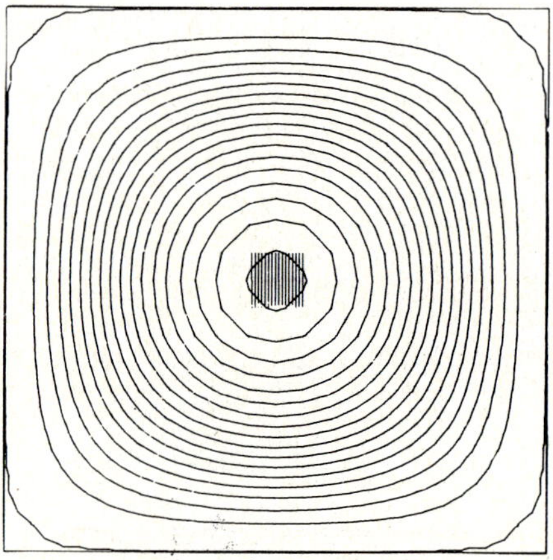

Fig. 11.5. Time 0.040.

PARAMETERS

PLASTICITY THRESHOLD	2.5
VISCOSITY	1.0
EXTERNAL FORCE	F1(X1,X2,T) = 300 × (X2 - .5)
	F2(X1,X2,T) = -300 × (X1 - .5)

BOUNDARY CONDITIONS

NORMAL COMPONENT OF VELOCITY	0.0
TANGENTIAL COMPONENT OF VELOCITY	0.0

INITIAL CONDITION

INITIAL VELOCITY	0.0

- STREAM FUNCTION CONTOURS -

||||||| RIGID ZONES

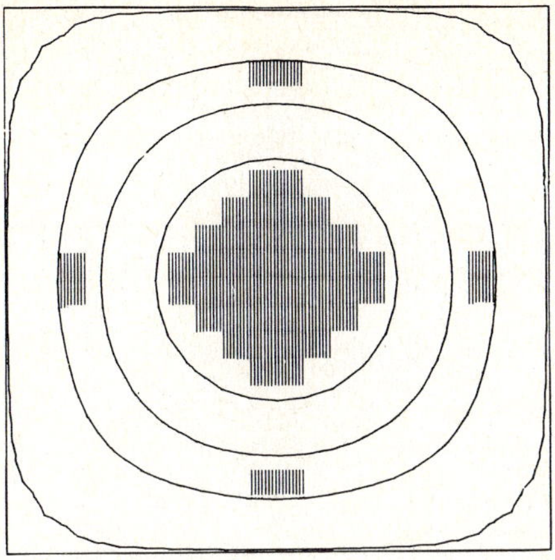

Fig. 11.6. Time 0.005.

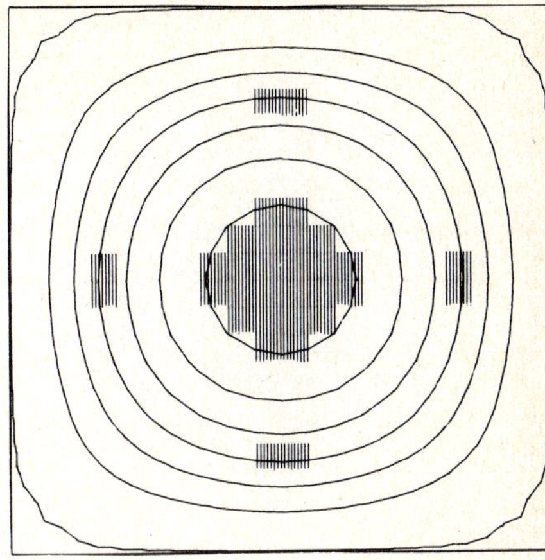

Fig. 11.7. Time 0.010.

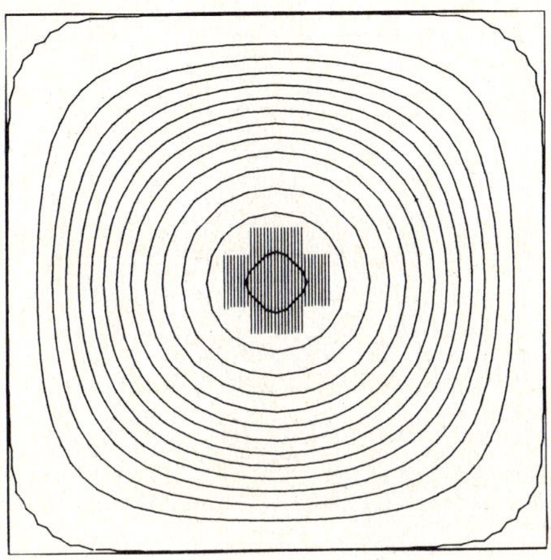

Fig. 11.8. Time 0.040.

PARAMETERS

PLASTICITY THRESHOLD	5.0
VISCOSITY	1.0
EXTERNAL FORCE	F1(X1,X2,T) = 300 × (X2 - .5)
	F2(X1,X2,T) = -300 × (X1 - .5)

BOUNDARY CONDITIONS

NORMAL COMPONENT OF VELOCITY	0.0
TANGENTIAL COMPONENT OF VELOCITY	0.0

INITIAL CONDITION

INITIAL VELOCITY	0.0

- STREAM FUNCTION CONTOURS -

▥ RIGID ZONES

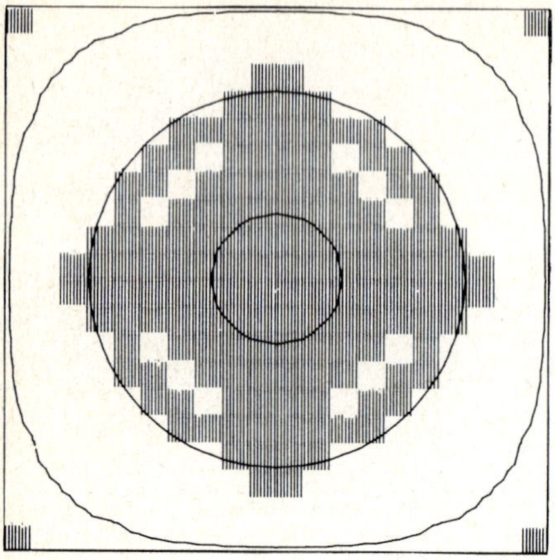

Fig. 11.9. Time 0.005.

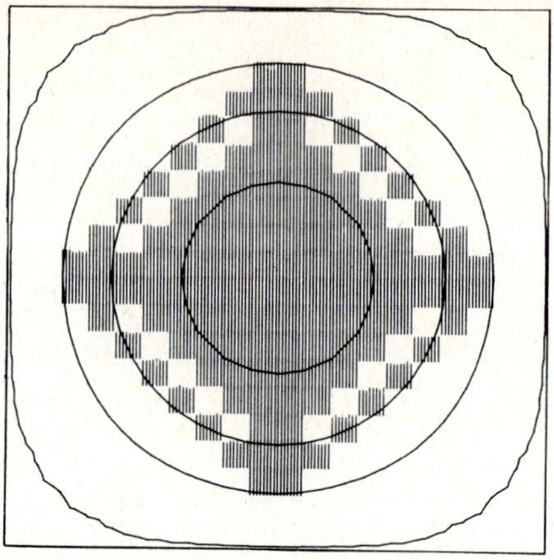

Fig. 11.10. Time 0.010.

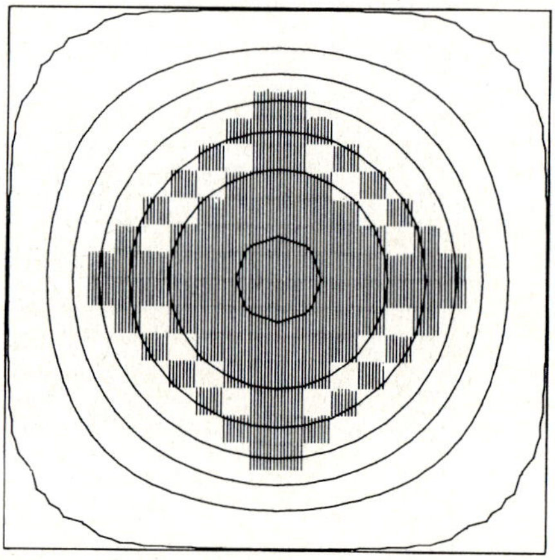

Fig. 11.11. Time 0.040.

PARAMETERS

PLASTICITY THRESHOLD	10.0
VISCOSITY	1.0
EXTERNAL VELOCITY	F1(X1,X2,T) = 300 × (X2 - .5)
	F2(X1,X2,T) = -300 × (X1 - .5)

BOUNDARY CONDITIONS

NORMAL COMPONENT OF VELOCITY	0.0
TANGENTIAL COMPONENT OF VELOCITY	0.0

INITIAL CONDITION

INITIAL VELOCITY	0.0

\- STREAM FUNCTION CONTOURS -

||||||| RIGID ZONES

PARAMETERS

PLASTICITY THRESHOLD	G
VISCOSITY	1.0
EXTERNAL FORCE	F1(X1,X2,T) = 300 × (X2 - .5)
	F2(X1,X2,T) = - 300 × (X1 - .5)

BOUNDARY CONDITIONS

NORMAL COMPONENT OF VELOCITY	0.0
TANGENTIAL COMPONENT OF VELOCITY	0.0

INITIAL CONDITION

INITIAL VELOCITY	0.0

$A(U(T)) = \text{VISCOSITY} \times (\text{GRAD}(U(X,T)), \text{GRAD}(U(X,T)))_{L^2(D)}$ INCREASING IN T

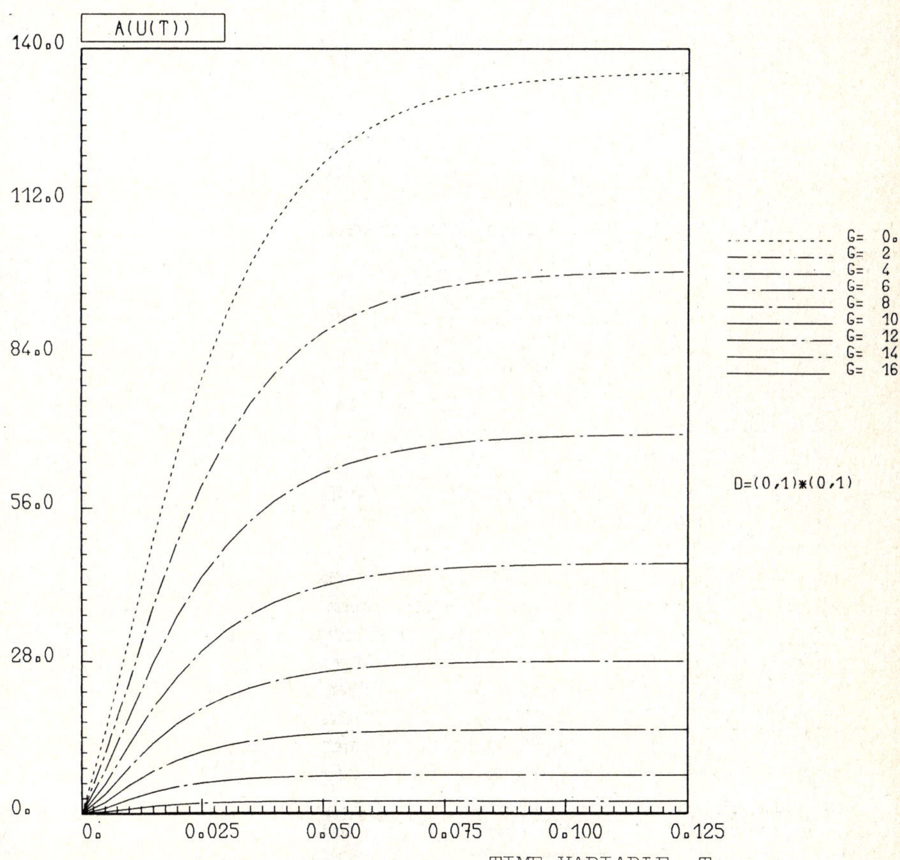

Fig. 11.12.

PARAMETERS

PLASTICITY THRESHOLD	G
VISCOSITY	1.0
EXTERNAL FORCE	F1(X1,X2,T) = 300 × (X2 − .5)
	F2(X1,X2,T) = −300 × (X1 − .5)

BOUNDARY CONDITIONS

NORMAL COMPONENT OF VELOCITY	0.0
TANGENTIAL COMPONENT OF VELOCITY	0.0

INITIAL CONDITION

INITIAL VELOCITY	0.0

$J(U(T)) = (|GRAD(U(X,T))|, 1.)_{L^2(D)}$ INCREASING IN T

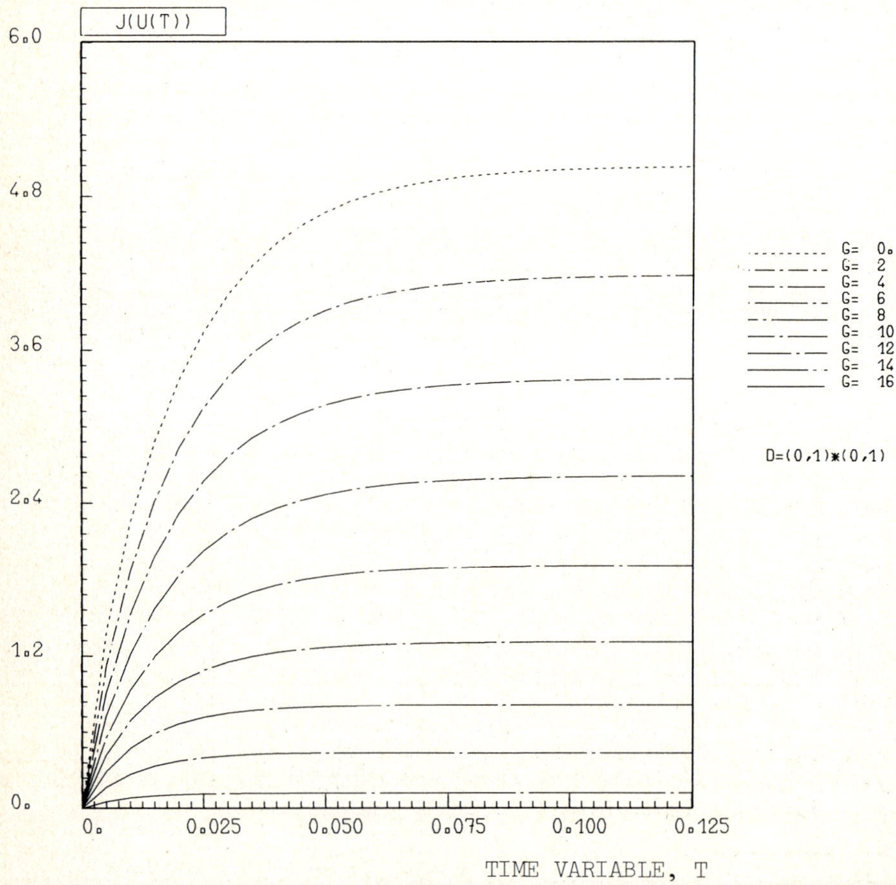

Fig. 11.13.

We observe that, in keeping with physical reasoning, *the domain of rigidity increases with* g, which is again confirmed by Figure 11.3 in which g is even larger.

Figure 11.4 presents a case where $f \neq 0$, the boundary conditions being zero. We note that in Figure 11.5, which refers to a time greater than that in Figure 11.4, the zone of rigidity decreases as t increases.

The *following conjecture* is supported by the figures and appears reasonable from a physical viewpoint: *let* u *be the solution of Problem I with* $f(t) = f$ *independent of* t *and with* $u_0 = 0$; *then the zone of rigidity decreases* (1) *when* t *increases, to "converge" towards the zone of rigidity of the steady problem.*

Figures 11.6, 11.7, and 11.8 show a case analogous to those of Figures 11.4 and 11.5, with a larger g, and hence more extensive zones of rigidity. The same is true of Figures 11.9, 11.10, 11.11.

Figures 11.12 and 11.13 represent graphs of the functions $t \to a(u_g(t))$ and $t \to J(u_g(t))$ for various values of g.

We note that $a(u_g(t))$ and $J(u_g(t))$ tend towards an equilibrium value, are increasing functions of time, and for fixed t, are increasing functions of g. It would be very interesting to obtain a proof of these facts, which appear very plausible from a physical point of view.

Further numerical results are given in D. Bégis, *loc. cit.*

12. DISCUSSION

The adaptation of the results of Raviart /1/ to the investigation of time-dependent inequalities was outlined in Lions /5/, and then extended and subjected to a detailed study in Trémolières /4/ and Viaud /1/; we also cite Trémolières /4/ for a more detailed investigation than that presented here for parabolic inequalities of the first type and for inequalities of the second order in t, and to Viaud /1/ regarding parabolic inequalities; this latter work also contains an investigation of the various "delay" inequalities (for the case of delay partial differential equations see M. Artola /1/, and for a study of the numerical approximation, Reverdy /1/). We also refer the reader to Bourgat /1/.

A numerical investigation of the inequalities governing the flow of a Bingham fluid (see Duvaut-Lions /1/, Cioranescu-Margolis /1/, /2/) has been undertaken by M. Fortin /1/ (we refer to the works of this author for a more complete investigation than that presented in Section 11 of the present chapter). Whilst the Bingham inequalities themselves contain the Navier-Stokes equations *as a special case*, it would not be possible for our investigation

---

(1) In the sense of the inclusion of the sets.

also to include as special cases all the (innumerable!) works relating to the numerical approximation of the Navier-Stokes equations; in this connection we refer to the work of Témam /2/ and the bibliography contained therein. Numerous problems remain to be solved in the numerical study of the flow of Bingham fluids. *For example*, the proof of the convergence of the approximate solutions u_h to the solution u has been outlined in the text; *does the "fluid-rigid" boundary for* u_h *similarly converge towards the free boundary for* u? It should be noted, moreover, that this type of (open) problem arises for *all* the problems of this chapter. An adaptation of the method due to Baiocchi /2/ to the parabolic case of the first type, for convex sets K of the type $\{v \mid v \leqslant \psi$ on $\Omega\}$, should be possible but has not been carried out. Also yet to be investigated are boundary layer phenomena (for the subject of singular perturbations in inequalities, which has not been considered here, we refer to Lions /8/).

We have not considered here the methods of splitting and decomposition which are introduced in Lions-Témam /1/ (see in particular Bensoussan-Lions-Témam /1/) and which constitute extensions of the classical methods of fractional steps to inequalities (see Yanenko /1/, Marchouk /1/ and the bibliographies therein).

The numerical approximation of various other problems of time-dependent inequalities is possible using the methods described in this chapter. Examples are given in Bégis /1/. For coupled systems, such as those arising in thermo-elasticity or magnetohydrodynamics (see Duvaut-Lions /3/) which involve "energy" methods, the results of this chapter can be extended. This is not the case for the problem of Bingham fluids whose viscosity depends on the temperature; see Duvaut-Lions /2/; for this problem, the convergence of a suitable approximation is open to question (the method used in Duvaut-Lions to demonstrate the existence of a solution is in fact not a constructive method). Other problems of coupled time-dependent inequalities are given in Lions /6/, in particular for problems of endo- or exothermic saturation; many questions remain unresolved with regard to this work; a partially heuristic numerical analysis of these problems is carried out in Marrocco /2/.

Other problems of time-dependent inequalities are encountered in problems of Stefan type (see G. Duvaut /1/) and similarly with regard to first-order hyperbolic operators ("Maxwell operators"; cf. the problems of insulation breakdown in antennas in Duvaut-Lions /1/, Chapter 7); this latter topic raises a general problem which we have not considered here; it would be interesting, in particular, to see to what extent it is possible to extend the results of Lesaint /1/ on first-order hyperbolic equations to "first-order hyperbolic inequalities".

Similarly we have the problem of the *optimal control* of systems governed by time-dependent inequalities. We refer to Kernevez /1/, Yvon /1/, Mignot /1/, Saguez /1/.

Questions of *error estimation* remain largely open; it would be very interesting to see to what extent the estimates of Douglas and Dupont /1/, and Raviart /2/ can be extended to inequalities.

Time-dependent inequalities with growth conditions at infinity are encountered in problems of optimal control and games theory (see Bensoussan-Lions /1/, A. Friedman /1/, /2/); their numerical investigation has not been attempted here. For the case of variational inequalities in which the convex set and/or the nondifferentiable functional depend on time, there is no special *numerical* difficulty; on the other hand there are serious theoretical difficulties when the convex sets depend "irregularly" on time; this is the case of the obstacle problem

$$\frac{\partial u}{\partial t} - \Delta u - f \geqslant 0, u - \psi \geqslant 0 ,$$

$$\left(\frac{\partial u}{\partial t} - \Delta u - f\right)(u - \psi) = 0 ,$$

$u = 0$ on the boundary, and for $t = 0$

where $\psi = \psi(x, t)$ is an "irregular" function of x and t.

In this case Mignot and Puel /1/ have proved the *existence of a (weak) minimum solution*, this minimum solution being the limit of penalised or semi-discretised solutions.

REFERENCES

ABADIE, J.(Ed.) /1/ *Nonlinear programming*, North Holland (1967).
/2/ *Integer and nonlinear programming*, North Holland (1970).

ABADIE, J., CARPENTIER, J. /1/ Generalisation of the Wolfe reduced gradient method to the case of nonlinear constraints. In *Optimisation*, R. Fletcher, Ed., Acad. Press, London and N.Y. (1969), pp. 37-49.

ABLOW, C.M., BRIGHAM, G. /1/ An analogue solution of programming problems. *Oper. Research*, Vol. 3 (1955), pp. 388-394.

ALTMAN, M. /1/ Generalised gradient methods of minimising a functional. *Bull. Acad. Pol. Sci.*, Vol. 14 (1966), pp. 313-318.

ANNIN, B.D. /1/ Existence and uniqueness of the solution of the elastic-plastic torsion of a cylindrical bar of oval cross-section. *J. of Appl. Math. Mech.* 29 (1965), pp. 1038-1047.

ANTOSIEWICZ, H.A., RHEINBOLDT, W.C. /1/ Conjugate direction methods and the method of steepest descent, in *A Survey of Numerical Analysis*, J.Todd, Ed., McGraw Hill, N.Y. (1962), pp. 501-512.

ARGYRIS, J.H. /1/ Elasto-plastic matrix displacement analysis of 3-D continua. *J. Royal Aero. Soc.*, 69 (1965), pp. 633-635.
/2/ Elasto-plastic analysis of three-dimensional media. *Acta Technica Academia Sci. Hungaricae*, vol. 54 (1966), pp. 219-238.
/3/ Continua and discontinua. *Proc. First Conference Matrix Methods in Struct. Mech.*, Wright-Patterson Air Force Base, Ohio, Oct. 1965.

ARGYRIS, J.H., BALMER, H., DOLTSINIS, J., WILLIAM, K.J. /1/ Finite element analysis of thermo-mechanical problems, *Third Conference Matrix Methods Struct. Mech.*, Wright-Patterson Air Force Base, Ohio, Oct. 1971.

ARGYRIS, J.H., SCHARPF, D.W. /1/ Methods of elasto-plastic analysis. *Proc. of the ISD-ISSC Symposium on Finite Element Techniques*. Institut für Statik und Dynamik der Luft und Raumfahrt-Konstruktione, Stüttgart (1969).

ARGYRIS, J.H., SCHARPF, D.W., SPOONER, J.B. /1/ The elasto-plastic calculation of general structures and continua. *Proc. 3rd Conference on Dimensioning*, Budapest (1969), pp. 345-384.
/2/ Die elasto-plastische Berechnung von allgemeinen Tragwerken und Kontinua. *Ingenieur Archiv.*, Vol. 37 (May 1969), pp. 326-352.

ARROW, K.J., HURWICZ, L., UZAWA, H. /1/ *Studies in linear and nonlinear programming*. Stanford University Press, Stanford, California (1958).

ARROW, K.J., HURWICZ, L. /1/ Gradient method for concave programming, in *Studies in linear and nonlinear programming*, Arrow, K.J., Hurwicz, L., Uzawa, H., Ed. Stanford University Press (1958), pp. 117-132.

ARTOLA, M. /1/ Sur les perturbations des équations d'évolution. Applications à des problèmes de retard. *Annales E.N.S.*, 2 (1969), pp. 137-253.

ATTOUCH, H., DAMLAMIAN, A. /1/ Problèmes d'évolution dans les Hilbert et Applications. *J. Math. Pures Appli.*, 54 (1975), pp. 53-74.

AUBIN, J.P., /1/ Approximation of variational inequalities, in *Functional Analysis and Optimisation*. Caianiello, Ed., Acad. Press (1966), pp. 7-14.

/2/ *Approximation of elliptic boundary value problems*. Wiley (1972).

AUSLENDER, A. /1/ Une méthode générale pour la décomposition et la minimisation de fonctions non différentiables. *C.R. Acad. Sci.*, Paris, vol. 271 (1970), pp. 1078-1081.

/2/ Méthodes numériques pour la décomposition et la minimisation de fonctions non différentiables. *Numer. Math.*, 18 (1972), pp. 213-223.

/3/ Résolution numérique d'inégalités variationnelles. *C.R. Acad. Sci.*, vol. 276 (1973), pp. 1063-1066.

AUSLENDER, A., GOURGAND, M., GUILLET, A. /1/ Résolution numérique d'inégalités variationnelles. *C.R. Acad. Sci.*, Paris, vol. 279 (1974), pp. 341-344.

BAIOCCHI, C. /1/ Su un problema di frontiera libera connesso a questioni di idraulica. *Annali Mat. Pura ed Appli.*, 92 (1972), pp. 107-127.

/2/ Estimations d'erreurs dans L^∞ pour les inéquations à obstacle. In *Mathematical Aspects of Finite Element Methods*, I. Galligani, E. Magenes, Eds., Lecture Notes in Math., 606, Springer-Verlag, Berlin, (1977), pp. 27-34.

BAIOCCHI, C., COMINCIOLI, V., MAGENES, E., POZZI, G.A. /1/ Free boundary problems in the theory of fluid flow through porous media. Existence and uniqueness theorems. *Annali di Mat. Pura ed Appli.*, XCVII (1973), pp. 1-82.

BALAKRISHNAN, A.V., NEUSTADT, L.W. (Ed.) /1/ *Computing methods in optimisation problems*. Acad. Press, New York (1964).

BAYADA, G. /1/ *Inéquations variationnelles elliptiques avec conditions aux limites périodiques. Application à la résolution de l'équation de Reynold*. Third-Cycle Thesis, Université de Lyon (1972).

BECKMAN, F.S. /1/ The solution of linear equations by the conjugate gradient method. In *Mathematical Methods for Digital Computers*, Ralston, A. and Wilf, H.S., Ed., Wiley (1960), pp. 62-72.

BEGIS, D. /1/ *Analyse numérique de l'écoulement d'un fluide de Bingham.* Third-Cycle Thesis, Université de Paris VI (1972).

/2/ *Résolution numérique de problèmes de convection et de transfert de chaleur.* LABORIA Report No. 15 (1973).

/3/ *Etude numérique du comportement d'un fluide de Bingham.* LABORIA Report No. 42 (1973).

BEGIS, D., GLOWINSKI, R. /1/ Dual numerical techniques for some variational problems involving biharmonic operators, in *Techniques of Optimisation.* A.V. Balakrishnan, Ed., Acad. Press, New York (1972), pp. 159-174.

BENSOUSSAN, A., LIONS, J.L. /1/ Problèmes de temps d'arrêt optimal et inéquations variationnelles paraboliques. *Journal of Applicable Analysis*, Vol. 3 (1973), pp. 267-294.

/2/ Book in preparation on "*Quasi-variational inequalities*".

BENSOUSSAN, A., GOURSAT, M., LIONS, J.L. /1/ Contrôle impulsionnel et inéquations quasi variationnelles stationnaires. *C.R. Acad. Sci.*, Paris (9 May 1973), vol. 276, Série A, pp. 1279-1284.

BENSOUSSAN, A., LIONS, J.L., TEMAM, R. /1/ Sur les méthodes de décomposition, de décentralisation et de coordinations et applications. *Cahier de l'IRIA*, No. 11 (1972), pp. 5-189.

BERGE, C. /1/ *Espaces topologiques*, Dunod, Paris (1959).

BIDAUT, M.J. /1/ *Théorème d'existence et d'existence "en général" pour des systèmes régis par des équations aux dérivées partielles non linéaires.* Thesis, Université de Paris VI (1973).

BLUM, E.K. /1/ A convergent gradient procedure in pre-Hilbert spaces. *Pacific J. of Mathematics*, Vol. 18, No. 1 (1966).

BORSENBERGER, A. /1/ Internal IRIA report (1973).

BOSSAVIT, A. /1/ Une méthode de décomposition de l'opérateur biharmonique. Note H.I. 585/2, Electricité de France (1971).

BOURGAT, J.F. /1/ *Analyse numérique du problème de la torsion élasto-plastique.* Third-Cycle Thesis, Université de Paris VI (1971).

/2/ Numerical study of a dual iterative method for solving a finite element approximation of the biharmonic equation. *Computer Meth. Appl. Mech. Eng.*, 9 (1976), pp. 203-218.

BOURGAT, J.F., DUVAUT, G. /1/ Numerical analysis of flow with or without wake past a symmetric two-dimensional profile with or without incidence. *Int. J. Numer. Meth. Eng.*, 11 (1977), pp. 975-993.

BOURGAT, J.F., TRÉMOLIÈRES, R. /1/ Analyse numérique du refroidissement d'un cylindre radio-actif. LABORIA Report No. 17 (1973).

BOX, M.J., DAVIES, O., SWANN, W.H. /1/ *Nonlinear optimisation techniques.* I.C.I. Monograph, Oliver and Boyd, Edinburgh (1969).

BRAM, J., SAATY, T.C. /1/ *Nonlinear Mathematics.* McGraw Hill (1964).

BRAMBLE, J.H., SCHATZ, A.M. /1/ Rayleigh-Ritz-Galerkin methods for Dirichlet's problem using subspaces without boundary conditions. *Comm. Pure and Applied Math.*, 23 (1970), pp. 115-175.

BREZIS, H. /1/ Equations et inéquations non linéaires dans les espaces vectoriels en dualité. *Annales de l'Institut Fourier*, 18 (1968), pp. 115-175.

/2/ Problèmes unilatéraux. *J. de Math. Pures et Appliquées*, 51 (1972), pp. 1-168.

/3/ *Opérateurs maximaux monotones et semi-groupes de contraction dans les espaces de Hilbert.* North-Holland/Elsevier (1973).

/4/ Un problème d'évolution avec contraintes unilatérale dépendant du temps. *C.R. Acad. Sci.*, Paris, vol. 274 (1972), pp. 310-312.

/5/ Multiplicateur de Lagrange en torsion élasto-plastique. *Archive Rat. Mech. Anal.*, 49 (1972), pp. 32-40.

/6/ Monotonicity methods in Hilbert spaces and some applications to nonlinear partial differential equations, in *Contributions to Nonlinear Functional Analysis.* E. Zarantonello, Ed., Acad. Press (1971), pp. 101-156.

BREZIS, H., DUVAUT, G. /1/ Ecoulement avec sillage autour d'un profil symétrique sans incidence. *C.R. Acad. Sci.*, vol. 276 (1973), pp. 875-878.

BREZIS, H., LIONS, J.L. /1/ Sur certains problèmes unilatéraux hyperboliques. *C.R. Acad. Sci.*, Paris, vol. 264 (1967), pp. 928-931.

BREZIS, H., SIBONY, M. /1/ Equivalence de deux inéquations variationnelles et applications. *Archive Rat. Mech. Anal.*, 41 (1971), pp. 254-265.

/2/ Méthodes d'approximation et d'itération pour les opérateurs monotones. *Archive Rat. Mech. Anal.*, 28 (1968), pp. 59-82.

BREZIS, H., STAMPACCHIA, G. /1/ Une nouvelle méthode pour l'étude d'écoulements stationnaires. *C.R. Acad. Sci.*, Paris, vol. 276 (1973), pp. 129-132.

/2/ Sur la régularité de la solution d'inéquations elliptiques. *Bull. Soc. Mathématique de France*, 96 (1968), pp. 153-180.

BROWDER, F.E. /1/ The fixed-point theory of multi-valued problems. *Math. Annalen*, 1977 (1968), pp. 283-301.

BUI-TRONG-LIEU, HUARD, P. /1/ La méthode des centres dans un espace topologique. *Num. Math.*, 8 (1966), pp. 56-67.

CANON, M., CULLUM, C., POLAK, E. /1/ Constrained minimisation problems in finite dimensional spaces. *SIAM J. on Control*, Vol. 4, No. 3 (1966), pp. 528-547.

CAPURSO, M. /1/ Principi di minimo per la soluzione incrementale dei problemi elasto-plastici. *Rendi Conti Acad. Naz. Lincei*, (1969), XLVI; I, pp. 417-425; II, pp. 552-560.

/2/ On the incremental solution of elasto-plastic continua in the range of large displacements. *Meccanica*, Vol. V, 1 (1970), pp. 1-9.

/3/ A general method for the incremental solution of elasto-plastic problems. *Meccanica*, IV (1969), 4, pp. 267-280.

CAPURSO, M., MAIER, G. /1/ Incremental elasto-plastic analysis and quadratic optimisation. *Meccanica*, V (1970), pp. 1-10.

CAROLL, C.W. /1/ *An operator research approach to the economic optimisation of a kraft pulping process*, Ph.D. Dissertation, Institute of Paper Chemistry, Appleton, Wis. (1959).

CAUCHY, A. /1/ Méthodes générales pour la résolution des systèmes d'équations simultanées. *C.R. Acad. Sci.*, Paris, Vol. 25 (1847), pp. 536-538.

CEA, J. /1/ Private communication.

/2/ *Optimisation. Théorie et algorithmes.* Dunod, Paris (1971).

/3/ Approximation variationnelle des problèmes aux limites. *Annales de l'Institut Fourier*, 14 (1964), pp. 345-444.

CEA, J., GLOWINSKI, R. /1/ Méthodes numériques pour l'écoulement laminaire d'un fluide rigide visco-plastique incompressible. *Int. J. of Comp. Math.*, B, Vol. 3 (1972), pp. 225-255.

/2/ Sur des méthodes d'optimisation par relaxation. *Revue Française d'Automatique, Informatique, Recherche Opérationnelle* (Dec. 1973), R-3, pp. 5-32.

CEA, J., GLOWINSKI, R., NEDELEC, J.C. /1/ Application des méthodes d'optimisation, de différences et d'éléments finis à l'analyse numérique de la torsion élasto-plastique d'une barre cylindrique, in *Approximation et méthodes itératives de résolution d'inéquations variationnelles et de problèmes non linéaires.* Cahier de l'IRIA, No. 12 (1974), pp. 7-138.

/2/ Minimisation de fonctionnelles non différentiables, in *Conference on Applications of Numerical Analysis.* Morris Ed., Lecture Notes in Math., B 228, Springer-Verlag (1971), pp. 19-38.

CIARLET, P.G., GLOWINSKI, R. /1/ Sur la résolution numérique du problème de Dirichlet pour l'opérateur biharmonique. *C.R. Acad. Sci.*, Paris, Série A, vol. 279 (1974), pp. 239-241.

/2/ Dual iterative techniques for solving a finite element approximation of the biharmonic equation. *Comp. Meth. Appl. Mech. Eng.*, 5 (1975), pp. 277-295.

CIARLET, P.G., RAVIART, P.A. /1/ General Lagrange and Hermite interpolation in $\mathbf{R}^n$ with applications to finite element methods. *Archive Rat. Mech. Anal.* 46 (1972), pp. 177-179.

/2/ Interpolation theory over curved element with applications to finite element methods. *Comp. Meth. in Appli. Mech. and Eng.* 1 (1972), pp. 217-249.

/3/ A mixed finite element method for the biharmonic equation, in *Mathematical aspects of finite elements in partial differential equations.* C. de Boor, Ed., Acad. Press, New York (1974), pp. 125-145.

CIARLET, P.G., WAGSCHAL, C. /1/ Multipoint Taylor formulas and applications to the finite element method. *Num. Math.*, 17 (1971), pp. 84-100.

CIORANESCU, D., MARGOLIS, B. /1/ Ecoulement d'un fluide de Bingham dans un domaine non-cylindrique avec singularité, *J. Functional Analysis*, 13 (1973), 2, pp. 125-131.

COMINCIOLI, V. /1/ Metodi di rilassamento per la minimizzazione in uno spazio prodotto. *L.A.N.*, *C.N.R.*, 20 (1971), Pavia.

COMINCIOLI, V., GUERRI, L., VOLPI, G. /1/ Analisi numerica di un problema di frontiera libera connesso col moto di un fluido attraverso un mezzo poroso. *Publication No. 17 of the Numerical Analysis Laboratory of the University of Pavia* (1971).

COURANT, R. /1/ Variational method for the solution of problems of equilibrium and vibrations. *Bull. Amer. Math. Society*, Vol. 49 (1943) pp. 1-23.

CROCKETT, J.B., CHERNOFF, H. /1/ Gradient methods of maximisation. *Pacific J. of Math.*, Vol. 5 (1955), pp. 33-50.

CROUZEIX, M. /1/ Etude d'une méthode de linéarisation. Résolution numérique des équations de Stokes stationnaires. Applications aux équations de Navier-Stokes stationnaires, in *Approximations et méthodes itératives de résolution d'inéquations variationnelles et de problèmes non linéaires*. Cahier de l'IRIA, No. 12, pp. 139-244 (1974).

CROUZEIX, M., RAVIART, P.A. /1/ Conforming and nonconforming finite element methods for solving the stationary Stokes equations. I. *Revue Française d'Automatique, d'Informatique et de Recherche Opérationnelle*, R-3 (1973), pp. 33-76.

CRYER, C.W. /1/ The solution of a quadratic programming problem using systematic over-relaxation. *SIAM J. on Control*, Vol. 9, No. 3 (1971), pp. 385-392.

/2/ The method of Christopherson for solving free-boundary problems. *Math. Comp.*, 25 (1971), pp. 435-443.

CURRY, B. /1/ The method of steepest descent for nonlinear minimisation problems. *Quart. of Applied Math.*, Vol. 2 (1944), pp. 258-261.

DANIEL, J.W. /1/ The conjugate gradient method for linear and nonlinear operator equations. *SIAM. J. Num. Analysis*, Vol. 4 (1967), pp. 10-26.

/2/ *The approximate minimisation of functionals.* Prentice Hall (1970).

DAVIDON, W.D. /1/ Variable metric method for minimisation. *A.G.G. Res. and Dev. Report.* ANL-5990 (1959).

DEMYANOV, V.F. /1/ Minimisation of functions on convex bounded sets. *Kibernetika*, Vol. 1, No. 6 (1965), pp. 65-74.

DOUGLAS, J., DUPONT, T. /1/ Galerkin methods for parabolic equations. *SIAM J. Num. Analysis* (1970), pp. 575-626.

DURAND, E. /1/ *Solutions numériques des équations algébriques.* Volumes 1 and 2, Masson, Paris (1960).

DUVAUT, G. /1/ Résolution d'un problème de Stefan. *C.R. Ac. Sci.* Paris, Série A, Vol. 276 (1973), pp. 1461-1463.

DUVAUT, G., LIONS, J.L. /1/ *Les inéquations en mécanique et en physique.* Dunod, Paris (1972).

/2/ Transfert de chaleur dans un fluide de Bingham dont la viscosité dépend de la température. *J. of Funct. Analysis*, 10 (1972).

/3/ Inéquations en thermo-élasticité et magnéto-hydrodynamique. *Archive Rat. Mech. Analysis*, 46 (1972), pp. 241-279.

/4/ Problèmes unilatéraux dans la théorie de la flexion forte des plaques. (I) Le cas stationnaire, *Journal de mécanique*, Vol. B, No. 1 (March 1974).

/5/ Problèmes unilatéraux dans la théorie de la flexion forte des plaques. (II) Le cas d'évolution, *Journal de mécanique*, Vol. B, No. 1 (June 1974).

EKELAND, I. /1/ Sur le contrôle optimal de systèmes gouvernés par des équations elliptiques. *J. of Funct. Analysis*, 9, No. 1 (1972), pp. 1-62.

EKELAND, I., TEMAM, R. /1/ *Analyse convexe et problèmes variationnels*. Dunod, Gauthier-Villars, Paris (1974).

ELKIN, R.M. /1/ *Convergence theorems for Gauss-Seidel and other minimisation algorithms*. Technical Report 68-59, University of Maryland, Comp. Sci. Center, College Park, Maryland (1968).

EHRLICH, L.W. /1/ Solving the biharmonic equation as coupled finite difference equations. *SIAM J. Num. Anal.*, 8 (1971), pp. 278-287.

FALK, R.S. /1/ *Approximate solutions of some variational inequalities with order of convergence estimates*. Ph. D. Thesis, Cornell University (1971).

/2/ Error estimates for the approximation of a class of variational inequalities. *Math. of Comp.*, Vol. 28, 128 (1974), pp. 963-971.

FENCHEL, W. /1/ On conjugate convex functions. *Canadian J. of Math.*, 1 (1949), pp. 73-77.

FIACCO, A., McCORMICK, G. /1/ *Nonlinear programming: unconstrained minimisation techniques*. Wiley, New York (1968).

FICHERA, G. /1/ Problemi elastostatici con vincoli unilaterali; il problema di Signorini con ambigue condizioni al contorno. *Mem. Accad. Naz. dei Lincei*, 8 (7) (1964), pp. 91-140.

FLETCHER, R. /1/ *Optimisation*. Academic Press, London (1969).

FLETCHER, R., POWELL, M. /1/ A rapidly convergent method for minimisation. *Comp. J.*, Vol. 6 (1963), pp. 163-168.

FLETCHER, R., REEVES, C.M. /1/ Functional minimisation by conjugate gradients. *Comp. Journal*, Vol. 7 (1964), pp. 149-153.

FORTIN, M. /1/ *Calcul numérique des écoulements des fluides de Bingham et des fluides newtoniens incompressibles par la méthode des éléments finis*. Thesis, Université de Paris VI (1972).

FORTIN, M., GLOWINSKI, R., MARROCCO, A. /1/ To appear.

FRAIEJS DE VEUBEKE, B. /1/ Displacement and equilibrium models in finite element method. In *Stress Analysis*, Chap. 9, O.C. Zienckiewicz and G.S. Molistek, Ed., Wiley (1965).

FRANK, M., WOLFE, P. /1/ An algorithm for quadratic programming. *Naval Res. Log. Quart.*, 3 (1956), pp. 95-110.

FREMOND, M. /1/ Solide reposant sur un sol stratifié. *Bull. du Laboratoire des Ponts et Chaussées*, April 1972, pp. 85-95.

/2/ *Etude des structures visco-élastiques stratifiées soumises à des charges harmoniques et de solides élastiques reposant sur des structures.* Thesis, Université de Paris VI (1971).

FRIEDMAN, A. /1/ Stochastic games and variational inequalities. *Archive Rat. Mech. and Anal.*, Vol. 51, No. 5 (1973), pp. 321-346.

/2/ Regularity theorems for variational inequalities in unbounded domains and applications to stopping-time problems. *Archive Rat. Mech. and Anal.*, Vol. 52, 2 (1973), pp. 134-160.

FUSCIARDI, A., MOSCO, U., SCARPINI, F., SCHIAFFINO, A. /1/ A dual method for the numerical solution of some variational inequalities. *J. Math. Anal. Appl.*, 40 (1972), pp. 471-493.

GABAY, D., MERCIER, B. /1/ A dual algorithm for the solution of nonlinear variational problems via finite element approximation. *Comp. and Math. with Applic.*, Vol. 2 (1976), pp. 17-40.

GLOWINSKI, R. /1/ Méthodes numériques pour l'écoulement d'un fluide rigide visco-plastique incompressible, in *Proc. of the 2nd Int. Conference on Numerical Methods in Fluid Dynamics*, M. Holt, Ed., Lecture Notes in Physics, Vol. 8, Springer Verlag (1971), pp. 385-394.

/2/ Sur l'écoulement d'un fluide de Bingham dans une conduite cylindrique. *J. de Mécanique*, Vol. 13, No. 4, Dec. 1973, pp. 601-621.

/3/ Méthodes itératives duales pour la minimisation de fonctionnelles convexes, in *Constructive aspects of functional Analysis*, G. Geymonat, Ed., Cremonese, Rome (1973), pp. 263-290.

/4/ *La méthode de relaxation.* Rendiconti di Matematica, 14 Università di Roma (1971).

/5/ Sur la minimisation par surrelaxation avec projection, de fonctionnelles quadratiques dans les espaces de Hilbert, *C.R. Acad. Sci.*, Paris, Vol. 276 (1973), pp. 1421-1423.

/6/ Approximations externes, par éléments finis de Lagrange d'ordre un et deux, du problème de Dirichlet pour l'opérateur biharmonique. Méthodes itératives de résolution des problèmes approchés, in *Topics in Numerical Analysis*, J.J.H. Miller, Ed., Acad. Press, London (1973), pp. 123-171.

/7/ *Numerical methods for some nonlinear elliptic boundary value problems.* Notes du Séminaire de Mathématiques supérieures de l'Université de Montréal, été 1975 (to appear)

/8/ Internal IRIA report (1975) (see also Sec. 5.2.5 of Glowinski, R., Pironneau, O. /1A/).

/9/ Introduction to the approximation of elliptic variational inequalities. Report 76006, *Laboratoire d'Analyse Numérique*, LA 189, *Univ. Paris* VI (1976).

GLOWINSKI, R., LANCHON, H. /1/ Torsion élasto-plastique d'une barre cylindrique de section multi-connexe. *J. de Mécanique*, 12 (1973), pp. 151-171.

GLOWINSKI, R., MARROCCO, A. /1/ Sur l'approximation par éléments finis d'ordre un et la résolution par pénalisation-dualité d'une classe de problèmes de Dirichlet non linéaires. *C.R. Acad. Sci.*, Paris, Vol. 178, Série A (June 1974), pp. 1649-1652.

/2/ Sur l'approximation par éléments finis d'ordre un et la résolution par pénalisation-dualité d'une classe de problèmes de Dirichlet non linéaires. *Revue Francaise d'Automatique, d'Informatique et de Recherche Opérationnelle*, R-2 (1975), pp. 41-76.

GODUNOV, S.K., PROKOPOV, G.L. /1/ Variational approach to the solution of large linear systems arising from strongly elliptic problems. *Publication of the Appl. Math. Inst. of the Academy of Sciences of the U.S.S.R.*, Moscow (1968) (in Russian).

GOLDFARB, D. /1/ Extension of Davidon's variable metric method to maximisation under linear inequality and equality constraints. *SIAM J. Appl. Math.*, Vol. 17, No. 4 (1969), pp. 739-764.

GOLDFARB, D., LAPIDUS, L. /1/ Conjugate gradient method for nonlinear programming problems with linear constraints. *I. and E.C. Fundamental*, Vol. 7 (1968), pp. 142-151.

GOLDSTEIN, A.A. /1/ Cauchy's method of minimisation. *Num. Math.*, 4 (1962), pp. 146-150.

/2/ Convex programming in Hilbert spaces. *Bull. Amer. Math. Society*, Vol. 70 (1964), pp. 709-710.

GOURSAT, M. /1/ Internal IRIA report (1975).

/2/ *Analyse numérique de problèmes d'élasto-plasticité et visco-plasticitié.* Third-Cycle Thesis, Université de Paris VI (1971).

GRAVES, R.L., WOLFE, P. (Ed.) /1/ *Recent Advances in Mathematical Programming.* McGraw Hill, New York (1963).

GREENSTADT, J.L. /1/ On the relative efficiencies of gradient methods. *Math. of Computation*, Vol. 21 (1967), pp. 360-367.

HADLEY, G. /1/ *Nonlinear and dynamic programming.* Addison-Wiley (1964).

HAUGAZEAU, Y. /1/ Sur les inéquations variationnelles. *C.R. Acade. Sci.*, Paris (1967).

/2/ Thèse d'Etat, Paris (1968).

HERAKOVITCH, C.T. and HODGE, P.G. /1/ Elastic-plastic torsion of hollow bars by quadratic programming. *Int. Journal Mech. Sciences*, Pergamon Press (1969), 11, pp. 53-63.

HESTENES, M.R. /1/ The conjugate gradient method for solving linear systems. *Proc. Symposium in Applied Math.*, Vol. 6, Numerical Analysis, pp. 83-102, McGraw Hill, New York (1956).

/2/ Multiplier and gradient methods, *J.O.T.A.*, 4, No. 5 (1969), pp. 303-320.

HESTENES, M.R., STIEFEL, E. /1/ Methods of conjugate gradient for solving linear systems. *J. Res. Nat. Bureau of Standards*, Vol. 49 (1952), pp. 409-436.

HODGE, P.G., HERAKOVITCH, C.T., STOUT, R.E. /1/ On numerical comparisons in elastic-plastic torsion. *Transactions of the A.S.M.E.* (1962), pp. 454-459.

HÖRMANDER, L. /1/ Sur la fonction d'appui des ensembles convexes dans un espace localement convexe. *Arkiv für Mat.*, 3 (1954), pp. 181-186.

HUARD, P. /1/ Resolution of mathematical programming with nonlinear constraints by the method of centers. *Nonlinear programming.* Abadie, J., Ed., North Holland (1967).

HUET, D. /1/ Perturbations singulières d'inéquations variationnelles, *C.R. Acad. Sci.*, Paris, 276A, (1968), pp. 932-934.

JOLY, J.C. /1/ Thesis, Faculté des Sciences de Grenoble (1970).

KALKER, J.J. /1/ Variational Principles of Contact Elastostatics, *J. Inst. Math. Appl.*, 20 (1977), pp. 199-219.

/2/ A minimum principle for the law of dry friction with application to elastic cylinders in rolling contact. *J. Appl. Mech.*, 38, E (1971), pp. 875-880 and 881-887.

KERNEVEZ, J.P. /1/ Systèmes bienzymatiques structurés et systèmes gouvernés par des inéquations variationnelles, *Report No.* 4, *Dpt. Bio-Math.*, Université Technologique de Compiègne (1974).

KOITER, W.I. /1/ General theorems for elastic plastic solids. *Progress in solid Mechanics*, Vol. 2, North-Holland, P.C. (1960), pp. 165-221.

KOWALIK, J., OSBORNE, M.R. /1/ *Methods for unconstrained optimisation problems*, Elsevier, New York (1968).

KUHN, H.W., TUCKER, A.W. /1/ *Nonlinear programming*, *Proc. Second Berkeley Symposium on Math. Stat. and Prob.*, University of California Press (1951), pp. 481-492.

KUNZI, M.P., KRELLE, W., OETTLI, W. /1/ *Nonlinear programming*. Blaisdell (1966).

KUNZI, H.P., TZSCHACH, H.G., ZEHNDER, C.A. /1/ *Numerical Methods of Mathematical Optimisation*. Acad. Press (1968).

KY-FAN /1/ Sur un théorème minimax. *C.R. Acad. Sci.*, Paris, vol. 259 (1964), pp. 3925-3928.

LANCHON, H. /1/ Solution du problème de torsion élasto-plastique d'une barre cylindrique de section quelconque. *C.R. Acad. Sci.*, Paris, vol. 269 (1969), pp. 791-794.

/2/ *Torsion élasto-plastique d'un arbre cylindrique de section simplement ou multiplement connexe*. Thesis, Université de Paris VI, 1972.

LASDON, L.S. /1/ *Optimisation theory for large systems*. McMillan Book Co., London, 1970.

LAURENT, P.J. /1/ *Approximation et optimisation*. Hermann, Paris, 1972.

LAVAINE, F. /1/ Méthodes de décomposition rapidement convergentes pour des inéquations variationnelles. Third-Cycle Thesis, Université de Paris-Sud, 1975.

LEMONNIER, P. /1/ Internal IRIA report (1971).

LERAY, J., LIONS, J.L. /1/ Quelques résultats de Visik sur les problèmes elliptiques non linéaires par les méthodes de Minty-Browder. *Bull. Soc. Math. de France*, 93 (1965), pp. 97-107.

LEROY, D. /1/ Internal IRIA report (1971).

LESAINT, P. /1/ Finite element methods for symmetric hyperbolic equations. *Num. Math.*, 21 (1973), pp. 244-255.

LEVITIN, E.S., POLYAK, B.T. /1/ Constrained minimisation methods. *U.S.S.R. Comput. Math. Math-Physics*, Vol. 6 (1966), pp. 1-50.

/2/ Convergence of minimising sequences in conditional extremum problems. *Soviet Math. Dokl.*, Vol. 7 (1966), pp. 764-767.

LEWY, H., STAMPACCHIA, G. /1/ On the regularity of the solution of variational inequality. *Comm. Pure Appl. Math.*, 22 (1969), pp. 153-188.

/2/ On the smoothness of superharmonics which solve a minimum problem. *Journal d'analyse mathématique*, 23 (1970), pp. 227-236.

LIONS, J.L. /1/ *Quelques méthodes de résolution des problèmes aux limites non linéaires.* Dunod, Gauthier-Villars, Paris, 1969.

/2/ *Contrôle optimal des systèmes gouvernés par des équations aux dérivées partielles.* Dunod, Gauthier-Villars, Paris, 1968.

/3/ *Cours d'analyse numérique.* Faculté des Sciences, Paris, 1961.

/4/ Sur un nouveau type de problème non linéaire pour opérateurs hyperboliques du deuxième ordre. *Séminaire J. Leray*, Collège de France, 1965-1966, II, pp. 17-33.

/5/ Sur l'approximation de la solution d'inéquations d'évolution. *C.R. Acad. Sci.*, Paris, vol. 263 (1966), pp. 55-57.

/6/ Sur l'analyse numérique de problèmes d'inéquations couplées. Rome, *Istituto di Alta Matematica* (1972).

/7/ Approximation numérique des inéquations d'évolution, in *Constructive Aspects of Functional Analysis*, C.I.M.E., 1971, G. Geymonat, Ed., Cremonese (1973), pp. 295-361.

/8/ *Méthodes de perturbations singulières dans les problèmes aux limites et le contrôle optimal.* Lecture Notes in Math., Springer, B 323, 1973.

LIONS, J.L., MAGENES, E. /1/ *Problèmes aux limites non homogènes.* Vol. 1, Dunod, Paris (1968).

LIONS, J.L., STAMPACCHIA, G. /1/ Variational inequalities. *Comm. Pure Appl. Math.*, XX (1967), pp. 493-519.

LIONS, J.L., TEMAM, R. /1/ Une méthode d'éclatement des opérateurs et des contraintes en calcul des variations. *C.R. Acad. Sci.*, Paris, vol. 268 (1966), pp. 563-565.

LOOTSMA, F.A. /1/ Boundary properties of penalty functions for constrained minimisation. Thesis, Eindhoven, 1970.

LUENBERGER, D.G. /1/ *Optimisation by vector space methods.* Wiley, 1969.

/2/ The conjugate residual method for constrained minimisation problems. *SIAM J. Num. Anal.*, Vol. 7, No. 3 (1970).

MAIER, G. /1/ Quadratic programming and theory of elastic perfectly plastic structures. *Meccanica*, IV (1968), pp. 1-9.

/2/ Some theorems for plastic strain rates and plastic strains. *Journal de Mécanique*, 8 (1969), pp. 5-19.

MAGENES, E., STAMPACCHIA, G. /1/ I problemi al contorno per le equazioni differenziali di tipo ellittico. *Ann. Scuola Normale Superiore*, Pisa, 12 (1958), pp. 247-358.

MANDEL, J. /1/ *Cours de mécanique des milieux continus*. Vol. 1: *Mécanique des fluides*, Vol. 2: *Mécanique des solides*. Gauthier-Villars, Paris (1966).

MANDELBROJT, S. /1/ Sur les fonctions convexes. *C.R. Acad. Sci.*, Paris, vol. 209 (1939), pp. 977-978.

MANGASARIAN, O.L. /1/ *Nonlinear programming*. McGraw Hill, New York, 1969.

MARCHOUK, G.I. /1/ *Méthodes de mathématiques numériques*. Novosibirsk, 1972.

MARROCCO, A. /1/ *Application de la méthode de pénalisation à la résolution d'un problème d'élasto-plasticité*. Third-Cycle Thesis, Paris, 1970.

/2/ Résolution numérique de problèmes de saturation endo- ou exo-thermique. IRIA Report INF. 7203/72003, Jan. 1972.

MARZULLI /1/ Risoluzione alle differenze finite di equazioni alle derivate parzialli di tipo ellittico con condizioni sú un contorno libero. *Calcolo*, *Suppl*. 1, 5 (1968), pp. 1-22.

MERCIER, B. /1/ Approximation par éléments finis et résolution par pénalisation-dualité d'un problème d'élasto-plasticité. *C.R. Acad. Sci.*, Paris, vol. 280, Série A (1975), pp. 287-290.

/2/ Internal IRIA report (1971).

MEZLYAKOV /1/ On a relaxation method for solving systems of linear inequalities. *J. of Math. and Phys.*, Vol. 2, No. 3 (1962).

MIELLOU, J.C. /1/ Méthodes de Jacobi, Gauss-Seidel, sous/sur-relaxation par blocs appliquées à une classe de problèmes non linéaires. *C.R. Acad. Sci.*, Paris, Vol. 273, Série A (1971), pp. 1257-1259.

MIGNOT, F., PUEL, J.P. /1/ Inéquations d'évolution paraboliques avec convexe dépendant du temps: application aux I.Q.V. *Archiv. Rat. Mech. Anal.* (1976).

MINTY, G. /1/ Monotone (nonlinear) operators in Hilbert space. *Duke Math. J.*, 29 (1962), pp. 341-346.

MOREAU, J.J. /1/ Fonctionnelles convexes. *Séminaire J. Leray*, Collège de France, 1966.

/2/ Fonctions de résistance et fonctions de dissipation. *Séminaire d'analyse convexe*, Montpellier (1971), No. 6.

/3/ Sur l'évolution d'un système élasto-visco-plastique. *C.R. Acad. Sci.*, Paris, vol. 273 (1971), pp. 118-121.

/4/ Rafle par un convexe variable. *Séminaire d'analyse convexe*, Montpellier (1971).

MORICE, P. /1/ Calcul parallèle et décomposition dans la résolution d'équations aux derivées partielles de type elliptique. *IRIA Report* INF. 7214/72017 (1972).

MOSCO, U. /1/ Dual variational inequalities. *J. Math. Anal. and Appl.*, 39 (1972), pp. 202-206.

/2/ Convergence of convex sets and of solutions of variational inequalities. *Advances in Math.*, 3 (4) (1969), pp. 510-585.

MOSCO, U., STRANG, G. /1/ One sided approximations and variational inequalities. *Bull. A.M.S.*, 80 (1974), pp. 308-312.

MOSSOLOV, P.P., MIASNIKOV, V.P. /1/ Variational methods in the theory of fluidity of a viscous-plastic medium. *P.M.M.*, 29 (1965), pp. 468-492.

MOTZKIN, SCHOENBERG /1/ The relaxation method for linear inequalities. *Canadian J. of Math.*, Vol. 6, 1954.

NAYAK, G., ZIENKIEWICZ, O.C. /1/ Note on the "alpha"-constant stiffness method for the analysis of nonlinear problems. *Int. J. of Num. Meth. Eng.*, 4 (1972), pp. 579-582.

NECAS, J. /1/ *Les méthodes directes dans la théorie des équations aux dérivées partielles.* Masson, Paris, 1967.

ORTEGA, J., RHEINBOLDT, W.C. /1/ Monotone iterations for nonlinear equations with applications to Gauss-Seidel methods. *SIAM J. Num. Anal.*, Vol. 4 (1967), pp. 171-190.

/2/ On a class of approximate iterative processes. *Archive Rat. Mech. Anal.*, Vol. 3 (1967), pp. 352-356.

/3/ *Iterative solution of nonlinear equations in several variables.* Acad. Press, 1970.

ORTEGA, J., ROCKOFF, M. /1/ Nonlinear difference equation and Gauss-Seidel type iterative methods. *SIAM J. Num. Anal.*, Vol. 3, No. 3 (1966), pp. 497-513.

PEARSON, J.D. /1/ On variable metric methods of minimisation. *Comput. J.*, Vol. 12 (1969), pp. 171-178.

PERALBA, J.C. /1/ Un problème d'évolution relatif à un opérateur sous-differérentiel dépendant du temps. *Séminaire d'Analyse convexe*, Montpellier, 1972.

PETRISCHYN, W.V. /1/ On the generalised over-relaxation method for operator equations. *Proc. Amer. Math. Society*, 14 (1963), pp. 917-921.

/2/ Operator equations in Hilbert space. *J. Math. Anal. Appl.*, Vol. 29, 3 (1970), pp. 558-568.

POLAK, E. /1/ *Computational methods in optimisation.* Academic Press, 1971.

POLYAK, B.T. /1/ Gradient methods for the minimisation of functionals. *Zh. fych. mat.*, 3, No. 4 (1963), pp. 643-653.

/2/ A general method for solving extremum problems. *Soviet Math. Dokl.*, Vol. 8, No. 3 (1967), pp. 593-597.

/3/ Convergence of minimising sequences in conditional extremum problems. *Doklady*, Vol. 168, No. 5 (1966), pp. 764-766.

/4/ Existence theorems and convergence of minimising sequences in extremum problems with restrictions. *Soviet Math. Dokl.*, Vol. 7 (1966), pp. 72-75.

POWELL, M.J.D. /1/ A method for nonlinear optimisation in minimisation problems, in *Optimisation*, R. Fletcher, Ed., Acad. Press, New York, 1969.

PRAGER, W. /1/ *Introduction to mechanics of continua*, Ginn and Co. (1961).

/2/ *Problèmes de plasticité théorique*, Dunod, 1958.

/3/ Plastic design and thermal stresses. *British Welding Journal* (1956), pp. 355-359.

/4/ On ideal locking materials. *Transaction of the Society of Rheology*, Vol. 1 (1957), pp. 169-175.

/5/ Unilateral constraints in mechanics of continuum. *Atti del Simp. Lagrangian Accad. Sc. Torino*, pp. 1-11.

/6/ Linearisation in visco-plasticity, *Osterreiches Ingenieur-Archive*, XV (1961), pp. 152-157.

/7/ Composite stress-strain relations for elasto-plastic solids. *Proc. IUTAM Symp.* Vienna (1966), pp. 315-325.

/8/ On elastic, perfectly locking materials. *Proc. Int. Conf. Applied Mech.* (1964), pp. 538-544.

PRAGER, W., HODGE, P.G. /1/ *Theory of perfectly plastic solids.* Wiley, 1961.

RAVIART, P.A. /1/ Sur l'approximation de certaines équations d'évolution linéaires et non linéaires. *J. Math. Pures et Appl.*, 46 (1967), pp. 11-183.

/2/ The use of numerical integration in finite element methods for solving parabolic equations, in *Topics in numerical analysis*, J.J.H. Miller, Ed., Acad. Press, London (1973), pp. 233-264.

REVERDY, J. /1/ Sur une méthode de Runge-Kutta pour des équations à retard. *Compt. Rend. Acad. Sci.* Paris, 286A, (1978), pp. 417-420.

ROCKAFELLAR, A. T. /1/ Convex functions, monotone operators and variational inequalities. In *NATO Course*, Venice (1968), Oderisi Ed. (1969), pp. 231-247.

/2/ Duality and stability in extremum problems involving convex functions. *Pacific J. Math.*, 21 (1962), pp. 167-187.

/3/ *Convex analysis*, Princeton University Press, Princeton, N.J. (1970).

/4/ The multiplier method of Hestenes and Powell applied to convex programming. *J. Opt. Th. and Appl.*, 12 (1973).

/5/ Augmented Lagrange multiplier functions and duality in non-convex programming. *SIAM J. on Control*, Vol. 12, No. 2 (1974), pp. 268-285.

ROSEN, J.B. /1/ The gradient projection method for nonlinear programming. Part I, linear constraints. *SIAM J.*, Vol. 8, No. 1 (1960), pp. 181-217.

/2/ The gradient projection method for nonlinear programming. Part II, nonlinear constraints. *SIAM J.*, Vol. 9, No. 4 (1961), pp. 514-532.

/3/ Approximate solution and error bounds for quasi-linear boundary value problems. *SIAM J. Num. Anal.*, 7 (1970), pp. 80-103.

SHAW, F.S. /1/ The torsion of solid and hollow prisms in the elastic and plastic range by relaxation methods. *Report ACA-11* (1944).

SCHECHTER, S. /1/ Iterations method for nonlinear problems. *Trans. Amer. Math. Society*, 104 (1962), pp. 601-612.

/2/ Minimisation of convex functions by relaxation, Ch. 7 of *Integer and nonlinear programming*, Abadie Ed., North-Holland (1970), pp. 177-189.

/3/ Relaxation method for convex problems. *SIAM J. of Num. Anal.*, Vol. 5 (1968), pp. 601-612.

SCHWARTZ, L. /1/ *Théorie des distributions.* Hermann, Paris (1967).

SIBONY, M. /1/ *Méthodes itératives pour les équations et inéquations aux dérivées partielles non linéaires de type monotone.* Calcolo 2 (1970), pp. 65-183.

SIGNORINI, A. /1/ Sopra alcune questioni di Elastostatica. *Atti della Società Italiana per il Progresso della Scienza*, 1933.

SION, M. /1/ Existence de cols pour les fonctions quasi convexes et semi-continues. *C.R. Acad. Sci.*, Paris, Vol. 244 (1954), pp. 2120-2123.

/2/ On general minimax theorems. *Pacific J. Math.*, 8 (1958), pp. 171-176.

SMITH, J. /1/ The coupled equation approach to the numerical solution of the biharmonic equation by finite differences, I. *SIAM J. Num. Anal.*, 5 (1968), pp. 323-339.

/2/ The coupled equation approach to the numerical solution of the biharmonic equation by finite differences, II. *SIAM J. Num. Anal.*, 7 (1970), pp. 104-111.

/3/ On the approximate solution of the first boundary value problem for $\nabla^4 u = f$. *SIAM J. Num. Anal.*, 10 (1973), pp. 967-982.

SOBOLEV, S.L. /1/ *Applications de l'analyse fonctionnelle aux équations de la physique mathématique.* Leningrad, 1950.

SOUTHWELL /1/ *Relaxation methods in theoretical physics*, Vol. 1, Clarendon Press, Oxford, 1946.

SPANG, H.A. /1/ A review of minimising techniques for nonlinear functions. *SIAM Review*, Vol. 4, No. 4 (1962), pp. 343-365.

STIEFEL, E. /1/ Recent developments in relaxation techniques. *Proceedings Int. Congress Appl. Math.*, Vol. 1 (1954), pp. 384-391.

STOUT, R.B., HODGE, P.G. /1/ Elastic-plastic torsion of hollow cylinders. *Int. J. Mech. Sci.*, 12 (1970), pp. 91-108.

STRANG, G., FIX, G. /1/ *An analysis of the finite element method.* Prentice Hall, New York, 1973.

TARTAR, L., TEMAM, R. /1/ Private communication.

TCHERNOUSLO, F.L., BANICHOUK, N.V. /1/ *Problèmes variationnels en mécanique et en contrôle*, Moscow, 1973.

TEMAM, R. /1/ Solutions généralisées de certaines équations du type hypersurfaces minima. *Archive Rat. Mech. Anal.*, 44 (1971), pp. 121-156.

/2/ *Navier-Stokes equations.* North-Holland, Amsterdam, 1977.

/3/ Sur la stabilité et la convergence de la méthode des pas fractionnaires. *Annali di Mat. Pura e Appl.*, 4, LXXIV (1968), pp. 191-380.

THOMASSET, T. /1/ Internal IRIA report, 1972.

/2/ *Etude d'une méthode d'éléments finis de degré cinq. Application aux problèmes de plaques et d'écoulement de fluides.* Third-Cycle Thesis, Université de Paris XI (1974).

TING, T.W. /1/ Elastic-plastic torsion problem III. *Archive Rat. Mech. Anal.*, 34 (1969), pp. 228-243.

/2/ Elastic-plastic torsion of convex cylindrical bars. *J. Math. Mech*, 19 (1969), pp. 531-551.

/3/ Elastic-plastic torsion of simply-connected cylindrical bars. *Indiana Univers. Math. J.*, Vol. 20, No. 11 (1971), pp. 1047-1076.

TRÉMOLIÈRES, R. /1/ *La méthode des centres à troncature variable.* Third-Cycle Thesis, Paris (1968).

/2/ Optimisation par la technique des séries divergentes. *Internal IRIA report*, (1969).

/3/ Optimisation non linéaire avec contraintes. *IRIA report* (1970).

/4/ *Inéquations variationnelles: existence, approximations, résolution.* Thesis, Université de Paris VI (1972).

UZAWA, H. /1/ Iterative methods for concave programming, in *Studies in linear and nonlinear programming*, Arrow, K.J., Hurwicz, L., Uzawa, H., Ed., Stanford University Press (1958), pp. 154-165.

VAINBERG, M.M. /1/ Le problème de la minimisation des fonctionnelles non linéaires, in *Problems in nonlinear analysis.* Proceedings of the C.I.M.E., G. Prodi, Ed., Crémonèse (1971), pp. 465-574.

VARGA, R.S. /1/ *Matrix iterative analysis.* Prentice Hall, Englewood Cliffs, New Jersey, 1962.

VARAYIA, P.P. /1/ Nonlinear programming in Banach spaces. *SIAM J. Applied Math.*, Vol. 15, No. 2, 1967.

/2/ *Notes on optimisation.* Van Nostrand, Rheinbold, 1972.

VIAUD, D. /1/ *Approximation numérique des problèmes d'asservissement et d'asservissement retardé.* Thesis, Université de Paris VI, 1973.

WOLFE, P. /1/ Methods of nonlinear programming, in *Recent advances in mathematical programming*, R.L. Graves and P. Wolfe, Ed., McGraw Hill, New York (1963), pp. 67-86.

YAMADA, Y. /1/ Recent developments in matrix displacement method for elastic problems in Japan, in *Recent Advances in Matrix Methods of Structural Analysis and Design*, R.H. Gallagher, Y. Yamada, J.T. Oden, Ed., University of Alabama Press (1971), pp. 283-316.

YANENKO, N.N. /1/ *Méthodes de pas fractionnaires*. A. Colin, 1968.

YVON, J.P. /1/ *Etude de quelques problèmes de contrôle pour des systèmes distribués*. Thesis, Université de Paris VI (1973).

/2/ Contrôle optimal de systèmes gouvernés par des inéquations variationnelles. *LABORIA Report* No. 53 (1974).

ZANGWILL, W.I. /1/ *Nonlinear programming: a unified approach*. Prentice Hall, 1969.

ZIENKIEWICZ, O.C. /1/ *The finite element method in structural and continuum mechanics*. McGraw Hill (1967).

/2/ *The finite element method in Engineering Science*. McGraw Hill (1978).

ZIENKIEWICZ, O.C., CORMEAU, J.C. /1/ Visco-plasticity solution by finite element process. *Archive of Mechanics*, 24 (1972), pp. 773-889, Warsaw.

ZIENKIEWICZ, O.C. VALLIAPAN, S. /1/ Analysis of real structures for creep plasticity and other complex constitutive laws, in *Structure, Solid Mechanics and Engineering Design*. McTe'eni, Ed., Wiley, Interscience (1971), pp. 27-48.

ZIENKIEWICZ, O.C., VALLIAPAN, S., KING, J.P. /1/ Elasto-plastic solutions of engineering problems, initial stress finite element approach. *Int. J. for Num. Methods in Engineering*, 1 (1969), pp. 75-100.

ZLAMAL /1/ On the finite element method. *Num. Math.*, 12 (1968), pp. 394-409.

/2/ On some finite element procedures for solving second order boundary value problems. *Num. Math.*, 14 (1969), pp. 42-48.

ZOUTENDIJK, G. /1/ Nonlinear programming: a numerical survey. *SIAM J. on Control*, Vol. 4, No. 1 (1966), pp. 194-210.

Appendix 1

FURTHER DISCUSSION OF STEADY-STATE INEQUALITIES

1. SYNOPSIS

In the following, we shall now consider in a rather more systematic manner the investigation of the *existence*, *uniqueness* and *approximation* properties of steady-state inequalities of the type

$$(1.1A)\quad \begin{cases} \textit{Find } u \in K \textit{ such that} \\ a(u,v-u) \geq L(v-u) \ \forall v \in K \end{cases}$$

and

$$(1.2A)\quad \begin{cases} \textit{Find } u \in V \textit{ such that} \\ a(u,v-u)+j(v)-j(u) \geq L(v-u) \quad \forall v \in V. \end{cases}$$

In Sections 3 and 4 we shall consider the approximation, using *conforming* and *nonconforming* finite elements, of the obstacle problem

$$(1.3A)\quad \begin{cases} \int_\Omega \nabla u \cdot \nabla (v-u)\,dx \geq \int_\Omega f(v-u)\,dx \quad \forall v \in K, \\ u \in K \end{cases}$$

with

$$(1.4A)\qquad K = \{v \in H^1(\Omega), \quad v \geq \psi \quad \text{a.e. on} \quad \Omega,\ v|_\Gamma = g\}\ .$$

In Section 5 we give a brief outline of the work of Bogomolnii-Eskin-Zuchowiskii relating to a stamp problem formulated as a variational inequality of type (1.1A), with the form $a(\cdot,\cdot)$ relating to a pseudo-differential operator. In Section 6 we give an example of nonlinear elliptic equations, the investigation of which can be reduced to that of variational inequalities of type

(1.2A). Finally, in Section 7, we give a brief introduction to the subject of *steady quasi-variational inequalities* (Q.V.I.).

2. EXISTENCE, UNIQUENESS AND APPROXIMATION RESULTS FOR PROBLEMS (1.1A) and (1.2A).

The results presented in this section were already known at the time when the French edition was being prepared, but they had not then been made explicit.

2.1 The functional setting

We shall assume throughout this section that

- V is a real Hilbert space with inner product $(\cdot,\cdot)$ and norm $\|\cdot\|$.

- $a : V \times V \to \mathbb{R}$ is a continuous coercive bilinear form (sometimes referred to as V-elliptic); there thus exists an $\alpha > 0$ such that

  $$a(v,v) \geq \alpha\|v\|^2 \quad \forall v \in V.$$

  It is not assumed a priori that $a(\cdot,\cdot)$ is symmetric.

- $L : V \to \mathbb{R}$ is a continuous bilinear form.
- K is a convex closed nonempty set in V.
- $j : V \to \bar{\mathbb{R}}$ $(=\mathbb{R} \cup \{+\infty\} \cup \{-\infty\})$ is a functional which is convex, lower semi-continuous (l.s.c.), and proper (this last property implying that $j(v) > -\infty \ \forall v \in V$, and that $j \not\equiv +\infty$).

2.2 Existence and uniqueness results for (1.1A) and (1.2A)

It was shown in Chapter 1, Section 3.1, that under the assumptions regarding V, K, a and L defined in Section 2.1, we have existence and uniqueness of a solution of problem (1.1A). We recall that the method used (due to Lions-Stampacchia /1/) reduces (1.1A) to a *fixed point* problem. We shall show below, by a similar process, that (1.2A) admits one and only one solution under the assumptions of Section 2.1. We thus have

Theorem 2.1: *Under the assumptions of* Section 2.1 *on* V, a, L *and* j, *problem* (1.2A) *admits one and only one solution.*

Proof: 1) *Uniqueness*

Let u_1 and u_2 be two solutions. We then have

$$a(u_1, v-u_1)+j(v)-j(u_1) \geq L(v-u_1) \quad \forall v \in V, \tag{2.1A}$$

$$a(u_2, v-u_2)+j(v)-j(u_2) \geq L(v-u_2) \quad \forall v \in V. \tag{2.2A}$$

We note that $j(u_i)$, i=1,2 is finite since, $j(\cdot)$ being *proper*, we have $j(u_i) > -\infty$ and moreover that (2.1A), (2.2A) imply, for i=1,2, that

$$j(u_i) \leq a(u_i, v-u_i)+j(v)-L(v-u_i) \quad \forall v \in V. \tag{2.3A}$$

Since $j \not\equiv +\infty$ there exists $v_o \in V$ such that $j(v_o) < +\infty$; putting $v=v_o$ in (2.3A) we hence deduce that $j(u_i) < +\infty$. Under these conditions we can put $v=u_2$ in (2.1A), $v=u_1$ in (2.2A) and add, giving

$$\alpha \| u_2 - u_1 \|^2 \leq 0 \Leftrightarrow u_2 = u_1 .$$

2) *Existence*

We proceed as in Lions-Stampacchia, loc. cit., using a fixed-point method. Let $u \in V$ (u arbitrary for the moment); we then define w as being the solution (if one exists) of the problem

$$(\pi_\rho(u)) \quad \begin{cases} \textit{Find } w \in V \textit{ such that} \\ (w, v-w)+\rho j(v)-\rho j(w) \geq (u, v-w)+\rho L(v-w)-\rho a(u, v-w) \quad \forall v \in V, \end{cases}$$

with $\rho > 0$. Since the vector u is given, $(\pi_\rho(u))$ is a problem of type (1.2A). We shall assume for the moment that $(\pi_\rho(u))$ admits a solution (which is necessarily unique).

We denote by $w = S_\rho(u)$ the mapping $u \to w$. If w_1 and w_2 correspond to u_1, u_2 respectively, we have

$$(w_1,w_2-w_1)+\rho j(w_2)-\rho j(w_1) \geq (u_1,w_2-w_1)+\rho L(w_2-w_1)-\rho a(u_1,w_2-w_1)$$

$$(w_2,w_1-w_2)+\rho j(w_1)-\rho j(w_2) \geq (u_2,w_1-w_2)+\rho L(w_1-w_2)-\rho a(u_2,w_1-w_2)$$

which, by addition, gives

$$\|w_2-w_1\|^2 \leq (u_2-u_1,w_2-w_1)-\rho a(u_2-u_1,w_2-w_1). \tag{2.3A}$$

We have (see Chapter 1, Section 3.1) the existence of a unique $\mathcal{A} \in \mathcal{L}(V,V)$ such that

$$a(u,v) = (\mathcal{A}u,v) \quad \forall u,v \in V.$$

We then deduce from (2.3A) that

$$\|w_2-w_1\|^2 \leq ((I-\rho\mathcal{A})(u_2-u_1),w_2-w_1) \leq \|I-\rho\mathcal{A}\| \, \|u_2-u_1\| \, \|w_2-w_1\|$$

or

$$\|w_2-w_1\| \leq \|I-\rho\mathcal{A}\| \, \|u_2-u_1\| . \tag{2.4A}$$

Taking account of the V-ellipticity of $a(\cdot,\cdot)$, it is clear that

$$0 < \rho < \frac{2\alpha}{\|\mathcal{A}\|^2} \tag{2.5A}$$

implies

$$\|I-\rho\mathcal{A}\| < 1 .$$

We have thus shown that if ρ satisfies (2.5A), then S_ρ is *strictly* and *uniformly* a *contraction mapping*; the *fixed-point* problem

$$u = S_\rho(u) \tag{2.6A}$$

then admits one and only one solution which satisfies, since u is a solution of $(\pi_\rho(u))$

$$\begin{cases} a(u,v-u)+j(v)-j(u) \geq L(v-u) & \forall v \in V, \\ u \in V. \end{cases}$$ ■

We have thus proved Theorem 2.1 up to the proof of

Lemma 2.1: *The problem* $(\pi_\rho(u))$ *admits one and only one solution.*

Proof: Since the bilinear form on the left-hand side of the inequality in $(\pi_\rho(u))$ is equal to $(\cdot,\cdot)$ it is *symmetric*. It then follows (see Chapter 1, Section 2.1, Céa /2/, Lions /2/, Ekeland-Témam /1/) that $(\pi_\rho(u))$ is equivalent to the minimisation problem

$$(2.7A)\qquad \begin{cases} \textit{Find } w \in V \textit{ such that} \\ J_\rho(u;w) \le J_\rho(u;v) \quad \forall v \in V \end{cases}$$

with

$$J_\rho(u;v) = \frac{1}{2}\|v\|^2 + \rho j(v) - (u,v) + \rho a(u,v) - \rho L(v).$$

The functional $v \to J_\rho(u;v)$ is *strictly convex* and l.s.c. . Moreover it satisfies

$$(2.8A)\qquad \lim_{\|v\|\to+\infty} J_\rho(u;v) = +\infty .$$

The property (2.8A) results from the fact that since $j(\cdot)$ (and therefore $\rho j(\cdot)$) is convex, proper and l.s.c., there exist (see Ekeland-Témam /1/, for example) $\lambda : V \to \mathbb{R}$, linear and continuous, and $\mu \in \mathbb{R}$, such that

$$(2.9A)\qquad j(v) \ge \lambda(v) + \mu \quad \forall v \in V.$$

In view of (2.9A) it can be easily shown that (2.8A) holds.

From the above properties of $J_\rho(u;\cdot)$ we have (from a standard result of convex analysis for which we again cite the three above references) the *existence* and *uniqueness* of a solution of (2.7A), and therefore of $(\pi_\rho(u))$. ■

Some applications of Theorem 2.1 were given in Chapters 4 and 5; further examples are given in Section 6.

2.3 On the interior approximation of (1.1A) and (1.2A)

In this section we shall reconsider Sections 4.3 and 4.4 of Chapter 1, weakening certain assumptions and expanding certain remarks (in particular Remark 4.1).

The assumptions on V, a, L, K and j are those of Section 2.1.

2.3.1 *Approximation of* (1.1A)

With $h \to 0$ we approximate V by the family $(V_h)_h$ where, $\forall h$, V_h is a closed subspace of V (in practice $\dim V_h < +\infty$). We then approximate K by the family $(K_h)_h$ where the K_h are, $\forall h$, nonempty, closed convex subsets of V_h; we do not assume a priori that $K_h \subset K$. We do assume however that $(K_h)_h$ satisfies the following two properties:

$$(P_1)\quad \begin{cases} \textit{If } (v_h)_h \textit{ is such that } v_h \in K_h \ \forall h, \textit{ then the cluster} \\ \textit{points of } (v_h)_h \textit{ for the weak topology of V belong to K.} \end{cases}$$

$$(P_2)\quad \begin{cases} \textit{There exist } \chi \subset K, \ \overline{\chi} = K \textit{ and } r_h : \chi \to K_h \textit{ such that} \\ \lim_{h\to 0} \| r_h v - v \| = 0 \quad \forall\, v \in \chi \end{cases}$$

Remark 2.1: It results from (P_1) that $\bigcap_h K_h \subset K$; moreover we note that if $K_h \subset K \ \forall h$ then (P_1) is *automatically satisfied.* ■

We thus approximate (1.1A) by

$$(2.10\text{A})\quad \begin{cases} \textit{Find } u_h \in K_h \textit{ such that} \\ a(u_h, v_h - u_h) \geq L(v_h - u_h) \quad \forall\, v_h \in K_h \ ; \end{cases}$$

it is clear that the approximate problem (2.10A) admits one and only one solution.

Remark 2.2: In the majority of cases, it is necessary to approximate $a(\cdot,\cdot)$ and $L(\cdot)$ by $a_h(\cdot,\cdot)$ and $L_h(\cdot)$ (generally by a *numerical integration* procedure in normal applications). The problems posed by the approximation of $a(\cdot,\cdot)$ and $L(\cdot)$ are those encountered in the case of approximating linear variational equations; we therefore refer to Ciarlet /1A/ and Strang-Fix /1/ for a detailed analysis of these problems (in the context of finite element approximations). ■

With regard to the convergence of $(u_h)_h$ when $h \to 0$, we have

Theorem 2.2: *Let* u *be the solution of* (1.1A) *and* u_h *that of* (2.10A); *under the above assumptions on* V, K, a *and* L, *and if* (P_1), (P_2) *are satisfied, we have*

(2.11A) $$\lim_{h\to 0} \|u_h - u\| = 0.$$

Proof: This is a variant of that of Theorem 4.2 of Chapter 1, Section 4.4.

1) A priori estimates for u_h

We deduce from (2.10A) that

$$a(u_h,u_h) \le a(u_h,v_h) - L(v_h-u_h) \quad \forall v_h \in K_h$$

which implies that

(2.12A) $$\alpha\|u_h\|^2 \le \|\mathcal{A}\|\,\|u_h\|\,\|v_h\| + \|L\|_*\,\|u_h\| + \|L\|_*\,\|v_h\| \quad \forall v_h \in K_h$$

where $\|L\|_* = \sup\limits_{v\in V-\{0\}} \dfrac{|L(v)|}{\|v\|}$ and $\mathcal{A}$ is the unique operator from $\mathcal{L}(V,V)$ determined by $a(u,v) = (\mathcal{A}u,v)$ $\forall u,v \in V$. It then results from (2.12A) and (P_2) that

(2.13A) $$\alpha\|u_h\|^2 \le \|\mathcal{A}\|\,\|u_h\|\,\|r_hv\| + \|L\|_*\,\|u_h\| + \|L\|_*\,\|r_hv\| \quad \forall v \in \chi.$$

Let $v_0 \in \chi$; we deduce from (P_2) (with C_i denoting various constants which depend on v_0 but not on h)

$$\|r_hv_0\| \le C_0$$

which implies

(2.14A) $$\alpha\|u_h\|^2 \le C_1\|u_h\| + C_2 \quad , \forall h$$

from which we finally deduce that $(u_h)_h$ is *bounded* in V.

2) Convergence of $(u_h)_h$

We can extract from $(u_h)_h$ a subsequence - also denoted by $(u_h)_h$ - such that

(2.15A) $$\lim_{h\to 0} u_h = u^* \text{ weakly in } V;$$

(P_1) then implies

(2.16A) $$u^* \in K.$$

In view of (2.15A) and (P_2) we obtain by proceeding to the limit in $a(u_h,u_h) \le a(u_h,r_hv)-L(r_hv-u_h) \quad \forall v \in \chi$,

(2.17A) $$\liminf_{h\to 0} a(u_h,u_h) \le a(u^*,v)-L(v-u^*) \quad \forall v \in \chi .$$

Moreover we have

$$\alpha\|u_h-u^*\|^2 \le a(u_h-u^*,u_h-u^*) = a(u_h,u_h)+a(u^*,u^*)-a(u_h,u^*)-a(u^*,u_h) ,$$

giving, in the limit,

(2.18A) $$\liminf_{h\to 0} a(u_h,u_h) \ge a(u^*,u^*)$$

which is a well-known property of *weak* l.s.c..

It then follows from (2.17A), (2.18A) that

(2.19A) $$a(u^*,u^*) \le a(u^*,v)-L(v-u^*) \quad \forall v \in \chi .$$

Since $\overline{\chi} = K$, (2.19A) in fact holds for all $v \in K$. Hence, from (2.16A) we have:

$$\begin{cases} a(u^*,v-u^*) \ge L(v-u^*) & \forall v \in K, \\ u^* \in K. \end{cases}$$

We have thus shown that u^* is a solution of (1.1A); in view of the uniqueness we have $u^* = u$ and the *entire* family $(u_h)_h$ converges weakly to u.

3) <u>Strong convergence of</u> $(u_h)_h$.

We have

(2.20A) $$\begin{aligned}\alpha\|u_h-u\|^2 &\le a(u_h-u,u_h-u) = a(u_h,u_h)+a(u,u)-a(u_h,u)-a(u,u_h) \\ &\le a(u_h,r_hv)-L(r_hv-u_h)+a(u,u)-a(u_h,u)-a(u,u_h) \quad \forall v \in \chi .\end{aligned}$$

In view of (P_2) and the fact that $\lim_{h\to 0} u_h = u$ weakly in V, we obtain by proceeding to the limit in (2.20A)

$$\begin{aligned}\alpha \limsup_{h\to 0} \|u_h-u\|^2 &\le a(u,v)-L(v-u)-a(u,u) \\ &= a(u,v-u)-L(v-u) \quad \forall v \in \chi .\end{aligned}$$

Since $\overline{\chi} = K$ we also have (by continuity)

(2.21A) $$\alpha \limsup \|u_h - u\|^2 \le a(u,v-u) - L(v-u) \quad \forall v \in K.$$

Taking v=u in (2.21A) we obtain

$$\limsup_{h \to 0} \|u_h - u\|^2 \le 0$$

which implies $\lim_{h \to 0} \|u_h - u\| = 0$, i.e. *strong convergence.* ∎

Some applications of Theorem 2.2 were given in Chapters 3 and 4.

2.3.2 *Approximation of* (1.2A)

Preliminary remark: In this section we shall assume that

(2.22A) $j(\cdot)$ *is continuous on* V.

In Section 6 we shall examine a class of inequalities of the type (1.2A) for which (2.22A) does not generally hold, but which can be investigated using methods very similar to those described below. ∎

Assuming again that $h \to 0$, we approximate V by $(V_h)_h$ where, $\forall h$, V_h is a closed subspace of V (finite dimensional in practical applications). We assume that $(V_h)_h$ satisfies

$$(P_1) \quad \begin{cases} \text{There exists } \mathcal{V} \subset V, \text{ with } \overline{\mathcal{V}} = V, \text{ and } r_h : \mathcal{V} \to V_h \text{ such that} \\ \lim_{h \to 0} \|r_h v - v\| = 0 \quad \forall v \in \mathcal{V}. \end{cases}$$

We then assume that the functional $j(\cdot)$ is approximated by a family such that

$$(2.23A) \quad \begin{cases} j_h : V_h \to \overline{\mathbb{R}}, \\ j_h \text{ is convex, l.s.c., and uniformly proper in } h; \end{cases}$$

the last property implies the existence of $\lambda \in V'$ and $\mu \in \mathbb{R}$ such that

(2.24A) $$j_h(v_h) \geq \lambda(v_h)+\mu \quad \forall v_h \in V_h, \quad \forall h .$$

We shall similarly assume that the following are satisfied

$$(P_2) \quad \begin{cases} \textit{If } v_h \to v \textit{ weakly in } V \textit{ then} \\ \liminf_{h \to 0} j_h(r_h v) \geq j(v) , \end{cases}$$

$$(P_3) \quad \lim_{h \to 0} j_h(r_h v) = j(v) \quad \forall v \in \mathcal{V} .$$

Remark 2.3: In all the applications of which we are aware, it has always been possible, if $j(\cdot)$ is *continuous* on V, to construct $j_h(\cdot)$, *continuous* on V_h and satisfying (2.24A), (P_2), (P_3).

Remark 2.4: If $j_h(v_h) = j(v_h) \quad \forall v_h \in V_h, \quad \forall h$, then (2.24A), (P_2), (P_3) are automatically satisfied. ∎

We thus approximate (1.2A) by

(2.25A) $$\begin{cases} \textit{Find } u_h \in V_h \textit{ such that} \\ a(u_h, v_h-u_h)+j_h(v_h)-j_h(u_h) \geq L(v_h-u_h) \quad \forall v_h \in V_h. \end{cases}$$

The approximate problem (2.25A) admits one and only one solution. Remark 2.2 of Section 2.3.1 also holds for problem (2.25A).

With regard to the convergence of $(u_h)_h$ when $h \to 0$, we have

Theorem 2.3: *Let* u *be the solution of* (1.2A), u_h *that of* (2.25A); *under the above assumptions on* V, a, L and j, *and if* (P_1), (2.23A), (2.24A), (P_2), (P_3) *are satisfied, we have*

(2.26A) $$\lim_{h \to 0} \|u_h-u\| = 0 ,$$

(2.27A) $$\lim_{h \to 0} j_h(u_h) = j(u).$$

Proof: This is a variant of that of Theorem 2.2 of Section 2.3.1.

1) A priori estimates for u_h

It follows from (2.25A) that

$$a(u_h,u_h)+j_h(u_h) \le a(u_h,v_h)+j_h(v_h)-L(v_h-u_h) \quad \forall v_h \in V_h ,$$

which, in view of (2.24A), implies

$$(2.28A) \quad \begin{cases} \alpha\|u_h\|^2 - \|\lambda\|_* \|u_h\| + \mu \\ \le \|a\| \|u_h\| \|v_h\| + j_h(v_h) + \|L\|_* (\|v_h\| + \|u_h\|). \end{cases}$$

It then follows from (P_1) and (2.28A) that

$$(2.29A) \quad \begin{cases} \alpha\|u_h\|^2 + \mu \le \|a\| \|u_h\| \|r_h v\| + j_h(r_h v) + \|L\|_* \|r_h v\| \\ \qquad + (\|L\|_* + \|\lambda\|_*) \|u_h\| \quad \forall v \in \mathcal{V} . \end{cases}$$

Let $v_o \in \mathcal{V}$; the C_i which appear below denote various quantities which depend on v_o , but not on h; we have, taking account of (P_2), (P_3),

$$(2.30A) \quad \|r_h v_o\| \le C_1 , \ |j_h(r_h v_o)| \le C_2 \quad \forall h.$$

If we take $v = v_o$ in (2.29A), we deduce from (2.30A)

$$\alpha\|u_h\|^2 \le C_3 \|u_h\| + C_4 ,$$

which implies that u_h is bounded in V.

2) Weak convergence of $(u_h)_h$.

Since the sequence $(u_h)_h$ is bounded in V, we can extract from it a subsequence, also denoted by $(u_h)_h$, such that

$$(2.31A) \quad \lim_{h\to 0} u_h = u^* \ \textit{weakly in } V.$$

Moreover we deduce from (2.25A) that

$$(2.32A) \quad a(u_h,u_h)+j_h(u_h) \le a(u_h,r_h v)+j_h(r_h v)-L(r_h v-u_h) \quad \forall v \in \mathcal{V} .$$

Using (2.31A), (P_1), (P_2), (P_3) we have in the limit in (2.32A), $\forall v \in \mathcal{V}$:

$$\text{(2.33A)} \quad \begin{cases} a(u^*,u^*)+j(u^*) \le \liminf a(u_h,u_h)+\liminf j_h(u_h) \\ \le \liminf (a(u_h,u_h)+ j_h(u_h)) \le a(u^*,v)+j(v)-L(v-u^*) \end{cases}$$

Using the *density* of $\mathcal{V}$ in V and the *continuity* of $j(\cdot)$, we deduce from (2.33A)

$$\text{(2.34A)} \quad \begin{cases} a(u^*,v-u^*)+j(v)-j(u^*) \ge L(v-u^*) \quad \forall\, v \in V, \\ u^* \in V\ , \end{cases}$$

which implies $u^* = u$ and the convergence of the *entire* sequence $(u_h)_h$ towards u.

3) <u>Strong convergence of</u> $(u_h)_h$

The relation

$$\alpha\|u_h-u\|^2+j_h(u_h) \le a(u_h-u,u_h-u)+j_h(u_h)$$

combined with (2.25A) implies that, $\forall\, v \in \mathcal{V}$, we have

$$\text{(2.35A)} \quad \begin{cases} \alpha\|u_h-u\|^2+j_h(u_h) \le a(u_h,u_h)+a(u,u)-a(u_h,u)-a(u,u_h)+j_h(u_h) \\ \le a(u_h,r_hv)+j_h(r_hv)-L(r_hv-u_h)+a(u,u)-a(u_h,u)-a(u,u_h). \end{cases}$$

In view of (P_1), (P_2), (P_3) and from the weak convergence result given above, we have, in the limit in (2.35A), $\forall\, v \in \mathcal{V}$

$$\text{(2.36A)} \quad \begin{cases} j(u) \le \liminf j_h(u_h) \le \liminf(\alpha\|u_h-u\|^2+j_h(u_h)) \\ \le \limsup(\alpha\|u_h-u\|^2+ j_h(u_h)) \le a(u,v-u)+j(v)-L(v-u) \end{cases}$$

Using the *density* of $\mathcal{V}$ in V and the *continuity* of $j(\cdot)$, we deduce from (2.36A) that

$$
(2.37A) \quad \begin{cases} j(u) \leq \liminf j_h(u_h) \leq \liminf(\alpha\|u_h-u\|^2 + j_h(u_h)) \leq \\ \limsup(\alpha\|u_h-u\|^2 + j_h(u_h)) \leq a(u,v-u)+j(v)-L(v-u) \quad \forall v \in V. \end{cases}
$$

Taking v=u in (2.37A) we obtain

$$
j(u) = \liminf j_h(u_h) = \lim(\alpha\|u_h-u\|^2 + j_h(u_h))
$$

which clearly implies (2.26A), (2.27A). ■

Some applications of the above Theorem are given in Chapters 4 and 5.

3. THE OBSTACLE PROBLEM. (I) GENERAL REMARKS. CONFORMING APPROXIMATIONS

3.1 Synopsis

In Chapter 1, Section 3.6, Example 3.5, we considered the special case of the obstacle problem (1.3A) corresponding to $\psi = 0$, $g = 0$. A variant of this special case which arises in lubrication theory was also considered in Chapter 2, Section 5. In the present section, which follows Glowinski /1A/, we shall b particularly concerned with the approximation of (1.3A) by means of *conforming finite element* methods, and we shall generalise the result of Chapter 1, Section 4.7, relating to the special case $\psi = 0$, $g = 0$. The approximation of (1.3A), using *nonconforming* finite elements of mixed type, will be considered in Section 4. It is appropriate to point out that variational problems of type (1.3A), (1.4A) (*the obstacle problem*), although particularly simple, provide a good mathematical model in several important applications (see Sections 3.2 and 3.3 below). Moreover, amongst variational inequality problems they are those for which the most detailed theoretical and numerical results have been obtained.

3.2 Formulation of the problem. Physical interpretation

Let Ω be a bounded domain in $\mathbb{R}^N$, with regular boundary $\Gamma = \partial\Omega$. With $x = \{x_i\}_{i=1}^N$, $\nabla = \{\frac{\partial}{\partial x_i}\}_{i=1}^N$, we consider the following *obstacle problem*, which has already been formulated in (1.3A), (1.4A),

(3.1A) $$\begin{cases} \textit{Find } u \in K \textit{ such that} \\ \int_\Omega \nabla u \cdot \nabla(v-u)\,dx \geq \int_\Omega f(v-u)\,dx \quad \forall v \in K, \end{cases}$$

where, in (3.1A), $f \in L^2(\Omega)$ and K is defined by

(3.2A) $$K = \{v \in H^1(\Omega),\ v \geq \psi \text{ a.e. on } \Omega,\ v|_\Gamma = g\}$$

where ψ and g are given functions.

Physical interpretation: We assume $\Omega \subset \mathbb{R}^2$; a standard interpretation of (3.1A), (3.2A) is then that its solution, u, represents the *vertical displacements*, assumed *small*, of a *horizontal elastic membrane* Ω under the effect of a *distribution of vertical forces* of intensity f (f is a surface density of vertical forces). This membrane is fixed along its boundary Γ ($u = g$), and is *constrained to lie above an obstacle* with height given by ψ ($u \geq \psi$) ; Figure 3.1 below describes this phenomenon geometrically:

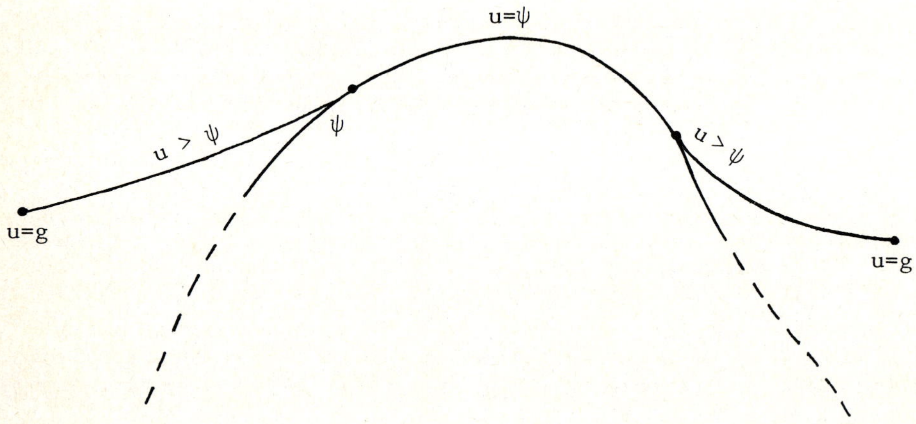

Fig. 3.1

3.3 <u>Other phenomena related to the obstacle problem</u>

Variational inequalities of the same type as (3.1A), (3.2A) but possibly with different boundary conditions and/or a non-symmetric bilinear form) arise in the following applications:

- *Lubrication phenomena*; see for example Cryer /1/, /2/, Marzulli /1/ and Chapter 2, Section 5 of the present book for solution by finite difference methods and some further references. For the modelling itself see Capriz /1A/.

- *Seepage of liquids in porous media*; see in particular Baiocchi /1/, /2/, /1A/, /2A/, Comincioli /1A/, Baiocchi-Brezzi-Comincioli /1A/, Baiocchi - Comincioli - Magènes-Pozzi /1/, /1A/, Cryer-Fetter /1A/, Baiocchi-Capelo /1A/, and also their associated bibliographies.

- *Two-dimensional potential flow of perfect fluids*; see Brézis-Stampacchia /1/, /1A/, Brézis /1A/, Ciavaldini-Pogu-Tournemine /1A/, Roux /1A/, and also the references therein.

- *Wake problems*; see Brézis-Duvaut /1/, Bourgat-Duvaut /1/.

The above list is far from exhaustive; numerous other applications also exist in biomathematics, mathematical economics, and in semi-conductor physics (see Hunt-Nassif /1A/, etc.).

3.4 Interpretation of (3.1A), (3.2A) as a free boundary problem.

Let u be the solution of (3.1A), (3.2A); we then define

$$\Omega^+ = \{x \mid x \in \Omega,\ u(x) > \psi(x)\} ,$$

$$\Omega^0 = \{x \mid x \in \Omega\ ,\ u(x) = \psi(x)\} ,$$

$$\gamma = \partial\Omega^+ \cap \partial\Omega^0 ,$$

and finally

$$u_+ = u\big|_{\Omega^+} ,\ u_0 = u\big|_{\Omega^0} .$$

A classical formulation of (3.1A), (3.2A) is then:

Find u and γ (*the free boundary*) such that

(3.3A) $-\Delta u = f$ on Ω^+,

(3.4A) $u = \psi$ on Ω^0,

(3.5A) $u = g$ on Γ,

$$u_+|_\gamma = u_o|_\gamma \; . \qquad (3.6A)$$

The physical interpretation of (3.3A) to (3.6A) is as follows: (3.3A) means that the membrane is *strictly above* the obstacle and has a *purely elastic* behaviour; (3.4A) means that on Ω^o the membrane is on contact with the obstacle; (3.6A) is a *transmission condition* at the free boundary. In fact (3.3A) - (3.6A) are not sufficient to characterise u and γ and it is necessary to introduce supplementary conditions concerning the behaviour of u on γ or alternatively the global regularity of u; for example, if ψ is sufficiently regular, we could require the "continuity" of ∇u on γ (more specifically we could impose $\nabla u \in H^1(\Omega) \times H^1(\Omega)$).

Remark 3.1: In Chapters 1 - 6 we have already investigated other variational inequality problems, interpreted as free-boundary problems.

3.5 Existence, uniqueness and regularity of the solution of (3.1A),(3.2A).

With regard to the existence and uniqueness of a solution of (3.1A), (3.2A), we can easily prove

Theorem 3.1: *Suppose that* Γ *is regular, and that* $\psi \in H^1(\Omega)$, $g \in H^{1/2}(\Gamma)$ *with* $\psi|_\Gamma \leq g$ *a.e. on* Γ; *then* (3.1A), (3.2A) *admits one and only one solution.*

Remark 3.2: We already have existence and uniqueness for functions f and ψ which are much less regular than those introduced in Theorem 3.1. ■

With regard to the regularity of u we recall the results below, which are due to Brézis-Stampacchia /2/

$$(3.7A) \quad \begin{cases} \text{If } \Gamma \textit{ is sufficiently regular and if for } p \in]1,+\infty[\\ \textit{we have } f \in L^p(\Omega) \cap (H^1(\Omega))' \;, \psi \in W^{2,p}(\Omega) \;, \\ g = \tilde{g}|_\Gamma \textit{ with } \tilde{g} \in W^{2,p}(\Omega), \textit{ then } u \in W^{2,p}(\Omega). \end{cases}$$

3.6 Finite-element approximations of problem (3.1A), (3.2A). (I) Piecewise-linear approximations

In this section we shall consider the approximation of (3.1A), (3.2A) *by a first-order conforming finite-element method* (i.e. using *continuous*, *piecewise-affine* approximations). Approximation by second-order conforming finite elements will be considered in Section 3.7.

also Falk /1A/). We have, from Green's identity

$$(3.18A)\quad a(u,v) = \int_\Omega \nabla u.\nabla v\, dx = -\int_\Omega \Delta uv\, dx + \int_\Gamma \frac{\partial u}{\partial n} v\, d\Gamma \quad \forall\, v \in H^1(\Omega).$$

It then follows from $v_h - u_h \in H_0^1(\Omega)\ \forall v_h \in K_h$ and (3.18A) that

$$(3.19A)\quad a(u,v_h-u_h) - \int_\Omega f(v_h-u_h)\,dx = \int_\Omega (-\Delta u-f)(v_h-u_h)\,dx \quad \forall\, v_h \in K_h.$$

Let π_h be the V_h-interpolation operator on Σ_h, i.e. the operator defined by

$$\begin{cases} \pi_h : H^1(\Omega) \cap C^0(\bar{\Omega}) \to V_h\,, \\ \forall v \in H^1(\Omega) \cap C^0(\bar{\Omega})\,,\ \forall P \in \Sigma_h \text{ we have } \pi_h v(P) = v(P). \end{cases}$$

We have $\Omega \subset \mathbb{R}^2$, and hence $H^2(\Omega) \subset C^0(\bar{\Omega})$ with continuous injection; the conditions $u \in H^2(\Omega)$, $u \geq \psi$ on Ω, $u = g$ on Γ, then imply $\pi_h u \in K_h$.

We take $v_h = \pi_h u$ in (3.15A), (3.19A) and hence obtain

$$(3.20A)\quad a(u_h-u,u_h-u) \leq a(u_h-u,\pi_h u-u) + \int_\Omega (-\Delta u-f)(\pi_h u-u_h)\,dx\ .$$

We note that

$$(3.21A)\quad \pi_h u-u_h = (\pi_h u-u)+(u-\psi)+(\psi-\pi_h\psi)+(\pi_h\psi-u_h).$$

Let $w = -\Delta u - f$; we have $w \in L^2(\Omega)$ and we deduce from (3.21A)

$$(3.22A)\quad \begin{cases} \int_\Omega w(\pi_h u-u_h)\,dx = \int_\Omega w(\pi_h u-u)\,dx + \int_\Omega w(u-\psi)\,dx \\ \quad + \int_\Omega w(\psi-\pi_h\psi)\,dx + \int_\Omega w(\pi_h\psi-u_h)\,dx. \end{cases}$$

From Lemma 3.1 we have $w \geq 0$ a.e. and $w(u-\psi) = 0$ a.e.; moreover since $u_h \in K_h$ we have $\pi_h\psi - u_h \leq 0$ on $\bar{\Omega}$. It then follows from (3.22A) that

$$(3.23A)\quad \int_\Omega w(\pi_h u-u_h)\,dx \leq \|w\|_{L^2(\Omega)} \left(\|\pi_h - u\|_{L^2(\Omega)} + \|\pi_h\psi-\psi\|_{L^2(\Omega)} \right).$$

From the *condition on the angles of* $\mathcal{C}_h$ stated in the Theorem, we have, since $u,\psi \in H^2(\Omega)$

$$(3.24A)\quad \|\pi_h u-u\|_{H^1(\Omega)} \leq Ch\,\|u\|_{H^2(\Omega)}\,,\quad \|\pi_h u-u\|_{L^2(\Omega)} \leq Ch^2\,\|u\|_{H^2(\Omega)}\,,$$

(3.25A) $\|\pi_h\psi-\psi\|_{H^1(\Omega)} \le Ch\|\psi\|_{H^2(\Omega)}$, $\|\pi_h\psi-\psi\|_{L^2(\Omega)} \le Ch^2\|\psi\|_{H^2(\Omega)}$

where, in (3.24A), (3.25A), C denotes various quantities which are independent of h,u,ψ.

We write $|v|_{1,\Omega} = (\int_\Omega |\nabla v|^2 dx)^{1/2}$ and $\|v\|_{1,\Omega} = \|v\|_{H^1(\Omega)}$; it then results from (3.20A), (3.23A)-(3.25A) that

(3.26A) $|u_h-u|_{1,\Omega} = O(h)$.

In order to estimate $\|u_h-u\|_{1,\Omega}$ we note that, since Ω is bounded, we have

(3.27A) $\|v\|_{1,\Omega} \le C|v|_{1,\Omega} \quad \forall v \in H^1_o(\Omega)$, C *independent of* v.

It then follows from (3.26A) and from $u_h-\pi_h u \in H^1_o(\Omega)$ that

$$\text{(3.28A)}\quad \begin{cases} \|u_h-u\|_{1,\Omega} \le \|u_h-\pi_h u\|_{1,\Omega} + \|\pi_h u-u\|_{1,\Omega} \le \\ \le C|u_h-\pi_h u|_{1,\Omega} + \|\pi_h u-u\|_{1,\Omega} \le \\ \le C|u_h-u|_{1,\Omega} + C|u-\pi_h u|_{1,\Omega} + \|\pi_h u-u\|_{1,\Omega}\,. \end{cases}$$

The estimate (3.17A) is then a clear consequence of (3.24A), (3.26A), (3.28A).

Comments: Concerning the approximation of the obstacle problem by first-order finite elements, the O(h) estimate was obtained by Falk /1/, and then by Mosco-Strang /1/ (see also Chapter 1, Section 4.7). In Falk /2/, /1A/ the validity of this estimate was shown under quite general assumptions on Ω (the results of Falk /1A/ were considered further in Ciarlet /1A/). The problem of obtaining L^2 estimates of optimal order (i.e. $O(h^2)$) of u_h-u via a generalisation of the Aubin-Nitsche method has not yet been completely resolved; for incomplete results in this direction we refer to Natterer /1A/ and Mosco /1A/.

To conclude our discussion of piecewise-affine approximations, it should be pointed out that using adequate assumptions Baiocchi /2/ (resp. Nitsche /1A/) have been able to obtain for the obstacle problem $\|u_h-u\|_{L^\infty(\Omega)} = O(h^{2-\varepsilon})$, $\varepsilon > 0$ arbitrarily small (resp. $\|u_h-u\|_{L^\infty(\Omega)} = O(h^2|\text{Log } h|)$).

3.7 Finite-element approximations of problem (3.1A), (3.2A) (II) Piecewise-quadratic approximations

With Σ_h and $\mathring{\Sigma}_h$ as in Section 3.6, we define

$$\Sigma_h' = \{P \in \overline{\Omega},\ P \text{ the midpoint of a side of a triangle } T \in \mathcal{T}_h\},$$

$$\mathring{\Sigma}_h' = \{P \in \Sigma_h',\ P \notin \Gamma\},$$

$$\Sigma_h'' = \Sigma_h \cup \Sigma_h',\quad \mathring{\Sigma}_h'' = \mathring{\Sigma}_h \cup \mathring{\Sigma}_h'.$$

We then approximate $H^1(\Omega)$ and K by

$$\text{(3.29A)}\quad V_h = \{v \in C^0(\overline{\Omega}),\ v_h|_T \in P_2\ \ \forall T \in \mathcal{T}_h\},$$

$$\text{(3.30A)}_1\quad K_h^1 = \{v_h \in V_h,\ v_h(P) \geq \psi(P)\ \ \forall P \in \mathring{\Sigma}_h'',\ v_h(P) = g(P)\ \ \forall P \in \Sigma_h'' \cap \Gamma\}$$

$$\text{(3.30A)}_2\quad K_h^2 = \{v_h \in V_h,\ v_h(P) \geq \psi(P)\ \ \forall P \in \mathring{\Sigma}_h',\ v_h(P)=g(P)\ \ \forall P \in \Sigma_h'' \cap \Gamma\}.$$

We remark that in K_h^2 the condition $v_h(P) \geq \psi(P)$ is only required at the midpoints of the sides.

Finally we approximate the obstacle problem (3.1A), (3.2A) by

$$\text{(3.31A)}_i\quad \begin{cases} \textit{Find } u_h^i \in K_h^i \textit{ such that} \\ \displaystyle\int_\Omega \nabla u_h^i \cdot \nabla (v_h - u_h^i)\,dx \geq \int_\Omega f(v_h - u_h^i)\,dx \quad \forall v_h \in K_h^i, \end{cases}$$

where i = 1, 2.

If $\psi|_\Gamma \leq g$ we have $K_h^i \neq \emptyset$ and thus:

Proposition 3.2: *The approximate problems* $(3.31A)_i$, i = 1,2, *admit one and only one solution.*

As regards the convergence of the approximate solutions, we refer to Glowinski /1A/ where $\lim_{h\to 0} \|u_h^i - u\|_{H^1(\Omega)} = 0$ is proved, provided we assume that the condition on the angles of $\mathcal{T}_h$ stated in Theorem 3.2 is satisfied. The investigation of the approximation error is much more complicated than in the case of piecewise-affine approximations and we therefore refer to B.H.R. /1A/ where, using suitable assumptions on f, u, ψ, g and the free boundary, it is shown that $\forall i=1,2$ we have

$$\|u_h^i - u\|_{H^1(\Omega)} = O(h^{3/2-\varepsilon}),\quad \varepsilon > 0 \text{ arbitrarily small},$$

if the above condition on the angles of $\mathcal{T}_h$ is satisfied.

4. THE OBSTACLE PROBLEM. (II) NON-CONFORMING APPROXIMATIONS USING MIXED FINITE ELEMENTS

4.1 Synopsis

In this section which follows B.H.R. /2A/ (see also Hlavachek /1A/), we consider the approximation of the obstacle problem (3.1A), (3.2A) by a *mixed finite element method, via a dual formulation of* (3.1A), (3.2A).

4.2 A dual formulation of the obstacle problem (3.1A), (3.2A).

It is clear that (3.1A), (3.2A) is equivalent to

$$\min_{\{v,q\}\in\mathcal{K}} \left\{ \frac{1}{2}\int_\Omega |q|^2 dx - \int_\Omega fv\, dx \right\}, \tag{4.1A}$$

where

$$\begin{cases} \mathcal{K} = \{\{v,q\} \in H^1(\Omega) \times (L^2(\Omega))^N,\ q=\nabla v\ ,\ v=g \ \text{ on } \Gamma, \\ v \geq \psi \ \text{ a.e. on } \Omega\}. \end{cases} \tag{4.2A}$$

We associate with (4.1A), (4.2A) *the Lagrangian*

$$\mathcal{L}(v,q,\mu) = \frac{1}{2}\int_\Omega |q|^2 dx - \int_\Omega fv\, dx + \int_\Omega \mu\cdot(\nabla v - q)\, dx\ , \tag{4.3A}$$

and we consider the *dual problem* (see Chapters 1 and 2 for the concepts of duality) of (4.1A), (4.2A) relative to $\mathcal{L}$, namely

$$\max_{\mu\in(L^2(\Omega))^N}\ \min_{\{v,q\}\in\tilde{\mathcal{K}}} \mathcal{L}(v,q,\mu) \tag{4.4A}$$

where

$$\tilde{\mathcal{K}} = \{\{v,q\} \in H^1(\Omega)\times(L^2(\Omega))^N, v=g \ \text{ on } \Gamma,\ v \geq \psi \ \text{ a.e. on } \Omega\}. \tag{4.5A}$$

It can be shown quite easily that:

Proposition 4.1: *The dual problem* (4.4A) *is equivalent to*

$$\min_{q\in C} \left\{\frac{1}{2}\int_\Omega |q|^2 dx + \int_\Omega \psi\ \nabla\cdot q\ dx - \int_\Gamma gq\cdot n\ d\Gamma\right\} \tag{4.6A}$$

where

$$C = \{q \in H(\mathrm{div},\Omega),\ \nabla\cdot q+f \leq 0 \ \textit{a.e. on } \Omega\} \tag{4.7A}$$

(4.8A) $$H(\text{div},\Omega) = \{q \in (L^2(\Omega))^N, \ \nabla\cdot q \in L^2(\Omega)\} .$$

Conversely (4.6A), (4.7A) *admits a unique solution* p *such that* p = ∇u *where* u *is the solution of the obstacle problem* (3.1A), (3.2A).

Remark 4.1: In the above, n denotes a unit vector normal to Γ, pointing outwards from Ω. We recall that if $q \in H(\text{div},\Omega)$, then $q\cdot n$ exists in $H^{-\frac{1}{2}}(\Gamma)$. Strictly, the term $\int_\Gamma g\, q\cdot n\, d\Gamma$ should be replaced by $\langle g, q\cdot n\rangle$ where $\langle\cdot,\cdot\rangle$ denotes the bilinear form of the duality between $H^{\frac{1}{2}}(\Gamma)$ and $H^{-\frac{1}{2}}(\Gamma)$.

Remark 4.2: Let

(4.9A) $$\Lambda = \{\mu \in L^2(\Omega), \ \mu \geq 0 \text{ a.e. on } \Omega\} .$$

We then have the following equivalence

(4.10A) $$q \in C \Leftrightarrow \begin{cases} q \in H(\text{div},\Omega), \\ \displaystyle\int_\Omega (\nabla\cdot q + f)\mu \, dx \leq 0 \quad \forall \mu \in \Lambda. \end{cases}$$

4.3 An approximation of the dual problem (4.6A), (4.7A) by mixed finite elements

We once again assume that Ω is a *bounded polygonal* domain in $\mathbb{R}^2$, and $\mathcal{T}_h$ is a *triangulation* of Ω. We then approximate $H(\text{div},\Omega)$, $L^2(\Omega)$, Λ, C respectively, by

(4.11A) $$\begin{cases} H_h = \{q_h \in H(\text{div},\Omega), \ q_h|_T \in P_{k+1}\times P_{k+1} \ \text{and} \\ \nabla\cdot q_h|_T \in P_k \ \ \forall T \in \mathcal{T}_h, \ q_h\cdot n|_S \in P_k \ \ \forall S \ \text{side of a} \ T \in \mathcal{T}_h\} \end{cases}$$

(n is a unit vector normal to S),

(4.12A) $$L_h = \{\mu_h \in L^2(\Omega), \ \mu_h|_T \in P_k \ \ \forall T \in \mathcal{T}_h\} ,$$

(4.13A) $$\Lambda_h = \Lambda \cap L_h = \{\mu_h \in L_h, \ \mu_h \geq 0 \text{ a.e. on } \Omega\} ,$$

(4.14A) $$C_h = \{q_h \in H_h, \ \int_\Omega (\nabla\cdot q_h + f)\mu_h \, dx \leq 0 \ \ \forall \mu_h \in \Lambda_h\} .$$

We then approximate the dual problem (4.6A), (4.7A) by

$$
\text{(4.15A)}\quad \begin{cases} \textit{Find } p_h \in C_h \textit{ such that } \forall q_h \in C_h \textit{ we have} \\ \displaystyle\int_\Omega p_h \cdot (q_h - p_h)\,dx + \int_\Omega \psi \nabla \cdot (q_h - p_h)\,dx \ \geq \int_\Gamma g(q_h - p_h)\cdot n \, d\Gamma \ . \end{cases}
$$

It is clear that (4.15A) admits one and only one solution.

Let λ_h be the *Kuhn-Tucker vector* (see Rockafellar /3/) relating to (4.14A); problem (4.15A) is thus equivalent to the *variational system*

$$
\text{(4.16A)}\quad \begin{cases} \textit{Find } \{p_h, \lambda_h\} \in H_h \times \Lambda_h \textit{ such that} \\ \displaystyle\int_\Omega p_h \cdot q_h dx + \int_\Omega (\lambda_h + \psi) \nabla \cdot q_h dx = \int_\Gamma g \ q_h \cdot n \ d\Gamma \quad \forall q_h \in H_h \ , \\ \displaystyle\int_\Omega (\nabla \cdot p_h + f)(\mu_h - \lambda_h)\,dx \leq 0 \quad \forall \mu_h \in \Lambda_h . \end{cases}
$$

It results from B.H.R. /2A/ that (4.16A) admits one and only one solution, with p_h the solution of (4.15A).

As regards the convergence of p_h, λ_h when $h \to 0$, the following results are proved in B.H.R. /2A/.

Theorem 4.1: *We assume that when* $h \to 0$ *the angles of* $\mathcal{C}_h$ *are bounded below by* $\theta_0 > 0$, *independent of* h. *Then*

(i) *If* $k = 0$ *and if* $f \in L^2(\Omega)$, $u, \psi \in H^2(\Omega)$ *we have*

$$
\text{(4.17A)}\qquad \|p_h - \nabla u\|_{L^2(\Omega)} = \|\lambda_h - (u - \psi)\|_{L^2(\Omega)} = O(h) \ ,
$$

(ii) *If* $k = 1$, *if* $f \in L^\infty(\Omega)$, $u, \psi \in W^{2,\infty}(\Omega)$ *and if*
$u \in W^{s,\alpha}(\Omega) \quad \forall \alpha \in]1, +\infty[$ *and* $s < 2 + \frac{1}{p}$,

then

$$
\text{(4.18A)}\qquad \|p_h - \nabla u\|_{L^2(\Omega)} = \|\lambda_h - (u - \psi)\|_{L^2(\Omega)} = O(h^{3/2-\varepsilon})
$$

with $\varepsilon > 0$ *arbitrarily small.* ■

Remark 4.3: If $k = 0$ and if $q_h \in H_h$, then $q_h|_T = \{q_{1T}, q_{2T}\}$ must *necessarily* satisfy (see J.M. Thomas, private communication):

$$
\text{(4.19A)}\quad \begin{cases} q_{1T}(x) = q_{1T} + b_T x_1 \ , \\ q_{2T}(x) = q_{2T} + b_T x_2 \end{cases}
$$

(where $x = \{x_1, x_2\}$). ■

Remark 4.4: The numerical solution of (4.15A), (4.16A) would appear to be more complicated than that of the approximate problems of Section 3 corresponding to *conforming finite-element approximations* of the obstacle problem (3.1A), (3.2A). ■

5. ON A STAMP PROBLEM WHICH LEADS TO A VARIATIONAL INEQUALITY FOR A PSEUDO-DIFFERENTIAL OPERATOR

This section follows the work of G.I. Eskin /1A/ and of A. Bogomolnii, G. Eskin and S. Zuchowizkii /1A/.

5.1 Statement of the problem

Consider a stamp with an external surface represented by

$$z = \phi(x)$$

where $x \in \mathbb{R}^2$, ϕ being a regular function with

$$\text{(5.1A)} \qquad \begin{cases} \phi(0) = 0 \ , \ \dfrac{\partial\phi}{\partial x_1}(0) = \dfrac{\partial\phi}{\partial x_2}(0) = 0 \ , \\ \phi(x) > 0 \text{ if } x \neq 0. \end{cases}$$

The stamp is applied to the elastic half space $z < 0$. Let $\delta(>0)$ be the displacement imposed on the stamp in the direction parallel to the z-axis. Let G be the contact domain, which is not known a priori. $w(x)$ denotes the displacement normal to the plane $z = 0$ and $\sigma(x)$ the normal component of the stress tensor. As in classical theory, w and σ are related by

$$\text{(5.2A)} \qquad w = k \frac{1}{|x|} * \sigma \quad , \ k > 0 \ , \quad |x| = \sqrt{x_1^2 + x_2^2} \ .$$

The conditions on $z = 0$ differ according to whether one is inside G or outside G: in G we have

$$\text{(5.3A)} \qquad w(x) = \phi(x) - \delta \quad \text{and} \quad \sigma(x) \le 0 \ ,$$

and outside G we have:

$$\text{(5.4A)} \qquad w \le \phi(x) - \delta \quad \text{and} \quad \sigma(x) = 0 \ .$$

If we introduce the notation (in accordance with the usual notation for this subject)

(5.5A) $$u = -\sigma$$

then

(5.6A) $$\begin{cases} u \geq 0 , \\ k \dfrac{1}{|x|} \star u - (\delta - \phi) \geq 0 , \\ u[k \dfrac{1}{|x|} \star u - (\delta - \phi)] = 0 \text{ in } \mathbb{R}^2 , \end{cases}$$

which is an "obstacle-type" problem - except that the second-order elliptic operator which normally appears has here been replaced by *the convolution operator*

(5.7A) $$Au = k \frac{1}{|x|} \star u . \quad \blacksquare$$

We note that problem (5.6A) can be formulated for a *bounded open domain* (at least with a stamp of "reasonable shape"). In fact, if $u \geq 0$, then $Au \geq 0$ and thus $Au > \delta - \phi$ when $\delta - \phi(x) < 0$; thus, using the last of conditions (5.6A), $u = 0$ when $\delta - \phi(x) < 0$. It is thus sufficient to consider the problem analogous to (5.6A) *in an open domain* Ω *containing the region* "$\delta - \phi(x) < 0$" in its interior, which is possible with Ω *bounded* if the stamp $z = \phi(x)$ is "reasonable" !

We thus put

(5.8A) $$\delta - \phi = f .$$

We finally have the problem: *Find* u *in* Ω, *a bounded open domain in* $\mathbb{R}^2$, *with*

(5.9A) $$u \geq 0 , \quad Au - f \geq 0 , \quad u(Au - f) = 0 \quad \textit{in } \Omega$$

where f *is given, with* $f < 0$ *in the neighbourhood of the boundary of* Ω, *and where in* (5.7A) u *is understood to be extended by zero outside* Ω. ■

5.2 Functional formulation

We denote by $\mathfrak{F}$ the Fourier transform; formally, we have:

$$\mathfrak{F}\phi = \hat{\phi}, \ \hat{\phi}(\xi) = \int_{\mathbb{R}^2} e^{-2\pi i x\xi} \phi(x)\,dx .$$

We note that

(5.10A) $$\mathfrak{F}\left(\frac{1}{|x|}\right) = \frac{1}{|\xi|}$$

so that

$$\mathfrak{F}(Au) = k \frac{1}{|\xi|} \hat{u} . \tag{5.11A}$$

Consequently, if ϕ and $\psi \in \mathcal{D}(\Omega)$, we have

$$(A\phi,\psi) = \int_{\Omega} (A\phi)\psi \, dx = k \int_{\mathbb{R}^2} \frac{1}{|\xi|} \hat{\phi} \, \hat{\psi} \, d\xi . \tag{5.12A}$$

It is then natural to introduce:

$$\begin{cases} E = \text{completion of } \mathcal{D}(\Omega) \text{ for the } \textit{norm} \\ \|\phi\|_E = \left(\int_{\mathbb{R}^2} \frac{1}{|\xi|} |\hat{\phi}(\xi)|^2 d\xi\right)^{1/2} ; \end{cases} \tag{5.13A}$$

this space is one of a family of spaces of 'fractional Sobolev' type, investigated elsewhere.

More generally, for s > 0, we introduce (see L. Hormander and J.L. Lions /1A/):

$$\begin{cases} \hat{\mathcal{D}}^s(\Omega) = \text{completion of } \mathcal{D}(\Omega) \text{ for the norm} \\ \left(\int_{\mathbb{R}^2} |\xi|^{2s} |\hat{\phi}(\xi)|^2 d\xi\right)^{1/2}. \end{cases} \tag{5.14A}$$

We must take note of the fact that it is not always possible to identify this space with a subspace of distributions on Ω. However

$$\hat{\mathcal{D}}^{1/2}(\mathbb{R}^2) \text{ is identified with a subspace of } \mathcal{D}'(\mathbb{R}^2) \tag{5.15A}$$

(and, consequently, $\hat{\mathcal{D}}^{1/2}(\Omega)$ is identified, *for any* Ω, *bounded or not*, with a subspace of $\mathcal{D}'(\Omega)$).

To prove (5.15A), we consider a Cauchy sequence ϕ_n for $\hat{\mathcal{D}}^{1/2}(\mathbb{R}^2)$; if $\psi \in \mathcal{D}(\mathbb{R}^2)$, we have:

$$\int_{\mathbb{R}^2} \phi_n \overline{\psi} \, dx = \int_{\mathbb{R}^2} |\xi|^{1/2} \hat{\phi}_n (|\xi|^{-1/2} \hat{\overline{\psi}}) d\xi ;$$

since $|\xi|^{1/2} \hat{\phi}_n$ is a Cauchy sequence in $L^2(\mathbb{R}^2)$ and $|\xi|^{-1/2} \hat{\psi}$ is in $L^2(\mathbb{R}^2)$ we thus see that $\int_{\mathbb{R}^2} \phi_n \overline{\psi} \, dx$ converges.

We note that

$$(5.16\text{A})\qquad \begin{cases} E = \text{the dual of the space } \hat{\mathcal{D}}^{1/2}(\Omega) \text{ (a space which} \\ \text{is generally denoted by } \hat{\mathcal{D}}^{-1/2}(\Omega)\text{)} \end{cases}$$

and from (5.11A) we see that

$$(5.17\text{A})\qquad A \text{ is an isomorphism from } \hat{\mathcal{D}}^{-1/2}(\Omega) \text{ onto } \hat{\mathcal{D}}^{+1/2}(\Omega).$$

If (f,ϕ) denotes the scalar product between $\hat{\mathcal{D}}^{-1/2}(\Omega)$ and $\hat{\mathcal{D}}^{1/2}(\Omega)$, we have:

$$(5.18\text{A})\qquad (A\phi,\phi) = k\|\phi\|_E^2 .$$

We introduce

$$(5.19\text{A})\qquad K = \{v \mid v \in E \ , \ v \geq 0 \text{ in } \Omega \ ^{(1)}\}$$

which defines a nonempty closed convex set in E. Problem (5.9) is thus formulated: *find* $u \in K$ *such that*

$$(5.20\text{A})\qquad (Au, v-u) \geq (f, v-u) \quad \forall v \in K$$

where f *is given in* $\hat{\mathcal{D}}^{1/2}(\Omega)$.

From (5.18A) this problem comes within the general theory of Variational Inequalities. *It admits a unique solution.*

5.3 Finite element approximation

In concept, any finite element method (with a suitable approximation of K) is applicable and *convergent*. The difficulty lies in the feasibility of *calculating the coefficients*. To this end the following method is proposed in Bogomolnii, Eskin and Zuchowizkii /1A/.

We consider a partitioning of $\mathbb{R}^2$ into squares of side h, with edges parallel to the coordinate axes. We put

$$(5.21\text{A})\qquad \phi_o(x) = (1-|x_1|)(1-|x_2|)$$

in the square $|x_1| < 1$, $|x_2| < 1$, and $\phi_o = 0$ outside the square. We then consider the *basis functions*:

$$(5.22\text{A})\qquad \phi_p(x) = \phi_o\left(\frac{x_1}{h} - p_1, \frac{x_2}{h} - p_2\right) , \ p = \{p_1, p_2\} \in \mathbb{Z}^2 .$$

---

(1) In the sense of distributions on Ω

We assume that Ω is approximated by the union of squares $|x_i| < h$ and their translations, and we approximate K by combinations of the ϕ_p corresponding to coefficients ≥ 0. This amounts to calculating

$$(5.23A) \qquad (A\phi_p, \phi_q) = a_{p,q} .$$

In fact $a_{p,q}$ depends only on p-q:

$$(5.24A) \qquad a_{p,q} = a_\rho \, , \; \rho = p-q$$

where, after a simple calculation

$$(5.25A) \qquad a_\rho = \frac{kh^3}{2\pi} \iint_{\mathbb{R}^2} e^{i\rho\xi} \frac{1}{|\xi|} \left(\frac{\sin \frac{\xi_1}{2}}{\frac{\xi_1}{2}} \; \frac{\sin \frac{\xi_2}{2}}{\frac{\xi_2}{2}} \right)^4 d\xi \, .$$

It is possible to calculate a_ρ using *an asymptotic formula*

$$(5.26A) \qquad a_\rho = kh^3 \sum_{|\alpha| \leq M} (-1)^{|\alpha|} \phi_\alpha \, D^{2\alpha}\left(\frac{1}{|\rho|}\right) + 0\left(\frac{1}{|\rho|^{2M+2}}\right)$$

where we consider ρ as a continuous variable and then take $\rho \in \mathbb{Z}^2$ in the result, and where the ϕ_α are the Taylor coefficients of the function

$$\left(\frac{\sin \frac{\xi_1}{2}}{\frac{\xi_1}{2}} \; \frac{\sin \frac{\xi_2}{2}}{\frac{\xi_2}{2}} \right)^4 .$$

We refer to Bogomolnii et al, loc. cit., for numerical results obtained using this technique.

6. SOLUTION OF NONLINEAR DIRICHLET PROBLEMS BY REDUCTION TO VARIATIONAL INEQUALITIES

6.1 Synopsis

In this section we show that the methodology of variational inequalities can also be useful in the solution of *equations*, a fact which has been used by Puel /1A/. We shall illustrate this using a family of *nonlinear Dirichlet problems*, concentrating in particular on their approximation by means of a *finite element method*. We shall omit most of the proofs, referring for these to Chan-Glowinski /1A/, Glowinski /2A/.

6.2 The continuous problem

Let Ω be a bounded domain in $\mathbb{R}^N$ $(N \geq 1)$ with a regular boundary $\Gamma = \partial\Omega$; with $V = H_0^1(\Omega)$ we consider:

- $L : V \to \mathbb{R}$, linear and continuous, i.e. $L(v) = \langle f,v\rangle \quad \forall v \in V$, where $f \in V' = H^{-1}(\Omega)$ (V' is the dual of V and $\langle\cdot,\cdot\rangle$ is the bilinear form of the duality between V' and V).

- $a : V \times V \to \mathbb{R}$ bilinear, continuous, coercive; we thus have the existence of an $\alpha > 0$ such that

$$a(v,v) \geq \alpha\|v\|_V^2 \quad \forall v \in V \tag{6.1A}$$

where

$$\|v\|_V = \left(\int_\Omega |\nabla v|^2\, dx\right)^{1/2}.$$

We do not assume a priori that $a(\cdot,\cdot)$ is symmetric.

- $\phi : \mathbb{R} \to \mathbb{R}$, $\phi \in C^0(\mathbb{R})$, ϕ non-decreasing (i.e. $\phi(t') \geq \phi(t) \quad \forall t,t' \in \mathbb{R},\ t' \geq t$), $\phi(0) = 0$.

We then consider the *nonlinear variational equation*

$$\begin{cases} \textit{Find } u \in V \textit{ such that } \phi(u) \in L^1(\Omega) \cap V' \textit{ and} \\ a(u,v) + \langle\phi(u),v\rangle = \langle f,v\rangle \quad \forall v \in V. \end{cases} \tag{6.2A}$$

Let $A \in \mathcal{L}(V,V')$ such that $a(v,u) = \langle Av,w\rangle \quad \forall v,w \in V$; it is clear that (6.2A) *is equivalent* to the *nonlinear Dirichlet problem*

$$\begin{cases} Au + \phi(u) = f, \\ u \in V,\ \phi(u) \in L^1(\Omega). \end{cases} \tag{6.3A}$$

Example: Let $a_0 \in L^\infty(\Omega)$ such that

$$a_0(x) \geq \alpha > 0 \text{ a.e. on } \Omega \tag{6.4A}$$

and let β be a *constant vector* in $\mathbb{R}^N$.

We define $a(\cdot,\cdot)$ *bilinear* and *continuous* on $H_0^1(\Omega) \times H_0^1(\Omega)$ by

$$a(u,v) = \int_\Omega a_0 \nabla u\cdot\nabla v\, dx + \int_\Omega \beta\cdot\nabla u\, v\, dx .$$

We have

$$\int_\Omega \beta\cdot\nabla v\ v\ dx = 0 \quad \forall v \in H_0^1(\Omega)$$

which, combined with (6.4A), implies

$$a(v,v) \geq \int_\Omega |\nabla v|^2 dx \quad \forall v \in H_0^1(\Omega) \ ,$$

i.e. the *coercivity* of $a(\cdot,\cdot)$.

6.3 Existence and uniqueness results for (6.2A), (6.3A).

6.3.1 *Introduction. A variational inequality associated with* (6.2A), (6.3A).

If $N=1$ we have $H_0^1(\Omega) \subset C^0(\bar{\Omega})$ with continuous injection and the proof of the existence and uniqueness of a solution of (6.2A), (6.3A) is then almost immediate. If $N \geq 2$, as we shall assume in the following, the essential difficulty is precisely the fact that $H_0^1(\Omega) \not\subset C^0(\bar{\Omega})$.

Remark 6.1: The analysis which follows can also be applied to problems in which $V = H^1(\Omega)$, or in which V is a closed subspace of $H^1(\Omega)$. ■

Let $\Phi : \mathbb{R} \rightarrow \mathbb{R}$ be defined by

$$\Phi(t) = \int_0^t \phi(\tau)d\tau \ ;$$

Φ is clearly C^1, convex and ≥ 0. We then define $j(\cdot)$ by

$$j(v) = \int_\Omega \Phi(v)\ dx. \tag{6.5A}$$

It is easily shown (see Chan-Glowinski /1A/, Glowinski /2A/) that:

Proposition 6.1: *The functional* $j(\cdot)$ *is convex, proper and l.s.c. on* $L^2(\Omega)$. ■

Assume that $a(\cdot,\cdot)$ is *symmetric*; it would then be natural to associate with (6.2A), (6.3A) the following *calculus-of-variations* problem:

$$\text{(6.6A)} \quad \begin{cases} \textit{Find } u \in V \textit{ such that} \\ J(u) \leq J(v)\ \forall v \in V \end{cases}$$

where

$$(6.7A) \qquad J(v) = \frac{1}{2}\, a(v,v)+j(v)-L(v).$$

If $a(\cdot,\cdot)$ is non-symmetric the above approach is not feasible, at least directly, but, on the other hand, we can associate with (6.2A), (6.3A) the *Elliptic Variational Inequality* (of type (1.2A)):

$$(6.8A) \qquad \begin{cases} \textit{Find } u \in V \textit{ such that} \\ a(u,v-u)+j(v)-j(u) \geq L(v-u) \quad \forall v \in V \end{cases}$$

which is in fact a generalisation of (6.6A), (6.7A).

As regards the existence and uniqueness of a solution of (6.8A), in view of Proposition 6.1 we can apply Theorem 2.1 of Section 2.2, which gives:

Theorem 6.1: *Under the above assumptions on* a, L, ϕ, *problem* (6.8A) *admits one and only one solution.*

6.3.2 *Equivalence between* (6.2A), (6.3A) and (6.8A).

It is possible to prove:

Theorem 6.2: *The solution of* (6.8A) *is necessarily a solution of* (6.2A), (6.3A). *Conversely, any solution of* (6.2A), (6.3A) *is a solution of* (6.8A).

The above theorem implies that (6.2A), (6.3A) admits one and only one solution. Another consequence of the above equivalence is that, from the point of view of the *numerical solution*, we can either work directly on (6.2A), (6.3A), or alternatively consider problem (6.8A); in fact, from the point of view of the *approximation*, it is in our opinion simpler to work with (6.8A).

6.4 The finite element approximation of (6.2A), (6.3A) and (6.8A).

6.4.1 *Synopsis*

We shall now approximate problem (6.2A), (6.3A) by working directly on the equivalent formulation (6.8A); however, we should point out that the considerations of Section 2.3.2 cannot be applied directly since assumption (2.22A) (i.e. $j(\cdot)$ continuous on V) is not satisfied.

6.4.2 *Triangulation of Ω. Definition of the approximate problem.*

We assume that Ω is a polygonal [1] bounded domain in $\mathbb{R}^2$. We introduce a family $(\mathcal{T}_h)_h$ of triangulations of Ω which satisfy the assumptions of Section 3.6 and we approximate $V = H_0^1(\Omega)$ by

$$V_h = \{v_h | v_h \in C^0(\bar{\Omega}),\ v_h|_\Gamma = 0\ ,\ v_h|_T \in P_1\ \ \forall T \in \mathcal{T}_h\}.$$

It is then natural to approximate (6.2A), (6.3A) and (6.8A), respectively, by:

$$\text{(6.9A)}\quad \begin{cases} \textit{Find } u_h \in V_h \textit{ such that} \\ a(u_h, v_h) + \int_\Omega \phi(u_h) v_h\, dx = L(v_h) \quad \forall v_h \in V_h \end{cases}$$

and

$$\text{(6.10A)}\quad \begin{cases} \textit{Find } u_h \in V_h \textit{ such that} \\ a(u_h, v_h - u_h) + j(v_h) - j(u_h) \geq L(v_h - u_h) \quad \forall v_h \in V_h . \end{cases}$$

It is clear that these two *finite-dimensional* problems are *equivalent*.

From a practical point of view it is not generally possible to use (6.9A) and (6.10A) as this requires the calculation of integrals which are impossible to evaluate exactly. It is thus necessary to modify (6.9A), (6.10A) via the use of numerical integration procedures. In the following we shall be content to approximate $j(\cdot)$. (Remark 2.2 of Section 2.3.1 also applies in this context). Using the notation of Figure 6.1 below, we approximate $j(\cdot)$ by $j_h(\cdot)$ defined by

$$\text{(6.11A)}\qquad j_h(v_h) = \frac{1}{3} \sum_{T \in \mathcal{T}_h} \text{meas.}(T) \sum_{i=1}^{3} \Phi(v_h(M_{iT})) \quad \forall v_h \in V_h .$$

In fact we could also consider $j_h(v_h)$ as being the *exact integral* of a piecewise-constant function; more precisely, if we assume that the set Σ_h of nodes of $\mathcal{T}_h$ is ordered from 1 to $N_h = \text{Card}\,(\Sigma_h)$, then with each $M_i \in \Sigma_h$ we associate the set Ω_i (shown hatched in Figure 6.2) obtained by joining the centres of the triangles with common vertex M_i to the midpoints of the edges which meet at M_i.

---

(1) This assumption is not essential.

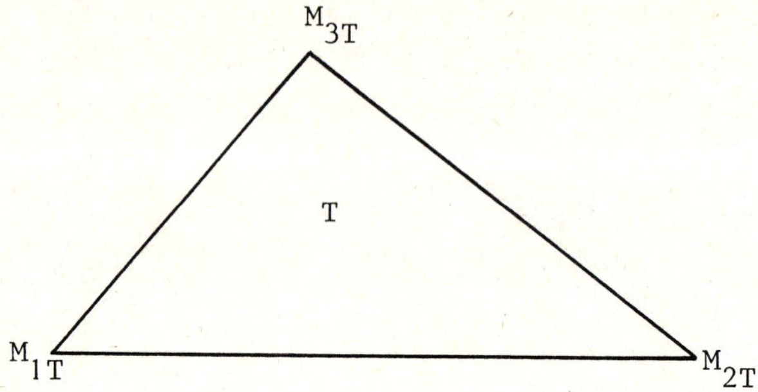

Fig. 6.1

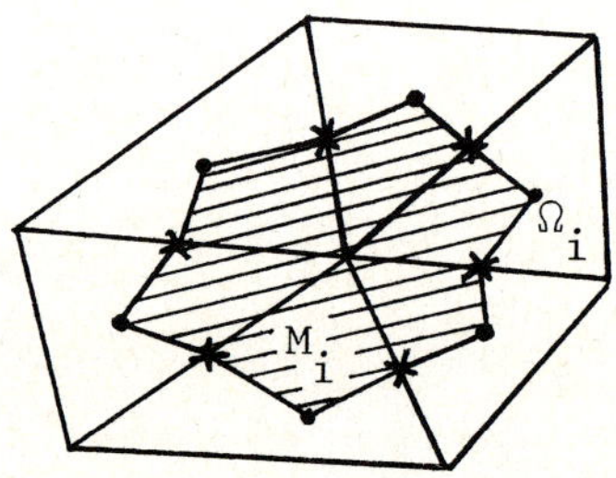

Fig. 6.2

(If $M_i \in \Gamma$, the modification required in Figure 6.2 is trivial).
We then define a space L_h of piecewise-constant functions by:

(6.12A) $$L_h = \{\mu_h | \mu_h = \sum_{i=1}^{N_h} \mu_i\chi_i \ , \ \mu_i \in \mathbb{R} \quad \forall\, i=1,\ldots N_h\} \quad ,$$

where χ_i is the characteristic function of Ω_i, i.e.

$$\chi_i(x) = 1 \text{ if } x \in \Omega_i \ ,$$

$$\chi_i(x) = 0 \text{ if } x \notin \Omega_i \ .$$

We next define $q_h : C^o(\overline{\Omega}) \cap H_o^1(\Omega) \to L_h$ by

(6.13A) $$q_h v = \sum_{i=1}^{N_h} v(M_i)\chi_i \ .$$

It then follows from (6.11A) - (6.13A) that

(6.14A) $$j_h(v_h) = \int_\Omega \Phi(q_h v_h)\,dx \quad \forall v_h \in V_h ,$$

and hence that

(6.15A) $$j_h(v_h) = j(q_h v_h) \quad \forall v_h \in V_h .$$

Finally, we approximate (6.2A), (6.3A) and (6.8A), respectively, by:

(6.16A) $$\begin{cases} \textit{Find } u_h \in V_h \textit{ such that} \\ a(u_h,v_h) + \int_\Omega \phi(q_h u_h) q_h v_h\,dx = L(v_h) \quad \forall v_h \in V_h \end{cases}$$

and

(6.17A) $$\begin{cases} \textit{Find } u_h \in V_h \textit{ such that} \\ a(u_h,v_h-u_h)+j_h(v_h)-j_h(u_h) \geq L(v_h-u_h) \quad \forall v_h \in V_h . \end{cases}$$

It can then easily be shown that (6.16A) *and* (6.17A) *are equivalent and admit one and only one solution.*

6.4.3. *Four lemmas*

With the aim of proving the convergence of $(u_h)_h$ to u when $h \to 0$, we shall use the four lemmas below, which are proved in Chan-Glowinski /1A/ and Glowinski /2A/:

<u>Lemma 1.6</u>: *Let* $v \in H_0^1(\Omega)$; *for* $n \in N$ *we define* $\tau_n v$ *by*

(6.18A) $$\tau_n v = \inf(n, \sup(-n,v)).$$

We then have

$$\tau_n v \in H_0^1(\Omega) \cap L^\infty(\Omega) ,$$

$$\lim_{n\to+\infty} \tau_n v = v \quad \textit{strongly in } H_0^1(\Omega)$$

$$\lim_{n\to+\infty} j(\tau_n v) = j(v)$$

(if $\Phi(v) \notin L^1(\Omega)$ *then* $\lim_{n\to+\infty} j(\tau_n v) = +\infty$).

Lemma 6.2: *The solution* u *of* (6.8A) (*and* (6.2A), (6.3A)) *is characterised by*

$$(6.19\text{A})\quad \begin{cases} u \in H_0^1(\Omega)\ ,\ \Phi(u) \in L^1(\Omega), \\ a(u,v-u)+j(v)-j(u) \geq L(v-u) \quad \forall v \in H_0^1(\Omega) \cap L^\infty(\Omega). \end{cases}$$

Lemma 6.3: *The space* $\mathcal{D}(\Omega)$ *is dense in* $H_0^1(\Omega) \cap L^\infty(\Omega)$, *this latter space being provided with the strong topology of* V *and the weak* topology of* $L^\infty(\Omega)$.

Lemma 6.4: *For all* p *such that* $1 \leq p \leq +\infty$ *we have*

$$(6.20\text{A})\quad \|q_h v_h - v_h\|_{L^p(\Omega)} \leq \frac{2}{3} h \|\nabla v_h\|_{L^p(\Omega)\times L^p(\Omega)} \qquad \forall\, v_h \in V_h\ . \blacksquare$$

Lemmas 6.1, 6.2, 6.3 are direct consequences of classical results on the *truncation* of functions of $H^1(\Omega)$ (and $H_0^1(\Omega)$), for which we refer to Stampacchia /1A/.

6.4.4 *Convergence of the approximate solutions*

Regarding the convergence of the approximate solutions when $h \to 0$, we have

Theorem 6.3: *We assume that when* $h \to 0$ *the angles of* $\mathcal{T}_h$ *are bounded below by a constant* $\theta_0 > 0$. *We then have*

$$(6.21\text{A})\quad \lim_{h\to 0} \|u_h - u\|_V = 0,$$

$$(6.22\text{A})\quad \lim_{h\to 0} j_h(u_h) = j(u)$$

where u *is the solution of* (6.2A), (6.3A) *and* (6.8A) *and* u_h *is that of* (6.16A) *and* (6.17A).

Proof: We proceed as in Chan-Glowinski, loc. cit., Glowinski /2A/; the proof is along the same general lines as that of Theorem 2.3, Section 2.3.2.

1) *A priori estimates for* u_h

We take $v_h = 0$ in (6.17A); we then deduce from (6.1A) and the fact that $j_h(0) = 0$

$$(6.23\text{A})\quad \|u_h\|_V \leq \frac{1}{\alpha} \|L\|_* \ ,$$

(6.24A) $0 \leq \int_\Omega \Phi(q_h u_h)\,dx \leq \frac{1}{\alpha}\|L\|_*^2$

where

$$\|L\|_* = \sup_{v\in V-\{0\}} \frac{|L(v)|}{\|v\|_V}.$$

2) Weak convergence of $(u_h)_h$.

It results from (6.23A) and from the *compactness* of the injection from $V = H_0^1(\Omega)$ into $L^2(\Omega)$ that we can extract from $(u_h)_h$ a subsequence – also denoted by $(u_h)_h$ – such that

(6.25A) $\lim_{h\to 0} u_h = u^*$ *weakly in* V

(6.26A) $\lim_{h\to 0} u_h = u^*$ *strongly in* $L^2(\Omega)$

(6.27A) $\lim_{h\to 0} u_h = u^*$ *a.e. on* Ω.

In view of (6.23A) – (6.26A) and Lemma 6.4 (with $p = 2$) we similarly have

(6.28A) $\lim_{h\to 0} q_h u_h = u^*$ *strongly in* $L^2(\Omega)$.

We thus have, up to a new extracted subsequence

(6.29A) $\lim_{h\to 0} q_h u_h = u^*$ *a.e. on* Ω

(6.30A) $\lim_{h\to 0} \Phi(q_h u_h) = \phi(u^*)$ *a.e. on* Ω.

Let $v \in \mathcal{D}(\Omega)$; it then results from e.g. Ciarlet /1A/ that under the assumption on the angles of $\mathcal{T}_h$ given in Theorem 6.3, we have

(6.31A) $\|r_h v - v\|_{W^{1,\infty}(\Omega)} \leq Ch\,\|v\|_{W^{2,\infty}(\Omega)} \quad \forall v \in \mathcal{D}(\Omega)$,

(6.32A) $\|r_h v - v\|_{L^\infty(\Omega)} \leq Ch^2\,\|v\|_{W^{2,\infty}(\Omega)} \quad \forall v \in \mathcal{D}(\Omega)$,

where C denotes various quantities which are independent of v and h, and where r_h is the *interpolation operator* on $\mathcal{T}_h$ defined by

$$\begin{cases} \forall v \in H^1_o(\Omega) \cap C^o(\overline{\Omega}) \text{ we have } r_h v \in V_h \text{ and} \\ r_h v(P) = v(P) \quad \forall P \in \Sigma_h . \end{cases}$$

Moreover, Lemma 6.4 (with p = + ∞) implies

$$\text{(6.33A)} \quad \lim_{h\to 0} \| q_h r_h v - v \|_{L^\infty(\Omega)} = 0 \quad \forall v \in \mathcal{D}(\Omega).$$

Taking $v_h = r_h v$ in (6.17A) we then obtain

$$\text{(6.34A)} \quad \begin{cases} a(u_h,u_h) + \int_\Omega \Phi(q_h u_h)\,dx \le a(u_h, r_h v) + \int_\Omega \Phi(q_h r_h v)\,dx - \\ - L(r_h v - u_h) \quad \forall v \in \mathcal{D}(\Omega) . \end{cases}$$

From (6.25A), (6.26A) and Proposition 6.1, we deduce

$$\text{(6.35A)} \quad a(u^*,u^*)+j(u^*) \le \liminf_{h \to 0} \, (a(u_h,u_h)+j(q_h u_h)).$$

In view of (6.24A), (6.30A) we can apply *Fatou's Lemma* to $\{\phi(q_h u_h)\}_h$ from which we obtain:

$$\text{(6.36A)} \quad \phi(u^*) \in L^1(\Omega).$$

Moreover, we have

$$\lim_{h\to 0} \int_\Omega \Phi(q_h r_h v)\,dx = \int_\Omega \Phi(v)\,dx = j(v) \quad \forall v \in \mathcal{D}(\Omega) ,$$

which, with (6.35A), implies, in the limit in (6.34A)

$$\text{(6.37A)} \quad a(u^*,u^*)+j(u^*) \le a(u^*,v)+j(v)-L(v-u^*) \quad \forall v \in \mathcal{D}(\Omega).$$

It then follows from (6.36A), (6.37A) that u^* satisfies

$$\text{(6.38A)} \quad \begin{cases} u^* \in V, \ \Phi(u^*) \in L^1(\Omega), \\ a(u^*, v-u^*)+j(v)-j(u^*) \ge L(v-u^*) \quad \forall v \in \mathcal{D}(\Omega). \end{cases}$$

Let $v \in V \cap L^\infty(\Omega)$; we deduce from Lemma 6.3 the existence of a sequence $\{v_n\}_n$ such that $v_n \in \mathcal{D}(\Omega)$ $\forall n$ and

$$\text{(6.39A)} \quad \lim_{n\to+\infty} v_n = v \textit{ strongly in } V,$$

$$\text{(6.40A)} \quad \lim_{n\to+\infty} v_n = v \textit{ weakly}^* \textit{ in } L^\infty(\Omega).$$

From (6.38A) we deduce

$$(6.41A) \quad \begin{cases} u^* \in V, \ \Phi(u^*) \in L^1(\Omega) \\ a(u^*,v_n-u^*)+j(v_n)-j(u^*) \geq L(v_n-u^*) \ \ \forall\, n. \end{cases}$$

Moreover (6.39A) implies

$$(6.42A) \quad \lim_{n\to+\infty} a(u^*,v_n-u^*) = a(u^*,v-u) \ ,$$

$$(6.43A) \quad \lim_{n\to+\infty} L(v_n-u^*) = L(v-u^*) \ ,$$

and, up to an extracted subsequence

$$(6.44A) \quad \lim_{n\to+\infty} v_n = v \ \textit{a.e. on } \Omega,$$

which implies

$$(6.45A) \quad \lim_{n\to+\infty} \Phi(v_n) = \Phi(v) \ \textit{a.e. on } \Omega.$$

In view of (6.40) we have

$$\|v_n\|_{L^\infty(\Omega)} \leq \text{const.} \ \ \forall\, n$$

and hence $\forall n$:

$$(6.46A) \quad 0 \leq \Phi(v_n) \leq \text{const.} \ \textit{a.e. on } \Omega.$$

In view of (6.45A), (6.46A) we can apply the *Lebesgue dominated convergence theorem*, which gives:

$$\lim_{n\to+\infty} j(v_n) = \lim_{n\to+\infty} \int_\Omega \Phi(v_n)dx = \int_\Omega \Phi(v)dx = j(v).$$

In the limit in (6.41A) we thus have

$$(6.47A) \quad \begin{cases} u^* \in V, \ \Phi(u^*) \in L^1(\Omega), \\ a(u^*,v-u^*)+j(v)-j(u^*) \geq L(v-u^*) \ \ \forall\, v \in V \cap L^\infty(\Omega) \ . \end{cases}$$

From Lemma 6.2, (6.47A) implies $u^* = u$ where u is the solution of (6.2A), (6.3A) and (6.8A). In view of the uniqueness of u it is the *entire* sequence $(u_h)_h$ which converges *weakly* to u.

3) Strong convergence of $(u_h)_h$.

From (6.17A) and from the *coercivity* of $a(\cdot,\cdot)$, it results that $\forall v \in \mathcal{D}(\Omega)$:

$$(6.48A)\quad \begin{cases} \alpha \|u_h - u\|_V^2 + j_h(u_h) \le a(u_h-u, u_h-u) + j_h(u_h) = \\ = a(u,u) - a(u_h,u) - a(u,u_h) + a(u_h,u_h) + j_h(u_h) \le \\ \le a(u,u) - a(u_h,u) - a(u,u_h) + a(u_h, r_h v) + j_h(r_h v) - L(r_h v - u_h). \end{cases}$$

Using the various convergence results from the second part of the proof, we have, in the limit in (6.48A)

$$(6.49A)\quad \begin{cases} j(u) \le \liminf j_h(u_h) \le \liminf(\alpha \|u_h - u\|_V^2 + j_h(u_h)) \le \\ \le \limsup(\alpha \|u_h - u\|_V^2 + j_h(u_h)) \le \\ \le a(u, v-u) + j(v) - L(v-u) \quad \forall v \in \mathcal{D}(\Omega). \end{cases}$$

Using, as above, the *density* of $\mathcal{D}(\Omega)$ in $V \cap L^\infty(\Omega)$, it can be shown that (6.49A) also holds for all the $v \in V \cap L^\infty(\Omega)$; with τ_n defined by (6.18A) we then have:

$$(6.50A)\quad \begin{cases} j(u) \le \liminf j_h(u_h) \le \liminf(\alpha \|u_h - u\|_V^2 + j_h(u_h)) \le \\ \le \limsup(\alpha \|u_h - u\|_V^2 + j_h(u_h)) \le \\ \le a(u, \tau_n v - u) + j(\tau_n v) - L(\tau_n v - u) \quad \forall v \in V, \forall n. \end{cases}$$

In view of the properties of τ_n (see Lemma 6.1) we have, in the limit in (6.50A) when $n \to +\infty$:

$$(6.51A)\quad \begin{cases} j(u) \le \liminf(\alpha \|u_h - u\|_V^2 + j_h(u_h)) \le \\ \le \limsup(\alpha \|u_h - u\|_V^2 = j_h(u_h)) \le \\ \le a(u, v-u) + j(v) - L(v-u) \quad \forall v \in V. \end{cases}$$

Taking v = u in (6.51A) we thus deduce

$$\lim_{h\to 0} j_h(u_h) = j(u) \ ,$$

$$\lim_{h\to 0} \| u_h - u \|_V = 0 \ ,$$

which proves the Theorem. ■

Descriptions of various iterative methods which can be used to solve the equivalent approximate problems (6.16A), (6.17A) can be found in Chan-Glowinski, loc. cit., which also contains the results of various numerical tests.

7. INTRODUCTION TO NUMERICAL ALGORITHMS FOR QUASI-VARIATIONAL INEQUALITIES [1]

7.1 Quasi-variational inequalities

Consider, in an open bounded [2] domain Ω, an operator

$$(7.1A) \qquad A = -\sum \frac{\partial}{\partial x_i}\left(a_{ij}(x)\frac{\partial}{\partial x_j}\right) + \sum a_i \frac{\partial}{\partial x_i} + a_o$$

with the *assumption of coercivity* on $H^1(\Omega)$ [3]:

$$(7.2A) \qquad a(v,v) \geq \alpha_o \|v\|^2 \quad \forall v \in H^1(\Omega)$$

where

$$(7.3A) \qquad a(u,v) = \sum \int_\Omega a_{ij} \frac{\partial u}{\partial x_j}\frac{\partial v}{\partial x_i}\, dx + \sum \int_\Omega a_i \frac{\partial u}{\partial x_i}\, v\, dx + \int_\Omega a_o uv\, dx.$$

We also introduce a *nonlinear operator*

$$\psi \to M(\psi)$$

from $L^\infty_+(\Omega) \to L^\infty_+(\Omega)$ where $L^\infty_+(\Omega)$ denotes the set of functions (a.e. ≥ 0) in $L^\infty(\Omega)$; we assume that this operator M has the

(1) Abbreviated to Q.V.I.

(2) In order to clarify the concepts.

(3) This assumption can be weakened and we can consider a Hilbert space V with $H^1_o(\Omega) \subseteq V \subseteq H^1(\Omega)$.

following properties:

$$(7.4A)\quad \begin{cases} M \text{ is } \textit{positive increasing} \text{ i.e.} \\ 0 \le \psi_1 \le \psi_2 \Leftrightarrow 0 \le M(\psi_1) \le M(\psi_2) \ , \end{cases}$$

$$(7.5A)\quad M(0) \ge k > 0 \ , \ k = \text{constant}.$$

Under these conditions, there exists *one and only one function*, u, *which is a solution of the Q.V.I.*

$$(7.6A)\quad \begin{cases} a(u,v-u) \ge (f,v-u) \quad \forall v \text{ with } 0 \le v \le M(u) \ , \\ 0 \le u \le M(u) \ . \end{cases}$$

In (7.6A), f is given in $L^\infty_+(\Omega)$.

Q.V.I's have been introduced in Bensoussan-Lions /1A/,/2A/,/3A/, /4A/in the *steady-state* and *time-dependent* (1) cases, *with the operator* M *given by*

$$(7.7A)\quad M(\psi)(x) = k + \inf_{\xi_i \ge 0} \psi(x+\xi) \ , \forall x \in \Omega$$

where in (7.7A) $\psi(x + \xi)$ is taken for the ξ such that $\xi_i \ge 0$ *and* $x + \xi \in \Omega$; the motivation for this work resides *in impulsive stochastic control* (2) (3).

L. Tartar /1A/ has made the observation that, *for a* (non-constructive) *existence proof*, the *assumption that* M *is increasing* is sufficient, and Th. Laetsch /1A/ has proved the *uniqueness* of the solution when the supplementary condition (7.5A) is met.

Remark 7.1: If in (7.6A) M(u) *were known*, we would then have a variational inequality and, in fact, an *obstacle problem*; however, since M(u) is not known, the problem is *implicit* and we have adopted the terminology Q.V.I.

---

(1) Not considered here.

(2) A systematic investigation of this is described in A. Bensoussan, J.L. Lions, *Contrôle Impulsionnel Stochastique*, Dunod-Gauthiers-Villars, to appear.

(3) Impulsive control leads to numerous operators M which are different from (7.7A) but which are still similar in nature.

Synopsis

In the following we shall be concerned solely with a *constructive* method for approximating the solution.

7.2 Iterative scheme

The *iterative scheme* to be introduced is completely standard; starting with $u^o \geq 0$ (suitably chosen; we shall return to this point later) we successively define u^1, u^2.... by the *solution of the variational inequalities (obstacle problem)*

$$\text{(7.8A)} \quad \begin{cases} a(u^n, v-u^n) \geq (f, v-u^n) \quad \forall v \text{ with } 0 \leq v \leq M(u^{n-1}), \\ 0 \leq u^n \leq M(u^{n-1}). \end{cases}$$

This is the scheme introduced in Bensoussan, Goursat and Lions /1/. Various numerical experiments, motivated by practical applications, notably *stock control* (see Goursat and Maarek /1A/), show a *very rapid convergence* of u^n to the solution. In the case where M is given by (7.7A) J.L. Menaldi /1A/ has proved convergence as 1/n by probabilistic methods. However, there is in fact a *very simple general result giving exponential convergence*, due to B. Hanouzet and J.L. Joly /1A/, which we shall now present (see also Cortey-Dumont /1A/).

We start by introducing the function $\bar{u}$, the solution of

$$\text{(7.9A)} \quad a(\bar{u}, v) = (f, v) \quad \forall v \in H^1(\Omega).$$

We have:

$$\text{(7.10A)} \quad \bar{u} \geq 0 \text{ and } \bar{u} \in L^\infty(\Omega).$$

We use the iterative scheme (7.8A) taking u^o such that

$$\text{(7.11A)} \quad 0 \leq u^o \leq \bar{u} .$$

We introduce λ with

$$\text{(7.12A)} \quad 0 < \lambda < 1 \ , \quad \lambda\bar{u} \leq M(0) \text{ a.e. on } \Omega.$$

Remark 7.2: From (7.5A) and (7.10A) *there exists* a λ satisfying (7.12A). In order to obtain a more accurate result (see (7.14A) below) it is advantageous to *choose λ as large as possible.*

Theorem 7.1: *We assume that the operator* M *satisfies* (7.4A), (7.5A) *and is concave in the sense*

$$(7.13A)\quad \begin{cases} 0 \le \psi_1, \psi_2 \implies \\ M((1-\theta)\psi_1 + \theta\psi_2) \ge (1-\theta)M(\psi_1) + \theta M(\psi_2), \\ \textit{if } 0 \le \theta \le 1. \end{cases}$$

We use the iterative scheme (7.8A) *starting from* u^0 *with* (7.11A), *and choose* λ *satisfying* (7.12A). *Then*

$$(7.14A)\quad \|u^n - \bar{u}\|_{L^\infty(\Omega)} \le C(1-\lambda)^n,\ C = \text{constant}.$$

Remark 7.3: If we can choose $\lambda = 1$ satisfying (7.12A), i.e. $\bar{u} \le M(0)$, *then* $\bar{u}$ *is a solution of* the Q.V.I. since we then have $a(\bar{u}, v-\bar{u}) = (f, v-\bar{u})\ \forall v$ and $0 \le \bar{u} \le M(0) \le M(\bar{u})$. In this case the Theorem is trivial.

Proof.

1) We introduce the following notation: $S(\psi)$ *denotes the solution of the V.I. with obstacle* $M(\psi)$, i.e.

$$(7.15A)\quad \begin{cases} a(S(\psi), v-S(\psi)) \ge (f, v-S(\psi))\ \forall v \quad \text{with } 0 \le v \le M(\psi) \\ S(\psi) \le M(\psi). \end{cases}$$

Note that $\bar{u}$ is the solution of the V.I. *with obstacle* $+\infty$ and that the value of the solution of the V.I. with obstacle *increases if the obstacle increases*; thus

$$(7.16A)\quad S(\psi) \le S(\infty) = \bar{u},\ \forall \psi.$$

2) Since $A(\lambda\bar{u}) = \lambda A\bar{u} = \lambda f \le f$, we see that $\lambda\bar{u}$ can be considered as the solution of the obstacle problem when the obstacle is $\lambda\bar{u}$, so that $\lambda\bar{u} \le M(0)$ implies

$$(7.17A)\quad \lambda\bar{u} \le S(0).$$

3) For all v, w with $0 \le v, w \le \bar{u}$, we can choose α, β with $0 \le \alpha, \beta \le 1$ such that

$$(7.18A)\quad -\alpha w \le v - w \le \beta v;$$

we shall now show that (7.18A) implies

$$(7.19A)\quad -(1-\lambda)\alpha\ S(w) \le S(v)-S(w) \le (1-\lambda)\beta\ S(v)\ .$$

In fact, the second inequality (7.18A) can be written

$$(7.20A)\quad w \ge (1-\beta)v = (1-\beta)v + \beta 0\ .$$

However, *the concavity of* M *implies the concavity of* S, i.e.

$$(7.21A)\quad S((1-\theta)\psi_1+\theta\psi_2) \ge (1-\theta)S(\psi_1) + \theta S(\psi_2)$$

so that (7.20A) implies

$$\begin{cases} S(w) \ge (1-\beta)S(v)+\beta\ S(0) \ge (\text{by } (7.17A))\,(1-\beta)S(v)+\beta\lambda\bar{u} \ge \\ \quad \ge (\text{by } (7.16A))\ (1-\beta)S(v)+\beta\lambda S(v) \end{cases}$$

which leads to the second inequality (7.19A).

The first inequality (7.18A) can be written $v \ge (1-\alpha)w + \alpha 0$ and hence

$$S(v) \ge (1-\alpha)S(w)+\alpha S(0) \ge (1-\alpha)S(w)+\alpha\lambda S(w)$$

giving the first inequality (7.19A).

4) We now use (7.18A) (7.19A) by recurrence; we note that $S(u) = u$; taking $w = u^o$, we choose α and β in such a way that

$$-\alpha u^o \le u-u^o \le \beta u\ ;$$

from this we deduce that

$$-(1-\lambda)\alpha S(u^o) \le S(u)-S(u^o) \le (1-\lambda)\beta\ S(u)$$

i.e.

$$-(1-\lambda)\alpha u^1 \le u-u^1 \le (1-\lambda)\beta u.$$

We now apply (7.18A) (7.19A) with $v = u$, $w = u^1$; α has become $(1 - \lambda)\alpha$ and β has become $(1 - \lambda)\beta$; from this we deduce

$$-(1-\lambda)^2\alpha\ S(u^1) \le S(u)-S(u^1) \le (1-\lambda)^2\beta\ S(u)$$

and so on by recurrence, thus

$$(7.22A)\quad -(1-\lambda)^n\alpha\ u^n \le u-u^n \le (1-\lambda)^n\beta\ u$$

giving, *in particular*, (7.14A).

Remark 7.4: We started with $0 \le u^o \le \bar{u}$. It is sufficient to start with $u^o \le$ upper solution, which is chosen to be a *sufficiently large constant*.

Remark 7.5: From a numerical point of view, we shall thus use (7.8A), each V.I. being solved by the methods described elsewhere in the book. A few iterations are sufficient.

Remark 7.6: It has been observed by C. Baiocchi /3A/, (and this observation has given rise to a large number of investigations) that Q.V.I.'s allow the solution of free-boundary problems which arise in the seepage of water in non-rectangular dams. In this case, assumption (7.5A) *is not* satisfied; nevertheless, from the numerical point of view, we observe a 'very rapid' convergence of scheme (7.8A), and this would seem to indicate that Theorem 7.1 could be extended still further.

Appendix 2

FURTHER DISCUSSION OF OPTIMISATION ALGORITHMS

1. SYNOPSIS

In this appendix we shall present a fairly succinct discussion, supplementary to Chapter 2, on *optimisation algorithms*. In Section 2 we describe a *constrained block-overrelaxation algorithm* and present a convergence result.

In Section 3 we return to the duality algorithms of Chapter 2, Section 4, in order to point out some new results on the *convergence of the multipliers*.

Section 4 is devoted to *complementarity* methods and Section 5 to *conjugate gradient* methods for the minimisation of *quadratic functionals over the products of intervals*.

Section 6 describes *alternating direction* algorithms which can be applied to the solution of certain nonlinear variational problems.

In Section 7 we show the relationship which exists between the *augmented Lagrangian* methods mentioned in Chapter 2, Section 6, Chapter 3, Section 10 and Chapter 5, Section 9 and the *alternating direction* methods; for further details on augmented Lagrangian methods and their application to the solution of linear and nonlinear variational problems the reader is referred to the monograph currently under preparation by Fortin-Glowinski /1A/ (and also the corresponding bibliography); see also Marrocco /1A/.

2. THE METHOD OF BLOCK OVERRELAXATION WITH PROJECTION

This section supplements Chapter 2, Section 1.4 on the *minimisation* of finite-dimensional quadratic functionals using *overrelaxation with projection*. We shall describe a *block* variant of algorithm (1.41), (1.42), which is also valid in *infinite-dimensional Hilbert spaces*, and give a convergence result. We follow the presentation of Cea-Glowinski /2/ (see also Glowinski /4/, /5/, Comincioli /1/, Cottle-Golub-Sacher /1A/).

2.1 Statement of the problem

Let V be a finite- or infinite-dimensional real vector space. We assume that V is of the form $V = \prod_{i=1}^{N} V_i$ where the V_i are Hilbert spaces; if $v \in V$ we then have

$$v = \{v_1, \ldots v_N\} \text{ with } v_i \in V_i \ \forall i=1,\ldots N.$$

We denote by $((\cdot,\cdot))_i$ and $\|\cdot\|_i$ the *inner product* and the *norm* associated with V_i; V is then a Hilbert space for the inner product and norm defined by

$$((u,v)) = \sum_{i=1}^{N} ((u_i,v_i))_i \ , \ \|v\| = \Big(\sum_{i=1}^{N} \|v_i\|_i^2\Big)^{1/2}. \tag{2.1A}$$

Let K be a *nonempty closed convex* subset of V, of the form

$$K = \prod_{i=1}^{N} K_i \tag{2.2A}$$

where i = 1, ... N, and the K_i are (nonempty) closed convex subsets of V_i.

We next consider $J : V \to \mathbb{R}$ defined by

$$J(v) = \frac{1}{2}a(v,v) - ((f,v)), \tag{2.3A}$$

where $f \in V$ and $a: V \times V \to \mathbb{R}$ is a *bilinear* form which is *continuous*, *symmetric* and *coercive* (i.e. $\exists \alpha > 0$ such that $a(v,v) \geq \alpha\|v\|^2 \ \forall v \in V$).

It then follows from the above properties that the minimisation problem

$$\begin{cases} \textit{Find } u \in K \textit{ such that} \\ J(u) \leq J(v) \quad \forall v \in K \end{cases} \tag{2.4A}$$

admits one and only one solution characterised by

$$\begin{cases} u \in K, \\ a(u,v-u) \geq ((f,v-u)) \quad \forall v \in K. \end{cases} \tag{2.5A}$$

2.2 Description of the algorithm

We note that

$$J(v) = J(v_1, \dots v_N) = \frac{1}{2} \sum_{1 \le i, j \le N} a_{ij}(v_i, v_j) - \sum_{i=1}^{N} ((f_i, v_i)) \tag{2.6A}$$

where $f_i \in V_i$ and where the a_{ij} are continuous bilinear forms on $V_i \times V_j$, the a_{ii} being symmetric and coercive (with $a_{ii}(v_i, v_i) \ge \alpha \|v_i\|_i^2 \quad \forall v_i \in V_i$) and, more generally

$$a_{ij} = a_{ji}^* \; (\Longleftrightarrow a_{ij}(v_i, v_j) = a_{ji}(v_j, v_i) \;\; \forall v_i \in V_i, v_j \in V_j). \tag{2.7A}$$

By using *Riesz's theorem* we deduce from (2.7A) the existence of operators $A_{ij} \in \mathcal{L}(V_j, V_i)$ such that

$$a_{ij}(v_i, v_j) = ((v_i, A_{ij} v_j))_i \; , \; A_{ij} = A_{ji}^* \; , \tag{2.8A}$$

where the A_{ii} (which are self-adjoint) are isomorphisms from V_i onto V_i. As the forms a_{ii} are *symmetric* and *coercive*, they define a Hilbert structure on V_i, whose norm $|||\cdot|||_i$ defined by

$$|||v_i|||_i^2 = a_{ii}(v_i, v_i) = ((A_{ii} v_i, v_i))_i \tag{2.9A}$$

is equivalent to the norm $\|\cdot\|_i$; we shall denote by P_i the *orthogonal projection operator* from V_i onto K_i according to the norm $|||\cdot|||_i$. Now let us consider N positive scalars $\omega_1, \dots \omega_N$; to solve (2.4A) we use the following generalisation of algorithm (1.41), (1.42) of Chapter 2:

$$u^0 = \{u_1^0, \dots u_N^0\} \text{ given arbitrarily in } K \tag{2.10A}$$

then for $n \ge 0$ *successively determine* u_i^{n+1}, $i = 1, \dots, N$, *by*

$$u_i^{n+1} = P_i(u_i^n - \omega_i A_{ii}^{-1}(\sum_{j<i} A_{ij} u_j^{n+1} + \sum_{j \ge i} A_{ij} u_j^n - f_i)). \tag{2.11A}$$

Remark 2.1: A variational description of algorithm (2.10A) (2.11A) is given in Comincioli /1/.

2.3 Convergence of algorithm (2.10A), (2.11A)

The following theorem is proved in Cea-Glowinski /2/:

Theorem 2.1: *If* $0 < \omega_i < 2$, $\forall i = 1, \dots N$, *we have for the sequence* $\{u^n\}_n$ *defined by* (2.10A), (2.11A)

$$\lim_{n\to+\infty} \|u^n-u\| = 0 ,$$

where u *is the solution of* (2.4A).

2.4 Various remarks

Remark 2.2: In the case in which dim V < + ∞, with $K = V = \prod_{i=1}^{N} V_i$, and in which $\forall i=1,\ldots N$ we have $\omega_i = \omega > 1$ (resp. $\omega = 1$, $\omega < 1$), algorithm (2.10A), (2.11A) coincides with the standard *block overrelaxation* (resp. *relaxation*, *underrelaxation*) algorithm associated with the solution of the linear system Au = f (see Varga /1/, Young /1A/) with $A \in \mathcal{L}(V,V)$ defined by $a(u,v) = ((Au,v))\ \forall\, u,v \in V$.

Remark 2.3: In the case in which $V = \mathbb{R}^N$, we can identify the above operator A with an N×N symmetric positive-definite matrix.

If $K = \prod_{i=1}^{N} K_i$ with $K_i = \{v_i \mid a_i \le v_i \le b_i\}$, then algorithm (2.10A), (2.11A) reduces to algorithm (1.41), (1.42) of Chapter 2, Section 1.4.

Remark 2.4: An application of algorithm (2.10A), (2.11A) was described in Chapter 3, Section 8.3. Further applications may be found in Marrocco /1A/, Fortin-Glowinski /1A/ and Glowinski-Marrocco /1A/, in connection with the use of *augmented Lagrangian* methods.

3. DUALITY METHODS - A FURTHER DISCUSSION

We shall now supplement Theorems 4.1, 4.2 and Remark 4.7 of Chapter 2, Section 4 (1) by describing the behaviour of the sequences $\{p^n\}_n$ defined by algorithms (4.12), (4.13) and (4.31), (4.32).

To this end we use the following lemma (see Opial /1A/):

Opial's Lemma: *Let* L *be a Hilbert space*, X *a subset of* L *and* $\{p^n\}_n$ *a sequence in* L *such that for all* $p \in X$ *the sequence* $\{\|p^n-p\|_L\}_n$ *is convergent. If every subsequence* $\{p^{n_k}\}_k$ *which converges weakly has its limit in* X *then* $\{p^n\}_n$ *converges to an element of* X.

We can then supplement Theorem 4.1 of Chapter 2, Section 4 by

(1) The notation of which we retain.

Theorem 3.1: *Assume that the conditions for the application of Theorem 4.1 of Chapter 2, Section 4 are satisfied; if the condition*

$$(3.1A)\qquad 0 < \rho_n = \rho < \frac{2\alpha}{C_1^2}$$

is satisfied we then have for the sequence $\{p^n\}_n$ *defined by* (4.12), (4.13)

$$(3.2A)\qquad \lim_{n\to+\infty} \|p^{n+1}-p^n\|_L = 0,$$

$$(3.3A)\qquad \lim_{n\to+\infty} p_n = p \text{ weakly in } L$$

where p *is such that* $\{u,p\}$ *is a saddle point of* $\mathcal{L}$ *on* $M\times\Lambda$.

Proof: With the Lagrangian $\mathcal{L}(\cdot,\cdot)$ defined by

$$(3.4A)\qquad \mathcal{L}(v,q) = J_o(v)+(q,\phi(v))_L$$

(see Chapter 2, Section 4.3, relation (4.9)) we consider the differences $\mathcal{L}(u^{n+1},p^{n+1})-\mathcal{L}(u^n,p^n)$. We have

$$(3.5A)\qquad \begin{cases} \mathcal{L}(u^{n+1},p^{n+1})-\mathcal{L}(u^n,p^n) = \\ = J_o(u^{n+1})-J_o(u^n)+(p^{n+1},\phi(u^{n+1}))_L-(p^n,\phi(u^n))_L \geq \\ \geq (J_o'(u^n),u^{n+1}-u^n)+(p^{n+1},\phi(u^{n+1}))_L-(p^n,\phi(u^n))_L = \\ = (J_o'(u^n),u^{n+1}-u^n)+(p^n,\phi(u^{n+1})-\phi(u^n))_L + \\ \quad + (p^{n+1}-p^n,\phi(u^{n+1}))_L . \end{cases}$$

Since the vector u^n is a solution of the minimisation problem

$$(3.6A)\qquad \min_{v\in M} \mathcal{L}(v,p^n)$$

we have

$$(3.7A)\qquad (J_o'(u^n),v-u^n)+(p^n,\phi(v)-\phi(u^n))_L \geq 0 \quad \forall v \in M$$

and hence

$$(J_o'(u^n),u^{n+1}-u^n)+(p^n,\phi(u^{n+1})-\phi(u^n))_L \geq 0$$

which in conjunction with (3.5A) implies

$$\mathcal{L}(u^{n+1},p^{n+1})-\mathcal{L}(u^n,p^n) \geq (p^{n+1}-p^n,\phi(u^{n+1}))_L \ ,$$

or alternatively

$$\text{(3.8A)} \quad \begin{cases} \mathcal{L}(u^{n+1},p^{n+1})-\mathcal{L}(u^n,p^n) \geq \\ \quad (p^{n+1}-p^n,\phi(u^{n+1})-\phi(u^n))_L+(p^{n+1}-p^n,\phi(u^n))_L. \end{cases}$$

From the properties of the projections, we have

$$(q-p^{n+1},p^n+\rho\phi(u^n)-p^{n+1})_L \leq 0 \quad \forall q \in \Lambda \ ,$$

and hence in particular $(p^n-p^{n+1},p^n+\rho\phi(u^n)-p^{n+1})_L \leq 0$, from which we deduce

$$\text{(3.9A)} \qquad ||p^{n+1}-p^n||_L^2 \leq \rho(\phi(u^n),p^{n+1}-p^n)_L.$$

Taking (3.9A) into account, (3.8A) implies

$$\text{(3.10A)} \quad \begin{cases} \mathcal{L}(u^{n+1},p^{n+1})-\mathcal{L}(u^n,p^n) \geq \\ \quad (p^{n+1}-p^n,\phi(u^{n+1})-\phi(u^n))_L + \frac{1}{\rho}||p^{n+1}-p^n||_L^2 \ . \end{cases}$$

It now remains to estimate $(p^{n+1}-p^n,\phi(u^{n+1})-\phi(u^n))_L$; we have

$$p^{n+1} = P_\Lambda(p^n+\rho\phi(u^n)) \ ,$$

$$p^{n+2} = P_\Lambda(p^{n+1}+\rho\phi(u^{n+1})),$$

and hence, since the projection is *contractive*

$$\text{(3.11A)} \quad \begin{cases} ||p^{n+2}-p^{n+1}||_L^2 \leq ||p^{n+1}-p^n||_L^2+2\rho(p^{n+1}-p^n,\phi(u^{n+1})-\phi(u^n))_L + \\ \qquad \rho^2 \, ||\phi(u^{n+1})-\phi(u^n)||_L^2 \ , \end{cases}$$

or alternatively (ϕ being *Lipschitz continuous* with constant C_1)

$$(3.12A)\quad \begin{cases} (p^{n+1}-p^n,\phi(u^{n+1})-\phi(u^n))_L \geq \\ \frac{1}{2\rho}(\|p^{n+2}-p^{n+1}\|_L^2-\|p^{n+1}-p^n\|_L^2)-\frac{\rho}{2}\|\phi(u^{n+1})-\phi(u^n)\|_L^2 \geq \\ \frac{1}{2\rho}(\|p^{n+2}-p^{n+1}\|_L^2-\|p^{n+1}-p^n\|_L^2)-\frac{\rho}{2}c_1^2\|u^{n+1}-u^n\|_V^2 . \end{cases}$$

By addition we then deduce from (3.10A), (3.12A)

$$\mathcal{L}(u^{n+1},p^{n+1})-\mathcal{L}(u^n,p^n) \geq \frac{1}{2\rho}\|p^{n+1}-p^n\|_L^2-\frac{\rho}{2}c_1^2\|u^{n+1}-u^n\|_V^2 ,$$

and hence by summation and $\forall n \geq 0$

$$(3.13A)\quad \begin{cases} \mathcal{L}(u^{n+1},p^{n+1})-\mathcal{L}(u^o,p^o) \geq \\ \frac{1}{2\rho}\sum_{j=0}^{n}\|p^{j+1}-p^j\|_L^2-\frac{\rho}{2}c_1^2\sum_{j=0}^{n}\|u^{j+1}-u^j\|_V^2 . \end{cases}$$

We shall now show that under condition (3.1A) the sequence $\{\sum_{j=0}^{n}\|u^{j+1}-u^j\|_V^2\}_n$ is convergent: in fact from (3.7A) we have

$$(J_o'(u^n),u^{n+1}-u^n)_V+(p^n,\phi(u^{n+1})-\phi(u^n))_L \geq 0$$

$$(J_o'(u^{n+1}),u^n-u^{n+1})_V+(p^{n+1},\phi(u^n)-\phi(u^{n+1}))_L \geq 0$$

and hence, by addition

$$(3.14A)\quad \begin{aligned} (p^n-p^{n+1},\phi(u^{n+1})-\phi(u^n))_L &\geq (J_o'(u^{n+1})-J_o'(u^n),u^{n+1}-u^n) \\ &\geq \alpha\|u^{n+1}-u^n\|_V^2 \end{aligned}$$

which in conjunction with (3.11A) implies

$$\|p^{n+1}-p^n\|_L^2-\|p^{n+2}-p^{n+1}\|_L^2 \geq \rho(2\alpha-c_1\rho)\|u^{n+1}-u^n\|_V^2 ,$$

and hence (by summation)

(3.15A) $$\|p^1-p^o\|_L^2 - \|p^{n+2}-p^{n+1}\|_L^2 \geq \rho(2\alpha-C_1\rho) \sum_{j=0}^{n} \|u^{j+1}-u^j\|_V^2 \,.$$

The proof of Theorem 4.1 in Chapter 2, Section 4.3 shows that under condition (3.1A) the sequence $\{p^n\}_n$ is bounded in L; (3.15A) then implies that the sequence $\{\sum_{j=0}^{n} \|u^{j+1}-u^j\|_V^2\}_n$ is convergent. In order to prove the convergence of the sequence $\{\sum_{j=0}^{n} \|p^{j+1}-p^j\|_L^2\}_n$, and hence that $\lim_{n\to+\infty} \|p^{n+1}-p^n\|_L = 0$, it remains to show that the term on the left-hand side of the inequality (3.13A) is bounded; since the sequence $\{u^n\}_n$ (resp. $\{p^n\}_n$) is convergent (resp. bounded) this follows simply from the continuity of $\mathcal{L}(\cdot\,,\cdot)$ and the linearity of $q \to (q,\phi(v))_L$. We have thus proved (3.2A).

We shall now use *Opial's Lemma* in order to prove (3.3A); let $u \in M$ be the solution of the *primal problem*

$$\operatorname*{Inf}_{v\in M}\ \operatorname*{Sup}_{q\in\Lambda} \mathcal{L}(v,q)$$

and let

$$X = \{p \in \Lambda\ ,\ \{u,p\}\ \textit{is a saddle point of } \mathcal{L} \textit{ on } M\times\Lambda\}.$$

The proof of Theorem 4.1 in Chapter 2, Section 4 shows that if condition (3.1A) is satisfied we have convergence of the sequence $\{\|p^n-p\|_L\}_n \quad \forall p \in X$. In order to be able to apply Opial's Lemma, it therefore remains to prove that *any* weakly convergent subsequence $\{p^{n_k}\}_k$ extracted from $\{p^n\}_n$ *has its limit in* X. We know that u^{n_k}, p^{n_k}, p^{n_k+1} satisfy

(3.16A) $$\begin{cases} (J_o'(u^{n_k}),v-u^{n_k})+(p^{n_k},\phi(v)-\phi(u^{n_k}))_L \geq 0 \quad \forall v \in M, \\ u^{n_k} \in M, \end{cases}$$

(3.17A) $$\begin{cases} (q-p^{n_k+1},p^{n_k}+\rho\phi(u^{n_k})-p^{n_k+1})_L \leq 0 \quad \forall q \in \Lambda\ , \\ p^{n_k} \in \Lambda\ . \end{cases}$$

Under condition (3.1A) we have $\lim_{k\to+\infty} u^{n_k} = u$ *strongly* in V, with $u \in M$ since M is *closed* in V; similarly we have $\lim_{k\to+\infty} p^{n_k} = p^*$ *weakly* in L, $\lim_{k\to+\infty} (p^{n_k+1}-p^{n_k})=0$ *strongly* in L (from (3.2A)) with $p^* \in \Lambda$ since Λ is convex and closed in L.

In view of these properties, we have in the limit in (3.16A), (3.17A)

$$\begin{cases} (J_o'(u), v-u) + (p^*, \phi(v) - \phi(u))_L \geq 0 \quad \forall v \in M, \\ u \in M, \end{cases}$$

$$\begin{cases} (q-p^*, \phi(u)) \leq 0 \quad \forall q \in \Lambda \ , \\ p^* \in \Lambda \ , \end{cases}$$

which shows that $\{u, p^*\}$ is a saddle-point of $\mathcal{L}$ on $M \times \Lambda$, and hence $p^* \in X$. We can thus apply Opial's Lemma to prove (3.3A), thereby proving Theorem 3.1. ■

Remark 3.1: The above results also hold for the sequence $\{u^n, p^n\}_n$ defined by algorithm (4.31), (4.32) of Chapter 2, Section 4.4. ■

4. INTRODUCTION TO COMPLEMENTARITY METHODS

4.1 General remarks. Synopsis.

This fourth section constitutes an introduction to *complementarity methods*, which have been widely used in recent years for the numerical solution of variational inequality problems of the *obstacle problem* type (see Appendix 1, Section 3).

The model problem for *linear complementarity* is as follows:

Given a matrix A *of type* N × N *and* $b \in \mathbb{R}^N$, *seek* $u \in \mathbb{R}^N$ *such that*

$$\text{(4.1A)} \qquad \begin{cases} Au-b \geq 0, \\ u \geq 0, \\ (Au-b, u) = 0 \ ; \end{cases}$$

in (4.1A), $(\cdot,\cdot)$ denotes the ordinary Euclidean inner product of $\mathbb{R}^N$, and in general

$$v = \{v_1, \dots v_N\} \geq 0 \Leftrightarrow v_i \geq 0 \quad \forall i=1,\dots N,$$

$$w \geq v \Leftrightarrow w-v \geq 0.$$

Hereafter we shall use the notation $\mathbb{R}_+^N = \{v \in \mathbb{R}^N, \ v \geq 0\}$.

It is clear that (4.1A) is *equivalent* to the *variational inequality* problem

$$(4.2\text{A}) \qquad \begin{cases} u \in \mathbb{R}_+^N \\ (Au, v-u) \geq (b, v-u) \quad \forall v \in \mathbb{R}_+^N \ . \end{cases}$$

If A is symmetric, (4.2A) is a necessary condition (and sufficient if A is positive semi-definite) of optimality for the minimisation problem

$$(4.3\text{A}) \qquad \begin{cases} J(u) \leq J(v) \quad \forall v \in \mathbb{R}_+^N \\ u \in \mathbb{R}_+^N \end{cases}$$

with

$$(4.4\text{A}) \qquad J(v) = \frac{1}{2} (Av, v) - (b, v).$$

With the above problems we shall associate

a) <u>The least element problem</u>

$$(4.5\text{A}) \qquad \begin{cases} \textit{Find } u \in \hat{K} \textit{ such that} \\ u \leq v \quad \forall v \in \hat{K} \end{cases}$$

where

$$(4.6\text{A}) \qquad \hat{K} = \{v \in \mathbb{R}_+^N \ , \ Av-b \geq 0\}$$

b) <u>The family of linear programming problems</u>

$$(4.8\text{A}) \qquad \begin{cases} \textit{Find } \lambda \in \hat{K}^* \textit{ such that} \\ (b, \lambda) \geq (b, \mu) \quad \forall \mu \in \hat{K}^* \end{cases}$$

where

$$\hat{K}^* = \{\mu \in \mathbb{R}_+^N, \ A^t\mu - p \le 0\} . \tag{4.9A}$$

Sufficient conditions on A *and* p for all the above problems to be equivalent are given in Cryer-Dempster /1A/ (see also the bibliography therein); if this equivalence can be established it is then possible to solve the variational inequality (4.2A) by *pivoting* methods inspired by the *simplex method* of linear programming.

We shall show in Section 4.2 (which is inspired by Cryer-Dempster, *loc. cit.*) that the model problem for *variational inequalities* of *obstacle* type

$$\begin{cases} u \in K \\ \int_\Omega \nabla u \cdot \nabla (v-u)\, dx \ge \int_\Omega f(v-u)\, dx \quad \forall v \in K \end{cases} \tag{4.10A}$$

where

$$K = \{v \in H_0^1(\Omega), \ v \ge 0 \ \text{a.e.}\} \tag{4.11A}$$

is equivalent to a linear complementarity problem (formally of the same type as (4.1A)) which is itself equivalent to an (*infinite-dimensional*) *linear programming* problem.

A discussion of complementarity methods, together with a number of references, will be given in Section 4.3.

4.2 The obstacle problem from the point of view of complementarity methods

We follow the account of Cryer-Dempster /1A/.

4.2.1 *Formulation of the problem. Standard results.*

Let Ω be a bounded open set in $\mathbb{R}^N$ with regular boundary Γ; we consider the *variational inequality* problem

$$\begin{cases} u \in K, \\ \int_\Omega \nabla u \cdot \nabla (v-u)\, dx \ \ge \langle f, v-u \rangle \quad \forall v \in K \end{cases} \tag{4.12A}$$

where, in (4.12A), K is given by (4.11A), $<\cdot,\cdot>$ denotes the duality between $H^{-1}(\Omega)$ and $H_o^1(\Omega)$ and $f \in H^{-1}(\Omega)$.

We know that (4.12A) (which was investigated earlier in Chapter 1, Sections 3.6 and 4.7 and in Appendix 1, Sections 3, 4) admits one and only one solution characterised by

$$(4.13A)\quad \begin{cases} J(u) \le J(v) \quad \forall v \in K, \\ u \in K \end{cases}$$

(where $J(v) = \frac{1}{2}\int_\Omega |\nabla v|^2 dx - <f,v>$) and also, K being a *convex cone with vertex* 0

$$(4.14A)\quad \begin{cases} \int_\Omega \nabla u \cdot \nabla v \, dx \ge <f,v> \quad \forall v \in K, \\ \int_\Omega \nabla u \cdot \nabla u \, dx = <f,u> \quad , \\ u \in K \; ; \end{cases}$$

(4.14A) can also be written

$$(4.15A)\quad \begin{cases} <-\Delta u - f, v> \ \ge 0 \quad \forall v \in K, \\ <-\Delta u - f, u> \ = 0 \ , \\ u \in K. \end{cases}$$

If $f \in L^2(\Omega)$, which implies that $u \in H^2(\Omega) \cap H_o^1(\Omega)$, and hence that $-\Delta u - f \in L^2(\Omega)$, we deduce from (4.15A)

$$(4.16A)\quad \begin{cases} -\Delta u - f \ge 0 \text{ a.e. on } \Omega \\ (-\Delta u - f)u = 0 \text{ a.e. on } \Omega \\ u \in K. \end{cases}$$

Conversely, it follows from (4.15A) that the solution u of (4.16A) is a solution of (4.12A).

It is clear that the *infinite-dimensional* problems (4.15A), (4.16A) are formally of the same type as (4.1A) and should thus also be considered as *linear complementarity* problems (in $H_o^1(\Omega)$).

Remark 4.1: The above properties extend to variational inequalities which are more complicated than (4.12A) and which possibly relate to *nonlinear* elliptic operators.

4.2.2 *Other properties associated with* (4.12A)

In view of Section 4.1 it is natural to associate the following problems with the equivalent problems (4.12A), (4.13A), (4.15A):

a) The least element problem

$$\text{(4.17A)} \quad \begin{cases} \textit{Find } u \in \hat{K} \textit{ such that} \\ \\ u \le v \text{ a.e. } \textit{on } \Omega, \ \forall v \in \hat{K} \end{cases}$$

where

$$\text{(4.18A)} \quad \begin{cases} \hat{K} = \{v \in K, \ -\Delta v - f \ge 0\} \\ \\ \quad = \{v \in K, \ \int_\Omega \nabla v \cdot \nabla w \, dx \ \ge \langle f,w\rangle \ \ \forall w \in K\} \end{cases}$$

b) The family of linear problems

$$\text{(4.19A)} \quad \begin{cases} \textit{Find } u \in \hat{K} \textit{ such that} \\ \\ \langle p,u\rangle \le \langle p,v\rangle \quad \forall v \in \hat{K} \end{cases}$$

where $p \in H^{-1}(\Omega)$ is *given*. Here again we therefore have a family of linear programming problems with parameter p. The *dual problem* to (4.19A) is given by

$$\text{(4.20A)} \quad \begin{cases} \textit{Find } \lambda \in \hat{K}^* \textit{ such that} \\ \\ \langle f,\lambda\rangle \ \ge \ \langle f,\mu\rangle \quad \forall \mu \in \hat{K}^* \end{cases}$$

$$\text{(4.21A)} \quad \hat{K}^* = \{\mu \in K, \ -\Delta\mu - p \le 0\} = \{\mu \in K, \ \int_\Omega \nabla\mu \cdot \nabla v \, dx \ \le \langle p,v\rangle \ \ \forall v \in K\} .$$

4.2.3 *The equivalence of problems* (4.12A), (4.13A), (4.15A) (4.17A), (4.19A)

By using a suitable positivity assumption on p, we shall now show that the above problems are equivalent. This will result from the following lemma, due to Stampacchia /1A/, /2A/;

Lemma 4.1: *If* $v_1, v_2 \in \hat{K}$ *then*

$$w = \inf(v_1, v_2) \; \left(= \frac{v_1+v_2}{2} - \frac{|v_1-v_2|}{2} \right) \quad \textit{also belongs to } K.$$

and also from

Lemma 4.2: *There is equivalence between problems* (4.12A), (4.13A), (4.15A) *and the variational inequality problem*

$$\text{(4.22A)} \qquad \begin{cases} u \in \hat{K} \\ \displaystyle\int_\Omega \nabla u \cdot \nabla(v-u)\,dx \geq \langle f, v-u \rangle \quad \forall v \in \hat{K}\,. \end{cases}$$

The proof of Lemma 4.2 is immediate.

We are now in a position to prove

Theorem 4.1: *There is equivalence between problems* (4.12A), (4.13A), (4.15A) *and the least element problem* (4.17A). *If, in addition, we have*

$$\text{(4.23A)} \qquad \langle p, v \rangle > 0 \quad \forall v \in K - \{0\}$$

then the linear programming problem (4.19A) *is equivalent to* (4.12A), (4.13A), (4.15A), (4.17A).

Proof: 1) The equivalence between (4.12A), (4.13A), (4.15A) has already been proved elsewhere. We shall now prove the equivalence between (4.15A) and (4.17A). Let u be the solution of (4.15A); we then have $u \in \hat{K}$, and hence (see Lemma 4.1):

$$w = \inf(u, v) \in \hat{K} \quad \forall v \in \hat{K}\,,$$

which implies, since $u - w \geq 0$ (i.e. $u - w \in K$),

$$\text{(4.24A)} \qquad \langle -\Delta w - f, u - w \rangle \geq 0.$$

We also have, since u is the solution of (4.15A) (and hence of (4.12A)) and since $w \in \hat{K} \subset K$

$$\text{(4.25A)} \qquad \langle -\Delta u - f, w - u \rangle \geq 0.$$

By adding (4.24A) and (4.25A) we thus have

$$0 \geq \langle -\Delta(w-u), w-u \rangle = \int_\Omega |\nabla(w-u)|^2 dx$$

and hence

$$\forall v \in \hat{K}\ ,\ u = w = \inf(v,u)$$

which implies $u \leq v \quad \forall v \in \hat{K}$ i.e. u is a solution of (4.17A). Conversely if u is a solution of (4.17A) we have

$$-\Delta u - f \geq 0 \quad \text{since} \quad u \in \hat{K}$$

and

$$v-u \geq 0 \quad \forall v \in \hat{K}\ ,$$

and hence

$$\begin{cases} \int_\Omega \nabla u \cdot \nabla(v-u)dx - \langle f, v-u \rangle = \langle -\Delta u - f, v-u \rangle \geq 0 \quad \forall v \in \hat{K}\ , \\ u \in \hat{K}, \end{cases}$$

which shows, from Lemma 4.2, that u is a solution of (4.12A), (4.15A).

2) If u is a solution of (4.17A) we have $v-u \geq 0 \quad \forall v \in \hat{K}$; if p satisfies (4.23) we thus have

$$\langle p, v-u \rangle \geq 0 \quad \forall v \in \hat{K}$$

which shows that (4.17A) implies (4.19A).

Conversely, if u is a solution of (4.19A) we have

$$\forall v \in \hat{K}\ ,\ w = \inf(u,v) \in \hat{K}$$

and hence

$$\langle p, w-u \rangle \geq 0. \tag{4.26A}$$

However (4.23A) implies, since $u-w \geq 0$ (i.e. $u-w \in K$)

$$\langle p, u-w \rangle \geq 0. \tag{4.27A}$$

By adding (4.26A), (4.27A) we obtain

$$\begin{cases} \langle p, u-w \rangle = 0\ , \\ u-w \geq 0 \end{cases}$$

and hence w = u, from (4.23A).

We thus have $u \in \hat{K}$, $u = \inf(u,v) \quad \forall v \in \hat{K}$, and hence $v \geq u \quad \forall v \in \hat{K}$, which proves that u is a solution of (4.17A). The theorem is thus completely proved.

Remark 4.2: There is no difficulty in finding $p \in H^{-1}(\Omega)$ such that $\langle p,v\rangle > 0 \quad \forall v \in K-\{0\}$. In fact it is sufficient to define p by

$$(4.28A) \qquad \langle p,v\rangle = C\int_\Omega v\,dx \quad \forall v \in H_0^1(\Omega)$$

where C is a *positive constant*.

4.3 Discussion and references

4.3.1 *Discussion*

From Section 4.2 we see that it is possible to reduce the task of solving the obstacle problem (4.12A) to that of solving an equivalent linear programming problem. From a practical point of view a discretisation is performed using *finite differences* or *finite elements* and the discrete analogues of problems (4.12A), (4.13A), (4.15A), (4.17A), (4.19A) are considered; in particular the discretised form of problem (4.19A) can be solved by a *pivoting* method of the simplex type. In this case the equivalence Theorem 4.1 results from the fact that the operator $-\Delta$ associated with Dirichlet boundary conditions obeys the *maximum principle* (this also holds for more complicated second-order operators and for other forms of boundary condition; see Stampacchia /1A/, /2A/; in order that the equivalence properties of Theorem 4.1 can be carried over to the discrete case it is necessary that the operator approximating $-\Delta$ obey a *discrete maximum principle* (see Ciarlet-Raviart /1A/). In the case of the obstacle problem (4.12A) with $\Omega \subset \mathbb{R}^2$, discretised by a *triangular finite-element method*, the discrete maximum principle will be satisfied if

(i) we use a first-order finite-element approximation (i.e. globally continuous and piecewise affine)

(ii) all the angles of the triangulation are $\leq \frac{\pi}{2}$.

If some of the angles are obtuse and/or finite elements of order ≥ 2 are used, the discrete maximum principle no longer applies.

We should also point out that a certain number of operators which are of fundamental importance in practical applications, such as the *biharmonic operator* $\psi \to \Delta^2\psi$ with the boundary conditions $\psi|_\Gamma = g_1$, $\frac{\partial\psi}{\partial n}\big|_\Gamma = g_2$, and similarly the *linear elasticity operator* do not satisfy a maximum principle.

In view of the above remarks, it would appear that these complementarity methods by definition have a fairly limited range of application, at least as far as the solution of variational inequalities is concerned. A further discussion of the above topics can be found in Cryer-Dempster /1A/.

4.3.2 *References*

Complementarity methods have been investigated by numerous authors, particularly with regard to the solution of variational inequalities of the obstacle problem type.

To attempt to give a list of all the references relating to this class of methods would be out of the question, and we therefore restrict ourselves to mentioning only the references below, the bibliographies of which are also worth consulting. Considering only works which are oriented towards the solution of free-boundary problems of the obstacle problem type, we may mention among others: Cottle-Sacher /1A/, Cottle-Golub-Sacher /1A/, Cottle /1A/, /2A/, Mosco-Scarpini /1A/, Scarpini /1A/ and of course Cryer-Dempster, *loc. cit.*

5. MINIMISATION OF QUADRATIC FUNCTIONALS OVER THE PRODUCTS OF INTERVALS, USING CONJUGATE GRADIENT METHODS

5.1 Synopsis

In this fifth section we shall supplement the investigations of Chapter 2, Section 2.3 (see also the discussion in Chapter 2, Section 6) with regard to the conjugate gradient method. We consider the model problem in $\mathbb{R}^N$

$$\text{(5.1A)} \quad \begin{cases} \textit{Find } u \in K \textit{ such that} \\ J(u) \leq J(v) \quad \forall\, v \in K, \end{cases}$$

where

$$\text{(5.2A)} \quad K = \prod_{i=1}^{N} K_i, \quad K_i = [a_i, b_i] \subset \mathbb{R}$$

$(a_i, b_i$ *finite or otherwise*),

$$\text{(5.3A)} \quad J(v) = \frac{1}{2}(Av,v) - (b,v) \;,$$

where A is an $N \times N$ symmetric positive-definite matrix and $b \in \mathbb{R}^N$; as usual (in this book) $(\cdot,\cdot)$ denotes the canonical Euclidean inner product of $\mathbb{R}^N$ (and more generally of $\mathbb{R}^p$). In Section 5.2 we shall describe an algorithm of *conjugate gradient* type, which allows an efficient solution of (5.1A) - (5.3A), and does so *in a finite number of iterations* (i.e. *as in the constraint-free case* of Chapter 2, Section 2.3).

5.2 Description of the method. Convergence results

In this section we have followed the account of D.P. O'Leary /1A/ to which we refer for further information and numerical examples.

5.2.1 *General remarks*

Let $u = \{u_1,\ldots u_N\}$ be the solution of (5.1A) - (5.3A), and let $\lambda = \{\lambda_1,\ldots\lambda_N\}$ be defined by

$$\text{(5.4A)} \qquad \lambda = Au - b ;$$

it can easily be shown that u *is characterised* by

$$\text{(5.5A)} \qquad \begin{cases} \forall\ i=1,\ldots N, \textit{ we have} \\ \lambda_i \geq 0 \textit{ if } u_i = a_i,\ \lambda_i \leq 0 \textit{ if } u_i = b_i , \\ \lambda_i = 0 \textit{ if } a_i < u_i < b_i . \end{cases}$$

Remark 5.1: The above characterisation contains λ as a Kuhn-Tucker multiplier (see e.g. Rockafellar /3, Section 28/) for the problem (5.1A) - (5.3A). ■

Remark 5.2: If $\forall i = 1,\ldots N$, we have $a_i = 0$, $b_i = +\infty$ then the above characterisation reduces to the linear complementarity problem

$$\text{(5.6A)} \qquad \begin{cases} \textit{Find } u \in \mathbf{R}^N \textit{ such that} \\ (Au-b,u) = 0, \\ Au-b \in \mathbf{R}^N_+ ,\ u \in \mathbf{R}^N_+ \end{cases}$$

the solution of which was the subject of Section 4 of this Appendix.

We shall describe Polyak's /1A/ algorithm in detail in Section 5.2.2, but we first shall give a broad outline of this algorithm and attempt to extricate its general principles. The algorithm generates a sequence $\{u^n\}_{n\geq 0}$ such that $\forall\ n,\ \ u^n \in K$, and we adjust $\lambda^n (= Au^n - b)$ in such a way as to satisfy (5.5A).

Given $u^o \in K$, for $n \geq 0$ we then construct $u^n \in K$, and λ^n satisfying (5.5A) "as well as possible", using a method which is itself iterative. With regard to the outer iteration loop, with u^n known we first define

$$I^n \subset \{1,\ldots N\}$$

as being the set of indices i for which $\{u_i^n, \lambda_i^n\}$ is consistent with (5.5A), i.e.

$$I^n = \{i \mid u_i^n = a_i \text{ and } \lambda_i > 0\} \cup \{i \mid u_i^n = b_i \text{ and } \lambda_i < 0\}\ ; \tag{5.7A}$$

with I^n and J^n $(=\{1,\ldots N\} - I^n)$ we associate the vectors $u_I^n = \{u_i\}_{i \in I^n}$ and $u_J^n = \{u_i\}_{i \in J^n}$ (we similarly define b_I^n, b_J^n, λ_I^n, λ_J^n). We then reorder the indices in such a way that $I^n = \{1,\ldots \text{Card}(I^n)\}$ and $J^n = \{1+\text{Card}(I^n),\ldots N\}$. In view of the above partitioning of $\{1,\ldots N\}$ and the corresponding decomposition of $\mathbb{R}^N$ (which implies $\mathbb{R}^N = \mathbb{R}^{\text{Card}(I^n)} \oplus \mathbb{R}^{\text{Card}(J^n)}$) we can write the relation

$$Au^n - b = \lambda^n \tag{5.8A}$$

in the form

$$\begin{pmatrix} A_{II}^n & A_{IJ}^n \\ A_{JI}^n & A_{JJ}^n \end{pmatrix} \begin{pmatrix} u_I^n \\ u_J^n \end{pmatrix} - \begin{pmatrix} b_I^n \\ b_J^n \end{pmatrix} = \begin{pmatrix} \lambda_I^n \\ \lambda_J^n \end{pmatrix}, \tag{5.9A}$$

where, in (5.9A), A_{II}^n and A_{JJ}^n are square, symmetric, positive-definite matrices.

The basic idea behind Polyak's method lies in modifying u_J^n (u_I^n remaining fixed) so as to attempt to make λ_J vanish; this leads to the linear system

$$A_{JJ}^n v_J = b_J^n - A_{JI}^n u_I^n\ . \tag{5.10A}$$

Since the matrix A_{JJ}^n is *symmetric and positive definite* we can apply the *conjugate gradient method* of Chapter 2, Section 2.3 for the solution of (5.10A); we can thus solve (5.10A) *exactly in a finite number of iterations (of the inner iteration loop for the overall algorithm).*

In practice, it is unlikely that the exact solution of (5.10A) will satisfy the bounds

(5.11A) $$a_i \leq v_i \leq b_i$$

$\forall i \in J^n$; we therefore have to be content with solving (5.10A) "as well as possible", whilst keeping the interval constraints (5.11A) satisfied. In order to do this we modify the conjugate gradient method; if during the *descent phase* of the conjugate gradient method (see (2.15), (2.16) of Chapter 2, Section 2.3) some of the bounds (5.11A) are violated, *we reduce the descent step* (denoted by λ in (2.15) of Chapter 2, Section 2.3) by *as little as possible* to ensure that the bounds (5.11A) stay satisfied; this results in some of the bounds being satisfied with equality; the corresponding indices are then added to I^n and the inner iteration loop is resumed with a new decomposition of A and of the vectors.

When the modified conjugate gradient algorithm has converged, we know that the variables, now denoted by u_i^{n+1}, of which the index i has remained in J^n, satisfy (5.11A) and that the corresponding scalars λ_i are zero.

Using the modified conjugate gradient method, we have thus constructed a vector $u^{n+1} \in K$; if $I^{n+1} = I^n$ then u^{n+1} satisfies (5.5A) and we therefore have $u^{n+1} = u$, where u is the solution of (5.1A); if $I^{n+1} \neq I^n$ the inner iteration loop is repeated with I^n replaced by I^{n+1}.

5.2.2 Detailed description of Polyak's algorithm

We follow D.P. O'Leary's /1A/ description; the various phases of the algorithm are

A. Initialisation of the overall algorithm

(5.12A) $u^o \in K$, *given arbitrarily*,

(5.13A) $n = 0$,

(5.14A) $I = \{1,2,\ldots N\}$.

B. Outer iteration loop

For $n \geq 0$, *set*

(5.15A) $n \leftarrow n+1,\ u^n = u^{n-1},\ \lambda^n = Au^n - b,$

(5.16A) $I^{n-1} = I,$

and then define I^n *by* (5.7A).

If $I^n = I^{n-1}$ *then* $u^n = u$ *the solution of* (5.1A); *if* $I^n \neq I^{n-1}$ *set*

(5.17A) $I = I^n$

and start the inner interation loop.

C. Inner iteration loop

C.1 Decomposition of A and of the vectors u^n,b. Notation

Let $J = \{1,\ldots N\} - I$; *decompose and reorder* A *and* u^n,b *relative to the partitioning* $I+J = \{1,\ldots.N\}$, *giving*

$$u^n \to \begin{pmatrix} u_I^n \\ u_J^n \end{pmatrix},\ b \to \begin{pmatrix} b_I^n \\ b_J^n \end{pmatrix},\ A \to \begin{pmatrix} A_{II} & A_{JI}^t \\ A_{JI} & A_{JJ} \end{pmatrix},$$

with A_{JJ} *symmetric and positive-definite.*

We denote by $\{z^q\}$, $\{w^q\}$, $\{r^q\}$ *respectively, the sequence of approximate solutions generated by the modified conjugate gradient algorithm (inner algorithm), the sequence of descent directions and the sequence of residuals*

$$r^q = b_J^n - A_{JJ}z^q - A_{JI}u_I^n .$$

C.2 Initialisation of the inner algorithm

Set

(5.18A) $q = 0$

and

(5.19A) $z^o = u_J^n ,$

(5.20A) $w^o = r^o = b_J^n - A_{JJ}z^o - A_{JI}u_I^n .$

C.3 Calculation of z^{q+1} *and* r^{q+1}

First determine the optimal descent step in the direction of w^q, *namely*

(5.21A)
$$\rho_1^q = \frac{(r^q, w^q)}{(A_{JJ}w^q, w^q)} = \frac{(r^q, r^q)}{(A_{JJ}w^q, w^q)} ,$$

and then the maximal descent step for which the bounds (5.11A) *are still satisfied, namely*

(5.22A)
$$\rho_2^q = \min\left(\min_{\substack{j\in J \\ w_j^q<0}} \frac{a_j - z_j^q}{w_j^q} , \min_{\substack{j\in J \\ w_j^q>0}} \frac{b_j - z_j^q}{w_j^q} \right) .$$

The descent step finally used is then

(5.23A)
$$\rho^q = \min(\rho_1^q, \rho_2^q) ,$$

and hence z^{q+1}, r^{q+1} *are given by*

(5.24A)
$$z^{q+1} = z^q + \rho^q w^q ,$$

(5.25A)
$$r^{q+1} = r^q - \rho^q A_{JJ} w^q .$$

C.4 Termination criterion for the inner iteration loop

—*If* $r^{q+1} = 0$, *put* $u_J^n = z^{q+1}$ *and restart the inner iteration loop.*

—*If* $\{j | z_j^{q+1} = a_j \text{ or } b_j\} = \emptyset$ *go to* C.5.

—*Otherwise, put* $u_j^n = z^{q+1}$ and $I = \{i | u_i^n = a_i \text{ or } b_i\}$. *If* $I = \{1,\ldots N\}$ *restart the outer iteration loop. Otherwise restart the inner iteration loop.*

C.5 Calculation of the new descent direction

Define

(5.26A) $$\gamma_q = -\frac{(A_{JJ}w^q, r^{q+1})}{(w^q, A_{JJ}w^q)} = \frac{(r^{q+1}, r^{q+1})}{(r^q, r^q)} ,$$

giving the new descent direction (A_{JJ}*-conjugate to the previous directions*)

(5.27A) $$w^{q+1} = r^{q+1} + \gamma_q w^q .$$

Then replace q *by* q + 1 *and go to* C.3. ■

Remark 5.3: We refer to O'Leary, *loc. cit.*, for a number of variants of the above algorithm, and also practical details of its implementation on a computer.

5.2.3 *Convergence results*

In O'Leary, *loc. cit.*, the following theorem is proved:

Theorem 5.1: *Polyak's algorithm of Section* 5.2.2 *converges to the solution of problem* (5.1A) *in a finite number of iterations.*

5.3 Discussion

A variant (called the S.N. method) of Polyak's algorithm of Section 5.2 is given in Trémolieres /4, Chap. 7/. More specifically, regarding the application of Polyak's algorithm to the solution of *variational inequality* problems (after suitable discretisation using finite differences or finite elements) we refer to O'Leary, *loc. cit.*, and O'Leary-Yang /1A/ where numerical applications to the *elasto-plastic torsion* problem of Chapter 3 can be found, in the equivalent formulation (2.4) of Chapter 3, Section 2.2 (see also Chapter 3, Section 3).

6. ALTERNATING DIRECTION METHODS. APPLICATION TO THE SOLUTION OF NONLINEAR VARIATIONAL PROBLEMS

In this section we follow the description of P.L. Lions - B. Mercier /1A/; the notation is that of H. Brézis /3/, to which we also refer for the various definitions and properties connected with the concept of monotonicity used in this section.

6.1 Formulation of the problem. Synopsis

Let H be a real Hilbert space of which the inner product and associated norm are denoted respectively by $(\cdot,\cdot)$ and $|\cdot|$. We consider the equation

$$0 \in C(u) \tag{6.1A}$$

where C is a *monotone* operator on H, which may possibly be *multivalued* [1] (in which case, if $v \in H$, $C(v)$ is a subset of H, possibly reduced to a single point or empty).

Of particular interest is the solution of (6.1A) when

$$\begin{cases} C = A+B\ , \\ A,B,C \ \textit{are maximal monotone.} \end{cases} \tag{6.2A}$$

Let $\lambda > 0$; to solve (6.1A) using the decomposition (6.2A), we consider (as in P.L. Lions-B. Mercier, *loc. cit.*) the following two algorithms:

First algorithm:

$$u^o \ \textit{given} \tag{6.3A}$$

and for $n \geq 0$, u^n *known*

$$u^{n+1} = (I+\lambda B)^{-1}(I-\lambda A)(I+\lambda A)^{-1}(I-\lambda B)u^n. \tag{6.4A}$$

Second algorithm:

$$u^o \ \textit{given} \tag{6.5A}$$

and for $n \geq 0$, u^n *known*

$$u^{n+1} = (I+\lambda B)^{-1}[(I+\lambda A)^{-1}(I-\lambda B)+\lambda B]\, u^n. \tag{6.6A}$$

Remark 6.1: We assume (for simplicity) that A,B,C are single-valued [2] operators and we put

---

Translator's notes: (1) The terminology *multivoque* is sometimes used.

(2) The terminology *univoque* is sometimes used.

(6.7A) $$u^{n+1/2} = (I+\lambda A)^{-1}(I-\lambda B)\, u^n \, .$$

Equation (6.7A) is equivalent to $(I+\lambda A)u^{n+1/2}+\lambda B(u^n) = u^n$, which allows us to rewrite the first algorithm in the form

(6.8A) u^o *given*

and for $n \geq 0$, u^n *known*

(6.9A) $$u^{n+1/2} + \lambda A(u^{n+1/2}) = u^n - \lambda B(u^n) \, ,$$

(6.10A) $$u^{n+1} + \lambda B(u^{n+1}) = u^{n+1/2} - \lambda A(u^{n+1/2}) \, ;$$

likewise the second algorithm is equivalent to

(6.11A) u^o *given*

and for $n \geq 0$, u^n *known*

(6.12A) $$u^{n+1/2}+\lambda A(u^{n+1/2}) = u^n-\lambda B(u^n),$$

(6.13A) $$u^{n+1}+\lambda B(u^{n+1}) = u^n- \lambda A(u^{n+1/2}).$$

In view of (6.8A)-(6.10A) (resp. (6.11A)-(6.13A)) algorithm (6.3A), (6.4A) (resp. (6.5A), (6.6A)) has the appearance of an alternating direction method of the Peaceman-Rachford /1A/ type (resp. Douglas-Rachford /1A/ type). ■

In Section 6.2 we shall give some information on the convergence of (6.3A), (6.4A) and (6.5A),(6.6A) (referring to P.L. Lions-B. Mercier, *loc. cit.*, and Gabay /1A/ for the proofs). In Section 6.3 we shall briefly describe the application of (6.3A), (6.4A) and (6.5A), (6.6A) to the solution of the obstacle problem of Appendix 1, Section 3.

For further details regarding algorithms (6.3A), (6.4A) and (6.5A), (6.6A), and also for their implementation, we refer the reader to the two references cited above, in which various generalisations are also given.

6.2 Convergence of algorithms (6.3A), (6.4A) and (6.5A), (6.6A)

We follow P.L. Lions-B. Mercier /1A, Section 1/ to which we refer for the proofs of the following results.

6.2.1 *Assumptions and additional notation. Initialisation*

We recall that A,B,C are maximal (Assumption (6.2A)). We denote by D(A) the *domain* of A (i.e. $D(A) = \{v \in H,\ A(v) \subset H,\ A(v) \neq \emptyset\}$), and by R(A) the *image* of A (i.e. $R(A) = \{v \in H,\ \exists w \in H$ such that $v \in A(w)\}$). We recall that A is monotone if

$$(y-z,u-v) \geq 0 \ , \ \forall u \in H,\ y \in A(u), \quad \forall v \in H,\ z \in A(v) \ ;$$

moreover the statement that A is *maximal monotone* is equivalent to saying that the *resolvent* $J_A^\lambda = (I+\lambda A)^{-1}$ (a *single-valued* operator) is a *contraction defined on the whole of* H.

In this present section we shall make the assumption that (6.1A) admits at least one solution, i.e.

$$\text{(6.14A)} \quad \left\{ \begin{array}{l} \textit{There exists } u \in H,\ a \in A(u),\ b \in B(u), \\ \\ \textit{such that } a+b = 0. \end{array} \right.$$

If A and B are multi-valued, it is appropriate to describe algorithms (6.3A), (6.4A) and (6.5A), (6.6A) in rather greater detail, in particular with regard to their initialisation. To initialise (6.3A), (6.4A) and (6.5A), (6.6A) we take $u^o \in D(B)$, then $b^o \in B(u^o)$, and we put

$$\text{(6.15A)} \qquad v^o = u^o + \lambda b^o \Longrightarrow u^o = J_B^\lambda v^o.$$

We next define the sequence $\{v^n\}_{n\geq 0}$ by

<u>*First algorithm*</u> ((6.3A, (6.4A))

$$\text{(6.16A)} \qquad v^{n+1} = (2J_A^\lambda - I)(2J_B^\lambda - I)v^n \ ;$$

<u>*Second algorithm*</u> ((6.5A), (6.6A))

$$\text{(6.17A)} \qquad v^{n+1} = J_A^\lambda(2J_B^\lambda - I)v^n + (I - J_B^\lambda)v^n.$$

In both cases we obtain the sequence $\{u^n\}_{n\geq 0}$ (from (6.3A), (6.4A) or (6.5A), (6.6A)), starting from $\{v^n\}_{n\geq 0}$, by

(6.18A) $u^n = J_B^\lambda v^n$.

6.2.2 *Convergence of algorithm* (6.3A), (6.4A).

We put

$$v = u+\lambda b \ , \ w = u+\lambda a,$$

$$w^n = 2u^n - v^n \ , \ b^n = \frac{v^n - u^n}{\lambda} \ , \ a^n = \frac{w^n - w^{n+1}}{2\lambda} \ .$$

We then have (see P.L. Lions-B. Mercier /1A, Section 1/):

Proposition 6.1: *Assume* (6.14A) *to be satisfied; we then have*

(6.19A) *the sequences* u^n, v^n, w^n, a^n, b^n *are bounded*

(6.20A) $\lim_{n\to+\infty} (b^n - b, u^n - u) = 0$,

(6.21A) $\lim_{n\to+\infty} \left(a^n - a, \frac{v^{n+1} + w^n}{2} - u\right) = 0.$ ■

The above proposition results in the convergence properties given in:

Corollary 6.1: *If* B *is single-valued and satisfies*

(6.22A) *For any sequence* $\{x_n\}$, $x_n \in D(B)$ $\forall n$, *such that* $\{Bx_n\}$ *is bounded,* $x_n \to \bar{x}$ *weakly and* $\lim_{n\to+\infty} (Bx_n - Bx, x_n - x) = 0$, *we have* $x = \bar{x}$

then we have

(6.23A) $\lim_{n\to+\infty} u^n = u$ *weakly in* H,

where u *is the solution of* (6.1A) (*unique, from* (6.22A)).

Remark 6.2: Property (6.22A) is satisfied in the following cases (see P.L. Lions-B. Mercier, *loc. cit.*, for the proofs):

(i) B is *coercive* (also termed H-elliptic), i.e. $\exists \alpha > 0$ such that

(6.24A) $$(B(y)-B(x),y-x) \geq \alpha|y-x|^2 \quad \forall x,y \in D(B).$$

In this case we actually have a stronger condition that (6.23A), since (6.20A) and (6.24A) imply the *strong convergence* of $\{u^n\}$ to u.

(ii) B^{-1} is *coercive*, i.e. $\exists\beta > 0$ such that

(6.25A) $$(B(y)-B(x),y-x) \geq \beta|B(y)-B(x)|^2 \quad \forall x,y \in D(B).$$

In this case, in addition to (6.23A), we also have the *strong convergence* of $\{Bu^n\}$ to Bu, from (6.20A) and (6.25A).

(iii) B is *strictly monotone* and *weakly closed*, i.e.

— $\forall x,y \in D(B)$ such that $(B(y)-B(x),y-x) = 0$ we have $y=x$,

— if $\{x_n\}$ is such that $x_n \in D(B)$ $\forall n$, $\lim_{n\to+\infty} x_n = x$ weakly in H and if there exists $y_n \in Bx_n$ with $\lim_{n\to+\infty} y_n = y$ weakly in H, then $x \in D(B)$ and $y = Bx$.

Remark 6.3: If B is linear, or if J_B is compact, then B is weakly closed.

Remark 6.4: If in Corollary 6.1 we assume that A rather than B satisfies (6.22A), then $\{\frac{v^{n+1}+w^n}{2}\}_n$ converges weakly to the unique solution of (6.1A). In fact everything said in Remark 6.2 with regard to $\{u^n\}_n$, also holds for $\{\frac{v^{n+1}+w^n}{2}\}_n$.

6.2.3 *Convergence of algorithm* (6.5A), (6.6A)

The notation is that of Sections 6.2.1 and 6.2.2; with regard to the convergence of (6.5A), (6.6A), the following theorem is proved in P.L. Lions - B. Mercier /1A, Section 1.3/:

Theorem 6.1: *If* (6.14A) *is satisfied, then as* $n \to +\infty$ *the sequence* $\{v^n\}$ *generated by* (6.15A), (6.17A) *converges weakly to* $v \in H$ *such that* $u = J_B^\lambda v$ *is a solution of* (6.1A). *Moreover for the sequence* $\{u^n\}$ *defined by* $u^n = J_B^\lambda v^n$, *we have the following convergence properties*:

- *If* B *is linear*, u^n *converges weakly to a solution of* (6.1A).

- *If* A *and* B *are odd* (i.e.
  $A(x) = -A(-x) \quad \forall x \in D(A), \quad B(x) = -B(-x) \quad \forall x \in D(B)$) *then* $\{u^n\}$ *converges strongly to a solution of* (6.1A).

- *If* A + B *is maximal monotone the weak cluster points of* $\{u^n\}$ *are solutions of* (6.1A).

The following propositions are also proved in P.L. Lions-B. Mercier, *loc. cit.*

Proposition 6.2: *If* J_B^λ *is weakly closed and if* (6.14A) *is satisfied, then* u^n *converges weakly to a solution of* (6.1A).

Proposition 6.3: *If* J_A^λ *is compact and if* (6.14A) *is satisfied, then* u^n *converges strongly to a solution of* (6.1A).

Proposition 6.4: *Assume that* B *is coercive and Lipschitz continuous*, i.e. $\exists \alpha$ *and* $M > 0$ *such that* $\forall x_1, x_2 \in H$ *we have*

$$|B(x_2)-B(x_1)| \leq M|x_2-x_1| , \tag{6.26A}$$

$$(B(x_2)-B(x_1), x_2-x_1) \geq \alpha |x_2-x_1|^2 . \tag{6.27A}$$

Under these conditions there exists a constant C *such that*

$$|v^n - v| \leq Ck^n , \quad |u^n - u| \leq Ck^n , \tag{6.28A}$$

where $u = J_B^\lambda v$ *is the unique solution of* (6.1A) *and where*

$$k = \left(1 - \frac{2\lambda\alpha}{(1+\lambda M)^2}\right)^{1/2} . \blacksquare$$

Remark 6.5: The estimates (6.28A) also hold for algorithm (6.3A), (6.4A) if B satisfies (6.26A), (6.27A). Both algorithms then have *linear convergence*.

Remark 6.6: We remark that algorithm (6.5A), (6.6A) has been investigated by J. Lieutaud /1A/ who has established its convergence under less restrictive assumptions on A and B (which are assumed *single-valued*, however).

Remark 6.7: In Section 7 we shall show the relationships which exist between the above alternating direction methods and augmented Lagrangian methods.

6.3 Application to the solution of the obstacle problem

We follow P.L. Lions-B. Mercier /1A, Section 3/: let Ω be a bounded open domain in $\mathbb{R}^2$, with regular boundary, and $f \in L^2(\Omega)$. The variational inequality

$$\text{(6.29A)} \quad \begin{cases} \textit{Find} \quad u \in H_0^1(\Omega) \cap K \ \textit{such that} \\ \int_\Omega \nabla u \cdot \nabla(v-u)\,dx \geq \int_\Omega f(v-u)\,dx \quad \forall v \in H_0^1(\Omega) \cap K, \end{cases}$$

where

$$\text{(6.30A)} \qquad K = \{v \in L^2(\Omega),\ v(x) \geq 0 \quad a.e.\ on\ \Omega\},$$

admits one and only one solution and we have $u \in H^2(\Omega)$. The solution u of (6.29A) is also the solution of the (multivoque (*)) equation in $L^2(\Omega)$:

$$\text{(6.31A)} \qquad 0 \in -\Delta u + f + \partial I_K(u) \ ,$$

where ∂I_K is the *sub-differential* (see Brézis /3/, Ekeland-Temam /1/ for this concept) of the *indicator functional* of the convex set K defined by (6.30A). Since the convex set K is closed and nonempty, I_K is convex, proper and l.s.c; it then follows that ∂I_K is maximal monotone. We put $H = L^2(\Omega)$, $A = \partial I_K$ and we define B by

$$B(v) = -\Delta v + f \ \text{with} \ D(B) = H^2(\Omega) \cap H_0^1(\Omega).$$

The domain Ω being bounded, we have *coercivity* of B since $\forall v_1, v_2 \in D(B)$

$$(B(v_2)-B(v_1), v_2-v_1) = \int_\Omega |\nabla(v_2-v_1)|^2 dx \geq \gamma_0 |v_2-v_1|^2$$

where $\gamma_0 > 0$ is the smallest eigenvalue of $-\Delta$ on $H^2(\Omega) \cap H_0^1(\Omega)$.

Moreover we have $J_A^\lambda = P_K$ where P_K is the projection operator onto K in the $L^2(\Omega)$ norm, namely

$$(P_K v)(x) = \sup(0, v(x)), \quad a.e.\ on\ \Omega\ .$$

(*) *Translator's Note*: The terminology *multivoque equation* refers to an equation related to a multivalued operator.

P_K is thus in fact a *truncation* operator.

We can thus apply algorithms (6.3A), (6.4A) and (6.5A), (6.6A) for the solution of (6.29A), (6.31A); at each iteration we have to carry out a truncation operation, which is numerically trivial, and to solve a homogeneous Dirichlet problem relating to the operator $I-\lambda\Delta$, which is a standard numerical process and nowadays poses few problems (at least if $\Omega \subset \mathbb{R}^2$).

We refer to P.L. Lions-B. Mercier, *loc. cit.*, for a comparison of the performance of the above two alternating direction algorithms applied to the solution of (6.29A), (6.31A) (after a suitable discretisation); it would appear that for this example the algorithm (6.3A), (6.4A) is twice as fast as (6.5A), (6.6A). In contrast, however, examples are given in Chan-Glowinski /1A/ relating to the nonlinear Dirichlet problems of Appendix 1, Section 6, for which algorithm (6.5A), (6.6A) appears much more efficient than (6.3A), (6.4A).

7. RELATIONSHIP BETWEEN THE ALTERNATING DIRECTION METHODS AND AUGMENTED LAGRANGIAN METHODS

In this section we shall supplement Remark 6.7 of Section 6.2; we shall show that the *alternating direction* methods of Section 6 are in fact special cases of a more general family of algorithms related to the *duality algorithms* of Chapter 2, Section 4 and to the augmented Lagrangian methods mentioned in Chapter 2, Section 6, Chapter 3, Section 10 and Chapter 5, Section 9. For further details we refer the reader to Bourgat-Dumay-Glowinski /1A, Section 6.3/ and Gabay /1A/.

7.1 A model problem in Hilbert space

The notation is that of Section 6; for simplicity we consider a particular model problem of the type (6.1A). We therefore let H be a Hilbert space on $\mathbb{R}$ with inner product and norm denoted by $(\cdot,\cdot)$ and $|\cdot|$, respectively. We consider the problem

$$C(u) = 0, \tag{7.1A}$$

where $C : H \to H$ is a single-valued operator. We assume (again for simplicity) that C is the derivative (in the Frechet or Gateaux sense) of a functional J (i.e. $C = J'$), *strictly convex*, such that

$$\lim_{\|v\| \to +\infty} J(v) = +\infty .$$

It follows from the properties of J that C is strictly monotone and that the minimisation problem

(7.2A)
$$\begin{cases} \textit{Find } u \in H \textit{ such that} \\ J(u) \leq J(v) \quad \forall v \in H, \end{cases}$$

admits one and only one solution characterised by (7.1A), which consequently admits a unique solution.

We assume that

(7.3A)
$$J = J_A + J_B,$$

where J_A, J_B are also convex, differentiable functionals; we write:

(7.4A)
$$A = J'_A \ , \ B = J'_B \ ;$$

A,B are *monotone* operators and we have A + B = C.

7.2 Decomposition-coordination of (7.1A), (7.2A) using the augmented Lagrangian

The problems (7.1A), (7.2A) are clearly equivalent to the minimisation problem

(7.5A)
$$\begin{cases} \textit{Find } \{u,p\} \in \mathcal{H} \textit{ such that } \forall \{v,q\} \in \mathcal{H} \\ j(u,p) \leq j(v,q) \ , \end{cases}$$

where

(7.6A)
$$j(v,q) = J_A(v) + J_B(q),$$

(7.7A)
$$\mathcal{H} = \{\{v,q\} \in H \times H, \ v-q = 0\}.$$

We shall handle the linear equality constraint v-q = 0 by *penalty and Lagrange multiplier*; to this end we introduce (with r > 0) the *augmented Lagrangian* $\mathcal{L}_r : H \times H \times H \to \mathbb{R}$ defined by

(7.8A)
$$\mathcal{L}_r(v,q,\mu) = j(v,q) + \frac{r}{2}\|v-q\|^2 + (\mu, v-q) \ .$$

The following proposition may readily be proved (see e.g. Fortin-Glowinski /2A/, Glowinski /2A/)

<u>Proposition 7.1</u>: *Under the above assumptions on* C, J, J_A, J_B *the augmented Lagrangian* $\mathcal{L}_r$ *admits a unique saddle point* $\{u,p,\lambda\}$ *on* $H \times H \times H$ *and we have* $p = u$, $\lambda = B(u) = -A(u)$, *where* u *is the solution of* (7.1A), (7.2A).

7.3 <u>Solution of (7.1A), (7.2A) via duality algorithms for $\mathcal{L}_r$</u>

In view of Proposition 7.1 it is natural to envisage solving (7.1A), (7.2A) via the determination of the saddle points of $\mathcal{L}_r$, using the duality algorithms of Chapter 2, Section 4; we shall restrict our attention to Uzawa's algorithm (4.12), (4.13) of Chapter 2, Section 4.3; we thus have:

$$\lambda^o \in H \text{ given} \tag{7.9A}$$

then for $n \geq 0$, *assuming that* λ^n *is known, determine* u^n, p^n and λ^n *by*

$$\begin{cases} \text{Find } \{u^n,p^n\} \in H \times H \text{ such that } \forall \{v,q\} \in H \times H \\ \text{we have} \\ \mathcal{L}_r(u^n,p^n,\lambda^n) \leq \mathcal{L}_r(v,q,\lambda^n) \end{cases}, \tag{7.10A}$$

$$\lambda^{n+1} = \lambda^n + \rho(u^n - p^n) \ , \ \rho > 0 \ ; \tag{7.11A}$$

Rendering (7.10A) explicit, we have

$$\begin{cases} ru^n + A(u^n) - rp^n = -\lambda^n, \\ \\ rp^n + B(p^n) - ru^n = \lambda^n. \end{cases} \tag{7.10A)'$$

The essential difficulty in the application of (7.9A)-(7.11A) is thus the solution of the system of two coupled equations (7.10A)'. As pointed out previously in Chapter 3, Section 10 and Chapter 5, Section 9 in a similar context, it is possible to envisage using *block* variants of the *relaxation* methods of Chapter 2, Section 1 (the convergence of which has been established by Cea-Glowinski /2/ under quite general assumptions) to solve (7.10A)'; in the case in which only a *single* relaxation iteration is performed, naturally using the results of the previous iteration as initial conditions, we are led to the following algorithm (introduced by Glowinski-Marrocco /2/):

(7.12A) p^o, λ^1 *given in* H,

then for $n \geq 1$, *assuming that* p^{n-1} *and* λ^n *are known, determine* u^n, p^n *and* λ^{n+1} *by*

(7.13A) $ru^n + A(u^n) = rp^{n-1} - \lambda^n,$

(7.14A) $rp^n + B(p^n) = ru^n + \lambda^n,$

(7.15A) $\lambda^{n+1} = \lambda^n + \rho(u^n - p^n).$

A variant (due to Gabay /1A/, to which we also refer for the convergence results) is given by

(7.16A) p^o, λ^1 *given in* H,

then for $n \geq 1$, *assuming that* p^{n-1} *and* λ^n *are known, determine* u^n, p^n *and* λ^{n+1} *by*

(7.17A) $ru^n + A(u^n) = rp^{n-1} - \lambda^n,$

(7.18A) $\lambda^{n+1/2} = \lambda^n + \rho(u^n - p^{n-1}),$

(7.19A) $rp^n + B(p^n) = ru^n + \lambda^{n+1/2},$

(7.20A) $\lambda^{n+1} = \lambda^{n+1/2} + \rho(u^n - p^n).$

Under "very reasonable" monotonicity and continuity assumptions for A and B, it is shown in Fortin-Glowinski /2A/, Glowinski /2A/ that for any $p^o, \lambda^1 \in H$ and if

(7.21A) $\rho \in]0, \frac{1+\sqrt{5}}{2} r[$

we have

(7.22A)
$$\begin{cases} \lim_{n\to+\infty} \|u^n-u\| = 0\,, \\ \lim_{n\to+\infty} \|p^n-u\| = 0, \\ \lim_{n\to+\infty} \lambda^n = \lambda \textit{ weakly in } H. \end{cases}$$

If J_B is *linear* (or *affine*), it is proved in Gabay-Mercier /1/ that (7.22A) also holds if $\rho \in]0,2r[$.

Remark 7.1: In fact the convergence proofs (based on energy methods) given in the above references also hold if A,B satisfy suitable monotonicity and continuity properties, without necessarily being the derivatives of convex functionals (and possibly if they are multivalued).

7.4 An "alternating direction" interpretation of algorithms (7.12A)-(7.15A) *and* (7.16A)-(7.20A)

The case of algorithm (7.12A)-(7.15A):

We assume that $\rho = r$; it then follows from (7.14A), (7.15A) that

$$\lambda^{n+1} = B(p^n) \text{ (and hence } \lambda^n = B(p^{n-1})\text{)}\,,$$

which combined with (7.13A) implies

(7.23A) $$ru^n + A(u^n) = rp^{n-1}-B(p^{n-1}).$$

We have, from (7.14A), $ru^n + \lambda^n = rp^n+B(p^n)$, and hence, by using this in (7.13A)

(7.24A) $$rp^n+B(p^n) = rp^{n-1}-A(u^n).$$

Putting $p^{n-1/2} = u^n$ and $r = \frac{1}{\lambda}$, we finally deduce from (7.23A), (7.24A)

(7.25A) $$p^{n+1/2}+ \lambda A(p^{n+1/2}) = p^n-\lambda B(p^n)\,,$$

(7.26A) $$p^{n+1}+\lambda B(p^{n+1}) = p^{n}-\lambda A(p^{n+1/2}).$$

Relations (7.25A), (7.26A) show that for $\rho = r$, (7.12A)-(7.15A) is in fact equivalent to the *Douglas-Rachford Alternating Direction algorithm* of Section 6.1, Remark 6.1.

The case of algorithm (7.16A)-(7.20A):

Here again we assume that $\rho = r$; it can then readily be deduced from (7.17A)-(7.20A) that

$$\lambda^{n+1/2} = -A(u^n),\ \lambda^{n+1} = B(p^n)\ (\text{hence}\ \ \lambda^n = B(p^{n-1})),$$

and using this in (7.17A), (7.19A) gives

(7.27A) $$ru^n+A(u^n) = rp^{n-1}-B(p^{n-1}),$$

(7.28A) $$rp^n+B(p^n) = ru^n - A(u^n).$$

Putting $p^{n-1/2} = u^n$ and $r = \frac{1}{\lambda}$ we finally deduce from (7.27A), (7.28A)

(7.29A) $$rp^{n+1/2}+A(p^{n+1/2}) = rp^n-B(p^n),$$

(7.30A) $$rp^{n+1}+B(p^{n+1}) = rp^{n+1/2}-A(p^{n+1/2}).$$

It follows from (7.29A), (7.30A) that, if $\rho = r$, (7.16A)-(7.20A) is in fact equivalent to the *Peaceman-Rachford Alternating Direction algorithm* of Section 6.1, Remark 6.1.

Remark 7.1: To the best of our knowledge, the relationships which exist between the *alternating direction* and *augmented Lagrangian* methods were first demonstrated in Chan-Glowinski /1A/, /2A/, in connection with the numerical solution of the nonlinear Dirichlet problems of Appendix 1, Section 6.

Appendix 3

FURTHER DISCUSSION OF THE NUMERICAL ANALYSIS OF THE ELASTO-PLASTIC TORSION PROBLEM

1. SYNOPSIS

The aim of this appendix is to supplement Chapter 3 relating to the numerical analysis of the *elasto-plastic torsion* problem; we shall, in fact, restrict our attention to several supplementary results concerning the finite-element approximation of this problem and its *iterative solution*. More specifically, we shall return in Section 2 to the approximation using *finite elements of order one* (i.e. *piecewise affine*), previously investigated in Chapter 3, Sections 4 and 6; it will be shown that under reasonable assumptions we "almost" have $\|u_h-u\|_{H_0^1(\Omega)} = O(h^{1/2})$ (in fact we have $\|u_h-u\|_{H_0^1(\Omega)} = O(h)$ if $\Omega \subset \mathbb{R}$). In Section 3, which follows Falk-Mercier /1A/, it will be shown that by using an equivalent formulation it is possible to obtain an approximation error of optimal order, even if $\Omega \subset \mathbb{R}^2$. Finally, in Section 4, we shall give some additional information on the iterative solution of the above problem.

Let us conclude this first section by recalling (see Chapter 3, Section 1) that the problem under consideration is defined by

$$(1.1A)\qquad \begin{cases} \textit{Find } u \in K \textit{ such that} \\ \\ a(u,v-u) \geq L(v-u) \quad \forall v \in K \end{cases}$$

where

$$(1.2A)\qquad K = \{v \in H_0^1(\Omega),\ |\nabla v| \leq 1 \ \textit{a.e.}\},$$

$$(1.3A)\qquad a(v,w) = \int_\Omega \nabla v \cdot \nabla w \, dx \quad \forall v,w \in H^1(\Omega),$$

$$(1.4A)\qquad L(v) = \langle f,v\rangle \quad \forall v \in H_0^1(\Omega)\ ,\ f \in H^{-1}(\Omega) = (H_0^1(\Omega))'\ .$$

2. THE FINITE-ELEMENT APPROXIMATION OF PROBLEM (1.1A). (I) ERROR ESTIMATES FOR PIECEWISE-LINEAR APPROXIMATIONS

This section, which follows Glowinski /2A, Chapter 2, Section 3/ will supplement the results of Chapter 3, Section 6.2.

2.1 One-dimensional case

Suppose that $\Omega =]0,1[$ and that in (1.4A) we have $f \in L^2(\Omega)$. Problem (1.1A) can then be written

$$\text{(2.1A)} \quad \begin{cases} \textit{Find } u \in K = \{v \in H_0^1(0,1),\ |\frac{dv}{dx}| \leq 1 \text{ a.e.}\} \textit{ such that} \\ \int_0^1 \frac{du}{dx}\left(\frac{dv}{dx} - \frac{du}{dx}\right) dx \geq \int_0^1 f(v-u)\,dx \quad \forall v \in K. \end{cases}$$

Let N be an integer > 0 and let $h = \frac{1}{N}$; we consider $x_i = ih$ for $i = 0,1,\ldots N$ and

$$e_i = [x_{i-1}, x_i],\ i=1,2,\ldots N.$$

We then approximate $H_0^1(0,1)$ and K respectively by

$$\text{(2.2A)} \qquad V_{oh} = \{v_h \in C^o[0,1],\ v_h(0)=v_h(1)=0,$$
$$v_h|_{e_i} \in P_1,\ i=1,2,\ldots N\}$$

and

$$\text{(2.3A)} \qquad K_h = K \cap V_{oh} = \{v_h \in V_{oh},\ |v_h(x_i)-v_h(x_{i-1})| \leq h \ \textit{ for }\ i=1,2,\ldots N\}$$

(with, as usual, P_1 = space of polynomials of degree ≤ 1).

The approximate problem is then defined by

$$\text{(2.4A)} \quad \begin{cases} \textit{Find } u_h \in K_h \textit{ such that} \\ \int_0^1 \frac{du_h}{dx}\left(\frac{dv_h}{dx} - \frac{du_h}{dx}\right) dx \geq \int_0^1 f(v_h-u_h)\,dx \quad \forall v_h \in K_h. \end{cases}$$

Problem (2.4A) clearly admits *a unique solution*. Regarding the approximation error $\|u_h-u\|_{H_0^1(0,1)}$ we have

Theorem 2.1: *Let* u *and* u_h *be the respective solutions of* (2.1A) *and* (2.4A). *If* $f \in L^2(0,1)$ *we then have*

$$\|u_h-u\|_{H_0^1(0,1)} = O(h). \tag{2.5A}$$

Proof: Since $u_h \in K_h \subset K$ it results from (2.1A) that

$$a(u,u_h-u) \geq \int_0^1 f(u_h-u)\,dx. \tag{2.6A}$$

We deduce, by adding (2.4A) and (2.6A), that $\forall v_h \in K_h$

$$a(u_h-u,u_h-u) \leq a(v_h-u,u_h-u)+a(u,u_h-u) - \int_0^1 f(v_h-u)\,dx$$

and hence $\forall v_h \in K_h$

$$\begin{cases} \frac{1}{2}\|u_h-u\|^2_{H_0^1(0,1)} \leq \\ \frac{1}{2}\|v_h-u\|^2_{H_0^1(0,1)} + \int_0^1 \frac{du}{dx}\left(\frac{dv_h}{dx}-\frac{du}{dx}\right) - \int_0^1 f(v_h-u)\,dx \ ; \end{cases} \tag{2.7A}$$

since $f \in L^2(0,1)$ implies that $u \in H^2(0,1) \cap K$ (see Chapter 3, Section 2.1) we have

$$\int_0^1 \frac{du}{dx}\frac{d}{dx}(v_h-u)\,dx = \int_0^1 \left(-\frac{d^2u}{dx^2}\right)(v_h-u)\,dx \tag{2.8A}$$

$$\leq \left\|\frac{d^2u}{dx^2}\right\|_{L^2} \|v_h-u\|_{L^2} .$$

We also have (the proof being left as an exercise for the reader):

$$\left\|\frac{d^2u}{dx^2}\right\|_{L^2} \leq \|f\|_{L^2} ,$$

which combined with (2.7A), (2.8A) implies $\forall v_h \in K_h$

$$\frac{1}{2}\|u_h-u\|^2_{H_0^1(0,1)} \leq \frac{1}{2}\|v_h-u\|^2_{H_0^1(0,1)} + 2\|f\|_{L^2}\|v_h-u\|_{L^2} . \tag{2.9A}$$

Let $v \in K$; we define the *linear interpolate* r_hv by

$$\begin{cases} r_hv \in V_{oh} , \\ (r_hv)(x_i) = v(x_i) \ , \ i=0,1,\ldots N. \end{cases} \tag{2.10A}$$

We have

$$\frac{d}{dx}(r_h v)\Big|_{e_i} = \frac{v(x_i)-v(x_{i-1})}{h} = \frac{1}{h}\int_{x_{i-1}}^{x_i} \frac{dv}{dx}\,dx ,$$

and hence

$$\left|\frac{d}{dx}(r_h v)\right|_{e_i} \le 1 \textit{ since } \left|\frac{dv}{dx}\right| \le 1 \textit{ a.e. on }]0,1[\; ;$$

we thus have

(2.11A) $$r_h v \in K_h \quad \forall v \in K.$$

Replacing v_h by $r_h u$ in (2.9A) we then have

(2.12A) $$\frac{1}{2}\|u_h-u\|^2_{H^1_o(0,1)} \le \frac{1}{2}\|r_h u-u\|^2_{H^1_o(0,1)} + 2\|f\|_{L^2}\|r_h u-u\|_{L^2} .$$

The *regularity* property $u \in H^2(0,1)$ and

$$\left\|\frac{d^2u}{dx^2}\right\|_{L^2(0,1)} \le \|f\|_{L^2(0,1)}$$

imply

(2.13A) $$\|r_h u-u\|_{H^1_o(0,1)} \le Ch\|u\|_{H^2(0,1)} \le Ch\|f\|_{L^2(0,1)} ,$$

(2.14A) $$\|r_h u-u\|_{L^2(0,1)} \le Ch^2\|u\|_{H^2(0,1)} \le Ch^2\|f\|_{L^2(0,1)} ,$$

where C denotes various constants independent of u and h. The estimate (2.5A) then follows trivially from (2.12A)-(2.14A). ∎

2.2 Two-dimensional case

Suppose that Ω is a bounded convex polygonal domain in $\mathbb{R}^2$ with boundary Γ, and that $f \in L^p(\Omega)$ with $p > 2$ (a reasonable assumption since for applications in mechanics we have f = const.). We use the notation of Appendix 1, Section 3.6; we thus approximate $H^1_o(\Omega)$ and K respectively by

(2.15A) $$V_{oh} = \{v_h \in C^o(\bar{\Omega}), v_h|_\Gamma = 0, \; v_h|_T \in P_1 \quad \forall T \in \mathcal{T}_h\}$$

and

$$K_h = K \cap V_{oh} , \tag{2.16A}$$

giving *the approximate problem*

$$\text{(2.17A)} \quad \begin{cases} \textit{Find } u_h \in K_h \textit{ such that} \\ \displaystyle\int_\Omega \nabla u_h \cdot \nabla (v_h - u_h)\, dx \geq \int_\Omega f(v_h - u_h)\, dx \quad \forall v_h \in K_h . \end{cases}$$

Problem (2.17A) admits a unique solution, and with regard to the convergence of u_h to u we have

Theorem 2.2: *Suppose that the angles of* $\mathcal{T}_h$ *are bounded below by* $\theta_o > 0$ *as* $h \to 0$. *Then under the above assumptions on* Ω *and* f *we have*

$$\| u_h - u \|_{H^1_o(\Omega)} = O(h^{1/2-1/p}), \tag{2.18A}$$

where u and u_h are the respective solutions of (1.1A) and (2.17A).

Proof: It results from Chapter 3, Section 2.1 that

$$u \in W^{2,p}(\Omega) . \tag{2.19A}$$

By proceeding as in the proof of Theorem 2.1, and using the fact that $K_h \subset K$ we obtain

$$\text{(2.20A)} \quad \begin{cases} \frac{1}{2} \| u_h - u \|^2_{H^1_o(\Omega)} \leq \frac{1}{2} \| v_h - u \|^2_{H^1_o(\Omega)} + a(u, v_h - u) - \displaystyle\int_\Omega f(v_h - u)\, dx \\ \leq \frac{1}{2} \| v_h - u \|^2_{H^1_o(\Omega)} - \displaystyle\int_\Omega (-\Delta u - f)(v_h - u)\, dx \quad \forall v_h \in K_h . \end{cases}$$

Using the *Hölder inequality* we deduce from (2.20A)

$$\text{(2.21A)} \quad \begin{cases} \frac{1}{2} \| u_h - u \|^2_{H^1_o(\Omega)} \leq \frac{1}{2} \| v_h - u \|^2_{H^1_o(\Omega)} \\ \qquad\qquad + \{ \| \Delta u \|_{L^p(\Omega)} + \| f \|_{L^p(\Omega)} \} \| v_h - u \|_{L^{p'}(\Omega)} \\ \forall v_h \in K_h, \textit{ with } \frac{1}{p} + \frac{1}{p'} = 1 \ (\text{i.e. } p' = \frac{p}{p-1}). \end{cases}$$

Let q be such that $1 \leq q \leq +\infty$; then suppose that $\mathcal{T}_h$ satisfies the assumptions of Theorem 2.2 and that $p > 2$. If

$W^{2,p}(T) \subset W^{1,q}(T)$ (with *continuous injection*), then it results from Ciarlet /1A/ and from *Sobolev's embedding theorem* ($W^{2,p}(T) \subset W^{1,\infty}(T) \subset C^0(T)$, with *continuous injection*) that $\forall T \in \mathcal{C}_h$ and $\forall v \in W^{2,p}(T)$ we have

$$\|\nabla(\pi_T v - v)\|_{L^q(T)\times L^q(T)} \leq C h_T^{1+2\left(\frac{1}{q}-\frac{1}{p}\right)} \|v\|_{W^{2,p}(T)} \tag{2.22A}$$

In (2.22A), $\pi_T v$ is the *linear interpolate* of v at the three vertices of T, h_T is the length of the longest side of T, and C is a constant independent of T and v. Let $v \in W^{2,p}(\Omega)$ and let $\pi_h : H_0^1(\Omega) \cap C^0(\bar{\Omega}) \to V_{oh}$ be defined by

$$\pi_h v \in V_{oh} \quad \forall v \in H_0^1(\Omega) \cap C^0(\bar{\Omega}) \ ,$$

$$(\pi_h v)(P) = v(P) \quad \forall\ P \in \mathring{\Sigma}_h \ ;$$

since $p > 2$ implies $W^{2,p}(\Omega) \subset C^0(\bar{\Omega})$, with *continuous injection*, we can define $\pi_h v$(1) but *in contrast to the one-dimensional case we have in general*

$$\pi_h v \notin K_h \quad \textit{if} \quad v \in W^{2,p}(\Omega) \cap K.$$

Since $W^{2,p}(\Omega) \subset W^{1,\infty}(\Omega)$ with continuous injection if $p > 2$, it follows from (2.22A) that *almost everywhere* on Ω we have

$$|\nabla(\pi_h v - v)(x)| \leq r h^{1-2/p} \|v\|_{W^{2,p}(\Omega)} \qquad \forall v \in W^{2,p}(\Omega)$$

from which we deduce that

$$\begin{cases} \textit{almost everywhere on } \Omega \textit{ we have} \\ |\nabla(\pi_h v)(x)| \leq 1 + r h^{1-2/p} \|v\|_{W^{2,p}(\Omega)} \qquad \forall v \in K \cap W^{2,p}(\Omega) \ ; \end{cases} \tag{2.23A}$$

the constant r in (2.23A) is independent of v and h. We then define $r_h : H_0^1(\Omega) \cap W^{2,p}(\Omega) \to V_{oh}$ by

$$r_h v = \frac{\pi_h v}{1 + r h^{1-2/p} \|v\|_{W^{2,p}(\Omega)}} \ ; \tag{2.24A}$$

---

(1) We can, in fact, define $\pi_h v$ $\forall v \in K$ since $K \subset W^{1,\infty}(\Omega) \subset C^0(\bar{\Omega})$.

it then follows from (2.23A), (2.24A) that

(2.25A) $$r_h v \in K_h \quad \forall v \in W^{2,p}(\Omega) \cap K.$$

Since $u \in W^{2,p}(\Omega) \cap K$, it follows from (2.25A) that we can take $v_h = r_h u$ in (2.21A), giving

(2.26A) $$\frac{1}{2}\|u_h-u\|^2_{H^1_o(\Omega)} \le \frac{1}{2}\|r_h u-u\|^2_{H^1_o(\Omega)} + \{\|\Delta u\|_{L^p} + \|f\|_{L^p}\}\|r_h u-u\|_{L^{p'}} .$$

We have

$$r_h u-u = \frac{\pi_h u-u-rh^{1-2/p}\|u\|_{W^{2,p}(\Omega)}u}{1+rh^{1-2/p}\|u\|_{W^{2,p}(\Omega)}} ,$$

which implies

(2.27A) $$\|r_h u-u\|_{H^1_o(\Omega)} \le \|\pi_h u-u\|_{H^1_o(\Omega)} + rh^{1-2/p}\|u\|_{W^{2,p}(\Omega)}\|u\|_{H^1_o(\Omega)} ,$$

(2.28A) $$\|r_h u-u\|_{L^{p'}(\Omega)} \le \|\pi_h u-u\|_{L^{p'}(\Omega)} + rh^{1-2/p}\|u\|_{W^{2,p}(\Omega)}\|u\|_{L^{p'}(\Omega)} .$$

Since $p > 2$ we have $L^p(\Omega) \subset L^{p'}(\Omega)$, with continuous injection; it then results from Strang-Fix /1/ and Ciarlet /1A/ that under the assumptions on $\mathcal{T}_h$ stated in the Theorem we have

(2.29A) $$\|\pi_h u-u\|_{H^1_o(\Omega)} \le Ch\|u\|_{W^{2,p}(\Omega)} ,$$

(2.30A) $$\|\pi_h u-u\|_{L^{p'}(\Omega)} \le Ch^2\|u\|_{W^{2,p}(\Omega)} ,$$

with, in (2.29A), (2.30A), C independent of h and u. Estimate (2.18A) then results trivially from (2.26A)-(2.30A). ■

Remark 2.1: It follows from Theorem 2.2 that if f = constant (which is the case for applications in mechanics) and if Ω is a convex polygonal domain, then we have an approximation error which is "practically" of $O(\sqrt{h})$ since $u \in W^{2,p}(\Omega)\ \forall p\ (<+\infty)$.

3. FINITE-ELEMENT APPROXIMATION OF PROBLEM (1.1A). (II) OPTIMAL ORDER ESTIMATES THROUGH THE USE OF AN EQUIVALENT FORMULATION

In this section, which follows the account of Falk-Mercier /1A/, we show that the use of an equivalent formulation enables us to obtain finite-element approximations which lead to error estimates of *optimal order*.

3.1 Equivalent formulation of problem (1.1A)

The following equivalence lemma forms the fundamental result of this third section:

Lemma 3.1: *Suppose that* Ω *is a simply-connected bounded domain in* $\mathbb{R}^2$; *there is then equivalence between problem* (1.1A) *and the variational problem* [1]

$$\text{(3.1A)} \quad \begin{cases} \text{Find } p \in \Lambda \cap H \text{ such that} \\ \int_\Omega p\cdot(q-p)\,dx \geq \int_\Omega \phi\cdot(q-p)\,dx \quad \forall q \in \Lambda \cap H, \end{cases}$$

where $\phi = \{\phi_1, \phi_2\}$ *is an (arbitrary) solution of*

$$\text{(3.2A)} \qquad f = \frac{\partial \phi_1}{\partial x_2} - \frac{\partial \phi_2}{\partial x_1},$$

and where

$$\text{(3.3A)} \qquad \Lambda = \{q \in L^2(\Omega) \times L^2(\Omega),\ |q| \leq 1 \quad a.e. \text{ on } \Omega\}$$

$$\text{(3.4A)} \qquad H = \{q \in L^2(\Omega) \times L^2(\Omega)\ ,\ \int_\Omega q\cdot\nabla w\,dx = 0 \quad \forall w \in H^1(\Omega)\}.$$

The solutions u *of* (1.1A) *and* p *of* (3.1A) *are connected by the relation*

$$\text{(3.5A)} \qquad p = \{\frac{\partial u}{\partial x_2}\ ,\ -\frac{\partial u}{\partial x_1}\}.$$

---

(1) In all that follows, $y\cdot z = \sum_{i=1}^{2} x_i y_i \quad \forall\ y,z \in \mathbb{R}^2$, $y = \{y_1, y_2\}$, $z = \{z_1, z_2\}$; also $|y| = (y\cdot y)^{1/2}$.

Proof: (i) Let u be the solution of (1.1A) with f satisfying (3.2A) and with $v \in H_o^1(\Omega)$; we use the notation

$$(3.6A) \qquad p = \{\frac{\partial u}{\partial x_2}, -\frac{\partial u}{\partial x_1}\}, \quad q = \{\frac{\partial v}{\partial x_2}, -\frac{\partial v}{\partial x_1}\}.$$

Clearly we have

$$(3.7A) \qquad v \in K \Longrightarrow q \in \Lambda ,$$

$$(3.8A) \qquad \int_\Omega fv\, dx = \int_\Omega \phi \cdot q\, dx \quad \forall\ v \in H_o^1(\Omega),$$

$$(3.9A) \qquad \int_\Omega \nabla u \cdot \nabla v\, dx = \int_\Omega p \cdot q\, dx \quad \forall v \in H_o^1(\Omega).$$

Moreover we have, $\forall w \in H^1(\Omega)$,

$$(3.10A) \qquad \int_\Omega q \cdot \nabla w\, dx = -\int_\Omega \nabla \cdot q\, w\, dx \quad \forall q = \{\frac{\partial v}{\partial x_2}, -\frac{\partial v}{\partial x_1}\},\ v \in \mathcal{D}(\Omega) ;$$

by the density of $\mathcal{D}(\Omega)$ in $H_o^1(\Omega)$, (3.10A) also holds $\forall q = \{\frac{\partial v}{\partial x_2}, -\frac{\partial v}{\partial x_1}\}$, $v \in H_o^1(\Omega)$, so that

$$(3.11A) \qquad v \in H_o^1(\Omega) \Longrightarrow q = \{\frac{\partial v}{\partial x_2}, -\frac{\partial v}{\partial x_1}\} \in H.$$

By using (3.6A)-(3.9A), (3.11A) we can easily deduce from (1.1A) that $p = \{\frac{\partial u}{\partial x_2}, -\frac{\partial u}{\partial x_1}\}$ is the solution of (3.1A). ■

(ii) It now remains to prove the converse; let p be the solution of (3.1A) and let $q \in \Lambda \cap H$. The condition $q \in H$ is in fact equivalent (from Green's formula) to

$$(3.12A) \qquad \nabla \cdot q = 0 \ \textit{a.e. in}\ \Omega,$$

$$(3.13A) \qquad q \cdot n = 0 \ \textit{on}\ \Gamma,$$

where $n = \{n_1, n_2\}$ is the unit *normal* pointing outwards from Ω; the vector $\tau = \{n_2, -n_1\}$ is thus tangential to Γ.

Conditions (3.12A), (3.13A) and the simple connectivity of Ω imply the existence of $v \in H^1(\Omega)$, defined to within an additive constant, such that

$$(3.14A) \qquad q = \{\frac{\partial v}{\partial x_2}, -\frac{\partial v}{\partial x_1}\},$$

$$(3.15A) \qquad \frac{\partial v}{\partial \tau} = 0 \text{ on } \Gamma \Leftrightarrow v = \text{constant on } \Gamma.$$

We can thus take $v = 0$ on Γ, which combined with $v \in H^1(\Omega)$ implies $v \in H^1_o(\Omega)$; in view of the above we then clearly have

$$(3.16A) \qquad q \in \Lambda \cap H \Leftrightarrow v \in K.$$

Since the above relations also hold for p, we deduce from (3.1A), (3.2A), (3.14A) the existence of a **u** which satisfies

$$\begin{cases} u \in K \\ \int_\Omega \nabla u \cdot \nabla (v-u)\,dx \geq \int_\Omega f(v-u)\,dx \quad \forall\, v \in K, \end{cases}$$

i.e. u is the (unique) solution of (1.1A). ■

Remark 3.1: It results from Duvaut-Lions /1, Chapter 5/ that (3.1A) is in fact the natural mechanical formulation for the problem of the elasto-plastic torsion of a cylindrical bar under the physical assumptions of Chapter 3, Section 1.2.

Remark 3.2: If f = const. = C (which corresponds to the physical situation), it follows immediately that $\phi = \{\phi_1, \phi_2\}$ defined by

$$(3.17A) \qquad \phi_1(x) = \lambda C x_2,\ \phi_2(x) = (\lambda-1) C x_1$$

satisfies condition (3.2A), $\forall \lambda \in \mathbb{R}$. If f *is not constant*, we (uniquely) determine $u_o \in H^1_o(\Omega)$ from

$$\begin{cases} \Delta u_o = f \text{ in } \Omega, \\ u_o = 0 \text{ on } \Gamma; \end{cases}$$

putting $\phi_1 = \frac{\partial u_o}{\partial x_2}$, $\phi_2 = -\frac{\partial u_o}{\partial x_1}$, it is clear that the function ϕ thus defined satisfies (3.2A).

Remark 3.3: The *regularity* results relating to problem (1.1A) (proved in Brézis-Stampacchia /2/ and quoted in Chapter 3, Section 2.1) and relation (3.5A) imply that if Γ is sufficiently regular, then we have for the solution p of problem (3.1A)

$$(3.18A) \qquad f \in L^2(\Omega) \Longrightarrow p \in H^1(\Omega) \times H^1(\Omega).$$

Remark 3.4: This remark concerns the existence of a *Lagrange multiplier*, associated with the condition $p \in H$, for problem (3.1A): it follows from (3.1A), (3.4A) that a natural Lagrangian associated with problem (3.1A) is $\mathcal{L} : (L^2(\Omega))^2 \times H^1(\Omega) \to \mathbb{R}$ defined by

$$(3.19A) \qquad \mathcal{L}(q,w) = \frac{1}{2}\int_\Omega |q|^2 dx - \int_\Omega \phi \cdot q \; dx - \int_\Omega q \cdot \nabla w \; dx ,$$

which leads to the saddle-point problem

$$(3.20A) \qquad \begin{cases} \textit{Find } \{p,\chi\} \in \Lambda \times H^1(\Omega) \textit{ such that} \\ \mathcal{L}(p,w) \le \mathcal{L}(p,\chi) \le \mathcal{L}(q,\chi) \quad \forall \{q,w\} \in \Lambda \times H^1(\Omega). \end{cases}$$

We then have

<u>Proposition 3.1</u>: *There is equivalence between* (3.20A) *and*

$$(3.21A) \qquad \begin{cases} \{p,\chi\} \in (\Lambda \cap H) \times H^1(\Omega), \\ \int_\Omega p \cdot (q-p)\,dx \ge \int_\Omega (\phi + \nabla\chi)\cdot(q-p)\,dx \quad \forall q \in \Lambda ; \end{cases}$$

moreover if $\{p,\chi\}$ *is a solution of* (3.20), (3.21A), *then* p *is the solution of problem* (3.1A).

Proof: (i) (3.20A) $\Longrightarrow$ (3.21A) : From the left-hand inequality in (3.20A) we deduce that

$$\int_\Omega p \cdot \nabla(w-\chi)\,dx \ge 0 \quad \forall w \in H^1(\Omega),$$

and hence

$$(3.22A) \qquad \int_\Omega p \cdot \nabla w \; dx = 0 \quad \forall w \in H^1(\Omega) \Longleftrightarrow p \in H ;$$

the right-hand inequality in (3.20A) implies (as is standard)

$$\text{(3.23A)} \qquad \lim_{\substack{t\to 0\\ t>0}} \frac{\mathcal{L}(p+t(q-p),\chi)-\mathcal{L}(p,\chi)}{t} \geq 0 \quad \forall q \in \Lambda .$$

Now

$$\lim_{\substack{t\to 0\\ t\neq 0}} \frac{\mathcal{L}(p+t(q-p),\chi)-\mathcal{L}(p,\chi)}{t}$$

$$= \int_\Omega p\cdot(q-p)\,dx - \int_\Omega (\phi+\nabla\chi)\cdot(q-p)\,dx$$

which with (3.23A) implies the inequality in (3.21A); in view of (3.22A) we have thus proved (3.20A) $\Longrightarrow$ (3.21A).

(ii) (3.21A) $\Longrightarrow$ (3.20A) : This is immediate since on the one hand

$$p\in H \Longrightarrow \mathcal{L}(p,w) = \frac{1}{2}\int_\Omega |q|^2 dx - \int_\Omega \phi\cdot q \;dx \quad \forall w \in H^1(\Omega),$$

and hence

$$\text{(3.24A)} \qquad \mathcal{L}(p,w) = \mathcal{L}(p,\chi) \quad \forall\; w \in H^1(\Omega).$$

On the other hand we have, by the *convexity* of $q \to \mathcal{L}(q,\chi)$,

$$\begin{cases} \mathcal{L}(q,\chi)-\mathcal{L}(p,\chi) = \frac{1}{2}\,|q|^2 - \frac{1}{2}\,|p|^2 - \int_\Omega (\phi+\nabla\chi)\cdot(q-p)\,dx \\ \geq \int_\Omega p\cdot(q-p)\,dx - \int_\Omega (\phi+\nabla\chi)\cdot(q-p)\,dx \quad \forall\; p,q \in L^2(\Omega)\times L^2(\Omega), \end{cases}$$

which combined with (3.21A) implies

$$\text{(3.25A)} \qquad \mathcal{L}(p,\chi) \leq \mathcal{L}(q,\chi) \quad \forall q \in \Lambda .$$

The double inequality in (3.20A) results from (3.24A), (3.25A).

(iii) It is moreover, immediate that (3.21A) implies that p satisfies (3.1A); it is sufficient to restrict q to $\Lambda \cap H$ in (3.21A) and to note that

$$p,q \in H \implies \int_\Omega \nabla\chi\cdot(q-p)\,dx = 0. \blacksquare$$

The existence of a multiplier χ satisfying (3.20A), (3.21A) is a tricky problem: however, it is shown in Duvaut-Lions /1/, using the results of Brézis /5/, that such a multiplier $\chi \in H^1(\Omega)$ exists in the case in which f = const. (the case in mechanics); it should be noted (see Duvaut-Lions, loc. cit.) that, in mechanics, χ is interpreted as a *displacement*. We note that the multiplier χ discussed above is defined only to within an arbitrary additive constant.

3.2 Approximation of problem (3.1A) by a finite-element method

We shall assume in the following that Ω is a bounded convex polygonal domain in $\mathbb{R}^2$.

3.2.1 *Definition of the approximate problem*

Using notation similar to that of Appendix 1, Section 3.6, we approximate $H^1(\Omega)$, $L^2(\Omega) \times L^2(\Omega)$, H and Λ respectively by:

(3.26A) $$V_h = \{w_h \in C^0(\bar{\Omega}),\ w_h|_T \in P_1 \quad \forall T \in \mathcal{C}_h\},$$

(3.27A) $$L_h = \{q_h \in L^2(\Omega) \times L^2(\Omega),\ \ q_h|_T \in P_0 \times P_0 \quad \forall\, T \in \mathcal{C}_h\}\,^{(1)},$$

(3.28A) $$H_h = \{q_h \in L_h,\ \int_\Omega q_h\cdot\nabla w_h\,dx = 0 \quad \forall w_h \in V_h\},$$

(3.29A) $$\Lambda_h = \Lambda \cap L_h = \{q_h \in L_h,\ \ |q_h| \le 1 \ \textit{a.e. on}\ \Omega\}.$$

We then approximate problem (3.1A) by

(3.30A) $$\begin{cases} \textit{Find } p_h \in \Lambda_h \cap H_h \textit{ such that} \\ \int_\Omega p_h\cdot(q_h-p_h)\,dx \ge \int_\Omega \phi\cdot(q_h-p_h)\,dx \quad \forall q_h \in \Lambda_h \cap H_h. \end{cases}$$

---

(1) P_0: space of *constant* real-valued functions.

It is easy to prove the following

<u>Proposition 3.2</u>: *Problem* (3.30A) *admits a unique solution.*

3.2.2 *Investigation of the convergence*

We shall assume in the following that $f \in L^2(\Omega)$; since Ω is convex and bounded, this implies (see Chapter 3, Section 2.1) that the solution u of (1.1A) satisfies $u \in H^2(\Omega)$, and hence for the solution p of (3.1A)

$$(3.31A)\qquad \begin{cases} p \in H^1(\Omega) \times H^1(\Omega), \\ (\textit{with}\quad \|p\|_{H^1(\Omega)\times H^1(\Omega)} \leq C\,\|f\|_{L^2(\Omega)}\ ) ; \end{cases}$$

similarly if ϕ is obtained from f using the relations of Remark 3.2, we have

$$(3.32A)\qquad \phi \in H^1(\Omega) \times H^1(\Omega)$$

with

$$(3.33A)\qquad \|\phi\|_{H^1(\Omega)\times H^1(\Omega)} \leq C\,\|f\|_{L^2(\Omega)}\quad ;$$

in (3.31A), (3.33A), C denotes constants.

We also assume the existence of a multiplier $\chi \in H^1(\Omega)$ such that $\{p,\chi\}$ satisfies (3.20A), (3.21A).

In order to investigate the convergence, we shall need the following results:

<u>Proposition 3.3</u>: *Let* $\pi_h : L^2(\Omega) \times L^2(\Omega) \to L_h$ *be the orthogonal projection operator onto* L_h *(in the norm* $L^2(\Omega) \times L^2(\Omega)$*); we then have the following properties*:

$$(3.34A)\qquad \nabla w_h \in L_h \quad \forall w_h \in V_h\ ,$$

$$(3.35A)\qquad \pi_h q \in \Lambda_h \quad \forall q \in \Lambda,$$

(3.36A) $$\int_\Omega (\pi_h q - q)\cdot z_h \, dx = 0 \quad \forall q \in L^2(\Omega)\times L^2(\Omega), \quad \forall z_h \in L_h,$$

(3.37A) $$\pi_h q \in H_h \quad \forall q \in H.$$

Proof: Properties (3.34A)-(3.36A) are almost immediate; we shall now prove (3.37A):

We have $V_h \subset H^1(\Omega)$, giving

(3.38A) $$\int_\Omega q\cdot \nabla w_h dx = 0 \quad \forall q \in H, \quad \forall w_h \in V_h .$$

Since $\nabla w_h \in L_h$ it follows from (3.36A) and (3.38A) that if $q \in H$ then

$$\int_\Omega \pi_h q\cdot\nabla w_h dx = \int_\Omega q\cdot\nabla w_h dx = 0 \quad \forall w_h \in V_h \Rightarrow \pi_h q \in H_h . \blacksquare$$

Moreover, by using the results of Ciarlet /1A/, for example, we may prove the following proposition regarding π_h:

Proposition 3.4: *We have* $\forall\, q \in H^1(\Omega)\times H^1(\Omega)$

(3.39A) $$\|\pi_h q - q\|_{L^2(\Omega)\times L^2(\Omega)} \le Ch\, \|q\|_{H^1(\Omega)\times H^1(\Omega)} ,$$

with C *independent of* h *and* q.

Regarding the convergence of p_h to p, we have

Theorem 3.1 (Falk-Mercier /1A/): *Under the above assumptions on* Ω, f, χ, *(the assumptions on* $\mathcal{T}_h$ *being those of Theorem* 2.2) *we have the (optimal order) error estimate*

(3.40A) $$\|p_h - p\|_{L^2(\Omega)\times L^2(\Omega)} = O(h) ,$$

where, in (3.40A), p *is the solution of problem* (3.1A) *and* p_h *that of the approximate problem* (3.30A).

Proof: We have, from the definition of the approximate problem

$$(3.41A)\qquad \int_\Omega p_h\cdot(q_h-p_h)\,dx \ge \int_\Omega \phi\cdot(q_h-p_h)\,dx \quad \forall q_h \in \Lambda_h \cap H_h,$$

and also

$$(3.42A)\qquad p_h \in \Lambda_h \cap H_h \subset \Lambda\ .$$

In view of (3.42A) we can take $q = p_h$ in (3.21A), giving

$$(3.43A)\qquad \int_\Omega p\cdot(p_h-p)\,dx \ge \int_\Omega (\phi+\nabla\chi)\cdot(p_h-p)\,dx.$$

By adding (3.41A) and (3.43A), we obtain after some straightforward manipulation, $\forall q_h \in \Lambda_h \cap H_h$, $\forall w_h \in V_h$:

$$(3.44A)\qquad \begin{cases} \|p_h-p\|^2_{L^2(\Omega)\times L^2(\Omega)} \le \int_\Omega p_h\cdot(q_h-p)\,dx - \int_\Omega \phi\cdot(q_h-p)\,dx \\ - \int_\Omega \nabla(\chi-w_h)\cdot(p_h-p)\,dx - \int_\Omega \nabla w_h\cdot(p_h-p)\,dx \quad . \end{cases}$$

In the following we shall use the notation $|q|_o = \|q\|_{L^2(\Omega)\times L^2(\Omega)}$ and $|q|_1 = \|q\|_{H^1(\Omega)\times H^1(\Omega)}$; we may then deduce from (3.44A) (via *Schwarz's inequality* and the relation $2|x||y| \le x^2+y^2 \ \ \forall x,y \in \mathbb{R}$), $\forall q_h \in \Lambda_h \cap H_h$, $\forall w_h \in V_h$:

$$(3.45A)\qquad \begin{cases} \frac{1}{2}\,|p_h-p|^2_o \le \left|\int_\Omega p_h\cdot(q_h-p)\,dx\right| + \left|\int_\Omega \phi\cdot(q_h-p)\,dx\right| \\ + \left|\int_\Omega \nabla w_h\cdot(p_h-p)\,dx\right| + \frac{1}{2}\,|\nabla(\chi-w_h)|^2_o\ . \end{cases}$$

Moreover we have $p_h \in H_h$, $p \in H$, giving $\forall w_h \in V_h$ $(\subset H^1(\Omega))$,

$$\int_\Omega \nabla w_h\cdot(p_h-p)\,dx = 0\ ,$$

which, together with (3.45A), implies

$$(3.46\text{A})\quad \begin{cases} \frac{1}{2}|p_h-p|_0^2 \le \left|\int_\Omega p_h\cdot(q_h-p)\,dx\right| + \left|\int_\Omega \phi\cdot(q_h-p)\,dx\right| \\ \qquad + \frac{1}{2}|\nabla(\chi-w_h)|_0^2 \qquad \forall\, q_h \in \Lambda_h \cap H_h, \quad \forall w_h \in V_h\ . \end{cases}$$

We then approximate p and χ by $\pi_h p$ and $r_h\chi$ where $r_h\ :\ C^0(\bar\Omega) \to V_h$ is the (linear) V_h-*interpolation* operator on $\mathcal{T}_h$, i.e.

$$r_h v \in V_h \quad \forall v \in C^0(\bar\Omega),$$

$$(r_h v)(P) = v(P) \quad \forall\ P \quad \text{vertex of} \quad T \in \mathcal{T}_h\ ;$$

it is perfectly acceptable to define $r_h\chi$, since $\chi \in H^2(\Omega) \subset C^0(\bar\Omega)$ in the two-dimensional case under consideration. We then have

$$(3.47\text{A})\qquad r_h\chi \in V_h\ ;$$

similarly we have (see Proposition 3.3)

$$(3.48\text{A})\qquad \pi_h p \in \Lambda_h \cap H_h\ ,$$

$$(3.49\text{A})\qquad \int_\Omega p_h\cdot(\pi_h p-p)\ dx = 0\ .$$

Moreover, since $p \in H^1(\Omega) \times H^1(\Omega)$ and $\chi \in H^2(\Omega)$ (see (3.39A) and Strang-Fix /1/, Ciarlet /1A/ for the following estimates), we have

$$(3.50\text{A})\qquad |\pi_h p-p|_0 \le Ch|p|_1\,,$$

$$(3.51\text{A})\qquad |\nabla(r_h\chi-\chi)|_0 \le Ch\|\chi\|_{H^2(\Omega)}\quad ,$$

where in (3.50A), (3.51A), C denotes constants which are independent of h, p, and χ.

In view of (3.47A), (3.48A) we can, in (3.46A), take $q_h = \pi_h p$ and $w_h = r_h\chi$; we then deduce from (3.46A), (3.49A), (3.51A)

$$\frac{1}{2}|p_h-p|_o^2 \le \left|\int_\Omega \phi\cdot(\pi_h p-p)\,dx\right| + C^2h^2\|\chi\|^2_{H^2(\Omega)} . \tag{3.52A}$$

From Proposition 3.3, relation (3.36A), we have

$$\left\{\begin{aligned}\int_\Omega \phi\cdot(\pi_h p-p)\,dx &= \int_\Omega(\phi-\pi_h\phi)\cdot(\pi_h p-p)\,dx + \int_\Omega \pi_h\phi\cdot(\pi_h p-p)\,dx\\ &= \int_\Omega(\phi-\pi_h\phi)\cdot(\pi_h p-p)\,dx ,\end{aligned}\right.$$

and hence

$$\left|\int_\Omega \phi\cdot(\pi_h p-p)\,dx\right| \le |\phi-\pi_h\phi|_o|\pi_h p-p|_o ; \tag{3.53A}$$

we then deduce from (3.31A)-(3.33A), (3.39A) and from (3.50A), (3.52A), (3.53A)

$$|p_h-p|_o^2 \le C^2h^2(\|f\|^2_{L^2(\Omega)} + \|\chi\|^2_{H^2(\Omega)}) ,$$

which proves the estimate (3.40A) and renders it more precise. ■

4. FURTHER DISCUSSION OF ITERATIVE METHODS FOR SOLVING THE ELASTO-PLASTIC TORSION PROBLEM

4.1 General discussion. Synopsis

In Chapter 3, various methods based on the formulation (1.1A) were dèscribed for the iterative solution of the elasto-plastic torsion problem; without going into greater detail (for which we refer to Marrocco /1A/, Gabay-Mercier /1/, Fortin-Glowinski /1A/, /2A/, Glowinski /2A/ and Glowinski-Marrocco /1A/) we can state that the most efficient calculation methods currently available are based on algorithm (10.5)-(10.7) of Chapter 3, Section 10 (*of the augmented Lagrangian type*); this algorithm ([1]) can be generalised in a particularly simple manner to cases in which the cross-section of the bar is *multiply-connected*.

We shall conclude this fourth section (and this appendix) by investigating, in Sections 4.2 and 4.3, the *iterative solution* of

([1]) Together with its variants for which cite the above references.

(1.1A) *via the equivalent formulation* (3.1A), using *duality*-type algorithms based on the formulation (3.20A); in Section 4.2 we describe a Uzawa-type algorithm and in Section 4.3 a variant of conjugate-gradient type.

4.2 Solution of problem (3.1A) by a duality method

4.2.1 *General discussion. Description of the algorithm*

Given the particular structure of problem (3.1A), we shall restrict our attention to a duality algorithm of the same type as those considered in Chapter 2, Section 4.3 (see also Appendix 2, Section 3). Using the duality results in Remark 3.4 of Section 3.1 of this appendix, we shall reduce the solution of (3.1A) (and therefore that of (1.1A)) to the solution of the saddle-point problem (3.20A), which as we recall is

$$\text{(4.1A)}\qquad \begin{cases} \textit{Find } \{p,\chi\} \in \Lambda \times H^1(\Omega) \textit{ such that} \\ \mathcal{L}(p,w) \le \mathcal{L}(p,\chi) \le \mathcal{L}(q,\chi) \quad \forall \{q,w\} \in \Lambda \times H^1(\Omega) \end{cases}$$

with

$$\text{(4.2A)}\qquad \mathcal{L}(q,w) = \frac{1}{2}\int_\Omega |q|^2 dx - \int_\Omega \phi\cdot q \, dx - \int_\Omega \nabla w\cdot q \, dx ,$$

which is justified since p *is in this case the solution of* (3.1A).

In view of Chapter 2, Section 4.3, this leads to the following (Uzawa-type) algorithm:

$$\text{(4.3A)}\qquad \chi^o \in H^1(\Omega) \textit{ given } (\chi^o = 0 \textit{ for example}),$$

then for $n \ge 0$, $\chi^n \in H^1(\Omega)$ *known*, *determine* $p^n \in \Lambda$ *and* $\chi^{n+1} \in H^1(\Omega)$ *using*

$$\text{(4.4A)}\qquad \begin{cases} p^n \in \Lambda , \\ \mathcal{L}(p^n,\chi^n) \le \mathcal{L}(q,\chi^n) \quad \forall q \in \Lambda, \end{cases}$$

$$(4.5A)\quad \begin{cases} \chi^{n+1} \in H^1(\Omega), \\ (\chi^{n+1}, w)_{H^1(\Omega)} = (\chi^n, w)_{H^1(\Omega)} - \rho\int_\Omega \nabla w \cdot p^n dx \quad \forall w \in H^1(\Omega) \end{cases}$$

with in (4.5A), $\rho > 0$.

The practical implementation of (4.3A)-(4.5A) presents no major difficulties; in effect (4.4A) is equivalent to

$$(4.6A)\qquad p^n = P_\Lambda(\phi + \nabla\chi^n),$$

where $P_\Lambda : (L^2(\Omega))^2 \to (L^2(\Omega))^2$ is the *orthogonal projection operator* onto Λ (in the norm $L^2(\Omega) \times L^2(\Omega)$); we then have

$$(4.7A)\qquad p^n = \frac{\phi + \nabla\chi^n}{\sup(1, |\phi + \nabla\chi^n|)}.$$

Regarding (4.6A), if we define $(\cdot,\cdot)_{H^1(\Omega)}$ by [1]

$$(4.8A)\qquad (v,w)_{H^1(\Omega)} = \int_\Omega (\nabla v \cdot \nabla w + vw)\,dx \quad \forall v, w \in H^1(\Omega) \quad [2]$$

then the determination of χ^{n+1} from χ^n and p^n (via (4.5A)) reduces to the solution of a *Neumann problem* for the operator $-\Delta + I$; this problem is well-posed since the linear mapping

$$w \to (\chi^n, w)_{H^1(\Omega)} - \rho\int_\Omega \nabla w \cdot p^n dx$$

is obviously continuous on $H^1(\Omega)$. It can thus be seen that the implementation of (4.3A)-(4.5A) (in practice, its finite-dimensional variants) is straightforward provided that efficient programs for the numerical solution of linear elliptic problems of order two are available.

---

(1) other choices are possible

(2) and hence $\|\nabla w\|_{L^2 \times L^2} \le \|w\|_{H^1} \quad \forall\, w \in H^1(\Omega)$.

4.2.2 *Convergence of algorithm* (4.3A) - (4.5A)

Regarding the convergence of algorithm (4.3A)-(4.5A) we have

Theorem 4.1: *Suppose that the saddle-point problem* (4.1A), (4.2A) *admits a solution* $\{p,\chi\}$ *and that* $(\cdot,\cdot)_{H^1(\Omega)}$ *is defined by* (4.8A); *then if*

$$0<\rho<2 \tag{4.9A}$$

we have for the sequence $\{p^n\}_{n\geq 0}$ *defined by algorithm* (4.3A)-(4.5A)

$$\lim_{n\to+\infty} \|p^n-p\|_{L^2(\Omega)\times L^2(\Omega)} = 0 \quad \forall\, \chi^o \in H^1(\Omega) \tag{4.10A}$$

where, *in* (4.10A), p *is the solution of* (3.1A).

Proof: This follows the same general lines as that of Theorem 4.1 in Chapter 2, Section 4.3; despite this we shall give the proof in full in order to demonstrate the convergence condition (4.9A).

Let $\{p,\chi\}$ be a solution of (4.1A), (4.2A); then (see Proposition 3.1) p must necessarily be a solution of (3.1A). In view of the fact that $p \in H$, we have

$$\int_\Omega p\cdot\nabla w \,dx = 0 \quad \forall w \in H^1(\Omega)$$

and hence

$$(\chi,w)_{H^1(\Omega)} = (\chi,w)_{H^1(\Omega)} - \rho\int_\Omega p\cdot\nabla w\,dx \quad \forall w\in H^1(\Omega),$$

which, by differencing with (4.5A) and writing

$$\bar{p}^n = p^n-p\ ,\quad \bar{\chi}^n = \chi^n-\chi\ ,$$

leads to

$$(\bar{\chi}^{n+1},w)_{H^1(\Omega)} = (\bar{\chi}^n,w)_{H^1(\Omega)} - \rho\int_\Omega \bar{p}^n\cdot\nabla w\,dx \quad \forall w\in H^1(\Omega). \tag{4.11A}$$

Moreover, by *convexity* and by taking account of (4.4A) and the right-hand inequality in (4.1A), we have

(4.12A) $$\int_\Omega p^n\cdot(p-p^n)\,dx \geq \int_\Omega(\phi+\nabla\chi^n)\cdot(p-p^n)\,dx\ ,$$

(4.13A) $$\int_\Omega p\cdot(p^n-p)\,dx \geq \int_\Omega(\phi+\nabla\chi)\cdot(p^n-p)\,dx\ ;$$

we deduce, by adding (4.12A) and (4.13A)

(4.14A) $$\|\bar{p}^n\|^2_{L^2(\Omega)\times L^2(\Omega)} \leq \int_\Omega \bar{p}^n\cdot\nabla\bar{\chi}^n dx\ .$$

At this stage it is convenient to introduce $\bar{y}^n \in H^1(\Omega)$, defined (uniquely) by

(4.15A) $$\begin{cases} \bar{y}^n \in H^1(\Omega)\,, \\ (\bar{y}^n,w)_{H^1(\Omega)} = \int_\Omega \bar{p}^n\cdot\nabla w\ dx \quad \forall\ w\in H^1(\Omega)\,. \end{cases}$$

We then have, from (4.11A), (4.15A)

$$\bar{\chi}^{n+1} = \bar{\chi}^n - \rho\bar{y}^n\ ,$$

giving

$$\|\bar{\chi}^{n+1}\|^2_{H^1(\Omega)} = \|\bar{\chi}^n\|^2_{H^1(\Omega)} - 2\rho(\bar{\chi}^n,\bar{y}^n)_{H^1(\Omega)} + \rho^2\|\bar{y}^n\|^2_{H^1(\Omega)},$$

or alternatively

(4.16A) $$\|\bar{\chi}^n\|^2_{H^1(\Omega)} - \|\bar{\chi}^{n+1}\|^2_{H^1(\Omega)} = 2\rho(\bar{\chi}^n,\bar{y}^n)_{H^1(\Omega)} - \rho^2\|\bar{y}^n\|^2_{H^1(\Omega)}.$$

By putting $w = \bar{\chi}^n$ in (4.15A) we deduce from (4.14A),(4.16A)

(4.17A) $$\|\bar{\chi}^n\|^2_{H^1(\Omega)} - \|\bar{\chi}^{n+1}\|^2_{H^1(\Omega)} \geq 2\rho\|\bar{p}^n\|^2 - \rho^2\|\bar{y}^n\|^2_{H^1(\Omega)}\ ;$$

by now putting $w = \bar{y}^n$ in (4.15A) we thereby deduce, using Schwarz's inequality and (4.8A)

$$\begin{cases} \|\bar{y}^n\|^2_{H^1(\Omega)} = \int_\Omega \bar{p}^n\cdot\nabla\bar{y}^n\ dx \leq \|\bar{p}^n\|_{L^2\times L^2}\|\nabla\bar{y}^n\|_{L^2\times L^2} \\ \qquad\qquad \leq \|\bar{p}^n\|_{L^2\times L^2}\|\bar{y}^n\|_{H^1} \end{cases}$$

and hence

$$\|\bar{y}^n\|_{H^1(\Omega)} \le \|\bar{p}^n\|_{L^2(\Omega)\times L^2(\Omega)} . \tag{4.18A}$$

It then follows from (4.17A), (4.18A) that

$$\|\bar{\chi}^n\|^2_{H^1(\Omega)} - \|\bar{\chi}^{n+1}\|^2_{H^1(\Omega)} \ge \rho(2-\rho)\,\|\bar{p}^n\|^2_{L^2(\Omega)\times L^2(\Omega)} \tag{4.19A}$$

It is clear, in view of (4.19A), that condition (4.9A) implies (4.10A) since $\bar{p}^n = p^n - p$.

Remark 4.1: We deduce from (4.5A), (4.10A) and from the fact that $p \in H$ that under the assumptions for which Theorem 4.1 above applies, we have

$$\lim_{n\to+\infty} \|\chi^{n+1}-\chi^n\|_{H^1(\Omega)} = 0. \tag{4.20A}$$

In fact we know a great deal more about the behaviour of the sequence $\{\chi^n\}_{n\ge 0}$, since, by applying Theorem 3.1 of Appendix 2, Section 3, it can be shown that under the above assumptions we have

$$\lim_{n\to+\infty} \chi^n = \chi^* \text{ weakly in } H^1(\Omega) \tag{4.21A}$$

where χ^* ($\in H^1(\Omega)$) is such that $\{p,\chi^*\}$ is a solution of the saddle-point problem (4.1A), (4.2A).

We note (see Lions-Magenes /1/) that (4.21A) implies

$$\lim_{n\to+\infty} \|\chi^n-\chi^*\|_{H^s(\Omega)} = 0 \quad \forall\, s<1, \tag{4.22A}$$

and hence, in particular

$$\lim_{n\to+\infty} \|\chi^n-\chi^*\|_{L^2(\Omega)} = 0.$$

Remark 4.2: The above can easily be "extended" to the approximate problem (3.30A) via the solution of the associated saddle-point problem

(4.23A)
$$\begin{cases} \text{Find } \{p_h, \chi_h\} \in \Lambda_h \times V_h \text{ such that} \\ \mathcal{L}(p_h, w_h) \leq \mathcal{L}(p_h, \chi_h) \leq \mathcal{L}(q_h, \chi_h) \quad \forall \{q_h, w_h\} \in \Lambda_h \times V_h \ ; \end{cases}$$

in particular, the convergence condition (4.9A) still holds for finite-dimensional variants of algorithm (4.3A)-(4.5A) (provided, of course, that $(\cdot,\cdot)_{H^1(\Omega)}$ is still defined by (4.8A)).

We note that (4.23A) always admits a saddle point; in fact (3.30A) admits a (unique) solution and the relations defining H_h are *linear* and are *finite* in number (this number being equal to $\dim V_h$); this is sufficient to ensure (see Rockafellar /3/) the existence of a solution to the saddle-point problem (4.23A) in the *finite-dimensional* case. ■

4.3 On conjugate-gradient-type variants of algorithm (4.3A)-(4.5A)

4.3.1 *General discussion*

It is often advantageous [1] to attempt to *accelerate the convergence* of gradient-type algorithms by using a variant of *conjugate-gradient* type in the sense of Chapter 2, Section 2.3 (see also Appendix 2, Section 5); this is particularly true for algorithms of the Uzawa type, since these can often be interpreted as gradient algorithms applied to a problem which is the *dual* of the original problem (the *primal problem*).

The conjugate-gradient algorithm (2.23)-(2.25) of Chapter 2, Section 2.3 was applied to the solution of the problem

(4.24A)
$$\begin{cases} J(u) \leq J(v) \quad \forall v \in \mathbb{R}^N, \\ u \in \mathbb{R}^N, \end{cases}$$

with

(4.25A)
$$J(v) = \frac{1}{2}(Av, v)_{\mathbb{R}^N} - (b, v)_{\mathbb{R}^N} \ ;$$

---

(1) At the expense of a certain increase in the complexity of the programming.

we also mentioned a variant which can be applied to non-quadratic cases (in fact this variant is known as the *Fletcher-Reeves* method, see Polak /1/). The above algorithms can in fact be generalised to much more complicated situations, both finite- and infinite-dimensional (see Daniel /1/, /2/, Polak /1/ and the additional references given in Section 4.3.3); in particular, as we shall see in Section 4.3.2, we can define conjugate-gradient-type variants of algorithm (4.3A)-(4.5A) (and of its finite-dimensional variants associated with the solution of the approximate problem (3.30A)).

4.3.2 *Dual functional. Description of conjugate-gradient-type variants of algorithm* (4.3A)-(4.5A)

Let $\mathcal{L} : (L^2(\Omega))^2 \times H^1(\Omega) \to \mathbb{R}$ be the Lagrangian defined by (4.2A): it is convenient to consider the functional $\mathcal{M} : H^1(\Omega) \to \mathbb{R}$ defined by

$$\mathcal{M}(w) = - \underset{q\in\Lambda}{\text{Min}}\ \mathcal{L}(q,w). \tag{4.26A}$$

If we denote by p(w) the solution of the problem

$$\underset{q\in\Lambda}{\text{Min}}\ \mathcal{L}(q,w) \tag{4.27A}$$

we have

$$p(w) = \frac{\phi+\nabla w}{\sup(1,|\phi+\nabla w|)} \tag{4.28A}$$

and hence

$$\left\{\begin{aligned}\mathcal{M}(w) &= -\mathcal{L}(p(w),w)\\ &= \int_\Omega \frac{|\phi+\nabla w|^2}{\sup(1,|\phi+\nabla w|)}\,dx - \frac{1}{2}\int_\Omega \frac{|\phi+\nabla w|^2}{\sup(1,|\phi+\nabla w|^2)}\,dx .\end{aligned}\right. \tag{4.29A}$$

We deduce from (4.29A) that

$$\mathcal{M}(w) = \int_\Omega M(w)\,dx \tag{4.30A}$$

with the integrand M(•) in (4.30A) defined by

(4.31A) $$M(w)(x) = \begin{cases} \frac{1}{2}|\phi(x)+\nabla w(x)|^2 & \textit{if} \quad |\phi(x)+\nabla w(x)| < 1, \\ |\phi(x)+\nabla w(x)| - \frac{1}{2} & \textit{if} \quad |\phi(x)+\nabla w(x)| \geq 1. \end{cases}$$

Starting from (4.26A)-(4.28A) and (4.30A), (4.31A), it is quite easy to prove:

<u>Proposition 4.1</u>. *The functional* $\mathcal{M} : H^1(\Omega) \to \mathbb{R}$ *defined by* (4.26A) *is convex, Lipschitz continuous and Gateaux differentiable; its Gateaux differential* $\mathcal{M}'$ *is also Lipschitz continuous, and is defined by*

(4.32A) $$\langle \mathcal{M}'(w), \eta \rangle = \int_\Omega p(w)\cdot\nabla\eta \, dx \quad \forall w, \eta \in H^1(\Omega) \;;$$

where, in (4.32A), $p(w)$ *is defined by* (4.27A), (4.28A) *and where* $\langle\cdot,\cdot\rangle$ *denotes the bilinear form of the duality between* $(H^1(\Omega))'$ *and* $H^1(\Omega)$. ■

Remark 4.3: It follows from (4.26A) that the problem

(4.33A) $$\begin{cases} \textit{Find } \chi \in H^1(\Omega) \textit{ such that} \\ \mathcal{M}(\chi) \leq \mathcal{M}(w) \quad \forall w \in H^1(\Omega) \end{cases}$$

is the dual of problem (3.1A) associated with the Lagrangian $\mathcal{L}$; moreover, algorithm (4.3A)-(4.5A) is simply a (fixed-step) gradient method applied to the solution of the dual problem (4.33A).

Problem (4.33A) is an *infinite-dimensional* minimisation problem, *non-quadratic* but *unconstrained*; the generalisation of the conjugate-gradient algorithms of Chapter 2, Section 2.3 to this problem leads to the following algorithms:

Stage 0: *Initialisation*

(4.34A) $$\chi^0 \in H^1(\Omega), \textit{ given}$$

then determine g^0 *by*

(4.35A) $$\begin{cases} g^0 \in H^1(\Omega), \\ (g^0, w)_{H^1(\Omega)} = \langle \mathcal{M}'(\chi^0), w \rangle \quad \forall\, w \in H^1(\Omega) \end{cases}$$

and put

$$z^o = g^o. \tag{4.36A}$$

Next, for $n \geq 0$ *and assuming that* χ^n, g^n, z^n *are known determine* χ^{n+1}, g^{n+1}, z^{n+1} *using*

Stage 1: Descent

Solve the one-variable minimisation problem

$$\begin{cases} \text{Find } \lambda^n \in \mathbb{R} \text{ such that} \\ \mathcal{M}(\chi^n - \lambda^n z^n) \leq \mathcal{M}(\chi^n - \lambda z^n) \quad \forall \lambda \in \mathbb{R} \end{cases} \tag{4.37A}$$

and put

$$\chi^{n+1} = \chi^n - \lambda^n z^n . \tag{4.38A}$$

Stage 2: Construction of the new descent direction

Define g^{n+1} *by*

$$\begin{cases} g^{n+1} \in H^1(\Omega), \\ (g^{n+1}, w)_{H^1(\Omega)} = \langle \mathcal{M}'(\chi^{n+1}), w \rangle \quad \forall w \in H^1(\Omega), \end{cases} \tag{4.39A}$$

and then γ^{n+1} *by*

$$\gamma^{n+1} = \frac{(g^{n+1}, g^{n+1})_{H^1(\Omega)}}{(g^n, g^n)_{H^1(\dot{\Omega})}} = \frac{\|g^{n+1}\|^2_{H^1(\Omega)}}{\|g^n\|^2_{H^1(\Omega)}} \tag{4.40A$_1$}$$

(or alternatively

$$\gamma^{n+1} = \frac{(g^{n+1}, g^{n+1} - g^n)_{H^1(\Omega)}}{(g^n, g^n)_{H^1(\Omega)}} \quad) \tag{4.40A$_2$}$$

and finally

(4.41A) $z^{n+1} = g^{n+1} + \gamma^{n+1} z^n$.

The next step, of course, is

(4.42A) $n = n+1$, *go to* (4.37A). ■

The above algorithms call for a number of comments:

Remark 4.4: The conjugate-gradient algorithm (4.34A)-(4.39A), $(4.40A)_1$, (4.41A), (4.42A) (resp. (4.34A)-(4.39A), $(4.40A)_2$, (4.42A) is of *Fletcher-Reeves* (resp. *Polak-Ribière*) type (see Polak /1/ for the *finite-dimensional* convergence of these algorithms). If $\mathcal{M}$ is *quadratic* it can be shown that *the two algorithms coincide*; however it appears that methods of the Polak-Ribière type in general perform better than those of the Fletcher-Reeves type (for an explanation of this phenomenon, see Powell /1A/). ■

Remark 4.5: The two non-trivial stages in the implementation of the two algorithms above are:

(i) The solution of the *one-variable* minimisation problem (4.37A); this problem can be solved (without calculating derivatives) using methods of *dichotomy* or *Fibonacci* type, etc... (see Polak /1/, Wilde-Beightler /1A/, Brent /1A/); it is also possible to use *Newton* or *secant* type methods. We should point out that it is not computationally expensive to evaluate $\mathcal{M}(w)$, $\forall w \in H^1(\Omega)$, from relations (4.28A), (4.29A).

(ii) The calculation of g^{n+1} from χ^{n+1} requires the solution of the Neumann problem (4.39A); we note that in view of (4.32A) we have, in (4.39A),

(4.43A) $$\langle \mathcal{M}'(\chi^{n+1}), w\rangle = \int_\Omega p^{n+1} \cdot \nabla w \, dx \quad \forall w \in H^1(\Omega),$$

where in (4.43A) we have put (see (4.28A))

(4.44A) $$p^{n+1} = p(\chi^{n+1}) = \frac{\phi + \nabla\chi^{n+1}}{\sup(1, |\phi + \nabla\chi^{n+1}|)} \,. ■$$

Remark 4.6: In view of the above remarks it is clear that, as for algorithm (4.3A)-(4.5A), an efficient method for solving Neumann problems will be a fundamental tool for the implementation of the two conjugate-gradient algorithms above; it will in fact be finite-dimensional variants of these two algorithms which will be applied to problem (3.30A), and under these conditions it will be advisable to factorise once and for all the (symmetric, positive definite) matrix associated with the bilinear form $V_h \times V_h \to \mathbb{R}$ defined by

$$\{v_h, w_h\} \to (v_h, w_h)_{H^1(\Omega)} \quad ; \tag{4.45A}$$

to this end we shall employ either a *Gauss* or a *Cholesky factorisation*. The above also holds, of course, for finite-dimensional variants of algorithm (4.3A)-(4.5A). ∎

Remark 4.7: The convergence of the two algorithms (4.34A)-(4.42A) is a tricky problem, and to the best of our knowledge is still an open question. ∎

4.3.3 *Discussion. Conclusion.*

The algorithm (4.3A)-(4.5A) (and its finite-dimensional variants) falls within the class of *gradient methods with auxiliary operator*, previously encountered in Chapter 1, Section 3.1 and Chapter 5, Section 6.4.

As regards the two algorithms of Section 4.3.2 above, these are *conjugate gradient methods with auxiliary operator* (in the specialist literature these are usually referred to as conjugate gradient algorithms with *preconditioning*); such algorithms have been used in particular for the numerical solution of problems involving linear or non-linear partial differential equations, some of which (see Bristeau-Glowinski-Periaux-Perrier-Pironneau-Poirier /1A/, Bristeau-Glowinski-Periaux-Perrier-Pironneau /1A/, Periaux /1A/ and Le Tallec /1A/) are much more complicated than those considered in this current section. In addition to the above references, further details may be found in Bartels-Daniel /1A/, Axelsson /1A/, Douglas-Dupont /1A/, Concus-Golub-O'Leary /1A/, Glowinski-Pironneau /1A/, and their respective bibliographies.

It would certainly appear that the algorithms discussed in this section, combined with the finite-element approximation methods (due to Falk-Mercier /1A/) of Section 3, provide methods which are efficient and relatively easy to program for solving, via formulation (3.1A), the problem of the elasto-plastic torsion of a cylindrical bar with simply-connected cross-section under the physical assumptions of Chapter 3, Section 1.2.

Appendix 4

FURTHER DISCUSSION OF BOUNDARY UNILATERAL PROBLEMS AND ELLIPTIC VARIATIONAL INEQUALITIES OF ORDER 4. APPLICATION TO FLUID MECHANICS.

1. SYNOPSIS

This appendix supplements Chapter 4 on *boundary unilateral problems* and on *elliptic variational inequalities of order* 4. In Section 2 we consider the approximation, using both *conforming* and *non-conforming finite elements*, of the unilateral problem

$$(1.1A)\qquad \begin{cases} u \in K, \\ \int_\Omega [\nabla u \cdot \nabla(v-u) + u(v-u)]dx \geq \int_\Omega f(v-u)dx \quad \forall v \in K \end{cases}$$

with ($\Gamma = \partial\Omega$)

$$(1.2A)\qquad K = \{v \in H^1(\Omega),\ v \geq g \ \ \textit{a.e. on } \Gamma\}.$$

In Section 3 we return to the method of mixed finite elements for fourth-order problems, which formed the subject of Section 4.5 of Chapter 4; we supplement the approximation results and describe an application of this method to the solution of the problem of the (fourth order) elliptic variational inequality

$$(1.3A)\qquad \begin{cases} u \in K, \\ \int_\Omega \Delta u\ \Delta(v-u)dx \geq \int_\Omega f(v-u)dx \quad \forall v \in K, \end{cases}$$

with

$$(1.4A)\qquad K = \{v \in H_0^2(\Omega),\ \alpha \leq \Delta v \leq \beta \quad \textit{a.e. in } \Omega\}.$$

In Section 4 we show that the techniques used for solving fourth-order elliptic variational inequalities facilitate the numerical treatment of certain problems in fluid mechanics; as an example we have taken the *transonic potential* flow of a *perfect compressible fluid*; further examples will be given in Appendix 6.

Finally in Section 5 we give a supplementary bibliography relating to the subjects contained in this appendix, or to closely related subjects.

2. CONFORMING AND NON-CONFORMING APPROXIMATIONS OF THE BOUNDARY UNILATERAL PROBLEM (1.1A), (1.2A).

The numerical analysis of problem (1.1A), (1.2A)(1) using *conforming finite elements of order one and two* was the subject of Chapter 4, Section 3, in which various iterative methods for solving the continuous problem and the corresponding approximate problems were also investigated. In the following, we supplement the above results by those of Brezzi-Hager-Raviart(2) /1A/, /2A/ regarding the finite-element approximation of (1.1A), (1.2A) and in particular the problem of obtaining *optimal order* error estimates for the approximation. In Section 2.1 (resp. 2.2) we consider the approximation of (1.1A), (1.2A) by *first-order conforming* (resp. *mixed-type non-conforming*) *finite elements*.

2.1. Approximation of the unilateral problem (1.1A), (1.2A) by first-order conforming finite elements.

We suppose in what follows that Ω is a bounded polygonal domain in $\mathbb{R}^2$ and that $f \in L^p(\Omega)$ (with $p > 1$), $g \in C^0(\Gamma)$.

2.1.1. *The approximate problem.*

We use the notation of Appendix 1, Section 3.6, namely

$$\Sigma_h = \{P \in \bar{\Omega},\ P \textit{ a vertex of } T \in \mathcal{T}_h\}, \tag{2.1A}$$

$$V_h = \{v_h \in C^0(\bar{\Omega}),\ v_h|_T \in P_1 \quad \forall\, T \in \mathcal{T}_h\} . \tag{2.2A}$$

We then approximate the convex set K (defined in (1.2A)) by

$$K_h = \{v_h \in V_h,\ v_h(P) \geq g(P) \quad \forall P \in \Sigma_h \cap \Gamma\} , \tag{2.3A}$$

and then problem (1.1A) by

$$\begin{cases} u_h \in K_h, \\ \displaystyle\int_\Omega [\nabla u_h \cdot \nabla(v_h - u_h) + u_h(v_h - u_h)]dx \geq \int_\Omega f(v_h - u_h)dx \quad \forall v_h \in K_h . \end{cases} \tag{2.4A}$$

---

(1) With g = 0 in (1.2A).
(2) Abbreviated hereafter to B.H.R.

It is then clear that

Proposition 2.1 : *The approximate problem* (2.3A), (2.4A) *admits one and only one solution.*

2.1.2. *Convergence of the approximate solutions. Error estimates of optimal order.*

As regards the convergence of the approximate solutions when $h \to 0$, we refer to Chapter 4, Section 3.5 for the case in which the solution u is not very regular (and $g = 0$). In the following we shall estimate the approximation error $\|u_h - u\|_{H^1(\Omega)}$ using a certain number of (reasonable) regularity assumptions on u, g *and* the free boundary (i.e. the boundary between $\Gamma_o = \{P \in \Gamma,\ u(P) = g(P)\}$ and $\Gamma_+ = \{P \in \Gamma,\ u(P) > g(P)\}$).

In order to obtain these error estimates we shall make use of

Lemma 2.1: *Suppose that* $u \in H^2(\Omega)$; *then*

(2.5A) $$-\Delta u + u = f \quad a.e.\ in\ \Omega,$$

(2.6A) $$\frac{\partial u}{\partial n} \geq 0 \quad a.e.\ on\ \Gamma,$$

(2.7A) $$\frac{\partial u}{\partial n}(u-g) = 0 \quad a.e.\ on\ \Gamma.$$

Proof: See Chapter 1, Section 1.1.

Lemma 2.2: *Let* u *and* u_h *be the respective solutions of* (1.1A), (1.2A) *and* (2.3A), (2.4A). *We then have (with*

$$a(v,w) = \int_\Omega [\nabla v \cdot \nabla w + vw]\, dx \quad \forall\, v,w \in H^1(\Omega)\ ), \quad \forall\ v_h \in K_h:$$

(2.8A) $$a(u_h-u, u_h-u) \leq a(u_h-u, v_h-u) + a(u, v_h-u_h) - \int_\Omega f(v_h-u_h)\,dx.$$

Proof: See that of Lemma 3.2 of Appendix 1, Section 3.6.

We then have

Theorem 2.1: *Suppose that the solution* u *of* (1.1A), (1.2A) *satisfies* $u \in W^{2,p}(\Omega)$ *with* $p > 2$, *that* $g \in W^{1,\infty}(\Gamma)$ *and that the set of points on the free boundary* (i.e. $\overline{\Gamma}_o \cap \overline{\Gamma}_+$) *is finite. If the angles of* $\mathcal{T}_h$ *are bounded below by* $\theta_o > 0$, *independent of* h, *we then have*

(2.9A) $$\|u_h - u\|_{H^1(\Omega)} = O(h)$$

where u_h *is the solution of the approximate problem* (2.3A), (2.4A).

Proof: We follow B.H.R. /1A, Lemma 6.1/. From Green's formula and since $u \in W^{2,p}(\Omega)$, we have, $\forall\ v \in H^1(\Omega)$,

$$a(u,v) = \int_\Omega [\nabla u \cdot \nabla v + uv]\,dx = \int_\Omega (-\Delta u + u)\,v\,dx + \int_\Gamma \frac{\partial u}{\partial n} v\,d\Gamma$$

and hence $\forall\ v_h \in K_h$

$$\text{(2.10A)} \quad \begin{cases} a(u,v_h-u_h) - \displaystyle\int_\Omega f(v_h-u_h)\,dx \\ = \displaystyle\int_\Omega (-\Delta u+u-f)(v_h-u_h)\,dx + \int_\Gamma \frac{\partial u}{\partial n}(v_h-u_h)\,d\Gamma. \end{cases}$$

Let π_h be the V_h-interpolation operator on Σ_h (see Appendix 1, Section 3.6); we have $\Omega \subset \mathbb{R}^2$, and hence $W^{2,p}(\Omega) \subset C^1(\bar{\Omega})$, with continuous injection, if $p > 2$; the conditions $u \in W^{2,p}(\Omega)$ and $u \geq g$ on Γ then imply that $\pi_h u \in K_h$.
We take $v_h = \pi_h u$ in (2.8A), (2.10A) giving

$$\begin{cases} a(u_h-u,u_h-u) \leq a(u_h-u,\pi_h u-u) \\ + \displaystyle\int_\Omega (-\Delta u+u-f)(\pi_h u-u_h)\,dx + \int_\Gamma \frac{\partial u}{\partial n}(\pi_h u-u_h)\,d\Gamma, \end{cases}$$

which implies, in view of (2.5A)

$$\text{(2.11A)} \quad a(u_h-u,u_h-u) \leq a(u_h-u,\pi_h u-u) + \int_\Gamma \frac{\partial u}{\partial n}(\pi_h u-u_h)\,d\Gamma.$$

It can easily be deduced from (2.11A) that

$$\text{(2.12A)} \quad \frac{1}{2}\|u_h-u\|^2_{H^1(\Omega)} \leq \frac{1}{2}\|\pi_h u-u\|^2_{H^1(\Omega)} + \int_\Gamma \frac{\partial u}{\partial n}(\pi_h u-u_h)\,d\Gamma.$$

It results from Ciarlet /1A/, for example, that $u \in W^{2,p}(\Omega)$ with $p > 2$ implies

$$\text{(2.13A)} \quad \|\pi_h u-u\|_{H^1(\Omega)} = O(h)$$

under the assumptions on $\mathcal{T}_h$ stated in the theorem.

In view of (2.12A), (2.13A) it will suffice to establish

$$\text{(2.14A)} \quad \int_\Gamma \frac{\partial u}{\partial n}(\pi_h u-u_h)\,d\Gamma \leq Ch^2$$

to prove the estimate (2.9A).

Let $\tilde{g} \in C^o(\bar{\Omega})$ be a lifting of g and let $\pi_h \tilde{g}$ be its V_h-interpolate on Σ_h; we denote by g_h the trace of $\pi_h \tilde{g}$ on Γ. This function g_h is in fact the interpolate [1] of g from the values taken on $\Gamma \cap \Sigma_h$. We then have

$$\text{(2.15A)} \qquad g_h \le u_h \ \textit{on}\ \Gamma,$$

which combined with (2.6A) implies

$$\text{(2.16A)} \qquad \int_\Gamma \frac{\partial u}{\partial n}(g_h - u_h)\, d\Gamma \le 0.$$

It then follows from (2.7A), (2.16A) that

$$\text{(2.17A)} \qquad \begin{cases} \displaystyle\int_\Gamma \frac{\partial u}{\partial n}(\pi_h u - u_h)\, d\Gamma = \int_\Gamma \frac{\partial u}{\partial n}[(\pi_h u - g_h) - (u-g)]\, d\Gamma \\ \displaystyle\qquad + \int_\Gamma \frac{\partial u}{\partial n}(u-g)\, d\Gamma + \int_\Gamma \frac{\partial u}{\partial n}(g_h - u_h)\, d\Gamma \\ \displaystyle\le \int_\Gamma \frac{\partial u}{\partial n}[(\pi_h u - g_h) - (u-g)]\, d\Gamma. \end{cases}$$

Let Γ_{oh} (resp. Γ_{+h}) be the subset of Γ which is the union of the sides of the triangles of $\mathcal{T}_h$ contained in Γ_o (resp. $\overline{\Gamma_+}$). We denote by s_h the generic element of the set of sides of the triangles of $\mathcal{T}_h$ lying on Γ; if $s_h \subset \Gamma_{+h}$, then $\frac{\partial u}{\partial n} = 0$ (see (2.7A)); if $s_h \subset \Gamma_{oh}$ then u=g on s_h, which implies $g_h = \pi_h u$ on s_h. From these relations we deduce

$$\text{(2.18A)} \qquad \int_{s_h} \frac{\partial u}{\partial n}[(\pi_h u - g_h) - (u-g)]\, d\Gamma = 0 \quad \forall\, s_h \subset \Gamma_{oh} \cup \Gamma_{+h}.$$

In view of the assumption on the free boundary stated in the theorem, we have

(2.19A) *the number of sides* $s_h \not\subset \Gamma_{oh} \cup \Gamma_{+h}$ *is less than or equal to the number of points on the free boundary.*

If $s_h \not\subset \Gamma_{oh} \cup \Gamma_{+h}$ there exists $P \in s_h$ such that u(P) = g(P). Since g and $u|_\Gamma$ belong to $W^{1,\infty}(\Gamma)$, we have (from Taylor's theorem and certain standard interpolation results for which we refer to Ciarlet /1A/)

---

(1) piecewise linear on Γ

(2.20A) $\|u-g\|_{L^{\infty}(s_h)} = O(h)$,

(2.21A) $\|\pi_h u-g_h\|_{L^{\infty}(s_h)} = O(h)$.

Since $u \in W^{2,p}(\Omega)$ with $p > 2$ implies $u \in C^1(\overline{\Omega})$ we have

(2.22A) $\frac{\partial u}{\partial n}\Big|_{\Gamma} \in L^{\infty}(\Gamma)$.

It then follows from (2.20A)-(2.22A) and from $\int_{s_h} d\Gamma = O(h)$ that

(2.23A) $\int_{s_h} \frac{\partial u}{\partial n} [(\pi_h u-g_h)-(u-g)]\, d\Gamma = O(h^2) \quad \forall\, s_h \not\subset \Gamma_{oh} \cup \Gamma_{+h}$.

The required relation (2.14A) then results trivially from (2.17A), (2.18A), (2.19A), (2.23A).

Remark 2.1: It is shown in B.H.R. /1A/ that the estimate (2.9A) also holds if Ω is a convex domain (possibly not polygonal) and the triangulation $\mathcal{C}_h$ and the approximate convex set K_h are suitably chosen.

2.2. Approximation of the boundary unilateral problem (1.1A), (1.2A) using non-conforming finite elements of mixed type.

In this section, which follows B.H.R. /2A/, we consider the approximation of the *boundary unilateral problem* (1.1A), (1.2A) using a *mixed finite-element* method *via a dual formulation* of (1.1A), (1.2A); we adopt the notation of Appendix 1, Section 4 and we assume that $g \in H^{1/2}(\Gamma)$.

2.2.1. *A dual formulation of the boundary unilateral problem* (1.1A), (1.2A).

It is clear that problem (1.1A), (1.2A) is equivalent to

(2.24A) $\underset{\{v,q\}\in \mathcal{K}}{\text{Min}} \left\{\frac{1}{2}\int_{\Omega}(|q|^2+|v|^2)\,dx - \int_{\Omega} fv\, dx\right\}$,

where

(2.25A) $\mathcal{K} = \{\{v,q\} \in H^1(\Omega) \times (L^2(\Omega))^N,\ \nabla v = q\ ,\ v \geq g$ *a.e. on* $\Gamma\}$.

With (2.24A), (2.25A) we associate the *Lagrangian*

$$\mathcal{L}(v,q,\mu) = \frac{1}{2}\int_\Omega (|q|^2+|v|^2)\,dx - \int_\Omega fv\,dx + \int_\Omega \mu\cdot(\nabla v-q)\,dx, \tag{2.26A}$$

and we consider the *dual problem* of (1.1A), (1.2A) relative to $\mathcal{L}$, namely

$$\underset{\mu\in(L^2(\Omega))^N}{\text{Max}}\ \underset{\{v,q\}\in\tilde{\mathcal{K}}}{\text{Min}}\ \mathcal{L}(v,q,\mu) \tag{2.27A}$$

where

$$\tilde{\mathcal{K}} = \{\{v,q\}\in H^1(\Omega)\times(L^2(\Omega))^N,\ v\geq g\ \textit{a.e. on}\ \Gamma\}. \tag{2.28A}$$

It is quite easy to prove:

Proposition 2.2: *The dual problem* (2.27A) *is equivalent to*

$$\underset{q\in C}{\text{Min}}\ \{\frac{1}{2}\int_\Omega(|q|^2+|\nabla\cdot q|^2)\,dx + \int_\Omega f\nabla\cdot q\,dx - \int_\Gamma g\,q\cdot n\,d\Gamma \tag{2.29A}$$

where

$$C = \{q\in H(\text{div},\Omega),\ q\cdot n\geq 0\ \textit{on}\ \Gamma\}, \tag{2.30A}$$

$$H(\text{div},\Omega) = \{q\in(L^2(\Omega))^N,\ \nabla\cdot q\in L^2(\Omega)\}. \tag{2.31A}$$

Conversely (2.29A), (2.30A) *admits a unique solution* p *such that* $p = \nabla u$ *where* u *is the solution of the boundary unilateral problem* (1.1A), (1.2A). ■

Remark 2.2: Remark 4.1 of Appendix 1, Section 4.2 also holds for (2.29A), (2.30A).

Remark 2.3: Let

$$\Lambda = \{\mu\in H^{1/2}(\Gamma),\ \mu\leq 0\ \textit{a.e. on}\ \Gamma\}. \tag{2.32A}$$

We then have the following equivalence

$$q\in C \Leftrightarrow \begin{cases} q\in H(\text{div},\Omega), \\ \langle\mu,q\cdot n\rangle \leq 0\quad \forall\mu\in\Lambda \end{cases}, \tag{2.33A}$$

where $<\cdot,\cdot>$ denotes the bilinear form of the duality between $H^{1/2}(\Gamma)$ and $H^{-1/2}(\Gamma)$.

The following saddle-point result can then easily be proved:

Theorem 2.2: *Suppose that the solution* u *of* (1.1A), (1.2A) *belongs to* $H^2(\Omega)$; *then let*

$$\mathcal{L}(q,\mu) = \frac{1}{2}\int_\Omega (|q|^2+|\nabla\cdot q|^2)\,dx + <\mu-g,q\cdot n> + \int_\Omega f\,\nabla\cdot q\,dx \tag{2.34A}$$

be the Lagrangian associated with (2.29A), (2.30A); $\mathcal{L}$ *admits a unique saddle point* $\{p,\lambda\}$ *on* $H(\mathrm{div},\Omega)\times\Lambda$ *such that*

$$p = \nabla u \quad \textit{in } \Omega \tag{2.35A}$$

$$\lambda = g-u \quad \textit{on } \Gamma. \tag{2.36A}$$

Moreover $\{p,\lambda\}$ *is characterised by*

$$\{p,\lambda\} \in H(\mathrm{div},\Omega)\times\Lambda\ , \tag{2.37A}$$

$$\int_\Omega [p\cdot q+\nabla\cdot p\ \nabla\cdot q]\,dx + <\lambda-g,q\cdot n> = -\int_\Omega f\,\nabla\cdot q\,dx \quad \forall q\in H(\mathrm{div},\Omega), \tag{2.38A}$$

$$<\mu-\lambda,p\cdot n> \le 0 \quad \forall\mu\in\Lambda\ . \tag{2.39A}$$

2.2.2. *An approximation to the dual problem* (2.29A),(2.30A) *using mixed finite elements.*

Once again we assume that Ω is a bounded polygonal domain in $\mathbb{R}^2$ and that $\mathcal{T}_h$ is a triangulation of Ω. We then approximate $H(\mathrm{div},\Omega)$, $H^{1/2}(\Gamma)$, Λ, C respectively by

$$\begin{cases} H_h = \{q_h\in H(\mathrm{div},\Omega),\ q_h|_T\in P_{k+1}\times P_{k+1} \ \ \textit{and} \\ \nabla\cdot q_h|_T\in P_k \ \ \forall T\in\mathcal{T}_h,\ q_h\cdot n|_S\in P_k \ \ \forall S \textit{ a side of } T\in\mathcal{T}_h\} \end{cases} \tag{2.40A}$$

(n is a unit vector, normal to S)

$$L_h = \{\mu_h\in L^2(\Gamma),\ \mu_h|_S\in P_k \ \ \forall S\subset\Gamma,\ S \textit{ a side of } T\in\mathcal{T}_h\}\ , \tag{2.41A}$$

$$\Lambda_h = \{\mu_h\in L_h\ ,\ \mu_h\le 0 \ \textit{a.e. on } \Gamma\}, \tag{2.42A}$$

(2.43A) $$C_h = \{q_h \in H_h, \int_\Gamma q_h \cdot n\, \mu_h \, d\Gamma \le 0 \quad \forall \mu_h \in \Lambda_h\} .$$

We then approximate the dual problem (2.29A), (2.30A) by

(2.44A) $$\begin{cases} \textit{Find } p_h \in C_h \textit{ such that } \forall q_h \in C_h \textit{ we have} \\ \int_\Omega [p_h \cdot (q_h - p_h) + \nabla \cdot p_h \, \nabla \cdot (q_h - p_h)] dx \\ \ge \int_\Gamma g(q_h - p_h) \cdot n \, d\Gamma - \int_\Omega f \nabla \cdot (q_h - p_h) dx , \end{cases}$$

which is the discrete analogue of (2.29A).

It is clear that (2.44A) *admits one and only one solution.*

Let λ_h be the Kuhn-Tucker vector associated with (2.43A); the approximate dual problem (2.44A) is then equivalent to the variational system

(2.45A) $$\begin{cases} \textit{Find } \{p_h, \lambda_h\} \in H_h \times \Lambda_h \textit{ such that } \int_\Omega [p_h \cdot q_h + \nabla \cdot p_h \, \nabla \cdot q_h] dx \\ \qquad + \int_\Gamma (\lambda_h - g) q_h \cdot n \, d\Gamma = - \int_\Omega f \, \nabla \cdot q_h \, dx \quad \forall q_h \in H_h, \\ \int_\Gamma p_h \cdot n(\mu_h - \lambda_h) \, d\Gamma \le 0 \quad \forall \mu_h \in \Lambda_h , \end{cases}$$

which is the discrete analogue of (2.37A)-(2.39A), p_h being the solution of (2.44A).

Regarding the convergence of p_h, λ_h when $h \to 0$, the following results are proved in B.H.R. /2A/:

<u>Theorem 2.3</u>: *Suppose that when* $h \to 0$ *the angles of* $\mathcal{T}_h$ *are bounded below by* $\theta_o > 0$, *independent of* h. *Then if* $k = 0$, *if* $u \in H^2(\Omega) \cap W^{1,\infty}(\Omega)$, $g \in W^{1,\infty}(\Gamma)$, $f \in H^1(\Omega) \cap L^\infty(\Omega)$ *and if the set of points on the free boundary* (i.e. $\overline{\Gamma}_o \cap \overline{\Gamma}_+$) *is finite, we have*

(2.46A) $$\| p_h - \nabla u \|_{H(div,\Omega)} = O(h)$$

(2.47A) $$\| \lambda_h - (g-u) \|_{L^2(\Gamma)} = O(h) .$$

B.H.R. /2A/ includes an investigation of the case in which Ω is not polygonal. Furthermore, Remarks 4.3 and 4.4 of Appendix 1, Section 4.3 also hold for (2.44A), (2.45A).

---

(1) $$\|q\|_{H(div,\Omega)} = \left(\int_\Omega (|q|^2 + |\nabla \cdot q|^2) dx\right)^{1/2}$$

3. FURTHER DISCUSSION ON THE APPROXIMATION OF FOURTH-ORDER VARIATIONAL PROBLEMS USING MIXED FINITE-ELEMENT METHODS

3.1 Synopsis

The aim of this section is to supplement Section 4 of Chapter 4 concerning the approximation of certain fourth-order variational problems. Section 3.2 gives some information on the convergence of the mixed finite-element method of Chapter 4, Section 4.5; in Section 3.3 we briefly describe certain results of Glowinski-Pironneau /1A/ concerning the solution of the approximate biharmonic problem and which generalise those of Chapter 4, Section 4.6.

Finally, in Section 3.4, we apply the above methods to the solution of the fourth-order variational inequality problem defined by (1.3A), (1.4A). The notation is that of Chapter 4.

3.2 Further discussion of the convergence of the mixed finite-element method of Chapter 4, Section 4.5.

Let Ω be a bounded domain in $\mathbb{R}^2$ with regular boundary; the homogeneous Dirichlet problem for Δ^2 is then defined by

$$\text{(3.1A)} \quad \begin{cases} \Delta^2 u = f \ \textit{in } \Omega, \\ u = \dfrac{\partial u}{\partial n} = 0 \ \textit{on } \Gamma. \end{cases}$$

In Chapter 4, Section 4.5 we described an approximation using mixed finite elements *of order* k, for which we stated the convergence result

$$\text{(3.2A)} \quad \|u_h - u\|_{H^1(\Omega)} + \|p_h - (-\Delta u)\|_{L^2(\Omega)} = O(h^{k-1})$$

due to Ciarlet-Raviart /3/; the estimate (3.2A) supposes that $u \in H^{k+2}(\Omega)$ and makes various regularity assumptions on the family $(\mathcal{C}_h)_h$, for which we refer to Chapter 4, Section 4.5.

In fact it results from the works of Scholz /1A/, /2A/, Rannacher /1A/, /2A/ and Girault-Raviart /1A/ that under the same assumptions on $(\mathcal{C}_h)_h$ and with suitable (and reasonable) regularity assumptions on u, we have $\forall\, k \geq 1$:

$$\text{(3.3A)} \quad \|u_h - u\|_{H^1(\Omega)} = O(h^{k-\varepsilon}) \quad (\textit{with } \varepsilon = 0 \textit{ if } k \geq 2)$$

(3.4A) $$\|u_h-u\|_{L^2(\Omega)} = O(h^{k+1-\varepsilon}) \text{ with } \varepsilon = 0 \text{ if } k \geq 2 ,$$

(3.5A) $$\|p_h-(-\Delta u)\|_{L^2(\Omega)} = O(h^{k-1/2-\varepsilon}) \text{ with } \varepsilon>0 \text{ arbitrarily small.}$$

Indeed, it is even proved in Scholz /3A/ that for $k \geq 1$ we have [1]

(3.6A) $$\begin{cases} \forall\, \Omega_o \subset\subset \Omega \\ \|p_h-(-\Delta u)\|_{L^2(\Omega_o)} = O(h^{k-\varepsilon}). \end{cases}$$

We remark that the estimates (3.3A), (3.4A), (3.6A) are of *optimal* or *quasi-optimal order*.

3.3 Discussion supplementing Chapter 4, Section 4.6 on the solution of the approximate biharmonic problem (4.72)

3.3.1. *Review of the approximate biharmonic problem*

In Chapter 4, Section 4.5.2 we defined the following discrete spaces

(3.7A) $$V_h = \{v_h | v_h \in C^o(\bar{\Omega}),\ v_h|_T \in P_k \quad \forall T \in \mathcal{C}_h\},$$

(3.8A) $$V_{oh} = \{v_h | v_h \in V_h,\ v_h|_\Gamma = 0\} = V_h \cap H_o^1(\Omega),$$

and also

(3.9A) $$\begin{cases} M_h : \text{complement of } V_{oh} \text{ in } V_h, \\ v_h \in M_h \Longrightarrow v_h|_T = 0 \quad \forall T \in \mathcal{C}_h,\ \partial T \cap \Gamma = \emptyset\ ; \end{cases}$$

M_h is defined uniquely by (3.9A).

We next consider

---

(1) $\Omega_o \subset\subset \Omega \Leftrightarrow \bar{\Omega}_o \subset \Omega$.

(3.10A) $$\begin{cases} W_h = \{\{v_h,q_h\} \mid \{v_h,q_h\} \in V_{oh} \times V_h \ , \ \int_\Omega \nabla v_h \cdot \nabla \mu_h \, dx \\ \quad = \int_\Omega q_h \mu_h dx \quad \forall \mu_h \in V_h\} \ . \end{cases}$$

The mixed finite-element formulation of problem (3.1A) is then

(3.11A) $$\begin{cases} \textit{Find } \{u_h,p_h\} \in W_h \textit{ such that} \\ j(u_h,p_h) \le j(v_h,q_h) \quad \forall \{v_h,q_h\} \in W_h \end{cases}$$

where

(3.12A) $$j(v_h,q_h) = \frac{1}{2}\int_\Omega |q_h|^2 dx - \int_\Omega f \ v_h \ dx.$$

Problem (3.11A) admits one and only one solution.

3.3.2. *Solution of* (3.11A) *by reduction to a variational problem in* M_h.

We have $V_h = V_{oh} \oplus M_h$; let $\{ u_h,p_h\}$ be the solution of (3.11A) and let λ_h be the component of p_h in M_h, i.e.

(3.13A) $$p_h = (p_h-\lambda_h) + \lambda_h \ ; \ p_h-\lambda_h \in V_{oh} \ , \ \lambda_h \in M_h.$$

The following theorem then results from Ciarlet-Glowinski /2/:

<u>Theorem 3.1</u>: *Let* $\{u_h,p_h\}$ *be the solution of* (3.11A) *and let* λ_h *be the component of* p_h *in* M_h; $\{u_h,p_h,\lambda_h\}$ *is then the unique element of* $V_{oh} \times V_h \times M_h$ *such that*

(3.14A) $$\int_\Omega \nabla p_h \cdot \nabla v_h dx = \int_\Omega f \ v_h \ dx \quad \forall v_h \in V_{oh}, \ p_h-\lambda_h \in V_{oh},$$

(3.15A) $$\int_\Omega \nabla u_h \cdot \nabla v_h dx = \int_\Omega p_h v_h dx \quad \forall v_h \in V_{oh} \ , \ u_h \in V_{oh} \ ,$$

(3.16A) $$\int_\Omega \nabla u_h \cdot \nabla \mu_h dx = \int_\Omega p_h \mu_h \ dx \quad \forall \ \mu_h \in M_h \ . \ \blacksquare$$

Let $\lambda_h \in M_h$ and ω_h, ψ_h be the respective solutions of the approximate Dirichlet problems

(3.17A) $$\int_\Omega \nabla \omega_h \cdot \nabla v_h dx = 0 \quad \forall \ v_h \in V_{oh} \ , \ \omega_h \in V_h, \ \omega_h-\lambda_h \in V_{oh}$$

$$\text{(3.18A)} \qquad \int_\Omega \nabla\psi_h \cdot \nabla v_h \, dx = \int_\Omega \omega_h v_h \, dx \quad \forall v_h \in V_{oh}, \; \psi_h \in V_{oh}.$$

We then define the bilinear form $a_h : M_h \times M_h \to \mathbb{R}$ by

$$\text{(3.19A)} \qquad a_h(\lambda_h, \mu_h) = \int_\Omega \omega_h \mu_h \, dx - \int_\Omega \nabla\psi_h \cdot \nabla\mu_h \, dx \quad \forall \mu_h \in M_h .$$

The following lemma is proved in Glowinski-Pironneau /1A/:

Lemma 3.1: *The bilinear form* $a_h(\cdot,\cdot)$ *is symmetric and positive-definite.* ∎

Now let u_{oh} and p_{oh} be the solutions of the two approximate Dirichlet problems

$$\text{(3.20A)} \qquad \int_\Omega \nabla p_{oh} \cdot \nabla v_h \, dx = \int_\Omega f \, v_h \, dx \quad \forall v_h \in V_{oh}, \; p_{oh} \in V_{oh},$$

$$\text{(3.21A)} \qquad \int_\Omega \nabla u_{oh} \cdot \nabla v_h \, dx = \int_\Omega p_{oh} v_h \, dx \quad \forall v_h \in V_{oh}, \; u_{oh} \in V_{oh};$$

the fundamental result concerning the solution of (3.11A), via M_h, is then

Theorem 3.2: *Let* $\{u_h, p_h\}$ *be the solution of* (3.11A) *and let* λ_h *be the component of* p_h *in* M_h; λ_h *is then the unique solution of the linear variational problem*

$$\text{(3.22A)} \qquad \begin{cases} \text{Find } \lambda_h \in M_h \text{ such that} \\ a_h(\lambda_h, \mu_h) = \int_\Omega \nabla u_{oh} \cdot \nabla\mu_h \, dx - \int_\Omega p_{oh} \mu_h \, dx \end{cases}$$

which is equivalent to a linear system with a symmetric positive-definite matrix.

It can then easily be shown that algorithm (4.77)-(4.80) of Chapter 4, Section 4.6.1 is actually a *fixed-step gradient algorithm* applied to the solution of problem (3.22A); Glowinski-Pironneau /1A/ gives [1] some algorithms for solving (3.11A), via (3.22A), which are *more efficient* than the above algorithm, and in particular some algorithms of the *conjugate-gradient* type, which are no more complicated to implement, each iteration requiring essentially the solution of two approximate Dirichlet

---

(1) See also Vidrascu /1A/ for further details and numerous numerical experiments.

problems for $-\Delta$.

3.4 Application to the numerical solution of problem (1.3A) (1.4A).

In this section we shall consider the numerical solution of the problem of the (fourth order) Elliptic Variational Inequality (1.3A), (1.4A).

3.4.1. *Formulation of the continuous problem. Regularity results.*

Let Ω be a bounded domain in $\mathbb{R}^2$ with regular boundary Γ, let $f \in H^{-2}(\Omega) = (H_0^2(\Omega))'$ and let $\alpha, \beta \in \mathbb{R}$ be such that $\alpha \leq \beta$.

We then consider the variational problem:

$$(3.23A) \qquad \begin{cases} \text{Find } u \in K \text{ such that} \\ \int_\Omega \Delta u \, \Delta(v-u) \, dx \geq \langle f, v-u \rangle \quad \forall v \in K \end{cases}$$

where

$$(3.24A) \qquad K = \{v \in H_0^2(\Omega) \; , \; \alpha \leq \Delta v \leq \beta \; \text{ a.e. on } \Omega\}$$

and where $\langle\cdot,\cdot\rangle$ denotes the standard bilinear form of the duality between $H^{-2}(\Omega)$ and $H_0^2(\Omega)$.

Remark 3.1: We have (cf. Green's formula)

$$(3.25A) \qquad \int_\Omega \Delta v \, dx = \int_\Gamma \frac{\partial v}{\partial n} \, d\Gamma = 0 \quad \forall v \in H_0^2(\Omega).$$

Hence it follows that $K = \emptyset$ if $\alpha > 0$ or $\beta < 0$; it also follows from (3.25A) that $K = \{0\}$ if $\alpha = 0$ or $\beta = 0$. We shall therefore assume hereafter that

$$(3.26A) \qquad \alpha < 0 < \beta.$$

Remark 3.2: There is *equivalence* between (3.23A) and the minimisation problem

$$(3.27A) \qquad \operatorname*{Min}_{v \in K} \left\{ \frac{1}{2} \int_\Omega |\Delta v|^2 \, dx - \langle f, v \rangle \right\} .$$

The mapping $v \to \|\Delta v\|_{L^2(\Omega)}$ defines a norm in $H_0^2(\Omega)$ equivalent to that induced by $H^2(\Omega)$; moreover if $K \neq \emptyset$ then K is necessarily closed in $H_0^2(\Omega)$. From these properties it is easy to deduce:

Proposition 3.1: *If* $K \neq \emptyset$, *then problem* (3.23A) *admits one and only one solution.*

Regarding the regularity of the solution, it is shown in Brézis-Stampacchia /2A/ (see also Torelli /1A/) that

(3.28A) $$f \in L^p(\Omega) \text{ with } p > 2 \Longrightarrow u \in W^{3,p}_{loc}(\Omega).$$

We refer to Caffarelli-Friedman-Torelli /1A/ for a detailed investigation of the free boundary associated with problem (3.23A) (3.27A), i.e. the set $\partial\Omega_o \cap \partial\Omega_+$ where

$$\Omega_o = \{x \in \Omega,\ \Delta v(x) = \alpha \text{ or } \Delta v(x) = \beta\},$$

$$\Omega_+ = \{x \in \Omega,\ \alpha < \Delta v(x) < \beta\}.$$

3.4.2. *A density lemma.*

In order to investigate the convergence of the finite-element approximations of Section 3.4.4 we shall require

Lemma 3.2: *Suppose that* $\alpha < 0 < \beta$ *and that* Ω *is a star-shaped* [1] *domain; let* $\chi = K \cap \mathcal{D}(\Omega)$; *we then have*

(3.29A) $$\overline{\chi} = K.$$

Proof: We can always assume that Ω is a *star-shaped domain with respect to the origin* $\{0,0\}$. Let $v \in K$ and $\varepsilon > 0$. We define $\tilde{v} \in H^2(\mathbb{R}^2)$ by

(3.30A) $$\tilde{v}(x) = \begin{cases} v(x) \text{ if } x \in \Omega, \\ 0 \text{ if } x \notin \Omega. \end{cases}$$

We have

(3.31A) $$\alpha \leq \Delta\tilde{v} \leq \beta \text{ a.e. in } \mathbb{R}^2.$$

We next define $\tilde{v}_\varepsilon \in H^2(\mathbb{R}^2)$ by

(3.32A) $$\tilde{v}_\varepsilon(x) = \tilde{v}((1+\varepsilon)x) \quad \forall x \in \mathbb{R}^2.$$

We note that $\tilde{v}_\varepsilon$ *has compact support in* Ω.

---

(1) All convex domains are star-shaped.

We next define $v_\varepsilon \in H_0^2(\Omega)$ by

(3.33A) $$v_\varepsilon = \frac{1}{(1+\varepsilon)^2} \tilde{v}_\varepsilon|_\Omega \cdot$$

We then have

(3.34A) $$v_\varepsilon \in H_0^2(\Omega),$$

(3.35A) $$\lim_{\varepsilon\to 0} v_\varepsilon = v \textit{ strongly in } H_0^2(\Omega)$$

(3.36A) $$\alpha \le \Delta v_\varepsilon \le \beta \textit{ a.e. in } \Omega.$$

From (3.34A)-(3.36A) we deduce that the set

$$\chi_1 = \{v \in K, \quad v \textit{ has compact support in } \Omega\}$$

is *dense* in K.

Let $v \in \chi_1$ and let $\{\rho_n\}_n$ be a *regularising sequence* (see Remark 3.3 of Chapter 3, Section 3.3 for the definition of such sequences); we then define $\tilde{v} \in H^2(\mathbb{R}^2)$ by (3.30A) and $\{\tilde{v}_n\}_n$, $\tilde{v}_n \in \mathcal{D}(\mathbb{R}^2)$ $\forall n$, by

(3.37A) $$\tilde{v}_n = \tilde{v} * \rho_n.$$

We then have

(3.38A) $$\lim_{n\to+\infty} \tilde{v}_n = \tilde{v} \textit{ strongly in } H^2(\mathbb{R}^2).$$

Moreover we have

(3.39A) $$\Delta\tilde{v}_n = \Delta\tilde{v} * \rho_n ,$$

or

(3.40A) $$\Delta\tilde{v}_n(x) = \int_{\mathbb{R}^2} \Delta\tilde{v}(y)\rho_n(x-y)\,dy \qquad \forall\, x \in \mathbb{R}^2.$$

We deduce from (3.40A) and from $\alpha \le \Delta\tilde{v} \le \beta$ *a.e.*, that $\forall x \in \mathbb{R}^2$ we have

$$(3.41A)\qquad \alpha = \alpha\int_{\mathbb{R}^2} \rho_n(x-y)\,dy \le \Delta\tilde{v}_n(x) \le \beta\int_{\mathbb{R}^2} \rho_n(x-y)\,dy = \beta.$$

For n *sufficiently large* we have

$$(3.42A)\qquad \mathrm{supp}(\tilde{v}_n) \subset \Omega \quad ;$$

if we define v_n by $v_n = \tilde{v}_n|_\Omega$, it follows from (3.38A), (3.41A), (3.42A) that *for* n *sufficiently large* we have

$$(3.43A)\qquad v_n \in \mathcal{D}(\Omega),$$

$$(3.44A)\qquad \lim_{n\to+\infty} v_n = v \textit{ strongly in } H_0^2(\Omega)$$

$$(3.45A)\qquad \alpha \le \Delta v_n \le \beta \quad \textit{on } \Omega.$$

In view of the *density* of χ_1 in K, relations (3.43A)-(3.45A) imply $\overline{\chi} = K$.

3.4.3. *Solution of* (3.23A), (3.27A) *by a duality algorithm.*

By analogy with the various obstacle-type problems encountered in this book it is natural to associate with problem (3.23A), (3.27A) the *Lagrangian* $\mathcal{L} : H^2(\Omega) \times (L^2(\Omega))^2 \to \mathbb{R}$ defined by

$$(3.46A)\qquad \begin{cases} \mathcal{L}(v,\mu) = \frac{1}{2}\int_\Omega |\Delta v|^2 dx - \langle f,v\rangle \\ \qquad + \int_\Omega \mu_1(\Delta v-\beta)\,dx + \int_\Omega \mu_2(\alpha-\Delta v)\,dx\,, \end{cases}$$

where $\mu = \{\mu_1,\mu_2\}$.

We next define

$$(3.47A)\qquad L_+^2(\Omega) = \{q \in L^2(\Omega),\ q(x) \ge 0 \textit{ a.e. in } \Omega\}$$

and

$$(3.48A)\qquad \Lambda = L_+^2(\Omega) \times L_+^2(\Omega).$$

We then have

Theorem 3.3: *Let* $\{u,\lambda\} \in H_o^2(\Omega) \times \Lambda$ *be a saddle point of* $\mathcal{L}$ *on* $H_o^2(\Omega) \times \Lambda$; *we then have*

(3.49A) u *is the solution of* (3.23A), (3.27A)

and

$$\int_\Omega (\Delta u-\beta)(\mu_1-\lambda_1)\,dx \le 0 \quad \forall \mu_1 \in L_+^2(\Omega), \tag{3.50A}$$

$$\int_\Omega (\alpha-\Delta u)(\mu_2-\lambda_2)\,dx \le 0 \quad \forall \mu_2 \in L_+^2(\Omega). \tag{3.51A}$$

Proof: From the definition of a saddle point, we have

$$\begin{cases} \{u,\lambda\} \in H_o^2(\Omega) \times \Lambda \text{ and } \forall \{v,\mu\} \in H_o^2(\Omega) \times \Lambda \\ \mathcal{L}(u,\mu) \le \mathcal{L}(u,\lambda) \le \mathcal{L}(v,\lambda). \end{cases} \tag{3.52A}$$

Relations (3.50A), (3.51A) are immediate consequences of (3.52A) and imply

$$\alpha \le \Delta u \le \beta \text{ a.e. on } \Omega, \tag{3.53A}$$

and

$$\int_\Omega \lambda_1 (\Delta u-\beta)\,dx = 0 , \int_\Omega \lambda_2 (\alpha-\Delta u)\,dx = 0 . \tag{3.54A}$$

We put

$$J(v) = \frac{1}{2}\int_\Omega |\Delta v|^2 dx - \langle f,v\rangle ; \tag{3.55A}$$

it follows from (3.54A) that

$$\mathcal{L}(u,\lambda) = J(u). \tag{3.56A}$$

We have (see (3.52A), (3.56A))

(3.57A) $$J(u) \le J(v) + \int_\Omega \lambda_1 (\Delta v - \beta)\, dx + \int_\Omega \lambda_2 (\alpha - \Delta v)\, dx \quad \forall v \in H_0^2(\Omega).$$

If $v \in K$ we have (since $\lambda \in \Lambda$)

$$\int_\Omega \lambda_1 (\Delta v - \beta)\, dx + \int_\Omega \lambda_2 (\alpha - \Delta v)\, dx \le 0 \quad \forall v \in K,$$

and hence, in view of (3.57A),

$$J(u) \le J(v) \quad \forall v \in K$$

which proves (3.49A). ■

Remark 3.3: In view of Theorem 3.3, $\lambda = \{\lambda_1, \lambda_2\} \in \Lambda$ appears as a Kuhn-Tucker vector associated with problem (3.23A), (3.27A).

Moreover it follows from (3.54A) and $\lambda \in \Lambda$ that we have

(3.58A) $$\lambda_1 (\Delta u - \beta) = \lambda_2 (\alpha - \Delta u) = 0 \text{ a.e. on } \Omega.$$

Remark 3.4: The existence of a multiplier $\lambda \in \Lambda$ is a tricky problem; this would not be the case if we were to seek a multiplier associated with K in $(L^\infty(\Omega))'_+ \times (L^\infty(\Omega))'_+$.

We refer to Remark 9.2 of Chapter 3, Section 9.1.4 for an analogous discussion concerning the elasto-plastic torsion problem. Naturally the above difficulties vanish for the finite-dimensional variants of problem (3.23A), (3.27A). ■

In view of Theorem 3.3 we can solve problem (3.23A), (3.27A) using the duality algorithms of Chapter 2, Section 4; if we apply algorithm (4.12), (4.13) of Chapter 2, Section 4.3 (with $\rho_n = \rho$) we are led to

(3.59A) $$\lambda^0 = \{\lambda_1^0, \lambda_2^0\} \in \Lambda, \text{ given arbitrarily,}$$

then for $n \ge 0$, $\lambda^n \in \Lambda$ *known, determine* u^n *and* λ^{n+1} *from*

(3.60A)
$$\begin{cases} u^n \in H_o^2(\Omega), \\ \int_\Omega \Delta u^n \, \Delta v \, dx = \langle f,v\rangle + \int_\Omega (\lambda_2^n - \lambda_1^n)\Delta v \, dx \quad \forall\, v \in H_o^2(\Omega), \end{cases}$$

(3.61A)
$$\begin{cases} \lambda_1^{n+1} = [\lambda_1^n + \rho(\Delta u^n - \beta)]^+ , \\ \lambda_2^{n+1} = [\lambda_2^n + \rho(\alpha - \Delta u^n)]^+ . \end{cases}$$

We note that (3.60A) is equivalent to the solution in $H_o^2(\Omega)$ of the *biharmonic problem*

(3.62A)
$$\Delta^2 u^n = f + \Delta(\lambda_2^n - \lambda_1^n) \quad \textit{in } \Omega,$$

(3.63A)
$$u^n = \frac{\partial u^n}{\partial n} = 0 \quad \textit{on } \Gamma.$$

Concerning the convergence of (3.59A)-(3.61A) we have

Theorem 3.4: *Suppose that* $\mathcal{L}$ *admits a saddle point on* $H_o^2(\Omega) \times \Lambda$; *then* $\forall\, \lambda^o \in \Lambda$ *and if*

(3.64A)
$$0 < \rho < 1 ,$$

we have

(3.65A)
$$\lim_{n\to+\infty} u^n = u \quad \textit{strongly in } H_o^2(\Omega)$$

(3.66A)
$$\lim_{n\to+\infty} \lambda^n = \lambda^* \quad \textit{weakly in } L^2(\Omega) \times L^2(\Omega)$$

where u *is the solution of problem* (3.23A), (3.27A) *and where* λ^* ($\in \Lambda$) *is such that* $\{u,\lambda^*\}$ *is a saddle point of* $\mathcal{L}$ *on* $H_o^2(\Omega) \times \Lambda$.

Proof: Let $\{u,\lambda\}$ be a saddle point of $\mathcal{L}$ on $H_o^2(\Omega) \times \Lambda$; relations (3.50A), (3.51A) imply

$$\text{(3.67A)}\qquad \begin{cases} \lambda_1 = [\lambda_1 + \rho(\Delta u - \beta)]^+ , \\ \lambda_2 = [\lambda_2 + \rho(\alpha - \Delta u)]^+ ; \end{cases}$$

relation (3.52A) implies

$$\text{(3.68A)}\qquad \begin{cases} u \in H_0^2(\Omega) \\ \int_\Omega \Delta u\, \Delta v\, dx = \langle f,v\rangle + \int_\Omega (\lambda_2 - \lambda_1)\Delta v\, dx \quad \forall\, v \in H_0^2(\Omega). \end{cases}$$

We put $\bar{\lambda}^n = \lambda^n - \lambda$, $\bar{u}^n = u^n - u$; since the mapping $q \to q^+ = \sup(0,q)$ is a contraction from $L^2(\Omega) \to L_+^2(\Omega)$, then by subtracting (3.67A) from (3.61A) (with $|q| = \|q\|_{L^2(\Omega)}$) we have

$$|\bar{\lambda}_1^{n+1}|^2 \le |\bar{\lambda}_1^n|^2 + 2\rho \int_\Omega \bar{\lambda}_1^n\, \Delta\bar{u}^n dx + \rho^2 |\Delta\bar{u}^n|^2 ,$$

$$|\bar{\lambda}_2^{n+1}|^2 \le |\bar{\lambda}_2^n|^2 - 2\rho \int_\Omega \bar{\lambda}_2^n\, \Delta\bar{u}^n dx + \rho^2 |\Delta\bar{u}^n|^2 ,$$

and hence by addition

$$\text{(3.69A)}\qquad |\bar{\lambda}^n|^2 - |\bar{\lambda}^{n+1}|^2 \ge 2\rho \int_\Omega (\bar{\lambda}_2^n - \bar{\lambda}_1^n)\Delta\bar{u}^n dx - 2\rho^2 |\Delta\bar{u}^n|^2 .$$

By subtracting (3.68A) from (3.60A) and putting $v = \bar{u}^n$, we deduce

$$|\Delta\bar{u}^n|^2 = \int_\Omega (\bar{\lambda}_2^n - \bar{\lambda}_1^n)\Delta\bar{u}^n\, dx ,$$

which in combination with (3.69A) implies

$$\text{(3.70A)}\qquad |\bar{\lambda}^n|^2 - |\bar{\lambda}^{n+1}|^2 \ge 2\rho(1-\rho)|\Delta\bar{u}^n|^2 .$$

If $\rho \in]0,1[$ the inequality (3.70A) clearly implies (3.65A); as regards (3.66A), this follows from Theorem 3.1 of Appendix 2, Section 3.

3.4.4 *Approximation of the variational problem* (1.3A), (1.4A) *using a mixed finite-element method*

We assume in the following that $f \in L^2(\Omega)$; with the notation of Chapter 4, Section 4.3 (1) it can easily be shown that there is equivalence between (1.3A), (1.4A) and

$$(3.71A)\quad \begin{cases} \text{Find } \{u,p\} \in \mathcal{K} \text{ such that} \\ j(u,p) \leq j(v,q) \quad \forall \{v,q\} \in \mathcal{K} \end{cases}$$

where

$$(3.72A)\quad j(v,q) = \frac{1}{2}\int_\Omega |q|^2 dx - \int_\Omega fv\, dx$$

$$(3.73A)\quad \mathcal{K} = \{\{v,q\} \in W,\ -\beta \leq q \leq -\alpha \text{ a.e. on } \Omega\}^{(2)}.$$

Hereafter we assume that Ω is a *polygonal* domain in $\mathbb{R}^2$, and we confine our attention to *triangular finite elements of order* 2; in view of the formulation (3.71A)-(3.73A) it is natural to approximate problem (1.3A), (1.4A) using the mixed finite-element method of Chapter 4, Section 4.5, which in the notation of Chapter 4, Section 4.5.2 and of Section 3.3.1 of this appendix is

$$(3.74A)\quad \begin{cases} \text{Find } \{u_h,p_h\} \in \mathcal{K}_h \text{ such that} \\ j(u_h,p_h) \leq j(v_h,q_h) \quad \forall \{v_h,q_h\} \in \mathcal{K}_h , \end{cases}$$

where

$$(3.75A)\quad \mathcal{K}_h = \{\{v_h,q_h\} \in W_h,\ -\beta \leq q_h(P) \leq -\alpha \quad \forall P \in \Sigma_h\}$$

with

---

(1) See (4.52), in particular, for the definition of W.
(2) We recall that $p = -\Delta u$.

(3.76A) $\Sigma_h = \{P | P \in \bar{\Omega},\ P$ *a vertex of* $T \in \mathcal{C}_h$ *or the midpoint of a side of* $T \in \mathcal{C}_h\}$.

It can easily be shown that *if* $\alpha < 0 < \beta$, *the approximate problem admits one and only one solution.*

3.4.5 *Convergence of the approximate solutions*

We shall assume in the following that Ω is a *bounded, polygonal, convex* domain in $\mathbb{R}^2$. In order to prove the convergence of the approximate solutions $\{u_h,p_h\}$ to the solution $\{u,-\Delta u\}$ of (3.71A) when $h \to 0$, we shall require the following lemmas:

Lemma 3.3: *Let* $(\{v_h,q_h\})_h$ *be such that*

(3.77A) $$\{v_h,q_h\} \in \mathcal{K}_h \quad \forall\, h,$$

(3.78A) $$\lim_{h\to 0} \{v_h,q_h\} = \{v,q\} \text{ weakly in } H_0^1(\Omega) \times L^2(\Omega).$$

Then if, when $h \to 0$, *the angles of* $\mathcal{C}_h$ *remain bounded below by* $\theta_0 > 0$ *independent of* h, *we have*

(3.79A) $$\{v,q\} \in \mathcal{K} .$$

Proof: Since $\{v_h,q_h\} \in W_h$, we have

(3.80A) $$\int_\Omega \nabla v_h \cdot \nabla \mu_h dx = \int_\Omega q_h \mu_h dx \quad \forall\, \mu_h \in V_h .$$

Let $\mu \in H^1(\Omega)$ and let μ_h be the projection of μ on V_h in the $H^1(\Omega)$ norm; we then have .

$$\begin{cases} \mu_h \in V_h, \\ (\mu_h,w_h)_{H^1(\Omega)} = (\mu,w_h)_{H^1(\Omega)} & \forall w_h \in V_h . \end{cases}$$

Under the assumptions on $\mathcal{C}_h$ stated in the lemma, we have

$$\lim_{h\to 0} \|\mu_h - \mu\|_{H^1(\Omega)} = 0.$$

From this we deduce in the limit in (3.80A) that

$$\int_\Omega \nabla v \cdot \nabla \mu \, dx = \int_\Omega q \, \mu \, dx \quad \forall \mu \in H^1(\Omega),$$

which since $\{v,q\} \in H_0^1(\Omega) \times L^2(\Omega)$ implies that

(3.81A) $\{v,q\} \in W.$

We shall now show that $\{v,q\} \in \mathcal{K}$. Let $\phi \in C^0(\bar{\Omega}), \phi \geq 0$; we define

(3.82A) $$s_h \phi = \sum_{T \in \mathcal{T}_h} \phi(G_T) \chi_T$$

where, in (3.82A)

G_T *is the centroid of the triangle* T,

and

χ_T *is the characteristic function of the triangle* T.

We have (*uniform continuity of* ϕ *on* $\bar{\Omega}$)

(3.83A) $$\lim_{h \to 0} \|\phi - s_h \phi\|_{L^\infty(\Omega)} = 0 .$$

Moreover we have

(3.84A) $$\int_\Omega (\beta + q_h) s_h \phi \, dx = \frac{1}{3} \sum_{T \in \mathcal{T}_h} \text{meas.}(T) \phi(G_T) \sum_{i=1}^{3} (\beta + q_h(m_{iT})),$$

where, in (3.84A), the m_{iT} are the midpoints of the sides of the triangle T; from the definition of $\mathcal{K}_h$ (see (3.75A)) we have in (3.84A)

$$\beta + q_h(m_{iT}) \geq 0 \quad \forall i=1,2,3, \quad \forall T \in \mathcal{T}_h, \quad \forall \{v_h, q_h\} \in \mathcal{K}_h ,$$

giving, since $\phi \geq 0$,

(3.85A) $$\int_\Omega (\beta + q_h) s_h \phi \, dx \geq 0 \quad \forall \phi \in C^0(\bar{\Omega}), \phi \geq 0.$$

We have (3.83A) and $\lim_{h \to 0} q_h = q$ *weakly* in $L^2(\Omega)$, and hence in the limit of (3.85A) we obtain:

$$\int_\Omega (\beta+q)\,\phi\,dx \geq 0 \qquad \forall\,\phi \in C^0(\bar{\Omega}),\ \phi \geq 0\ ,$$

which is equivalent to

$$-\beta \leq q \ \textit{a.e. on}\ \Omega;$$

similarly it can be shown that

$$q \leq -\alpha \ \textit{a.e. on}\ \Omega,$$

giving $\{v,q\} \in \mathcal{K}$ (in view of (3.81A)).

Let $\chi = K \cap \mathcal{D}(\Omega)$; it follows from Lemma 3.2 that $\bar{\chi} = K$. We define $\chi^* = \bigcup_{0\leq\lambda<1} \lambda\chi$; it is clear that

$$\bar{\chi}^* = K. \tag{3.86A}$$

If we define $\mathcal{K}^* \subset \mathcal{K}$ by

$$\mathcal{K}^* = \{\{v,q\} \in W,\ v \in \chi^*\},$$

then it follows from (3.86A) and the definition of W that

$$\bar{\mathcal{K}}^* = \mathcal{K}\,. \tag{3.87A}$$

Let $\{v,q\} \in \mathcal{K}^*$; we define $\{r_h v,\ r_h q\} \in W_h$ by

$$\begin{cases} \{r_h v, r_h q\} \in W_h \ \textit{and} \ \forall\{v_h,q_h\} \in W_h \ \textit{we have} \\ \frac{1}{2}\int_\Omega |r_h q|^2 dx - \int_\Omega \Delta^2 v\, r_h v\, dx \leq \frac{1}{2}\int_\Omega |q_h|^2 dx - \int_\Omega \Delta^2 v\, v_h dx\,. \end{cases} \tag{3.88A}$$

We then have

Lemma 3.4: *Suppose that the assumptions on* $\mathcal{T}_h$ *of Lemma 3.3 are satisfied, along with condition* (4.73) *of Chapter* 4, *Section* 4.5.4. *Then we have* $\forall\,\{v,q\} \in \mathcal{K}^*$

$$\{r_h v, r_h q\} \in \mathcal{K}_h \ \textit{for}\ h\ \textit{sufficiently small}, \tag{3.89A}$$

(3.90A) $$\lim_{h\to 0} \{r_h v, r_h q\} = \{v,q\} \text{ strongly in } H_o^1(\Omega)\times L^2(\Omega) .$$

Proof: Problem (3.88A) is the special case of problem (3.11A), (3.12A) which corresponds to $f = \Delta^2 v$, i.e. the approximate problem associated with the biharmonic problem (3.1A) when $f = \Delta^2 v$; the solution of (3.1A) is trivially equal to v when $f = \Delta^2 v$. We have $\{v,q\} \in \mathcal{K}^*$, giving

(3.91A) $$v \in \mathcal{D}(\Omega),$$

(3.92A) $$-\beta < q = -\Delta v < -\alpha \quad on\ \Omega;$$

it then follows from Rannacher /2A/, and from (3.91A), that under the assumptions on $\mathcal{T}_h$ stated in the lemma we have (with C independent of v and h)

(3.93A) $$\|r_h v-v\|_{W^{1,\infty}(\Omega)} + h\,\|r_h q-q\|_{L^\infty(\Omega)} \leq Ch^2|\text{Log } h|^{3/2}\|v\|_{W^{4,\infty}(\Omega)}$$

which proves (3.90A); moreover we deduce from (3.93A) that

$$\|r_h q-q\|_{L^\infty(\Omega)} \leq Ch|\text{Log } h|^{3/2}\,\|v\|_{W^{4,\infty}(\Omega)}$$

and hence, in view of (3.92A)

(3.94A) $$-\beta < r_h q < -\alpha \quad \textit{for } h \textit{ sufficiently small.}$$

Property (3.89A) then results from (3.75A), (3.88A) and (3.94A). ∎

In view of the two preceding lemmas, we have

Theorem 3.5: *With the same assumptions on* $(\mathcal{T}_h)_h$ *as in Lemma* 3.4, *we have*

(3.95A) $$\lim_{h\to 0} \{u_h, p_h\} = \{u,-\Delta u\} \text{ strongly in } H_o^1(\Omega)\times L^2(\Omega)$$

where in (3.95A) $\{u_h,p_h\}$ *is the solution of the approximate problem* (3.74A) *and* u *that of problem* (1.3A), (1.4A).

Proof: The proof follows the same general lines as that of Theorem 2.2 of Appendix 1, Section 2.3.

1) *A priori estimates for* $\{u_h,p_h\}$.

In what follows, C denotes various constants; since $\alpha < 0 < \beta$ we have $\{0,0\} \in \mathcal{K}_h$, which, on putting $\{v_h,q_h\} = \{0,0\}$ in (3.74A), gives

$$\frac{1}{2}\int_\Omega |p_h|^2 dx - \int_\Omega f u_h \, dx \le 0. \tag{3.96A}$$

From (3.96A) we deduce that

$$\|p_h\|^2_{L^2(\Omega)} \le C \|f\|_{L^2(\Omega)} \|u_h\|_{H^1_0(\Omega)}. \tag{3.97A}$$

From the definition of W_h (see (3.10A)) we have

$$\int_\Omega \nabla u_h \cdot \nabla \mu_h dx = \int_\Omega p_h \mu_h dx \quad \forall \mu_h \in V_h. \tag{3.98A}$$

Since $u_h \in V_{oh} \subset V_h$ it is permissible to put $\mu_h = u_h$ in (3.98A), giving

$$\|u_h\|_{H^1_0(\Omega)} \le C \|p_h\|_{L^2(\Omega)},$$

which in combination with (3.97A) implies

$$\|p_h\|_{L^2(\Omega)} \le C \quad \forall h, \tag{3.99A}$$

$$\|u_h\|_{H^1_0(\Omega)} \le C \quad \forall h. \tag{3.100A}$$

2) *Weak convergence*

In view of (3.99A), (3.100A) we can extract from $(\{u_h,p_h\})_h$ a subsequence - also denoted by $(\{u_h,p_h\})_h$ - such that

$$\lim_{h\to 0} u_h = u^* \quad \textit{weakly in } H^1_0(\Omega) \tag{3.101A}$$

$$\lim_{h\to 0} p_h = p^* \quad \textit{strongly in } L^2(\Omega) \tag{3.102A}$$

Since $\{u_h,p_h\} \in \mathcal{K}_h \ \forall h$, it follows from Lemma 3.3 that

(3.103A) $\{u^*, p^*\} \in \mathcal{K}$.

Let $\{v,q\} \in \mathcal{K}^*$; it follows from Lemma 3.4, relation (3.89A), that for h sufficiently small we have $\{r_h v, r_h q\} \in \mathcal{K}_h$; we then deduce from (3.72A), (3.74A) that for h sufficiently small we have $\forall \{v,q\} \in \mathcal{K}^*$

$$\frac{1}{2}\int_\Omega |p_h|^2 dx - \int_\Omega f u_h dx \le \frac{1}{2}\int_\Omega |r_h q|^2 dx - \int_\Omega f\, r_h v\, dx. \tag{3.104A}$$

It then results from (3.90A) and from the (weak) lower semi-continuity of $z \to \|z\|^2_{L^2(\Omega)} : L^2(\Omega) \to \mathbb{R}$ that

$$j(u^*,p^*) \le j(v,q) \quad \forall \{v,q\} \in \mathcal{K}^*. \tag{3.105A}$$

In view of $\overline{\mathcal{K}^*} = \mathcal{K}$ and the *continuity* of $j(\cdot,\cdot)$ we deduce from (3.105A)

$$j(u^*,p^*) \le j(v,q) \quad \forall \{v,q\} \in \mathcal{K} \tag{3.106A}$$

which combined with (3.103A) shows that $\{u^*,p^*\}$ is the solution of (3.71A), i.e. $u^* = u$, $p^* = p = -\Delta u$ where u is the solution of (1.3A), (1.4A).

The *uniqueness* of u implies that the whole family $\{u_h, p_h\} \to \{u,p\}$ and not simply an extracted subsequence.

3) *Strong convergence*

Since $\{u_h, p_h\} \in W_h$ and $\{u,p\} \in W$ (with $p = -\Delta u$), we have

$$\int_\Omega |\nabla u_h|^2\, dx = \int_\Omega p_h u_h dx\ , \tag{3.107A}$$

$$\int_\Omega |\nabla u|^2 dx = \int_\Omega pu\ dx. \tag{3.108A}$$

It was shown earlier that

(3.109A) $\lim_{h\to 0} p_h = p$ *weakly in* $L^2(\Omega)$;

from the compactness of the canonical injection from $H_0^1(\Omega)$ into $L^2(\Omega)$ we also have

(3.110A) $\lim_{h\to 0} u_h = u$ *strongly in* $L^2(\Omega)$.

We then deduce from (3.107A)-(3.110A) that

$$\lim_{h\to 0} \int_\Omega |\nabla u_h|^2\, dx = \lim_{h\to 0} \int_\Omega p_h u_h dx = \int_\Omega pu\, dx = \int_\Omega |\nabla u|^2\, dx\ ,$$

which, in combination with the *weak convergence* of u_h to u in $H_0^1(\Omega)$, implies

(3.111A) $\lim_{h\to 0} \| u_h - u \|_{H_0^1(\Omega)} = 0.$

It now remains to demonstrate the strong convergence of p_h to p in $L^2(\Omega)$; it follows from (3.74A) and from (3.89A) that for h sufficiently small we have, $\forall \{v,q\} \in \mathcal{H}^*$

(3.112A) $\frac{1}{2}\int_\Omega |p_h|^2\, dx \le \int_\Omega fu_h + \frac{1}{2}\int_\Omega |r_h q|^2\, dx - \int_\Omega fr_h v\, dx.$

It results from (3.90A) that, proceeding to the limit in (3.112A) we have, $\forall \{v,q\} \in \mathcal{H}^*$

(3.113A)
$$\begin{cases} \frac{1}{2}\int_\Omega |p|^2 dx \le \frac{1}{2} \liminf_{h\to 0} \int_\Omega |p_h|^2 dx \le \frac{1}{2} \limsup_{h\to 0} \int_\Omega |p_h|^2\, dx \\ \le \frac{1}{2}\int_\Omega |q|^2 dx - \int_\Omega f(v-u)\, dx\ . \end{cases}$$

By density and continuity, (3.113A) also holds $\forall \{v,q\} \in \mathcal{H}$; we can thus put $\{v,q\} = \{u,p\}$ in (3.113A) giving

$$\lim_{h\to 0} \|p_h\|_{L^2(\Omega)} = \|p\|_{L^2(\Omega)}$$

which combined with (3.109A) proves that $\lim_{h\to 0} \|p_h - p\|_{L^2(\Omega)} = 0$

and conclude the proof of the theorem. ■

Remark 3.5: In view of the proof of Lemma 3.3, if in (3.75A) we merely imposed the condition

$$-\beta \leq q_h(P) \leq -\alpha$$

at the midpoints of the sides of the $T \in \mathcal{T}_h$, the convergence results of Theorem 3.5 would still apply. ■

Remark 3.6: An investigation of other methods for approximating problem (1.3A), (1.4A) by finite elements can be found in Glowinski-Marini-Vidrascu /1A/. ■

3.4.6 *Numerical solution of the approximate problem*

In order to solve the approximate problem (3.74A) we can use, as we do below, a finite-dimensional variant of algorithm (3.59A)-(3.61A) of Section 3.4.3; first we define

(3.114A) $$V_h^+ = \{q_h \in V_h,\ q_h(P) \geq 0 \quad \forall P \in \Sigma_h\}$$

where, in (3.114A), Σ_h is as defined in Section 3.4.4 by (3.76A) (note that $q_h \in V_h^+ \not\Rightarrow q_h \geq 0$ on $\bar{\Omega}$ since $k = 2$). We next define

(3.115A) $$\Lambda_h = V_h^+ \times V_h^+ ,$$

and then a new inner product on V_h, namely $(\cdot,\cdot)_h$, by

$$(y_h, z_h)_h = \frac{1}{6} \sum_{P \in \Sigma_h} m_P y_h(P) z_h(P) \quad \forall y_h, z_h \in V_h ,$$

where m_P is the area of the domain consisting of the union of those triangles of $\mathcal{T}_h$ which have P as a common vertex or as the midpoint of a common side. Using the inner product $(\cdot,\cdot)_h$, we define the *symmetric positive-definite* operator S_h: $V_h \rightarrow V_h$, by [1]

(1) S_h is in fact an approximation to the identity operator.

$$\begin{cases} S_h y_h \in V_h & \forall y_h \in V_h , \\ \int_\Omega S_h y_h z_h dx = (y_h, z_h)_h & \forall z_h \in V_h . \end{cases}$$

The discrete analogue of algorithm (3.59A)-(3.61A) is then

(3.116A) $\lambda_h^o = \{\lambda_{1h}^o, \lambda_{1h}^o\} \in \Lambda_h$ *given,*

then with $\lambda_h^n \in \Lambda_h$ *known, determine* $u_h^n \in V_{oh}$, $p_h^n \in V_h$ *and* $\lambda_h^{n+1} \in \Lambda_h$ *from*

(3.117A) $$\begin{cases} \{u_h^n, p_h^n\} \in W_h , \\ \int_\Omega p_h^n q_h \, dx = \int_\Omega f v_h dx + (\lambda_{1h}^n - \lambda_{2h}^n, q_h)_h & \forall \{v_h, q_h\} \in W_h , \end{cases}$$

(3.118A) $$\lambda_{1h}^{n+1}(P) = [\lambda_{1h}^n(P) - \rho(p_h^n(P)+\beta)]^+ \quad \forall P \in \Sigma_h ,$$

(3.119A) $$\lambda_{2h}^{n+1}(P) = [\lambda_{2h}^n(P) + \rho(p_h^n(P)+\alpha)]^+ \quad \forall P \in \Sigma_h .$$

Regarding the *convergence* of algorithm (3.116A)-(3.119A), it can be shown that there exists a $\delta_h > 0$ such that $\forall \lambda_h^o \in \Lambda_h$ and for all ρ satisfying

(3.120A) $$0 < \rho < \delta_h$$

we have

(3.121A) $$\lim_{n \to +\infty} \{u_h^n, p_h^n\} = \{u_h, p_h\}$$

where $\{u_h, p_h\}$ is the solution of (3.74A). It can also be shown that $\lim_{h \to 0} \delta_h = 1$.

Remark 3.7: The problem (3.117A) is simply an approximate mixed formulation of the biharmonic problem (3.60A); a formulation equivalent to (3.117A) is given by

$$
(3.122\mathrm{A})\quad
\begin{cases}
\{u_h^n, p_h^n\} \in V_{oh} \times V_h, \\
\int_\Omega \nabla p_h^n \cdot \nabla v_h \, dx = \int_\Omega f v_h \, dx + \int_\Omega \nabla S_h(\lambda_{2h}^n - \lambda_{1h}^n) \cdot \nabla v_h \, dx \quad \forall v_h \in V_{oh}, \\
\int_\Omega \nabla u_h^n \cdot \nabla q_h \, dx = \int_\Omega p_h^n q_h \, dx \quad \forall q_h \in V_h .
\end{cases}
$$

For solving the approximate biharmonic problem (3.117A), (3.122A) at each iteration, we refer to the discussion in Section 3.3 of this appendix and to Chapter 4, Section 4.6; see also Glowinski-Pironneau /1A/, Vidrascu /1A/, and Glowinski-Marini-Vidrascu /1A/.

3.4.7 *Application to the numerical solution of a model problem*

We shall now consider the particular case of the problem (1.3A), (1.4A) in which $\Omega =]0,1[\times]0,1[$, f = const., $\alpha = -1$, $\beta = +1$. The triangulation $\mathcal{T}_h$ used to define the approximation with finite elements of order two is shown in Figure 1.10 of Chapter 4, Section 1.7, and consists of 128 triangles and 289 nodes; we thus have $\dim V_h = 289$, $\dim V_{oh} = 225$. In implementing algorithm (3.116A)-(3.119A), we have used $\lambda_h^o = \{0,0\}$; furthermore, the optimum value for ρ, for various values of f, is close to 0.8. For $f = 100$, Figures 3.1-3.4 show the *contours* of p_h^n, corresponding to $n = 0$, 20, 40 and 240; the contours of u_h are shown in Figure 3.5 (convergence is in fact virtually complete after 50 iterations).

Figures 3.6-3.10 show the contours of p_h^n corresponding to $n = 0$, 20, 40, 60 and 180, for $f = 1000$; the contours of u_h are indicated in Figure 3.11.

In the above figures the regions where p_h ($\simeq -\Delta u$) attains its limiting values are clearly visible; p_h is equal to + 1 in the central region and - 1 in the four regions in contact with the edges. However, it should be noted that the contours of p_h depend quite appreciably on $\mathcal{T}_h$, especially near the free boundaries.

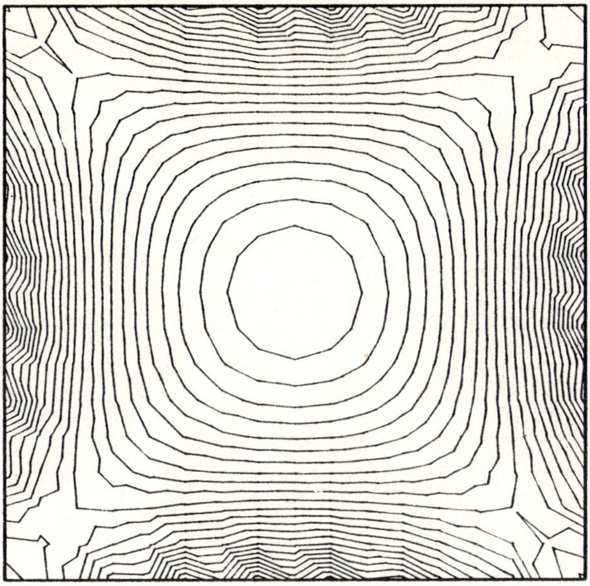

Fig. 3.1
(f = 100, *contours of* p_h^o)

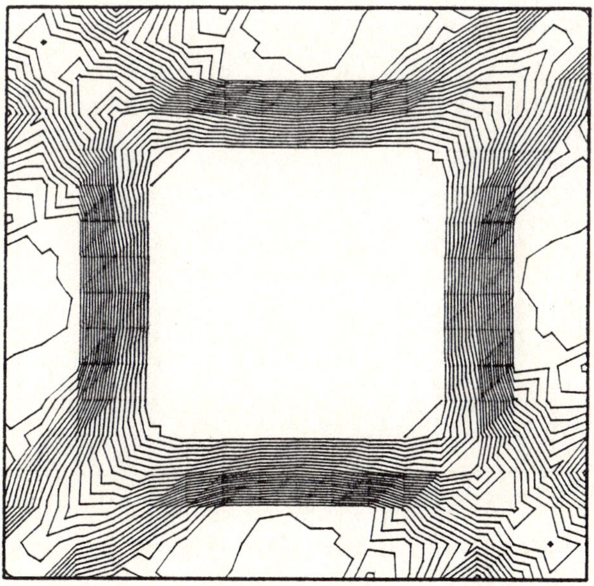

Fig. 3.2
(f = 100, *contours of* p_h^{20})

Fig. 3.3

(f = 100, *contours of* p_h^{40})

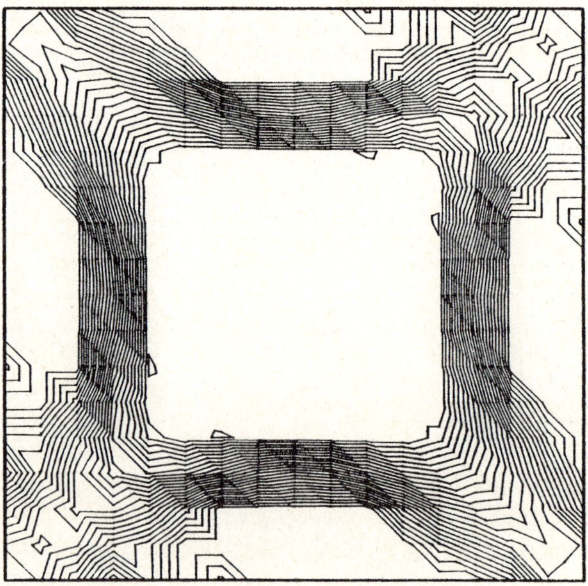

Fig. 3.4

(f = 100, *contours of* p_h^{240})

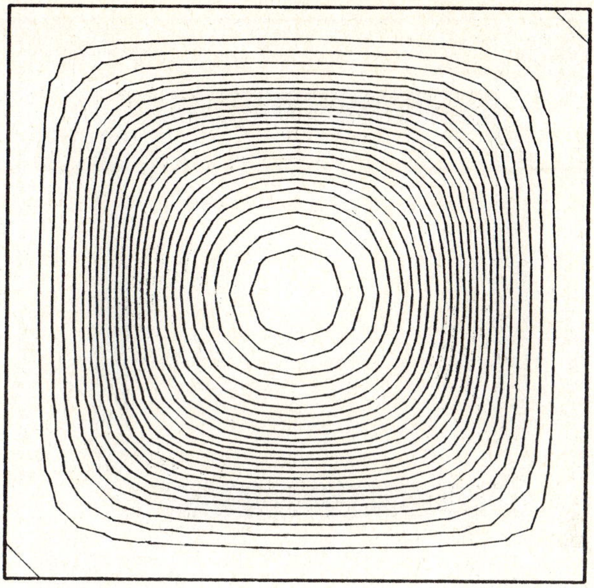

Fig. 3.5

(f = 100, *contours of* u_h)

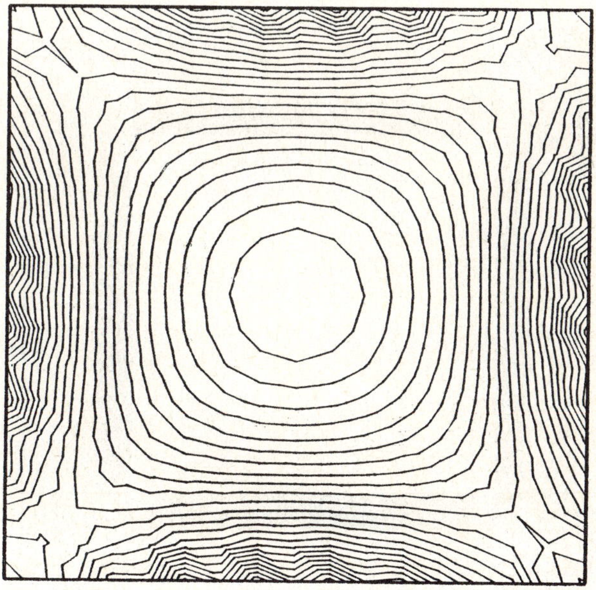

Fig. 3.6

(f = 1000, *contours of* p_h^o)

Fig. 3.7
(f = 1000, *contours of* p_h^{20})

Fig. 3.8
(f = 1000, *contours of* p_h^{40})

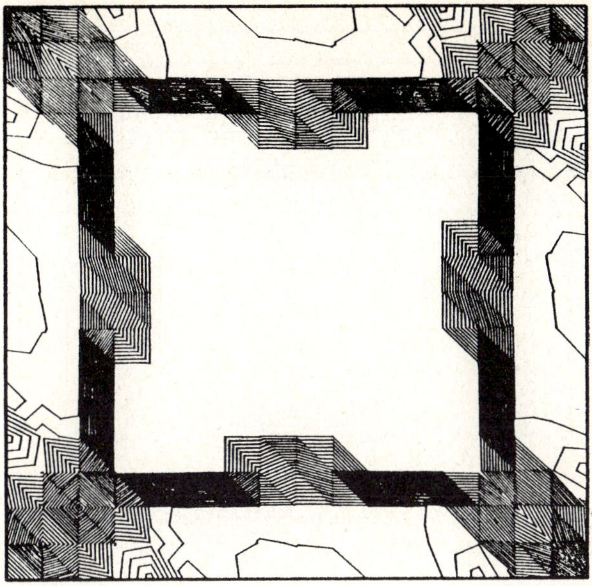

Fig. 3.9

(f = 1000, *contours of* p_h^{60})

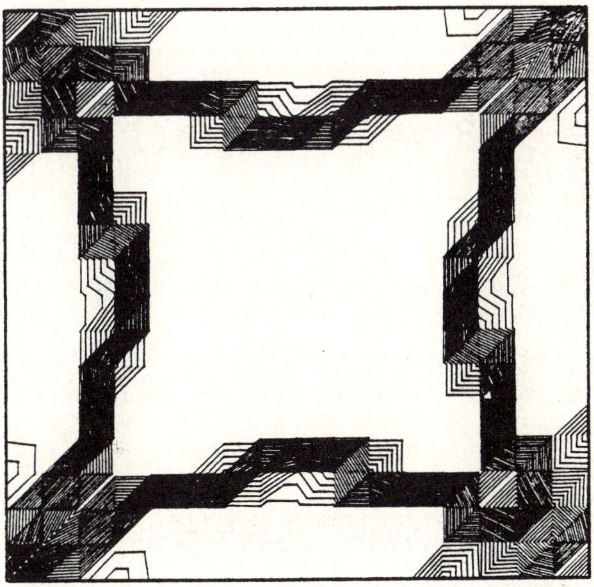

Fig. 3.10

(f = 1000, *contours of* p_h^{180})

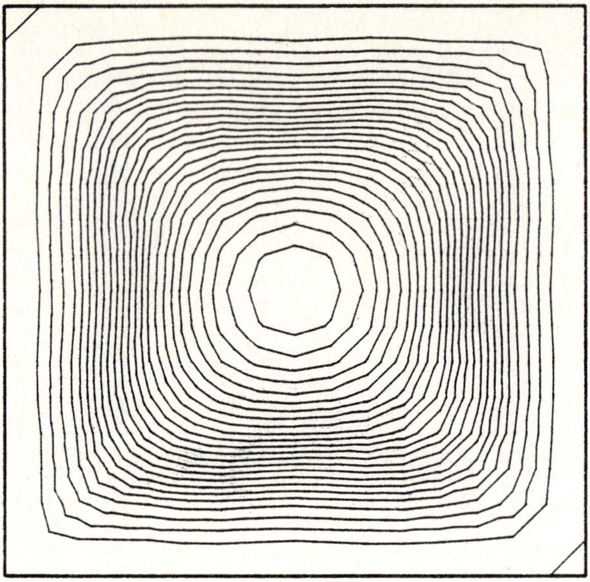

Fig. 3.11

(f = 1000, *contours of* u_h)

4. NUMERICAL SIMULATION OF THE TRANSONIC POTENTIAL FLOW OF IDEAL COMPRESSIBLE FLUIDS USING VARIATIONAL INEQUALITY METHODS

4.1 Synopsis

As we have already remarked in Appendix 1, Section 6, the methodology of variational inequalities can provide an effective tool for the solution of problems initially formulated in a more traditional manner. A first example has already been given in Appendix 1, Section 6; in the present section we shall consider another example which is much more important from the applications point of view, but much more difficult mathematically, namely the numerical simulation of the *transonic potential flow of perfect compressible fluids*. Within the limited scope of this appendix, it would be out of the question to attempt to provide a detailed account of this subject; we shall therefore be content to give only an outline of the material considered in more depth in Bristeau-Glowinski-Periaux-Perrier-Pironneau-Poirier [1] /1A/, /2A/ (which also contains a substantial bibliography on the subject.

---

(1) Abbreviated hereafter to B.G.4P.

4.2 Mathematical formulation

4.2.1 *The equation of continuity*

If Ω is the flow domain and Γ its boundary, it is shown in Landau-Lifchitz /1A/ that the flow satisfies

$$\nabla\cdot\rho\vec{u} = 0 \text{ in } \Omega \tag{4.1A}$$

with

$$\rho = \rho_o\left(1 - \frac{|\vec{u}|^2}{\frac{\gamma+1}{\gamma-1}C_*^2}\right)^{\frac{1}{\gamma-1}} \tag{4.2A}$$

$$\vec{u} = \nabla\phi\ , \tag{4.3A}$$

where, in (4.1A)-(4.3A)

a) ϕ is the *velocity potential*
b) ρ is the *density* of the fluid
c) γ is the ratio of specific heats (γ = 1.4 in air)
d) C_* is the *critical speed of sound.*

4.2.2 *Boundary conditions*

In the case of an *aerofoil* such as B (see Figure 4.1), if we assume that the flow is uniform at infinity and tangential to the aerofoil surface Γ_B, we have

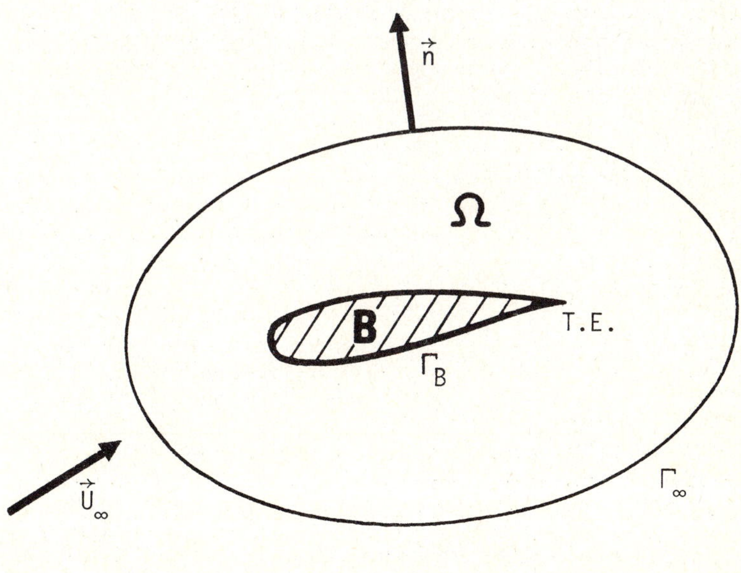

Fig. 4.1

(4.4A) $\vec{u} = \vec{u}_\infty$ *at infinity*

(4.5A) $\frac{\partial\phi}{\partial n} = 0$ *on* Γ_B.

In practice we bound Ω by Γ_∞ as shown in Figure 4.1 and in place of (4.4A) we use

(4.4A)' $\frac{\partial\phi}{\partial n}\Big|_{\Gamma_\infty} = \vec{u}_\infty\cdot\vec{n}$.

It turns out that for the above conditions the potential ϕ is determined only to within an arbitrary additive constant; to remedy this situation we can prescribe a value of ϕ at some point in Ω; for example we can use

(4.6A) $\phi = 0$ *at the trailing edge* T.E. *of* B.

4.2.3 *Lifting aerofoils and the Kutta-Joukowsky condition*

The Kutta-Joukowsky condition is not particular to transonic flows, as it also applies in the flow of incompressible perfect fluids and in the subsonic flow of compressible perfect fluids. We refer to B.G. 4P /1A/ and Periaux /1A/ for the numerical treatment of the Kutta-Joukowsky condition in the case of two- and three-dimensional flows.

4.2.4 *Shock-jump relations*

The flow must satisfy the *Rankine-Hugoniot* relations across a shock, namely

(4.7A) $(\rho\vec{u}\cdot\vec{n})_+ = (\rho\vec{u}\cdot\vec{n})_-$ (where $\vec{n}$ is *normal* to the shock line or surface)

(4.8A) *the tangential component of the velocity is continuous.*

A suitable variational formulation of (4.1A)-(4.3A) will take (4.7A)-(4.8A) into account automatically.

4.2.5 *Entropy condition*

This can be formulated as follows (see Landau-Lifchitz /1A/ for more details):

(4.9A) *There cannot be an increase in the flow velocity across a shock as there would then be an associated decrease in the entropy, which is physically impossible.*

The numerical implementation of (4.9A) will be discussed in Section 4.5.

Remark 4.1: Since the boundary between the subsonic region and the supersonic region is unknown, the problem to be solved is in fact a free-boundary problem; the numerical treatment of this problem will take this into account.

4.3 Least-squares formulation of the continuous problem

In this section we shall not consider the practical implementation of (4.9A); we shall only discuss the variational formulation of (4.1A)-(4.5A), (4.7A), (4.8A) and of an associated non-linear least-squares formulation.

4.3.1 *A variational formulation of the equation of continuity*

We consider for simplicity the situation shown in Figure 4.2, representing a *symmetric flow*, *subsonic at infinity*, around a symmetric aerofoil; since the flow is symmetric, the Kutta-Joukowsky condition is *automatically* satisfied and the aerofoïl is non-lifting.

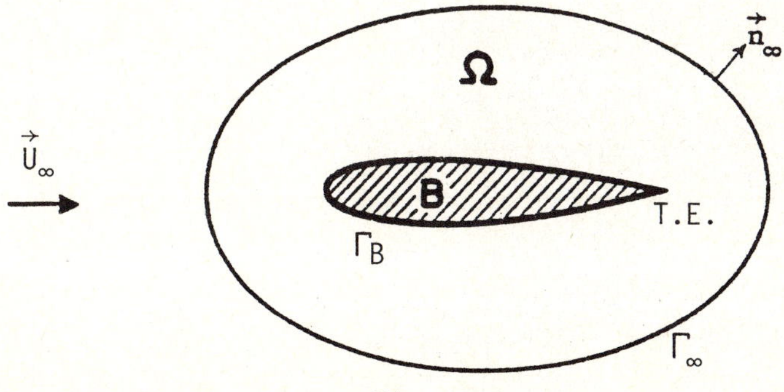

Fig. 4.2.

For practicality (although other approaches are possible) we imbed the aerofoil in a "large" domain; using the notation of Section 4.2, the continuity equation and the boundary conditions are

$$(4.10A) \qquad \nabla\cdot\rho(\phi)\nabla\phi = 0 \ \textit{in} \ \Omega$$

with

$$(4.11A) \qquad \rho(\phi) = \rho_0\Big(1 - \frac{|\nabla\phi|^2}{\frac{\gamma+1}{\gamma-1}\,C_*^2}\Big)^{\frac{1}{\gamma-1}} ,$$

and

$$(4.12A) \qquad \rho\,\frac{\partial\phi}{\partial n} = 0 \ \textit{on} \ \Gamma_B,\ \rho\,\frac{\partial\phi}{\partial n} = \rho_\infty\vec{u}_\infty\cdot\vec{n}_\infty \ \textit{on} \ \Gamma_\infty ;$$

on $\Gamma(=\Gamma_B\cup\Gamma_\infty)$ we define g by

$$(4.13A) \qquad g=0 \ \textit{on} \ \Gamma_B,\ g = \rho_\infty\vec{u}_\infty\cdot\vec{n}_\infty \ \textit{on} \ \Gamma_\infty .$$

We clearly have

$$(4.14A) \qquad \rho\,\frac{\partial\phi}{\partial n} = g \ \textit{on} \ \Gamma \ \textit{and} \ \int_\Gamma g\,d\Gamma = 0.$$

An equivalent variational formulation of (4.10A), (4.14A) is

$$(4.15A) \qquad \begin{cases} \int_\Omega \rho(\phi)\nabla\phi\cdot\nabla v\,dx = \int_\Gamma gv\,d\Gamma \qquad \forall\, v\in H^1(\Omega), \\ \phi\in W^{1,\infty}(\Omega)/\mathbb{R}. \end{cases}$$

The function ϕ can be determined only to within an arbitrary constant.

Remark 4.2: The space $W^{1,\infty}(\Omega)$ is a natural choice for ϕ since *physical flows* require (among other properties) a *positive density* ρ; therefore from (4.11A) ϕ has to satisfy

$$|\nabla\phi| \le \delta < \left(\frac{\gamma+1}{\gamma-1}\right)^{1/2} C_* \quad \textit{a.e. on } \Omega. \tag{4.16A}$$

4.3.2 *A least-squares formulation of* (4.15A)

For a *genuine transonic flow*, problem (4.15A) *is not equivalent* to a standard problem in the *Calculus of Variations* (as is the case for purely subsonic flows); to remedy this situation and - in some sense - "*convexify*" the problem under consideration, we introduce a nonlinear least-squares formulation of the transonic flow problem (4.15A), as follows:

Let X *be a set of feasible transonic flow solutions; the least-squares problem is then*

$$\operatorname*{Min}_{\xi \in X} J(\xi), \tag{4.17A}$$

with

$$J(\xi) = \frac{1}{2}\int_\Omega |\nabla y(\xi)|^2 dx , \tag{4.18A}$$

where, in (4.18A) $y(\xi)$ (= y) *is a solution of the state equation*

$$\begin{cases} \textit{Find } y \in H^1(\Omega)/\mathbb{R} \textit{ such that} \\ \displaystyle\int_\Omega \nabla y \cdot \nabla v \, dx = \int_\Omega \rho(\xi)\nabla\xi\cdot\nabla v \, dx - \int_\Gamma gv \, d\Gamma \quad \forall v \in H^1(\Omega). \end{cases} \tag{4.19A}$$

If the solutions of the transonic flow problem exist, then these are solutions of the least-squares problem and *give the value zero* to the objective function J.

4.4 Finite-element approximation and least-squares-conjugate-gradient solution of the approximate problems

We shall consider here only two-dimensional problems, although the following methods can also be (and have been) applied to three-dimensional problems.

4.4.1 *Finite-element approximation of the nonlinear variational equation* (4.15A)

We again consider the non-lifting situation of Section 4.3.1. The flow region is imbedded in a large bounded domain Ω, which we approximate by a polygonal domain Ω_h. With $\mathcal{C}_h$ a standard triangulation of Ω_h, we approximate $H^1(\Omega)$ (and in fact $W^{1,p}(\Omega)$, $\forall p \geq 1$) by

(4.20A) $$H_h^1 = \{v_h | v_h \in C^o(\overline{\Omega}_h),\ v_h|_T \in P_1 \ \forall T \in \mathcal{C}_h\}$$

where, in (4.20A), P_1 is the space of polynomials in two variables of degree ≤ 1. We prescribe the value zero for the potential at the trailing edge; this leads to

(4.21A) $$V_h = \{v_h \in H_h^1,\ v_h\ (\text{T.E.}) = 0\} \quad ;$$

we clearly have

(4.22A) $$\dim H_h^1 = 1 + \dim V_h = \textit{number of vertices of } \mathcal{C}_h.$$

We then approximate the variational equation (4.15A) by

(4.23A) $$\begin{cases} \textit{Find } \phi_h \in V_h \textit{ such that} \\ \int_\Omega \rho(\phi_h)\nabla\phi_h \cdot \nabla v_h \, dx = \int_\Omega g_h v_h d\Gamma \quad \forall v_h \in V_h \end{cases}$$

where, in (4.23A), g_h is an approximation of the function g in (4.13A).

The above discrete variational formulation implies that $\frac{\partial\phi}{\partial n}\Big|_\Gamma = g$ is automatically satisfied "approximately".

Let $\mathcal{B}_h = \{w_i\}_{i=1}^{N_h}$ be a vector basis of V_h; then (4.23A) is equivalent to the *nonlinear finite-dimensional system*

(4.24A) $$\begin{cases} \phi_h = \sum_{j=1}^{N_h} \phi_j w_j \ , \\ \int_\Omega \rho(\phi_h)\nabla\phi_h \cdot \nabla w_i \, dx = \int_\Gamma g_h w_i d\Gamma \quad \forall i=1,\ldots N_h \ , \end{cases}$$

where $\phi_j = \phi(P_j)$ $\forall j=1,\ldots N_h$, assuming that $\{P_j\}_{j=1}^{N_h}$ is the set of the vertices of $\mathcal{T}_h$ different from T.E.

With the above choice of H_h^1 and V_h, there is no problem with the numerical integration, since in (4.23A), (4.24A), $\nabla\phi_h$, ∇v_h, ∇w_i (and therefore $\rho(\phi_h)$) are piecewise constant.

4.4.2 *Least-squares formulation of the discrete problem* (4.23A).

Combining the results of Sections 4.3.2 and 4.4.1, we introduce the following least-squares formulation of the approximate problem (4.23A)

(4.25A)
$$\min_{\xi_h \in X_h} J_h(\xi_h)$$

where, in (4.25A), X_h is the set of feasible discrete solutions, and where

(4.26A)
$$J_h(\xi_h) = \frac{1}{2}\int_\Omega |\nabla y_h(\xi_h)|^2 dx ,$$

with $y_h (= y_h(\xi_h))$ the solution of the discrete variational state equation

(4.27A)
$$\begin{cases} \textit{Find } y_h \in V_h \textit{ such that} \\ \int_\Omega \nabla y_h \cdot \nabla v_h dx = \int_\Omega \rho(\xi_h)\nabla\xi_h \cdot \nabla v_h dx - \int_\Gamma g_h v_h d\Gamma \quad \forall v_h \in V_h . \end{cases}$$

4.3.3 *Conjugate-gradient solution of the least-squares problem* (4.25A)-(4.27A)

We follow B.G. 4P /1A/, /2A/ and Periaux /1A/; a preconditioned conjugate-gradient algorithm for solving (4.25A)-(4.27A) - with $X_h = V_h$ - is

<u>Step 0:</u> <u>Initialisation</u>

(4.28A)
$$\phi_h^o \in V_h \quad \textit{given},$$

then compute g_h^o *from*

$$(4.29A)\qquad \begin{cases} g_h^o \in V_h \\ \int_\Omega \nabla g_h^o \cdot \nabla v_h \, dx = \langle J'(\phi_h^o), v_h \rangle \quad \forall v_h \in V_h , \end{cases}$$

and set

$$(4.30A)\qquad z_h^o = g_h^o .$$

Then for $n \geq 0$, *assuming* ϕ_h^n, g_h^n, z_h^n *known, compute* $\phi_h^{n+1}, g_h^{n+1}, z_h^{n+1}$ *by*

Step 1: Descent

$$(4.31A)\qquad \text{Compute } \lambda^n = \operatorname{Arg}\min_{\lambda \in \mathbb{R}} J_h(\phi_h^n - \lambda z_h^n) ,$$

$$(4.32A)\qquad \phi_h^{n+1} = \phi_h^n - \lambda^n z_h^n .$$

Step 2: Construction of the new descent direction

Define g_h^{n+1} *by*

$$(4.33A)\qquad \begin{cases} g_h^{n+1} \in V_h \\ \int_\Omega \nabla g_h^{n+1} \cdot \nabla v_h \, dx = \langle J_h'(\phi_h^{n+1}), v_h \rangle \quad \forall v_h \in V_h , \end{cases}$$

then

$$(4.34A)\qquad \gamma_{n+1} = \frac{\int_\Omega \nabla g_h^{n+1} \cdot \nabla (g_h^{n+1} - g_h^n) \, dx}{\int_\Omega |\nabla g_h^n|^2 dx} ,$$

(4.35A) $$z_h^{n+1} = g_h^{n+1} + \gamma_{n+1} z_h^n ,$$

$n = n+1$, *go to* (4.31A).

The two non-trivial steps of algorithms (4.28A)-(4.35A) are:

(i) the solution of the *single-variable* minimisation problem (4.31A); part (i) of Remark 4.5 in Appendix 3, Section 4.3.2 still holds for (4.31A). It should be noted that each evaluation of $J_h(\xi_h)$, for a given argument ξ_h, requires the solution of the linear approximate Neumann problem (4.27A) to obtain the corresponding y_h.

(ii) the calculation of g_h^{n+1} from ϕ_h^{n+1}, which requires the solution of two linear approximate Neumann problems (namely (4.27A) with $\xi_h = \phi_h^{n+1}$ and (4.33A)).

Calculation of $J_h'(\xi_h^n)$ *and* g_h^n: In view of the importance of step (ii), we shall now describe in detail the calculation of $J_h'(\xi_h^n)$ and g_h^n (supposing for simplicity that $\rho_o = 1$).

We have by differentiation

(4.36A) $$\langle J_h'(\xi_h), \delta\xi_h \rangle = \int_\Omega \nabla y_h \cdot \nabla \delta y_h \, dx ,$$

where δy_h is, from (4.27A), the solution of

(4.37A) $$\begin{cases} \delta y_h \in V_h \quad \textit{and} \ \forall v_h \in V_h \ \textit{we have} \\ \int_\Omega \nabla \delta y_h \cdot \nabla v_h \, dx = \int_\Omega \rho(\xi_h) \nabla \delta\xi_h \cdot \nabla v_h dx + \int_\Omega \delta\rho(\xi_h) \nabla \xi_h \cdot \nabla v_h dx . \end{cases}$$

Since $\rho(\xi_h) = \rho_o(1-K|\nabla\xi_h|^2)^\alpha$ with $K = \frac{1}{C_*^2}\frac{\gamma-1}{\gamma+1}$ and $\alpha = 1/(\gamma-1)$, we have

(4.38A) $$\delta\rho(\xi_h) = -2K\alpha(1-K|\nabla\xi_h|^2)^{\alpha-1} \nabla\xi_h \cdot \nabla\delta\xi_h .$$

It follows from (4.36A)-(4.38A) that

$$\left\{\begin{aligned}&\int_\Omega \nabla y_h \cdot \nabla \delta y_h dx = \int_\Omega \rho(\xi_h) \nabla y_h . \nabla \delta \xi_h\, dx - 2K\alpha \int_\Omega (\rho(\xi_h))^{2-\gamma} \times \\ &\qquad \times \nabla \xi_h \cdot \nabla y_h \nabla \xi_h \cdot \nabla \delta \xi_h dx\,.\end{aligned}\right. \tag{4.39A}$$

In fact $\langle J_h'(\xi_h), \eta_h \rangle$ can be identified with the linear functional

$$\eta_h \to \int_\Omega \rho(\xi_h) \nabla y_h \cdot \nabla \eta_h dx - 2K\alpha \int_\Omega (\rho(\xi_h))^{2-\gamma} \nabla \xi_h \cdot \nabla y_h \nabla \xi_h \cdot \nabla \eta_h dx. \tag{4.40A}$$

It is quite easy to obtain g_h^{n+1} from ϕ_h^{n+1}, using (4.33A), (4.40A).

Remark 4.3: An efficient discrete Poisson solver will form a fundamental part of the method if the above algorithm (4.28A)-(4.35A) is used.

4.4.4 *Numerical solution of a test problem*

We shall now apply the above methods to the numerical simulation of the flow around a disc, the flow being uniform and subsonic at infinity. Owing to the symmetry of the flow, the Kutta-Joukowsky condition is satisfied automatically. If $\|\vec{u}_\infty\|$ is sufficiently small the flow is purely subsonic and a very good solution is obtained in a very small number of iterations of algorithm (4.28A)-(4.35A) ($\simeq$ 5 iterations). For greater values of $\|\vec{u}_\infty\|$ a supersonic pocket appears, and if the computed [1] Mach number distribution on the surface of the body is plotted, we obtain the distribution in Figure 4.3 showing a rarefaction (or expansion) shock ; the latter cannot exist physically. The correct Mach number distribution is shown in Figure 4.4.

---

[1] The convergence is still very fast ($\simeq$ 10 iterations)

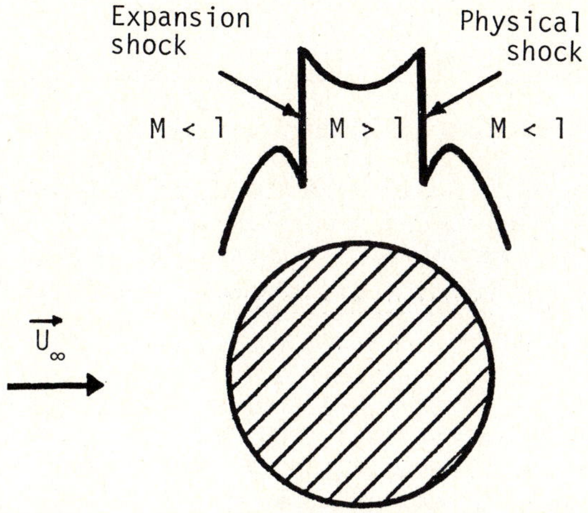

Expansion shock
Physical shock
M < 1
M > 1
M < 1
$\vec{U}_\infty$

Fig. 4.3

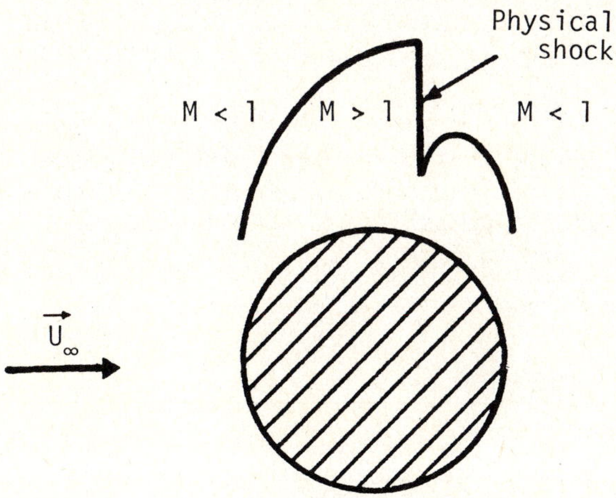

Physical shock
M < 1
M > 1
M < 1
$\vec{U}_\infty$

Fig. 4.4

From the above tests it is apparent that we have to incorporate in our numerical procedure some device (in fact some dissipation) aimed at suppressing these non-physical shocks which violate the *entropy condition* (4.9A).

The numerical implementation of (4.9A) is discussed below in Section 4.5.

4.5 Numerical implementation of the entropy condition

4.5.1 *General discussion. Synopsis.*

Several methods based on *penalisation* and/or *artificial viscosity* have been discussed and numerically tested in B.G. 4P /1A/; we introduce in Section 4.5.2 an *interior penalty method* which is effective in suppressing rarefaction shocks and gives a good approximation of the solution in the neighbourhood of physical shocks.

If we were to go back to the example of Section 4.4.4 and try to represent the corresponding least-squares functional $\xi_h \to J_h(\xi_h)$, it would most probably look like the graph in Figure 4.5 (if we assume $\|\vec{u}_\infty\|$ sufficiently large). This feeling about Figure 4.5 is based on the fact that ξ_h is a very powerful attractor for almost any iterative method, and in particular for algorithm (4.28A)-(4.35A); in contrast the physical solution seems to lack these attractor properties (except possibly in a very small neighbourhood of this solution). Based on these observations, one possible idea consists of replacing the continuous curve in Figure 4.5 by the discontinuous one, corresponding to a functional $\xi_h \to J_h^*(\xi_h)$, taking very large values for the non-physical solution $\underline{\xi}_h$ and taking its minimal value close to $\hat{\xi}_h$.

4.5.2 *An interior penalty method with truncation*

4.5.2.1 The physical motivation

It is known from Landau-Lifchitz /1A/ that in the case of a *weak shock* we have, *following the streamlines*,

$$[\mathcal{S}] = O([v]^3), \tag{4.41A}$$

where in (4.41A), $[\mathcal{S}]$ (resp. $[v]$) denotes the *jump in entropy* (resp. the *jump in velocity*) across the shock, with the sign of $[\mathcal{S}]$ opposite to that of $[v]$. Now for a physical shock we must have

(4.42A) $[\mathcal{S}] \geq 0$;

Combining (4.41A), (4.42A) we shall use as *entropy condition*

(4.43A) $[v]^3 \leq 0$.

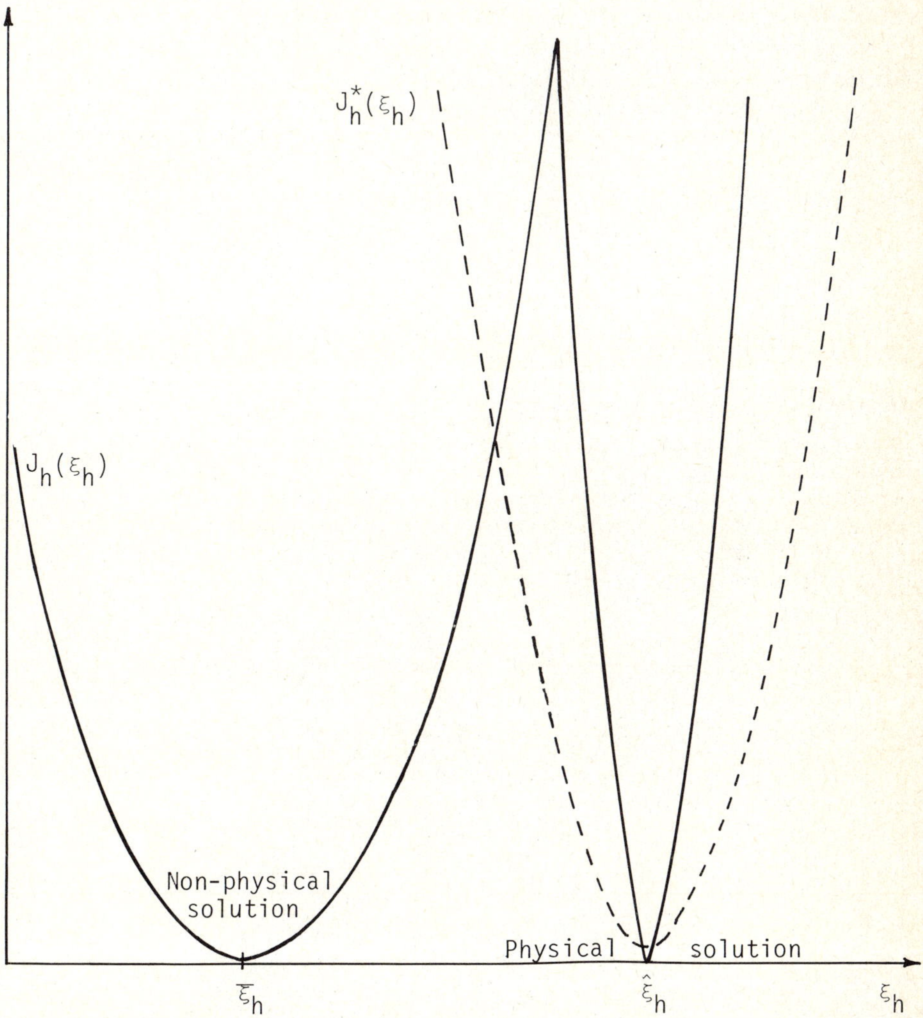

Fig. 4.5

This choice is based on the fact that our finite element method gives us direct access to velocity variables.

4.5.2.2 Finite element implementation of (4.43A) using an interior penalty method

Consider two adjacent triangles, like those shown in Fig. 4.6 (in fact the following method can be (and has been) applied to three-dimensional problems); we denote by $\vec{n}$ the unit vector of the normal to their common side AB, directed from T_1 into T_2. Since we are using linear elements, we have

$$\text{(4.44A)} \qquad \vec{v}_h|_{T_j} = \nabla\phi_h|_{T_j}, \quad j=1,2 .$$

Moreover since we are using C^0-conforming elements the component along AB of the velocity $\vec{v}_h = \nabla\phi_h$ is *continuous*, unlike the normal component which is clearly discontinuous; from these considerations the discontinuities of $\vec{v}_h$ are supported by the local normal components of the velocity. From these observations and in view of (4.44A) we introduce the functional E_h defined as follows

$$\text{(4.45A)} \qquad E_h(\phi_h) = \sum_i \sigma_i \left(\left[\frac{\partial\phi_h}{\partial n_i} \right]^{3+} \right)^2$$

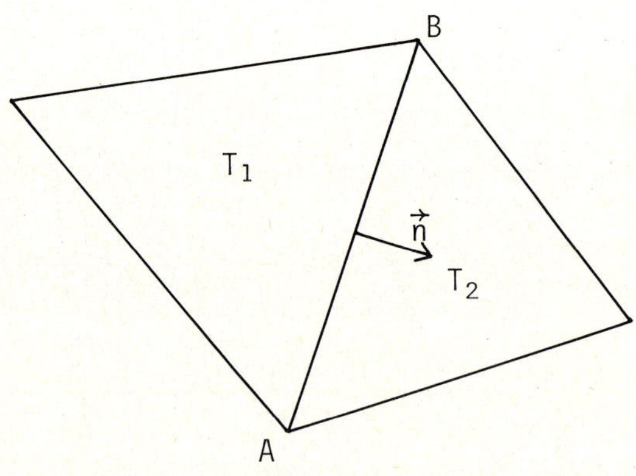

Fig. 4.6

where, in (4.45A), the summation is carried out over the common sides of all the pairs of adjacent triangles in the triangulation $\mathcal{T}_h$, σ_i being a coefficient which takes account of the local size of the triangulation, and where the symbolic notation t^+ is defined as follows:

(4.46A) $$t^+ = \max (0;t) \quad \forall\, t \in \mathbb{R} .$$

Remark 4.4: We observe that (4.45A) is in fact very similar to the integral over Ω of the sixth power of a truncated second-order derivative; for homogeneity this suggests we take

(4.47A) $$\sigma_i = d_i^{\alpha}$$

where, in (4.47A), d_i is the length of the common side and where $\alpha = -5$ for one-dimensional problems, $\alpha = -4$ for two-dimensional problems and $\alpha = -3$ for three-dimensional problems. ■

Combining the above functional E_h and the approximate problem (4.25A)-(4.27A) of Section 4.4.2, we obtain the following discrete problem, which incorporates a dissipative device aimed at suppressing non-physical shocks

(4.48A) $$\underset{\xi_h \in X_h}{\mathrm{Min}} \; J_{rh}(\xi_h) ,$$

where, in (4.48A), X_h is the set of feasible discrete solutions and where

(4.49A) $$J_{rh}(\xi_h) = \frac{1}{2} \int_{\Omega} |\nabla y_h(\xi_h)|^2 \, dx + rE_h(\xi_h) ,$$

with $r > 0$ and y_h $(=y_h(\xi_h))$ the solution of the discrete variational state equation

(4.50A) $$\begin{cases} \textit{Find } y_h \in V_h \textit{ such that} \\ \int_{\Omega} \nabla y_h \cdot \nabla v_h dx = \int_{\Omega} \rho(\xi_h) \nabla \xi_h \cdot \nabla v_h dx - \int_{\Gamma} g_h v_h d\Gamma \quad \forall v_h \in V_h . \end{cases}$$

To solve (4.48A)-(4.50A) we can use a variant of the conjugate gradient algorithm (4.28A)-(4.35A) (with J_h replaced by J_{rh}).

The numerical results produced by the above methods are fairly good, the sonic transition from the subsonic region to the supersonic region being very well approximated, (non-physical "expansion shocks" having been suppressed); furthermore the computed compression shocks are located accurately and are very sharp; we refer to Section 4.6 and also to B.G. 4P /2A/ and Periaux /1A/, in which numerical results produced by the above methods are presented and discussed.

A practical and very important problem concerns the proper choice of r, since for a given aerofoil the optimal value of r seems to be a complicated and sensitive function of $\|\vec{u}_\infty\|$, of the angle of attack, etc.; we shall discuss in Section 4.5.2.3 a technique which we have recently discovered, which seems to be very effective in removing (or at least reducing) this sensitivity to the choice of r.

Remark 4.5: We may supplement Remark 4.4 as follows: since $E_h(\xi_h)$ appears like the integral over Ω of the sixth power of a discrete truncated second-order derivative of ξ_h, we can, by differentiation, associate with E_h a discrete fourth-order nonlinear operator. Since E_h is a *convex functional* its differential is a *monotone operator*. From these properties the addition of rE_h to J_h may be interpreted as a *non-linear fourth-order regularisation process*.

Remark 4.6: We may justify the title of Section 4.5.2 by noting that the functional E_h which we have introduced to regularise our problem is in fact a (more complicated) variant of the *interior penalty functionals* discussed in Douglas-Dupont /2A/ and Wheeler /1A/.

Remark 4.7: We have performed computations using exponents other than 2 and 3 in (4.45A); these two exponents in fact appear to be the optimal combination, based on the quality of the computed solutions.

4.5.2.3 A nonlinearly weighted interior penalty method

From the comments made in Section 4.5.2.2 regarding the sensitivity to the choice of r in the definition of J_{rh} - which is clearly related to the strong nonlinearity of our problem - it is very tempting to introduce local control of the regularisation process associated with $\left(\left[\frac{\partial\phi_h}{\partial n}\right]^{3+}\right)^2$. To achieve this goal, we have introduced in the above functional E_h (defined in (4.45A)) a *nonlinear weighting* directly related to the local value of the density; more precisely, instead of using E_h we use $\tilde{E}_h$ defined by

(4.51A) $$\tilde{E}_h(\phi_h) = \sum_i \frac{\sigma_i}{\rho_i^{2n(\gamma-1)}} \left(\left[\frac{\partial \phi_h}{\partial n_i} \right]^{3+} \right)^2$$

where, in (4.51A), $\frac{1}{\rho_i^\beta}$ is a symbolic notation defined either by

$(4.52A)_1$ $$\frac{1}{\rho_i^\beta} = \frac{1}{2} \left(\frac{1}{\rho_{i1}^\beta} + \frac{1}{\rho_{i2}^\beta} \right)$$

or by

$(4.52A)_2$ $$\frac{1}{\rho_i^\beta} = \frac{1}{2^\beta} \left(\frac{1}{\rho_{i1}} + \frac{1}{\rho_{i2}} \right)^\beta$$

with

(4.53A) $$\rho_{ij} = \left(1 - \frac{|(\nabla\phi_h)_{ij}|^2}{\frac{\gamma+1}{\gamma-1} c_*^2} \right)^{\frac{1}{\gamma-1}}$$

where in (4.53A)

$$(\nabla\phi_h)_{ij} = \nabla\phi_h|_{T_{ij}} ,$$

T_{i1}, T_{i2} being the adjacent triangles of $\mathcal{T}_h$ which have the ith side of the triangulation in common.

Remark 4.8: Our computer experiments indicate that a "good" value for n in (4.51A) is n = 2. ■

Replacing E_h by $\tilde{E}_h$, we obtain a variant of the approximate problem (4.48A)-(4.50A) which can be solved by the same type of iterative methods; computer experiments with $\tilde{E}_h$ have shown that for a NACA 0012 aerofoil the same value of r was optimal (or nearly optimal) for the following conditions

Angle of attack (degrees)	M_∞
6	0.6
1	0.78
0	0.8
0	0.85

Table 4.1

The corresponding results are described in Section 4.6.

4.5.2.4 An interior penalty method using density jumps

The interior penalty method discussed in Section 4.5.2.3 combines two effects:

(i) It penalises expansion shocks via $[v]^+$

(ii) By virtue of the nonlinear weighting that we have introduced, the regularisation effect is amplified in regions where the Mach number is high, since in these regions ρ is "small".

It is therefore natural to look for a variant of (4.51A) which combines these two effects more closely; we start by noting that for a physical shock we must have (still following the streamlines)

(4.54A) $$[\frac{1}{\rho}] \leq 0.$$

This jump condition (4.54A) leads to the following variant of the entropy functionals E_h and $\tilde{E}_h$ of Sections 4.5.2.2 and 4.5.2.3 (the notation of which has been retained)

(4.55A) $$R_h = \sum_i \sigma_i \left(\left(\frac{w_i}{C_*}\left[-\frac{1}{\rho^\beta}\right]_i\right)^+\right)^{2n} .$$

The notation in (4.55A) is self-explanatory, except for w_i which is defined by

(4.56A) $$w_i = \frac{1}{2}\sum_{j=1}^{2}\left(\frac{\partial\phi_h}{\partial n_i}\right)_j ,$$

the indices j = 1,2 corresponding to the two adjacent triangles of $\mathcal{T}_h$ which have the ith side of the triangulation in common.

It is then quite easy to formulate an approximate problem in which E_h or $\tilde{E}_h$ is replaced by R_h; moreover the same type of iterative methods can be applied to this new approximate problem. Numerical experiments have yet to be carried out to check the validity of this new approach and also to determine an optimal choice for the two exponents β and n in (4.55A).

4.5.2.5 Further comments

The interior penalty methods with truncation that we have discussed in the above sections have been directly inspired by some of the methods used for the numerical treatment of variational inequalities (cf. Chapter 2, Section 3 and also Glowinski /2A/, /3A/). In the present case the problem to be solved definitely lacks those monotonicity properties which are so useful in the theory and approximation of variational inequalities; however the least-squares formulation and also the formulation of an *entropy condition* as a set of inequality relations suggest very clearly a methodology founded on techniques which have proved successful for simpler types of inequality problems.

4.6 Numerical experiments

In this section we shall present some of the numerical results obtained using the above methods. The results of Section 4.6.1 relate to a NACA 0012 aerofoil, and those of Section 4.6.2 to the flow around a two-component aerofoil.

4.6.1 *Simulation of flows around a NACA 0012 aerofoil*

As a first example we consider the flow around a NACA 0012 aerofoil at various angles of attack and various freestream Mach numbers. The corresponding *pressure distributions* are shown in Figures 4.7-4.11, in which the *isomach lines in the supersonic region* are also shown. We observe that the physical shocks are quite well-defined and also that the computed transition from the subsonic region to the supersonic region is smooth and shock-free, implying that the entropy condition has been satisfied. The above numerical results lie very close to those obtained by various authors using finite difference methods (see particularly Jameson /1A/).

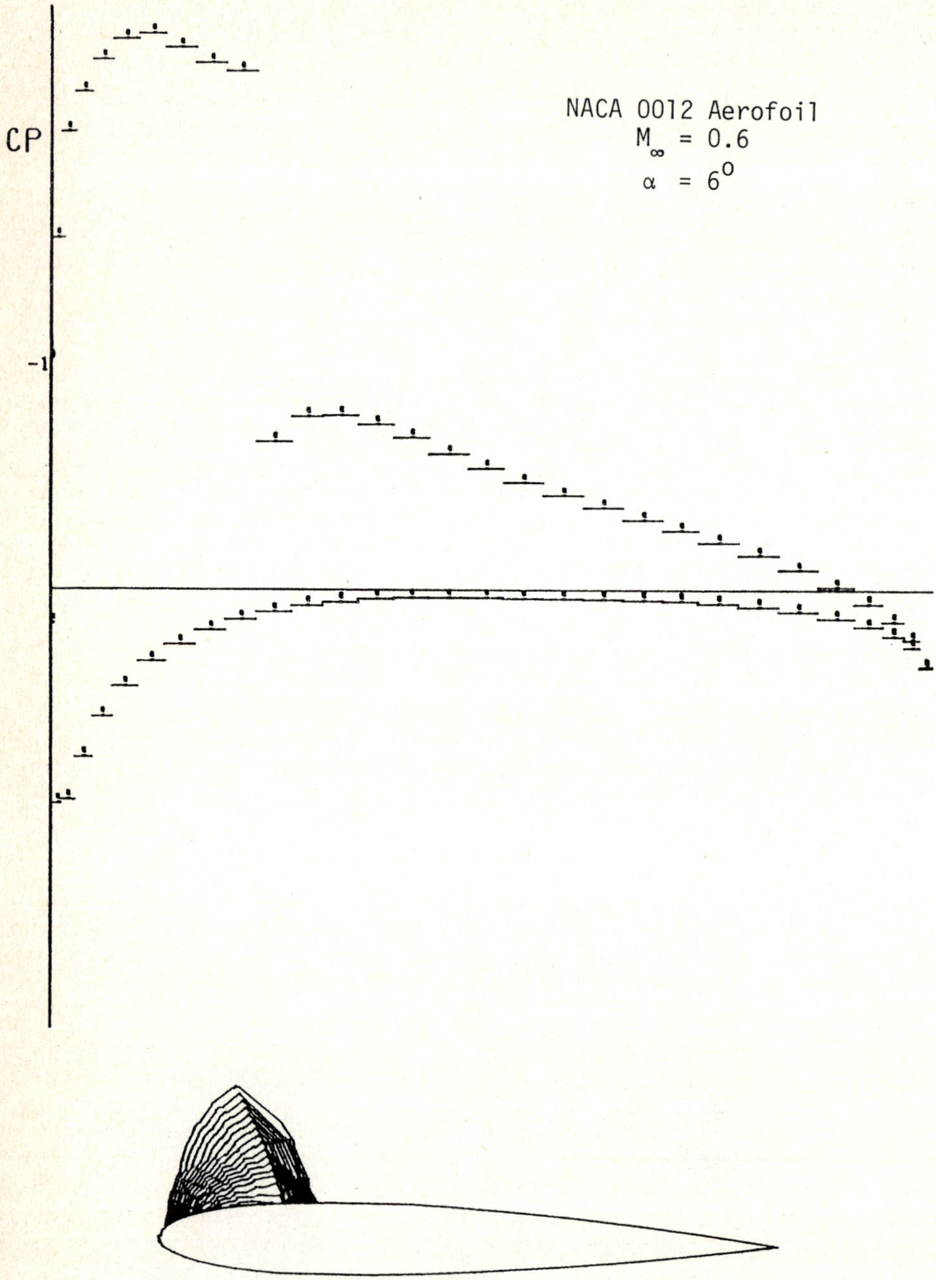

Fig. 4.7

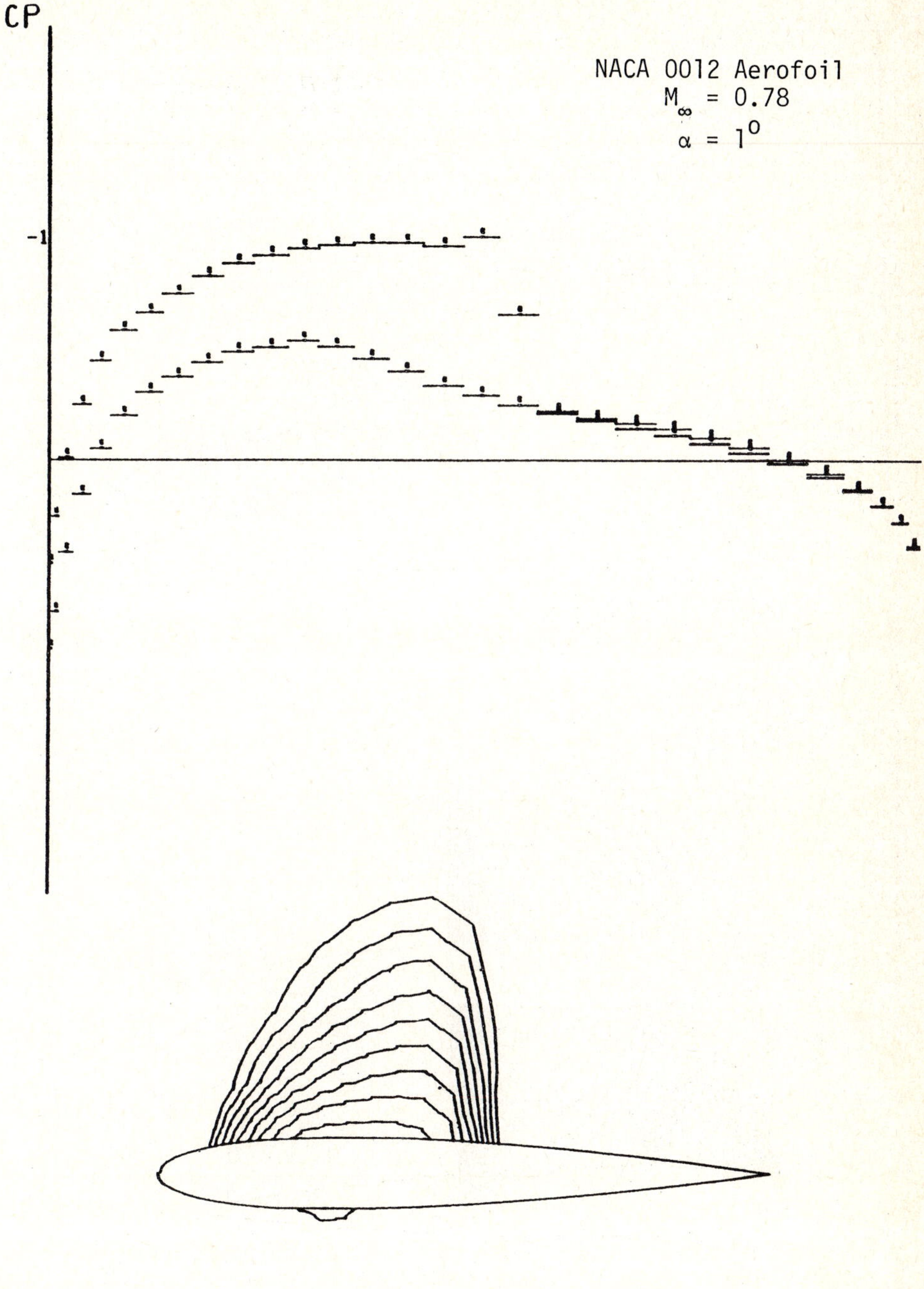

Fig. 4.8

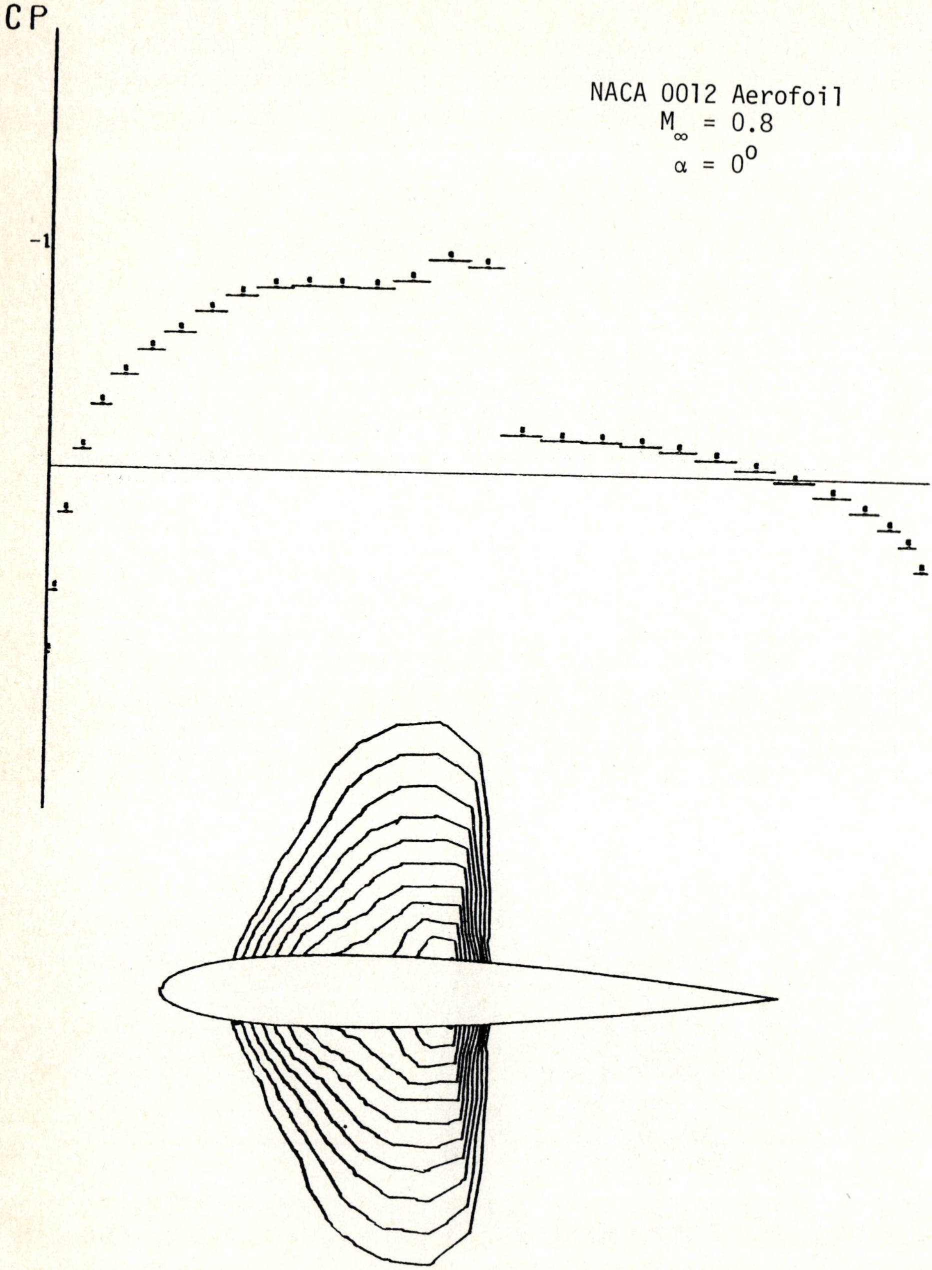

Fig. 4.9

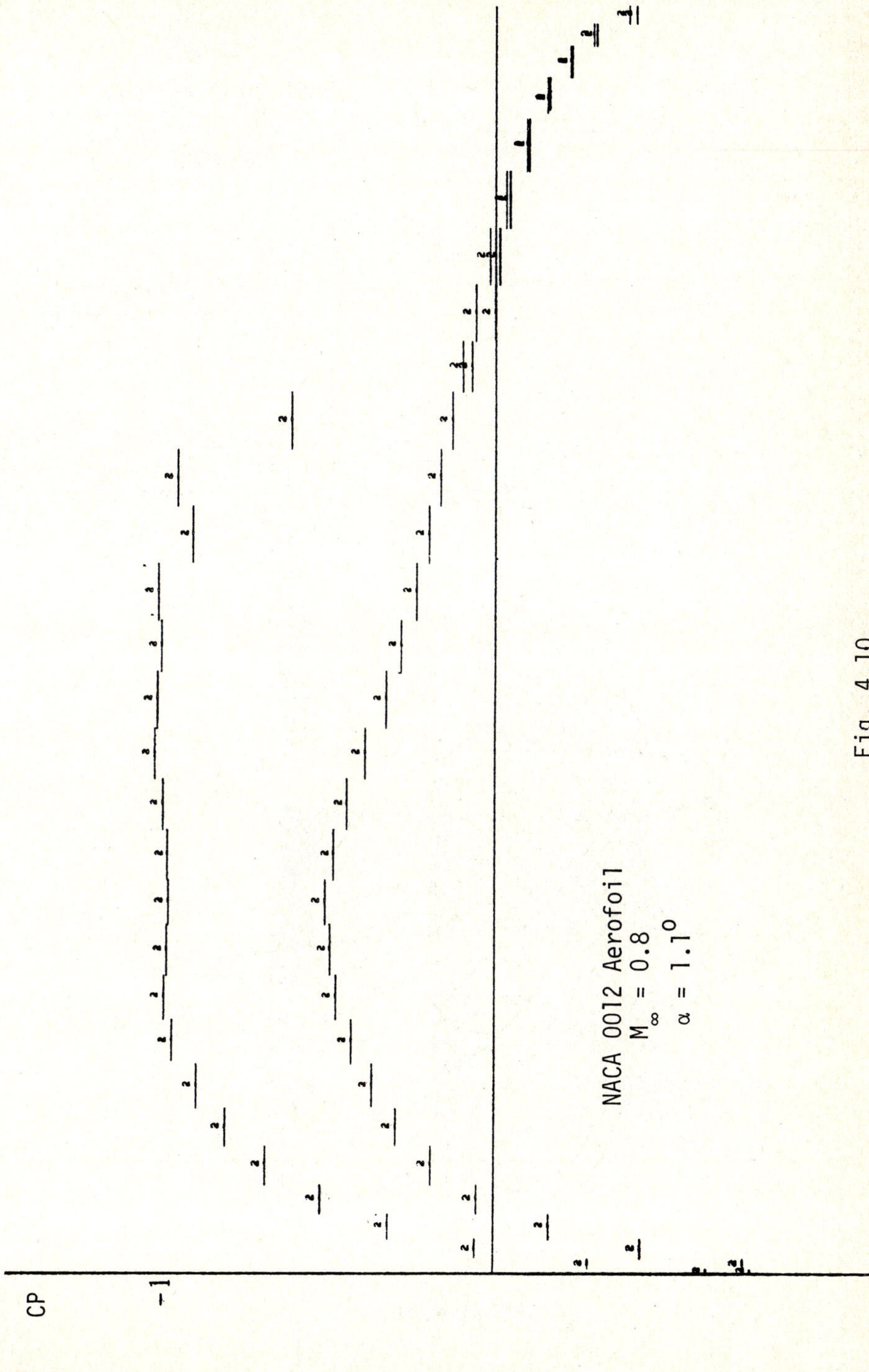

Fig. 4.10

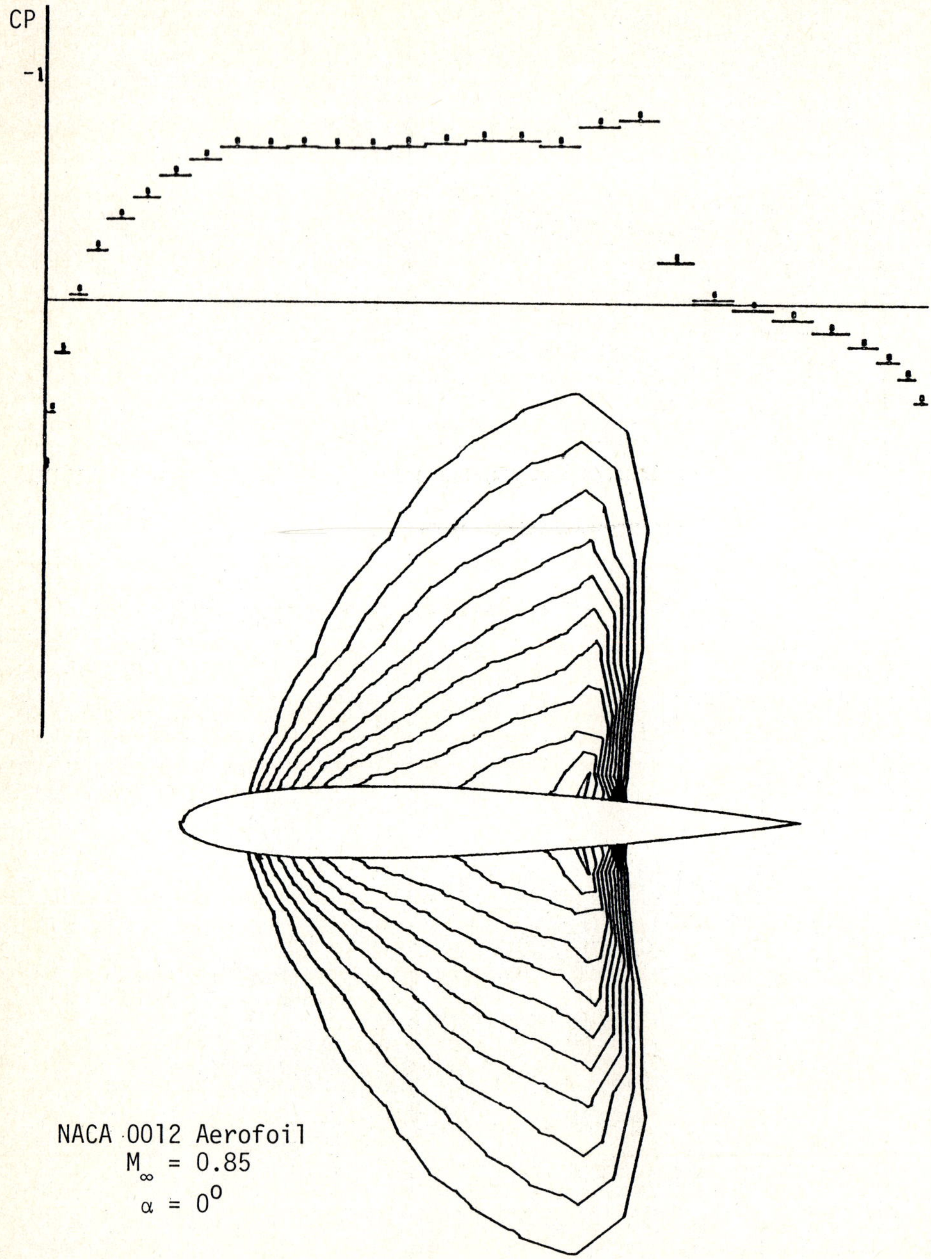

Fig. 4.11

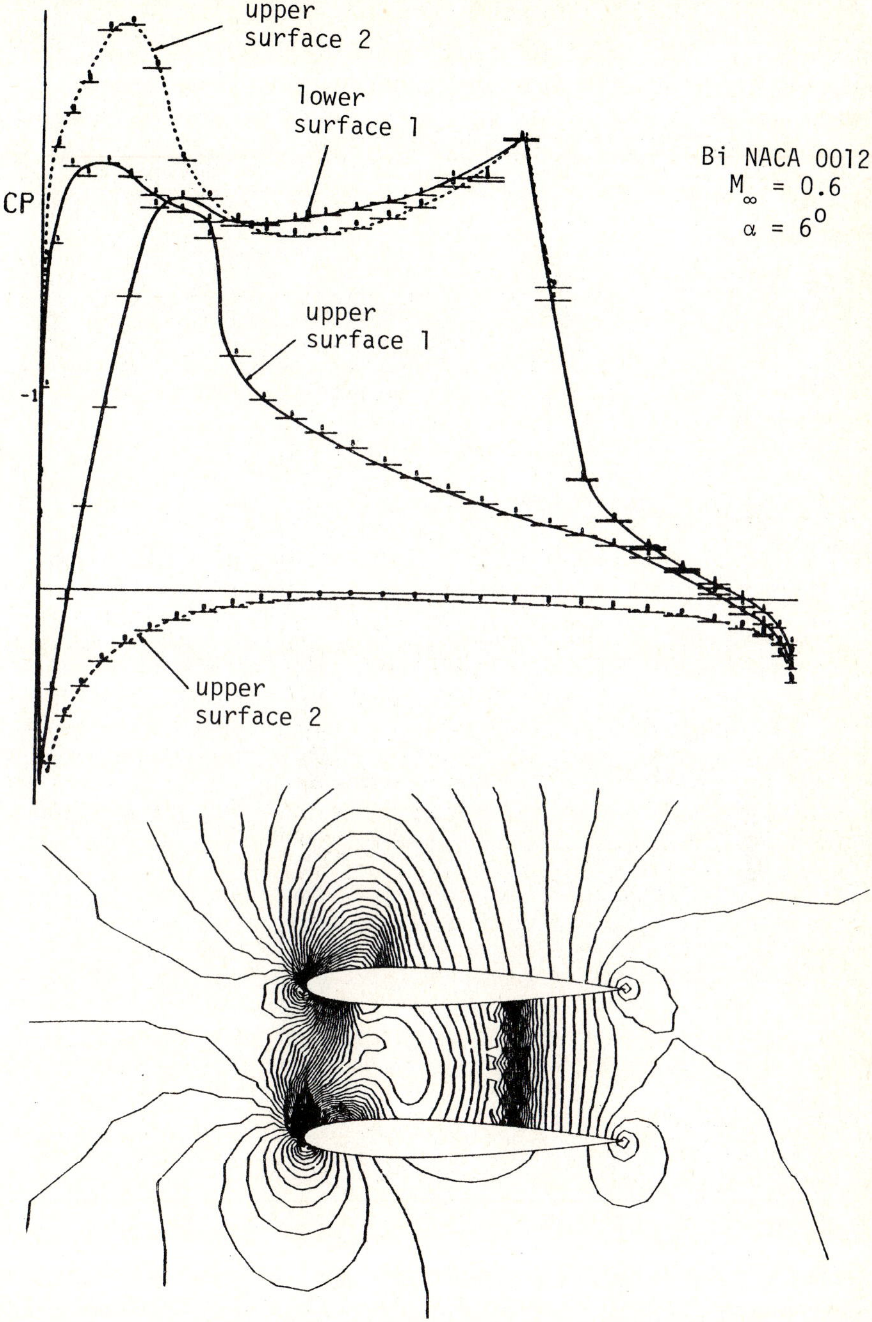
upper
surface 2
lower
surface 1
Bi NACA 0012
$M_\infty = 0.6$
$\alpha = 6^o$
CP
upper
surface 1
-1
upper
surface 2

Fig. 4.12

4.6.2 *Flow around a two-component aerofoil*

The two-component aerofoil investigated is shown on Figure 4.12; each component is a NACA 0012 aerofoil (the upper one being component No. 1). The pressure distribution and the isomach lines are shown on Figure 4.12. We observe that the region between the two aerofoils acts as a nozzle; we also observe two supersonic regions, one between the two aerofoils and one adjacent to the upper surface of body No. 1.

4.6.3 *Concluding remark*

It appears from the above numerical results that the method used leads to sharp shocks and to a smooth transition from the subsonic to the supersonic region.

5. SUPPLEMENTARY BIBLIOGRAPHY

In Section 2 of this appendix we have already given various references relating to the approximation of second-order boundary-unilateral problems using the method of conforming or non-conforming finite elements; it is also appropriate to mention, amongst other references on this and related subjects: Hlavacek /2A/, /3A/, Haslinger /1A/, Hlavacek-Lovisek /1A/, Mosco /1A/, Scarpini-Vivaldi /1A/, Kawohl /1A/, Johnson /1A/ and also the monograph (1) of Kikuchi-Oden /1A/ on the numerical analysis of *contact problems*. Regarding the numerical analysis of fourth-order variational inequalities, we cite Mercier /1A/, /2A/, Brezzi-Johnson-Mercier /1A/, Haslinger /2A/.

The numerical simulation of transonic flows has given rise to a large number of works, and we shall therefore merely cite the references on this subject contained in B.G. 4P /1A/, /2A/, Jameson /1A/, Periaux /1A/ and Hunt /1A/.

---

(1) The substantial bibliography of this work will also reward investigation.

Appendix 5

FURTHER DISCUSSION OF THE NUMERICAL ANALYSIS OF THE STEADY FLOW OF A BINGHAM FLUID IN A CYLINDRICAL DUCT

1. SYNOPSIS

The aim of this Appendix is to supplement Chapter 5 on the Numerical Analysis of the *problem of the steady flow of a Bingham fluid in a cylindrical duct*. We shall give some supplementary results on the *finite-element approximation* of this problem and on its *iterative solution*.

In Section 2 we shall return to approximation by *finite elements of order one* (i.e. *piecewise affine*), previously investigated in Chapter 5, Sections 3 and 5; we shall show that under reasonable assumptions we "almost" have $\|u_h - u\|_{H_0^1(\Omega)} = O(h)$ (1). In Section 3, which follows Falk-Mercier /1A/, it will be shown that the use of an equivalent formulation allows an approximation to be obtained which is of *optimal* order, even if $\Omega \subset \mathbb{R}^2$. Finally, in Section 4, we give supplementary information on the iterative solution of the above problems; in particular we describe conjugate-gradient methods similar to those in Appendix 3, Section 4.3.

To conclude this first section we recall (see Chapter 5, Section 1) that the problem under consideration is defined by

$$(1.1A)\quad \begin{cases} \textit{Find } u \in H_0^1(\Omega) \textit{ such that} \\ a(u,v-u)+gj(v)-gj(u) \geq L(v-u) \quad \forall v \in H_0^1(\Omega) \end{cases}$$

where

$$(1.2A)\quad j(v) = \int_\Omega |\nabla v| \, dx \quad \forall\, v \in H_0^1(\Omega),$$

$$(1.3A)\quad a(v,w) = \mu\int_\Omega \nabla v \cdot \nabla w \, dx \quad \forall\, v,w \in H_0^1(\Omega),$$

$$(1.4A)\quad L(v) = \langle f,v\rangle \quad \forall v \in H_0^1(\Omega), \quad f \in H^{-1}(\Omega) = (H_0^1(\Omega))',$$

---

(1) We actually have $\|u_h - u\|_{H_0^1(\Omega)} = O(h)$ if $\Omega \subset \mathbb{R}$.

and g and μ are *positive constants*.

2. FINITE-ELEMENT APPROXIMATION OF PROBLEM (1.1A). (I) ERROR ESTIMATES FOR PIECEWISE-LINEAR APPROXIMATIONS

In this section, which in part follows Glowinski /1A/, /2A, Chapter 2, Section 6.7/ and /3A/, we shall supplement the results of Chapter 5, Section 5.1.

2.1 One-dimensional case

We take $\Omega =]0,1[$ and assume that in (1.4A) we have $f \in L^2(\Omega)$. Problem (1.1A) can then be written

$$\text{(2.1A)} \quad \begin{cases} \textit{Find } u \in H_0^1(0,1) \textit{ such that } \forall v \in H_0^1(0,1) \\ \mu \int_0^1 \frac{du}{dx}\left(\frac{dv}{dx} - \frac{du}{dx}\right) dx + g\int_0^1 \left|\frac{dv}{dx}\right| dx - g\int_0^1 \left|\frac{du}{dx}\right| dx \\ \qquad \geq \int_0^1 f(v-u)\,dx. \end{cases}$$

Let N be an integer >0 and let $h = \frac{1}{N}$; we consider $x_i = ih$ for $i=0,1,\ldots N$ and

$$e_i = [x_{i-1}, x_i], \quad i=1,2,\ldots N.$$

We then approximate $H_0^1(0,1)$ by

$$\text{(2.2A)} \qquad V_{oh} = \{v_h \in C^0[0,1],\ v_h(0)=v_h(1)=0,\ v_h|_{e_i} \in P_1,\ i=1,2,\ldots N\}. \quad (^1)$$

The *approximate problem* is then defined by

$$\text{(2.3A)} \quad \begin{cases} \textit{Find } u_h \in V_{oh} \textit{ such that } \forall v_h \in K_h \\ \mu\int_0^1 \frac{d}{dx}u_h \frac{d}{dx}(v_h-u_h)\,dx + gj(v_h) - gj(u_h) \geq \int_0^1 f(v_h-u_h)\,dx. \end{cases}$$

---

(1) P_1 = space of polynomials in one variable of degree ≤ 1.

Problem (2.3A) admits a unique solution. Concerning the approximation error, we have

Theorem 2.1: *Let* u *and* u_h *be the respective solutions of* (2.1A) *and* (2.3A). *If* $f \in L^2(0,1)$ *then we have*

$$\|u_h - u\|_{H_0^1(0,1)} = O(h). \tag{2.4A}$$

Proof: We recall that $V_{oh} \subset H_0^1(0,1) \subset C^0[0,1]$; let $r_h u$ be the V_{oh}-interpolate of u, i.e. let

$$\begin{cases} r_h u \in V_{oh} \\ r_h u(x_i) = u(x_i) \quad \forall x_i, \quad i=1,2,\dots N. \end{cases} \tag{2.5A}$$

Proceeding as in Chapter 5, Section 5.1 it can readily be shown that

$$a(u_h-u,u_h-u) \le a(u_h,r_hu-u)+g(j(r_hu)-j(u)) - \int_0^1 f(r_hu-u)\,dx\ ; \tag{2.6A}$$

we deduce from (2.6A) that

$$a(u_h-u,u_h-u) \le a(u_h-u,r_hu-u)+g(j(r_hu)-j(u))+a(u,r_hu-u) - \int_0^1 f(r_hu-u)\,dx,$$

or in a more explicit form (after using Schwarz's inequality in $H_0^1(0,1)$ and from the fact that $2|x||y| \le x^2+y^2 \quad \forall x,y \in \mathbb{R}$):

$$\begin{cases} \frac{\mu}{2}\|u_h-u\|^2_{H_0^1(0,1)} \le \frac{\mu}{2}\|r_hu-u\|^2_{H_0^1(0,1)} + g(j(r_hu)-j(u)) + \\ \qquad + \mu\int_0^1 \frac{du}{dx}\frac{d}{dx}(r_hu-u)\,dx - \int_0^1 f(r_hu-u)\,dx. \end{cases} \tag{2.7A}$$

The condition $f \in L^2(0,1)$ implies (see Chapter 5, Section 2.1 and Brézis /6/) that

$$u \in H^2(0,1) \cap H_0^1(0,1)\ , \tag{2.8A}$$

with in fact

$$\left\|\frac{d^2u}{dx^2}\right\|_{L^2(0,1)} \leq \frac{1}{\mu}\|f\|_{L^2(0,1)} \quad ; \tag{2.9A}$$

the proof of this is left as an exercise for the reader.

By integrating by parts, we deduce from (2.7A)-(2.9A)

$$\left\{\begin{aligned} \frac{\mu}{2}\|u_h-u\|^2_{H^1_0(0,1)} &\leq \frac{\mu}{2}\|r_hu-u\|^2_{H^1_0(0,1)} + g(j(r_hu)-j(u)) \\ &\quad + \int_0^1 \left(-\mu\frac{d^2u}{dx^2} - f\right)(r_hu-u)\,dx \\ &\leq \frac{\mu}{2}\|r_hu-u\|^2_{H^1_0(0,1)} + g(j(r_hu)-j(u)) \\ &\quad + 2\|f\|_{L^2(0,1)}\|r_hu-u\|_{L^2(0,1)} . \end{aligned}\right. \tag{2.10A}$$

In view of (2.8A) we have

$$\|r_hu-u\|_{H^1_0(0,1)} \leq Ch\|u\|_{H^2(0,1)} \quad , \tag{2.11A}$$

$$\|r_hu-u\|_{L^2(0,1)} \leq Ch^2\|u\|_{H^2(0,1)} \quad . \tag{2.12A}$$

The estimate (2.4A) will thus follow from (2.10A)-(2.12A) if we can succeed in proving that $j(r_hu) \leq j(u) \quad \forall h$; we have

$$j(r_hu) = \int_0^1 \left|\frac{d}{dx}(r_hu)\right|dx = \sum_{i=1}^N \int_{x_{i-1}}^{x_i} \left|\frac{d}{dx}(r_hu)\right| dx. \tag{2.13A}$$

However, on $]x_{i-1},x_i[$ we have

$$\frac{d}{dx}(r_hu) = \frac{u(x_i)-u(x_{i-1})}{h} = \frac{1}{h}\int_{x_{i-1}}^{x_i} \frac{du}{d\xi}\,d\xi \quad . \tag{2.14A}$$

We deduce from (2.14A) that, *on* $]x_{i-1},x_i[$,

(2.15A) $$\left|\frac{d}{dx}(r_h u)\right| = \frac{1}{h}\left|\int_{x_{i-1}}^{x_i} \frac{du}{d\xi}\, d\xi\right| \le \frac{1}{h}\int_{x_{i-1}}^{x_i} \left|\frac{du}{d\xi}\right| d\xi$$

and hence by integrating the inequality (2.15A)

$$j(r_h u) = \int_0^1 \left|\frac{d}{dx}(r_h u)\right| dx \le \sum_{i=1}^{N}\int_{x_{i-1}}^{x_i} \left|\frac{du}{d\xi}\right| d\xi = \int_0^1 \left|\frac{du}{d\xi}\right| d\xi = j(u),$$

which is exactly the inequality we were trying to prove.

2.2 Two-dimensional case

If $\Omega \subset \mathbb{R}^2$, then a somewhat more complicated argument (which among other things makes use of the characterisation (3.48)-(3.51) of Chapter 1, Section 3.4 and (7.1), (7.2) of Chapter 5, Section 7.1) shows that if suitable regularity assumptions [1] for u and for the *free boundary* [2] are satisfied, then for the first-order finite-element approximation of Chapter 5, Section 3, and using the same assumptions on $\mathcal{C}_h$ as in Theorem 5.1 of Chapter 5, Section 5.1, we have

(2.16A) $$\|u_h - u\|_{H_0^1(\Omega)} = O(h|\mathrm{Log}\ h|^{1/2})$$

or "almost" O(h); we refer to Glowinski /1A/, /2A/, /3A/ for the derivation of the estimate (2.16A).

3. FINITE-ELEMENT APPROXIMATIONS OF PROBLEM (1.1A). (II) OPTIMAL-ORDER ERROR ESTIMATES THROUGH THE USE OF AN EQUIVALENT FORMULATION

In this section we follow the account of Falk-Mercier /1A/; we show that the use of an equivalent formulation enables us to obtain finite-element approximations which lead to error estimates *of optimal order*. In view of the analogies which exist with Section 3

(1) satisfied in particular if Ω is a disc and if f = const.

(2) i.e. the interface between $\Omega^o = \{x \in \Omega,\ |\nabla u(x)| = 0\}$ and $\Omega^+ = \{x \in \Omega,\ |\nabla u(x)| > 0\}$.

of Appendix 3, many of the results will be stated without proof.

3.1 Equivalent formulation of problem (1.1A)

The following equivalence lemma forms the fundamental result of this third section; its proof is a variation of that of Lemma 3.1 of Appendix 3, Section 3.1, the notation of which is retained.

Lemma 3.1: *Suppose that* Ω *is a simply-connected bounded domain in* $\mathbb{R}^2$; *there is then equivalence between problem* (1.1A) *and the variational problem*

$$(3.1A)\quad \begin{cases} \text{Find } p \in H \text{ such that } \forall q \in H, \\ \mu\int_\Omega p\cdot(q-p)\,dx + g\int_\Omega |q|\,dx - g\int_\Omega |p|\,dx \geq \int_\Omega \phi\cdot(q-p)\,dx \end{cases}$$

where $\phi = \{\phi_1, \phi_2\}$ *is an (arbitrary) solution of*

$$(3.2A)\quad f = \frac{\partial\phi_1}{\partial x_2} - \frac{\partial\phi_2}{\partial x_1},$$

$$(3.3A)\quad H = \{q \in L^2(\Omega)\times L^2(\Omega),\ \int_\Omega q\cdot\nabla w\,dx = 0\quad \forall w \in H^1(\Omega)\}.$$

The solutions u *of* (1.1A) *and* p *of* (3.1A) *are connected by the relation*

$$(3.4A)\quad p = \{\frac{\partial u}{\partial x_2}, -\frac{\partial u}{\partial x_1}\}. \ \blacksquare$$

Remark 3.1: We consider the vector $\underset{\sim}{U} = \{0,0,u\}$ in $\mathbb{R}^3$ and $\underset{\sim}{\nabla} = \{\frac{\partial}{\partial x_1}, \frac{\partial}{\partial x_2}, \frac{\partial}{\partial x_3}\}$; we then have $\underset{\sim}{\omega} = \underset{\sim}{\nabla}\times\underset{\sim}{U} = \{\frac{\partial u}{\partial x_2}, -\frac{\partial u}{\partial x_1}, 0\}$.

In view of (3.4A) we can therefore regard p as being a *vector-valued vorticity function* associated with the velocity field defined by u.

Remark 3.2: Remark 3.2 of Appendix 3, Section 3.1 relating to the construction of ϕ from f is also applicable here. ■

Remark 3.3: The regularity results relating to problem (1.1A) (proved in Brézis /6/ and referred to in Chapter 5, Section 2.1), together with relation (3.4A), imply that if Γ is sufficiently regular, then for the solution p of problem (3.1A) we have

$$f \in L^2(\Omega) \Longrightarrow p \in (H^1(\Omega))^2 \ . \qquad (3.5A)$$ ■

Remark 3.4: This remark concerns the existence of *Lagrange multipliers* associated with the condition $p \in H$ for problem (3.1A); it follows from (3.1A), (3.3A) that a natural Lagrangian associated with problem (3.1A) is $\mathcal{L} : L^2(\Omega))^2 \times H^1(\Omega) \to \mathbb{R}$, defined by

$$\mathcal{L}(q,w) = \frac{\mu}{2}\int_\Omega |q|^2 dx + g\int_\Omega |q|\,dx - \int_\Omega \phi\cdot q\,dx - \int_\Omega q\cdot\nabla w\,dx\ , \qquad (3.6A)$$

which leads to the saddle-point problem

$$\begin{cases} \textit{Find } \{p,\chi\} \in (L^2(\Omega))^2 \times H^1(\Omega) \textit{ such that} \\ \mathcal{L}(p,w) \le \mathcal{L}(p,\chi) \le \mathcal{L}(q,\chi) \quad \forall \{q,w\} \in (L^2(\Omega))^2 \times H^1(\Omega)\ . \end{cases} \qquad (3.7A)$$

We then have the following proposition, the proof of which is an immediate variant of that of Proposition 3.1 in Appendix 3, Section 3.1:

Proposition 3.1: *There is equivalence between* (3.7A) *and*

$$\begin{cases} \{p,\chi\} \in H\times H^1(\Omega), \\ \mu\int_\Omega p\cdot(q-p)\,dx + g\left(\int_\Omega |q|\,dx - \int_\Omega |p|\,dx\right) \\ \qquad \ge \int_\Omega (\phi+\nabla\chi)\cdot(q-p)\,dx \quad \forall q \in (L^2(\Omega))^2; \end{cases} \qquad (3.8A)$$

moreover if $\{p,\chi\}$ *is a solution of* (3.7A), (3.8A), *then* p *is the solution of problem* (3.1A).

In contrast to problems (3.20A), (3.21A) of Appendix 3, Section 3.1, which are formally similar, the existence of χ such that $\{p,\chi\}$ is a solution of problems (3.7A), (3.8A) poses no major difficulty. Hence we have

Proposition 3.2: *Using the notation of Proposition* 3.1 *above, we have the existence of* $\chi \in H^1(\Omega)$, *such that* $\{p,\chi\}$ *is a solution of problems* (3.7A), (3.8A).

Proof: If we assume the existence of χ, then χ is a *solution of the dual problem*

$$\text{(3.9A)} \qquad \underset{w\in H^1(\Omega)}{\text{Max}} \ \underset{q\in(L^2(\Omega))^2}{\text{Min}} \mathcal{L}(q,w) \ ;$$

the converse is also true.

We denote by p(w) the (unique) solution of the problem

$$\text{(3.10A)} \qquad \underset{q\in(L^2(\Omega))^2}{\text{Min}} \mathcal{L}(q,w) \ ;$$

it can easily be shown that [1]

$$\text{(3.11A)} \qquad p(w) = \frac{1}{\mu}(|\phi+\nabla w|-g)^+ \frac{\phi+\nabla w}{|\phi+\nabla w|} \ .$$

If we define $\mathcal{M} : H^1(\Omega) \to \mathbb{R}$ by

$$\text{(3.12A)} \qquad \mathcal{M}(w) = -\mathcal{L}(p(w),w),$$

then we may deduce from (3.6A), (3.9A)-(3.12A) that the dual problem (3.9A) admits the formulation

$$\text{(3.13A)} \qquad \underset{w\in H^1(\Omega)}{\text{Min}} \ \mathcal{M}(w)$$

with

---

[1] we use the notation $t^+ = \sup(0,t), \ \forall\, t\in\mathbb{R}$.

(3.14A) $$\mathcal{M}(w) = \frac{1}{2\mu} \|(|\phi+\nabla w|-g)^+\|^2_{L^2(\Omega)} .$$

It follows from (3.14A) that $\mathcal{M}$ is *convex* (*but not strictly convex*) and *continuous* on $H^1(\Omega)$; moreover we have

(3.15A) $$\lim_{\|w\|_{\dot{H}^1(\Omega)} \to +\infty} \mathcal{M}(w) = +\infty,$$

where, in (3.15A), we have used the notation $\dot{H}^1(\Omega) = H^1(\Omega)/\mathbb{R}$. The above properties of $\mathcal{M}$ imply the existence of solutions of the dual problem (3.9A), to within an additive constant.

3.2 Approximation of problem (3.1A) by a finite-element method

We assume hereafter that Ω is a bounded, convex, polygonal domain in $\mathbb{R}^2$.

3.2.1 *Definition of the approximate problem*

Using the notation of Appendix 3, Section 3.1, we approximate (3.1A) by

(3.16A) $$\begin{cases} \textit{Find } p_h \in H_h \textit{ such that } \forall q_h \in H_h \\ \mu\int_\Omega p_h\cdot(q_h-p_h)+g\left(\int_\Omega |q_h|\,dx - \int_\Omega |p_h|\,dx\right) \geq \int_\Omega \phi\cdot(q_h-p_h)\,dx. \end{cases}$$

It can readily be proved that

Proposition 3.3: *Problem* (3.16A) *admits a unique solution.*

3.2.2 *Investigation of the convergence*

We shall assume hereafter that $f \in L^2(\Omega)$; with Ω *convex* and *bounded*, this implies (see Brézis /6/) that the solution u of (1.1A) satisfies $u \in H^2(\Omega)$ with

(3.17A) $$\|u\|_{H^2(\Omega)} \leq \frac{\gamma_o}{\mu} \|f\|_{L^2(\Omega)}$$

where γ_o depends only on Ω. We deduce from (3.4A), (3.17A) that the solution p of (3.1A) satisfies

$$\text{(3.18A)} \quad \begin{cases} p \in H^1(\Omega) \times H^1(\Omega) \\ \text{with } \|p\|_{(H^1(\Omega))^2} \le \frac{C_o}{\mu} \|f\|_{L^2(\Omega)} . \end{cases}$$

Similarly if ϕ is obtained from f using the relations of Remark 3.2 in Appendix 3, Section 3.1, then

$$\text{(3.19A)} \quad \phi \in (H^1(\Omega))^2$$

with

$$\text{(3.20A)} \quad \|\phi\|_{(H^1(\Omega))^2} \le C \|f\|_{L^2(\Omega)} .$$

We also assume the existence of a multiplier $\chi \in H^2(\Omega)$ such that $\{p,\chi\}$ satisfies (3.7A), (3.8A).

Hereafter we shall use π_h to denote the orthogonal projection operator onto L_h, in the norm $((L^2(\Omega))^2$; we refer to Propositions 3.3 and 3.4 of Appendix 3, Section 3.2.2 for the various properties of π_h, which are useful in investigating the convergence of the solution p_h of (3.16A) to the solution p of (3.1A). More specifically, with regard to this convergence we have

Theorem 3.1 (Falk-Mercier /1A/): *Under the above assumptions on* Ω, f, χ *and with the same assumptions on* $\mathcal{T}_h$ *as in Theorem* 3.1 *of Appendix* 3, *Section* 3.2.2, *we have the optimal-order error estimate*

$$\text{(3.21A)} \quad \|p_h - p\|_{(L^2(\Omega))^2} = O(h),$$

where, in (3.21A), p is the solution of problem (3.1A) and p_h that of the approximate problem (3.16A).

Proof: Proceeding as in the proof of Theorem 3.1 of Appendix 3, Section 3.2.2, we obtain the estimate

$$
(3.22A)\quad \begin{cases} \frac{\mu}{2}\|p_h-p\|_{(L^2(\Omega))^2} \\ \\ \leq c^2h^2(\|f\|^2_{L^2(\Omega)}+\|\chi\|^2_{H^2(\Omega)})+g(\int_\Omega |\pi_h p|\,dx - \int_\Omega |p|\,dx). \end{cases}
$$

The theorem will be proved if

$$
(3.23A)\qquad \int_\Omega |\pi_h p|\,dx \leq \int_\Omega |p|\,dx \quad \forall h.
$$

We have

$$
(3.24A)\qquad \int_\Omega |\pi_h p|\,dx = \sum_{T\in \mathcal{C}_h} \int_T |\pi_h p|\,dx ,
$$

but

$$
(3.25A)\qquad \pi_h p|_T = \frac{1}{m(T)} \int_T p\,dx
$$

where, in (3.25A), $m(T)$ = measure of T. We thus deduce

$$
(3.26A)\qquad |(\pi_h p|_T)| = |\pi_h p||_T \leq \frac{1}{m(T)} \int_T |p|\,dx .
$$

The inequality (3.23A) then results trivially from (3.24A), (3.26A).

4. SUPPLEMENTARY INFORMATION ON ITERATIVE METHODS FOR SOLVING THE PROBLEM OF THE STEADY FLOW OF A BINGHAM FLUID IN A CYLINDRICAL DUCT

4.1 General discussion. Synopsis

In Chapter 5 we described various iterative methods for solving problem (1.1A) in its direct formulation; in this case, the most efficient calculation methods again seem to be those based on algorithm (9.6)-(9.8) of Chapter 5, Section 9 (of the augmented Lagrangian type); for further details we refer to Marrocco /1A/,

Gabay-Mercier /1/, Fortin-Glowinski /1A/, /2A/, Glowinski /2A/, Glowinski-Marrocco /1A/.

Furthermore, it is extremely likely that we could improve the speed of convergence of algorithm (6.93)-(6.96) of Chapter 5, Section 6.4 (which relates to the regularised problems (6.3)) by using a variant of the conjugate-gradient type (and thus with preconditioning in the sense of Appendix 3, Section 4.3.3); the algorithm used would then be similar to the algorithms (4.34A)-(4.42A) of Appendix 3, Section 4.3.2.

In Sections 4.2, 4.3 below, we shall conclude this fourth section with an investigation of the *iterative solution of* (1.1A) via the equivalent formulation (3.1A), using duality-type algorithms based on the formulation (3.7A); in Section 4.2 we describe a Uzawa-type algorithm, and variants of the conjugate-gradient type are described in Section 4.3.

4.2 Solution of problem (3.1A) by a duality method

The material discussed in this section is similar to that in Appendix 3, Section 4.2.

4.2.1 *Description of the algorithm*

In view of the duality results of Remark 3.4 of Section 3.1 of this appendix we shall reduce the solution of (3.1A) (and hence of (1.1A)) to that of the saddle-point problem (3.7A), namely

$$\text{(4.1A)} \quad \begin{cases} \textit{Find } \{p,\chi\} \in (L^2(\Omega))^2 \times H^1(\Omega) \textit{ such that} \\ \mathcal{L}(p,w) \le \mathcal{L}(p,\chi) \le \mathcal{L}(q,\chi) \quad \forall \{q,w\} \in (L^2(\Omega))^2 \times H^1(\Omega) \end{cases}$$

with

$$\text{(4.2A)} \qquad \mathcal{L}(q,w) = \frac{\mu}{2}\int_\Omega |q|^2 dx + g\int_\Omega |q|\,dx - \int_\Omega \phi\cdot q\,dx - \int_\Omega q\cdot\nabla w\,dx\ ,$$

which is justified since p *is then the solution of* (3.1A). By analogy with Appendix 3, Section 4.2.1, we are led to the following algorithm:

(4.3A) $\chi^o \in H^1(\Omega)$ *given* ($\chi^o = 0$ *for example*)

then for $n \geq 0$, $\chi^n \in H^1(\Omega)$ *known, determine* $p^n \in (L^2(\Omega))^2$ *and* $\chi^{n+1} \in H^1(\Omega)$ *by*

$$\text{(4.4A)} \quad \begin{cases} p^n \in (L^2(\Omega))^2, \\ \mathcal{L}(p^n,\chi^n) \leq \mathcal{L}(q,\chi^n) \quad \forall q \in (L^2(\Omega))^2, \end{cases}$$

$$\text{(4.5A)} \quad \begin{cases} \chi^{n+1} \in H^1(\Omega), \\ (\chi^{n+1},w)_{H^1(\Omega)} = (\chi^n,w)_{H^1(\Omega)} - \rho\int_\Omega \nabla w\cdot p^n dx \quad \forall w \in H^1(\Omega) \end{cases}$$

with, in (4.5A), $\rho > 0$.

The practical implementation of (4.3A)-(4.5A) does not pose any major difficulties; in fact (4.4A) is equivalent (see (3.11A) of Section 3.1) to

$$\text{(4.6A)} \quad p^n = p(\chi^n) = \frac{1}{\mu}\,(|\phi+\nabla\chi^n|-g)^+ \;\frac{\phi+\nabla\chi^n}{|\phi+\nabla\chi^n|}\,.$$

In connection with (4.5A), we refer to the comments in Appendix 3, Section 4.2.1.

4.2.2 *Convergence of algorithm* (4.3A)-(4.5A)

By proceeding as in Appendix 3, Section 4.2.2, the following theorem on the convergence of algorithm (4.3A)-(4.5A) can be proved.

Theorem 4.1: *Suppose that* $(\cdot,\cdot)_{H^1(\Omega)}$ *is defined by*

$$(v,w)_{H^1(\Omega)} = \int_\Omega (\nabla \mathbf{v}\cdot\nabla w+vw)\,dx\ ;$$

then if

(4.7A) $0 < \rho < 2\mu$

then for the sequence $\{p^n\}_{n\geq 0}$ *defined by algorithm* (4.3A)-(4.5A) *we have*

(4.8A) $$\lim_{n\to+\infty} \|p^n - p\|_{L^2(\Omega)\times L^2(\Omega)} = 0 \quad \forall \chi^o \in H^1(\Omega)$$

where, in (4.8A), p *is the solution of* (3.1A).

Remarks 4.1, 4.2 and 4.3 of Appendix 3, Section 4.2.2 and 4.3.2 also hold for algorithm (4.3A)-(4.5A).

4.3 On conjugate-gradient type variants of algorithm (4.3A)-(4.5A).

As previously mentioned in Appendix 3, Section 4.3.1 in relation to a similar problem, we might expect to accelerate the convergence of algorithm (4.3A)-(4.5A) by using a variant of conjugate-gradient type. In fact this possibility is justified by

Proposition 4.1: *The functional* $\mathcal{M} : H^1(\Omega) \to \mathbf{R}$ *defined by* (3.14A) *is convex, Lipschitz continuous and Gateaux-differentiable; its Gateaux differential* $\mathcal{M}'$ *is also Lipschitz continuous and is defined by*

(4.9A) $$\langle \mathcal{M}'(w), \eta \rangle = \int_\Omega p(w)\cdot\nabla\eta \, dx \quad \forall w, \eta \in H^1(\Omega)$$

with (see (3.11A))

(4.10A) $$p(w) = \frac{1}{\mu}\left(|\phi+\nabla w| - g\right)^+ \frac{\phi+\nabla w}{|\phi+\nabla w|} \quad ;$$

in (4.9A) $\langle\cdot,\cdot\rangle$ *denotes the bilinear form of the duality between* $(H^1(\Omega))'$ *and* $H^1(\Omega)$.

By proceeding as in Appendix 3, Section 4.3.2, we are led to the following conjugate-gradient algorithms:

Stage 0: Initialisation

(4.11A) $\chi^o \in H^1(\Omega)$, *given*,

then determine g^o *using*

(4.12A)
$$\begin{cases} g^o \in H^1(\Omega), \\ (g^o, w)_{H^1(\Omega)} = \langle \mathcal{M}'(\chi^o), w\rangle \quad \forall w \in H^1(\Omega), \end{cases}$$

and put

(4.13A) $z^o = g^o$.

Next, for $n \geq 0$ *and assuming that* χ^n, g^n, z^n *are known, determine* $\chi^{n+1}, g^{n+1}, z^{n+1}$ *using*

Stage 1: Descent

Solve the one-variable minimisation problem

(4.14A)
$$\begin{cases} \textit{Find } \lambda^n \in \mathbb{R} \textit{ such that} \\ \mathcal{M}(\chi^n - \lambda^n z^n) \leq \mathcal{M}(\chi^n - \lambda z^n) \quad \forall \lambda \in \mathbb{R} \end{cases}$$

and put

(4.15A) $\chi^{n+1} = \chi^n - \lambda^n z^n$.

Stage 2: Construction of the new direction of descent

Define g^{n+1} *by*

(4.16A)
$$\begin{cases} g^{n+1} \in H^1(\Omega), \\ (g^{n+1}, w)_{H^1(\Omega)} = \langle \mathcal{M}'(\chi^{n+1}), w\rangle \quad \forall w \in H^1(\Omega), \end{cases}$$

and then γ^{n+1} *by*

$$(4.17A)_1 \qquad \gamma^{n+1} = \frac{\|g^{n+1}\|^2_{H^1(\Omega)}}{\|g^n\|^2_{H^1(\Omega)}}$$

(*or alternatively by*

$$(4.17A)_2 \qquad \gamma^{n+1} = \frac{(g^{n+1}, g^{n+1}-g^n)_{H^1(\Omega)}}{(g^n, g^n)_{H^1(\Omega)}} \,)$$

and finally

$$(4.18A) \qquad z^{n+1} = g^{n+1} + \gamma^{n+1} z^n.$$

Naturally, the next step is

$$(4.19A) \qquad n = n+1, \text{ go to } (4.14A).$$

Remarks 4.4, 4.5 and 4.6 of Appendix 3, Section 4.3.2 also apply to the above algorithm (4.11A)-(4.19A); in particular g^{n+1} is determined from χ^{n+1} via (4.16A) and the following relations

$$(4.20A) \qquad \langle \mathcal{M}'(\chi^{n+1}), w\rangle = \int_\Omega p^{n+1} \cdot \nabla w \, dx \quad \forall w \in H^1(\Omega),$$

where

$$(4.21A) \qquad p^{n+1} = p(\chi^{n+1}) = \frac{1}{\mu} (|\phi + \nabla\chi^{n+1}| - g)^+ \frac{\phi + \nabla\chi^{n+1}}{|\phi + \nabla\chi^{n+1}|} \,.$$

Appendix 6

FURTHER DISCUSSION OF THE NUMERICAL ANALYSIS OF TIME-DEPENDENT VARIATIONAL INEQUALITIES

1. SYNOPSIS

This appendix is intended to provide a fairly brief supplement to Chapter 6 on the *numerical analysis of time-dependent inequalities*. In Section 2 we give a number of references concerning the approximation of time-dependent inequalities, mainly in the modelling of problems related to the mechanics of continuous media. Sections 3 and 4 supplement the discussion in Chapter 6, Section 11 concerning the *time-dependent flow of incompressible Bingham fluids*. Section 3 considers the case of the flow in a cylindrical duct, and thus forms an extension to Chapter 5 and Appendix 5 of this book. In Section 4 we return to the numerical investigation of Chapter 6, Section 11, concerning the *flow in a two-dimensional cavity*; by introducing a *stream function* we are led to a *time-dependent variational inequality* problem of *parabolic* type, which is of *fourth order with respect to the space variables*, and which we shall discretise (spatially) by a *method of mixed finite elements* similar to that discussed in Chapter 4, Section 4.5 and in Appendix 4, Section 3.

2. SUPPLEMENTARY BIBLIOGRAPHY

The numerical analysis of problems which arise in general from the *mechanics of continuous media* and which can be reduced to *time-dependent inequalities* has given rise to a number of works, among which we cite the following:

a) Johnson /2A/, Berger /1A/, Berger-Falk /1A/ concerning the approximation in space and time of the *model time-dependent obstacle problem*, which in the notation of Chapter 6, Sections 1 and 2 can be written

$$(2.1A)\quad \begin{cases} \textit{Find } u(t) \in K \textit{ a.e. on }]0,T[, \quad \textit{such that a.e. on }]0,T[\\ (\frac{\partial u}{\partial t}, v-u) + \int_\Omega \nabla u \cdot \nabla (v-u)\, dx \geq (f, v-u) \quad \forall v \in K, \\ u(0) = u_o , \end{cases}$$

where

(2.2A) $$K = \{v \in H_o^1(\Omega),\ v \geq \psi \ \ a.e.\ on\ \Omega\}$$

where $\psi \in H^1(\Omega) \cap C^o(\bar{\Omega})$ with $\psi \leq 0$ on $\partial\Omega$.

We assume that Ω is a bounded polygonal domain in $\mathbb{R}^2$ and that $(\mathcal{T}_h)_h$ is a standard family of triangulations of Ω ; we then, in the standard way, approximate $H_o^1(\Omega)$ and K by means of a *first-order finite-element method*, giving

(2.3A) $$V_{oh} = \{v_h \in C^o(\bar{\Omega}),\ v_h = 0\ on\ \partial\Omega,\ v_h|_T \in P_1 \quad \forall T \in \mathcal{T}_h\},$$

(2.4A) $$K_h = K \cap V_{oh} = \{v_h \in V_{oh},\ v_h(P) \geq \psi(P) \quad \forall P \in \Sigma_{oh}\}$$

where Σ_{oh} is the set of vertices of $\mathcal{T}_h$ interior to Ω.

This leads to the approximation of (2.1A) by - amongst other schemes - the following *implicit scheme* (with $\Delta t > 0$)

(2.5A) $$\begin{cases} \textit{with } u_h^n \textit{ known, determine } u_h^{n+1} \textit{ by} \\ \displaystyle\int_\Omega \left(\frac{u_h^{n+1} - u_h^n}{\Delta t}\right)(v_h - u_h^{n+1})\,dx + \int_\Omega \nabla u_h^{n+1} \cdot \nabla(v_h - u_h^{n+1})\,dx \\ \qquad \geq (f, v_h - u_h^{n+1}) \quad \forall v_h \in K_h, \\ u_h^{n+1} \in K_h\ , \quad n=0,1,2,\ldots\ ; \qquad u_h^o = u_{oh} \end{cases}$$

where $u_{oh} \in V_{oh}$ is an approximation of u_o.

It is shown in Berger-Falk /1A/ that under suitable (and reasonable) assumptions on $(\mathcal{T}_h)_h$, u_o, f, ψ, u and $u_{oh} - u_o$ we have

(2.6A) $$\| u_h^{\Delta t} - u \|_{L^2(0,T;H_o^1(\Omega)) \cap L^\infty(0,T;L^2(\Omega))} = O(h + |\Delta t|^{1-\varepsilon})$$

with, in (2.6A), $\varepsilon > 0$ arbitrarily small (we have $\varepsilon = 0$ if $\Omega \subset \mathbb{R}$) and

$$u_h^{\Delta t}(t) = \sum_{n \geq 0} u_h^n \varpi_n(t)$$

where ϖ_n is the characteristic function of the interval

$$]0,T[\ \cap\ ](n-\tfrac{1}{2})\Delta t,\ (n+\tfrac{1}{2})\Delta t[\ .$$

It is also shown in Berger-Falk, loc. cit., that the above estimate (2.6A) also holds for more explicit variants of scheme (2.5A), provided a *stability condition* of the type $\Delta t \le Ch^2$ is satisfied.

b) P.L. Lions – B. Mercier /1A/, which deals with the investigation of the convergence of implicit schemes of *alternating direction* type (A.D.I.), applicable to the numerical integration of a large number of *parabolic time-dependent inequalities*.

c) Baiocchi-Pozzi /1A/ which investigates the *finite-difference* approximation, in space and time, of a *free-boundary* problem with one space variable, after first reducing it to a parabolic variational inequality. This problem has its origin in *bio-mathematics* (diffusion-absorption of oxygen in tissues). This paper derives estimates of the approximation error, together with results on the convergence of the free boundary in the approximate problem towards that in the continuous problem.

d) Comincioli-Torelli /1A/ which investigates the *finite-element* approximation of *parabolic time-dependent inequalities* in the modelling of the unsteady flow due to the *seepage* of an incompressible fluid in a porous medium.

e) Johnson /3A/, /4A/, Laborde /1A/, Mercier /1A/ and Samuelsson-Froier /1A/ which investigate the *finite-element* approximation of time-dependent inequalities modelling nonlinear phenomena in the mechanics of continuous media (elasto-plasticity, visco-plasticity, hardening, etc...).

f) Lötstedt /1A/-/4A/ which investigates the numerical solution of *variational inequalities* which are of *second order* in t, in the modelling of *dynamic friction* and *contact* problems in the *Mechanics of Structures*. ■

We also refer to Glowinski /2A, Chapter 3/ which gives a description of various *time-discretisation schemes* for *parabolic inequalities* ("quasi" explicit, implicit, Crank-Nicolson, multi-step schemes, etc...). We conclude this section by mentioning the use of techniques for the numerical integration of time-dependent inequalities for solving *nonlinear vibration problems* in *incompressible* or *inextensible* media ([1]); we refer to Bourgat-Dumay-Glowinski /1A/ for an application in the study of *large-displacement* vibrations in *inextensible elastic pipelines*.

3. ON THE STEADY FLOW OF A BINGHAM FLUID IN A CYLINDRICAL DUCT. ASYMPTOTIC PROPERTIES OF THE CONTINUOUS AND DISCRETE PROBLEMS.

Following Glowinski /2A, Chapter 3, Section 6/ (see also Bristeau /1A/ and Bristeau-Glowinski /1A/) we consider the *time-dependent* problem associated with the elliptic variational inequality of Chapter 5 and Appendix 5, and study its *asymptotic properties* along with those of a related approximate problem.

3.1 Formulation of the problem. Existence and uniqueness results.

Let Ω be a *bounded domain* in $\mathbb{R}^2$ with a regular boundary Γ (Ω is the *cross section* of the duct). Using the notation of Chapters 5, 6 and Appendix 5, we consider

- $V = H_0^1(\Omega)$, $H = L^2(\Omega)$, $V' = H^{-1}(\Omega)$,
- $a(v,w) = \int_\Omega \nabla v \cdot \nabla w \, dx \quad \forall\, v,w \in H_0^1(\Omega)$,
- A time interval $[0,T]$, $0 < T < +\infty$,
- $f \in L^2(0,T;V')$, $u_0 \in H$,
- $j(v) = \int_\Omega |\nabla v| \, dx$,
- $\mu > 0$, $g > 0$

(see Chapter 5, Section 1.2 for the mechanical interpretation of μ, g).

([1]) The conditions for incompressibility or inextensibility are expressed as *nonlinear equality-type constraints* at each point of the deformable medium under consideration.

We then have the following theorem:

Theorem 3.1: *For every* $u_o \in H$ *and every* $f \in L^2(0,T;V')$, *the parabolic variational inequality*

$$\text{(3.1A)} \quad \begin{cases} \left(\frac{\partial u}{\partial t}, v-u\right) + \mu a(u,v-u) + gj(v) - gj(u) \geq (f,v-u) \quad \forall v \in V \\ \quad \textit{a.e. in } t \in]0,T[, \\ u(x,0) = u_o(x) \end{cases}$$

has a unique solution u *such that*

$$\text{(3.2A)} \quad \begin{cases} u \in L^2(0,T;V) \cap C^o([0,T];H), \\ \frac{\partial u}{\partial t} \in L^2(0,T;V'). \end{cases}$$

For a proof of this theorem see Duvaut-Lions /1, Chapter 6/.

3.2 On the asymptotic behaviour of the continuous solution

We assume that f is *independent* of t and that $f \in L^2(\Omega)$. We consider the following *steady-state* problem

$$\text{(3.3A)} \quad \begin{cases} \mu a(u,v-u) + gj(v) - gj(u) \geq \int_\Omega f(v-u)\,dx \quad \forall v \in V, \\ u \in V \end{cases}$$

the numerical solution of which was discussed in Chapter 5 and Appendix 5. It is proved in Duvaut-Lions /1, Chapter 6/ that

$$\text{(3.4A)} \qquad u \equiv 0 \quad \textit{if} \quad g\beta \geq \|f\|_{L^2(\Omega)}$$

where

$$\text{(3.5A)} \qquad \beta = \inf_{v \in V-\{0\}} \frac{j(v)}{\|v\|_{L^2(\Omega)}} > 0 .$$

We then have the following theorem:

Theorem 3.2: *Suppose that* $f \in L^2(\Omega)$ *with* $\|f\|_{L^2(\Omega)} < \beta g$; *then if* u *is the solution of* (3.1A) we have

$$u(t) = 0 \quad \textit{for} \quad t \geq \frac{1}{\lambda_o \mu} \operatorname{Log}\left(1+\lambda_o \mu \frac{\|u_o\|_{L^2(\Omega)}}{\beta g - \|f\|_{L^2(\Omega)}}\right) \tag{3.6A}$$

where λ_o *is the smallest eigenvalue of* $-\Delta$ *in* $H^1_o(\Omega)$ $(\lambda_o > 0)$.

Proof: We use $|\cdot|$ for the $L^2(\Omega)$-norm and $\|\cdot\|$ for the $H^1_o(\Omega)$-norm. Since $f \in L^\infty(\mathbb{R}_+; L^2(\Omega))$ it follows from Theorem 3.1 that the solution of (3.1A) is defined on the whole of $\mathbb{R}_+$.

We now observe that if $g\beta > |f|$, then *zero* is the unique solution of (3.3A); it then follows from Theorem 3.1 that if $u(t_o) = 0$ for some $t_o \geq 0$, then

$$u(t) = 0 \quad \forall t \geq t_o . \tag{3.7A}$$

Taking $v = 0$ and $v = 2u$ in (3.1A), we obtain

$$\left(\frac{\partial u}{\partial t}, u\right) + \mu a(u,u) + gj(u) = \int_\Omega fu \, dx \quad \textit{a.e. in } t. \tag{3.8A}$$

Since $v \in L^2(0,T;V)$, $\frac{dv}{dt} \in L^2(0,T;V')$ imply (this being a general result) that $t \to |v(t)|^2$ is *absolutely continuous* with $\frac{d}{dt}|v|^2 = 2\left(\frac{dv}{dt}, v\right)$, we obtain from (3.8A) that

$$\frac{1}{2}\frac{d}{dt}|u|^2 + \mu a(u,u) + gj(u) = \int_\Omega fu \, dx \leq |f|\,|u| \quad \textit{a.e. in } t. \tag{3.9A}$$

Since $a(v,v) \geq \lambda_o |v|^2 \ \forall v \in V$, and $j(v) \geq \beta |v| \ \forall v \in V$ (from (3.5A)) it follows from (3.9A) that

$$\frac{1}{2}\frac{d}{dt}|u|^2 + \mu\lambda_o |u|^2 + (g\beta - |f|)|u| \leq 0 \quad \textit{a.e. in } t \in \mathbb{R}_+. \tag{3.10A}$$

Suppose that $u(t) \neq 0 \ \forall t \geq 0$; then, since $t \to |u(t)|^2$ is *absolutely continuous* with $|u(t)| > 0$, this implies that $t \to |u(t)|$ is also *absolutely continuous*. Therefore from (3.10A) and from $\frac{1}{2}\frac{d}{dt}|u|^2 = |u|\frac{d}{dt}|u|$ we obtain:

(3.11A) $$\frac{d}{dt}|u(t)| + \mu\lambda_o|u(t)| + (g\beta - |f|) \le 0 \quad a.e. \ in \ t \in \mathbb{R}_+.$$

It follows from (3.11A) that

(3.12A) $$\frac{\frac{d}{dt}|u(t)|}{|u(t)| + \frac{g\beta - |f|}{\mu\lambda_o}} \le -\mu\lambda_o \quad a.e. \ in \ t \in \mathbb{R}_+.$$

Define γ by $\gamma = \frac{g\beta - |f|}{\mu\lambda_o}$; then $\gamma > 0$. It follows then by integrating (3.12A) that

(3.13A) $$|u(t)| + \gamma \le (|u_o| + \gamma)e^{-\mu\lambda_o t} \quad \forall\, t \in \mathbb{R}_+ \ ;$$

(3.13A) is clearly absurd for t sufficiently large. In fact we have $u(t) = 0$ if

$$\gamma \ge (|u_o| + \gamma)e^{-\mu\lambda_o t},$$

i.e.

(3.14A) $$t \ge \frac{1}{\lambda_o\mu} \operatorname{Log}\left(1 + \frac{\lambda_o\mu|u_o|}{g\beta - |f|}\right),$$

which completes the proof of the theorem.

Remark 3.1: Let $f \in L^2(\Omega)$, possibly with $|f| \ge g\beta$, and denote by u_∞ the solution of (3.3A); it can be then easily proved that

$$|u(t) - u_\infty| \le |u_o - u_\infty| e^{-\mu\lambda_o t}$$

where $u(t)$ is the solution of (3.1A).

3.3 Approximation of (3.1A). Asymptotic properties of the discrete solution.

We still suppose that $f \in L^2(\Omega)$; to approximate (3.1A) we proceed as follows: assuming that Ω is a polygonal domain, we approximate $V = H_o^1(\Omega)$ by

$$V_h = \{v_h \in C^o(\bar{\Omega}),\ v_h = 0 \ \ on \ \Gamma,\ v_h|_T \in P_1 \ \ \forall\, T \in \mathcal{T}_h\},$$

i.e. by means of a *first-order finite-element method*. We then approximate (3.1A) by the following *implicit scheme*

$$\text{(3.15A)}\quad \begin{cases} \textit{with } u_h^n \textit{ known, determine } u_h^{n+1} \textit{ by} \\ \int_\Omega \left(\frac{u_h^{n+1}-u_h^n}{\Delta t}\right)(v_h-u_h^{n+1})\,dx + \mu a(u_h^{n+1}, v_h-u_h^{n+1}) \\ \quad + gj(v_h)-gj(u_h^{n+1}) \geq \int_\Omega f_h(v_h-u_h^{n+1})\,dx \quad \forall v_h \in V_h , \\ u_h^{n+1} \in V_h , \\ n=0,1,2,\ldots ; \; u_h^o = u_{oh} . \end{cases}$$

We assume that $u_{oh} \in V_h \;\; \forall h$, and that

$$\text{(3.16A)}\qquad \lim_{h\to 0} \| u_{oh}-u_o\|_{L^2(\Omega)} = 0 .$$

We also assume that f is approximated by $\{f_h\}_h$ such that $\int_\Omega f_h v_h dx$ can be computed easily, and that

$$\text{(3.17A)}\qquad \lim_{h\to 0} \| f_h-f\|_{L^2(\Omega)} = 0.$$

Using the methods of Chapter 6, it is possible to prove the *convergence* (for some topology) of the *discrete solution* to the *continuous* one, as $h \to 0$, $\Delta t \to 0$, if we assume that the usual *angle condition* holds as $h \to 0$.

Concerning the asymptotic properties of the discrete solution, we have the following discrete variant of Theorem 3.2:

<u>Theorem 3.3</u>: *Suppose that* $|f| < \beta g$; *if* (3.16A), (3.17A) *hold and if* h *is sufficiently small*, we have

$$\text{(3.18A)}\qquad u_h^n = 0 \quad \textit{for } n \textit{ sufficiently large},$$

(*an estimate being given in* (3.26A) *below*).

Proof: Taking $v_h = 0$ and $v_h = 2u_h^{n+1}$ in (3.15A), we obtain

$$\text{(3.19A)} \quad \begin{cases} \int_\Omega \left(\frac{u_h^{n+1}-u_h^n}{\Delta t}\right) u_h^{n+1}\,dx + \mu a(u_h^{n+1},u_h^{n+1}) + gj(u_h^{n+1}) \\ = \int_\Omega f_h u_h^{n+1}\,dx \quad \forall n \geq 0 \ ; \end{cases}$$

using Schwarz's inequality in $L^2(\Omega)$, it follows from (3.19A) that $\forall n \geq 0$:

$$\text{(3.20A)} \quad \frac{|u_h^{n+1}|-|u_h^n|}{\Delta t}|u_h^{n+1}| + \mu\lambda_o |u_h^{n+1}|^2 + (g\beta - |f_h|)|u_h^{n+1}| \leq 0 .$$

Since $\lim_{h\to 0} |f_h - f| = 0$ we have

$$\text{(3.21A)} \quad g\beta - |f_h| > 0 \text{ for } h \text{ sufficiently small.}$$

It then follows from (3.20A), (3.21A) that

$$\text{(3.22A)} \quad \begin{cases} u_h^m = 0 \ \textit{implies} \ u_h^n = 0 \ \textit{for} \ n \geq m \ \textit{if} \ h \ \textit{is} \\ \textit{sufficiently small.} \end{cases}$$

Suppose that $u_h^n \neq 0 \ \forall n \geq 0$; then (3.20A) implies

$$\text{(3.23A)} \quad \frac{|u_h^{n+1}|-|u_h^n|}{\Delta t} + \mu\lambda_o |u_h^{n+1}| + g\beta - |f_h| \leq 0 \quad \forall n \geq 0.$$

We define γ_h by $\gamma_h = g\beta - |f_h|$; then

$$\text{(3.24A)} \quad \begin{cases} \gamma_h > 0 \ \textit{for} \ h \ \textit{sufficiently small}, \\ \lim_{h\to 0} \gamma_h = \gamma = g\beta - |f|. \end{cases}$$

It follows from (3.23A) that

$$\left(|u_h^{n+1}| + \frac{\gamma_h}{\mu\lambda_o}\right)(1+\mu\lambda_o\Delta t) \leq |u_h^n| + \frac{\gamma_h}{\mu\lambda_o} \quad \forall n \geq 0$$

which implies that

$$\text{(3.25A)} \quad \left(|u_h^n| + \frac{\gamma_h}{\mu\lambda_o}\right) \leq (1+\mu\lambda_o\Delta t)^{-n} \left(|u_h^o| + \frac{\gamma_h}{\mu\lambda_o}\right) .$$

Since $\gamma_h > 0$ for h sufficiently small, (3.25A) is *impossible* for n sufficiently large. More precisely, we will have $u_h^n = 0$ if

$$\frac{\gamma_h}{\mu\lambda_o} \geq (1+\mu\lambda_o\Delta t)^{-n}\left(|u_h^o| + \frac{\gamma_h}{\mu\lambda_o}\right),$$

which implies

$$\text{(3.26A)} \quad \begin{cases} \textit{If } h \textit{ is sufficiently small, then } u_h^n = 0 \\ \textit{if} \quad n \geq \dfrac{\operatorname{Log}\left(1+\mu\lambda_o \dfrac{|u_h^o|}{\gamma_h}\right)}{\operatorname{Log}(1+\mu\lambda_o\Delta t)} \end{cases} .$$

Relation (3.26A) makes the statement of Theorem 3.3 more precise. Moreover, in terms of *time*, (3.26A) implies that u_h^n is equal to zero if

$$\text{(3.27A)} \qquad n\Delta t \geq \Delta t \frac{\operatorname{Log}\left(1+\mu\lambda_o \dfrac{|u_h^o|}{\gamma_h}\right)}{\operatorname{Log}(1+\mu\lambda_o\Delta t)} .$$

We note that

$$\lim_{\substack{h\to 0\\ \Delta t\to 0}} \Delta t \frac{\operatorname{Log}\left(1+\mu\lambda_o \dfrac{|u_h^o|}{\gamma_h}\right)}{\operatorname{Log}(1+\mu\lambda_o\Delta t)} = \frac{1}{\mu\lambda_o}\operatorname{Log}\left(1+\lambda_o\mu\frac{|u^o|}{\gamma}\right).$$

Hence taking the limit in (3.27A) we obtain a further proof (since u_h^n converges to u in some topology) of the estimate (3.6A) given in the statement of Theorem 3.2.

Remark 3.2: Let u_h^∞ be the solution of the steady-state discrete problem associated with f_h (and (3.15A)), possibly with $|f_h| \geq \beta g$. We then have

$$\text{(3.28A)} \qquad |u_h^n - u_h^\infty| \leq (1+2\mu\lambda_o\Delta t)^{-n/2}\,|u_h^o - u_h^\infty| \quad \forall n \geq 0.$$

The above estimate (3.28A) is the discrete analogue of the estimate in Remark 3.1.

3.4 Further remarks

Remark 3.3: We can generalise Theorem 3.1 to the case of a Bingham flow in a two-dimensional bounded cavity.

Remark 3.4: Numerical examinations of the above asymptotic properties are performed in Glowinski /2/, Bristeau /1A/, Bristeau-Glowinski /1A/ and Begis /3/, and show them to be consistent with the theoretical predictions; see also Begis /1A/ in which numerical simulations of two-dimensional Bingham flows in a cavity are performed using the method described in Section 4 below.

4. NUMERICAL SIMULATION OF THE FLOW OF A BINGHAM FLUID IN A TWO-DIMENSIONAL CAVITY, USING A STREAM FUNCTION METHOD

4.1 Synopsis

The object of this section, which in part follows Begis /1A/, is to supplement Chapter 6, Section 11, on the numerical analysis of the flow of Bingham fluids. After first reviewing the initial formulation of the problem (i.e. that of Chapter 6, Section 11) in Section 4.2, we show in Section 4.3 that the introduction of a *stream function* enables the problem to be reduced to a *parabolic variational inequality*, which is of *fourth order with respect to the space variables*.

In Sections 4.4 and 4.5 we examine the *approximation* of *steady-state* and *time-dependent* problems using *mixed finite-element methods* (for the *spatial approximation*) and *finite differences* (for the *time approximation*). Next, it is shown in Section 4.6 that the methods of *decomposition-coordination* using an *augmented Lagrangian*, mentioned in Appendix 2, Section 7 (see also Chapter 3, Section 10 and Chapter 5, Section 9), allow the above approximate problems to be solved in a particularly simple manner.

Finally, in Section 4.7 we present some numerical results obtained using the above methods, which enable us to demonstate certain properties of the Bingham fluid flows under consideration.

4.2 Review of the velocity formulation of the Bingham flow problem

Let Ω be a *bounded domain* in $\mathbb{R}^2$, with regular boundary Γ. If $v = \{v_1, v_2\}$ is a function with values in $\mathbb{R}^2$, we put (using essentially the same notation as in Chapter 6, Section 11.1)

(4.1A) $$D_{ij}(v) = \frac{1}{2}\left(\frac{\partial v_i}{\partial x_j} + \frac{\partial v_j}{\partial x_i}\right) ,$$

(4.2A) $$D_{II}(v) = \frac{1}{2}\sum_{i,j=1}^{2} (D_{ij}(v))^2 ,$$

(4.3A) $$a(v,w) = 2 \sum_{i,j=1}^{2} \quad D_{ij}(v)D_{ij}(w)dx ,$$

(4.4A) $$j(v) = \int_\Omega (D_{II}(v))^{1/2}dx ,$$

(4.5A) $$V_o = \{v | v \in H_o^1(\Omega) \times H_o^1(\Omega),\ \nabla\cdot v = 0\} ,$$

(4.6A) $$H = \{v | v \in L^2(\Omega) \times L^2(\Omega),\ \nabla\cdot v = 0,\ v\cdot n = 0 \textit{ on } \Gamma\}.$$

If in addition $z \in H^{1/2}(\Gamma) \times H^{1/2}(\Gamma)$ with (1)

(4.7A) $$\int_\Gamma z\cdot n \, d\Gamma = 0 ,$$

we associate with z the affine space (which is *nonempty*, see Cattabriga /1A/):

(4.8A) $$V_z = \{v | v \in H^1(\Omega) \times H^1(\Omega),\ \nabla\cdot v = 0 ,\ v = z \textit{ on } \Gamma\}.$$

Hereafter we shall neglect the nonlinear terms associated with the trilinear form b(•,•,•) defined by (11.5) in Chapter 6, Section 11.1. We are then led (see Duvaut-Lions /1, Chapter 6/) to modelling the flow of a Bingham fluid in Ω, satisfying u = z on Γ, by

---

(1) n: unit normal vector pointing outwards from Ω.

(4.9A)
$$\begin{cases} \text{Find } u \in L^2(0,T;V_z) \cap L^\infty(0,T;H),\ u' \in L^2(0,T;V_0') \text{ such that} \\ (u'(t),v-u(t))+\nu a(u(t),v-u(t))+gj(v)-gj(u(t)) \\ \quad \geq (f(t),v-u(t)) \quad \forall v \in V_z, \text{ a.e. in } t, \\ u(0) = u_0 \in H,\ f \in L^2(0,T;V_0')\ . \end{cases}$$

We recall that in (4.9A)

ν is the *viscosity* of the fluid,
g is the *plasticity threshold*, (i.e. the *yield stress*),
f is the *density of the external forces*.

Duvaut-Lions /1, Chapter 6/ proves (for $z = 0$) the *existence and uniqueness* of a solution of problem (4.9A); the case $z \neq 0$, with z satisfying (4.7A), may be treated in similar fashion.

Remark 4.1: In the above we have assumed that $z(= u|_\Gamma)$ does not depend on t: in fact there is no extra difficulty in handling such a case numerically, using the following methods.

4.3 A stream-function formulation of the Bingham flow problem

In this section we make the following two simplifying assumptions

(i) Ω is *simply connected*,

(ii) $z = u|_\Gamma = 0$;

however, Remark 4.1 above still holds for situations in which (i) and/or (ii) are not satisfied.

In view of the fact that in this appendix we are restricting our attention to two-dimensional flows, it is natural to eliminate the condition $\nabla \cdot u = 0$ by introducing a *stream function*, defined (to within an additive constant) by

(4.10A) $$u_1 = \frac{\partial\psi}{\partial x_2}\ ,\ u_2 = -\frac{\partial\psi}{\partial x_1}\ .$$

The condition $u = 0$ on Γ implies

(4.11A) $\psi = \textit{constant on } \Gamma,$

(4.12A) $\dfrac{\partial\psi}{\partial n} = 0 \textit{ on } \Gamma;$

we shall take $\psi = 0$ on Γ (which fixes the constant mentioned above).

Let $v \in V_o$; with v we then associate $\phi \in H_o^2(\Omega)$ by

(4.13A) $v_1 = \dfrac{\partial\phi}{\partial x_2}\ , \ v_2 = -\dfrac{\partial\phi}{\partial x_1}\ ,$

(4.14A) $\phi = \dfrac{\partial\phi}{\partial n} = 0 \textit{ on } \Gamma.$

Substituting the functions (4.10A), (4.13A) reduces (4.9A) to the following *parabolic variational inequality* (of fourth order with respect to the space variables)

(4.15A)
$$\begin{cases} \textit{Find } \psi \in L^2(0,T;H_o^2(\Omega)) \cap L^\infty(0,T;H_o^1(\Omega)) \textit{ such that} \\ -\displaystyle\int_\Omega \psi'(t)\Delta(\phi-\psi(t))dx + \nu\tilde{a}(\psi(t),\phi-\psi(t)) + g\tilde{j}(\phi) - g\tilde{j}(\psi(t)) \\ \geq (\tilde{f}(t),\phi-\psi(t))\ ^{(1)} \qquad \forall\phi\in H_o^2(\Omega),\ \textit{a.e. in } t, \\ \psi(0) = \psi_o \in H_o^1(\Omega)\ , \end{cases}$$

where

(4.16A)
$$\begin{cases} \tilde{a}(\phi_1,\phi_2) = \displaystyle\int_\Omega [(2\frac{\partial^2\phi_1}{\partial x_1\partial x_2})(2\frac{\partial^2\phi_2}{\partial x_1\partial x_2}) \\ \qquad + (\dfrac{\partial^2\phi_1}{\partial x_2^2} - \dfrac{\partial^2\phi_1}{\partial x_1^2})(\dfrac{\partial^2\phi_2}{\partial x_2^2} - \dfrac{\partial^2\phi_2}{\partial x_1^2})]dx, \\ \forall\phi_1,\phi_2 \in H_o^2(\Omega), \end{cases}$$

(1) If, in (4.9A), $(f,v) = \int_\Omega f\cdot v\,dx$ then $\tilde{f} = \dfrac{\partial f_2}{\partial x_1} - \dfrac{\partial f_1}{\partial x_2}$.

(4.17A)
$$\begin{cases} \tilde{j}(\phi) = \int_\Omega [(2\frac{\partial^2\phi}{\partial x_1 \partial x_2})^2 + (\frac{\partial^2\phi}{\partial x_2^2} - \frac{\partial^2\phi}{\partial x_1^2})^2]^{1/2} dx \\ \forall\phi \in H_o^2(\Omega) . \end{cases}$$

Remark 4.2: In fact we have

(4.18A)
$$\tilde{a}(\phi_1,\phi_2) = \int_\Omega \Delta\phi_1 \Delta\phi_2 \, dx \quad \forall\, \phi_1,\phi_2 \in H_o^2(\Omega) \;;$$

in the following we shall use (4.16A) and (4.18A) simultaneously.

4.4 Approximation of the steady-state problem

4.4.1 *Synopsis. Formulation of the steady-state problem*

With a view to investigating the approximation of problem (4.15A) - by a mixed finite-element method - we shall first investigate the approximation of the corresponding steady-state problem, i.e. the fourth-order elliptic variational inequality [1]

(4.19A)
$$\begin{cases} \textit{Find } \psi \in H_o^2(\Omega) \textit{ such that} \\ \nu a(\psi,\phi-\psi)+gj(\phi)-gj(\psi) \geq (f,\phi-\psi) \quad \forall\phi \in H_o^2(\Omega) , \end{cases}$$

where a and j are defined by (4.16A), (4.17A); we remark that (4.19A) is equivalent to the minimisation problem

(4.20A)
$$\begin{cases} \textit{Find } \psi \in H_o^2(\Omega) \textit{ such that} \\ J(\psi) \leq J(\phi) \quad \forall\phi \in H_o^2(\Omega) \end{cases}$$

where, in (4.20A)

(4.21A)
$$J(\phi) = \frac{\nu}{2} a(\phi,\phi)+gj(\phi)-(f,\phi) .$$

---

[1] Hereafter we shall omit the symbol ~.

Hereafter we shall assume that $f \in H^{-1}(\Omega)$; however, there is no difficulty in considering the case in which

$$(f,\phi) = \int_\Omega f\, \Delta\phi\, dx \quad \forall \phi \in H_o^2(\Omega) \;\; ; \; f \in L^2(\Omega).$$

Since the form $a(\cdot,\cdot)$ is $H_o^2(\Omega)$-*elliptic* (i.e. *coercive*) from (4.18A), and since $j(\cdot)$ is *continuous* and *convex* on $H_o^2(\Omega)$, *problem* (4.19A), (4.20A) *admits a unique solution.*

4.2.2 *Approximation of* (4.19A),(4.20A) *by a method of mixed finite elements*

In the following we shall assume that Ω is a convex polygonal domain in $\mathbb{R}^2$; let $(\mathcal{C}_h)_h$ denote a standard family of triangulations of Ω. We shall approximate (4.19A), (4.20A) by a *method of mixed finite elements*, a variant (inspired by Miyoshi /1A/) of the mixed method considered in Chapter 4, Section 4 and Appendix 4, Section 3, of which we retain most of the notation. We thus have

$$V_h = \{v_h \in C^o(\bar{\Omega}),\; v_h|_T \in P_k \quad \forall\, T \in \mathcal{C}_h\}, \tag{4.22A}$$

$$V_{oh} = V_h \cap H_o^1(\Omega). \tag{4.23A}$$

Let $\phi \in H_o^2(\Omega)$; for $1 \le i, j \le 2$ we define

$$q_{ij} = q_{ji} = \frac{\partial^2\phi}{\partial x_i \partial x_j}\,. \tag{4.24A}$$

We then have $\forall\, \mu \in H^1(\Omega)$

$$\int_\Omega q_{ij}\, \mu\, dx = -\frac{1}{2}\int_\Omega \left(\frac{\partial\phi}{\partial x_i}\frac{\partial\mu}{\partial x_j} + \frac{\partial\phi}{\partial x_j}\frac{\partial\mu}{\partial x_i}\right) dx\,. \tag{4.25A}$$

Conversely if $\phi \in H_o^1(\Omega)$ and $q = \{q_{ij}\}_{1\le i\le j\le 2}$ satisfy (4.25A), then $\phi \in H_o^2(\Omega)$ and q and ϕ are related by (4.24A).

This leads to the introduction of the space $W_h \subset V_{oh} \times (V_h)^3$ defined by

(4.26A)
$$\left\{\begin{aligned} W_h &= \{\{\phi_h,q_h\}\in V_{oh}\times(V_h)^3,\ 2\int_\Omega q_{ijh}\mu_h dx \\ &= -\int_\Omega\left(\frac{\partial\phi_h}{\partial x_i}\frac{\partial\mu_h}{\partial x_j}+\frac{\partial\phi_h}{\partial x_j}\frac{\partial\mu_h}{\partial x_i}\right)dx \\ &\forall\mu_h\in V_h\ ,\ 1\le i\le j\le 2\}\ . \end{aligned}\right.$$

Let $\alpha,\beta\in\,]0,1[$ be such that $\alpha+\beta=1$; in view of (4.16A), (4.18A) we "approximate" the functional J of (4.20A), (4.21A) by

(4.27A)
$$\left\{\begin{aligned} J_h(\phi_h,q_h) &= \frac{\alpha\nu}{2}\int_\Omega(q_{11h}+q_{22h})^2dx \\ &+\frac{\beta\nu}{2}\int_\Omega[(2q_{12h})^2+(q_{22h}-q_{11h})^2]dx+gj_h(q_h)-(f,\phi_h) \end{aligned}\right.$$

where

(4.28A)
$$j_h(q_h)=\int_\Omega[(2q_{12h})^2+(q_{22h}-q_{11h})^2]^{1/2}dx\ .$$

We then approximate the problem (4.19A), (4.20A) by

(4.29A)
$$\left\{\begin{aligned} &\textit{Find } \{\psi_h,p_h\}\in W_h \textit{ such that} \\ &J_h(\psi_h,p_h)\le J_h(\phi_h,q_h)\quad\forall\{\phi_h,q_h\}\in W_h\ . \end{aligned}\right.$$

4.4.3 *Solvability of the approximate problem* (4.29A)

Regarding the solvability of the approximate problem (4.29A) we have

Theorem 4.1: *The approximate problem* (4.29A) *admits one and only one solution.*

Proof: The functional $j_h(\cdot)$ is *convex* and *continuous*; moreover if we put $\mu_h=\phi_h$ in the relations which define W_h (see (4.26A)), we have

(4.30A) $$\|\phi_h\|_{H_0^1(\Omega)} \le C|q_h| \quad \forall\{\phi_h,q_h\}\in W_h$$

where, in (4.30A), C is independent of $\{\phi_h,q_h\}$ and h, and where

$$|q_h| = \|q_h\|_{(L^2(\Omega))^3} = \Big(\sum_{1\le i\le j\le 2}\int_\Omega |q_{ijh}|^2\,dx\Big)^{1/2}.$$

In view of the above properties it is sufficient, in order to prove the existence of a solution, to show that

(4.31A) $$\begin{cases} \alpha\displaystyle\int_\Omega (q_{11h}+q_{22h})^2dx + \beta\int_\Omega [(2q_{12h})^2+(q_{22h}-q_{11h})^2]dx \\ \ge C_0|q_h|^2 \qquad \forall q_h\in(V_h)^3 \end{cases}$$

with $C_0 > 0$; it may then readily be verified that

(4.32A) $$\begin{cases} \alpha(\xi_1+\xi_3)^2+\beta(4\xi_2^2+(\xi_3-\xi_1)^2) \ge 2\,\mathrm{Min}(\alpha,\beta)(\xi_1^2+\xi_2^2+\xi_3^2), \\ \forall\{\xi_1,\xi_2,\xi_3\}\in\mathbb{R}^3, \end{cases}$$

which implies (4.31A) (with $C_0 = 2\,\mathrm{Min}(\alpha,\beta)$).

The iterative solution of (4.29A) will form the subject of Section 4.6 below.

4.4.4 *Convergence of the approximate solutions*

We shall consider only the case k = 2 (see Remark 4.3 below for k ≠ 2); regarding the convergence of the approximate solutions as h → 0, we have

Theorem 4.2: *Suppose that when* h → 0 *the angles of* $\mathcal{C}_h$ *remain bounded below, uniformly in* h, *by* $\theta_0 > 0$; *suppose also that condition* (4.73) *of Chapter* 4, *Section* 4.5.4 *is satisfied. We then have*

(4.33A) $$\lim_{h\to 0}\{\psi_h,p_h\} = \{\psi,p\} \textit{ strongly in } H_0^1(\Omega)\times(L^2(\Omega))^3$$

where $\{\psi_h, p_h\}$ *is the solution of the approximate problem* (4.29A), ψ *that of the continuous problem* (4.19A), (4.20), *and where*

$$p = \{p_{ij}\}_{1 \le i \le j \le 2} \quad \text{with} \quad p_{ij} = \frac{\partial^2 \psi}{\partial x_i \partial x_j}.$$

Proof: This is a variant of the proof of Theorem 2.3 of Appendix 1, Section 2.3.2 and that of Theorem 3.5 of Appendix 4, Section 3.4.5, so that we have:

1) A priori estimates for $\{\psi_h, p_h\}$

In the following, C denotes various constants. Since $\{0,0\} \in W_h$ we can take $\{\phi_h, q_h\} = \{0,0\}$ in (4.29A); in view of $j_h(q_h) \ge 0 \;\; \forall q_h \in (V_h)^3$ and (4.31A) we thus deduce

$$\frac{\nu C_o}{2} |p_h|^2 \le \|f\|_{H^{-1}(\Omega)} \|\psi_h\|_{H_o^1(\Omega)}. \tag{4.34A}$$

In combination with (4.30A), relation (4.34A) implies

$$\|\psi_h\|_{H_o^1(\Omega)} \le C \;\; \forall h, \tag{4.35A}$$

$$|p_h| \le C \;\; \forall h. \tag{4.36A}$$

2) Weak convergence

In view of (4.35A), (4.36A) we can extract a subsequence from $(\{\psi_h, p_h\})_h$ - also denoted by $(\{\psi_h, p_h\})_h$ - such that

$$\psi_h \to \psi^* \;\; \text{weakly in} \;\; H_o^1(\Omega) \tag{4.37A}$$

$$p_h \to p^* \;\; \text{weakly in} \;\; (L^2(\Omega))^3. \tag{4.38A}$$

It can be shown, using a variant of the proof of Lemma 3.3 of Appendix 4, Section 3.4.5, that

$$\psi^* \in H_o^2(\Omega), \tag{4.39A}$$

$$p_{ij}^* = \frac{\partial^2 \psi^*}{\partial x_i \partial x_j} \;\; \forall\, 1 \le i \le j \le 2. \tag{4.40A}$$

It remains to prove that $\psi^* = \psi$; we proceed as in the proof of Lemma 3.4 of Appendix 4, Section 3.4.5: let us suppose then that $\phi \in \mathcal{D}(\Omega)$; we define $\{r_h\phi, r_h q\} \in W_h$ by solving (using the mixed method of Chapter 4, Section 4) the approximate biharmonic problem

$$(4.41A)\quad \begin{cases} \{r_h\phi, \omega_h\} \in W_h^* \quad \textit{and} \quad \forall \{v_h, w_h\} \in W_h^* \quad \textit{we have} \\ \frac{1}{2}\int_\Omega |\omega_h|^2 dx - \int_\Omega \Delta^2\phi \; r_h\phi \; dx \le \frac{1}{2}\int_\Omega |w_h|^2 dx - \int_\Omega \Delta^2\phi \; v_h dx \end{cases}$$

where

$$W_h^* = \{\{v_h, w_h\} \in V_{oh} \times V_h, \int_\Omega \nabla v_h \cdot \nabla \mu_h dx = \int_\Omega w_h \mu_h \; dx \quad \forall \mu_h \in V_h\};$$

problem (4.41A) is well posed. The function $r_h\phi$ being known, we (*uniquely*) determine $r_h q \in (V_h)^3$ such that $\{r_h\phi, r_h q\} \in W_h$, from

$$(4.42A)\quad \begin{cases} (r_h q)_{ij} \in V_h \quad \forall\, 1 \le i \le j \le 2, \\ 2\int_\Omega (r_h q)_{ij}\mu_h dx = -\int_\Omega [\frac{\partial}{\partial x_i} r_h\phi \frac{\partial \mu_h}{\partial x_j} + \frac{\partial}{\partial x_j} r_h\phi \frac{\partial \mu_h}{\partial x_i}] dx \\ \forall \mu_h \in V_h. \end{cases}$$

It then follows from Rannacher /2A/ that under the assumptions on $(\mathcal{T}_h)_h$ stated in the theorem, we have

$$(4.43A)\quad r_h\phi \to \phi \quad \textit{strongly in} \quad H_o^1(\Omega).$$

$$(4.44A)\quad (r_h q)_{ij} \to \frac{\partial^2\phi}{\partial x_i \partial x_j} \quad \textit{strongly in} \quad L^2(\Omega), \quad \forall\, 1 \le i \le j \le 2.$$

Taking $\{\phi_h, q_h\} = \{r_h\phi, r_h q\}$ in (4.29A) we deduce from (4.37A)-(4.40A) and from (4.43A), (4.44A), by using the *weak lower semi-continuity* of convex continuous functionals, that

$$(4.45A)\quad \begin{cases} J(\psi^*) \le J(\phi) \quad \forall \phi \in \mathcal{D}(\Omega), \\ \psi^* \in H_o^2(\Omega). \end{cases}$$

From the *density* of $\mathcal{D}(\Omega)$ in $H_o^2(\Omega)$ and the *continuity* of $J(\cdot)$ on $H_o^2(\Omega)$, (4.45A) also holds $\forall \phi \in H_o^2(\Omega)$ which proves that

$\psi^* = \psi$, $p^*_{ij} = p_{ij} = \frac{\partial^2\phi}{\partial x_i \partial x_j}$; hence we have convergence of the entire family $(\{\psi_h, p_h\})_h$.

3) Strong convergence

Without going into details, it can be shown, by using a variant of the proof of Theorem 2.3 of Appendix 1, Section 2.3.2 and of Theorem 3.5 of Appendix 4, Section 3.4.5, that

(4.46A) $$\lim_{h\to 0} \psi_h = \psi \textit{ strongly in } H^1_o(\Omega)$$

(4.47A) $$\lim_{h\to 0} p_h = \{\frac{\partial^2\psi}{\partial x_i \partial x_j}\}_{1\le i,j\le 2} \textit{ strongly in } L^2(\Omega)$$

(4.48A) $$\lim_{h\to 0} J_h(\psi_h, p_h) = J(\psi)$$

(4.49A) $$\lim_{h\to 0} j_h(p_h) = j(\psi). \blacksquare$$

Remark 4.3: In the above we have assumed that k = 2; in fact, similar convergence results would be obtained for approximations based on finite elements of order $k \ge 3$, but given the *limited regularity* of the solutions ($\psi \notin H^4(\Omega) \cap H^2_0(\Omega)$, in general) the use of such high-order elements is not justified.

Regarding the case k = 1, it follows from Miyoshi /1A/ that the above convergence results also hold if $(\mathcal{C}_h)_h$ is generated by 3 families of parallel straight lines; in fact it is reasonable to conjecture that the estimates of Scholz /2A/, Girault-Raviart /1A/ (established for the mixed finite-element method of Chapter 4, Section 4) can be generalised to the Hermann-Miyoshi scheme (see Miyoshi /1A/); if this were the case, then Theorem 4.2 would hold for k = 1. (1)

4.4.5 *Approximation using numerical integration*

On the practical level, it is necessary to use a *numerical integration* process to approximate the functional $J_h(\cdot,\cdot)$ in (4.27A), (4.29A); we shall examine only the case k = 1. Let Σ_h be the set of vertices of $\mathcal{C}_h$; we approximate on V_h the inner product

(1) Very recent results of L. Reinhart show that this conjecture is correct.

induced by $L^2(\Omega)$, i.e. $\{\lambda_h,\mu_h\} \to \int_\Omega \lambda_h\mu_h \, dx$, by

$$(\lambda_h,\mu_h)_h = \frac{1}{3} \sum_{P \in \Sigma_h} m(P)\lambda_h(P)\mu_h(P) \tag{4.50A}$$

where m(P) is the sum of the areas of the triangles which have P as a common vertex. In view of (4.50A) we shall actually use in (4.29A) the functional $J_h(\cdot,\cdot)$ defined by

$$(4.51A)\quad \begin{cases} J_h(\phi_h,q_h) \\ = \frac{\alpha\nu}{2}(q_{11h}+q_{22h},q_{11h}+q_{22h})_h + \frac{\beta\nu}{2}[(2q_{12h},2q_{12h})_h \\ + (q_{22h}-q_{11h},q_{22h}-q_{11h})_h] + \frac{g}{3}\sum_{P\in\Sigma_h} m(P)[(2q_{12h}(P))^2 \\ + (q_{22h}(P)-q_{11h}(P))^2]^{1/2} - (f_h,\phi_h) \end{cases}$$

where f_h is an approximation of f.

Similarly, in place of using W_h defined by (4.26A), if k = 1 we shall use W_h defined by

$$(4.52A)\quad \begin{cases} W_h = \{\{\phi_h,q_h\} \in V_{oh} \times (V_h)^3,\ 2(q_{ijh},\mu_h)_h = -\int_\Omega [\frac{\partial\phi_h}{\partial x_i}\frac{\partial\mu_h}{\partial x_j} \\ \quad + \frac{\partial\phi_h}{\partial x_j}\frac{\partial\mu_h}{\partial x_i}]\, dx \quad \forall\, \mu_h \in V_h,\ 1 \le i \le j \le 2\}. \end{cases}$$

Using the relations in (4.52A), it is easy to express the $q_{ijh}(P)$, $\forall\, P \in \Sigma_h$, explicitly in terms of the values taken by ψ_h on Σ_h; we refer to Begis /1A/ for further details.

4.5 Approximation of the time-dependent problem (4.15A)

4.5.1 *Semi-discretisation in time*

Let $k = \Delta t (> 0)$ be a time step; we then approximate (4.15A) by the following *implicit scheme* (where $\psi^n \simeq \psi(nk)$ and in which the ~ have been omitted)

(4.53A)
$$\begin{cases} \textit{with } \psi^n \textit{ known, determine } \psi^{n+1} \textit{ from} \\ \int_\Omega \nabla\left(\frac{\psi^{n+1}-\psi^n}{k}\right)\cdot\nabla(\phi-\psi^{n+1})\,dx + \nu a(\psi^{n+1},\phi-\psi^{n+1}) \\ \quad + gj(\phi)-gj(\psi^{n+1}) \geq (f((n+1)k),\phi-\psi^{n+1}) \\ \forall\phi \in H_o^2(\Omega),\ \psi^{n+1} \in H_o^2(\Omega),\ n=0,1,\ldots\ ;\ \psi^o = \psi(0) = \psi_o\ . \end{cases}$$

Using the above semi-discrete scheme we have thus reduced the solution of (4.15A) to that of a sequence of elliptic variational inequalities, equivalent to the sequence of minimisation problems (for $n \geq 0$)

(4.54A)
$$\begin{cases} \textit{Find } \psi^{n+1} \in H_o^2(\Omega) \textit{ such that} \\ J_k^{n+1}(\psi^{n+1}) \leq J_k^{n+1}(\phi)\quad \forall\phi \in H_o^2(\Omega) \end{cases}$$

where

(4.55A)
$$\begin{cases} J_k^{n+1}(\phi) = \frac{1}{2k}\int_\Omega |\nabla\phi|^2\,dx + \frac{\nu}{2}\,a(\phi,\phi) + gj(\phi) \\ \qquad -(f((n+1)k),\phi) - \frac{1}{k}\int_\Omega \nabla\psi^n\cdot\nabla\phi\,dx\ . \end{cases}$$

The solution of (4.55A) by mixed finite elements will form the subject of Section 4.5.2.

4.5.2 *Full discretisation of* (4.15A)

The notation is that of Section 4.4.2; we approximate $\psi^o = \psi_o$ by $\psi_h^o \in V_{oh}$, and then the semi-discrete scheme (4.53A) by:

With the function $\psi_h^n \in V_{oh}$ *known, obtain* $\{\psi_h^{n+1}, p_h^{n+1}\}$ *by solving the following minimisation problem for* $n = 0, 1\ldots.$

(4.56A)
$$\begin{cases} \textit{Find } \{\psi_h^{n+1}, p_h^{n+1}\} \in W_h \textit{ such that} \\ J_{kh}^{n+1}(\psi_h^{n+1}, p_h^{n+1}) \leq J_{kh}^{n+1}(\phi_h, q_h)\quad \forall\{\phi_h, q_h\} \in W_h \end{cases}$$

where (with the notation of Section 4.4.2)

$$
(4.57\text{A})\quad \begin{cases} J_{kh}^{n+1}(\phi_h,q_h) = \dfrac{1}{2k}\displaystyle\int_\Omega |\nabla\phi_h|^2 dx + \dfrac{\alpha\nu}{2}\int_\Omega (q_{11h}+q_{22h})^2 dx \\[2mm] + \dfrac{\beta\nu}{2}\displaystyle\int_\Omega [(2q_{12h})^2 + (q_{22h}-q_{11h})^2] dx + gj_h(q_h) \\[2mm] - (f((n+1)k),\phi_h) - \dfrac{1}{k}\displaystyle\int_\Omega \nabla\psi_h^n\cdot\nabla\phi_h dx . \end{cases}
$$

It can easily be shown that problem (4.56A) admits one and only one solution; moreover the comments in Section 4.4.5 on the numerical integration also hold for problem (4.56A).

Using the techniques developed in Chapter 6, and under suitable regularity assumptions on ψ_o, f and ψ, it is possible to prove the convergence of $\{\psi_h^n\}$ to ψ as $h, k \to 0$, provided that the assumptions on $(\mathcal{E}_h)_h$ stated in Theorem 4.2 are still satisfied.

4.6 Solution of (4.19A), (4.54A) by augmented Lagrangian methods

4.6.1 *Synopsis*

In this section we shall show that it is possible to solve the steady-state problem (4.19A), or the sequence of problems (4.54A) (obtained by the *semi-discretisation in time* of the time-dependent problem (4.15A)), by augmented Lagrangian methods which are variants of algorithm (9.6)-(9.8) of Chapter 5, Section 9. We shall restrict our attention to the case which is continuous with respect to the space variables, but the extension to approximate problems in space and time does not present any particular difficulties (apart from the manipulation of a somewhat cumbersome formalism).

4.6.2 *The model problem. Introduction of an augmented Lagrangian*

Problems (4.19A) and (4.54A) lead us to consider the minimisation problem

$$
(4.58\text{A})\quad \begin{cases} \textit{Find } \psi \in H_o^2(\Omega) \textit{ such that} \\ J(\psi) \le J(\phi) \quad \forall\phi \in H_o^2(\Omega) \end{cases}
$$

where

$$(4.59A)\quad \begin{cases} J(\phi) = \frac{\gamma}{2}\int_\Omega |\nabla\phi|^2 dx + \frac{\nu}{2}\int_\Omega |\Delta\phi|^2 dx + g\int_\Omega [(2\,\frac{\partial^2\phi}{\partial x_1 \partial x_2})^2 \\ \quad + (\frac{\partial^2\phi}{\partial x_2^2} - \frac{\partial^2\phi}{\partial x_1^2})^2]^{1/2} dx - (f,\phi) \end{cases}$$

and $\gamma \geq 0$ ($\gamma = 0$ in the steady-state case, $\gamma = \frac{1}{k}$ if (4.58A) originates from (4.54A)). The essential difficulty in the solution of (4.58A) arises from the *non-differentiable* functional

$$\phi \to \int_\Omega [(2\,\frac{\partial^2\phi}{\partial x_1\partial x_2})^2 + (\frac{\partial^2\phi}{\partial x_2^2} - \frac{\partial^2\phi}{\partial x_1^2})^2]^{1/2}\, dx\ ;$$

hence, taking (with $q = \{q_1, q_2\}$)

$$W = \{\{\phi,q\} \in H_0^2(\Omega) \times (L^2(\Omega))^2,\ q_1 = 2\,\frac{\partial^2\phi}{\partial x_1\partial x_2},\ q_2 = \frac{\partial^2\phi}{\partial x_2^2} - \frac{\partial^2\phi}{\partial x_1^2}\},$$

it is clear that (4.58A) is equivalent to

$$(4.60A)\quad \begin{cases} \textit{Find } \{\psi,p\} \in W \textit{ such that} \\ j(\psi,p) \leq j(\phi,q) \quad \forall\{\phi,q\} \in W, \end{cases}$$

where (with $|q(x)| = \sqrt{q_1^2(x) + q_2^2(x)}$)

$$j(\phi,q) = \frac{\gamma}{2}\int_\Omega |\nabla\phi|^2 dx + \frac{\nu}{2}\int_\Omega |\Delta\phi|^2 dx + g\int_\Omega |q|\,dx - (f,\phi).$$

It is then natural to associate with (4.60A) the *augmented Lagrangian* $\mathcal{L}_r : H_0^2(\Omega) \times (L^2(\Omega))^2 \times (L^2(\Omega))^2$ defined (with $r > 0$ and $\mu = \{\mu_1, \mu_2\}$) by

$$\begin{cases} \mathcal{L}_r(\phi,q,\mu) = j(\phi,q) + \frac{r}{2}\int_\Omega |2\,\frac{\partial^2\phi}{\partial x_1\partial x_2} - q_1|^2\, dx \\ \quad + \frac{r}{2}\int_\Omega |(\frac{\partial^2\phi}{\partial x_2^2} - \frac{\partial^2\phi}{\partial x_1^2}) - q_2|^2 dx + \int_\Omega \mu_1 (2\,\frac{\partial^2\phi}{\partial x_1\partial x_2} - q_1)\,dx \\ \quad + \int_\Omega \mu_2 (\frac{\partial^2\phi}{\partial x_2^2} - \frac{\partial^2\phi}{\partial x_1^2} - q_2)\,dx - (f,\phi). \end{cases}$$

It can easily be shown that if $\{\psi,p,\lambda\}$ is a saddle point of

$\mathcal{L}_r$ on $H_0^2(\Omega) \times (L^2(\Omega))^2 \times (L^2(\Omega))^2$, then ψ is a solution of problem (4.58A) and

$$p = \{2 \frac{\partial^2\psi}{\partial x_1 \partial x_2}, \frac{\partial^2\phi}{\partial x_2^2} - \frac{\partial^2\phi}{\partial x_1^2}\}.$$

4.6.3 *An augmented-Lagrangian algorithm*

In view of Section 4.6.2 it is natural to solve (4.58A) by seeking saddle points of $\mathcal{L}_r$ on $H_0^2(\Omega) \times (L^2(\Omega))^2 \times (L^2(\Omega))^2$; by analogy with Chapter 2, Section 4, we are led to the following (Uzawa type) algorithm:

(4.61A) $\lambda^o \in (L^2(\Omega))^2$ *given*

then, for $n \geq 0$, $\lambda^n \in (L^2(\Omega))^2$ *known, determine* $\psi^n, p^n, \lambda^{n+1}$ *using*

$$\text{(4.62A)} \quad \begin{cases} \{\psi^n, p^n\} \in H_0^2(\Omega) \times (L^2(\Omega))^2 \text{ and} \\ \mathcal{L}_r(\psi^n, p^n, \lambda^n) \leq \mathcal{L}_r(\phi, q, \lambda^n) \quad \forall \{\phi, q\} \in H_0^2(\Omega) \times (L^2(\Omega))^2, \end{cases}$$

$$\text{(4.63A)} \quad \lambda_1^{n+1} = \lambda_1^n + \rho(2 \frac{\partial^2\psi^n}{\partial x_1 \partial x_2} - p_1^n),$$

$$\text{(4.64A)} \quad \lambda_2^{n+1} = \lambda_2^n + \rho(\frac{\partial^2\psi^n}{\partial x_2^2} - \frac{\partial^2\psi^n}{\partial x_1^2} - p_2^n).$$

It then follows from Fortin-Glowinski /1A, Chapter 3/, Glowinski /2A, Chapter 5/ and Theorem 3.1 of Appendix 2, Section 3 that regarding the convergence of (4.61A)-(4.64A) we have

<u>Theorem 4.3</u>: *Suppose that* $\mathcal{L}_r$ *admits a saddle point* $\{\psi, p, \lambda\}$ *on* $H_0^2(\Omega) \times (L^2(\Omega))^2 \times (L^2(\Omega))^2$; *then if*

(4.65A) $0 < \rho < 2r$

we have $\forall \lambda^o \in (L^2(\Omega))^2$,

(4.66A) $\lim_{n\to+\infty} \psi^n = \psi$ *strongly in* $H_0^2(\Omega)$

(4.67A) $\lim_{n\to+\infty} p^n = p$ *strongly in* $(L^2(\Omega))^2$

(4.68A) $$\lim_{n\to+\infty} \lambda^n = \lambda \quad \textit{weakly in } (L^2(\Omega))^2,$$

where λ *is such that* $\{\psi,p,\lambda\}$ *is a saddle point of* $\mathcal{L}_r$ *on* $H_0^2(\Omega) \times (L^2(\Omega))^2 \times (L^2(\Omega))^2$. ■

It is clear that the essential difficulty in the solution of (4.58A), via the intermediary of algorithm (4.61A)-(4.64A), lies in solving problem (4.62A) at each iteration; in view of the special structure of $\mathcal{L}_r$, we can solve this problem by a *block-overrelaxation* method of the type examined in Cea-Glowinski /2/; we refer to Fortin-Glowinski /1A, Chapter 3/, Glowinski /2A, Chapter 5/ and Bourgat-Dumay-Glowinski /1A/ for further details on the solution of problems of the type (4.62A) and on the optimal choice of the parameter r; the optimal choice for ρ appears (on the basis of a large number of numerical experiments) to lie very close to $\rho = r$.

4.6.4 *On variants of algorithm* (4.61A)-(4.64A)

If, when solving problem (4.62A) by block relaxation, we perform only a single inner iteration, we are led to the following variant of algorithm (4.61A)-(4.64A):

(4.69A) $$\{\psi^{-1},\lambda^0\} \in H_0^2(\Omega) \times (L^2(\Omega))^2 \textit{ given}$$

then, for $n \geq 0$, $\{\psi^{n-1},\lambda^n\}$ *known, successively determine* p^n,ψ^n,λ^{n+1} *from*

(4.70A) $$\begin{cases} p^n \in L^2(\Omega) \times L^2(\Omega), \\ \mathcal{L}_r(\psi^{n-1},p^n,\lambda^n) \leq \mathcal{L}_r(\psi^{n-1},q,\lambda^n) \quad \forall q \in L^2(\Omega) \times L^2(\Omega), \end{cases}$$

(4.71A) $$\begin{cases} \psi^n \in H_0^2(\Omega), \\ \mathcal{L}_r(\psi^n,p^n,\lambda^n) \leq \mathcal{L}_r(\phi,p^n,\lambda^n) \quad \forall\phi \in H_0^2(\Omega), \end{cases}$$

(4.72A) *Determine* λ^{n+1} *using* (4.63A), (4.64A).

The convergence results of Theorem 4.3 also hold (see Fortin-Glowinski, Glowinski, loc. cit.) if in place of (4.65A) we have

$$0 < \rho < \frac{1+\sqrt{5}}{2}\ r. \tag{4.73A}$$

It will be seen that the solution of problems (4.70A), (4.71A) does not pose any difficulties; in fact problem (4.71A) is equivalent to solving the following problem (of linear biharmonic type (1)):

$$\begin{cases} \psi^n \in H_0^2(\Omega) \quad \textit{and} \\ \gamma\int_\Omega \nabla\psi^n\cdot\nabla\phi\, dx + (\nu+r)\int_\Omega \Delta\psi^n\Delta\phi\, dx = (f,\phi) \\ + \int_\Omega (rp_1^n-\lambda_1^n)\left(2\,\frac{\partial^2\phi}{\partial x_1\partial x_2}\right)dx + \int_\Omega (rp_2^n-\lambda_2^n)\left(\frac{\partial^2\phi}{\partial x_2^2}-\frac{\partial^2\phi}{\partial x_1^2}\right)dx \\ \forall\,\phi \in H_0^2(\Omega) \end{cases} \tag{4.74A}$$

which *admits one and only one solution*. Regarding (4.70A), it can easily be shown that this problem admits a unique solution, given explicitly by

$$p^n = \frac{1}{r}\left(|X^n|-g\right)^+ \frac{X^n}{|X^n|}, \tag{4.75A}$$

where $X^n = \{X_1^n, X_2^n\}$ is defined by

$$X_1^n = \lambda_1^n + 2r\,\frac{\partial^2\psi^{n-1}}{\partial x_1\partial x_2}, \tag{4.76A}$$

$$X_2^n = \lambda_2^n + r\left(\frac{\partial^2\psi^{n-1}}{\partial x_2^2} - \frac{\partial^2\psi^{n-1}}{\partial x_1^2}\right), \tag{4.77A}$$

and where $|q| = \sqrt{q_1^2+q_2^2}$ if $q = \{q_1,q_2\}$.

Once again it would appear that the optimal choice for ρ is close to $\rho = r$; the optimal choice for r, however, is a more

---

(1) Relating to the operator $-\gamma\Delta + (\nu+r)\Delta^2$

complicated problem, since it is apparent that the optimal value for r is a function of - amongst other things - ν and g.

To conclude this section we should mention the variant (due to Gabay /1A/) of algorithm (4.69A)-(4.72A) in which, with (4.69A), (4.70A) unchanged, $\lambda^{n+\frac{1}{2}}$ is determined using

(4.78A) $$\lambda_1^{n+1/2} = \lambda_1^n + \rho\left(2\,\frac{\partial^2\psi^{n-1}}{\partial x_1\partial x_2} - p_1^n\right),$$

(4.79A) $$\lambda_2^{n+1/2} = \lambda_2^n + \rho\left(\frac{\partial^2\psi^{n-1}}{\partial x_2^2} - \frac{\partial^2\psi^{n-1}}{\partial x_1^2} - p_2^n\right),$$

and then ψ^n is determined using

(4.80A) $$\begin{cases} \psi^n \in H_o^2(\Omega), \\ \mathscr{L}_r(\psi^n, p^n, \lambda^{n+1/2}) \le \mathscr{L}_r(\phi, p^n, \lambda^{n+1/2}) \quad \forall\, \phi \in H_o^2(\Omega), \end{cases}$$

and finally λ^{n+1} is obtained from (4.63A), (4.64A) with $\lambda^{n+1/2}$ in place of λ^n. We remark that in the above algorithm the variables ϕ and q play a *symmetric* role; this is not the case in algorithm (4.69A)-(4.72A) in which the order chosen is governed by ellipticity considerations (for the details of which we refer to Fortin-Glowinski /1A, Chapter 3/).

Section 4.7 below presents the results of numerical experiments in which the above algorithms (actually their *finite-dimensional* variants) have been applied to the solution of problems (4.15A) and (4.19A).

4.7 Numerical experiments

In this section we shall describe certain numerical results of Begis /1A/, obtained by the methods described in the previous sections; the above reference also gives various comments and other numerical experiments.

4.7.1 *Formulation of the test problem*

Let $\Omega =]0,1[\times]0,1[$; we consider a family of Bingham fluid flows in Ω, for which

(4.81A) $\nu=1$,

(4.82A) $f=0$,

(4.83A)
$$\begin{cases} u|_\Gamma = z = \{z_1, z_2\} \text{ with } z_2 = 0 \text{ on } \Gamma, \\ z_1(0,x_2) = z_1(1,x_2) = 0 \quad \forall x_2 \in]0,1[\ , \\ z_1(x_1,1) = 0 \ \forall x_1 \in]0,1[\ , \ z_1(x_1,0) = 1 \ \forall x_1 \in]0,1[. \end{cases}$$

These are thus problems of the *sliding wall* type; we remark that $\int_\Gamma z \cdot n \, d\Gamma = 0$, but also that $z \notin H^{1/2}(\Gamma) \times H^{1/2}(\Gamma)$ (we have $z \in H^s(\Gamma) \times H^s(\Gamma)$ with $s < \frac{1}{2} - \varepsilon$, $\forall \varepsilon > 0$).

4.7.2 *Numerical results*

We refer to Bégis /1A/ for details of the actual implementation of the approximation and iterative solution methods described in the previous sections.

Steady-state cases: For various values of g, Figures 4.1-4.5 show the rigid (hatched) and visco-plastic (blank) regions, together with the streamlines. The growth of the rigid region with increasing g can be observed.

Time-dependent cases: For various values of g, we have examined the case in which the medium filling the cavity Ω is initially at rest (hence $u(x,0) = 0 \ \forall x \in \Omega \Leftrightarrow \psi(x,0) = 0 \ \forall x \in \Omega$), and is set in motion by the sliding (defined by (4.83A)) of the lower wall of Ω. When steady conditions have more or less been attained, we arrest the motion of the lower wall in order to examine the return to the corresponding steady state, which is $u = 0$ on $\Omega \Leftrightarrow \psi = 0$ on Ω.

For various values of g (including $g = 0$), Figure 4.6 shows the behaviour of

$$t \to \int_\Omega |\psi(x,t)|^2 dx \ ;$$

we observe that for g > 0 *the rigid state is attained after a finite time* which decreases as g increases; this is in agreement with mechanical intuition.

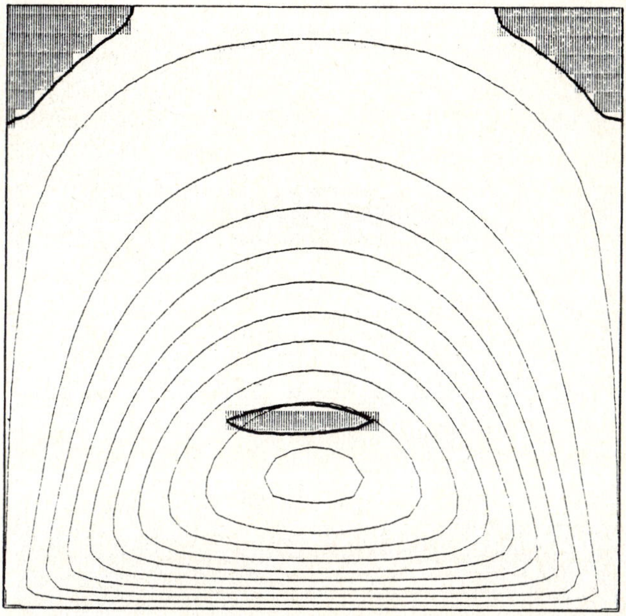

Fig. 4.1 (g = 0.5)

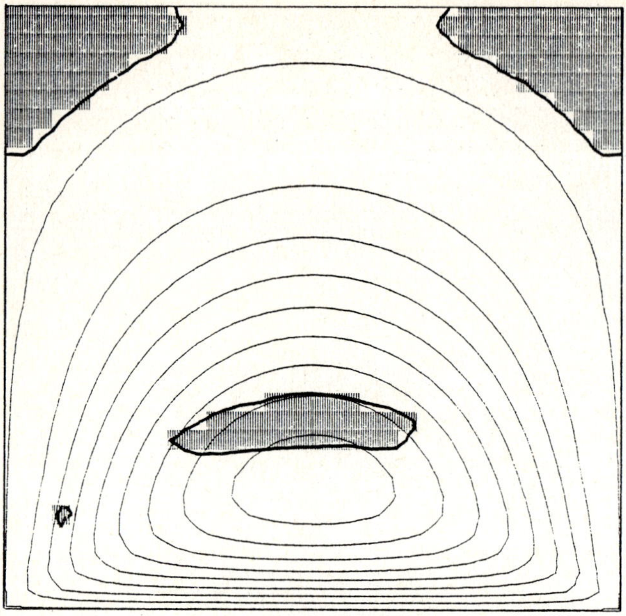

Fig. 4.2 (g = 1)

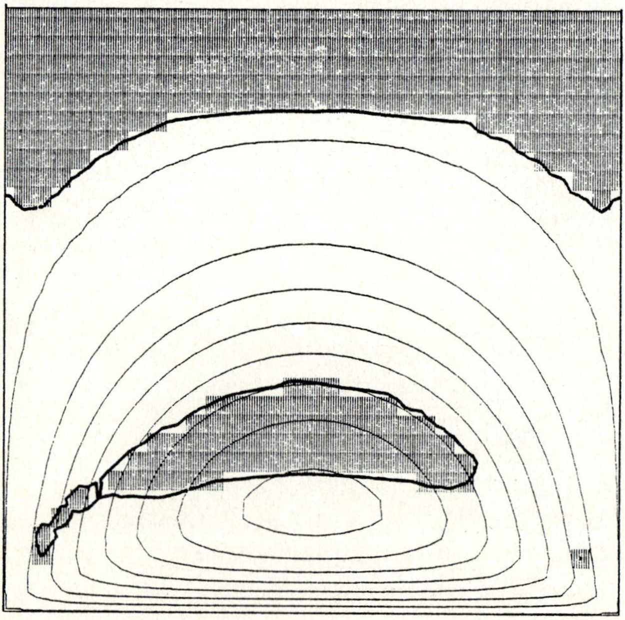

Fig. 4.3 (g = 2.5)

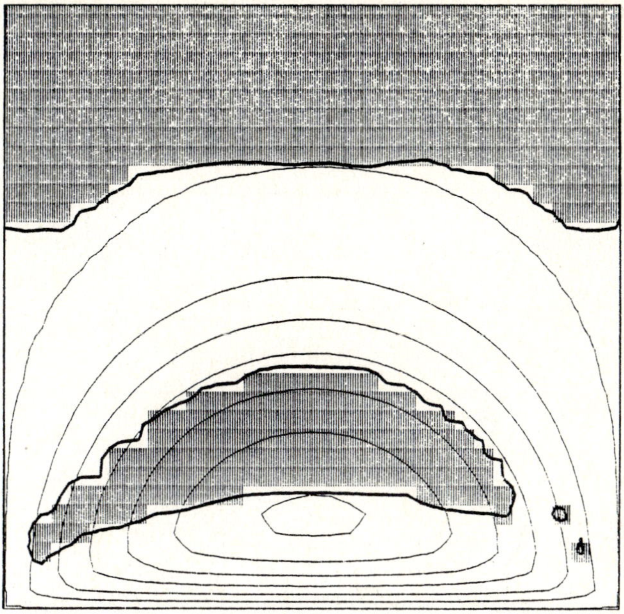

Fig. 4.4 (g = 5)

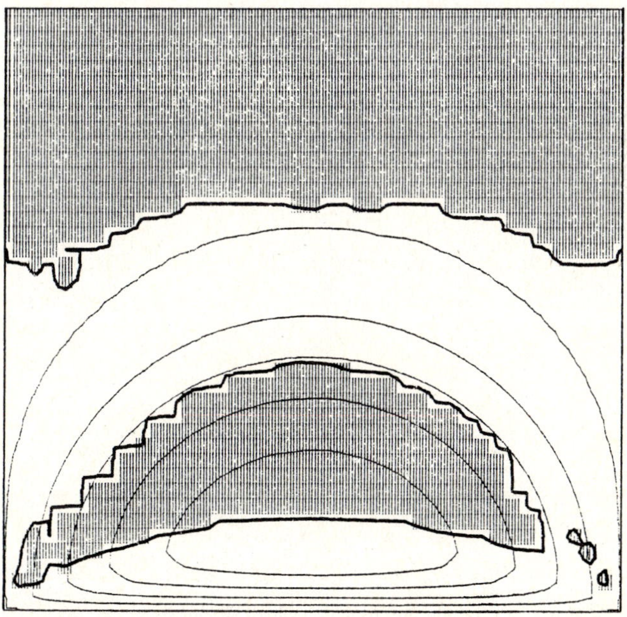

Fig. 4.5 (g = 10)

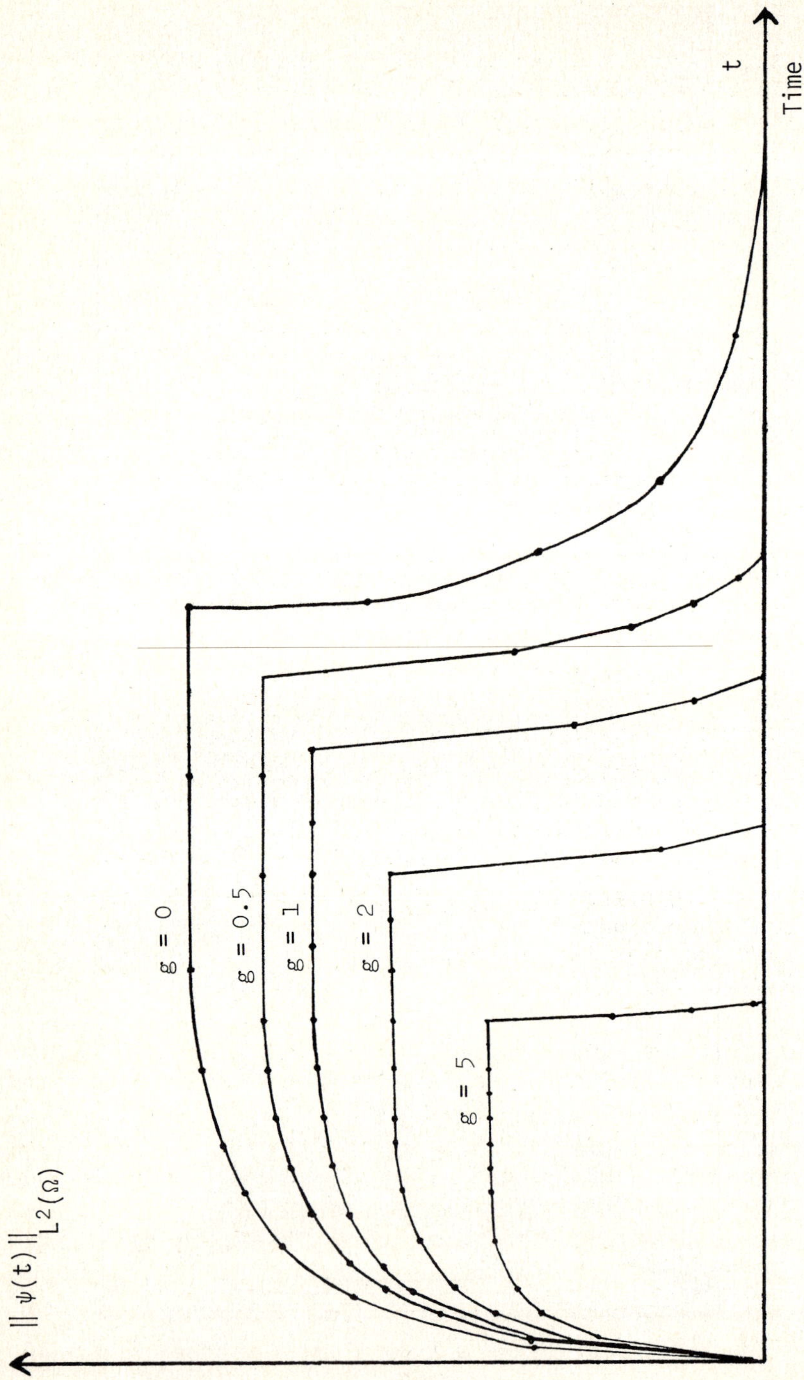

Fig. 4.6

BIBLIOGRAPHY OF THE APPENDICES

AXELSSON, O. /1A/ A class of iterative methods for finite element equations. *Comp. Meth. Appl. Mech. Eng.*, 9 (1976), 2, pp. 123-138.

BAIOCCHI, C. /1A/ Sur un problème à frontière libre traduisant le filtrage de liquides à travers des milieux poreux. *Comptes Rendus Acad. Sc. Paris*, 273A, (1971), pp. 1215-1217.

/2A/ Free boundary problems in the theory of fluid flow through porous media. *Proc. of the Int. Congress of Math.* (Vancouver 1974), 2, 1975, pp. 237-243.

/3A/ Studio di un problema quazi-variazionale connesso a problemi di frontiera libera. *Boll. U.M.I.* 11, (1975), 4 (suppl. fasc. 3) pp. 589-613.

BAIOCCHI, C., BREZZI, F., COMINCIOLI, V. /1A/ Free boundary problems in fluid flow through porous media. *2nd Int. Symp. Finite Element Methods in Flow Problems*, Santa Margharita, 1976, pp. 407-420.

BAIOCCHI, C., CAPELO, A. /1A/ *Dissequazioni Variazionali e Quasivariazionali. Applicazioni a problemi di frontiera libera.* Quaderni 4 e 7 dell' Unione Mathematica Italiana, Pitagor Editrice, Bologna 1978.

BAIOCCHI, C., COMINCIOLI, V., MAGENES, E., POZZI, G.A. /1A/ *Fluid flow through porous media; a new theoretical and numerical approach.* Publicazioni 69, L.A.N. - C.N.R., Pavia (1974).

BAIOCCHI, C., POZZI, G.A. /1A/ Error estimates and free-boundary convergence for a finite-difference discretisation of a parabolic variational inequality. *Rev. Franc. Autom. Info. Rech. Op., Anal. Num.*, 11, (1977), 4, pp. 315-340.

BARTELS, R., DANIEL, J.W. /1A/ A conjugate-gradient approach to nonlinear boundary-value problems in irregular regions; in *Conference on the Numerical Solution of Differential Equations, Dundee*, 1973, G.A. Watson (ed.), Lecture Notes in Math., 363, Springer-Verlag, Berlin, 1974.

BEGIS, D. /1A/ Analyse Numérique de l'écoulement d'un fluide viscoplastique de Bingham par une méthode de Lagrangien augmenté. *LABORIA Report* 355, 1979.

BENSOUSSAN, A., LIONS, J.L. /1A/ Nouvelle formulation de problèmes de contrôle impulsionnel et applications. *Comp. Rend. Acad. Sc. Paris*, 276A (1973), pp. 1189-1192.

/2A/ Nouvelles méthodes en contrôle impulsionnel. *Appl. Math. & Opt.*, 1, (1975), 4, pp. 289-312.

/3A/ Optimal impulse and continuous control: method of nonlinear quasi-variational inequalities. *Trudy Mat. Inst. Steklov*, 134 (1975), A.M.S. transl. pp. 7-25.

/4A/ Contrôle impulsionnel et inéquations quasi-variationnelles d'évolution. *Compt. Rend. Acad. Sci. Paris*, 276A, (1973), pp. 1279-1284.

BERGER, A.E. /1A/ The truncation method for the solution of a class of variational inequalities. *Rev. Franc. Autom. Inf. Rech. Op., Anal. Num.*, 10 (1976) 3, pp. 29-42.

BERGER, A.E., FALK, R.S. /1A/ An error estimate for the truncation method for the solution of parabolic obstacle variational inequalities. *Math. of Comp.*, 31 (1977), pp. 619-628.

BOGOMOLNII, A., ESKIN, G., ZUCHOWISKII, S. /1A/ Numerical solution of the stamp problem. *Comp. Meth. Appl. Mech. Eng.*, 15 (1978) 2, pp. 149-159.

BOURGAT, J.F., DUMAY, J.M., GLOWINSKI, R. /1A/ Large displacement calculations of flexible pipe lines by finite elements and nonlinear programming methods. *SIAM J. Sc. and Stat. Computing*, 1, (1980), 1, pp. 34-81.

BRENT, R. /1A/ *Algorithms for minimisation without derivatives.* Prentice Hall, Englewood Cliffs, N.J., 1973.

BREZIS, H. /1A/ A new method in the study of subsonic flows. In *Partial Differential Equations and Related Topics*, J. Goldstein (ed.), Lecture Notes in Math., 446, Springer-Verlag, Berlin, 1975, pp. 50-60.

BREZIS, H., STAMPACCHIA, G. /1A/ The hodograph method in fluid dynamics in the light of variational inequalities. *Arch. Rat. Mech. Anal.* 61, (1976), pp. 1-18.

/2A/ Remarks on some fourth-order variational inequalities, *Ann. Scu. Norm. Sup. Pisa*, 4, (1977), 4, pp. 363-371.

BREZZI, F., HAGER, W.W., RAVIART, P.A. /1A/ Error estimate for the finite-element solution of variational inequalities. (I) Primal theory; *Numerische Math.*, 18, (1977), pp. 431-443.

/2A/ Error estimates for the finite element solution of variational inequalities. (II) Mixed Methods, *Numerische Math.*, 31 (1978), pp. 1-16.

BREZZI, F., JOHNSON, C., MERCIER, B. /1A/ Analysis of a mixed finite-element method for elasto-plastic plates. *Math. of Comp.*, 31 (1977) 140, pp. 809-817.

BRISTEAU, M.O. /1A/ *Application de la méthode des éléments finis à la résolution d'inéquations variationnelles d'évolution de type Bingham.* Third-Cycle Thesis, Université Pierre et Marie Curie, Paris, 1975.

BRISTEAU, M.O., GLOWINSKI, R. /1A/ Finite-element analysis of the unsteady flow of a viscous-plastic fluid in a cylindrical pipe; in *Finite Element Methods in Flow Problems*, J.T. Oden, O.C. Zienkiewicz, R.H. Gallagher, C. Taylor (eds.), Univ. of Alabama Press, Huntsville, Alabama, 1974, pp. 471-488.

BRISTEAU, M.O., GLOWINSKI, R., PERIAUX, J., PERRIER, P., PIRONNEAU, O. /1A/ On the numerical solution of nonlinear problems in Fluid Dynamics by least-squares and finite-element methods. (I) Least-square formulations and conjugate-gradient solution of the continuous problems, *Comp. Meth. Appl. Mech. Eng.*, 17/18 (1979), pp. 619-657.

/2A/ Transonic flow simulations by finite elements and least-square methods; in *Proc. of the Third Conf. on Finite Elements in Flow Problems*, Calgary, Canada, June 1980.

BRISTEAU, M.O., GLOWINSKI, R., PERIAUX, J., PERRIER, P., PIRONNEAU, O., POIRIER, G. /1A/ Application of Optimal Control and Finite Element Methods to the Calculation of Transonic Flows and Incompressible Viscous Flows; in *Numerical Methods in Applied Fluid Dynamics*, B. Hunt (ed.), Acad. Press, London, 1980 (and also *LABORIA Report* 294 (1978)).

CAFFARELLI, L.A., FRIEDMAN, A., TORELLI, A. /1A/ The free boundary for a fourth-order variational inequality. *Pub.* 227, *Lab. di Anal. Num. del C.N.R.*, Pavia, 1979.

CAPRIZ, G. /1A/ Variational techniques for the analysis of a lubrication problem; in *Mathematical Aspects of Finite Element Methods*, I. Galligani, E. Magenes (Eds.), Lecture Notes in Math., 606, Springer-Verlag, Berlin, 1977, pp. 47-55.

CHAN, T.F., GLOWINSKI, R. /1A/ Numerical methods for solving some mildly nonlinear elliptic partial differential equations. *Stanford Report* STAN-CS-674, Comp. Sciences Dept., Stanford University, 1978.

/2A/ Numerical methods for a class of mildly nonlinear elliptic equations. *Atos do Decime Primeiro Coloquio Brasiliero de Matematica*, *Vol.* I, IMPA, Rio de Janeiro, 1978, pp. 279-318.

CIARLET, P.G. /1A/ *The finite element method for elliptic problems*, North-Holland, Amsterdam, 1978.

CIARLET, P.G. RAVIART, P.A. /1A/ Maximum principle and uniform convergence for the finite element method, *Comp. Meth. Appl. Mech. Eng.*, 2, (1973), pp. 17-31.

CIAVALDINI, J.F., POGU, M., TOURNEMINE, G. /1A/ Une méthode variationnelle non linéaire pour l'étude dans le plan physique d'écoulements compressibles subcritiques en atmosphère infinie. *Compt. Rend. Acad. Sc. Paris*, 281 A, (1975), pp. 1105-1108.

COMINCIOLI, V. /1A/ On some oblique derivative problems arising in the fluid flow in porous media. A theoretical and numerical approach. *Applied Math. and Optimisation* (1975), pp. 313-336.

COMINCIOLI, V., TORELLI, A. /1A/ A new numerical approach to a non-steady filtration problem, *Calcolo*, 16, (1979), 1, pp. 93-124.

CONCUS, P., GOLUB, G.H., O'LEARY, D.P. /1A/ Numerical solution of nonlinear partial differential equations by a generalised conjugate-gradient method. *Computing*, 19, (1977), pp. 321-340.

CORTEY-DUMONT, P. /1A/ Contribution à l'étude de l'approximation par la méthode des éléments finis d'inéquations quasi-variationnelles elliptiques. *Compt. Rend. Acad. Sc. Paris*, 288A (1979), p. 141-143.

COTTLE, R.W. /1A/ Computational experience with large-scale linear complementarity problems; in *Fixed Points: Algorithms and Applications*, S. Karamardian (ed.), Acad. Press, New-York, 1977, pp. 281-313.

/2A/ Numerical methods for complementarity problems in engineering and applied science; in *Computing Methods in Applied Sciences and Engineering*, 1977, 1, R. Glowinski, J.L. Lions (eds.), Lecture Notes in Math., Vol. 704, Springer-Verlag, Berlin, 1979, pp. 37-52.

COTTLE, R.W., SACHER, R.S. /1A/ On the solution of large, structured linear complementarity problems: the tridiagonal case, *Applied Math. Optimiz.*, 3, (1977) 4, pp. 321-340.

COTTLE, R.W., GOLUB, G.H., SACHER, R.S. /1A/ On the solution of large, structured linear complementarity problems: the block partitioned case. *Applied Math. Optimiz.* 4, (1978), 4, pp. 347-364.

CRYER, C.W., DEMPSTER, M.A.H. /1A/ Equivalence of linear complementarity problems and linear programs in vector lattice Hilbert spaces, *SIAM J. Control and Optimiz.*, 18 (1980) 1, pp. 76-90.

CRYER, C.W., FETTER, H. /1A/ The numerical solution of axi-symmetric free-boundary porous-flow well problems using variational inequalities. *M.R.C. Technical Summary Report* 1761, University of Wisconsin, Madison, 1977.

DOUGLAS, J., DUPONT, T. /1A/ Preconditioned conjugate gradient iteration applied to Galerkin methods for mildly nonlinear Dirichlet problems; in *Sparse Matrix Computations*, J.R. Bunch, D.J. Rose (eds.), Acad. Press, New-York, 1976, pp. 333-348.

/2A/ Interior penalty procedures for elliptic and parabolic Galerkin methods; in *Computing Methods in Applied Sciences*, R. Glowinski, J.L. Lions (eds.), Lecture Notes in Physics, Vol. 58, Springer-Verlag, Berlin, 1976, pp. 207-216.

DOUGLAS, J., RACHFORD, H.H. /1A/ On the numerical solution of the heat conduction problem in 2 and 3 space variables. *Trans. Amer. Math. Soc.*, 82 (1956), pp. 421-439.

ESKIN, G.I. /1A/ About a variational-difference method for solution of elliptic pseudo-differential equations. *Uspeki Mat. Nauk.*, 28, (1973), pp. 255-256.

FALK, R.S. /1A/ Approximation of an elliptic boundary value problem with unilateral constraints. *Rev. Franc. Autom. Rech. Op., Anal. Num.*, R2, (1975), pp. 5-12.

FALK, R.S., MERCIER, B. /1A/ Error estimates for elasto-plastic problems. *Rev. Fr. Auto. Inf. Rech. Op., Anal. Num.*, 11,(1977), pp. 135-144.

FORTIN, M., GLOWINSKI, R. /1A/ *Résolution numérique de problèmes aux limites par des méthodes de Lagrangien augmenté.* Book in preparation.

/2A/ Méthodes de décomposition-coordination par lagrangien augmenté. Chapter 3 of *Résolution numérique de problèmes aux limites par des méthodes de lagrangien augmenté*, M. Fortin, R. Glowinski (eds.) (to appear).

GABAY, D. /1A/ Résolution et décomposition des inéquations variationnelles par des méthodes de multiplicateurs; in *Thèse d'Etat*, Université Pierre et Marie Curie, Paris, 1979.

GIRAULT, V., RAVIART, P.A. /1A/ An Analysis of a Mixed Finite Element Method for the Navier-Stokes Equations. *Numerische Math.*, 33, (1979), pp. 235-271.

GLOWINSKI, R. /1A/ *Introduction to the approximation of variational inequalities.* Report 76006, Laboratoire d'Analyse Numérique, Université Pierre et Marie Curie, 1976.

/2A/ *Numerical Methods for Nonlinear Variational Problems.* Lecture Notes, Tata Institute, Bombay, and Springer-Verlag, Berlin, 1980.

/3A/ Finite elements and variational inequalities. Chapter 12 of *The Mathematics of Finite Elements and Applications, III, MAFELAP* 1978, J.R. Whiteman (ed.), Acad. Press, London, 1979, pp. 135-171.

/4A/ Sur l'approximation d'une inéquation variationnelle elliptique de type Bingham. *Rev. Franc. Autom. Inf. Rech, Op., Anal. Num.*, 10 (1976), pp. 13-30.

GLOWINSKI, R., MARINI, D., VIDRASCU, M. /1/ Finite-element approximations and iterative solution of a fourth-order variational inequality (to appear).

GLOWINSKI, R., MARRACCO, A. /1A/ Numerical solution of two-dimensional magnetostatic problems by augmented Lagrangian methods. *Comp. Meth. Appl. Mech. Eng.*, 12 (1977), pp. 33-46.

/2A/ Chapter 5 of *Résolution Numérique de Problèmes aux Limites par des Méthodes de Lagrangiens Augmentés*, M. Fortin, R. Glowinski (eds.) (to appear).

GLOWINSKI, R., PIRONNEAU, O. /1A/ Numerical methods for the first biharmonic equation and for the two-dimensional Stokes problem. *SIAM Review*, 21 (1979) 2, pp. 167-212.

GOURSAT, M., MAAREK, G. /1A/ Nouvelle approche des problèmes de gestion de stocks. Comparaison avec les méthodes classiques. *LABORIA Report* 148, March 1976.

HANOUZET, B., JOLY, J.L. /1A/ Convergence uniforme des itérés définissant la solution d'inéquations quasi-variationnelles et application à la régularité, *Num. Funct. Anal. and Optimiz.* 1, (1979), 4, pp. 399-414.

HASLINGER, J. /1A/ Finite-element analysis for unilateral problems with obstacles on the boundary. *Aplikace Matematiky*, 22, (1977) 3, pp. 180-188.

/2A/ On numerical solution of a variational inequality of the 4th order by finite element method. *Aplikace Matematiky*, 23, (1978) 5, pp. 334-345.

HLAVACEK, I. /1A/ Dual finite-element analysis for unilateral boundary value problems. *Aplikace Matematiky*, 22, (1977), pp. 14-51.

/2A/ Dual finite-element analysis for elliptic problems with obstacles on the boundary. *Aplikace Matematiky*, 22, (1977), pp. 244-255.

/3A/ Dual finite-element analysis for semi-coercive unilateral boundary value problems. *Aplikace Matematiky*, 23, (1978), pp. 52-71.

HLAVACEK, I., LOVISEK, J. /1A/ A finite-element analysis for the Signorini problem in plane elastostatics. *Aplikace Matematiky* 22, (1977), pp. 215-228.

HORMANDER, L., LIONS, J.L. /1A/ Sur la complétion par rapport à une intégrale de Dirichlet. *Math. Scand.*, 4 (1956), pp. 259-270.

HUNT, B. (ed.) /1A/ *Numerical Methods in Applied Fluid Dynamics*. Acad. Press, London, 1980.

HUNT, C., NASSIF, N.R. /1A/ On a variational inequality and its application in the theory of semi-conductors. *SIAM J. Num. Anal.*, 12, (1975), 6, pp. 938-950.

JAMESON, A. /1A/ Transonic flow calculations; in *Numerical Methods in Fluid Dynamics*, H.J. Wirz, J.J. Smolderen (eds.), McGraw-Hill, New-York, 1978, pp. 1-87.

JOHNSON, C. /1A/ An elasto-plastic contact problem. *Rev. Franc. Aut. Inf. Rech. Op., Anal. Num.*, 12, (1978), 1, pp. 59-74.

/2A/ A convergence estimate for an approximation of a parabolic variational inequality. *SIAM J. Num. Anal.*, 13 (1976), 4, pp. 599-606.

/3A/ A mixed finite-element method for plasticity with hardening. *SIAM J. Num. Anal.*, 14 (1977), 4, pp. 575-583.

/4A/ A finite-element method for consolidation of clay. *Comp. Meth. Appl. Mech. Eng.*, 16, (1978), 2, pp. 177-184.

KAWOHL, B. /1A/ On a mixed Signorini problem. *Num. Funct. Anal. and Optimiz.*, 1, (1979), 6, pp. 633-645.

KIKUCHI, N., ODEN, J.T. /1A/ *Contact problems in Elasticity.* TICOM Report 79-8, Univ. of Texas at Austin, July 1979.

LABORDE, P. /1A/ On visco-plasticity with hardening. *Num. Func. Anal. and Optimiz.*, 1, (1979), 3, pp. 315-339.

LAETSCH, T.H. /1A/ A uniqueness theorem for elliptic Q.V.I., *J. Functional Analysis*, 18, (1975), pp. 286-288.

LANDAU, L., LIFCHITZ, E. /1A/ *Mécanique des Fluides.* Mir, Moscow, 1953.

LE TALLEC, P. /1A/ *Simulation numérique d'écoulements visqueux incompressibles par des méthodes d'éléments finis mixtes.* Third-Cycle thesis, Université Pierre et Marie Curie, 1978.

LIEUTAUD, J. /1A/ *Approximation d'opérateurs par des méthodes de décomposition.* Thèse d'Etat, Université Paris VI, 1968.

LIONS, P.L., MERCIER, B. /1A/ Splitting algorithms for the sum of two nonlinear operators. *SIAM J. Num. Analysis*, 16, (1979), 6, pp. 964-979.

LÖTSTEDT, P. /1A/ Analysis of some difficulties encountered in the simulation of mechanical systems with constraints. Dpt. of Num. Anal. and Comp. Sci., *Royal Inst. of Tech., Stockholm*, TRITA-NA-7914, 1979.

/2A/ On a penalty function method for the simulation of mechanical systems subject to constraints. *Dpt. of Num. Anal. and Comp. Sci., Royal Inst. of Tech., Stockholm*, TRITA-NA-7919, 1979.

/3A/ A numerical method for the simulation of mechanical systems with unilateral constraints. *Dpt. of Num. Anal. and Comp. Sci., Royal Inst. of Tech., Stockholm*, TRITA-NA-7920, 1979.

/4A/ Interactive simulation of the progressive collapse of a building, revisited. *Dpt. of Num. Anal. and Comp. Sci., Royal Inst. of Tech., Stockholm*, TRITA-NA-7921, 1979.

MARROCCO, A. /1A/ Expériences numériques sur des problèmes non linéaires résolus par éléments finis et lagrangien augmenté. *LABORIA Report* 309, May 1978.

MENALDI, J.L. /1/ Sur le contrôle impulsionnel et l'I.Q.V. associée. *Compt. Rend. Acad. Sci. Paris*, 284A (1977), pp. 1499-1502.

MERCIER, B. /1A/ *Sur la Théorie et l'Analyse Numérique de Problèmes de Plasticité*. Thèse d'Etat, Université Pierre et Marie Curie, Paris, 1977.

/2A/ Computation of the limit load and elasto-plastic bending of thin plates. *Int. J. Num. Math. Eng.*, 14, (1979), pp. 235-250.

MIYOSHI, T. /1A/ A finite-element method for the solution of fourth-order partial differential equations. *Kunamoto J. Sci. (Math.)*, 9, (1973), pp. 87-116.

MOSCO, U. /1A/ Error estimates for some variational inequalities; in *Mathematical Aspects of Finite Element Methods*, I. Galligani, E. Magenes (eds.), Lecture Notes in Math., 606, Springer-Verlag, Berlin, 1977, pp. 224-236.

MOSCO, U., SCARPINI, F. /1A/ Complementarity systems and approximation of variational inequalities. *Rev. Franc. Autom. Inf. Rech. Op., Anal. Num.*, R-1 (1975), pp. 83-104.

NATTERER, F. /1A/ Optimale L_2-Konvergenz finiten Elemente bei Variationsungleichungen. *Bonn Math. Schr.*, 89, (1976), pp. 1-12.

NITSCHE, J. /1A/ L^{∞}-convergence of finite-element approximation; in *Mathematical Aspects of Finite Element Approximations*, I. Galligani, E. Magenes (eds.), Lecture Notes in Math., 606, Springer-Verlag, 1977, pp. 261-274.

O'LEARY, D.P. /1A/ A generalised conjugate-gradient algorithm for solving a class of quadratic programming problems. *Stanford Report*, *Comp. Science Dept.*, STAN-CA-77-638, Dec. 1977.

O'LEARY, D.P., YANG, W.H. /1A/ Elasto-plastic torsion by quadratic programming. *Comp. Meth. Appl. Mech. Eng.*, 16, (1978), 3, pp. 361-368.

OPIAL, Z. /1A/ Weak convergence of the successive approximations for non-expansive mappings in Banach spaces. *Bull. A.M.S.*, 73, (1967), pp. 591-597.

PEACEMAN, D.H., RACHFORD, H.H. /1A/ The numerical solution of parabolic elliptic differential equations. *J. Soc. Ind. Appl. Math.*, 3, (1955), pp. 28-41.

PERIAUX, J. /1A/ *Résolution de quelques problèmes non linéaires en aérodynamique par des méthodes d'éléments finis et de moindres carrés fonctionnels*. Third-Cycle thesis, Université Pierre et Marie Curie, 1979.

POLYAK, B.T. /1A/ The conjugate-gradient method in extremal problems. *U.S.S.R. Comp. Math. and Math. Phys.*, 9 (1969), pp. 94-112.

POWELL, M.J.D. /1A/ Restart procedure for the conjugate gradient method. *Math. Programming*, 12, (1977), pp. 148-162.

PUEL, J.P. /1A/ Existence, comportement à l'infini et stabilité dans certains problèmes quasilinéaires elliptiques et paraboliques d'ordre 2. *Annali Scuola Normale Superiore, Pisa, Cl. di Scienze, Serie IV*, 3, (1976), 1, pp. 89-119.

RANNACHER, R. /1A/ Punktweise Konvergenz der Methode der finiten Elemente heim Platten problem. *Manuscripta Math.*, 19, (1976), pp. 401-416.

/2A/ On nonconforming and mixed finite element methods for plate bending problems. The linear case. *Rev. Franc. Autom. Inf. Rech. Op., Anal. Num.*, 13, (1979), 4, pp. 369-387.

ROUX, J. /1A/ Résolution numérique d'un problème d'écoulement subsonique de fluide compressible. *Rev. Franc. Autom. Inf. Rech. Op., Anal. Num.* 11 (1977), pp. 197-208.

SAGUEZ, Ch. /1A/ *Contrôle de Systèmes à Frontière Libre.* Thèse d'Etat, Université Technologique de Compiègne, 1980.

SAMUELSSON, A., FRÖIER, M. /1A/ Finite elements in plasticity. A variational inequality approach. Chapter 9 of *The Mathematics of Finite Elements and Applications, III, MAFELAP* 1978, J.R. Whiteman (ed.), Acad. Press, London, 1979, pp. 105-115.

SCARPINI, F. /1A/ Some nonlinear complementarity systems algorithms and applications to unilateral boundary value problems. *Fasc.* 6, *Istituto Matematico G. Castelnuovo*, Universita degli studi di Roma, 1977-78.

SCARPINI, F., VIVALDI, M.A. /1A/ Error estimates for the approximation of some unilateral problems. *Rev. Franc. Autom. Inf. Rech. Op., Anal. Num.*, 11 (1977) 2, pp. 197-208.

SCHOLZ, R. /1A/ Approximation von Sattelpunkten mit finiten Elementen. *Math. Schriften*, 89, (1976), pp. 53-66.

/2A/ A mixed method for 4th order problems using linear finite elements. *Rev. Franc. Autom. Inf. Rech., Anal. Num.*, 12, (1978), 1, pp. 85-90.

/3A/ Interior error estimates for a mixed finite element method. *Num. Funct. Analysis Optimiz.*, 1, (1979), 4, pp. 415-429.

STAMPACCHIA, G. /1A/ *Equations Elliptiques du Second Ordre à Coefficients Discontinus.* Presse de l'Université de Montréal, 1965.

/2A/ Le problème de Dirichlet pour les équations elliptiques du second ordre à coefficients discontinus. *Ann. Inst. Fourier, Grenoble*, 15, (1965), pp. 189-258.

TARTAR, L. /1A/ Inéquations Quasi Variationnelles Abstraites. *Compt. Rend. Acad. Sc. Paris*, 278A, (1974), pp. 1193-1196.

TORELLI, A. /1A/ Some regularity results for a family of variational inequalities. *Ann. Scu. Norm. Sup. Pisa*, to appear.

VIDRASCU, M. /1A/ *Sur la résolution numérique du problème de Dirichlet pour l'opérateur biharmonique.* Third-cycle thesis, Université Pierre et Marie Curie, Paris, 1978.

WHEELER, M.F. /1A/ An elliptic collocation finite-element method with interior penalties. *SIAM J. Num. Anal.*, 15, (1978), 1, pp. 152-161.

WILDE, D.J., BEIGHTLER, C.S. /1A/ *Foundations of Optimisation.* Prentice Hall, Englewood Cliffs, N.J., 1967.

YOUNG, D.M. /1A/ *Iterative solution of large linear systems.* Academic Press, New York, 1971.

2